T5-DHH-775

Optimal Control of
Differential and Functional Equations

OPTIMAL CONTROL OF DIFFERENTIAL AND FUNCTIONAL EQUATIONS

J. Warga

DEPARTMENT OF MATHEMATICS
NORTHEASTERN UNIVERSITY
BOSTON, MASSACHUSETTS

ACADEMIC PRESS New York and London 1972

ACADEMIC PRESS, INC.
111 Fifth Avenue, New York, New York 10003

United Kingdom Edition published by
ACADEMIC PRESS, INC. (LONDON) LTD.
24/28 Oval Road, London NW1

LIBRARY OF CONGRESS CATALOG CARD NUMBER: 72-87229

AMS (MOS) Subject Classifications: 49A, 49B

PRINTED IN THE UNITED STATES OF AMERICA

TO THE MEMORY OF

Czarna, Artur, and Ruth Warga, victims of the Holocaust, and of Herman Warga who told their story. Z"L

Contents

Chapter II. **Functional Equations**

Part Two **OPTIMAL CONTROL**

Chapter III. **Basic Problems and Concepts, and Heuristic Considerations**

Chapter IV. **Original and Relaxed Control Functions**

Chapter V. **Control Problems Defined by Equations in Banach Spaces**

Chapter VI. **Optimal Control of Ordinary Differential Equations**

Chapter VII. **Optimal Control of Functional-Integral Equations in $C(T, \mathbb{R}^n)$**

Chapter VIII. **Optimal Control of Functional-Integral Equations in $L^p(T, \mathbb{R}^n)$**

Chapter IX. Conflicting Control Problems with Relaxed Adverse Controls

Chapter X. Conflicting Control Problems with Hyperrelaxed Adverse Controls

Preface

The purpose of this book is to present a mathematical theory of deterministic optimal control, with a special emphasis on problems involving functional-integral equations and functional restrictions. We address ourselves to three categories of readers: mature mathematicians, graduate students in analysis, and practitioners of optimal control whose primary interests and training are in science or engineering. In order to reach these last two categories of readers, we felt it necessary to devote Part One (comprising the first half of the book) to establishing the necessary analytical foundations. In theory at least, we only presuppose the knowledge of the real number system. In practice, some familiarity with at least the elementary concepts of topology, functional analysis, and measure theory appears essential.

In Part Two we study deterministic optimal control problems that are defined by specifying: a set $\mathscr{Y}$ of states (or state functions), a set $\mathscr{U}$ of "original" control functions $u : T \to R$, a set B of control parameters, a functional equation $y = F(y, u, b)$ in $\mathscr{Y} \times \mathscr{U} \times B$, a restriction $g_1(y, u, b) \in C_1$, and a cost functional $g_0(y, u, b)$. In addition to "original" solutions that are points of $\mathscr{Y} \times \mathscr{U} \times B$, we also consider "approximate" solutions that are sequences in $\mathscr{Y} \times \mathscr{U} \times B$ and "relaxed" solutions that are a form of weak, or extended, solutions.

We investigate original solutions because they present a challenge to the mathematician. We deal with approximate solutions because they represent the appropriate answer to the engineering control problem. Finally, we study relaxed solutions for a number of reasons: they yield a complete theory that encompasses both existence theorems and necessary conditions; they provide the means for constructing optimal approximate solutions; they properly model certain physical situations; and, in "normal" problems, they enable us to either determine the optimal original solutions or to prove that no such solutions exist.

We study the general optimal control problem (in Chapters IV and V) in a framework where (a) T and R are compact metric spaces and (T, Σ, μ) is a nonatomic positive Radon measure space, (b) $\mathscr{U}$ is a subset of the class of μ-measurable functions on T to R that satisfy a condition of the form $u(t) \in R^{\#}(t)$ $(t \in T)$, (c) C_1 is a subset of a topological vector space, and (d) $\mathscr{Y}$ is a Banach space. We then concentrate (in Chapters VI, VII, and VIII) on problems for which $\mathscr{Y} = C(T, \mathbb{R}^n)$ or $\mathscr{Y} = L^p(T, \Sigma, \mu, \mathbb{R}^n)$ and the equation $y = F(y, u, b)$ is a functional-integral equation of the form

$$y(t) = \int f(t, \tau, \xi(y)(\tau), u(\tau), b)\, \mu(d\tau).$$

Chapter VI deals with the optimal control of ordinary differential equations and, on occasions, the assumption that T and R are compact is considerably weakened. Chapter VII deals with functional-integral equations in $C(T, \mathbb{R}^n)$ (which include delay-differential and functional-differential equations as well as analogous integral equations). Chapter VIII is devoted to functional-integral equations in $L^p(T, \Sigma, \mu, \mathbb{R}^n)$.

Among the restrictions that we single out for particular attention are the unilateral restrictions $y(t) \in A(t)$ $(t \in T)$ and (in Chapters IX and X) restrictions involving conflicting and minimax controls.

A few remarks appear in order concerning the degree of preparation required for understanding Part Two and the order in which the various chapters can be read. The reader who is acquainted with the basic concepts of general topology, functional analysis, and the theory of measure and integration of scalar functions will be able to read Part Two after familiarizing himself with the definitions of

(1) the space $\mathscr{F}(S, \Sigma, \mu, X)$ (preliminary remarks in I.4.B and I.4.16);
(2) the space $L^p(S, \Sigma, \mu, \mathscr{X})$ (I.4.29);
(3) the Bochner integral (I.4.33);
(4) a μ-measurable set-valued mapping (preliminary remarks in I.7);
(5) Fréchet and directional derivatives and n-differentiable functions (preliminary remarks in II.3); and
(6) the Carathéodory functions (preliminary remarks in II.4).

He will have to refer frequently to various results presented in Chapters I and II, but will also have the option of postponing the study of these results and of the related material until he is motivated to do so by his own interest or by the needs of research.

The reading of Part Two need not proceed in a sequential manner. The reader will no doubt wish to begin with Chapter III which is descriptive and partly heuristic and whose main purpose is to survey some of the basic problems studied in the book and to motivate the rigorous arguments of the following chapters. He ought to follow with Chapter IV and Sections V.1–V.3 which provide the basis for the remainder of the book, and then with Section V.6 which presents a serious but reasonably simple application of this material. From this point on, each of the remaining sections of Chapter V and each of the Chapters VI–X can be studied independently, Chapters VI–VIII relying on Sections II.4–II.6, respectively.

The material of Chapter II had served as the basis for a course in functional equations that I taught at Northeastern University in the fall of 1969; and Section I.7, Chapter IV, and Sections V.1–V.3, V.6, and VI.3 had provided the material for a course on mathematical control theory in the Winter of 1969.

Except where otherwise indicated, the basic contents of Part Two, and to some extent of Sections II.5 and II.6, are believed to represent new results or an elaboration of my published work. References to related work are provided in the Notes that follow the pertinent chapters but these references should not be considered exhaustive or even reasonably complete.

I wish to express my appreciation to my wife Faye without whose help, encouragement, and sympathetic understanding this book could never have been completed.

Since this book is essentially self-contained, there are frequent references to results established in previous chapters or sections. Many of these results are quite familiar to all students of mathematics and, whenever possible, we have tried to refer to them not only by their appropriate numbers but also by the names most commonly associated with them in the mathematical literature. Thus we speak of Hölder's inequality I.5.13 (the number I referring to the chapter, 5 to the section, and 13 to the item in the section), the (Lebesgue) dominated convergence theorem I.4.35, Fubini's theorem I.4.45, etc. In some cases, we ascribe a well-known theorem to a mathematician who had established an important part or an important special case of it (e.g., Egoroff's theorem I.4.18 or Theorem IX.1.1 of von Neumann *et al.*), or whose arguments represent an essential ingredient in the proof (Saks' theorem I.4.10).

The choice of material for Part One was strictly governed by the requirements of the subsequent chapters. Every definition, lemma, or theorem in Part One is either directly or indirectly applied in Part Two.

PART ONE

Foundations

CHAPTER I

Analytical Foundations

I.1 Sets, Functions, Sequences

SETS

We denote *sets* (also called *collections*, *families*, and *classes*) by capital letters (Latin, script, or Greek) and the *empty set* by $\varnothing$. We write $a \in A$ if a is an *element* (or a *point*) of the set A (A *contains* a); $a_1, a_2, \ldots \in A$ if $a_1 \in A$ and $a_2 \in A, \ldots$ etc.; $a \notin A$ if a is not an element of A; $a = b$ if a and b are both names of the same element; $a \neq b$ if a and b designate different elements; $A \subset B$ or $B \supset A$ (A *is a subset* or a *subfamily* or a *subcollection* of B or A *is contained in* B or B *contains* A) if $a \in B$ whenever $a \in A$; and $A = B$ if $A \subset B$ and $B \subset A$. When there is no danger of confusion, the notation $a \in A$ may also mean "the object a that is an element of A" (as in "for every $a \in A$") and similarly for $A \subset B$. We often write $a \triangleq b$ instead of $a = b$ if this relation defines a or b.

If $S(x)$ is a statement asserting that an element x of some given set A has a particular property, we denote by $\{x \in A \mid S(x)\}$ or $\{x \mid S(x)\}$ the collection of the elements characterized by this property. We also write $\{a_1, a_2, \ldots\}$ for the set with elements $a_1, a_2, \ldots$.

The *union* of sets A and B, written $A \cup B$, is the set $\{x \mid x \in A \text{ or } x \in B\}$; the *intersection* of the sets A and B, written $A \cap B$, is the set $\{x \mid x \in A$ and $x \in B\}$; the *complement* of A in B (or the *complement of A* if B is specified), written $B \sim A$, is the set $\{b \in B \mid b \notin A\}$. The sets A and B are *disjoint* if $A \cap B = \varnothing$. If $\mathscr{T}$ is a collection of subsets of A, then the *union of* $\mathscr{T}$ is $\bigcup_{T\in\mathscr{T}} T \triangleq \{t \in A \mid t \in T \text{ for some } T \in \mathscr{T}\}$ and the *intersection of* $\mathscr{T}$ is $\bigcap_{T\in\mathscr{T}} T \triangleq \{t \in A \mid t \in T \text{ for every } T \in \mathscr{T}\}$. If the elements of $\mathscr{T}$ are distinguished by an index $\omega \in \Omega$, that is, if $\mathscr{T} = \{T_\omega \mid \omega \in \Omega\}$, then $\bigcup_{\omega\in\Omega} T_\omega \triangleq \bigcup_{T\in\mathscr{T}} T$ and $\bigcap_{\omega\in\Omega} T_\omega \triangleq \bigcap_{T\in\mathscr{T}} T$. If i and j are integers and $i \leqslant j$, then

$$\bigcup_{k=i}^{j} T_k \triangleq T_i \cup T_{i+1} \cup \cdots \cup T_j \quad \text{and} \quad \bigcap_{k=i}^{j} T_k \triangleq T_i \cap T_{i+1} \cap \cdots \cap T_j \,.$$

If $j < i$, then

$$\bigcup_{k=i}^{j} T_k \triangleq \varnothing \quad \text{and} \quad \bigcap_{k=i}^{j} T_k \triangleq A.$$

We also write $\bigcup_{k=i}^{\infty} T_k$ for $T_i \cup T_{i+1} \cup \cdots$ and $\bigcap_{k=i}^{\infty} T_k \triangleq T_i \cap T_{i+1} \cap \cdots$. If $\mathscr{T}$ is empty, then $\bigcup_{T\in\mathscr{T}} T \triangleq \varnothing$ and $\bigcap_{T\in\mathscr{T}} T \triangleq A$.

In order to avoid the excessive use of parentheses, we shall agree that, in expressions involving the symbols $\cap$, $\cup$, and $\sim$, the operations defined by these symbols are performed in the order indicated. The *Rules of de Morgan* (which we accept) assert that $A \sim \bigcup_{T\in\mathscr{T}} T = \bigcap_{T\in\mathscr{T}} (A \sim T)$ and $A \sim \bigcap_{T\in\mathscr{T}} T = \bigcup_{T\in\mathscr{T}} (A \sim T)$. A collection $\mathscr{T}$ of subsets of A is a *partition* of A if $T \cap S = \varnothing$ $(T, S \in \mathscr{T}, T \neq S)$ and $\bigcup_{T\in\mathscr{T}} T = A$. A collection $\mathscr{T}$ of subsets of A is a *covering* of a set $B \subset A$ and the elements of $\mathscr{T}$ *cover* B if $B \subset \bigcup_{T\in\mathscr{T}} T$.

If A is a given set and, for each $a \in A$, $S_1(a)$, $S_2(a)$, etc. are statements, we write $(S_1(a), S_2(a), \text{etc.})$ or $[S_1(a), S_2(a), \text{etc.}]$ to mean "for all choices of $a \in A$ for which $S_1(a)$ and $S_2(a)$, etc. are valid" [e.g., "the sets A_i $(i \in \{1, 2,...\}, i \text{ odd})$" means "the sets $A_1, A_3, ...$"]. We also write $x = a, b, c,...$ to mean $x \in \{a, b, c,...\}$.

Functions

An ordered couple of elements a and b is written (a, b); similarly, an ordered n-tuple of elements $a_1 ,..., a_n$ is written $(a_1 ,..., a_n)$. If $a_i \in A$ $(i = 1,..., n)$, then the ordered n-tuple $(a_1 ,..., a_n)$ is sometimes referred to as a *finite sequence in A*. If A and B are given sets and f is a rule that assigns to every element $a \in A$ an element $f(a) \in B$, we say that f is a *function* (or a *mapping*, or a *transformation*) *on A to B*, or *between A and B*, or with *domain A* and *range B*, or a *B-valued* function on A, or a function of a in A, and write $f: A \to B$, or $f(\cdot) : A \to B$, or $a \to f(a) : A \to B$, or $(f_a)_{a\in A}$ (respectively,

f, or $f(\cdot)$, or $a \to f(a)$, or (f_a) if A and B have been specified). Depending on the context, the notation $f: A \to B$ is read "f is a function on A to B" or "the function f on A to B."

When written $(f_a)_{a \in A}$ the function f is often called an *indexed family*. The symbol a appearing in the expressions $a \to f(a)$ or $(f_a)_{a \in A}$ is called a *dummy variable* or a *general point* of A and does not represent a (specific) point of A as in the expression $f(a)$ but rather a "place of substitution." Thus, $a \to f(5 + 2a) : \{1, 2,...\} \to \{1, 2,...\}$ represents the function $g : \{1, 2,...\} \to \{1, 2,...\}$ defined by the relation $g(a) \triangleq f(5 + 2a)$ $(a \in \{1, 2,...\})$. The same dummy variable can be represented by various symbols; thus the notations $a \to f(a) : A \to B$ and $\alpha \to f(\alpha) : A \to B$ describe the same function. We shall sometimes refer to a property of "the function $f(a)$ of a" meaning a property of $a \to f(a)$.

The set $\{(a, f(a)) \mid a \in A\}$ is the *graph* of the function $f: A \to B$. We call the *image of A under f* the set $f(A) \triangleq \{f(a) \mid a \in A\} \subset B$. If $C \subset B$, we call the *inverse image* (or the *preimage*) *of C under f* the set $f^{-1}(C) \triangleq \{a \in A \mid f(a) \in C)$. The elements of A are *arguments* of f and those of $f(A)$ are *values* of f. If $f(A)$ contains only one element, then f is a *constant function* or a *constant.* If the values of f are real numbers, then f is a *functional.*

If $f: A \to B$ and $g : B \to C$, then the function $h \triangleq g \circ f$, defined by $h(a) \triangleq g(f(a))$ for all $a \in A$, is the *composition* of g and f. We observe that, for $f_i : A_i \to A_{i+1}$ $(i = 1, 2, 3)$, the double compositions $f_3 \circ (f_2 \circ f_1)$ and $(f_3 \circ f_2) \circ f_1$ are both defined and represent the same function

$$a_1 \to f_3(f_2(f_1(a_1))) : A_1 \to A_4 .$$

A function $f: A \to B$ is *one-to-one* or an *injection* or *injective* if $f(a_1) = f(a_2)$ implies $a_1 = a_2$; it is *onto* or a *surjection* or *surjective* if $f(A) = B$. If f is both one-to-one and onto, also said to be *bijective* or a *one-to-one correspondence* or a *bijection*, then the set $f^{-1}(\{b\})$ contains a single point of A for every $b \in B$, and we denote this point by $f^{-1}(b)$. Then the rule f^{-1} is a function with domain B and range A which is itself bijective, and we call f^{-1} the *inverse* of f. The function $I_A : A \to A$, defined by $I_A(a) \triangleq a$ for every $a \in A$, is the *identity map* in A (also designated as I if A has been specified). We observe that if $f: A \to B$ is bijective, then $(f^{-1})^{-1} = f$, $f^{-1} \circ f = I_A$, and $f \circ f^{-1} = I_B$ and that, if $f: A \to B$ and $g : B \to A$ satisfy the relations $g \circ f = I_A$ and $f \circ g = I_B$, then both are bijective and $f = g^{-1}$ and $g = f^{-1}$. Furthermore, if $f: A_1 \to A_2$ and $g : A_2 \to A_3$ are both bijective, then $g \circ f$ is also bijective and $(g \circ f)^{-1} = f^{-1} \circ g^{-1}$. If there exists a one-to-one correspondence $f: A \to B$, then we say that A and B have the same *power* (are *equipotent*). We observe that if A and B have the same power and A and C have the same power, then B and C also have the same power.

Identification, Embedding, and Equivalence Classes

Let X and Y be two sets and f a given function on X to Y. If f has been previously specified, it is sometimes convenient to refer to an element x of X as if it were an element of Y in the sense that x is "a name" of the element $f(x)$ of Y. We say in such cases that we *identify* x and $f(x)$. If Y is contained in some set Z and f is a one-to-one correspondence, we may refer to X as a subset of Z (in the sense just described), in which case we say that we *embed* X in Z (or X is *embedded* in Z). If f is not a bijection, we may choose to refer to each set $f^{-1}(\{y\})$ $[\triangleq\{x \in X \mid f(x) = y\}]$ corresponding to some $y \in Y$ as an *equivalence class* and *identify* any two elements x_1 and x_2 in the same equivalence class in the sense that we consider both x_1 and x_2 to be "names" of this equivalence class. In particular, if X is a given set and $\mathscr{T}$ a partition of X, we may define a function $f: X \to \mathscr{T}$ by the relation $f(x) \triangleq T$ if $x \in T \in \mathscr{T}$. Then $T = f^{-1}(\{T\})$ for each $T \in \mathscr{T}$ and each T is an equivalence class whose elements can be identified.

An *equivalence relation* in a set X is a collection R of ordered couples (x, y), with $x, y \in X$, and such that $(y, x) \in R$ whenever $(x, y) \in R$; $(x, z) \in R$ whenever $(x, y) \in R$ and $(y, z) \in R$; and $(x, x) \in R$. If R is an equivalence relation in X and $\mathscr{P}(X)$ the collection of all subsets of X, we can define a function $f: X \to \mathscr{P}(X)$ by $f(x) \triangleq \{y \in X \mid (x, y) \in R\}$. We then verify that $f(x) = f(z)$ if $(x, z) \in R$ and $f(x) \cap f(z) = \varnothing$ otherwise. Thus the collection $\{f(x) \mid x \in X\}$ defines a partition of X and each $f(x)$ is an equivalence class whose elements can be identified.

Cartesian Product

We denote by $\mathscr{P}(A)$ the collection of all the subsets of a set A. If Ω is a set and $G(\cdot) : \Omega \to \mathscr{P}(A)$, we call the *cartesian product* of the indexed family $(G(\omega))_{\omega\in\Omega}$ of sets the set $\prod_{\omega\in\Omega} G(\omega)$ whose elements are functions $g : \Omega \to A$ such that $g(\omega) \in G(\omega)$ for all $\omega \in \Omega$. If $\Omega = \{1, 2,..., k\}$, $G(1) = P$, $G(2) = Q,..., G(k) = T$, we write $\prod_{i=1}^{k} G(i)$ or $P \times Q \times \cdots \times T$ for $\prod_{\omega\in\Omega} G(\omega)$ and observe that there exists a one-to-one correspondence between $P \times Q \times \cdots \times T$ and the set $\{(p, q,..., t) \mid p \in P, q \in Q,..., t \in T\}$ of ordered k-tuples, the function $\omega \to g(\omega)$ corresponding to the k-tuple $(g(1),..., g(k))$. It will cause no confusion in that case to identify g with the k-tuple $(g(1),..., g(k))$ which we thus consider as an element of $P \times Q \times \cdots \times T$. We write P^k for $\prod_{i=1}^{k} G(i)$ if $G(1) = \cdots = G(k) = P$, $\text{pr}_{\bar\omega} A \triangleq \{g(\bar\omega) \mid g \in A\}$ if $A \subset \prod_{\omega\in\Omega} G(\omega)$ and $\text{pr}_R A \triangleq \{r \in R \mid (p, q,..., r,..., t) \in A\}$ if

$$A \subset P \times Q \times \cdots \times R \times \cdots \times T.$$

We refer to $\text{pr}_{\bar\omega} A$ as the projection of A on $G(\bar\omega)$.

If $\mathscr{T}$ is a partition of Ω, then it follows from the definition of a function that there exists a one-to-one correspondence between $\prod_{\omega\in\Omega} G(\omega)$ and $\prod_{T\in\mathscr{T}}(\prod_{\omega\in T}G(\omega))$, and we identify the corresponding elements of each set. In this sense, we write, for a given $f: A_1 \times \cdots \times A_k \to B$, $f((a_1,..., a_k))$ or $f(a_1,..., a_k)$ for $f(g)$ if $(g(1),..., g(k)) \triangleq (a_1,..., a_k)$. Similarly, we write $f(a_1, a_2, a_3)$ for either $f(a_1, (a_2, a_3))$ or $f((a_1, a_2), a_3)$ and $f(\cdot, \cdot,..., \cdot)$ or $(a_1, a_2,..., a_k) \to f(a_1,..., a_k)$ for f.

Restrictions and Extensions

Let $A_1 \subset A$, $f: A \to B$, and $f_1: A_1 \to B$. We say that f_1 is the *restriction* of f to A_1, written $f \mid A_1$ or $f: A_1 \to B$, and f an *extension* of f_1 to A, if $f(a) = f_1(a)$ for all $a \in A_1$. When there is no danger of confusion, we use the same designation for two functions, one of which is a restriction of the other.

If $f: A_1 \times A_2 \times \cdots \times A_k \to B$ and $\bar{a}_i \in A_i$ $(i \neq j)$ are fixed, we denote by $a_j \to f(\bar{a}_1,..., \bar{a}_{j-1}, a_j, \bar{a}_{j+1},..., \bar{a}_k)$ or $f(\bar{a}_1,..., \bar{a}_{j-1}, \cdot, \bar{a}_{j+1},..., \bar{a}_k)$ the function on A_j to B that assigns to each $a_j \in A_j$ the value $f(\bar{a}_1,..., \bar{a}_{j-1}, a_j, \bar{a}_{j+1},..., \bar{a}_k)$. If, for fixed $\bar{a}_i$ $(i \neq j)$, the function $f(\bar{a}_1,..., \bar{a}_{j-1}, \cdot, \bar{a}_{j+1},..., \bar{a}_k)$ has a property P, we say that f has property P in a_j for $a_i = \bar{a}_i$ $(i \neq j)$.

Countable Sets and Sequences

We shall denote by $\mathbb{N}$ the set $\{1, 2, 3,...\}$ of all positive integers. Unless otherwise specified, the letters i, j, k, l, m, n will represent elements of $\mathbb{N}$. If a set A is empty or has the same power as $\{1, 2,..., k\}$ for some positive integer k, we say that A is *finite*; otherwise A is *infinite*. If a set A and $\mathbb{N}$ have the same power, we say that A is *countable* or *denumerable*. A set A is *at most denumerable* or *at most countable* if it is either finite or countable. If $k \in \mathbb{N}$ and the sets $A_1, A_2,..., A_k$ are denumerable, then $A_1 \times \cdots \times A_k$ is also denumerable which we can verify by constructing the one-to-one correspondence $f: \mathbb{N}^k \to \mathbb{N}$ defined as follows: If the positive integer x_i has the (unique) decimal representation $x_{i,1} + 10x_{i,2} + 10^2x_{i,3} + \cdots$ for $i = 1, 2,..., k$, then we let $f(x_1,..., x_k) \triangleq \sum_{i=1}^{k} x_{i,1}10^{i-1} + 10^k \sum_{i=1}^{k} x_{i,2}10^{i-1} + \cdots$. Similarly, we can show that the union of a finite or a denumerable collection of denumerable sets is denumerable, that the set of all rationals

$$\{\pm\, i/j \mid i \in \{0, 1, 2,...\}, j \in \mathbb{N}\}$$

is denumerable, and that any subset of a denumerable set is either finite or denumerable.

If A is any set and $a: \mathbb{N} \to A$, we say that $a(\cdot)$ is a *sequence in* A, and denote it by $(a_1, a_2,...)$ or as an indexed family $(a_i)_{i\in\mathbb{N}}$. We also write $(a_i)_i$ or (a_i) for $(a_i)_{i\in\mathbb{N}}$, the notation $(a_i)_i$ being used, in particular, if a_i is an expression involving other indices than i [For example $(2i + 5j)_i$ is the sequence $(2 + 5j, 4 + 5j, 6 + 5j,...)$ while $(2i + 5j)_j \triangleq (2i + 5, 2i + 10, 2i + 15,...)$]. If $i \to j_i : \mathbb{N} \to \mathbb{N}$, $j_{i+1} > j_i$, and $b_i \triangleq a_{j_i}$ for all $i \in \mathbb{N}$, we say that (b_i) is a subsequence of (a_i) and write $(b_i) \subset (a_i)$. We may also denote (b_i) as $(a_i)_{i \in K}$ if $K \triangleq (j_1, j_2,...)$. If $\mathscr{A} \triangleq (a_i)_{i \in K}$, then we write $b \in \mathscr{A}$ for $b \in \{a_i \mid i \in K\}$.

If (a_i) is a sequence, then to *extract* a subsequence $(a_i)_{i \in K}$ means to determine such a subsequence. If $((a_i^1), (a_i^2),...)$ is a sequence of sequences such that $(a_i^{k+1})_i \subset (a_i^k)_i$ for all $k \in \mathbb{N}$, then the sequence $(a_i^i)_i = (a_1^1, a_2^2,...)$ is the *diagonal subsequence of* $(a_i^1)_i$, $(a_i^2)_i$,... and, for every $k \in \mathbb{N}$,

$$(a_k^k, a_{k+1}^{k+1},...) \subset (a_i^k)_i .$$

The construction of the diagonal subsequence is the (Cantor) *diagonal process.*

If $S(i)$ is, for each $i \in \mathbb{N}$, a statement such that $S(1), S(2),..., S(i)$ imply $S(i + 1)$, and if $S(1)$ is true, then $S(i)$ must be true for all $i \in \mathbb{N}$. Indeed, if $S(j)$ is false for some $j \in \mathbb{N}$, let k be the smallest element of $\{1, 2,..., j\}$ for which $S(k)$ is false. Then $k > 1$ and $S(1), S(2),..., S(k - 1)$ are true, implying that $S(k)$ is true. Thus $S(j)$ cannot be false for any $j \in \mathbb{N}$. A *proof by induction* that $S(i)$ is true for all $i \in \mathbb{N}$ consists of two steps: (1) the proof that $S(1)$ is true, and (2) the proof that $S(1),..., S(i)$ imply $S(i + 1)$ for all $i \in \mathbb{N}$. To carry out step (2) is to *complete the induction.*

Zorn's Lemma

If A is any set, a *partial order relation* in A is a subset R of $A \times A$ such that $(a, c) \in R$ whenever $(a, b) \in R$ and $(b, c) \in R$, and $(a, a) \in R$ for all $a \in A$. We also write $a \leqslant b$ for $(a, b) \in R$. The couple (A, R) is then a *partially ordered set.* A *totally ordered* subset B of a partially ordered set (A, R) is a set B such that, for all $b_1 \in B$, $b_2 \in B$, we have either $b_1 \leqslant b_2$ or $b_2 \leqslant b_1$. An element x in a partially ordered set (A, R) is *maximal* if $x \leqslant a$ implies $a \leqslant x$; a subset B of A has an *upper bound* $x \in A$ if $b \leqslant x$ for all $b \in B$. We accept as an axiom *Zorn's lemma* asserting that if $A \neq \varnothing$ and (A, R) is a partially ordered set such that every totally ordered subset of it has an upper bound, then (A, R) has a maximal element. [Zorn's lemma can be deduced from the *Axiom of choice* asserting that the cartesian product $\prod_{\omega \in \Omega} G(\omega) \neq \varnothing$ if $G(\omega) \neq \varnothing$ for all $\omega \in \Omega$.]

1.2 Topology

A collection $\mathscr{T}$ of subsets of a set X is a *topology* in X (or a *topological structure of* X) if $\varnothing \in \mathscr{T}$, $X \in \mathscr{T}$, and $\mathscr{T}$ contains the union of every subcollection of $\mathscr{T}$ and the intersection of every finite subcollection of $\mathscr{T}$. A topology $\mathscr{T}_1$ is *weaker* or *smaller* than a topology $\mathscr{T}_2$, and $\mathscr{T}_2$ is *stronger* or *larger* than $\mathscr{T}_1$, if $\mathscr{T}_1 \subset \mathscr{T}_2$. Elements of a topology are called *open sets.* The couple $(X, \mathscr{T})$ is a *topological space*; alternately, X is called a "topological space" if a topology in X has been specified (X has been *topologized*).

If $\mathscr{U}$ is a collection of subsets of X, then there exists a unique topology $\mathscr{T}(\mathscr{U})$ in X containing $\mathscr{U}$ and weaker than any other topology containing $\mathscr{U}$; we construct $\mathscr{T}(\mathscr{U})$ by first forming $\mathscr{V}(\mathscr{U}) \triangleq \{\bigcap_{j=1}^{k} A_j \mid k \in \mathbb{N},\ A_j \in \mathscr{U}\}$ and then defining $\mathscr{T}(\mathscr{U})$ as the set of the unions of all subcollections of $\mathscr{V}(\mathscr{U})$. We refer to $\mathscr{U}$ as a *subbase* of $\mathscr{T}(\mathscr{U})$ and call a *base* of a topology $\mathscr{T}$ any collection $\mathscr{V} \subset \mathscr{T}$ such that every element of $\mathscr{T}$ is a union of some subcollection of $\mathscr{V}$. [Thus $\mathscr{V}(\mathscr{U})$ is a base of $\mathscr{T}(\mathscr{U})$].

If $(X, \mathscr{T})$ is a topological space and $A \subset X$, the collection $\mathscr{T}_A \triangleq \{A \cap B \mid B \in \mathscr{T}\}$ is a topology in A called the *relative topology* of A in $(X, \mathscr{T})$ or the *relativization of* $\mathscr{T}$ to A. Its elements are called *relatively open sets in* A. An *open neighborhood* of a point x (of a set A) is any open set containing x (containing A). A *neighborhood* of x (of A) is any set containing an open neighborhood of x (of A). A point x is a *limit point* of A if every neighborhood of x contains a point $y \in A \sim \{x\}$. A point x is a *limit* of a sequence (x_j), denoted $\lim_j x_j$ or $\lim_{j\to\infty} x_j$ (respectively limit of a sequence $(x_j)_{j\in J}$, denoted $\lim_{j\in J} x_j$ or $\lim x_j$ for $j \in J$) or, equivalently, (x_j) *converges* to x, written $x_j \to x$ or $x_j \to_j x$, if every neighborhood of x contains all the elements x_j except for a finite number; a sequence that converges to some point is *convergent*. We write $x = \lim_j x_j$ whenever x is one of possibly many limits of (x_j). Similarly, if $k \in \mathbb{N}$ and $(i_1, i_2, ..., i_k) \to x_{i_1,...,i_k} : \mathbb{N}^k \to X$, then $x_{i_1,...,i_k} \to x$ or $\lim_{i_1,...,i_k} x_{i_1,...,i_k} = x$ if every neighborhood of x contains all the points $x_{i_1,...,i_k}$ except for a finite number. A set A is *closed* if $X \sim A$ is open. The *closure* of a set A, denoted $\bar{A}$, is the set of all points $x \in X$ such that the intersection of A with every neighborhood of x is nonempty; it can be easily verified that $\bar{A}$ is closed for every $A \subset X$ and $\bar{A}$ is the intersection of all the closed sets that contain A. A point x is an *interior point* of A if some neighborhood of x is contained in A. We verify that a set is open if and only if every one of its points is interior. The *interior of* A, denoted A°, is the set of all the interior points of A (or, equivalently, the union of all the open subsets of A). The *boundary* of A, denoted ∂A, is the set $\bar{A} \sim A^\circ$. We say that A is a *dense* subset of B if $A \subset B \subset \bar{A}$. A set A is *sequentially closed* if every convergent sequence in A has its limits in A. The *sequential closure* of A, denoted seq cl(A), is the collection of all limits of convergent sequences in A. It is clear that seq cl$(A) \subset \bar{A}$.

A topological space $(X, \mathscr{T})$ is *separable* if X contains a finite or countable dense subset. In a topological space $(X, \mathscr{T})$ a subcollection $\mathscr{V}$ of $\mathscr{T}$ is an *open covering* of a set $B \subset X$ if $B \subset \bigcup_{V\in\mathscr{V}} V$. If $\mathscr{V}_1 \subset \mathscr{V}$ and both $\mathscr{V}_1$ and $\mathscr{V}$ are open coverings of a set B, then $\mathscr{V}_1$ is a *subcovering* of $\mathscr{V}$. A collection of sets has the *finite intersection property* if every finite subcollection has a nonempty intersection. A set $B \subset X$ is *compact* if every open covering $\mathscr{V}$ of B has a finite subcovering or (via the Rules of de Morgan) if every family of relatively closed subsets of B with the finite intersection property has a nonempty intersection. A set $B \subset X$ is *conditionally compact* if $\bar{B}$ is compact. The space $(X, \mathscr{T})$ and the topology $\mathscr{T}$ are *compact* if the set X is compact. A set A in the topological space X is *sequentially compact* if every sequence in A has a subsequence converging to some point in A. The space $(X, \mathscr{T})$ and the topology $\mathscr{T}$ are *sequentially compact* if the set X is sequentially compact. We verify that a closed subset of a compact set is compact and a sequentially closed subset of a sequentially compact set is sequentially compact.

A topological space $(X, \mathscr{T})$ and the topology $\mathscr{T}$ are *Hausdorff* if any two distinct points of X have disjoint neighborhoods. We verify that, for every point x in a Hausdorff space, the set $\{x\}$ is closed.

If $(X_\omega, \mathscr{T}_\omega)$ $(\omega \in \Omega)$ are topological spaces, then the *product topology* in $\prod_{\omega \in \Omega} X_\omega$ is the smallest topology containing sets of the form $\prod_{\omega \in \Omega} A_\omega$, where $A_\omega \in \mathscr{T}_\omega$ $(\omega \in \Omega)$ and $A_\omega = X_\omega$ except for a finite number of A_ω.

Metric Spaces

A function $d(\cdot, \cdot)$ on $X \times X$ to the set of real numbers is a *metric* (or a *distance*) function (for, or on, the set X) if

(1) $d(x, y) \geqslant 0$,

(2) $d(x, y) \leqslant d(x, z) + d(z, y)$,

(3) $d(x, y) = d(y, x)$, and

(4) $d(x, y) = 0$ if and only if $x = y$.

The set $S(x, a) \triangleq \{y \in X \mid d(x, y) < a\}$ is an *open ball* with *center* x and *radius* a, and the set $S^F(x, a) \triangleq \{y \in X \mid d(x, y) \leqslant a\}$ is a *closed ball* with *center* x and *radius* a. The topology $\mathscr{D}$ whose subbase is the collection of all open balls in X is the *metric topology*. We verify that the open balls actually form a base of $\mathscr{D}$. The topological space $(X, \mathscr{D})$ is called a *metric space*; alternately, X is called a metric space or *metric* if a metric function has been specified. Since $S(x, \frac{1}{2}d(x, y)) \cap S(y, \frac{1}{2}d(x, y)) = \varnothing$, every metric space is Hausdorff. The number $d(x, y)$ is the *distance* of x to y.

Let $d(\cdot, \cdot)$ and $\delta(\cdot, \cdot)$ be two metric functions for X with the following property: for every x and $\epsilon > 0$ there exists $\eta(x, \epsilon) > 0$ such that (a) $\delta(x, y) \leqslant \eta(x, \epsilon)$ implies $d(x, y) \leqslant \epsilon$, and (b) $d(x, y) \leqslant \eta(x, \epsilon)$ implies $\delta(x, y) \leqslant \epsilon$. We then verify that the metric topologies defined by $d(\cdot, \cdot)$ and

$\delta(\cdot, \cdot)$ coincide, and call $d(\cdot, \cdot)$ and $\delta(\cdot, \cdot)$ *equivalent metrics*. We can verify that if $d(\cdot, \cdot)$ is a metric function for X and $\delta(x, y)$ is the smaller of $d(x, y)$ and 1, then $\delta(x, y)$ is also a metric function for X, and that $d(\cdot, \cdot)$ and $\delta(\cdot, \cdot)$ are equivalent metrics.

We say that a topological space $(X, \mathscr{T})$ is *metrizable* if there exists a distance function $d(\cdot, \cdot)$ for X such that the corresponding metric topology coincides with $\mathscr{T}$. It is clear from our previous remark that we can choose this distance function so that $d(x, y) \leqslant 1$ for all x and y.

A sequence (x_j) in a metric space X is a *Cauchy sequence* if for every $\epsilon > 0$ there exists $k(\epsilon) \in \mathbb{N}$ such that $d(x_i, x_j) \leqslant \epsilon$ for $i, j \geqslant k(\epsilon)$. A function $(i, j) \to x_{i,j} : \mathbb{N} \times \mathbb{N} \to X$ is a *Cauchy double sequence* if for every $\epsilon > 0$ there exists $k(\epsilon) \in \mathbb{N}$ such that $d(x_{i_1,j_1}, x_{i_2,j_2}) \leqslant \epsilon$ for $i_1, i_2, j_1, j_2 \geqslant k(\epsilon)$.

A subset A of a metric space is *complete* if every Cauchy sequence in A has a limit in A. A metric space X is *complete* if its subset X is complete. Thus a closed subset of a complete metric space is complete.

A subset A of a metric space is *bounded* if there exists a real number c such that $d(x, y) \leqslant c$ for all $x, y \in A$. The set A is *totally bounded* if, for every $\epsilon > 0$, A can be covered by a finite collection of open balls with centers in A and radii ϵ.

The Metric Spaces $\mathbb{R}$ and $\overline{\mathbb{R}}$

We denote by $\mathbb{R}$ the metric space whose elements are real numbers (which we shall also refer to as *numbers*), and with the metric function defined by $d(x, y) \triangleq |x - y|$ (the absolute value of $x - y$). A number c is an *upper bound* (*lower bound*) of a nonempty subset A of $\mathbb{R}$ if $a \leqslant c$ ($a \geqslant c$) for all $a \in A$. We assume it known that the set of rationals $\{\pm i/j \mid i \in \{0, 1, 2, \ldots\}, j \in \mathbb{N}\}$ is dense in $\mathbb{R}$, that $\mathbb{R}$ is complete, and that, for every nonempty set $A \subset \mathbb{R}$ with an upper (lower) bound c, there exist a unique real number $\sup A$ ($\inf A$) and a sequence (a_j) in A such that $\lim_j a_j = \sup A$ ($\lim_j a_j = \inf A$) and $a \leqslant \sup A \leqslant c$ ($a \geqslant \inf A \geqslant c$) for all $a \in A$.

An *extended real number* is either a real number or one of the symbols $-\infty$ and ∞ (or $+\infty$). The set $\overline{\mathbb{R}}$ of extended real numbers is $\mathbb{R} \cup \{-\infty, \infty\}$. If we set $f(x) \triangleq x/(1 + |x|)$ for $x \in \mathbb{R}$, $f(\infty) \triangleq 1$, and $f(-\infty) \triangleq -1$, then we can verify that the function $(x, y) \to \delta(x, y) \triangleq |f(x) - f(y)| : \overline{\mathbb{R}} \times \overline{\mathbb{R}} \to \mathbb{R}$ is a metric function for $\overline{\mathbb{R}}$, and we choose the corresponding metric topology for $\overline{\mathbb{R}}$. We observe that the metric topology of $\mathbb{R}$ is the relativization of the metric topology of $\overline{\mathbb{R}}$ and that $\lim_j x_j = \infty$ ($\lim_j x_j = -\infty$) if and only if for each $n \in \mathbb{N}$ there exists $j(n) \in \mathbb{N}$ such that $x_j \geqslant n$ ($x_j \leqslant -n$) for all $j \geqslant j(n)$.

We consider $\overline{\mathbb{R}}$ to be totally ordered by the relation $\leqslant$, where $a \leqslant b$ is defined in the usual manner if $a, b \in \mathbb{R}$ and where $-\infty < a < \infty$ for all

$a \in \mathbb{R}$ ($a < b$ means $a \leqslant b$ and $a \neq b$). We write $b \geqslant a$ to mean $a \leqslant b$. For $x, y \in \overline{\mathbb{R}}$, we set

$$x + \infty = \infty + x \triangleq \infty \quad \text{if} \quad x \neq -\infty,$$
$$x - \infty = -\infty + x \triangleq -\infty \quad \text{if} \quad x \neq \infty,$$
$$a\infty \triangleq \infty \quad \text{if} \quad a > 0, \qquad a\infty \triangleq -\infty \quad \text{if} \quad a < 0,$$
$$x/\infty = x/(-\infty) \triangleq 0 \quad \text{if} \quad x \neq \infty \quad \text{and} \quad x \neq -\infty,$$

and

$$x - y \triangleq x + (-1) \cdot y.$$

The expressions $0/0$, $0 \cdot \infty$, $\infty - \infty$, ∞/∞, etc. remain undefined.

We say that elements of $\{-\infty, \infty\}$ are *infinite* and those of $\mathbb{R}$ *finite*. We write $|x|$ for x if $x \geqslant 0$ and for $-x$ if $x < 0$ and observe that $|x + y| \leqslant |x| + |y|$ $(x, y \in \overline{\mathbb{R}})$.

For $a, b \in \overline{\mathbb{R}}$ and $a \leqslant b$, we write

$$(a, b) \triangleq \{x \in \overline{\mathbb{R}} \mid a < x < b\}, \qquad (a, b] \triangleq \{x \in \overline{\mathbb{R}} \mid a < x \leqslant b\},$$

$$[a, b) \triangleq \{x \in \overline{\mathbb{R}} \mid a \leqslant x < b\}, \qquad [a, b] \triangleq \{x \in \overline{\mathbb{R}} \mid a \leqslant x \leqslant b\}.$$

We refer to these sets as *intervals* if they are nonempty and contained in $\mathbb{R}$, and then (a, b) is an *open interval* and $[a, b]$ a *closed interval*. A *subinterval* of a set T is an interval contained in T.

We say that $c \in \overline{\mathbb{R}}$ is an *upper bound* (*lower bound*) of a nonempty set $A \subset \overline{\mathbb{R}}$ if $a \leqslant c$ ($a \geqslant c$) for all $a \in A$. We then verify that for every nonempty set $A \subset \overline{\mathbb{R}}$ there exist unique extended real numbers $\sup A$ and $\inf A$ and sequences (a_j) and (b_j) in A such that $\lim_j a_j = \sup A$, $\lim_j b_j = \inf A$, and $d \leqslant \inf A \leqslant a \leqslant \sup A \leqslant c$ for every $a \in A$, every upper bound c, and every lower bound d. Furthermore, if A has an upper (lower) bound in $\mathbb{R}$, then $\sup A$ ($\inf A$) in $\overline{\mathbb{R}}$ coincides with $\sup(A \cap \mathbb{R})$ [$\inf(A \cap \mathbb{R})$] in $\mathbb{R}$, and otherwise $\sup A = \infty$ ($\inf A = -\infty$). We set $\sup \varnothing \triangleq -\infty$ and $\inf \varnothing \triangleq \infty$. If $\sup A \in (A \cap \mathbb{R})$ [$\inf A \in (A \cap \mathbb{R})$], we write $\max A$ ($\min A$) for $\sup A$ ($\inf A$). We also write Sup(Inf, Max, Min) for sup(inf, max, min), $\max(a_1, ..., a_k)$ for $\max\{a_1, ..., a_k\}$, and similarly for min. If X is any set, $B \subset X$ and $f : X \to \overline{\mathbb{R}}$, we write $\sup_{x \in B} f(x) \triangleq \sup f(B)$, $\inf_{x \in B} f(x) \triangleq \inf f(B)$. A function f on $A \subset \overline{\mathbb{R}}$ to $\overline{\mathbb{R}}$ is *increasing* (*nondecreasing*, *decreasing*, *nonincreasing*) if $f(x) < f(y)$ [$f(x) \leqslant f(y)$, $f(x) > f(y)$, $f(x) \geqslant f(y)$] whenever $x < y$. A function f is *monotonic* if it is either nondecreasing or nonincreasing.

If (x_j) (that is, the function $j \to x_j : \mathbb{N} \to \mathbb{R}$) is nondecreasing (nonincreasing) and $\{x_1, x_2, ...\}$ has an upper (lower) bound $c \in \mathbb{R}$, we verify that

$\lim_{i,j} | x_i - x_j | = 0$ and (x_j) is therefore convergent in the complete metric space $\mathbb{R}$.

For every sequence (x_j) in $\mathbb{R}$ we write $\limsup_j x_j \triangleq \inf_{k \in \mathbb{N}} \sup\{x_i \mid i \geqslant k\}$ and $\liminf_j x_j \triangleq \sup_{k \in \mathbb{N}} \inf\{x_i \mid i \geqslant k\}$.

Distance and Diameter

If X is a metric space, $A, B \subset X$, and $x \in X$, we set

$$d[A, x] = d[x, A] \triangleq \inf_{y \in A} d(x, y) \qquad \text{and} \qquad d[A, B] = d[B, A] \triangleq \inf_{z \in B} d[A, z].$$

(We write $d[A, x]$ instead of the often used notation $d(A, x)$ to prevent confusion with the set $\{d(y, x) \mid y \in A\}$; and similarly for $d[A, B]$). We refer to $d[x, A]$ as the *distance from x to A* and to $d[A, B]$ as the *distance from A* to B. We call the *diameter* of A and denote by $\operatorname{diam}(A)$ the extended real number $\sup_{x, y \in A} d(x, y)$.

Limits and Continuous Functions

Let $(X, \mathscr{T})$ and $(Y, \mathscr{V})$ be topological spaces, Z an arbitrary set, $f : Z \to Y$, and $g : Z \to X$. Then we write $\lim_{g(z) \to \bar{x}} f(z) = y$ or $\lim f(z) = y$ as $g(z) \to \bar{x}$ if for every neighborhood A of y there exists a neighborhood B of $\bar{x}$ such that $f(g^{-1}(B)) \subset A$. Now let $Z = X$ and $g(x) \triangleq x$. We write $\lim_{x \to \bar{x}} f(x) = y$ for $\lim f(z) = y$ as $g(z) \to \bar{x}$. We also write $\lim f(x) = y$ as $x \to \bar{x}$, $x \in X_1$ if for every neighborhood A of y there exists a neighborhood B of $\bar{x}$ such that $f(B \cap X_1) \subset A$; and we write $\lim_{x \to +0} f(x)$ for $\lim f(x)$ as $x \to 0$, $x \in (0, \infty)$ if $X \subset \mathbb{R}$. If $Y \subset \mathbb{R}$, then we set

$$\limsup_{x \to \bar{x}} f(x) \triangleq \inf\{\sup f(B) \mid \bar{x} \in B \in \mathscr{T}\}$$

and

$$\liminf_{x \to \bar{x}} f(x) = \sup\{\inf f(B) \mid \bar{x} \in B \in \mathscr{T}\}.$$

A function $f : X \to Y$ is *continuous at* $\bar{x}$ if $\lim_{x \to \bar{x}} f(x) = f(\bar{x})$. We say that f is *continuous* if it is continuous at $\bar{x}$ for every $\bar{x} \in X$. A function is *discontinuous* if it is not continuous.

It is easy to see that the following statements are equivalent: (a) f is continuous, (b) $f^{-1}(A)$ is open for every open $A \subset Y$, and (c) $f^{-1}(A)$ is closed for every closed $A \subset Y$. If $f : X \to Y$ is a bijection and both f and f^{-1} are continuous, then f is a *homeomorphism* (or a *topological isomorphism*) and X and Y are *homeomorphic*. A function $f : X \to \mathbb{R}$ is *upper semicontinuous at* $\bar{x}$ if $\limsup_{x \to \bar{x}} f(x) = f(\bar{x})$ and *lower semicontinuous at* $\bar{x}$ if

$$\liminf_{x \to \bar{x}} f(x) = f(\bar{x}).$$

We say that f is *upper semicontinuous* (*lower semicontinuous*) if it is upper semicontinuous (lower semicontinuous) at $\bar{x}$ for every $\bar{x} \in X$. We then verify that the set $\{x \in X \mid f(x) < a\}$ [the set $\{x \in X \mid f(x) > a\}$] is open for every $a \in \mathbb{R}$.

A function f on $T \triangleq [t_0, t_1]$ to a topological space Y is *piecewise continuous* respectively *piecewise constant* if there exists a finite partition of T into intervals $T_1, \ldots, T_k$ such that $f \mid T_i$ is continuous respectively constant for each i.

A function $f: X \to Y$ is *sequentially continuous at* $\bar{x}$ if the sequence $(f(x_j))$ converges to $f(\bar{x})$ whenever (x_j) converges to $\bar{x}$. The function f is *sequentially continuous* if it is sequentially continuous at $\bar{x}$ for all $\bar{x} \in X$.

If X is any set, Y a metric space, $f_j : X \to Y$ $(j \in \mathbb{N})$, and $f: X \to Y$, then $\lim_j f_j(\cdot) = f(\cdot)$ *uniformly* or, equivalently, $\lim_j f_j(x) = f(x)$ *uniformly for* $x \in X$ if for every $\epsilon > 0$ there exists $j(\epsilon) \in \mathbb{N}$ such that $d(f_j(x), f(x)) \leqslant \epsilon$ for all $x \in X$ and $j \geqslant j(\epsilon)$. If X and Y are any sets, Z a metric space, and $f: X \times Y \to Z$, we say that $\lim_j f(x_j, y) = f(x, y)$ *uniformly* for $y \in Y$ if $\lim_j f(x_j, \cdot) = f(x, \cdot)$ uniformly. If X and Y are metric, $\bar{x} \in X$, and $\mathscr{A}$ is a collection of functions $f: X \to Y$, we say that the functions belonging to $\mathscr{A}$ and the set $\mathscr{A}$ itself are *equicontinuous at* $\bar{x}$ if for every $\epsilon > 0$ there is $\delta(\epsilon, \bar{x}) > 0$ such that $d(f(x), f(\bar{x})) \leqslant \epsilon$ for all $f \in \mathscr{A}$ if $d(x, \bar{x}) \leqslant \delta(\epsilon, \bar{x})$. If $\mathscr{A}$ is equicontinuous at $\bar{x}$ for every $\bar{x} \in X$, we say that the functions belonging to $\mathscr{A}$ and the set $\mathscr{A}$ itself are *equicontinuous.*

A continuous function $f: X \to Y$ between metric spaces is *uniformly continuous* if for every $\epsilon > 0$ there exists $\delta > 0$ such that $d(x, \xi) \leqslant \delta$ implies $d(f(x), f(\xi)) \leqslant \epsilon$. (The metric functions d_X on X and d_Y on Y are both represented by $d(\cdot, \cdot)$ to simplify notation). A function $\Omega : (0, \infty) \to (0, \infty]$ is a *modulus of continuity* of a function $f: X \to Y$ between metric spaces if $d(f(x), f(\xi)) \leqslant \Omega(d(x, \xi))$ and $\lim_{h \to 0} \Omega(h) = 0$. A function f on a set X to a metric space Y is *bounded* if $f(X)$ is bounded. A collection K of functions on a set X to a metric space Y is *uniformly bounded* if there exists a bounded set $B \subset Y$ such that $f(X) \subset B$ $(f \in K)$.

A triplet (X, X', Φ) is a *compactification* of a topological space Y if X is a compact topological space, X' a dense subset of X, and $\Phi : Y \to X'$ a homeomorphism; (X, X', Φ) is a *metric compactification* of Y if the topology of X is metric. For example, $X = [-1, 1]$, $X' = (-1, 1)$, and $\Phi(y) = y/(|y| + 1)$ $(y \in \mathbb{R})$ define a metric compactification of $\mathbb{R}$.

I.2.1 ***Theorem*** (Lindelöf) If $(X, \mathscr{T})$ is a topological space and $\mathscr{T}$ has a countable base, then any open covering of a set $A \subset X$ has an at most countable subcovering.

▌PROOF Let $\mathscr{V}$ be an open covering of A and $\{B_1, B_2, \ldots\}$ a base for $\mathscr{T}$. For each $j \in \mathbb{N}$, we select $C_j \in \mathscr{V}$ such that $B_j \subset C_j$; if no such C_j exists, we set $C_j \triangleq \varnothing$. If $a \in C \in \mathscr{V}$, then there exists some $j \in \mathbb{N}$ such that $a \in B_j \subset C$; hence $B_j \subset C_j$ and $a \in C_j$. Thus $\{C_1, C_2, ...\}$ is an open covering of A. QED

I.2.2 ***Theorem*** If $(X, \mathscr{T})$ is a Hausdorff space, then a limit is unique (i.e., $\lim_j x_j = x$ and $\lim_j x_j = y$ implies $x = y$), a compact subset of X is closed, and a sequentially compact subset of X is sequentially closed.

▌PROOF Let $\lim_j x_j = x$ and $\lim_j x_j = y$. Then every neighborhood of x intersects every neighborhood of y; hence $x = y$ since X is Hausdorff. Now let $A \subset X$ be compact and $y \in X \sim A$. For every $a \in A$ there exist $U_a, V_a \in \mathscr{T}$ such that $a \in U_a, y \in V_a$, and $U_a \cap V_a = \varnothing$. Since $\bigcup_{a \in A} U_a$ covers the compact set A, there exists a finite subcovering $U_{a_1}, ..., U_{a_k}$. Then $V^y \triangleq \bigcap_{i=1}^k V_{a_i} \in \mathscr{T}$ and $V^y \cap A = \varnothing$, showing that y is an interior point of $X \sim A$. Thus $X \sim A$ is open and A is closed.

If A is sequentially compact, then A contains the (unique) limit of every convergent sequence in A; hence A is sequentially closed. QED

I.2.3 ***Theorem*** Let X, Y, and Z be topological spaces and $f: X \to Y$ and $g: Y \to Z$ continuous. Then $g \circ f: X \to Z$ is continuous.

▌PROOF For every $x \in X$ and every open neighborhood A of $g(f(x))$, the set $g^{-1}(A)$ is open in Y and $(g \circ f)^{-1}(A) = f^{-1}(g^{-1}(A))$ is open in X. QED

I.2.4 ***Theorem*** If X is a metric space, $\lim_j x_j = \bar{x}$ and $\lim_{i,j} x_{i,j} = \bar{y}$ in X, then (x_j) is a Cauchy sequence and $(x_{i,j})$ a Cauchy double sequence. If X is a complete metric space and $(x_{i,j})$ is a Cauchy double sequence, then $(x_{i,j})$ has a unique limit.

▌PROOF Let X be metric, $\lim_j x_j = \bar{x}$ and $\lim_{i,j} x_{i,j} = \bar{y}$. Then for every $\epsilon > 0$ there exists $k(\epsilon) \in \mathbb{N}$ such that $d(x_i, \bar{x}) \leqslant \epsilon$ and $d(x_{i,j}, \bar{y}) \leqslant \epsilon$ if $i, j \geqslant k(\epsilon)$. Thus

$$d(x_{i_1}, x_{i_2}) \leqslant d(x_{i_1}, \bar{x}) + d(x_{i_2}, \bar{x}) \leqslant 2\epsilon \qquad \text{if} \quad i_1, i_2 \geqslant k(\epsilon)$$

and

$$d(x_{i_1,j_1}, x_{i_2,j_2}) \leqslant d(x_{i_1,j_1}, \bar{y}) + d(x_{i_2,j_2}, \bar{y}) \leqslant 2\epsilon \qquad \text{if} \ i_1, j_1, i_2, j_2 \geqslant k(\epsilon).$$

This shows that (x_j) is a Cauchy sequence and $(x_{i,j})$ is a Cauchy double sequence.

Now assume that X is a complete metric space and $(x_{i,j})$ is a Cauchy double sequence. Then for every $m \in \mathbb{N}$ there exists $k(m) \in \mathbb{N}$ such that $d(x_{i_1,j_1}, x_{i_2,j_2}) \leqslant 1/m$ if $i_1, j_1, i_2, j_2 \geqslant k(m)$. We set $y_m \triangleq x_{k(m),k(m)}$ $(m \in \mathbb{N})$ and observe that $d(y_m, y_n) \leqslant 1/m$ if $n \geqslant m$. Thus (y_m) is a Cauchy sequence in X with a limit $\bar{y}$ and, for all $i, j \geqslant k(m)$,

$$d(x_{i,j}, \bar{y}) \leqslant d(x_{i,j}, x_{k(m),k(m)}) + d(x_{k(m),k(m)}, \bar{y}) \underset{m}{\longrightarrow} 0,$$

showing that $\lim_{i,j} x_{i,j} = \bar{y}$. QED

I.2.5 ***Theorem*** If X is a metric space and $A \subset X$, then the following statements are equivalent:

(1) A is compact;
(2) A is sequentially compact; and
(3) A is complete and totally bounded.

▌PROOF We assume that A is an infinite set, the theorem being obvious otherwise.

Step 1 Let A be compact. Then, by I.2.2, A is closed. If a sequence (x_j) in A has no convergent subsequence, then every point in A has a neighborhood containing only finitely many distinct x_j. Since A is compact and covered by such neighborhoods, it is covered a finite number of them, showing that only a finite number of x_j are distinct. Thus $x_{j_i} = x_{j_1}$ for $i = 1, 2,...$ and some subsequence (j_i) of $(1, 2,...)$, contradicting the assumption that (x_j) has no convergent subsequence. Thus (1) implies (2).

Now let (x_j) be a Cauchy sequence in A. Then, as just shown, there exist $\bar{x} \in A$ and $J \subset (1, 2,...)$ such that $\lim_{j \in J} x_j = \bar{x}$. Since (x_j) is a Cauchy sequence, for every $\epsilon > 0$ there exists $j(\epsilon)$ such that $d(x_i, \bar{x}) \leqslant \epsilon/2$ and $d(x_i, x_j) \leqslant \epsilon/2$ for all $j \geqslant j(\epsilon)$ and $i \in J$, $i \geqslant j(\epsilon)$; hence $d(x_j, \bar{x}) \leqslant d(x_j, x_i) + d(x_i, \bar{x}) \leqslant \epsilon$. Thus $\lim_j x_j = \bar{x}$ and this shows that $\bar{A} = A$ is complete. We also observe that A is totally bounded. Indeed, for every $\epsilon > 0$ the family $\{S(x, \epsilon) \mid x \in A\}$ is an open covering of the compact set A and admits therefore a finite sub-covering. Thus (1) implies (3).

Step 2 Now assume that (2) is true. If $\mathrm{diam}(A) = \infty$, then there exists a sequence (x_j) in A such that $d(x_i, x_j) \geqslant 1$ for all $i < j \in \mathbb{N}$, showing that no subsequence of (x_j) can converge, contrary to (2). Thus $\mathrm{diam}(A) < \infty$.

Now let $y_1, y_2 \in A$. We can successively choose y_i $(i = 3, 4,...)$ as follows: We set $d_i \triangleq \sup_{y \in A} \inf_{1 \leqslant k \leqslant i} d(y, y_k)$ $(i = 2, 3,...)$. Then we can determine some $y_{i+1} \in A$ such that $d(y_{i+1}, y_k) \geqslant d_i/2$ for $k = 1, 2,..., i$. Since A is sequentially compact, there exists a subsequence (y_{i_k}) that is convergent; hence, by I.2.4, a Cauchy sequence. Therefore

$$\lim_k d_{i_{k+1}-1}/2 \leqslant \lim_k d(y_{i_k}, y_{i_{k+1}}) = 0.$$

Since $d_{j+1} \leqslant d_j$ for all j, it follows that $\lim_j d_j = 0$. If $y \in A$, then

$$\inf_{1 \leqslant k \leqslant i} d(y, y_k) \leqslant d_i \xrightarrow[i \to \infty]{} 0.$$

It follows that $\{y_1, y_2, ...\}$ is a dense subset of A. This implies, in turn, that the collection $\{S(y_j, r) \mid j \in \mathbb{N}, r \text{ a rational number}\}$ is an at most countable base of the relative metric topology of A.

Now let $\{B_\omega' \mid \omega \in \Omega\}$ be an open covering of A. By Lindelöf's theorem I.2.1, there exists a countable subcovering $\{B_1, B_2, \ldots\} \triangleq \{B'_{\omega_1}, B'_{\omega_2}, \ldots\}$. If $A \sim \bigcup_{j=1}^k B_j \neq \varnothing$ for all $k \in \mathbb{N}$, then there exist sequences J in $\mathbb{N}$ and $(x_j)_{j \in J}$ in A, with

$$x_j \in B_j \sim \bigcup_{i=1}^{j-1} B_i \qquad (j \in J). \tag{4}$$

Since A is sequentially compact, there exist $\bar{x} \in A$ and $J_1 \subset J$ such that

$$\lim_{j \in J_1} x_j = \bar{x} \in A \subset \bigcup_{j=1}^{\infty} B_j . \tag{5}$$

Thus $\bar{x} \in B_k$ for some $k \in \mathbb{N}$, contradicting (4) and (5). It follows that $A \subset \bigcup_{j=1}^k B_j$ for some k; hence A is compact. Thus (2) implies (1) and, by Step 1, (2) implies (3).

Step 3 Now assume (3) is valid, and let (x_j) be a sequence in A. For each $k \in \mathbb{N}$, A can be covered by a finite collection of open balls of radius $1/k$. Then some subsequence $(x_j)_{j \in J_1}$ of (x_j) is contained in some open ball of radius 1, some $(x_j)_{j \in J_2}$ is contained in some open ball of radius $\frac{1}{2}$, with $J_2 \subset J_1$, etc. Let J be the diagonal subsequence of $J_1, J_2, \ldots$. Then we verify that $d(x_i, x_j) \leqslant 2/i$ for $j > i$. Thus $(x_j)_{j \in J}$ is a Cauchy sequence in the complete set A. Therefore (3) implies (2), and this shows that statements (1)–(3) are equivalent. QED

I.2.6 ***Theorem*** If X_i $(i = 1, 2, \ldots, k)$ are metric spaces with metric functions $d_i(\cdot, \cdot)$ and $X \triangleq \prod_{i=1}^k X_i$, then the product topology in X is the metric topology defined by the metric function

$$(x, y) \to d(x, y) \triangleq \sum_{i=1}^{k} d_i(x^i, y^i) \qquad \text{for } x \triangleq (x^1, \ldots, x^k) \text{ and } y \triangleq (y^1, \ldots, y^k).$$

If each X_i is complete (respectively separable, respectively compact), then X is complete (respectively separable, respectively compact).

▌PROOF It is clear that $d(\cdot, \cdot)$ is a metric function for X. To prove that the product topology in X is defined by $d(\cdot, \cdot)$, we must show that the open balls $\{y \mid d(x, y) < \epsilon\}$ $(x \in X, \epsilon > 0)$ form a subbase of the product topology. Let $x \triangleq (x^1, \ldots, x^k) \in \prod_{i=1}^k A_i$ and A_i be open in X_i. Then

$$S_i(x^i, \epsilon_i) \triangleq \{y^i \in X_i \mid d_i(x^i, y^i) < \epsilon_i\} \subset A_i \qquad \text{for some } \epsilon_i > 0 \quad (i = 1, \ldots, k).$$

Let $\epsilon \triangleq \min_{1 \leqslant i \leqslant k} \epsilon_i$. Then $y \triangleq (y^1, \ldots, y^k) \in X$ and $d(x, y) < \epsilon$ imply $d_i(x^i, y^i) < \epsilon_i$ for each i. Thus

$$S(x, \epsilon) \subset A_1 \times \cdots \times A_k .$$

Conversely, let $\epsilon > 0$ and $x \triangleq (x^1,..., x^k) \in X$. Then $d(x, y) < \epsilon$ if $d_i(x^i, y^i) < \epsilon/k$; hence

$$\prod_{i=1}^{k} S_i(x^i, \epsilon/k) \subset S(x, \epsilon).$$

This shows that the product topology and the topology defined by $d(\cdot, \cdot)$ coincide.

Now let each X_i be complete, and let $(x_j) \triangleq ((x_j^1,..., x_j^k))_j$ be a Cauchy sequence in X. Then $(x_j^i)_j$ is a Cauchy sequence in X_i for each i, and has a limit $\bar{x}^i$. It is easy to verify that $\lim_j x_j = (\bar{x}^1,..., \bar{x}^k)$.

If X_i is separable for each i, let $\{x_j^i \mid j \in J_i\}$ be a dense subset, where J_i is either a finite sequence or $J_i \subset (1, 2,...)$. Then $\{(x_{j_1}^1, x_{j_2}^2,..., x_{j_k}^k) \mid j_i \in J_i$ $(i = 1, 2,..., k)\}$ is a dense at most denumerable subset of X.

Finally, let each X_i be compact. By I.2.5, X_i is then sequentially compact. Let $(x_j) = ((x_j^1,..., x_j^k))$ be a sequence in X. Then there exist sequences J_i $(i = 1, 2,..., k)$ such that $(1, 2,...) \supset J_1 \supset J_2 \supset \cdots \supset J_k$ and $(x_j^i)_{j \in J_i}$ is convergent. It follows that $(x_j)_{j \in J_k}$ is convergent. Thus X is sequentially compact and, by I.2.5, X is compact. QED

I.2.7 ***Theorem*** Let X and Y be topological spaces. If $f: X \to Y$ is continuous at $\bar{x}$, then f is sequentially continuous at $\bar{x}$. If X is a metric space, then the converse is also true.

▌PROOF Let f be continuous at $\bar{x}$ and $\bar{x} = \lim_j x_j$. Then for every neighborhood U of $f(\bar{x})$ there is a neighborhood V of $\bar{x}$ such that $f(x) \in U$ if $x \in V$. Since only a finite number of x_j can be outside V, it follows that all but a finite number of $f(x_j)$ are inside U. Thus $\lim_j f(x_j) = f(\bar{x})$.

Now let X be metric and $\lim_j f(x_j) = f(\bar{x})$ if $\lim_j x_j = \bar{x}$. Assume that there exists a neighborhood U of $f(\bar{x})$ such that for every $i \in \mathbb{N}$ there exists $\xi_i \in S(\bar{x}, 1/i)$, with $f(\xi_i) \notin U$. Then $\lim_i \xi_i = \bar{x}$; hence $\lim_i f(\xi_i) = f(\bar{x})$ contrary to the assumption that $f(\xi_i) \notin U$ for all i. This shows that $\lim_{x \to \bar{x}} f(x) = f(\bar{x})$ and f is continuous at $\bar{x}$. QED

I.2.8 ***Theorem*** Let $[a_0, a_1] \subset \mathbb{R}$. Then

(1) $[a_0, a_1]$ is compact and sequentially compact;
(2) a relatively open subset of $[a_0, a_1]$ is a finite or a denumerable union of disjoint open intervals and a subset of $\{a_0, a_1\}$; and
(3) $[a_0, a_1]$ is not the union of two nonempty relatively open disjoint subsets.

▌PROOF *Proof of* (1) For every $\epsilon > 0$, let $n(\epsilon)$ be the smallest integer greater than or equal $2(a_1 - a_0)/\epsilon$, and

$$b_j \triangleq a_0 + \frac{j}{n(\epsilon)}(a_1 - a_0) \qquad [j = 1, 2,..., n(\epsilon) - 1].$$

Then the relatively open sets $(b_j - \epsilon, b_j + \epsilon) \cap [a_0, a_1]$ $[j = 1, \ldots, n(\epsilon) - 1]$ cover $[a_0, a_1]$. This shows that $[a_0, a_1]$ is totally bounded. Since the sets $(-\infty, a_0)$ and (a_1, ∞) are open, it follows that $[a_0, a_1] = \mathbb{R} \sim (-\infty, a_0) \cup (a_1, \infty)$ is a closed subset of the complete space $\mathbb{R}$; hence complete. Thus, by I.2.5, $[a_0, a_1]$ is compact and sequentially compact. This proves (1).

Proof of (2) Let B be a relatively open subset of $[a_0, a_1]$ and $x \in B$. We set $r(x) \triangleq \sup\{y \in B \mid [x, y] \subset B\}$ and $l(x) \triangleq \inf\{y \in B \mid [y, x] \subset B\}$. Since B is bounded, both $r(x)$ and $l(x)$ are in $\mathbb{R}$. If $r(x) \in B \sim \{a_0, a_1\}$, then $\big(r(x) - \epsilon, r(x) + \epsilon\big) \subset B$ for some $\epsilon > 0$, contradicting the definition of $r(x)$. Thus $r(x) \notin B \sim \{a_0, a_1\}$ and the same argument shows that $l(x) \notin B \sim \{a_0, a_1\}$; however $\big(l(x), r(x)\big) \subset B$. Furthermore, it is easy to verify that if $x_1, x_2 \in B$, then the sets $\big(l(x_1), r(x_1)\big)$ and $\big(l(x_2), r(x_2)\big)$ are either the same or disjoint. Thus B is the union of disjoint open intervals and a subset of $\{a_0, a_1\}$. For each $n \in \mathbb{N}$, no more than n such open intervals of length at least $(a_1 - a_0)/n$ can be contained in B, showing that B is a finite or denumerable collection of disjoint open intervals and a subset of $\{a_0, a_1\}$.

Proof of (3) If B_1 and B_2 are two nonempty disjoint relatively open subsets of $[a_0, a_1]$, then, by (2), there exist nonempty subsets J_1 and J_2 of $\mathbb{N}$, $C \subset \{a_0, a_1\}$, and nonempty disjoint open intervals (a_i^1, b_i^1) $(i \in J_1)$ and (a_j^2, b_j^2) $(j \in J_2)$ such that

$$B_1 \cup B_2 = \bigcup_{i \in J_1} (a_i^1, b_i^1) \bigcup_{j \in J_2} (a_j^2, b_j^2) \bigcup C.$$

For any $i \in J_1$, we cannot have $a_i^1 = a_0$ and $b_i^1 = a_1$ since then $(a_i^1, b_i^1) \cap (a_j^2, b_j^2) \neq \varnothing$ for every $j \in J_2$, which cannot be. Thus one of the points a_i^1 or b_i^1 belongs to (a_0, a_1) but not to $B_1 \cup B_2$. QED

I.2.9 ***Theorem*** Let X and Y be topological spaces and $f: X \to Y$ continuous. Then

(1) $f(X)$ is compact (respectively sequentially compact) if X is compact (respectively sequentially compact);
(2) if X is either compact or sequentially compact and $Y = \mathbb{R}$, then $\inf f(X)$ and $\sup f(X)$ both belong to $f(X)$;
(3) if $X = [a_0, a_1] \subset \mathbb{R}$ and $Y = \mathbb{R}$, then $f([a_0, a_1])$ is a closed interval; and
(4) if X is compact Hausdorff and f a bijection, then f is a homeomorphism.

▌PROOF *Proof of* (1) Let X be compact and $\mathscr{V}$ an open covering of $f(X)$. Then $\{f^{-1}(A) \mid A \in \mathscr{V}\}$ is an open covering of X and has a finite subcovering $\mathscr{T}_1$. It follows that $\{f(B) \mid B \in \mathscr{T}_1\}$ is a finite subcovering of $\mathscr{V}$. Thus $f(X)$ is compact. If X is sequentially compact, then every sequence (x_j) in X has a subsequence $(x_j)_{j\in J}$ converging to some x; hence, by I.2.7, $\lim_{j\in J} f(x_j) = f(x)$, showing that $f(X)$ is sequentially compact.

Proof of (2) Let X be either compact or sequentially compact and $Y = \mathbb{R}$. Then, by (1) and I.2.5, $f(X)$ is sequentially compact. If $\lim_j f(x_j) = \sup f(X)$, then $\lim_{j\in J} x_j = \bar{x}$ for some $J \subset (1, 2,...)$ and $\bar{x} \in X$ and $f(\bar{x}) = \sup f(X)$. A similar argument shows that $\inf f(X) \in f(X)$.

Proof of (3) By (2), there exist $a', a'' \in [a_0, a_1]$ such that $f(a') \leqslant f(x) \leqslant f(a'')$ for all x. Now assume that $b \in [f(a'), f(a'')]$ but $f(x) \neq b$ for all x. Then $[a_0, a_1] = f^{-1}((b, f(a''))) \cup f^{-1}((f(a'), b))$. Since f is continuous, this shows that $[a_0, a_1]$ is a union of two nonempty disjoint relatively open subsets, contrary to I.2.8(3).

Proof of (4) If the assumptions of (4) are satisfied, then every closed $A \subset X$ is compact and, by (1), $f(A)$ is a compact subset of a Hausdorff space; hence closed. This shows that f^{-1} is continuous. QED

I.2.10 ***Theorem*** If X is a metric space with the metric function $d(\cdot, \cdot)$ and $A \subset X$, then the functions $(x, y) \to d(x, y) : X \times X \to \mathbb{R}$ and $x \to d[x, A] : X \to \mathbb{R}$ are continuous.

▌PROOF If $d(x, \xi) < \epsilon_1$ and $d(y, \eta) < \epsilon_2$, then

$$d(x, y) - d(x, \xi) - d(y, \eta) \leqslant d(\xi, \eta) \leqslant d(x, y) + d(x, \xi) + d(y, \eta);$$

hence

$$| d(\xi, \eta) - d(x, y)| < \epsilon_1 + \epsilon_2 .$$

Thus $(x, y) \to d(x, y)$ is continuous. If $d(x, \xi) < \epsilon$, then, for all $y \in A$,

$$d[x, A] \leqslant d(x, y) \leqslant d(x, \xi) + d(\xi, y);$$

hence

$$d[x, A] < \epsilon + d[\xi, A]$$

and, interchanging x and ξ,

$$d[\xi, A] < \epsilon + d[x, A].$$

Thus $| d[x, A] - d[\xi, A]| < \epsilon$ and $x \to d[x, A]$ is continuous. QED

I.2.11 ***Theorem*** If X is a metric space and $A \subset X$, then $x \in \bar{A}$ or $d[x, A] = 0$ if and only if there exists a sequence (x_j) in A converging to x.

▌PROOF If $x \in \bar{A}$, then $S(x, 1/j)$ contains a point $x_j \in A$ for $j = 1, 2,...$. We have $\lim_j x_j = x$. Conversely, if $\lim_j x_j = x$, then every neighborhood of x contains some $x_j \in A$; hence $x \in \bar{A}$. It follows that $d[x, A] = 0$ if and only if $x \in \bar{A}$. QED

I.2.12 ***Theorem*** Let X and Y be metric spaces and $f: X \to Y$. If $\epsilon \to \delta(\epsilon) : (0, \infty) \to (0, \infty)$ is such that $d(f(x_1), f(x_2)) \leqslant \epsilon$ whenever $d(x_1, y_1) \leqslant \delta(\epsilon)$, then the function

$$h \to \Omega(h) \triangleq \sup\{d(f(x_1), f(x_2)) \mid d(x_1, x_2) \leqslant h\} : (0, \infty) \to (0, \infty]$$

is a modulus of continuity of f. Conversely, if f has a modulus of continuity, then f is uniformly continuous.

▌PROOF Let $\epsilon \to \delta(\epsilon)$ be as described. Then $\Omega(h) \leqslant \epsilon$ if $h \leqslant \delta(\epsilon)$ and $\epsilon > 0$, and

$$d(f(x_1), f(x_2)) \leqslant \Omega(d(x_1, x_2)) \qquad (x_1, x_2 \in X).$$

Thus Ω is a modulus of continuity of f.

Conversely, let $\Omega(h) > 0$ for all $h > 0$, $\lim_{h \to 0} \Omega(h) = 0$, and $d(f(x_1), f(x_2)) \leqslant \Omega(d(x_1, x_2))$. Then for every $\epsilon > 0$ there exists $\delta(\epsilon) > 0$ such that $\Omega(h) \leqslant \epsilon$ if $h \leqslant \delta(\epsilon)$; hence

$$d(f(x_1), f(x_2)) \leqslant \Omega(d(x_1, x_2)) \leqslant \epsilon \qquad \text{if} \quad d(x_1, x_2) \leqslant \delta(\epsilon). \quad \text{QED}$$

I.2.13 ***Theorem*** Let X and Y be metric spaces, Y complete, $A \subset X$, and $f: A \to Y$ uniformly continuous. Then there exists a unique continuous extension $\hat{f}$ of f to $\bar{A}$ and $\hat{f}$ is uniformly continuous.

▌PROOF Let $x \in \bar{A}$. By I.2.11, there exists a sequence (x_j) in A converging to x. Since f is uniformly continuous, for every $\epsilon > 0$ there exists $\delta(\epsilon) > 0$ such that $d(f(x_i), f(x_j)) \leqslant \epsilon$ if $d(x_i, x_j) \leqslant \delta(\epsilon)$. Thus $(f(x_j))$ is a Cauchy sequence in the complete metric space Y and has a limit $\hat{f}(x)$. If $\lim_j \xi_j = x$, then

$$d(f(\xi_j), \hat{f}(x)) \leqslant d(f(\xi_j), f(x_j)) + d(f(x_j), \hat{f}(x)) \underset{j}{\to} 0$$

since f is uniformly continuous. Thus $\hat{f}(x) = \lim_j f(x_j)$ for every sequence (x_j) converging to x; hence, by I.2.7, $\hat{f} : \bar{A} \to Y$ is continuous.

Now let $\epsilon > 0$, $x, \xi \in \bar{A}$, and $d(x, \xi) \leqslant \delta(\epsilon)/2$. Then, by I.2.11, there exist sequences (x_j) and (ξ_j) in A, with $\lim_j x_j = x$ and $\lim_j \xi_j = \xi$. By I.2.10, $d(x_j, \xi_j) \leqslant d(x, \xi) + \delta(\epsilon)/2 \leqslant \delta(\epsilon)$ for all sufficiently large j, say for $j \geqslant j_0$; hence $d(f(x_j), f(\xi_j)) \leqslant \epsilon$ for $j \geqslant j_0$ and therefore

$$d(\hat{f}(x), \hat{f}(\xi)) = \lim_j d(f(x_j), f(\xi_j)) \leqslant \epsilon.$$

Thus $\hat{f}$ is uniformly continuous. QED

I.2.14 ***Lemma*** Let X be a compact metric space and $\omega : X \times X \to \mathbb{R}$ such that $\omega(x, y) \leqslant \omega(x, z) + \omega(y, z)$ and $\lim_j \omega(x_j, x) = 0$ if $\lim_j x_j = x$.

Then there exists a function $\Omega : (0, \infty) \to (0, \infty]$ such that $\omega(x, y) \leqslant \Omega(d(x, y))$ and $\lim \Omega(h) = 0$ as $h \to 0$, $h > 0$.

PROOF Since $\omega(x, x) = 0$, we have $\omega(y, z) \geqslant 0$ for all $y, z \in X$. Next we show that for every $\epsilon > 0$ there exists $\delta(\epsilon) > 0$ such that $\omega(x, y) \leqslant \epsilon$ if $d(x, y) \leqslant \delta(\epsilon)$. Assume the contrary. Then there exist $\epsilon > 0$ and sequences (x_j) and (y_j) in X such that $\lim_j d(x_j, y_j) = 0$ and $\omega(x_j, y_j) > \epsilon$. By I.2.5, there exist convergent subsequences (x_{j_i}) and (y_{j_i}) which have a common limit $\bar{x}$ since $\lim_i d(x_{j_i}, y_{j_i}) = 0$. Thus

$$\epsilon < \omega(x_{j_i}, y_{j_i}) \leqslant \omega(x_{j_i}, \bar{x}) + \omega(y_{j_i}, \bar{x}) \to 0 \qquad \text{as} \quad i \to \infty,$$

which is a contradiction.

We now set $\Omega(h) \triangleq \sup\{\omega(x, y) \mid d(x, y) \leqslant h\}$ and deduce that $\omega(x, y) \leqslant \Omega(d(x, y))$ and $\Omega(h) \leqslant \epsilon$ if $h \leqslant \delta(\epsilon)$ and $\epsilon > 0$; hence $\lim \Omega(h) = 0$ as $h \to 0$, $h > 0$. QED

I.2.15 ***Theorem*** Let X be a topological space, Y a metric space, and $f : X \to Y$ continuous.

(1) If X is a compact metric space, then f is uniformly continuous.

(2) If W is a topological space, Z a compact metric space, $X = Z \times W$ with the product topology, and $\lim_i w_i = \bar{w}$ in W, then $\lim_i f(z, w_i) = f(z, \bar{w})$ uniformly for all $z \in Z$.

PROOF If X is a compact metric space, let $\omega(x, x') \triangleq d(f(x), f(x'))$. Then, by I.2.14, there exists a modulus of continuity of f; hence, by I.2.12, f is uniformly continuous. Thus proves (1).

Now let the conditions of (2) be satisfied and assume that there exist $\epsilon > 0$, $J \subset (1, 2, \ldots)$ and a sequence $(z_i)_{i \in J}$ in Z such that

$$d(f(z_i, w_i), f(z_i, \bar{w})) > \epsilon \qquad (i \in J). \tag{3}$$

Then, by I.2.5, there exist $J_1 \subset J$ and $\bar{z} \in Z$ such that $\lim_{i \in J_1} z_i = \bar{z}$ and we have

$$f(\bar{z}, \bar{w}) = \lim_{i \in J_1} f(z_i, \bar{w}) = \lim_{i \in J_1} f(z_i, w_i);$$

hence

$$\lim_{i \in J_1} d(f(z_i, w_i), f(z_i, \bar{w})) = 0,$$

contrary to (3). Thus $\lim_i f(z, w_i) = f(z, \bar{w})$ uniformly for $z \in Z$. QED

I.2.16 ***Theorem*** Let X and Y be metric spaces, $f_j : X \to Y$ ($j \in \mathbb{N}$) have a common modulus of continuity Ω, $f : X \to Y$, and $\lim_j f_j(x) = f(x)$ for each $x \in X$. Then Ω is a modulus of continuity of f.

▌PROOF Let $\epsilon > 0$ and $x_1, x_2 \in X$. Then, for some $k \in \mathbb{N}$, $d(f_k(x_i), f(x_i)) \leqslant \epsilon/2$ for $i = 1, 2$; hence

$$\begin{aligned} d(f(x_1), f(x_2)) &\leqslant d(f(x_1), f_k(x_1)) + d(f_k(x_1), f_k(x_2)) + d(f_k(x_2), f(x_2)) \\ &\leqslant \epsilon + \Omega(d(x_1, x_2)). \end{aligned}$$

Since ϵ is arbitrary, the conclusion follows. QED

I.2.17 ***Theorem*** If X is a compact metric space, then X is separable. If X is a separable metric space, then X has an at most countable base of topology and every nonempty subset A of X has an at most denumerable dense subset.

▌PROOF Let X be a compact metric space. Then, by I.2.5, X is totally bounded. Thus, for every $i \in \mathbb{N}$, there exist $k(i) \in \mathbb{N}$ and points $x_1^i, \ldots, x_{k(i)}^i$ such that the open balls $S(x_j^i, 1/i)$ $[j = 1, 2, \ldots, k(i)]$ cover X. It is now clear that the set $X_\infty \triangleq \{x_j^i \mid i \in \mathbb{N}, j \in \{1, 2, \ldots, k(i)\}\}$ is at most denumerable and dense in X.

Now let X be a separable metric space and $\{x_1, x_2, \ldots\}$ a dense subset. Then the collection $\{S(x_j, 1/i) \mid i, j \in \mathbb{N}\}$ is an at most countable base of topology for X. If $A \subset X$ and $A \neq \varnothing$, then the collection $\mathscr{T} \triangleq \{S(x_j, 1/i) \cap A \mid i, j \in \mathbb{N}\}$ is a base of the relative topology of A. If we choose a point a_T in every nonempty set $T \in \mathscr{T}$, then $\{a_T \mid T \neq \varnothing, T \in \mathscr{T}\}$ is a dense at most denumerable subset of A. QED

I.2.18 ***Theorem*** (Arzela) Let X be a separable metric space, Y a compact metric space, and K a collection of continuous functions $f : X \to Y$ with a common modulus of continuity Ω. Then for every sequence (f_i) in K there exist a uniformly continuous function $f_0 : X \to Y$ and a subsequence $(f_i)_{i \in J}$ of (f_i) such that $\lim_{i \in J} f_i(x) = f_0(x)$ $(x \in X)$ and Ω is a modulus of continuity for f_0.

▌PROOF We shall assume that X is infinite, the modifications being clear if X is finite. Let $\{x_1, x_2, \ldots\}$ be a dense subset of X. By I.2.5, the sequence $(f_i(x_1))$ in Y has a subsequence $(f_i(x_1))_{i \in J_1}$ converging to some $f_0(x_1) \in Y$. Similarly, $(f_i(x_2))_{i \in J_1}$ has a subsequence $(f_i(x_2))_{i \in J_2}$ converging to some $f_0(x_2)$. Continuing is this manner, we construct a sequence of sequences $((f_i)_{i \in J_1}, (f_i)_{i \in J_2}, \ldots)$ and their diagonal subsequence $(f_i)_{i \in J}$. We verify that $(f_i(x_k))_{i \in J}$ has a limit $f_0(x_k)$ for each $k \in \mathbb{N}$. It follows, by I.2.16, that $f_0 : \{x_1, x_2, \ldots\} \to Y$ has the modulus of continuity Ω and, by I.2.12 and

I.2.13, f_0 has a unique uniformly continuous extension to X (which we shall also designate as f_0). Now let $(x_j)_{j \in K}$ converge to x in X, and let $\epsilon > 0$. Then, for some $k \in K$ and all $j \in \{0, 1, 2, \ldots\}$, $d(f_j(x), f_j(x_k)) \leqslant \epsilon/3$ and, for some $i_0 \in \mathbb{N}$, $d(f_0(x_k), f_i(x_k)) \leqslant \epsilon/3$ for all $i \geqslant i_0$; hence

$$d(f_0(x), f_i(x)) \leqslant d(f_0(x), f_0(x_k)) + d(f_0(x_k), f_i(x_k)) + d(f_i(x_k), f_i(x)) \leqslant \epsilon \qquad \text{for all} \quad i \geqslant i_0 .$$

Since ϵ is arbitrary, we have $\lim_i f_i(x) = f_0(x)$. By I.2.16, Ω is a modulus of continuity for f_0. QED

I.2.19 ***Theorem*** Let X be a topological space, Y a complete metric space, and $f_j : X \to Y$ ($j \in \mathbb{N}$) such that $\lim_{i,j} \sup_{x \in X} d(f_i(x), f_j(x)) = 0$. Then there exists $f : X \to Y$ such that $\lim_j f_j(x) = f(x)$ uniformly for all $x \in X$. If f_j are continuous, then f is also continuous.

▌PROOF For each $x' \in X$, $(f_j(x'))$ is a Cauchy sequence in the complete space Y and it converges, therefore, to some $f(x')$. Now let $a_{ij} \triangleq \sup_{x \in X} d(f_i(x), f_j(x))$ and $\epsilon > 0$, and let $n \in \mathbb{N}$ be such that $a_{ij} \leqslant \epsilon$ for $i, j \geqslant n$. We have

$$\begin{aligned} d(f_j(x'), f(x')) &\leqslant a_{ij} + d(f_i(x'), f(x_i)) \\ &\leqslant \epsilon + d(f_i(x'), f(x')) \qquad \text{for} \quad i, j \geqslant n \quad \text{and all} \quad x' \in X. \end{aligned}$$

Now let j be fixed and $i \to \infty$. Then $d(f_j(x'), f(x')) \leqslant \epsilon$ for all $x' \in X$, showing that $\lim_j f_j(x) = f(x)$ uniformly for all $x \in X$. Now assume that f_j are continuous for all $j \in \mathbb{N}$, let $a > 0$ and $x' \in X$. Then there exists $j_0 \in \mathbb{N}$ such that $d(f_j(x), f(x)) < a/3$ for $j \geqslant j_0$ and all x. Since f_{j_0} is continuous, $A \triangleq f_{j_0}^{-1}(S(f_{j_0}(x'), a/3))$ is open. For each $x \in A$ we have

$$d(f(x), f(x')) \leqslant d(f(x), f_{j_0}(x)) + d(f_{j_0}(x), f_{j_0}(x')) + d(f_{j_0}(x'), f(x')) < a.$$

Thus $f^{-1}(S(f(x'), a))$ contains the open set A, from which it follows that f is continuous at x'. QED

I.3 Topological Vector Spaces

Vector Spaces

A *real vector space* $\mathscr{X}$ (or a *real linear space*), which we shall henceforth refer to as a *vector space*, is a set for which are defined functions $a(\cdot, \cdot) : \mathscr{X} \times \mathscr{X} \to \mathscr{X}$, $b(\cdot, \cdot) : \mathbb{R} \times \mathscr{X} \to \mathscr{X}$, and $c(\cdot) : \mathscr{X} \to \mathscr{X}$, written

$a(x, y) \triangleq x + y$, $b(\alpha, x) \triangleq \alpha x$, $c(x) \triangleq -x$, and with an element 0 (called the *origin*) satisfying the following conditions

$$x + y = y + x, \qquad (x + y) + z = x + (y + z),$$
$$x + (-x) = 0, \qquad x + 0 = x,$$
$$\alpha(x + y) = \alpha x + \alpha y, \qquad (\alpha + \beta)x = \alpha x + \beta x,$$
$$\alpha(\beta x) = (\alpha\beta)x, \qquad \text{and} \qquad 1 \cdot x = x.$$

The functions $a(\cdot, \cdot)$, $b(\cdot, \cdot)$, and $c(\cdot)$ are said to define the *algebraic structure* of $\mathscr{X}$. The element $(\cdots(((\alpha_1 x_1 + \alpha_2 x_2) + \alpha_3 x_3) + \cdots) + \alpha_k x_k)$ is written $\alpha_1 x_1 + \alpha_2 x_2 + \cdots + \alpha_k x_k$ or $\sum_{j=1}^k \alpha_j x_j$ and called a *linear combination* of $x_1, \ldots, x_k$. We verify that

$$\alpha_1 x_1 + \cdots + \alpha_k x_k = \alpha_{i_1} x_{i_1} + \cdots + \alpha_{i_k} x_{i_k}$$

for every *permutation* $(i_1, \ldots, i_k)$ of $(1, 2, \ldots, k)$ (that is, for every bijective mapping $j \to i_j : \{1, 2, \ldots, k\} \to \{1, 2, \ldots, k\}$), and that $0x = 0$. A point $x \in \mathscr{X}$ is a *linear combination of elements of* $A \subset \mathscr{X}$ if $x = \sum_{j=1}^k \alpha_j x_j$ for some $k \in \mathbb{N}$, $\alpha_j \in \mathbb{R}$, and $x_j \in A$. We observe that if $\sum_{j=1}^n \alpha_j x_j = 0$ and $\alpha_1 \neq 0$, then $x_1 = -\sum_{j=2}^n (\alpha_j/\alpha_1)\, x_j$. We write

$$x - y \triangleq x + (-y), \qquad A + B \triangleq \{a + b \mid a \in A, b \in B\},$$
$$\alpha A \triangleq \{\alpha a \mid a \in A\}, \qquad -A \triangleq (-1)A,$$
$$A - B \triangleq A + (-B), \qquad \text{and} \qquad A + x \triangleq A + \{x\}.$$

INDEPENDENT SETS, DIMENSION

A set A in a vector space $\mathscr{X}$ is *independent* (or *linearly independent*) if $k \in \mathbb{N}$, $\alpha_j \in \mathbb{R}$, $x_j \in A$ $(j = 1, \ldots, k)$, and $\sum_{j=1}^k \alpha_j x_j = 0$ imply $\alpha_j = 0$ $(j = 1, \ldots, k)$. A set A is *dependent* (or *linearly dependent*) if it is not independent. A subset $\mathscr{Y}$ of a vector space $\mathscr{X}$ is a *vector subspace* (or a *linear subspace*) of $\mathscr{X}$ if $0 \in \mathscr{Y}$ and the functions $(x, y) \to x + y$ and $(\alpha, x) \to \alpha x$ map $\mathscr{Y} \times \mathscr{Y}$ respectively $\mathbb{R} \times \mathscr{Y}$ into $\mathscr{Y}$. Then the set $\mathscr{Y}$ and the restrictions to $\mathscr{Y} \times \mathscr{Y}$, $\mathscr{Y}$ and $\mathbb{R} \times \mathscr{Y}$, respectively, of the functions $(x, y) \to x + y$, $x \to -x$, and $(\alpha, y) \to \alpha y$ define a vector space. The *span* of a set A in a vector space $\mathscr{X}$, written $\mathrm{sp}(A)$, is the intersection of all the vector subspaces of $\mathscr{X}$ containing A.

An independent set A in a vector space $\mathscr{X}$ is a *Hamel basis* for $\mathscr{X}$ if every $x \in \mathscr{X}$ is a linear combination of elements of A. A vector space $\mathscr{X}$ whose only element is 0 is said to be of dimension 0. We say that a vector space $\mathscr{X}$ is *finite-dimensional* if either $\mathscr{X} = \{0\}$ or $\mathscr{X}$ has a finite Hamel basis; otherwise we say that $\mathscr{X}$ is *infinite-dimensional.* By Theorem I.3.1 below, every Hamel basis in a finite-dimensional vector space has the same number of elements, called the *dimension* of $\mathscr{X}$ and written dim $\mathscr{X}$.

Vector Space of Functions and Multiplication of Functions

If S is any set, $\mathscr{X}$ a vector space, $\alpha \in \mathbb{R}$, $f : S \to \mathscr{X}$, and $g : S \to \mathscr{X}$, we define the functions $f + g : S \to \mathscr{X}$, $\alpha f : S \to \mathscr{X}$, $0 : S \to \mathscr{X}$, and $-f : S \to \mathscr{X}$ by the relations

$$(f + g)(x) \triangleq f(s) + g(s), \qquad (\alpha f)(s) \triangleq \alpha f(s),$$
$$0(s) \triangleq 0, \qquad \text{and} \qquad (-f)(s) \triangleq -f(s).$$

This defines the set of all functions on S to $\mathscr{X}$ as a vector space. If $h : S \to \mathbb{R}$ and $f : S \to \mathscr{X}$, we write hf for the function $s \to h(s)f(s) : S \to \mathscr{X}$.

Linear Operators

If $\mathscr{X}$ and $\mathscr{Y}$ are vector spaces, a function $T : \mathscr{X} \to \mathscr{Y}$ is a *linear operator* on $\mathscr{X}$ to $\mathscr{Y}$ (is *linear*) if $T(x + y) = T(x) + T(y)$ and $T(\alpha x) = \alpha T(x)$. We also write Tx for $T(x)$. A linear operator on $\mathscr{X}$ to $\mathbb{R}$ is a *linear functional on* $\mathscr{X}$. If T is a linear operator on $\mathscr{X}$ to $\mathscr{X}$, we say that T is a *linear operator in* $\mathscr{X}$. We verify that the set $\mathscr{L}(\mathscr{X}, \mathscr{Y})$ of all linear operators on $\mathscr{X}$ to $\mathscr{Y}$ is a vector space.

Isomorphism

Two vector spaces $\mathscr{X}$ and $\mathscr{Y}$ are *isomorphic* (*algebraically isomorphic*) if there exists a bijective $T \in \mathscr{L}(\mathscr{X}, \mathscr{Y})$. Then T is an (*algebraic*) *isomorphism* of $\mathscr{X}$ onto $\mathscr{Y}$.

Product Vector Spaces

If $\mathscr{X}_i$ $(i = 1,..., k)$ are vector spaces, we define $\mathscr{X} \triangleq \prod_{j=1}^{k} \mathscr{X}_j$ as a vector space by setting

$$(x_1 ,..., x_k) + (y_1 ,..., y_k) \triangleq (x_1 + y_1 ,..., x_k + y_k),$$
$$\alpha(x_1 ,..., x_k) \triangleq (\alpha x_1 ,..., \alpha x_k),$$
$$0 \triangleq (0,..., 0), \qquad \text{and} \qquad -(x_1 ,..., x_k) \triangleq (-x_1 ,..., -x_k).$$

Topological and Normed Vector Spaces, Banach Spaces

A *real topological vector space* $\mathscr{X}$ (which we shall henceforth refer to as a *topological vector space*) is a vector space with a Hausdorff topology such that the functions $(x, y) \to x + y : \mathscr{X} \times \mathscr{X} \to \mathscr{X}$ and $(\alpha, x) \to \alpha x : \mathbb{R} \times \mathscr{X} \to \mathscr{X}$

are continuous. A function $x \to |x| : \mathscr{X} \to \mathbb{R}$ on a real vector space $\mathscr{X}$ is a *seminorm* if $|x| \geqslant 0$, $|\alpha x| = |\alpha| |x|$ and $|x + y| \leqslant |x| + |y|$. A seminorm $|\cdot|$ is a *norm* if $|x| = 0$ implies $x = 0$. A (*real*) *normed vector space* $(\mathscr{X}, |\cdot|)$ is a vector space $\mathscr{X}$ with a norm $|\cdot|$ and the metric topology defined by the metric function $d(\cdot, \cdot)$, with $d(x, y) \triangleq |x - y|$. This metric topology is called the *norm topology* of $(\mathscr{X}, |\cdot|)$. If the norm $|\cdot|$ is specified, we may refer to $\mathscr{X}$ as a normed vector space. Two normed vector spaces $(\mathscr{X}, |\cdot|)$ and $(\mathscr{Y}, \|\cdot\|)$ are *isometrically isomorphic* if there exists an isomorphism T on $\mathscr{X}$ to $\mathscr{Y}$ such that $\| Tx \| = |x|$. A *Banach space* is a complete normed vector space. It follows easily from the properties of a norm that a normed vector space is a topological vector space. If $\mathscr{X}$ is a normed vector space, we call $S(0, 1) \triangleq \{x \in \mathscr{X} \mid |x| < 1\}$ the *open unit ball* in $\mathscr{X}$ and $S^F(0, 1) \triangleq \{x \in \mathscr{X} \mid |x| \leqslant 1\}$ the *closed unit* ball in $\mathscr{X}$.

$o(g(x))$

If X is any set, $\bar{x} \in X$, $\mathscr{Y}$ a topological vector space, $f : X \to \mathscr{Y}$ and $g : X \to \mathbb{R}$, we say that $f(x) = o(g(x))$ $(x \to \bar{x})$ if $\lim[1/g(x)]f(x) = 0$ as $x \to \bar{x}$, $x \in X \sim \{\bar{x}\}$. We write $f(x) = o(g(x))$ if $\bar{x}$ has been specified.

Product Spaces

We can verify that if $\mathscr{X}_i$ $(i = 1,..., k)$ are topological vector (normed, Banach) spaces, then the vector space $\mathscr{X} \triangleq \prod_{i=1}^k \mathscr{X}_i$ with the product topology (with the norm defined by $|(x_1, ..., x_k)| \triangleq |x_1| + \cdots + |x_k|$) is a topological vector (normed, Banach) space.

The Space $\mathbb{R}^k$

We verify that, with the definition of sum and product in $\mathbb{R}$ as that of real numbers, $\mathbb{R}$ is a vector space, and that the absolute value is a norm on $\mathbb{R}$ that yields the previously defined metric topology for $\mathbb{R}$. Since $\mathbb{R}$ is complete, it is a Banach space. We denote by $\mathbb{R}^k$ the Banach space $\mathbb{R} \times \cdots \times \mathbb{R}$ (k times). Thus, for $x \triangleq (x^1, ..., x^k) \in \mathbb{R}^k$, $|x| \triangleq |x^1| + \cdots + |x^k|$. It is often customary to use a different norm, called *euclidean*, on $\mathbb{R}^k$. This norm $|\cdot|_2$ is defined by $|x|_2 \triangleq |(x^1, ..., x^k)|_2 \triangleq [\sum_{i=1}^k (x^j)^2]^{1/2}$. However, the metric topologies defined by the two norms $|\cdot|$ and $|\cdot|_2$ can be easily shown to coincide.

We denote by $\mathbb{R}^0$ the vector space containing the single element 0.

Series

If $\mathscr{X}$ is a topological vector space and (x_j) a sequence in $\mathscr{X}$, we denote by $\sum_{j=1}^\infty x_j$ and call *series with terms* x_j *in* $\mathscr{X}$ the sequence $(\sum_{j=1}^i x_j)_i$. If

$K \triangleq (k_1, k_2, \ldots)$ is a sequence in some set A and $(x_{k_j})_j$ a sequence in $\mathscr{X}$, we write $\sum_{k\in K} x_k$ for $\sum_{j=1}^{\infty} x_{k_j}$. We write $\sum_{j=\alpha}^{\infty} x_j$ for $\sum_{i=1}^{\infty} x_{\alpha-1+i}$ if α is an integer and $\sum_{S(j)} x_j$ for $\sum_{j\in J} x_j$, where J is the increasing sequence of integers j for which $S(j)$ is true. The series $\sum_{j=1}^{\infty} x_j$ is *convergent* (*converges to* $\bar{x}$) if $(\sum_{j=1}^{i} x_j)_i$ is convergent (converges to $\bar{x}$) and, if this is the case, we write $\sum_{j=1}^{\infty} x_j = \bar{x}$.

In a Banach space $\mathscr{X}$, the series $\sum_{j=1}^{\infty} x_j$ is *absolutely convergent* if the series $\sum_{j=1}^{\infty} |x_j|$ in $\mathbb{R}$ is convergent. If $a_j \geqslant 0$ and $|x_j| \leqslant a_j$ $(j \in \mathbb{N})$, then the series $\sum_{j=1}^{\infty} x_j$ is *dominated* by $\sum_{j=1}^{\infty} a_j$.

If $b_j \in \bar{\mathbb{R}}$ and the sequence $(\sum_{j=1}^{i} b_j)_i$ is defined in $\bar{\mathbb{R}}$ and converges to some $\bar{b} \in \bar{\mathbb{R}}$, we say that the series $\sum_{j=1}^{\infty} b_j$ converges to $\bar{b}$ in $\bar{\mathbb{R}}$ and $\sum_{j=1}^{\infty} b_j = \bar{b}$ in $\bar{\mathbb{R}}$. We easily verify that if $\sum_{j=1}^{\infty} b_j$ converges to $\bar{b}$ in $\bar{\mathbb{R}}$ and $\bar{b} \in \mathbb{R}$ then the sequence $(\sum_{j=1}^{i} b_j)_i$ is in $\mathbb{R}$ and converges to $\bar{b}$ in $\mathbb{R}$.

If $\mathscr{X}$ is a Banach space and $x_{i,j} \in \mathscr{X}$ $(i, j \in \mathbb{N})$, then we denote by $\sum_{i,j=1}^{\infty} x_{i,j}$ and call *series with terms* $x_{i,j}$ in $\mathscr{X}$ the function

$$(\alpha, \beta) \to X_{\alpha,\beta} \triangleq \sum_{i=1}^{\alpha} \sum_{j=1}^{\beta} x_{i,j} : \mathbb{N} \times \mathbb{N} \to \mathscr{X}.$$

This series is *convergent* and *converges to* $\bar{x}$ if $\lim_{\alpha,\beta} X_{\alpha,\beta} = \bar{x}$ and, if this is the case, we write $\sum_{i,j=1}^{\infty} x_{i,j} = \bar{x}$. The series $\sum_{i,j=1}^{\infty} x_{i,j}$ is *absolutely convergent* if the series $\sum_{i,j=1}^{\infty} |x_{i,j}|$ in $\mathbb{R}$ is convergent. If $a_{i,j} \geqslant 0$ $(i, j \in \mathbb{N})$ and $|x_{i,j}| \leqslant a_{i,j}$, then the series $\sum_{i,j=1}^{\infty} x_{i,j}$ is *dominated* by $\sum_{i,j=1}^{\infty} a_{i,j}$. If $X_i \triangleq \sum_{j=1}^{\infty} x_{i,j}$ exists for all $i \in \mathbb{N}$, then we write $\sum_{i=1}^{\infty} \sum_{j=1}^{\infty} x_{i,j}$ for $\sum_{i=1}^{\infty} X_i$, and we similarly define $\sum_{j=1}^{\infty} \sum_{i=1}^{\infty} x_{i,j}$.

If $\mathscr{X}$ is a topological vector space and $x_j \in \mathscr{X}$ for all integers j, we write $\sum_{j=-\infty}^{\infty} x_j$ for both the double sequence $(\sum_{i=0}^{\alpha} x_i + \sum_{j=1}^{\beta} x_{-j})_{\alpha,\beta}$ and its limit (if it exists).

Continuous Linear Operators

In a topological vector space $\mathscr{X}$, a subset A is *bounded* if for every neighborhood G of 0 in $\mathscr{X}$ there exists $\epsilon > 0$ such that $\alpha A \subset G$ for all $\alpha \in [-\epsilon, \epsilon]$. If $\mathscr{X}$ and $\mathscr{Y}$ are topological vector spaces, then it is easy to see that $T \in \mathscr{L}(\mathscr{X}, \mathscr{Y})$ is continuous if and only if T is continuous at 0. By I.2.3, the collection $B(\mathscr{X}, \mathscr{Y})$ of all *continuous linear operators* on $\mathscr{X}$ to $\mathscr{Y}$ is a vector subspace of $\mathscr{L}(\mathscr{X}, \mathscr{Y})$. An element $T \in B(\mathscr{X}, \mathscr{Y})$ is a *compact operator* (or a *completely continuous operator*) if $\overline{T(X)}$ is compact whenever X is bounded. If $T_1, T_2 \in B(\mathscr{X}, \mathscr{X})$, the composition $T_1 \circ T_2$, written $T_1 T_2$ or $T_1 \cdot T_2$, clearly belongs to $\mathscr{L}(\mathscr{X}, \mathscr{X})$; hence, by I.2.3, it belongs to $B(\mathscr{X}, \mathscr{X})$. We write T^n for $T \cdot T \cdot \cdots \cdot T$ (n times). If $\mathscr{X}$ and $\mathscr{Y}$ are topological vector spaces and there exists a linear operator $T : \mathscr{X} \to \mathscr{Y}$ that is also a homeomorphism, we say that T is an *isomorphism* on $\mathscr{X}$ to $\mathscr{Y}$ and that $\mathscr{X}$ and $\mathscr{Y}$ are *isomorphic*.

Dual Space

We denote by $\mathscr{X}^*$, and call the *conjugate space* or the *topological dual* of $\mathscr{X}$, the space $B(\mathscr{X}, \mathbb{R})$. By Theorem I.3.6 below, if $\mathscr{X}$ is a normed vector space, then the function $T \to |T| \triangleq \sup_{|x| \leqslant 1} |Tx|$ is a norm in $\mathscr{X}^*$ which we refer to as *the norm* or the *strong norm* in $\mathscr{X}^*$ (to distinguish it from a *weak norm* $|\cdot|_w$ that we introduce in I.3.11 below); the corresponding metric topology is the (*strong*) *norm topology* of $\mathscr{X}^*$. The sets

$$\{T \in \mathscr{X}^* \mid |(T - T_1)x_j| < \epsilon \ (j = 1,..., k)\}$$

for $k \in \mathbb{N}$, $x_1,..., x_k \in \mathscr{X}$, $T_1 \in \mathscr{X}^*$ and $\epsilon > 0$ form a base of the *weak star topology* of $\mathscr{X}^*$ (also called the $\mathscr{X}$ *topology of* $\mathscr{X}^*$).

Conjugate of a Cartesian Product

If $\mathscr{X} \triangleq \prod_{i=1}^n \mathscr{X}_i$ and $\mathscr{X}_i$ are topological vector spaces, then an element l of $\mathscr{X}^*$ has the property that, for $x \triangleq (x^1,..., x^n) \in \mathscr{X}$,

$$l(x) = l(x^1, 0,..., 0) + l(0, x^2, 0,...) + \cdots + l(0,..., 0, 0, x^n).$$

If we denote by l_i the mapping $x^i \to l(0,..., 0, x^i, 0,..., 0) : \mathscr{X}_i \to \mathbb{R}$, then $l(x) = \sum_{j=1}^n l_j(x^j)$ and $l_i \in \mathscr{X}_i^*$. Conversely if $l_i \in \mathscr{X}_i^*$ and l is defined by $l(x^1,..., x^n) \triangleq \sum_{j=1}^n l_j(x^j)$, then $l \in \mathscr{X}^*$. Thus every element l of $\mathscr{X}^*$ corresponds uniquely to a n-tuple of elements $(l_1,..., l_n) \in \prod_{j=1}^n \mathscr{X}_j^*$ and this correspondence is clearly an algebraic isomorphism. If $\mathscr{X}_i$ are normed vector spaces, then

$$|l| \triangleq \sup \left\{ |l(x)| \;\middle|\; \sum_{j=1}^n |x^j| \leqslant 1 \right\} \leqslant \sum_{j=1}^n |l_j|$$

and

$$\sum_{j=1}^n |l_j| \triangleq \sum_{j=1}^n \sup\{|l_j(x^j)| \;\big|\; |x^j| \leqslant 1\} \leqslant \sum_{j=1}^n \sup \left\{ \sum_{k=1}^n |l_k(x^k)| \;\middle|\; \sum_{k=1}^n |x^k| \leqslant 1 \right\}$$
$$= n\,|l|.$$

Thus the isomorphism of $(\prod_{i=1}^n \mathscr{X}_i)^*$ and $\prod_{i=1}^n \mathscr{X}_i^*$ mentioned above is a homeomorphism if we choose the strong norm topologies for both sets. In particular, if $\mathscr{X}_i = \mathbb{R}$ for $i = 1,..., n$, then $(\mathbb{R}^n)^*$ and $(\mathbb{R}^*)^n$ are homeomorphic under this isomorphism. We observe that every $l \in \mathbb{R}^*$ corresponds uniquely to an element $\lambda \in \mathbb{R}$ such that $l(x) = \lambda x$, and conversely. Every element l of $(\mathbb{R}^n)^*$ is thus in a one-to-one correspondence with an element $\lambda \triangleq (\lambda^1,..., \lambda^n) \in \mathbb{R}^n$ and we have $l(x) = \sum_{j=1}^n \lambda^j x^j$ for each $x \triangleq (x^1,..., x^n) \in \mathbb{R}^n$. We write $\lambda \cdot x$ for $\sum_{j=1}^n \lambda^j x^j$ and identify l and λ.

Uniformly Continuous Functions

If X is any set, $\mathscr{Y}$ a topological vector space, $f: X \to \mathscr{Y}$, and $f_j : X \to \mathscr{Y}$ ($j \in \mathbb{N}$), then we say that $\lim_j f_j(\cdot) = f(\cdot)$ *uniformly*, or $\lim_j f_j = f$ *uniformly*, or $\lim_j f_j(x) = f(x)$ *uniformly* for $x \in X$ if for every neighborhood V of 0 in $\mathscr{Y}$ there exists an integer $j(V)$ such that $f_j(x) - f(x) \in V$ for all $x \in X$ and $j \geqslant j(V)$. If X is a metric space and $\mathscr{Y}$ a topological vector space, then a continuous function $f: X \to \mathscr{Y}$ is *uniformly continuous* if for every neighborhood V of 0 in $\mathscr{Y}$ there exists $\delta(V) > 0$ such that $f(x_1) - f(x_2) \in V$ if $d(x_1, x_2) \leqslant \delta(V)$.

Matrices

Let $n, m \in \mathbb{N}$ and $M \in \mathscr{L}(\mathbb{R}^n, \mathbb{R}^m)$. By I.3.3 below, there exists a unique array $[M] \triangleq (\alpha_{ij})$ $(i = 1,..., m; j = 1,..., n)$, with $\alpha_{i,j} \in \mathbb{R}$, such that $Mx = \sum_{j=1}^n x^j \alpha_j$ for all $x \triangleq (x^1,..., x^n) \in \mathbb{R}^n$, where $\alpha_j \triangleq (\alpha_{1,j},..., \alpha_{m,j})$ $(j = 1,..., n)$. We refer to the array $[M] \triangleq (\alpha_{i,j})$ as the *matrix* of M, to α_j as the jth *column* of $[M]$, to the point $\alpha^i \triangleq (\alpha_{i,1},..., \alpha_{i,n})$ as the ith *row* of $[M]$ (for $i = 1,..., m$), and to the elements $\alpha_{i,j}$ of the matrix as the *coefficients* of $[M]$. If a matrix has n columns and m rows, we refer to it as an $m \times n$ matrix.

It is easy to verify that if $M \in \mathscr{L}(\mathbb{R}^n, \mathbb{R}^m)$ and $N \in \mathscr{L}(\mathbb{R}^l, \mathbb{R}^n)$, then $M \circ N$, written MN, belongs to $\mathscr{L}(\mathbb{R}^l, \mathbb{R}^m)$. If $[M] \triangleq (\alpha_{i,j})$, $[N] \triangleq (\beta_{i,j})$ are the matrices of M, N, respectively, then we verify that $[M \circ N]$, written $[M] \cdot [N]$, is the matrix $(\gamma_{i,j})$ $(i = 1,..., m; j = 1,..., l)$, with $\gamma_{i,j} \triangleq \sum_{k=1}^n \alpha_{i,k}\beta_{k,j}$. If $M, N \in \mathscr{L}(\mathbb{R}^n, \mathbb{R}^m)$, $[M] \triangleq (\alpha_{i,j})$, and $[N] \triangleq (\beta_{i,j})$, then $[M + N]$, written $[M] + [N]$, is the matrix $(\alpha_{i,j} + \beta_{i,j})$.

By I.3.3 below, if $M \in \mathscr{L}(\mathbb{R}^n, \mathbb{R}^m)$, then M can have an inverse only if $m = n$. If $m = n$ and M^{-1} exists, we write $[M]^{-1}$ for $[M^{-1}]$ and observe that $[M]^{-1}[M] = [M]\,[M]^{-1} = [I]$, where $[I]$, called the *unit* $n \times n$ *matrix*, is the matrix of the identity operator in $\mathbb{R}^n$ and $[I] = (\delta_{i,j})$ $(i, j = 1,..., n)$, with $\delta_{i,j} \triangleq 0$ for $i \neq j$ and $\delta_{i,i} \triangleq 1$. We refer to a matrix with an equal number of rows and columns as a *square matrix*.

Since the composition of three functions f, g, h, if it is defined, is associative [that is, $(f \circ g) \circ h = f \circ (g \circ h)$], it follows that $([M] \cdot [N]) \cdot [L] = [M] \cdot ([N] \cdot [L])$ if $M \cdot N \cdot L$ is defined.

It is convenient to treat each row of a matrix (α_{ij}) $(i = 1,..., m; j = 1,..., n)$ as a $1 \times n$ matrix and each column of this matrix as an $m \times 1$ matrix. If $x \triangleq (x^1,..., x^n) \in \mathbb{R}^n$, we also denote by x the $n \times 1$ *column matrix* with $x^1,..., x^n$ as coefficients and by x^T the $1 \times n$ *row matrix* with $x^1,..., x^n$ as coefficients. It follows that if $M \in \mathscr{L}(\mathbb{R}^n, \mathbb{R}^m)$ and $x \in \mathbb{R}^n$, then $Mx = [M]x$. When there is no danger of confusion, we identify a 1×1 matrix with its single coefficient; then $x \cdot y = x^T y = y^T x$. If $A = (\alpha_{i,j})$ $(i = 1,..., m;$

$j = 1,\ldots, n$), we denote by A^T the matrix whose columns are the rows of A; that is, $A^T \triangleq (\gamma_{i,j})$ ($i = 1,\ldots, n; j = 1,\ldots, m$), with $\gamma_{i,j} = \alpha_{j,i}$.

As we often do in other cases where two spaces are isomorphic, we identify an element M of $\mathscr{L}(\mathbb{R}^n, \mathbb{R}^m)$ with its matrix $[M]$ and thus consider $\mathscr{L}(\mathbb{R}^n, \mathbb{R}^m)$ as consisting of matrices. In general, this will cause no confusion, and when necessary we shall specify whether we refer to an array or a linear operator.

The vector space $\mathscr{L}(\mathbb{R}^n, \mathbb{R}^m)$ has the "conventional" (strong) norm $| M | \triangleq \sup_{|x| \leqslant 1} | Mx |$. We shall find it more convenient on occasions to use a different norm $M \to \| M \|$, defined by $\| M \| \triangleq \sum_{i=1}^m \sum_{j=1}^n | \alpha_{i,j} |$, where $[M] \triangleq (\alpha_{i,j})$ is the matrix of M. It is easy to verify that $\| \cdot \|$ is a norm on $\mathscr{L}(\mathbb{R}^n, \mathbb{R}^m)$ and that $| Mx | \leqslant \| M \| \cdot | x |$ for all $x \in \mathbb{R}^n$. Thus $| M | \leqslant \| M \|$.

If $A : S \to \mathscr{L}(\mathbb{R}^n, \mathbb{R}^m)$, we refer to the function $s \to [A(s)]$ as the *matrix of* A and to the function $s \to \alpha_{i,j}(s)$ as the i, jth *coefficient* of A, where $[A(s)] \triangleq (\alpha_{i,j}(s))$ ($i = 1,\ldots, m; j = 1,\ldots, n$) for all $s \in S$.

I.3.1 *Theorem* Every Hamel basis in a finite-dimensional vector space $\mathscr{X}$ has the same number of elements, called dimension of $\mathscr{X}$, and written dim $\mathscr{X}$. If $B \triangleq \{y_1, \ldots, y_n\} \subset \mathscr{X}$ and $n > \dim \mathscr{X}$, then B is dependent.

▌PROOF Let $A \triangleq \{x_1, \ldots, x_k\}$ be a Hamel basis for $\mathscr{X}$. We shall prove that every set $B \triangleq \{y_1, \ldots, y_n\}$, with $n > k$, is dependent. Indeed, assume the contrary. Then we shall show that every set of relations of the form

$$(1.i) \qquad y_i + \sum_{j=i+1}^{n} \alpha_{i,j} y_j + \sum_{l=1}^{p} \beta_{i,l} x_{i_l} = 0 \qquad (i = 1,\ldots, q),$$

where $n \geqslant q > p \geqslant 1$, implies a similar set with p and q reduced by 1. Indeed, $\beta_{q,\delta} \neq 0$ for some δ since B is independent. Thus relation (1.q) shows that x_{i_δ} is a linear combination of $y_q, y_{q+1}, \ldots, y_n$ and x_{i_l} ($l \neq \delta$). Substituting this linear combination for x_{i_δ} in the relations (1.i) for $i < q$ yields $q - 1$ relations similar to (1.i) and involving $p - 1$ elements of A.

Since A is a Hamel basis, we have

$$y_i - \sum_{l=1}^{k} \beta_{i,l} x_l = 0 \qquad (i = 1,\ldots, n),$$

and our previous argument shows that we can successively eliminate all the x_l, and obtain relations of the form

$$\sum_{j=1}^{n} \alpha_{i,j} y_j = 0 \qquad (i = 1,\ldots, n - k),$$

where $\alpha_{i,i} = 1$. This implies that B is dependent, contrary to assumption.

Thus no Hamel basis, as an independent set, can have more elements than any finite Hamel basis. QED

I.3.2 ***Theorem*** Let $\mathscr{X}$ be a vector space.

(1) If $k \in \mathbb{N}$ and $\mathscr{X} = \text{sp}(\{x_1, ..., x_k\})$, then

$$\mathscr{X} = \left\{ \sum_{j=1}^{k} \alpha_j x_j \mid \alpha_j \in \mathbb{R} \ (j = 1, ..., k) \right\}$$

and $\dim \mathscr{X} \leqslant k$;
(2) if $\dim \mathscr{X} = n \in \mathbb{N}$, then every independent subset $\{x_1, ..., x_n\}$ of $\mathscr{X}$ is a Hamel basis;
(3) if $n \in \{0, 1, 2, ...\}$, then $\dim \mathbb{R}^n = n$; and
(4) if $n \in \mathbb{N}$ and $\{x_1, ..., x_n\}$ is a Hamel basis in $\mathscr{X}$, then for every $x \in \mathscr{X}$ there exists a unique $\alpha \triangleq (\alpha^1, ..., \alpha^n) \in \mathbb{R}^n$ such that $x = \sum_{j=1}^{n} \alpha^j x_j$.

▌PROOF *Proof of* (1) Let $k \in \mathbb{N}$ and $\mathscr{X} \triangleq \text{sp}(\{x_1, ..., x_k\})$. Then it is easy to verify that $\mathscr{X} = \{\sum_{j=1}^{k} \alpha_j x_j \mid \alpha_j \in \mathbb{R} \ (j = 1, ..., k)\}$. Now let $\{y_1, ..., y_l\}$ be the subset of $\{x_1, ..., x_k\}$ obtained by keeping only those x_i that are not 0 and such that $x_i \notin \text{sp}(\{x_1, ..., x_{i-1}\})$ $(i = 1, 2, ..., k)$. If $x_i = 0$ for all i, then $\mathscr{X} = \{0\}$ and $\dim \mathscr{X} = 0 < k$. Otherwise we have $1 \leqslant l \leqslant k$, $\text{sp}(\{y_1, ..., y_l\}) = \mathscr{X}$, and $\{y_1, ..., y_l\}$ is independent. Thus $\{y_1, ..., y_l\}$ is a Hamel basis and, by I.3.1, $\dim \mathscr{X} = l \leqslant k$.

Proof of (2) Let $\dim \mathscr{X} = n \in \mathbb{N}$, $\{x_1, ..., x_n\}$ be an independent subset of $\mathscr{X}$, and $x \in \mathscr{X}$. Then, by I.3.1, the set $\{x_1, ..., x_n, x\}$ is dependent and there exist $\alpha_1, ..., \alpha_{n+1} \in \mathbb{R}$, not all 0, and such that $\alpha_1 x_1 + \cdots + \alpha_n x_n + \alpha_{n+1} x = 0$. We must have $\alpha_{n+1} \neq 0$ (since $\{x_1, ..., x_n\}$ is independent) and therefore $x = \sum_{j=1}^{n} (-\alpha_j/\alpha_{n+1}) x_j$, showing that $\{x_1, ..., x_n\}$ is a Hamel basis.

Proof of (3) Let $n \in \mathbb{N}$ and $\delta_j \triangleq (\delta_j^{\,1}, ..., \delta_j^{\,n}) \in \mathbb{R}^n$ $(j = 1, 2, ..., n)$, where $\delta_j^{\,i} \triangleq 0$ if $i \neq j$, and $\delta_i^{\,i} \triangleq 1$. Then $\sum_{j=1}^{n} \alpha_j \delta_j = 0$ implies $\alpha_j = 0$ $(j = 1, ..., n)$, showing that $\{\delta_1, ..., \delta_n\}$ is independent. If $x \triangleq (x^1, ..., x^n) \in \mathbb{R}^n$, then $x = \sum_{j=1}^{n} x^j \delta_j$, showing that $\{\delta_1, ..., \delta_n\}$ is a Hamel basis and, by I.3.1, $\dim \mathbb{R}^n = n$. If $n = 0$, then $\dim \mathbb{R}^n = 0$ by definition.

Proof of (4) Let $n \in \mathbb{N}$, $\{x_1, ..., x_n\}$ be a Hamel basis in $\mathscr{X}$, and $x = \sum_{j=1}^{n} \alpha^j x_j = \sum_{j=1}^{n} \beta^j x_j$. Then $\sum_{j=1}^{n} (\alpha^j - \beta^j) x_j = 0$ and $\alpha^j - \beta^j = 0$.
QED

I.3.3 ***Theorem*** Let $n, m \in \mathbb{N}$. Then $\mathscr{L}(\mathbb{R}^n, \mathbb{R}^m) = B(\mathbb{R}^n, \mathbb{R}^m)$ and there exists a one-to-one correspondence between $\mathscr{L}(\mathbb{R}^n, \mathbb{R}^m)$ and the collection of all arrays $(\alpha_{i,j})$ $(i = 1, ..., m;\ j = 1, ..., n)$ (*matrices*) with $\alpha_{i,j} \in \mathbb{R}$. If $(\alpha_{i,j})$ corresponds to $T \in \mathscr{L}(\mathbb{R}^n, \mathbb{R}^m)$ and $\alpha_j \triangleq (\alpha_{1j}, ..., \alpha_{m,j})$ $(j = 1, ..., n)$, then $Tx = \sum_{j=1}^{n} x^j \alpha_j$ for all $x \triangleq (x^1, ..., x^n) \in \mathbb{R}^n$ and T has an inverse if and only if $m = n$ and the set $\{\alpha_1, ..., \alpha_n\}$ in $\mathbb{R}^m$ is independent, and then $T^{-1} \in \mathscr{L}(\mathbb{R}^n, \mathbb{R}^n)$.

▮ PROOF Let $D \triangleq \{\delta_1, ..., \delta_n\}$ and $E \triangleq \{\epsilon_1, ..., \epsilon_m\}$, with $\delta_j \triangleq (\delta_j^1, ..., \delta_j^n)$, $\epsilon_k \triangleq (\epsilon_k^1, ..., \epsilon_k^m)$, $\delta_j^j \triangleq 1$, $\epsilon_k^k \triangleq 1$, $\delta_j^i \triangleq 0$ for $i \neq j$, and $\epsilon_k^l \triangleq 0$ for $l \neq k$. Then D and E are Hamel bases for $\mathbb{R}^n$ and $\mathbb{R}^m$, respectively. Let $T \in \mathscr{L}(\mathbb{R}^n, \mathbb{R}^m)$ and $\alpha_j \triangleq (\alpha_{1,j}, ..., \alpha_{m,j}) \triangleq T\delta_j$. Then, for $x = \sum_{j=1}^n x^j \delta_j \in \mathbb{R}^n$, $Tx = \sum_{j=1}^n x^j \alpha_j$. Conversely, if $\alpha_1, ..., \alpha_n \in \mathbb{R}^m$, then a unique $T \in \mathscr{L}(\mathbb{R}^n, \mathbb{R}^m)$ is defined by

$$Tx \triangleq \sum_{j=1}^{n} x^j \alpha_j$$

for all $x \triangleq (x^1, ..., x^n) \in \mathbb{R}^n$, and we have $T\delta_j = \alpha_j$. If we set

$$c \triangleq \operatorname*{Max}_{1 \leqslant j \leqslant n} |\alpha_j|,$$

then $|Tx| \leqslant c \sum_{j=1}^n |x^j| = c|x|$. Thus, whenever $\lim_i x_i = \bar{x}$ in $\mathbb{R}^n$, we have $\lim_i |Tx_i - T\bar{x}| \leqslant c \lim_i |x_i - \bar{x}| = 0$; hence $\lim_i Tx_i = T\bar{x}$. It follows, by I.2.7, that T is continuous.

We next show that if T^{-1} exists, then $m = n$ and $\{\alpha_1, ..., \alpha_n\}$ is independent. Indeed, if T is one-to-one, then $Tx = \sum_{j=1}^n x^j \alpha_j = 0$ implies that $x = 0$; hence the set $\{\alpha_1, ..., \alpha_n\}$ is independent. If T is onto, then for every $y \in \mathbb{R}^m$ there exists $x \in \mathbb{R}^n$ such that $Tx = \sum_{j=1}^n x^j \alpha_j = y$; hence $\{\alpha_1, ..., \alpha_n\}$ is a Hamel basis for $\mathbb{R}^m$ and, by I.3.2(3), $n = m$.

Conversely, let $m = n$ and $\{\alpha_1, ..., \alpha_n\}$ be independent. Then, by I.3.2(2), $\{\alpha_1, ..., \alpha_n\}$ is a Hamel basis for $\mathbb{R}^n$ and, by I.3.2(4), T has an inverse T^{-1}. If $\beta_1, \beta_2 \in \mathbb{R}$ and $y_1, y_2 \in \mathbb{R}^n$, then

$$T(\beta_1 T^{-1}(y_1) + \beta_2 T^{-1}(y_2)) = \beta_1 TT^{-1}(y_1) + \beta_2 TT^{-1}(y_2) = \beta_1 y_1 + \beta_2 y_2,$$

showing that $T^{-1} \in \mathscr{L}(\mathbb{R}^n, \mathbb{R}^n)$. QED

I.3.4 ***Theorem*** Let $\mathscr{X}$ be a topological vector space and $\{x_1, ..., x_n\}$ a Hamel basis of $\mathscr{X}$. Then there exists a linear homeomorphism T on $\mathbb{R}^n$ to $\mathscr{X}$ defined by

$$T((\alpha^1, ..., \alpha^n)) \triangleq \sum_{j=1}^{n} \alpha^j x_j \qquad \text{for} \quad (\alpha^1, ..., \alpha^n) \in \mathbb{R}^n.$$

▮ PROOF Let $\{x_1, ..., x_n\}$ be a Hamel basis in $\mathscr{X}$ and $T : \mathbb{R}^n \to \mathscr{X}$ be defined as above. By I.3.2(4), T is bijective and T is clearly linear and continuous. As in I.3.3, T^{-1} is linear. It remains to show that T^{-1} is continuous.

Let $\epsilon \in (0, 1)$ and

$$D \triangleq \partial S(0, \epsilon) = \left\{ \alpha \triangleq (\alpha^1, ..., \alpha^n) \in \mathbb{R}^n \;\middle|\; |\alpha| \triangleq \sum_{j=1}^{n} |\alpha^j| = \epsilon \right\}.$$

Since, by I.2.8, $[-1, 1]$ is a compact subset of $\mathbb{R}$, $D \subset [-1, 1]^n$ and D is closed, it follows from I.2.6 that D is compact and, by I.2.2 and I.2.9, that

$T(D)$ is closed. Since $T^{-1}(0) = 0 \notin D$, $T(D)$ does not contain 0 and therefore $G \triangleq \mathscr{X} \sim T(D)$ is an open neighborhood of 0. Since $(\alpha, x) \to \alpha x : \mathbb{R} \times \mathscr{X} \to \mathscr{X}$ is continuous, there exist $\delta > 0$ and an open neighborhood U of 0 in $\mathscr{X}$ such that

$$\text{(1)} \qquad \beta U \subset G \quad \text{if} \quad |\beta| \leqslant \delta.$$

We now show that if $x \triangleq \sum_{j=1}^{n} \alpha^j x_j \in \delta U$, then $|\alpha| \triangleq |(\alpha^1,..., \alpha^n)| \triangleq \sum_{j=1}^{n} |\alpha^j| < \epsilon$. Indeed, assume the contrary. Then

$$0 < \gamma \triangleq \epsilon \Big/ \sum_{j=1}^{n} |\alpha^j| \leqslant 1.$$

Thus $\gamma x \in T(D)$ and, in view of (1), $\gamma x \in G$, which cannot be since $T(D) \cap G = \varnothing$.

Thus $T^{-1}(\delta U) \subset S(0, \epsilon)$. If $\lim_j y_j = 0$ in $\mathscr{X}$, then $y_j \in \delta U$ for sufficiently large j; hence $|T^{-1}(y_j)| < \epsilon$. Since $\epsilon > 0$ is arbitrary, this shows that T^{-1} is continuous at 0 and, being linear, T^{-1} is continuous. QED

I.3.5 ***Theorem*** Let $\mathscr{X}$ be a topological vector space. Then there exists a compact neighborhood of 0 in $\mathscr{X}$ if and only if $\dim \mathscr{X} < \infty$.

▌PROOF If $\mathscr{X} = \{0\}$, then our conclusion is trivially true. If $\dim \mathscr{X} = n \in \mathbb{N}$, then, by I.3.4, there exists a linear homeomorphism T of $\mathbb{R}^n$ to $\mathscr{X}$. Since $(-1, 1)$ and $[-1, 1]$ are, respectively, open and compact in $\mathbb{R}$, the sets $(-1, 1)^n$ and $[-1, 1]^n$ are, respectively, open and compact in $\mathbb{R}^n$ and their images under T are, respectively, open and compact in $\mathscr{X}$. Thus the set $T([-1, 1]^n)$ is a compact neighborhood of 0 in $\mathscr{X}$.

Now assume that there exists a compact neighborhood U of 0 in $\mathscr{X}$, and let V be any neighborhood of 0. The collection $\{x + \frac{1}{2}V \mid x \in U\}$ of neighborhoods covers U and, because U is compact, a finite subcollection $x_1 + \frac{1}{2}V,..., x_k + \frac{1}{2}V$ also covers U. Thus, for every $y \in U$, we have

$$\text{(1)} \qquad y = x_i + \tfrac{1}{2}z \quad \text{for some } i \in \{1, 2,..., k\} \text{ and } z \in V.$$

Now let W be any neighborhood of 0 in $\mathscr{X}$. Then there exist an open neighborhood V of 0 and $\delta > 0$ such that $\alpha_1 V + \alpha_2 V \subset W$ if $|\alpha_i| \leqslant \delta$. We can then determine $\beta \in (0, \delta]$ such that $\gamma x_i \in \delta V$ for $i = 1,..., k$ and $|\gamma| \leqslant \beta$. Relation (1) then shows that $\gamma U \subset W$ if $|\gamma| \leqslant \beta$.

Now consider relation (1) for $V = U$. We can apply this relation to z instead of y and substitute m times, obtaining

$$\text{(2)} \qquad y \in x_{i_1} + (1/2)\, x_{i_2} + \cdots + (1/2^m)\, x_{i_m} + (1/2^{m+1})U.$$

We observe that, for each $i \in \{1, 2,..., k\}$ and $J(i, \alpha) \triangleq \{j \in \{1, 2,..., \alpha\} \mid i_j = i\}$, the sequence $(\sum_{j \in J(i,\alpha)} 1/2^j)_\alpha$, being nondecreasing and bounded, converges to some a^i. Since for every neighborhood W of 0, $2^{-m-1}U \subset W$ for all sufficiently large m, we conclude from (2) that $y = \sum_{i=1}^k a^i x_i$. Thus every $y \in U$ is a linear combination of $x_1, ..., x_k$. Since $\lim_j (1/j)x = 0$ $(x \in \mathscr{X})$, to every $x \in \mathscr{X}$ there corresponds some $\beta \in \mathbb{R}$ such that $\beta x \in U$. We conclude that $\mathscr{X}$ is the span of $\{x_1, ..., x_k\}$ and, by I.3.2(1), $\dim \mathscr{X} \leqslant k$. QED

I.3.6 ***Theorem*** Let $\mathscr{X}$ and $\mathscr{Y}$ be normed vector spaces and $T \in \mathscr{L}(\mathscr{X}, \mathscr{Y})$. Then $T \in B(\mathscr{X}, \mathscr{Y})$ if and only if $|T| \triangleq \sup_{|x| \leqslant 1} |Tx| < \infty$, and $|\cdot|$ is a norm on $B(\mathscr{X}, \mathscr{Y})$. If $\mathscr{Y}$ is a Banach space, then $(B(\mathscr{X}, \mathscr{Y}), |\cdot|)$ is a Banach space.

PROOF Let $\mathscr{X}$ and $\mathscr{Y}$ be normed vector spaces, $T \in B(\mathscr{X}, \mathscr{Y})$, and (x_j) a sequence in $\mathscr{X}$ such that $|x_j| \leqslant 1$ and $\lim_j |Tx_j| = \infty$. If $y_j \triangleq |Tx_j|^{-1} x_j$ $(j \in \mathbb{N})$, then $\lim_j |y_j| = 0$ and, by I.2.7, $\lim_j |Ty_j| = |T \cdot 0| = 0$ which cannot be since $|Ty_j| = 1$. Thus $|T| \triangleq \sup_{|x| \leqslant 1} |Tx| < \infty$ for every $T \in B(\mathscr{X}, \mathscr{Y})$. Conversely, let $T \in \mathscr{L}(\mathscr{X}, \mathscr{Y})$ and $\sup_{|x| \leqslant 1} |Tx| = c < \infty$. If $\lim_j x_j = x$ in $\mathscr{X}$ and $x_j \neq x$, then

$$|Tx - Tx_j| = |x - x_j| \, |T(|x - x_j|^{-1}(x - x_j))| \leqslant c\,|x - x_j| \underset{j}{\to} 0;$$

hence $\lim_j Tx_j = Tx$ and, by I.2.7, T is continuous. Thus $B(\mathscr{X}, \mathscr{Y}) = \{T \in \mathscr{L}(\mathscr{X}, \mathscr{Y}) \mid |T| < \infty\}$, and we easily verify that $|\cdot|$ is a norm on $B(\mathscr{X}, \mathscr{Y})$.

Now assume that $\mathscr{Y}$ is a Banach space and let (T_j) be a Cauchy sequence in $(B(\mathscr{X}, \mathscr{Y}), |\cdot|)$. Then, for every $x \in \mathscr{X}$,

$$\lim_{i,j} |T_i x - T_j x| \leqslant \lim_{i,j} |T_i - T_j| \, |x| = 0;$$

hence $(T_j x)_j$ is a Cauchy sequence in the complete space $\mathscr{Y}$ with a limit $T(x)$. We verify that the function $x \to T(x)$ is a linear operator on $\mathscr{X}$ to $\mathscr{Y}$. We have, for every x with $|x| \leqslant 1$ and $i \leqslant j$, $|Tx - T_i x| \leqslant |Tx - T_j x| + |T_j - T_i|$. Since for any $\epsilon > 0$ we may choose i_0 sufficiently large so that $|T_j - T_i| \leqslant \epsilon$ for $j \geqslant i \geqslant i_0$, we have $|Tx - T_i x| \leqslant |Tx - T_j x| + \epsilon$ for $j \geqslant i \geqslant i_0$ and $|x| \leqslant 1$ and, since $\lim_j |Tx - T_j x| = 0$, we have $|Tx - T_i x| \leqslant \epsilon$ for $i \geqslant i_0$ and $|x| \leqslant 1$. This shows that

$$\sup_{|x| \leqslant 1} |Tx| \leqslant \sup_{|x| \leqslant 1} |T_{i_0} x| + \epsilon = |T_{i_0}| + \epsilon < \infty$$

and $\lim_i |T - T_i| = 0$, thus proving that $(B(\mathscr{X}, \mathscr{Y}), |\cdot|)$ is complete. QED

I.3.7 ***Theorem*** Let $\mathscr{X}$ be a Banach space and $T_1, T_2 \in B(\mathscr{X}, \mathscr{X})$. Then $T_1 T_2 \in B(\mathscr{X}, \mathscr{X})$ and $|T_1 T_2| \leqslant |T_1| \cdot |T_2|$. If $T, T_0, T_0^{-1} \in B(\mathscr{X}, \mathscr{X})$ and $|T - T_0| < |T_0^{-1}|^{-1}$, then T is bijective and $T^{-1} \in B(\mathscr{X}, \mathscr{X})$.

▌PROOF We have

$$|T_1T_2| = \sup_{|x|\leqslant 1} |T_1(T_2x)| \leqslant |T_1| \cdot \sup_{|x|\leqslant 1} |T_2x| = |T_1| \cdot |T_2|.$$

Thus, by I.3.6, $T_1 \cdot T_2 \in B(\mathscr{X}, \mathscr{X})$. Now let $I \triangleq I_{\mathscr{X}}$ be the identity operator in $\mathscr{X}$. We shall next show that $(I - K)^{-1} \in B(\mathscr{X}, \mathscr{X})$ for all $K \in B(\mathscr{X}, \mathscr{X})$ such that $|K| < 1$. Let $R_j \triangleq I + K + \cdots + K^j$ $(j \in \mathbb{N})$. Then

$$|R_i - R_j| = \left| \sum_{m=j+1}^{i} K^m \right| \leqslant \sum_{m=j+1}^{i} |K|^m$$

$$\leqslant |K|^{j+1} (1 - |K|)^{-1} \to 0 \qquad \text{as} \quad i, j \to \infty, \quad i \geqslant j.$$

Thus (R_j) is a Cauchy sequence in $B(\mathscr{X}, \mathscr{X})$ and, by I.3.6, has a limit $R \in B(\mathscr{X}, \mathscr{X})$. We have

$$\lim_j |R_j \cdot (I - K) - R \cdot (I - K)| \leqslant \lim_j |R_j - R| (1 + |K|) = 0$$

and

$$\lim_j |I - R_j(I - K)| = \lim_j |K^{j+1}| \leqslant \lim_j |K|^{j+1} = 0;$$

hence $\lim_j R_j \cdot (I - K) = R(I - K) = I$. We similarly show that $(I - K) \cdot R = I$. Thus $(I - K)^{-1} = R \in B(\mathscr{X}, \mathscr{X})$.

Now let $T, T_0, T_0^{-1} \in B(\mathscr{X}, \mathscr{X})$, $|T - T_0| < |T_0^{-1}|^{-1}$ and $K \triangleq -T_0^{-1}(T - T_0)$. Then $|K| < 1$ and $T_0^{-1}T = I - K$ has an inverse in $B(\mathscr{X}, \mathscr{X})$. Since $T_0^{-1} \in B(\mathscr{X}, \mathscr{X})$, it follows that $T = T_0(T_0^{-1}T)$ has an inverse $T^{-1} = (T_0^{-1}T)^{-1} T_0^{-1} \in B(\mathscr{X}, \mathscr{X})$. QED

I.3.8 ***Theorem*** (Hahn–Banach) Let $\mathscr{X}$ be a vector space, $\mathscr{Y}$ a subspace of $\mathscr{X}$, $p : \mathscr{X} \to \mathbb{R}$, and $f \in \mathscr{L}(\mathscr{Y}, \mathbb{R})$. If

$$p(x_1 + x_2) \leqslant p(x_1) + p(x_2) \quad \text{and} \quad p(\alpha x) = \alpha p(x) \quad (x, x_1, x_2 \in \mathscr{X}, \ \alpha \geqslant 0)$$

and

$$f(y) \leqslant p(y) \qquad (y \in \mathscr{Y}),$$

then there exists $\tilde{f} \in \mathscr{L}(\mathscr{X}, \mathbb{R})$ such that $f = \tilde{f} \mid \mathscr{Y}$ and $\tilde{f}(x) \leqslant p(x)$ for all $x \in \mathscr{X}$.

In particular, if $\mathscr{X}$ is a normed vector space, then every $f \in \mathscr{Y}^*$ can be extended to some $\tilde{f} \in \mathscr{X}^*$ in such a manner that $|\tilde{f}| = |f|$.

▌PROOF Let $\mathscr{F}$ be the set of all linear functionals g, each of which is defined on some subspace $\mathscr{X}_g$ of $\mathscr{X}$ containing $\mathscr{Y}$ and such that $g \mid \mathscr{Y} = f$ and $g(x) \leqslant p(x)$ $(x \in \mathscr{X}_g)$. Since $f \in \mathscr{F}$, $\mathscr{F}$ is nonempty. Let $(\mathscr{F}, \leqslant)$ be a partially ordered set, with $g_1 \leqslant g_2$ meaning that g_1 is a restriction of g_2.

If $\mathscr{G}$ is a totally ordered subset of $\mathscr{F}$, let $\mathscr{X}(\mathscr{G})$ be the union of the subspaces of $\mathscr{X}$ on which the elements of $\mathscr{G}$ are defined. Then for every $x \in \mathscr{X}(\mathscr{G})$ there exists some linear functional g whose domain contains x, and $g(x) = g'(x)$ for every other g' in $\mathscr{G}$ defined on x. Setting $\tilde{g}(x) \triangleq g(x)$ for every $x \in \mathscr{X}(\mathscr{G})$, we define a real-valued function $\tilde{g}$ on $\mathscr{X}(\mathscr{G})$ which must clearly be an element of $\mathscr{F}$. Thus $\mathscr{G}$ has an upper bound. It follows, by Zorn's lemma, that $\mathscr{F}$ has a maximal element h.

Now assume that there exists a point x_1 in $\mathscr{X}$ but not in the domain $\mathscr{X}_0$ of h, and let $\mathscr{X}_1 \triangleq \text{sp}(\mathscr{X}_0 \cup \{x_1\})$. If $x \in \mathscr{X}_1$, then $x = y + \alpha x_1$, with $y \in \mathscr{X}_0$, and we verify that α is unique.

We now observe that, for all $x, y \in \mathscr{X}_0$,

$$h(y) - h(x) = h(y - x) \leqslant p(y - x) \leqslant p(y + x_1) + p(-x_1 - x); \tag{1}$$

hence

$$-p(-x_1 - x) - h(x) \leqslant p(y + x_1) - h(y).$$

Since the left side of this inequality depends only on x and the right side only on y, there exists a constant c such that

$$-p(-x_1 - x) - h(x) \leqslant c \leqslant p(y + x_1) - h(y);$$

hence, setting $z = \alpha y$ for $\alpha \geqslant 0$ and $z = \alpha x$ for $\alpha < 0$, we deduce the relation

$$h(z) + \alpha c \leqslant p(z + \alpha x_1) \qquad (z \in \mathscr{X}_0, \quad \alpha \in \mathbb{R}). \tag{2}$$

Now let $\tilde{h}(z + \alpha x_1) \triangleq h(z) + \alpha c$ $(z \in \mathscr{X}_0, \alpha \in \mathbb{R})$. Then $\tilde{h}$ is a linear functional on $\mathscr{X}_1$, $h = \tilde{h} \mid \mathscr{X}_0$, and, by (2), $\tilde{h}(x) \leqslant p(x)$ for all $x \in \mathscr{X}_1$. Therefore $\tilde{h} \in \mathscr{F}$, contradicting the assumption that h was maximal. This shows that $\mathscr{X}_0 = \mathscr{X}$ and completes the proof of the first part of the theorem.

Now assume that $\mathscr{X}$ is a normed vector space and $f \in \mathscr{Y}^*$. Then, by I.3.6, $|f| \triangleq \sup\{|f(y)| \mid y \in \mathscr{Y}, |y| \leqslant 1\} < \infty$. We set $p(x) \triangleq |f|\,|x|$ $(x \in \mathscr{X})$ and observe that, by our previous argument, there exists a linear extension $\tilde{f}$ of f to $\mathscr{X}$ such that $\tilde{f}(x) \leqslant |f|\,|x|$ for all $x \in \mathscr{X}$. It follows that $\tilde{f}(-x) = -\tilde{f}(x) \leqslant |f|\,|x|$; hence $|\tilde{f}(x)| \leqslant |f|\,|x|$ and $|\tilde{f}| \leqslant |f|$. Since $\tilde{f} \mid \mathscr{Y} = f$, we must have $|\tilde{f}| = |f|$. Furthermore, by I.3.6, $\tilde{f} \in \mathscr{X}^*$. QED

I.3.9 ***Theorem*** Let X be a compact metric space, $\mathscr{Y}$ a topological vector space, and $f: X \to \mathscr{Y}$ continuous. Then f is uniformly continuous.

▌PROOF Let A be a neighborhood of the origin in $\mathscr{Y}$. Since the functions $(y_1, y_2) \to y_1 + y_2 : \mathscr{Y} \times \mathscr{Y} \to \mathscr{Y}$ and $(\alpha, x) \to \alpha x : \mathbb{R} \times \mathscr{Y} \to \mathscr{Y}$ are continuous, there exists an open neighborhood B of the origin such that $B + B \subset A$, and we may assume that $-B = B$ [replacing B by $(-B) \cap B$].

For every $x \in X$ there exists $\delta(x) > 0$ such that $f(x) - f(x') \in B$ if $d(x, x') \leqslant 2\delta(x)$. The collection $\{S(x, \delta(x)) \mid x \in X\}$ is an open covering of the compact set X and there exists, therefore, a finite subcovering $\{S(x_1, \delta(x_1)),..., S(x_k, \delta(x_k))\}$. Thus, for every $x \in X$ there exists $j \in \{1, 2,..., k\}$ such that $x \in S(x_j, \delta(x_j))$. Let $\eta = \mathrm{Min}_{1 \leqslant i \leqslant k} \delta(x_i)$. If $d(x, x') \leqslant \eta$, then

$$d(x', x_j) \leqslant d(x', x) + d(x, x_j) \leqslant \eta + \delta(x_j) \leqslant 2\delta(x_j);$$

hence

$$f(x) - f(x') = (f(x) - f(x_j)) + (f(x_j) - f(x')) \in B + B \subset A. \quad \text{QED}$$

I.3.10 ***Theorem*** Let $\mathscr{X}$ be a normed vector space, $U \triangleq \{l \in \mathscr{X}^* \mid |l| \leqslant 1\}$, $l_j \in U$ $(j \in \mathbb{N})$, $f: \mathscr{X} \to \mathbb{R}$ continuous, and $\lim_j l_j y = f(y)$ for all y in a dense subset $\mathscr{X}_1$ of $\mathscr{X}$. Then $\lim_j l_j x = f(x)$ for all $x \in \mathscr{X}$.

▌PROOF Let $x \in \mathscr{X}$ and $\epsilon > 0$. Then there are $j_0 \in \mathbb{N}$ and $y \in \mathscr{X}_1$ such that $|f(x) - f(y)| < \epsilon/3$, $|y - x| < \epsilon/3$, and $|f(y) - l_j y| < \epsilon/3$ for all $j \geqslant j_0$. Thus

$$|f(x) - l_j x| \leqslant |f(x) - f(y)| + |f(y) - l_j y| + |l_j(y - x)| < \epsilon$$

$$\text{for all} \quad j \geqslant j_0. \quad \text{QED}$$

I.3.11 ***Theorem*** (Bishop) Let $\mathscr{X}$ be a separable normed vector space and $U \triangleq \{l \in \mathscr{X}^* \mid |l| \triangleq \sup_{|x| \leqslant 1} |l(x)| \leqslant 1\}$. Then there exists a norm $|\cdot|_w$ (*weak norm*) on $\mathscr{X}^*$ such that the relativizations to U of the corresponding metric topology (*weak metric topology*) and of the weak star topology of $\mathscr{X}^*$ coincide, and $\lim_j |l - l_j|_w = 0$ for $l, l_j \in U$ if and only if $\lim_j l_j x = lx$ for each $x \in \mathscr{X}$.

▌PROOF Let $\mathscr{X}_\infty \triangleq \{x_1, x_2, ...\}$ be a dense subset of $\mathscr{X}$ and

$$|T|_w \triangleq \lim_n \sum_{j=1}^{n} 2^{-j} |Tx_j|/(1 + |x_j|) \qquad \text{for all} \quad T \in \mathscr{X}^*.$$

It is easy to verify that this limit exists and that $|T|_w \leqslant |T|$, $|T_1 + T_2|_w \leqslant |T_1|_w + |T_2|_w$, $|\alpha T|_w = |\alpha| \, |T|_w$, and $|T|_w \geqslant 0$. Furthermore, $|T|_w = 0$ implies $|Tx_j| = 0$ for all $j \in \mathbb{N}$; since both T and 0 are continuous extensions of $T \mid \mathscr{X}_\infty$ to $\mathscr{X} = \bar{\mathscr{X}}_\infty$, it follows from I.2.13 that $T = 0$. Thus $|\cdot|_w$ is a norm on $\mathscr{X}^*$.

Now let $\mathscr{T}_1$ be the relative weak star topology of U, $\mathscr{T}_2$ the relative $|\cdot|_w$-topology of U, and $T' \in A \in \mathscr{T}_1$. Then there exist $k \in \mathbb{N}$, $x^j \in \mathscr{X}$ $(j = 1,..., k)$ and $\epsilon > 0$ such that

$$\{T \in U \,\big|\, |(T - T')\, x^j| < \epsilon \ (j = 1,..., k)\} \subset A.$$

For each $j \in \{1, 2, ..., k\}$, we choose $x_{i_j} \in \mathscr{X}_\infty$ in such a manner that $| x^j - x_{i_j} | < \epsilon/4$ and we set $\alpha \triangleq \inf_{1 \leqslant j \leqslant k} 2^{-i_j}/(1 + | x_{i_j} |)$. Thus, if $T \in U$, $| T - T' |_w < \alpha\epsilon/2$ and $j \in \{1, 2, ..., k\}$, we have

$$|(T - T') x_{i_j} | \leqslant \alpha^{-1} | T - T' |_w < \epsilon/2$$

and

$$|(T - T')(x^j - x_{i_j})| \leqslant (| T | + | T' |) | x^j - x_{i_j} | < \epsilon/2;$$

hence

$$|(T - T') x^j | < \epsilon.$$

This shows that the set $\{T \in U \mid | T - T' |_w < \alpha\epsilon/2\} \in \mathscr{T}_2$ is contained in A and, since A is a union of such sets, $A \in \mathscr{T}_2$.

Conversely, let $T' \in B \in \mathscr{T}_2$. Then there exists $\epsilon > 0$ such that $\{T \in U \mid | T - T' |_w < \epsilon\} \subset B$ and we can choose $j_0 \in \mathbb{N}$ such that

$$\lim_n \sum_{j=j_0+1}^{n} 2^{-j} |(T - T') x_j |/(1 + | x_j |) \leqslant 2^{-j_0+1} < \epsilon/2.$$

We set $\alpha \triangleq \sup_{1 \leqslant j \leqslant j_0} 2^{-j}/(1 + | x_j |)$. Then, if $T \in U$ and $|(T - T') x_j | < \epsilon/(2\alpha j_0)$ $(j = 1, ..., j_0)$, we have

$$| T - T' |_w \leqslant \alpha \sum_{j=1}^{j_0} |(T - T') x_j | + \epsilon/2 < \epsilon.$$

This shows that every $T' \in B$ is contained in a set $\{T \in U \mid |(T - T') x^j | < \epsilon/(2\alpha j_0)$ $(j = 1, ..., j_0)\} \in \mathscr{T}_1$ which is itself contained in B. Thus $B \in \mathscr{T}_1$, completing the proof that $\mathscr{T}_1 = \mathscr{T}_2$.

Now let $l, l_j \in U$ $(j \in \mathbb{N})$. If $\lim_j | l - l_j |_w = 0$, then $lx_i = \lim_j l_j x_i$ for each $x_i \in \mathscr{X}_\infty$; hence, by I.3.10, $lx = \lim_j l_j x$ $(x \in \mathscr{X})$. Conversely, if $lx = \lim_j l_j x$ $(x \in \mathscr{X})$, then for every $\epsilon > 0$ we choose $i_0 \in \mathbb{N}$ such that

$$\lim_n \sum_{1=i_0+1}^{n} 2^{-i} |(l - l_j) x_i |/(1 + | x_i |) < 2^{-i_0+1} < \epsilon/2 \qquad (j \in \mathbb{N})$$

and $j_0 \in \mathbb{N}$ such that

$$\sum_{i=1}^{i_0} 2^{-i} |(l - l_j) x_i |/(1 + | x_i |) < \epsilon/2 \qquad \text{for all} \quad j \geqslant j_0 .$$

Then $| l - l_j |_w < \epsilon$ for all $j \geqslant j_0$; hence $\lim_j | l - l_j |_w = 0$. QED

I.3.12 ***Theorem*** (Alaoglu) Let $\mathscr{X}$ be a separable normed vector space, $| \cdot |_w$ a norm on $\mathscr{X}^*$ defined as in I.3.11 and $U \triangleq \{l \in \mathscr{X}^* \mid | l | \leqslant 1\}$. Then

U with the relative topology of $(\mathscr{X}^*, |\cdot|_w)$ is a compact and a sequentially compact metric space.

▌PROOF Let $\mathscr{X}_\infty \triangleq \{x_1, x_2, \ldots\}$ be the dense subset of $\mathscr{X}$ that defines $|\cdot|_w$ and (l_j) a sequence in U. For all $i, j \in \mathbb{N}$ we have $|l_j(x_i)| \leqslant |x_i|$. Thus the sequence $(l_j(x_1))_j$ has a subsequence $(l_j(x_1))_{j\in J_1}$ converging to some number $l(x_1)$. Similarly $(l_j(x_2))_{j\in J_1}$ has a subsequence $(l_j(x_2))_{j\in J_2}$ converging to some number $l(x_2)$. Continuing in this manner, we can construct a sequence of sequences $((l_j)_{j\in J_1}, (l_j)_{j\in J_2}, \ldots)$ and their diagonal subsequence $(l_j)_{j\in J}$, and we verify that $(l_j(x_i))_{j\in J}$ converges to a number $l(x_i)$ for each $i \in \mathbb{N}$. Since $|l_j(x_m) - l_j(x_n)| \leqslant |x_m - x_n|$ for all $j, m, n \in \mathbb{N}$, it follows that

$$|l(x_m) - l(x_n)| = \lim_{j\in J} |l_j(x_m) - l_j(x_n)| \leqslant |x_m - x_n| \qquad \text{for all} \quad m, n \in \mathbb{N}.$$

Thus $l \mid \mathscr{X}_\infty$ is uniformly continuous and, by I.2.13, has a uniformly continuous extension l to $\mathscr{X} = \bar{\mathscr{X}}_\infty$. By I.3.10, $\lim_{j\in J} l_j(x) = l(x)$ for each $x \in \mathscr{X}$, and it is easy to see that l is linear; hence, by I.3.11, $\lim_{j\in J} |l_j - l|_w = 0$ and $\lim_{j\in J} l_j = l$ in the relative topology of the metric space $(\mathscr{X}^*, |\cdot|_w)$. Thus U is a sequentially compact subset of $(\mathscr{X}^*, |\cdot|_w)$; hence, by I.2.5, it is compact. QED

I.3.13 ***Theorem*** (Riesz) Let $\mathscr{Y}$ be a normed vector space, I the identity operator in $\mathscr{Y}$, $T \in B(\mathscr{Y}, \mathscr{Y})$, T compact, and $I - T$ an injection. Then $I - T$ is a homeomorphism of $\mathscr{Y}$ onto $\mathscr{Y}$ and $(I - T)^{-1} - I$ is compact.

▌PROOF *Step* 1 Let $U \triangleq I - T$. We first show that if $\mathscr{X}_1$ and $\mathscr{X}_2$ are vector subspaces of $\mathscr{Y}$, the sets $\mathscr{X}_1$ and $\mathscr{X}_2$ closed, $\mathscr{X}_1 \subset \mathscr{X}_2$, $\mathscr{X}_1 \neq \mathscr{X}_2$, and $U(\mathscr{X}_2) \subset \mathscr{X}_1$, then there exists $a \in \mathscr{X}_2 \sim \mathscr{X}_1$ such that $|a| = 1$ and $|Ta - Tx| \geqslant \frac{1}{2}$ for each $x \in \mathscr{X}_1$. Indeed, there exist $b \in \mathscr{X}_2 \sim \mathscr{X}_1$ with $d[b, \mathscr{X}_1] = \alpha > 0$ and $c \in \mathscr{X}_1$ with $|b - c| \leqslant 2\alpha$. Let

$$a \triangleq \frac{1}{|b - c|}(b - c).$$

Then $|a| = 1$. For any $y \in \mathscr{X}_1$, we have

$$|a - y| = \frac{1}{|b - c|}|b - (c + |b - c|\, y)|.$$

Since $d[b, \mathscr{X}_1] = \alpha$, we have $|a - y| \geqslant \alpha / |b - c| \geqslant \frac{1}{2}$. Now let $x \in \mathscr{X}_1$. Then $Ta - Tx = (I - U)(a - x) = a - (x + U(a - x))$. Since $x + U(a - x) \in \mathscr{X}_1$, it follows that $|Ta - Tx| \geqslant \frac{1}{2}$.

Step 2 We now show that $U(A)$ is closed if A is closed. Let A be a closed subset of $\mathscr{Y}$ and $y \in \overline{U(A)}$. Then $y = \lim_j Ux_j$ for some sequence (x_j) in A.

If (x_j) is unbounded, we may assume (extracting a subsequence, if necessary) that $\lim_j | x_j | = \infty$. Let $z_j \triangleq (1/| x_j |) x_j$. Then

$$\lim_j Uz_j = \lim_j \frac{1}{| x_j |} Ux_j = \lim_j \frac{1}{| x_j |} y = 0$$

and $| z_j | = 1$. Since $Uz_j = z_j - Tz_j$ and some sequence $(Tz_j)_{j \in J}$ converges to a point a (because T is compact), we conclude that $\lim_{j \in J} (z_j - a) = \lim_{j \in J} Uz_j = 0$; hence $\lim_{j \in J} z_j = a$ and $Ua = \lim_{j \in J} Uz_j = 0$. Thus $a \in U^{-1}(\{0\}) = \{0\}$ contradicting $| a | = \lim_{j \in J} | z_j | = 1$. Therefore (x_j) is bounded and there exist $J \subset (1, 2,...)$ and some $b \in \mathscr{Y}$ such that

$$b = \lim_{j \in J} Tx_j = \lim_{j \in J} (x_j - Ux_j) = \lim_{j \in J} (x_j - y);$$

hence $b + y = \lim_j x_j \in A$. Since U is continuous, it follows that

$$y = \lim_{j \in J} Ux_j = U(b + y) \in U(A),$$

thus showing that $U(A) = \overline{U(A)}$.

Step 3 We next show that $\mathscr{Y} = U(\mathscr{Y})$. Let $T_k \triangleq I - U^k = I - (I - T)^k$ for $k \in \mathbb{N}$. We verify by induction on k that

$$(I - T)^k = \sum_{j=0}^{k} (-1)^j \binom{k}{j} T^j,$$

where

$$T^0 \triangleq I, \quad \binom{k}{j} \triangleq \frac{k!}{j!(k-j)!}, \quad 0! \triangleq 1, \quad \text{and} \quad j! \triangleq 1 \cdot 2 \cdot \cdots \cdot j \quad \text{for} \quad j \in \mathbb{N}.$$

Thus

$$I - T_k = (I - T)^k = I + T \sum_{j=1}^{k} (-1)^j \binom{k}{j} T^{j-1};$$

hence $T_k = TV$, where $V \triangleq -\sum_{j=1}^{k} (-1)^j \binom{k}{j} T^{j-1}$ and, by I.3.7, $| V | < \infty$. Thus, for any bounded set $A \subset \mathscr{Y}$, $\sup\{| V(y)| \mid y \in A\} \leqslant | V | \sup\{| x | \mid x \in A\}$; hence $T_k(A) = T(V(A))$ is conditionally compact. Thus $U^k = I - T_k$, where T_k is a compact operator in $\mathscr{Y}$.

Now let $F_k \triangleq U^k(\mathscr{Y})$ for $k = 0, 1, 2,...$. Then $F_{k+1} = U^k(F_1) \subset U^k(\mathscr{Y}) = F_k$ for all k. By Step 2, each F_k is closed. If $F_{k+1} \neq F_k$ for all k, then, for each fixed k, $\mathscr{X}_1 \triangleq F_{k+1}$ and $\mathscr{X}_2 \triangleq F_k$ are as in Step 1 because $U(F_k) = F_{k+1} \subset F_k$. Thus there exist $x_k \in F_k$ with $| x_k | = 1$ and $| Tx_k - Tx_j | \geqslant \frac{1}{2}$ for all $j > k$.

This contradicts the assumption that T is compact since the latter implies, by I.2.5, that $\overline{T(S^F(0, 1))}$ is sequentially compact. Thus $F_k = F_m$ for some (smallest) m and all $k > m$.

We next show that $m = 0$. We have $F_m \subset \mathscr{Y} = F_0$ and $U(F_m) = F_m$. If $F_m \neq F_0$, then $m > 0$ and there exists $z \in F_{m-1} \sim F_m \subset F_0 \sim F_m$. Since $U(z) \in F_m = U(F_m)$ there exists $w \in F_m$ such that $Uz = Uw$, implying $z = w$. This contradicts $z \notin F_m$. Thus $\mathscr{Y} = F_0 = F_1 = U(\mathscr{Y})$.

Step 4 We have shown in Step 3 that $\mathscr{Y} = U(\mathscr{Y})$; hence U is bijective and U^{-1} exists. Since $(U^{-1})^{-1}(A) = U(A)$ is closed for every closed A, U^{-1} is continuous. Thus U is a homeomorphism of $\mathscr{Y}$ onto $\mathscr{Y}$. Now let $R \triangleq (I - T)^{-1} - I$. Then $(I - T) \cdot (I + R) = I$; hence $R = T(I + R)$. By I.3.6, $I + R$ is bounded and therefore the set $(I + R)(A)$ is bounded whenever A is bounded, and $T(I + R)(A)$ is conditionally compact. We conclude that R is a compact operator. QED

I.3.14 ***Theorem*** Let $\mathscr{X}$ be a Banach space. Then

(1) the series $\sum_{j=1}^{\infty} x_j$ in $\mathscr{X}$ is convergent if it is dominated by a convergent series $\sum_{j=1}^{k} a_j$ (in particular, if it is absolutely convergent) or if $(\sum_{j=1}^{k} | x_j |)_k$ is bounded;

(2) $\lim_\alpha \lim_\beta A_{\alpha,\beta} = \lim_\beta \lim_\alpha A_{\alpha,\beta} = \lim_m A_{\alpha(m),\beta(m)} = \sup\{A_{\alpha,\beta} \mid \alpha, \beta \in \mathbb{N}\}$ in $\overline{\mathbb{R}}$ if $(A_{\alpha,\beta})_\alpha$ and $(A_{\alpha,\beta})_\beta$ are nondecreasing sequences in $\overline{\mathbb{R}}$ for each α and β and $m \to (\alpha(m), \beta(m)) : \mathbb{N} \to \mathbb{N} \times \mathbb{N}$ is a bijection; and

(3) the series $\sum_{i,j=1}^{\infty} x_{i,j}$ in $\mathscr{X}$ is convergent and

$$\sum_{i,j=1}^{\infty} x_{i,j} = \sum_{i=1}^{\infty} \sum_{j=1}^{\infty} x_{i,j} = \sum_{j=1}^{\infty} \sum_{i=1}^{\infty} x_{i,j} = \sum_{m=1}^{\infty} x_{i(m),j(m)}$$

if $\sum_{i,j=1}^{\infty} x_{i,j}$ is dominated by a convergent series $\sum_{i,j=1} a_{i,j}$ in $\mathbb{R}$ (in particular, if it is absolutely convergent) and $m \to (i(m), j(m)) : \mathbb{N} \to \mathbb{N} \times \mathbb{N}$ is a bijection. Furthermore, a series $\sum_{i,j=1}^{\infty} b_{i,j}$ with $b_{i,j} \geq 0$ is convergent in $\overline{\mathbb{R}}$.

▌PROOF *Proof of* (1) Assume that $\sum_{j=1}^{\infty} x_j$ is dominated by a convergent series $\sum_{j=1}^{\infty} a_j$. Then, by I.2.4, $(\sum_{j=1}^{i} a_j)_i$ is a Cauchy sequence and for every $\epsilon > 0$ there exists $j(\epsilon) \in \mathbb{N}$ such that $\sum_{j=i_1}^{i_2} a_j \leq \epsilon$ if $i_2 \geq i_1 \geq j(\epsilon)$. It follows then that

$$\left| \sum_{j=1}^{i_2} x_j - \sum_{j=1}^{i_1-1} x_j \right| \leq \sum_{j=i_1}^{i_2} | x_j | \leq \sum_{j=i_1}^{i_2} a_j \leq \epsilon,$$

showing that $(\sum_{j=1}^{i} x_j)_i$ is a Cauchy sequence in the complete metric space $\mathscr{X}$; hence convergent. If $(\sum_{j=1}^{k} | x_j |)_k$ is bounded, then, as a bounded nondecreasing sequence in $\mathbb{R}$, it must be a Cauchy sequence; hence convergent. Thus $\sum_{j=1}^{\infty} x_j$ is dominated by the convergent series $\sum_{j=1}^{\infty} | x_j |$. This proves (1).

Proof of (2) Now let the conditions of statement (2) be satisfied, and assume first that $s \triangleq \sup\{A_{\alpha,\beta} \mid \alpha, \beta \in \mathbb{N}\} < \infty$. Then, for each $\alpha \in \mathbb{N}$, $(A_{\alpha,\beta})_\beta$ is a nondecreasing bounded sequence in $\mathbb{R}$ and $\lim_\beta A_{\alpha,\beta} \triangleq s_\alpha = \sup_\beta A_{\alpha,\beta}$. We also observe that $s_{\alpha'} \leqslant s_{\alpha''} \leqslant s$ if $\alpha' \leqslant \alpha''$ and that for every $\epsilon > 0$ there exists $k(\epsilon) \in \mathbb{N}$ such that $0 \leqslant s - A_{\alpha,\beta} \leqslant \epsilon$ if $\alpha, \beta \geqslant k(\epsilon)$. Thus, if $\alpha, \beta \geqslant k(\epsilon)$, then $A_{\alpha,\beta} \leqslant s_\alpha \leqslant s$ and $0 \leqslant s - s_\alpha \leqslant s - A_{\alpha,\beta} \leqslant \epsilon$, showing that $\lim_\alpha s_\alpha = \lim_\alpha \lim_\beta A_{\alpha,\beta} = s$. A similar argument shows that $\lim_\beta \lim_\alpha A_{\alpha,\beta} = s$.

Now let $m \to (\alpha(m), \beta(m))$ be a bijection and $k(\epsilon)$ as just defined. Then there exists $m(\epsilon) \in \mathbb{N}$ such that $\alpha(m), \beta(m) \geqslant k(\epsilon)$ if $m \geqslant m(\epsilon)$ and $0 \leqslant s - A_{\alpha(m),\beta(m)} \leqslant \epsilon$ if $m \geqslant m(\epsilon)$. Thus $\lim_m A_{\alpha(m),\beta(m)} = s$. This proves (2) for the case where $s < \infty$.

If $s = \infty$, then there exists a sequence $((\alpha_j, \beta_j))_j$ in $\mathbb{N} \times \mathbb{N}$ such that $\lim A_{\alpha_j,\beta_j} = \infty$. It follows that for every $n \in \mathbb{N}$ there exists $j(n)$ such that $A_{\alpha_j,\beta_j} \geqslant n$ if $j \geqslant j(n)$. Then, for all $\alpha, \beta \geqslant \alpha_{j(n)}$, $s_\alpha \geqslant A_{\alpha,\beta} \geqslant n$, showing that $\lim_{\alpha,\beta} A_{\alpha,\beta} = \infty = s$ and $\lim_\alpha s_\alpha = \lim_\alpha \lim_\beta A_{\alpha,\beta} = \infty = s$. We similarly show that $\lim_\beta \lim_\alpha A_{\alpha,\beta} = \infty = s$. If we denote by $m'(n)$ a positive integer such that $\alpha(m), \beta(m) \geqslant \alpha_{j(n)}$ for $m \geqslant m'(n)$, then $A_{\alpha(m),\beta(m)} \geqslant n$ for $m \geqslant m'(n)$. Thus $\lim_m A_{\alpha(m),\beta(m)} = \infty = s$. This completes the proof of (2).

Proof of (3) Let the conditions of (3) be satisfied,

$$X_{\alpha,\beta} \triangleq \sum_{i=1}^{\alpha} \sum_{j=1}^{\beta} x_{i,j} \qquad \text{and} \qquad A_{\alpha,\beta} \triangleq \sum_{i=1}^{\alpha} \sum_{j=1}^{\beta} a_{i,j} \qquad (\alpha, \beta \in \mathbb{N}).$$

Since $A \triangleq \lim_{\alpha,\beta} A_{\alpha,\beta}$ exists in $\mathbb{R}$, for every $\epsilon > 0$ there exists $k(\epsilon) \in \mathbb{N}$ such that, for $\alpha, \beta \geqslant k(\epsilon)$, $A_{\alpha,\beta} - A_{k(\epsilon),k(\epsilon)} = \sum' a_{i,j} \leqslant \epsilon/2$, where $\sum'$ is the sum over the finite set of indices $\{1, 2,..., \alpha\} \times \{1, 2,..., \beta\} \sim \{1, 2,..., k(\epsilon)\}^2$. It follows that

$$| X_{\alpha,\beta} - X_{k(\epsilon),k(\epsilon)} | = \left| \sum{}' x_{i,j} \right| \leqslant \sum{}' | x_{i,j} | \leqslant \sum{}' a_{i,j} \leqslant \epsilon/2 \quad \text{for } \alpha, \beta \geqslant k(\epsilon)$$

and

$$| X_{\alpha,\beta} - X_{\alpha',\beta'} | \leqslant | X_{\alpha,\beta} - X_{k(\epsilon),k(\epsilon)} | + | X_{\alpha',\beta'} - X_{k(\epsilon),k(\epsilon)} | \leqslant \epsilon \qquad \text{for } \alpha', \beta' \geqslant k(\epsilon).$$

Thus, by I.2.4, the series $\sum_{i,j=1}^{\infty} x_{i,j}$ is convergent, $\sum_{i,j=1}^{\infty} x_{i,j} = \lim_{\alpha,\beta} X_{\alpha,\beta} \triangleq \bar{X} \in \mathscr{X}$, and

$$(4) \qquad | X_{\alpha,\beta} - \bar{X} | = \lim_{\alpha',\beta'} | X_{\alpha,\beta} - X_{\alpha',\beta'} | \leqslant \epsilon \qquad \text{if } \alpha, \beta \geqslant k(\epsilon).$$

For each $i \in \mathbb{N}$ the sequence $(\sum_{j=1}^{\beta} | x_{i,j} |)_\beta$ is bounded by $\sum_{i,j=1}^{\infty} a_{i,j}$; hence, by (1), the series $\sum_{j=1}^{\infty} x_{i,j}$ converges to some y_i in $\mathscr{X}$ and there exists

$j(\epsilon, i) \in \mathbb{N}$ such that $| y_i - \sum_{j=1}^{\beta} x_{i,j} | \leqslant \epsilon/2^i$ for $\beta \geqslant j(\epsilon, i)$. Then, for each $\alpha \in \mathbb{N}$,

$$\left| \sum_{i=1}^{\alpha} y_i - \sum_{i=1}^{\alpha} \sum_{j=1}^{\beta} x_{i,j} \right| \leqslant \epsilon \sum_{i=1}^{\alpha} 1/2^i \leqslant \epsilon$$

for $\beta \geqslant \beta(\epsilon, \alpha) \triangleq \sup\{j(\epsilon, 1),..., j(\epsilon, \alpha)\}$; hence, by (4),

$$\left| \sum_{i=1}^{\alpha} y_i - \bar{X} \right| \leqslant \left| \sum_{i=1}^{\alpha} y_i - X_{\alpha,\beta} \right| + | X_{\alpha,\beta} - \bar{X} | \leqslant 2\epsilon$$

$$\text{if} \quad \alpha \geqslant k(\epsilon) \quad \text{and} \quad \beta \geqslant \max\big(\beta(\epsilon, \alpha), k(\epsilon)\big).$$

Thus $\sum_{i=1}^{\infty} \sum_{j=1}^{\infty} x_{i,j} = \sum_{i,j=1}^{\infty} x_{i,j}$ and the same argument shows that $\sum_{j=1}^{\infty} \sum_{i=1}^{\infty} x_{i,j} = \sum_{i,j=1}^{\infty} x_{i,j}$.

Now let $m \to \big(i(m), j(m)\big) : \mathbb{N} \to \mathbb{N} \times \mathbb{N}$ be a bijection, $IJ(k) \triangleq \{\big(i(m), j(m)\big) \mid m \in \{1, 2,..., k\}\}$, and $I(k) \triangleq \{1, 2,..., k\}^2$. Then, for each $k \in \mathbb{N}$, there exist integers $l_1(k) \leqslant l_2(k)$ such that $\lim_k l_1(k) = \infty$ and $I\big(l_1(k)\big) \subset IJ(k) \subset I\big(l_2(k)\big)$. We have

$$\left| X_{l_2(k), l_2(k)} - \sum_{m=1}^{k} x_{i(m), j(m)} \right| \leqslant \sum{}'' a_{ij},$$

where $\sum''$ is the sum over $I\big(l_2(k)\big) \sim I\big(l_1(k)\big)$. Since $\lim_{\alpha,\beta} X_{\alpha,\beta}$ and $\lim_{\alpha,\beta} A_{\alpha,\beta}$ exist, it follows that $\sum'' \to 0$ as $k \to \infty$ and

$$\lim_k \sum_{m=1}^{k} x_{i(m), j(m)} = \lim_{\alpha,\beta} X_{\alpha,\beta} \triangleq \bar{X}.$$

Finally, if $b_{i,j} \geqslant 0$ $(i, j \in \mathbb{N})$, then it follows from (2) (setting $A_{\alpha,\beta} \triangleq \sum_{i=1}^{\alpha} \sum_{j=1}^{\beta} b_{i,j}$) that $\sum_{i,j=1}^{\infty} b_{i,j}$ exists in $\bar{\mathbb{R}}$. QED

I.3.15 *The Exponential and Logarithmic Functions*

DEFINITION OF $\exp(x)$

For any $x \in \mathbb{R}$ and $j \in \mathbb{N}$, let x^j represent the jth power of x (if there is no danger of confusing the exponent with a superscript), $0! \triangleq 1$ and $j! \triangleq 1 \cdot 2 \cdot 3 \cdot \cdots \cdot j$. If $y \in [0, 1)$, the series

$$\sum_{j=0}^{\infty} y^j = \lim_k \frac{1 - y^{k+1}}{1 - y} = (1 - y)^{-1}$$

is convergent. If $x \in \mathbb{R}$ and we denote by n the smallest integer greater than or equal to $|x| + 1$, then $j! \geqslant n^{j-n}$ for all $j \geqslant n$; therefore the series $\sum_{j \geqslant n} x^j/j!$ is dominated by the convergent series $n^n \sum_{j \geqslant n} (|x|/n)^j$. Thus, by I.3.14(1), the series

$$\exp(x) \triangleq \sum_{j=0}^{\infty} x^j/j! \tag{1}$$

is convergent for every $x \in \mathbb{R}$ and we refer to the function $\exp(\cdot) : \mathbb{R} \to \mathbb{R}$ as the *exponential* function.

THE FUNCTIONAL RELATION FOR $\exp(\cdot)$

We shall next show that

$$\exp(x + y) = \exp(x) \cdot \exp(y) \qquad (x, y \in \mathbb{R}). \tag{2}$$

Since $\sum_{j=0}^{\alpha} |x|^j/j! \leqslant \exp(|x|)$ for all $x \in \mathbb{R}$ and $\alpha \in \mathbb{N}$, we have

$$\sum_{i=1}^{\alpha} \sum_{j=1}^{\beta} |x|^i |y|^j/(i!\,j!) \leqslant \exp(|x|) \exp(|y|) \qquad (\alpha, \beta \in \mathbb{N})$$

and, by I.3.14(3) [enumerating the points (i, j) by first listing those with $i + j = 0$, then those with $i + j = 1, 2,\ldots$ etc.], we have

$$\exp(x) \cdot \exp(y) = \sum_{i=0}^{\infty} x^i/i! \sum_{j=0}^{\infty} y^j/j! = \sum_{m=0}^{\infty} \sum_{i=0}^{m} (x^i/i!) \cdot (y^{m-i}/(m-i)!).$$

We can verify, by induction on m, that

$$(x + y)^m = \sum_{i=0}^{m} [m!/(i!(m-i)!)] x^i y^{m-i}.$$

It follows thus that

$$\exp(x) \exp(y) = \sum_{m=0}^{\infty} (1/m!)(x + y)^m = \exp(x + y).$$

THE LOGARITHMIC FUNCTION

Since the terms of the series in (1) are increasing nonnegative functions of x for $x \geqslant 0$, we easily deduce that $\exp(\cdot)$ is nonnegative and increasing on $[0, \infty)$. Since, by (2), $\exp(x) \exp(-x) = \exp(0) = 1$, it follows that $\exp(\cdot)$ is nonnegative and increasing on $\mathbb{R}$. Since $\sum_{j=2}^{\infty} (1/j!)\, x^j \geqslant 0$ for $x \geqslant 0$, we have $\exp(x) \geqslant 1 + x$ for $x \geqslant 0$ and therefore $\lim_{x \to \infty} \exp(x) = \infty$ and $\lim_{x \to \infty} \exp(-x) = \lim_{x \to \infty} 1/\exp(x) = 0$. Thus $\exp(\cdot)$ maps $\mathbb{R}$ onto $(0, \infty)$ and, being an increasing function, it is an injection. Thus $\exp(\cdot) : \mathbb{R} \to (0, \infty)$ has an inverse $\exp^{-1} : (0, \infty) \to \mathbb{R}$, and we write $\log x$ for $\exp^{-1}(x)$ and refer to $\log(\cdot)$ as the *logarithmic* function.

Powers

If $a > 0$ and $x \in \mathbb{R}$, we write a^x for $\exp(x \log a)$ and observe that $a^{x+y} = a^x a^y$, $a^{bx} = (a^b)^x$, and $a^0 = 1$. We write e for $\exp(1)$ and observe that $\log e = 1$. Thus

$$e^x = \exp(x) \qquad (x \in \mathbb{R}).$$

An Inequality

We shall next show that, for all $\xi, \eta > 0$ and $\alpha \in [0, 1]$, we have

$$\xi^\alpha \eta^{1-\alpha} \leqslant \alpha\xi + (1 - \alpha)\eta. \tag{3}$$

We shall first prove that

$$e^{\alpha x+(1-\alpha)y} \leqslant \alpha e^x + (1 - \alpha)\, e^y \qquad \text{for all} \quad x, y \geqslant 0.$$

Indeed, for $j = 1, 2$ we have

$$[\alpha x + (1 - \alpha)\, y]^{j-1} \leqslant \alpha x^{j-1} + (1 - \alpha)\, y^{j-1}. \tag{4}$$

We shall assume, for purposes of induction, that this relation holds for some $j \geqslant 2$. We then multiply both sides of (4) by $\alpha x + (1 - \alpha)\, y$ to yield

$$[\alpha x + (1 - \alpha)\, y]^j \leqslant \alpha x^j + (1 - \alpha)\, y^j - \alpha(1 - \alpha)(x^j - x^{j-1}y + y^j - y^{j-1}x),$$

and observe that

$$x^j - x^{j-1}y + y^j - y^{j-1}x = (x - y)(x^{j-1} - y^{j-1}) = (x - y)^2 \sum_{k=0}^{j-2} x^k y^{j-2-k} \geqslant 0,$$

thus completing the induction and proving (4) for all $j \in \mathbb{N}$. It follows that

$$\begin{aligned} e^{\alpha x+(1-\alpha)y} &= \sum_{j=0}^{\infty} (1/j!)[\alpha x + (1 - \alpha)\, y]^j \leqslant \sum_{j=0}^{\infty} (1/j!][\alpha x^j + (1 - \alpha)\, y^j] \\ &= \alpha e^x + (1 - \alpha)\, e^y. \end{aligned} \tag{5}$$

We set $\xi = e^x$, $\eta = e^y$ in (5) to obtain

$$\xi^\alpha \eta^{1-\alpha} \leqslant \alpha\xi + (1 - \alpha)\eta \qquad \text{for all} \quad \xi, \eta \geqslant e;$$

hence, setting $z = \xi/\eta$,

$$z^\alpha \leqslant \alpha z + 1 - \alpha \qquad (z > 0)$$

If we again replace z by ξ/η for arbitrary positive ξ and η, we can now deduce relation (3) for all $\xi, \eta > 0$.

CONTINUITY OF $\exp(\cdot)$ AND $\log(\cdot)$

We observe that for each $x \in (-1, 1)$, we have

$$|\exp(x) - 1| = \left| \sum_{j=1}^{\infty} x^j/j! \right| \leqslant |x| \sum_{j=1}^{\infty} 1/j! = (e - 1)|x|;$$

hence $\lim_{x\to 0} \exp(x) = 1$ and, by (2),

$$\lim_{h\to 0} \exp(x + h) = \exp(x) \lim_{h\to 0} \exp(h) = \exp(x).$$

Thus $\exp(\cdot)$ is continuous.

Since $\exp(\cdot)$ is increasing, it follows that its restriction to any interval $[a_0, a_1]$ is a bijection of $[a_0, a_1]$ onto $[\exp(a_0), \exp(a_1)]$ and therefore, by I.2.9(4), $\log | [\exp(a_0), \exp(a_1)]$ is continuous. We conclude that $\log(\cdot)$ is continuous on $(0, \infty)$.

I.4 Measures, Measurable Functions, and Integrals

I.4.A Measures

Let S be an arbitrary set. A family Σ of subsets of S is a *field* (or a *field of sets*, or a *field in S*, or a *Boolean algebra of sets*) if Σ contains the empty set, the complement in S of every element of Σ, and the union of every finite subcollection of Σ. A field Σ is a *σ-field* if it contains the union of every denumerable subcollection or, equivalently, the union of every denumerable subcollection of disjoint sets.

If Σ is a field and $\mu : \Sigma \to \overline{\mathbb{R}}$, then μ is *additive* (or an *additive set function*) if $\mu(\varnothing) = 0$ and $\mu(A \cup B) = \mu(A) + \mu(B)$ whenever $A, B \in \Sigma$ and $A \cap B = \varnothing$. (Thus $\mu(A) + \mu(B)$ is defined in $\overline{\mathbb{R}}$.) It follows then by induction that $\mu(\bigcup_{j=1}^{k} A_j) = \sum_{j=1}^{k} \mu(A_k)$ if $(A_1, A_2, \ldots, A_k)$ is a finite sequence of disjoint elements of Σ. If, furthermore, $\mu(\bigcup_{j=1}^{\infty} A_j) = \sum_{j=1}^{\infty} \mu(A_j)$ for every sequence (A_j) of disjoint elements of Σ whose union is in Σ, then we say that μ is *countably additive* (on the field Σ).

If Σ is a σ-field in S and $\mu : \Sigma \to \overline{\mathbb{R}}$ is countably additive, then we say that μ is a *measure* (or a *measure in S*). An additive set function (respectively measure) μ is *finite* if its values are in $\mathbb{R}$.

It is easily verified that a linear combination $\sum_{j=1}^{k} \alpha_j \mu_j$ of finite additive set functions μ_j (respectively finite measures μ_j) is a finite additive set function (respectively a finite measure).

If Σ is a σ-field of subsets of S, we refer to the couple (S, Σ) as a *measurable space*. If $\mu : \Sigma \to \overline{\mathbb{R}}$ is a measure, then the triplet (S, Σ, μ) is a *measure space*.

We sometimes refer to S as a measure space if Σ and μ have been specified. A measure μ is *positive* if $\mu(E) \geqslant 0$ for all $E \in \Sigma$ and it is clear that, in that case, $\mu(A) \leqslant \mu(A) + \mu(B \sim A) = \mu(A \cup B)$ for all $A, B \in \Sigma$. A positive measure μ is a *probability measure* if $\mu(S) = 1$. A measure $\mu : \Sigma \to \overline{\mathbb{R}}$ is *supported on* A (has its *support in* A) if $\mu(B) = 0$ whenever $A \cap B = \varnothing$, $B \in \Sigma$.

If $\{\Sigma_\omega \mid \omega \in \Omega\}$ is a collection of fields of subsets of S, then we easily verify that $\bigcap_{\omega \in \Omega} \Sigma_\omega$ is a field. Thus, for every class $\mathscr{A}$ of subsets of S there exists a unique smallest field containing $\mathscr{A}$; namely, the intersection of all the fields containing $\mathscr{A}$ (of which the collection of all the subsets of S is one). The same argument shows that there exists a unique smallest σ-field containing $\mathscr{A}$.

If $(S, \mathscr{V})$ is a topological space, then the smallest σ-field containing $\mathscr{V}$ is called the *Borel field of sets*, and denoted by $\Sigma_{\text{Borel}}(S)$, and the elements of $\Sigma_{\text{Borel}}(S)$ are called *Borel sets*. A measure defined on $\Sigma_{\text{Borel}}(S)$ is called a *Borel measure* (as distinguished from *the Borel measure* on an interval defined in I.4.15 below). A Borel measure μ is the *Dirac measure* at $\bar{s} \in S$ if it is a probability measure and $\mu(\{\bar{s}\}) = 1$.

If Σ is a field and $\mu : \Sigma \to \overline{\mathbb{R}}$ an additive set function, then we call the *variation of* μ, and denote by $|\mu|$, the nonnegative function on Σ to $\overline{\mathbb{R}}$ defined for all $E \in \Sigma$ by

$$|\mu|(E) \triangleq \sup \left\{ \sum_{j=1}^{k} |\mu(E_j)| \;\middle|\; k \in \mathbb{N},\ E_1, E_2, \ldots, E_k \text{ disjoint elements of } \Sigma \text{ contained in } E \right\}.$$

We denote by μ^+ respectively μ^-, and call the *positive*, respectively *negative*, *variation* of μ, the function on Σ to $\overline{\mathbb{R}}$ defined (for finite measures μ) by

$$\mu^+(E) \triangleq \tfrac{1}{2}(|\mu|(E) + \mu(E)) \quad \text{respectively} \quad \mu^-(E) \triangleq \tfrac{1}{2}(|\mu|(E) - \mu(E)).$$

It is easy to verify that $|\mu| = \mu^+ = \mu$ if μ is positive.

If $(S, \mathscr{V})$ is a topological space, Σ a field in S, and $\mu : \Sigma \to \mathbb{R}$ additive, then μ is *regular* if for every $E \in \Sigma$ and $\epsilon > 0$ there exist $A, B \in \Sigma$ such that $\bar{A} \subset E \subset B^\circ$ and $|\mu|(B \sim A) \leqslant \epsilon$. In particular, if Σ contains both open and closed subsets of S, then μ is regular if for every $E \in \Sigma$ and $\epsilon > 0$ there exist an open $G\,(= B)$ and a closed $C\,(= A)$ such that $C \subset E \subset G$ and $|\mu|(G \sim C) \leqslant \epsilon$. A finite regular Borel measure is a *Radon measure*. We denote by $\text{frm}(S)$ the vector space of all Radon measures in S, by $\text{frm}^+(S)$ the set of all positive Radon measures in S, and by $rpm(S)$ the set of all Radon probability measures in S.

A measure $\mu : \Sigma \to \overline{\mathbb{R}}$ is *nonatomic* if for every $E \in \Sigma$ with $|\mu|(E) > 0$ there exists $A \in \Sigma$ with $0 < |\mu|(A) < |\mu|(E)$ and $A \subset E$.

If (S, Σ, μ) is a measure space, then Z is a *μ-null set* if there exists $A \in \Sigma$ such that $Z \subset A$ and $|\mu|(A) = 0$. The measure μ is *complete* if every μ-null set belongs to Σ. We say that a relation involving $s \in S$ is valid μ-a.e. (*μ-almost everywhere*) or holds for μ-a.a. $s \in S$ (*μ-almost all $s \in S$*) if there exists a μ-null set Z such that the relation holds for all $s \notin Z$. If $E \subset S$ and the relation holds for all $s \in E \sim Z$, then we say that it holds μ-a.e. in E or for μ-a.a. $s \in E$.

If (S, Σ, μ) and (S, Σ, λ) are measure spaces, then λ is *μ-continuous* (or *absolutely continuous with respect to* μ) if $\lambda(E) = 0$ whenever $|\mu|(E) = 0$.

A measure space (S, Σ, μ) is called *positive*, *finite*, *probability*, *regular*, etc. if the measure μ has the corresponding property.

I.4.1 ***Theorem*** Let $\mu : \Sigma \to \mathbb{R}$ be an additive set function and μ bounded. Then $|\mu|$, μ^+, and μ^- are bounded and each is an additive set function.

▌PROOF Let $|\mu(E)| \leqslant c < \infty$ for all $E \in \Sigma$. If $(E_1, \ldots, E_k)$ is a finite sequence of disjoint elements of Σ, let $J \triangleq \{j \in \{1, 2, \ldots, k\} \mid \mu(E_j) \geqslant 0\}$ and $J' \triangleq \{1, 2, \ldots, k\} \sim J$. Then

$$\sum_{j=1}^{k} |\mu(E_j)| = \sum_{j \in J} \mu(E_j) - \sum_{j \in J'} \mu(E_j) = \mu\left(\bigcup_{j \in J} E_j\right) - \mu\left(\bigcup_{j \in J'} E_j\right) \leqslant 2c.$$

Thus $|\mu|(E) \leqslant 2c$ for all $E \in \Sigma$, showing that $|\mu|$ is bounded; hence μ^+ and μ^- are also bounded.

It is clear that $|\mu|(\varnothing) = 0$. We shall next show that $|\mu|(\bigcup_{j=1}^{i} A_j) = \sum_{j=1}^{i} |\mu|(A_j)$ if $i \geqslant 2$ and $A_1, \ldots, A_i$ are disjoint elements of Σ. This will follow, by induction, for every $i \geqslant 2$ if we prove that it holds for $i = 2$. Let, therefore, $A, B \in \Sigma$ and $A \cap B = \varnothing$, and let $E_1, \ldots, E_k$ be disjoint subsets of $A \cup B$ belonging to Σ. We set $A_j \triangleq A \cap E_j$, $B_j \triangleq B \cap E_j$. Then $E_j = A_j \cup B_j$ and

$$\sum_{j=1}^{k} |\mu(E_j)| = \sum_{j=1}^{k} |\mu(A_j) + \mu(B_j)| \leqslant \sum_{j=1}^{k} |\mu(A_j)| + \sum_{j=1}^{k} |\mu(B_j)|$$

$$\leqslant |\mu|(A) + |\mu|(B),$$

implying

$$|\mu|(A \cup B) \leqslant |\mu|(A) + |\mu|(B). \tag{1}$$

On the other hand, for every $\epsilon > 0$ there exist disjoint sets $A_1, \ldots, A_m \subset A$ and $B_1, \ldots, B_n \subset B$, with $m, n \in \mathbb{N}$ and $A_i, B_j \in \Sigma$, such that

$$|\mu|(A) \leqslant \sum_{j=1}^{m} |\mu(A_j)| + \epsilon/2 \qquad \text{and} \qquad |\mu|(B) \leqslant \sum_{j=1}^{n} |\mu(B_j)| + \epsilon/2.$$

The sets $A_1, \ldots, A_m, B_1, \ldots, B_n$ are disjoint subsets of $A \cup B$ belonging to Σ; hence

$$| \mu | (A) + | \mu | (B) \leqslant \sum_{j=1}^{m} | \mu(A_j)| + \sum_{j=1}^{n} | \mu(B_j)| + \epsilon \leqslant | \mu | (A \cup B) + \epsilon.$$

Since $\epsilon > 0$ is arbitrary, this relation and (1) imply that

$$| \mu |(A \cup B) = | \mu |(A) + | \mu |(B).$$

Thus $| \mu |$ is a finite additive set function and it follows that $\mu^+ \triangleq \frac{1}{2}(| \mu | + \mu)$ and $\mu^- \triangleq \frac{1}{2}(| \mu | - \mu)$ are also finite additive set functions. QED

I.4.2 ***Theorem*** (Jordan decomposition) Let $\mu : \Sigma \to \mathbb{R}$ be bounded and an additive set function. Then, for all $E \in \Sigma$,

$$\mu^+(E) = \sup_{A \subset E} \mu(A) \geqslant 0, \qquad \mu^-(E) = -\inf_{A \subset E} \mu(A) \geqslant 0,$$

$$\mu(E) = \mu^+(E) - \mu^-(E), \quad | \mu |(E) = \mu^+(E) + \mu^-(E), \quad | \mu(E)| \leqslant | \mu |(E).$$

If μ is positive, then $\mu = \mu^+ = | \mu |$ and $\mu^- = 0$.

▌PROOF For all $A \subset E$, $A \in \Sigma$, we have

$$2\mu(A) = \mu(A) + \mu(E) - \mu(E \sim A) \leqslant \mu(E) + | \mu(A)| + | \mu(E \sim A)|$$
$$\leqslant \mu(E) + | \mu |(E);$$

hence

$$\sup_{A \subset E} \mu(A) \leqslant \mu^+(E). \tag{1}$$

Since, by I.4.1, $| \mu |$ is bounded, for every $\epsilon > 0$ there exist $k \in \mathbb{N}$ and disjoint subsets $A_1, \ldots, A_k$ of E in Σ such that $\bigcup_{j=1}^{k} A_j = E$ and

$$| \mu | (E) \leqslant \sum_{j=1}^{k} | \mu(A_j)| + \epsilon.$$

We let $J \triangleq \{j \in \mathbb{N} \mid \mu(A_j) \geqslant 0\}$ and observe that

$$\begin{aligned} 2\mu^+(E) &= | \mu | (E) + \mu(E) \\ &\leqslant 2 \sum_{j \in J} \mu(A_j) + \epsilon \\ &= 2\mu\Big(\bigcup_{j \in J} A_j\Big) + \epsilon \\ &\leqslant 2 \sup_{A \subset E} \mu(A) + \epsilon. \end{aligned}$$

This last relation, together with (1), shows that

$$\mu^+(E) = \sup_{A \subset E} \mu(A) \geqslant \mu(\varnothing) = 0.$$

We also observe that $\mu^- = (-\mu)^+$, yielding $\mu^-(E) = -\inf_{A \subset E} \mu(A) \geqslant 0$. The relations $\mu(E) = \mu^+(E) - \mu^-(E)$ and $|\mu|(E) = \mu^+(E) + \mu^-(E)$ follow from the definition of μ^+ and μ^- and imply that $|\mu(E)| \leqslant |\mu|(E)$. Finally, if μ is positive, then $\mu^-(E) = -\inf_{A \subset E} \mu(A) = 0$ for all $E \in \Sigma$; hence $\mu = \mu^+ = |\mu|$. QED

I.4.3 ***Theorem*** Let (S, Σ, μ) be a finite measure space. Then μ, $|\mu|$, μ^+, and μ^- are bounded finite measures and $\mu(E) \leqslant \mu^+(E) \leqslant |\mu|(E)$ for all $E \in \Sigma$.

▌PROOF If $A_1 \supset A_2 \supset \cdots$ and $A_j \in \Sigma$, then $\bigcap_{j=1}^{\infty} A_j$, $A_1 \sim A_2$, $A_2 \sim A_3$,... is a partition of A_1 into elements of Σ and, therefore,

$$\mu(A_1) = \mu\left(\bigcap_{j=1}^{\infty} A_j\right) + \sum_{i=1}^{\infty} \mu(A_i \sim A_{i+1})$$

$$= \mu\left(\bigcap_{j=1}^{\infty} A_j\right) + \lim_n \sum_{i=1}^{n} \mu(A_i \sim A_{i+1}) \in \mathbb{R};$$

hence $\lim_n \sum_{i=n}^{\infty} \mu(A_i \sim A_{i+1}) = 0$ and

$$\text{(1)} \qquad \lim_n \mu(A_n) = \mu\left(\bigcap_{j=1}^{\infty} A_j\right) + \lim_n \sum_{i=n}^{\infty} \mu(A_i \sim A_{i+1}) = \mu\left(\bigcap_{j=1}^{\infty} A_j\right) \in \mathbb{R}.$$

Now assume that $\sup_{E \subset S} \mu(E) = \infty$, and let $\mathscr{A}$ be the subcollection of Σ with the property that, for every $A \in \mathscr{A}$, $\sup_{B \subset A} \mu(B) = \infty$. Then there exist some $A \in \mathscr{A}$ and $c \in \mathbb{R}$ such that $\mu(B) \leqslant c$ if $B \in \mathscr{A}$ and $B \subset A$; indeed, otherwise we could find a sequence (A_j) in Σ such that $A_1 \supset A_2 \supset \cdots$ and $\lim_n \mu(A_n) = \infty$, contradicting (1). If A and c are as just defined, then there exists some $A_1 \subset A$ with $\mu(A_1) > c$ and, therefore, $A_1 \notin \mathscr{A}$. It follows that $A \sim A_1 \in \mathscr{A}$ and, therefore, $\mu(B) \leqslant c$ if $B \in \mathscr{A}$ and $B \subset A \sim A_1 \subset A$. Continuing in this manner, we determine sets A_1, A_2, A_3,... such that $\mu(A_j) > c$ and $A_{j+1} \subset A \sim \bigcup_{i=1}^{j} A_i$. Then

$$\mu\left(\bigcup_{j=1}^{\infty} A_j\right) = \sum_{j=1}^{\infty} \mu(A_j) > \sum_{j=1}^{\infty} c = \infty,$$

contradicting the assumption that μ is a finite measure.

Thus the assumption that $\sup_{E \subset S} \mu(E) = \infty$ is inadmissible. Applying the same argument to the finite measure $-\mu$ shows that

$$-\infty < -\sup_{E \subset S} (-\mu(E)) = \inf_{E \subset S} \mu(E).$$

Thus the finite measure μ is bounded and, by I.4.1, so are $|\mu|$, μ^+, and μ^-.

We shall next show that $|\mu|$ is a finite measure. By I.4.1, $|\mu|$ is a finite additive set function. Now let $E \in \Sigma$ and $A_1, A_2, \ldots$ be a denumerable partition of E into elements of Σ. Then, for each $k \in \mathbb{N}$, we have

$$\sum_{j=1}^{k} |\mu|(A_j) = |\mu|\left(\bigcup_{j=1}^{k} A_j\right) \leqslant |\mu|(E);$$

hence

$$\sum_{j=1}^{\infty} |\mu|(A_j) \leqslant |\mu|(E), \tag{2}$$

On the other hand, for every finite sequence $(B_1, \ldots, B_m)$ in Σ of disjoint subsets of E, we have (by I.3.14)

$$\begin{aligned}
\sum_{i=1}^{m} |\mu(B_i)| &= \sum_{i=1}^{m} \left| \mu\left(\bigcup_{j=1}^{\infty} B_i \cap A_j\right)\right| \\
&= \sum_{i=1}^{m} \left| \sum_{j=1}^{\infty} \mu(B_i \cap A_j)\right| \\
&\leqslant \sum_{i=1}^{m} \sum_{j=1}^{\infty} |\mu(B_i \cap A_j)| \\
&= \sum_{j=1}^{\infty} \sum_{i=1}^{m} |\mu(B_i \cap A_j)| \\
&\leqslant \sum_{j=1}^{\infty} |\mu|(A_j),
\end{aligned}$$

implying that $|\mu|(E) \leqslant \sum_{j=1}^{\infty} |\mu|(A_j)$ and, by (2), that

$$|\mu|(E) = \sum_{j=1}^{\infty} |\mu|(A_j).$$

Thus $|\mu|$ is a finite measure and so are the functions $\mu^+ \triangleq \frac{1}{2}(|\mu| + \mu)$ and $\mu^- \triangleq \frac{1}{2}(|\mu| - \mu)$. For each $E \in \Sigma$, we have

$$\mu(E) = \mu^+(E) - \mu^-(E) \leqslant \mu^+(E) \leqslant \mu^+(E) + \mu^-(E) = |\mu|(E). \quad \text{QED}$$

I.4.4 ***Lemma*** Let (S, Σ, μ) be a positive finite measure space and (E_j) a sequence in Σ. Then

$$\mu\left(\bigcup_{i=1}^{\infty} \bigcap_{j=i}^{\infty} E_j\right) = \lim_n \mu\left(\bigcap_{j=n}^{\infty} E_j\right) \leqslant \liminf_n \mu(E_n) \tag{1}$$

and

(2)
$$\mu\left(\bigcap_{i=1}^{\infty}\bigcup_{j=i}^{\infty}E_j\right)=\lim_n \mu\left(\bigcup_{j=n}^{\infty}E_j\right)\geqslant \limsup_n \mu(E_n).$$

PROOF We observe that if $B_1 \subset B_2 \subset \cdots$ and $B_i \in \Sigma$, then

(3)
$$\begin{aligned}\mu\left(\bigcup_{i=1}^{\infty}B_i\right)&=\mu(B_1\cup(B_2\sim B_1)\cup(B_3\sim B_2)\cup\cdots)\\&=\mu(B_1)+\sum_{i=1}^{\infty}\mu(B_{i+1}\sim B_i)\\&=\mu(B_1)+\lim_n\sum_{i=1}^{n-1}\mu(B_{i+1}\sim B_i)\\&=\lim_n \mu(B_n).\end{aligned}$$

If we set $B_i \triangleq \bigcap_{j=i}^{\infty} E_j$, then, for each $n \in \mathbb{N}$, $\mu(B_n) \leqslant \mu(E_n)$ which, together with (3), yields relation (1).

If $A_1 \supset A_2 \supset \cdots$ and $A_i \in \Sigma$, then we set $B_i \triangleq S \sim A_i$ and apply (3) to derive the relation

(4)
$$\begin{aligned}\mu\left(\bigcap_{i=1}^{\infty}A_i\right)&=\mu\left(\bigcap_{i=1}^{\infty}(S\sim B_i)\right)\\&=\mu\left(S\sim\bigcup_{i=1}^{\infty}B_i\right)\\&=\mu(S)-\mu\left(\bigcup_{i=1}^{\infty}B_i\right)\\&=\mu(S)-\lim_n\mu(B_n)=\lim_n\mu(A_n).\end{aligned}$$

If we set $A_i \triangleq \bigcup_{j=i}^{\infty} E_j$, then, for each $n \in \mathbb{N}$, $\mu(A_n) \geqslant \mu(E_n)$ which, together with (4), yields relation (2). QED

I.4.5 ***Theorem*** (Hahn decomposition) Let (S, Σ, μ) be a finite measure space. Then there exists a set $S_0 \in \Sigma$ such that

$$\mu^+(A)=\mu(S_0\cap A)\quad\text{and}\quad -\mu^-(A)=\mu((S\sim S_0)\cap A)\quad\text{for all } A\in\Sigma.$$

PROOF By I.4.3, $|\mu|$, μ^+, and μ^- are bounded positive finite measures and, by the Jordan decomposition theorem I.4.2, for every $j \in \mathbb{N}$ there exists $E_j \in \Sigma$ such that

$$\text{(1)} \qquad \mu^+(E_j) \geqslant \mu^+(E_j) - \mu^-(E_j) = \mu(E_j) \geqslant \mu^+(S) - 2^{-j};$$

hence

$$\text{(2)} \qquad \mu^+(S \sim E_j) = \mu^+(S) - \mu^+(E_j) \leqslant 2^{-j}.$$

Since $\mu = \mu^+ - \mu^-$, it follows from (1) that

$$\text{(3)} \qquad \mu^-(E_j) = \mu^+(E_j) - \mu(E_j) \leqslant \mu^+(S) - \mu(E_j) \leqslant 2^{-j}.$$

We now set $S_0 \triangleq S \sim \bigcup_{i=1}^{\infty} \bigcap_{j=i}^{\infty} (S \sim E_j) = \bigcap_{i=1}^{\infty} \bigcup_{j=i}^{\infty} E_j$. Then I.4.4(1) and (2) yield

$$\mu^+(S \sim S_0) = \mu^+ \left(\bigcup_{i=1}^{\infty} \bigcap_{j=i}^{\infty} (S \sim E_j) \right) \leqslant \liminf_n \mu^+(S \sim E_n) = 0;$$

hence $\mu^+(S \sim S_0) = 0$. Furthermore, by (3),

$$\mu^-(S_0) = \mu^- \left(\bigcap_{i=1}^{\infty} \bigcup_{j=i}^{\infty} E_j \right) \leqslant \mu^- \left(\bigcup_{j=n}^{\infty} E_j \right) \leqslant \sum_{j=n}^{\infty} \mu^-(E_j) \leqslant 2^{-n+1} \qquad (n \in \mathbb{N}).$$

Thus $\mu^+(S \sim S_0) = \mu^-(S_0) = 0$; hence $\mu^-(S_0 \cap A) = \mu^+((S - S_0) \cap A) = 0$ for all $A \in \Sigma$, yielding

$$\mu(S_0 \cap A) = \mu^+(S_0 \cap A) - \mu^-(S_0 \cap A) = \mu^+(S_0 \cap A) = \mu^+(A)$$

and

$$\begin{aligned} \mu((S \sim S_0) \cap A) &= \mu^+((S \sim S_0) \cap A) - \mu^-((S \sim S_0) \cap A) \\ &= -\mu^-((S \sim S_0) \cap A) = -\mu^-(A). \end{aligned}$$

QED

We shall henceforth refer to any set with the properties of S_0 in the Hahn decomposition I.4.5 as a *positive set for* μ and to its complement as a *negative set for* μ.

I.4.6 ***Theorem*** Let S be a topological space, Σ a field in S, and $\mu : \Sigma \to \mathbb{R}$ a bounded regular additive set function. Then $|\mu|$, μ^+, and μ^- are bounded regular additive set functions.

▌PROOF By I.4.1, $|\mu|$, μ^+, and μ^- are bounded and additive and $|\mu|$ is therefore regular by the definition of the regularity of μ. If $E, A, B \in \Sigma$, $\bar{A} \subset E \subset B^\circ$, and $|\mu|(B \sim A) = \mu^+(B \sim A) + \mu^-(B \sim A) \leqslant \epsilon$, then $\mu^+(B \sim A) \leqslant \epsilon$ and $\mu^-(B \sim A) \leqslant \epsilon$, showing that μ^+ and μ^- are regular.

QED

I.4.7 ***Theorem*** Let (S, Σ, μ) and (S, Σ, λ) be finite measure spaces and λ be μ-continuous. Then $\lim_i \lambda(A_i) = 0$ whenever $\lim_i |\mu|(A_i) = 0$.

▌PROOF Since $\lambda = \lambda^+ - \lambda^-$ and λ is μ-continuous if and only if it is $|\mu|$-continuous, we may assume that both μ and λ are finite positive measures. Let (A_i) be a sequence in Σ such that $\lim_i \mu(A_i) = 0$. If $(\lambda(A_i))_i$ does not converge to 0, then there exist $\epsilon > 0$ and a sequence $J \subset (1, 2, \ldots)$ such that $\lambda(A_i) > \epsilon$ and $\mu(A_i) \leqslant 2^{-i}$ for all $i \in J$. We shall assume that $A_1, A_2, \ldots$ were chosen so that $J = (1, 2, \ldots)$. Now let $E \triangleq \bigcap_{i=1}^{\infty} \bigcup_{j=i}^{\infty} A_j$. Then, for each $k \in \mathbb{N}$,

$$\mu(E) \leqslant \mu\left(\bigcup_{j=k}^{\infty} A_j\right) \leqslant \sum_{j=k}^{\infty} \mu(A_j) \leqslant \sum_{j=k}^{\infty} 2^{-j} = 2^{-k+1};$$

hence $\mu(E) = 0$ and therefore $\lambda(E) = 0$. By I.4.4(2), we have

$$\lambda(E) \geqslant \limsup_n \lambda(A_n) \geqslant \epsilon,$$

contradicting $\lambda(E) = 0$. QED

I.4.8 ***Theorem*** Let (S, Σ, μ) be a finite measure space, $\Sigma^* \triangleq \{E \cup Z \mid E \in \Sigma, Z \text{ is a } \mu\text{-null set}\}$, and $\mu^*(E \cup Z) \triangleq \mu(E)$ for all $E \in \Sigma$ and all μ-null sets Z. Then (S, Σ^*, μ^*) is a finite measure space, μ^* is complete and $\mu = \mu^* \mid \Sigma$.

▌PROOF Let $E_1, E_2, B_1, B_2 \in \Sigma$, $Z_1 \subset B_1$, $Z_2 \subset B_2$, $|\mu|(B_1) = |\mu|(B_2) = 0$, and $E_1 \cup Z_1 = E_2 \cup Z_2$. Then $E_1 \sim E_2 \subset B_2$ and $E_2 \sim E_1 \subset B_1$ implying, by I.4.2, that

$$|\mu(E_1) - \mu(E_1 \cap E_2)| = |\mu(E_1 \sim E_2)| \leqslant |\mu|(E_1 \sim E_2) \leqslant |\mu|(B_2) = 0.$$

Similarly, $|\mu(E_2) - \mu(E_1 \cap E_2)| = 0$, showing that $\mu(E_1) = \mu(E_1 \cap E_2) = \mu(E_2)$. Thus $\mu^*(A)$ is uniquely defined for all $A \in \Sigma^*$.

Now let $E_i, B_i \in \Sigma$, $|\mu|(B_i) = 0$, and $Z_i \subset B_i$ $(i \in \mathbb{N})$. Then, by I.4.3, $|\mu|(\bigcup_{i=1}^{\infty} B_i) = \sum_{i=1}^{\infty} |\mu|(B_i) = 0$ and therefore

$$\bigcup_{i=1}^{\infty} (E_i \cup Z_i) = \left(\bigcup_{i=1}^{\infty} E_i\right) \cup \left(\bigcup_{i=1}^{\infty} Z_i\right) \in \Sigma^*.$$

This shows that Σ^* contains the union of every denumerable subcollection of Σ^* and, since $\varnothing \in \Sigma^*$, Σ^* also contains the union of all of its finite subcollections.

Now, if $E, U \in \Sigma$, $|\mu|(U) = 0$, and $Z \subset U$, then $S \sim (E \cup Z) = (S \sim E) \cap (S \sim Z) \supset (S \sim E) \cap (S \sim U)$. Thus $S \sim (E \cup Z) = E_1 \cup Z_1$, where $E_1 \triangleq (S \sim E) \cap (S \sim U) \in \Sigma$ and $Z_1 \triangleq (S \sim E) \cap (U \sim Z) \subset U$, showing that Σ^* contains the complement of every one of its elements. Thus Σ^* is a σ-field.

If $A_1, A_2, \ldots$ are disjoint elements of Σ^*, with $A_i = E_i \cup Z_i$, $E_i \in \Sigma$ and Z_i μ-null, then $\bigcup_{i=1}^{\infty} Z_i$ is μ-null and, therefore,

$$\mu^*\left(\bigcup_{i=1}^{\infty} A_i\right) = \mu^*\left(\left(\bigcup_{i=1}^{\infty} E_i\right) \cup \left(\bigcup_{i=1}^{\infty} Z_i\right)\right)$$

$$= \mu\left(\bigcup_{i=1}^{\infty} E_i\right) = \sum_{i=1}^{\infty} \mu(E_i) = \sum_{i=1}^{\infty} \mu^*(E_i \cup Z_i).$$

Thus μ^* is a finite measure.

If A is μ^*-null, then $A \subset E \cup Z$, where $E \in \Sigma$, $|\mu|(E) = 0$, and Z is μ-null. Thus A is μ-null and, therefore, $A \in \Sigma^*$, showing that μ^* is complete. QED

The finite measure space (S, Σ^*, μ^*) defined in I.4.8 is referred to as the *Lebesgue extension* or the *Lebesgue completion* of (S, Σ, μ) and μ^* is the *Lebesgue extension* of μ. A set $A \in \Sigma^*$ is also called *μ-measurable*. It is clear that (S, Σ^*, μ^*) is its own Lebesgue extension and a set is μ-measurable if and only if it is μ^*-measurable. We shall usually denote by the same symbol both a measure and its Lebesgue extension.

I.4.9 ***Theorem*** Let S be a topological space, (S, Σ, μ_j) $(j \in \mathbb{N})$ regular positive finite measure spaces, and $\mu_j(S) \leqslant c < \infty$ $(j \in \mathbb{N})$ for some c. Then the function $\mu : \Sigma \to \mathbb{R}$, defined by the relation

$$\mu(A) \triangleq \sum_{j=1}^{\infty} 2^{-j}\mu_j(A) \qquad (A \in \Sigma),$$

is a regular positive finite measure on Σ.

▌PROOF The series defining $\mu(A)$ is clearly convergent and $\mu(A) \geqslant 0$ for all $A \in \Sigma$. Clearly $\mu(\varnothing) = 0$. Now let $A_1, A_2, \ldots \in \Sigma$ be disjoint and $\bigcup_{i=1}^{\infty} A_i = A$. Then, by I.3.14(3),

$$\mu(A) \triangleq \sum_{j=1}^{\infty} 2^{-j}\mu_j(A) = \sum_{j=1}^{\infty} 2^{-j} \sum_{i=1}^{\infty} \mu_j(A_i) = \sum_{i=1}^{\infty}\sum_{j=1}^{\infty} 2^{-j}\mu_j(A_i) = \sum_{i=1}^{\infty} \mu(A_i)$$

and $\mu(A) \leqslant c \cdot \sum_{j=1}^{\infty} 2^{-j} = c$ because all the terms in the series are non-negative. Thus μ is a positive finite measure. Finally, let $E \in \Sigma$ and $\epsilon > 0$. We can determine $k \in \mathbb{N}$ such that $\sum_{j=k+1}^{\infty} 2^{-j}\mu_j(C) \leqslant \epsilon/2$ $(C \in \Sigma)$. Then for each $j \in \{1, 2, \ldots k\}$ there exist $A_j, B_j \in \Sigma$ such that $\bar{A}_j \subset E \subset B_j^{\circ}$ and

$\mu_j(B_j \sim A_j) \leqslant \epsilon/2$. Let $A \triangleq \bigcup_{j=1}^{k} A_j$ and $B \triangleq \bigcap_{j=1}^{k} B_j$. Then $\bar{A} \subset E \subset B^\circ$ and $\mu_j(B \sim A) \leqslant \epsilon/2$ for $j = 1,..., k$. We have

$$\mu(B \sim A) = \sum_{j=1}^{\infty} 2^{-j}\mu_j(B \sim A) \leqslant \sum_{j=1}^{k} 2^{-j}\mu_j(B \sim A) + \epsilon/2$$

$$\leqslant \epsilon \sum_{j=1}^{k} 2^{-j-1} + \epsilon/2 \leqslant \epsilon.$$

Thus μ is regular. QED

I.4.10 ***Theorem*** (Saks) Let (S, Σ, μ) be a measure space, μ finite, positive, and nonatomic, and $M \in \Sigma$. Then there exists a function $B : [0, 1] \to \Sigma$ such that $B(\alpha) \subset B(\beta)$ if $0 \leqslant \alpha \leqslant \beta \leqslant 1$, $B(0) = \varnothing$, $B(1) = M$ and $\mu(B(\alpha)) = \alpha\mu(M)$ for all $\alpha \in [0, 1]$.

▌PROOF *Step* 1 Let $\mu(C_0) > 0$. We first show that for every $\epsilon > 0$ there exists $D \subset C_0$ with $0 < \mu(D) \leqslant \epsilon$. This will be the case if there exists a sequence (D_j) in Σ such that $D_j \subset C_0$, $\mu(D_j) > 0$ and $\lim_j \mu(D_j) = 0$. Indeed, since μ is nonatomic, there exists $C_1 \subset C_0$ such that $0 < \mu(C_1) < \mu(C_0)$ and, recursively, $C_2, C_3,...$ such that $C_{j+1} \subset C_j$ and $0 < \mu(C_{j+1}) < \mu(C_j)$. We observe that $(C_j \sim C_{j+1}) \cap (C_k \sim C_{k+1}) = \varnothing$ for $k \neq j$ and $\mu(\bigcup_{j=0}^{\infty} (C_j \sim C_{j+1})) \leqslant \mu(S) < \infty$; hence $D_j \triangleq C_j \sim C_{j+1} \subset C_0$, $\mu(D_j) > 0$, and $\lim_j \mu(D_j) = 0$.

Step 2 We next show that if $\mu(C) > 0$ and $\epsilon > 0$, then there exists a finite partition $C_1, C_2,..., C_k$ of C such that $0 < \mu(C_j) \leqslant \epsilon$. For every $A \in \Sigma$ we set $\Sigma_A \triangleq \{E \in \Sigma \mid E \subset A, \mu(E) \leqslant \epsilon\}$ and $s(A) \triangleq \sup \mu(\Sigma_A)$. By Step 1, we have $s(C) > 0$ and thus we can determine a set $C_1' \subset C$ with $s(C)/2 \leqslant \mu(C_1') \leqslant \epsilon$. We now choose recursively for $j = 1, 2,...$ a subset C'_{j+1} of $C \sim \bigcup_{i=1}^{j} C_i'$ satisfying the relation $\frac{1}{2}s(C \sim \bigcup_{i=1}^{j} C_i') \leqslant \mu(C'_{j+1}) \leqslant \epsilon$. Since the sets $C_1', C_2',...$ are disjoint and $\sum_{j=1}^{\infty} \mu(C_j') \leqslant \mu(C) < \infty$, we have $\lim_j \mu(C_j') = 0$; hence $s(C \sim \bigcup_{j=1}^{\infty} C_j') = 0$ implying $\mu(C \sim \bigcup_{j=1}^{\infty} C_j') = 0$ and $\mu(C) = \sum_{j=1}^{\infty} \mu(C_j')$. We choose $k \in \mathbb{N}$ so that $\sum_{j=k}^{\infty} \mu(C_j') \leqslant \epsilon$ and set $C_j \triangleq C_j'$ $(j = 1,..., k-1)$ and $C_k \triangleq C \sim \bigcup_{j=1}^{k-1} C_j'$.

Step 3 We now prove that for every $A \in \Sigma$ there exists $A' \subset A$ with $\mu(A') = \frac{1}{2}\mu(A)$. If $\mu(A) = 0$, then $A' \triangleq A$ yields $\mu(A') = \frac{1}{2}\mu(A)$. We therefore assume that $\mu(A) > 0$. By Step 2, if $\mu(C) > 0$ and $m \in \mathbb{N}$, we can determine a partition $C_1^m,..., C_{k(m)}^m$ of C with $0 < \mu(C_j^m) \leqslant 1/m$. Then the sets $D^m(j, C) \triangleq \bigcup_{i=1}^{j} C_i^m \subset C$ belong to Σ and are such that for every point $\theta \in [0, \mu(C)]$ there exists some $j(\theta) \in \{0, 1,..., k(m)\}$ with $\theta - 1/m < \mu(D^m(j(\theta), C)) \leqslant \theta$. We can therefore determine $D_1 \subset A$ with $\frac{1}{2}\mu(A) - 1 < \mu(D_1) \leqslant \frac{1}{2}\mu(A)$, then recursively, for $j = 1, 2,..., D_{j+1} \subset A \sim \bigcup_{i=1}^{j} D_i$ with $\frac{1}{2}\mu(A) - \sum_{i=1}^{j} \mu(D_i) - 1/j < \mu(D_{j+1}) \leqslant \frac{1}{2}\mu(A) - \sum_{i=1}^{j} \mu(D_i)$. It follows that $A' \triangleq \bigcup_{j=1}^{\infty} D_j \subset A$ and $\mu(A') = \frac{1}{2}\mu(A)$.

Step 4. We can now complete the proof. We set $B(0) = \varnothing$ and $B(1) = M$. By Step 3, there exists $B(\frac{1}{2}) \subset M$ with $\mu(B(\frac{1}{2})) = \frac{1}{2}\mu(M)$. Now assume, for purposes of induction, that $k \in \mathbb{N}$ and there exist Σ-measurable subsets $B(j2^{-k})$ of M $(j = 0, 1, \ldots, 2^k - 1)$ such that $B(j2^{-k}) \subset B((j+1)2^{-k})$ and $\mu(B(j2^{-k})) = j2^{-k}\mu(M)$. Then it follows from Step 3 that for each $j \in \{0, 1, \ldots, 2^k - 1\}$ there exists $B'_{k,j} \subset B((j+1)2^{-k}) \sim B(j2^{-k})$ with $\mu(B'_{k,j}) = 2^{-k-1}\mu(M)$. We set $B((2l+1)2^{-k-1}) \triangleq B(l2^{-k}) \cup B'_{k,l}$ and verify that $B(l2^{-k-1}) \subset B((l+1)2^{-k-1})$ and $\mu(B(l2^{-k-1})) = l2^{-k-1}\mu(M)$ for $l = 0, 1, \ldots, 2^{k+1} - 1$.

We thus define $B(j2^{-k})$ $(j = 0, 1, \ldots, 2^k)$ for $k = 0, 1, 2, \ldots$. If $0 < \alpha < 1$ we can form an increasing sequence $(j_i 2^{-k_i})$ in $\mathbb{R}$ with $\lim_i j_i 2^{-k_i} = \alpha$. We set

$$B(\alpha) = B(j_1 2^{-k_1}) + \bigcup_{i=1}^{\infty} \left(B(j_{i+1}2^{-k_{i+1}}) \sim B(j_i 2^{-k_i})\right)$$

and verify that the function $B(\cdot)$ satisfies the conditions of the theorem. QED

We shall next indicate how certain finite additive set functions defined on a field Σ are extended to finite measures on the smallest σ-field containing Σ. This is primarily aimed at defining the Borel and Lebesgue measures and will also be used in the proof of the Riesz representation theorem I.5.8.

I.4.11 ***Theorem*** (Alexandroff) Let S be a compact topological space, Σ a field in S, and $\mu : \Sigma \to \mathbb{R}$ additive, regular, and bounded. Then μ is countably additive on the field Σ.

▌PROOF We shall first show that $|\mu|$, which by I.4.1 is bounded and additive, is countably additive. Let (E_j) be a sequence of disjoint elements of Σ with $E \triangleq \bigcup_{j=1}^{\infty} E_j \in \Sigma$, and $\epsilon > 0$. Then there exist $A, B_j \in \Sigma$ $(j \in \mathbb{N})$ such that $\bar{A} \subset E$, $E_j \subset B_j^\circ$, $|\mu|(E \sim A) \leqslant \epsilon/2$, and $|\mu|(B_j \sim E_j) \leqslant \epsilon 2^{-j-1}$. The sets $B_1^\circ, B_2^\circ, \ldots$ are an open covering of the closed (and therefore compact) subset $\bar{A}$ of S and there exists, therefore, a finite subcovering, say $B_1^\circ, B_2^\circ, \ldots, B_k^\circ$. We have

$$\begin{aligned}
|\mu|(E) &= |\mu|(A) + |\mu|(E \sim A) \\
&\leqslant |\mu|(A) + \epsilon/2 \\
&\leqslant |\mu|\left(\bigcup_{j=1}^{k} B_j\right) + \epsilon/2 \\
&\leqslant \sum_{j=1}^{k} |\mu|(B_j) + \epsilon/2 \\
&\leqslant \sum_{j=1}^{\infty} |\mu|(E_j) + \epsilon.
\end{aligned}$$

On the other hand,

$$\sum_{j=1}^{m} |\mu|(E_j) = |\mu|\left(\bigcup_{j=1}^{m} E_j\right) \leqslant |\mu|(E) \qquad \text{for all} \quad m \in \mathbb{N}.$$

Since ϵ and m are arbitrary, these last two relations show that $|\mu|$ is countably additive.

By I.4.1, $|\mu|$ is bounded; hence $\sum_{j=1}^{\infty} |\mu|(E_j) = |\mu|(E) < \infty$ and $\lim_n \sum_{j=n+1}^{\infty} |\mu|(E_j) = 0$. It follows that

$$\begin{aligned}\lim_n \left| \mu(E) - \sum_{j=1}^{n} \mu(E_j) \right| &= \lim_n \left| \mu\left(\bigcup_{j=n+1}^{\infty} E_j\right) \right| \\ &\leqslant \lim_n |\mu| \left(\bigcup_{j=n+1}^{\infty} E_j\right) \\ &= \lim_n \sum_{j=n+1}^{\infty} |\mu|(E_j) = 0,\end{aligned}$$

showing that μ is countably additive. QED

Let $\mathscr{P}(S)$ be the collection of all the subsets of a set S. For every function $\nu : \mathscr{P}(S) \to \mathbb{R}$ we set

$$\mathscr{P}_\nu \triangleq \{E \in \mathscr{P}(S) \mid \nu(A) = \nu(A \cap E) + \nu(A \sim E)\,(A \in \mathscr{P}(S))\}.$$

A function $\nu : \mathscr{P}(S) \to \mathbb{R}$ is a *finite outer measure* if $\nu(\varnothing) = 0$, $\nu(A) \leqslant \nu(B)$ for $A \subset B$ and $\nu(\bigcup_{j=1}^{\infty} A_j) \leqslant \sum_{j=1}^{\infty} \nu(A_j)$.

I.4.12 ***Lemma*** Let Σ be a field in S and $\mu : \Sigma \to \mathbb{R}$ nonnegative and countably additive. For each $A \subset S$, let $\tilde{\mu}(A) \triangleq \inf\{\sum_{j=1}^{\infty} \mu(A_j) \mid A_j \in \Sigma, A \subset \bigcup_{j=1}^{\infty} A_j\}$. Then $\tilde{\mu}$ is a finite outer measure, $\Sigma \subset \mathscr{P}_{\tilde{\mu}}$ and $\tilde{\mu}(A) = \mu(A)$ for $A \in \Sigma$.

▌PROOF It is clear that $\tilde{\mu}$ is finite, $\tilde{\mu}(\varnothing) = 0$, and $\tilde{\mu}(A) \leqslant \tilde{\mu}(B)$ for $A \subset B$. Now let $(E^i)_i$ be a sequence in $\mathscr{P}(S)$. Then for each $\epsilon > 0$ and $i \in \mathbb{N}$ there exists a sequence $(A_j^i)_j$ in Σ such that $E^i \subset \bigcup_{j=1}^{\infty} A_j^i$ and $\tilde{\mu}(E^i) \geqslant \sum_{j=1}^{\infty} \mu(A_j^i) - \epsilon 2^{-i}$. Then, by the definition of $\tilde{\mu}$,

$$\tilde{\mu}\left(\bigcup_{i=1}^{\infty} E^i\right) \leqslant \sum_{i=1}^{\infty} \sum_{j=1}^{\infty} \mu(A_j^i) \leqslant \sum_{i=1}^{\infty} \tilde{\mu}(E^i) + \epsilon,$$

showing that $\tilde{\mu}$ is a finite outer measure.

Now let $E \in \Sigma$ and $A \in \mathscr{P}(S)$. By the definition of $\tilde{\mu}$, for every $\epsilon > 0$ there exists a sequence (E_j) in Σ such that $A \subset \bigcup_{j=1}^{\infty} E_j$ and $\tilde{\mu}(A) \geqslant \sum_{j=1}^{\infty} \mu(E_j) - \epsilon$; hence

$$\text{(1)} \qquad \tilde{\mu}(A \cap E) + \tilde{\mu}(A \sim E) \leqslant \tilde{\mu}\left(\bigcup_{j=1}^{\infty} E_j \cap E\right) + \tilde{\mu}\left(\bigcup_{j=1}^{\infty} (E_j \sim E)\right)$$

$$\leqslant \sum_{j=1}^{\infty} \mu(E_j \cap E) + \sum_{j=1}^{\infty} \mu(E_j \sim E)$$

$$= \sum_{j=1}^{\infty} \mu(E_j) \leqslant \tilde{\mu}(A) + \epsilon.$$

On the other hand, $\tilde{\mu}(A) = \tilde{\mu}(A \cap E \cup (A \sim E)) \leqslant \tilde{\mu}(A \cap E) + \tilde{\mu}(A \sim E)$, which together with (1) shows that $E \in \mathscr{P}_{\tilde{\mu}}$.

Finally, if $E \in \Sigma$ and (E_j) is a sequence in Σ with $E \subset \bigcup_{j=1}^{\infty} E_j$, we set $F_j \triangleq E_j \sim \bigcup_{i=1}^{j-1} E_i$ $(j \in \mathbb{N})$. Then $F_1, F_2, \ldots$ are disjoint elements of Σ and $E \subset \bigcup_{j=1}^{\infty} F_j$. It follows that

$$\mu(E) = \mu\left(\bigcup_{j=1}^{\infty} E \cap F_j\right) = \sum_{j=1}^{\infty} \mu(E \cap F_j) \leqslant \sum_{j=1}^{\infty} \mu(F_j) \leqslant \sum_{j=1}^{\infty} \mu(E_j);$$

hence $\mu(E) \leqslant \tilde{\mu}(E)$. By the definition of $\tilde{\mu}$, however, $\tilde{\mu}(E) \leqslant \mu(E)$ for $E \in \Sigma$, implying that $\mu(E) = \tilde{\mu}(E)$ for $E \in \Sigma$. QED

I.4.13 ***Theorem*** (Caratheodory) If $\nu : \mathscr{P}(S) \to \mathbb{R}$ and $\nu(\varnothing) = 0$, then $\mathscr{P}_\nu$ is a field and $\nu \mid \mathscr{P}_\nu$ is additive. If, furthermore, ν is a finite outer measure, then $\mathscr{P}_\nu$ is a σ-field and the restriction of ν to $\mathscr{P}_\nu$ is a finite measure.

▌PROOF We first assume that $\nu(\varnothing) = 0$ and observe that $\mathscr{P}_\nu$ contains $\varnothing$ and the complement in S of every one of its elements. We shall next show that $\mathscr{P}_\nu$ contains the intersection of any two of its elements. Indeed, if $E_1, E_2 \in \mathscr{P}_\nu$, then for every $A \in \mathscr{P}(S)$ we have

$$\nu(A) = \nu(A \cap E_1) + \nu(A \sim E_1)$$

and

$$\nu(A \cap E_1) = \nu(A \cap E_1 \cap E_2) + \nu(A \cap E_1 \sim E_2);$$

hence

$$\text{(1)} \qquad \nu(A) = \nu(A \cap E_1 \cap E_2) + \nu(A \cap E_1 \sim E_2) + \nu(A \sim E_1).$$

Similarly,

$$\begin{aligned} \nu(A \sim E_1 \cap E_2) &= \nu((A \sim E_1 \cap E_2) \cap E_1) + \nu((A \sim E_1 \cap E_2) \sim E_1) \\ &= \nu(A \cap E_1 \sim E_2) + \nu(A \sim E_1) \end{aligned}$$

which, combined with (1), yields

$$\nu(A) = \nu(A \cap E_1 \cap E_2) + \nu(A \sim E_1 \cap E_2)$$

Thus $E_1 \cap E_2 \in \mathscr{P}_\nu$; hence $E_1 \cup E_2 = S \sim (S \sim E_1) \cap (S \sim E_2) \in \mathscr{P}_\nu$, showing that $\mathscr{P}_\nu$ is a field.

We next observe that, if $A \subset S$, $E_1, E_2 \in \mathscr{P}_\nu$, $E_1 \cap E_2 = \varnothing$, and $B \triangleq A \cap (E_1 \cup E_2)$, then

$$\text{(2)} \qquad \nu(B) = \nu(B \cap E_1) + \nu(B \sim E_1) = \nu(A \cap E_1) + \nu(A \cap E_2).$$

This shows, setting $A = S$, that $\nu \mid \mathscr{P}_\nu$ is additive.

Now we assume that ν is a finite outer measure, (E_j) a sequence of disjoint elements of $\mathscr{P}_\nu$, $E \triangleq \bigcup_{j=1}^\infty E_j$, $A \in \mathscr{P}(S)$, and $k \in \mathbb{N}$. Since ν is a finite outer measure, we have, by (2),

$$\begin{aligned} \nu(A) &= \nu\left(A \cap \left(\bigcup_{j=1}^k E_j\right)\right) + \nu\left(A \sim \bigcup_{j=1}^k E_j\right) \\ &= \sum_{j=1}^k \nu(A \cap E_j) + \nu\left(A \sim \bigcup_{j=1}^k E_j\right) \\ &\geqslant \sum_{j=1}^k \nu(A \cap E_j) + \nu(A \sim E). \end{aligned}$$

Since k is arbitrary, we have $\nu(A) \geqslant \sum_{j=1}^\infty \nu(A \cap E_j) + \nu(A \sim E)$ and therefore

$$\begin{aligned} \nu(A \cap E) + \nu(A \sim E) \geqslant \nu(A) &\geqslant \sum_{j=1}^\infty \nu(A \cap E_j) + \nu(A \sim E) \\ &\geqslant \nu(A \cap E) + \nu(A \sim E). \end{aligned}$$

This shows that $E \in \mathscr{P}_\nu$ and, replacing A by E, that

$$\nu(E) = \sum_{j=1}^\infty \nu(E_j).$$

Thus $\nu \mid \mathscr{P}_\nu$ is a finite measure. QED

I.4.14 ***Theorem*** Let S be a compact topological space, Σ a field in S, $\mu : \Sigma \to \mathbb{R}$ additive, regular and bounded, and $\tilde{\Sigma}$ the smallest σ-field containing Σ. Then there exists a unique bounded regular measure $\tilde{\mu} : \tilde{\Sigma} \to \mathbb{R}$ that is an extension of μ to $\tilde{\Sigma}$.

▌PROOF We shall first assume that μ is nonnegative. By Alexandroff's theorem I.4.11, μ is countably additive on Σ. We define the function $\tilde{\mu}$ as in I.4.12 which implies that $\tilde{\mu}$ is a finite outer measure, $\Sigma \subset \mathscr{P}_{\tilde{\mu}}$ and $\mu = \tilde{\mu} \mid \Sigma$. By Caratheodory's theorem I.4.13, $\mathscr{P}_{\tilde{\mu}}$ is a σ-field (and contains, therefore, $\tilde{\Sigma}$) and $\tilde{\mu} \mid \mathscr{P}_{\tilde{\mu}}$ is a finite measure; hence $\tilde{\mu} \mid \tilde{\Sigma}$ is a positive finite measure and an extension of μ. By I.4.3, $\tilde{\mu}$ is bounded on $\tilde{\Sigma}$.

If $\nu : \tilde{\Sigma} \to \mathbb{R}$ is a positive measure and an extension of μ, we consider a set $E \in \tilde{\Sigma}$ and a sequence (E_j) in Σ such that $E \subset \bigcup_{j=1}^{\infty} E_j$. Then

$$\nu(E) \leqslant \nu\left(\bigcup_{j=1}^{\infty} E_j\right) \leqslant \sum_{j=1}^{\infty} \nu(E_j) = \sum_{j=1}^{\infty} \mu(E_j)$$

and, therefore, by the definition of $\tilde{\mu}$, $\nu(E) \leqslant \tilde{\mu}(E)$. Replacing E by $S \sim E$ yields $\nu(S \sim E) \leqslant \tilde{\mu}(S \sim E)$ and adding these two inequalities yields $\nu(E) + \nu(S \sim E) = \nu(S) = \tilde{\mu}(S) \leqslant \tilde{\mu}(E) + \tilde{\mu}(S \sim E) = \tilde{\mu}(S)$, showing that only the equality sign is permissible. Thus $\nu = \tilde{\mu} \mid \tilde{\Sigma}$ and $\tilde{\mu}$ is the unique non-negative countably additive extension of μ to $\tilde{\Sigma}$.

To prove that $\tilde{\mu} \mid \tilde{\Sigma}$ is regular, we observe that for every $E \in \tilde{\Sigma}$ and $\epsilon > 0$ there exists a sequence (E_j) in Σ such that $E \subset \bigcup_{j=1}^{\infty} E_j$ and $\tilde{\mu}(E) \geqslant \sum_{j=1}^{\infty} \tilde{\mu}(E_j) - \epsilon/2$; hence $\tilde{\mu}(\bigcup_{j=1}^{\infty} E_j \sim E) \leqslant \epsilon/2$. Since $\mu : \Sigma \to \mathbb{R}$ is regular, for every $j \in \mathbb{N}$ there exists $B_j \in \Sigma$ with $E_j \subset B_j^{\circ}$ and $\mu(B_j \sim E_j) \leqslant \epsilon 2^{-j-1}$. We set $B \triangleq \bigcup_{j=1}^{\infty} B_j$ and observe that $B \in \tilde{\Sigma}$, $E \subset \bigcup_{j=1}^{\infty} B_j^{\circ} \subset B^{\circ}$ and

$$0 \leqslant \tilde{\mu}(B \sim E) \leqslant \tilde{\mu}\left(B \sim \bigcup_{j=1}^{\infty} E_j\right) + \tilde{\mu}\left(\bigcup_{j=1}^{\infty} E_j \sim E\right)$$

$$\leqslant \sum_{j=1}^{\infty} \mu(B_j \sim E_j) + \epsilon/2 \leqslant \epsilon.$$

Similarly, with E replaced by $S \sim E$, there exists a set $C \in \tilde{\Sigma}$, with $S \sim E \subset C^{\circ}$ and $\tilde{\mu}(C \sim (S \sim E)) = \tilde{\mu}(E \sim (S \sim C)) \leqslant \epsilon$. Thus $\overline{(S \sim C)} \subset E \subset B^{\circ}$ and $\tilde{\mu}(B \sim (S \sim C)) = \tilde{\mu}(B \sim E) + \tilde{\mu}(E \sim (S \sim C)) \leqslant 2\epsilon$, showing that $\tilde{\mu}$ is regular.

Finally, we consider the case where μ is not nonnegative. By I.4.1, μ^+ and μ^- are additive and bounded, by I.4.2 they are nonnegative, and by I.4.6 they are regular. Thus each has a unique nonnegative extension to a measure μ_1 respectively μ_2 on $\tilde{\Sigma}$, and both μ_1 and μ_2 are bounded and regular. Therefore $\mu_1 - \mu_2$ is a bounded and regular extension of μ to a measure on $\tilde{\Sigma}$. If ν is another such extension, then ν^+ must be such an extension of μ^+ and ν^- of μ^-, and by our previous argument, $\nu^+ = \mu_1$ and $\nu^- = \mu_2$; hence $\nu = \mu_1 - \mu_2$. QED

I.4.15 ***The Borel and Lebesgue Measures in*** $\mathbb{R}$ Theorem I.4.14 can now serve to define the Borel and Lebesgue measures in $\mathbb{R}$. Let $t_0 < t_1$, $S \triangleq [t_0, t_1]$ and Σ be a collection of subsets of S of the following form: $\{t_0\} = [t_0, t_0]$, $(t', t'']$ for $t_0 \leqslant t' < t'' \leqslant t_1$, and finite unions of such intervals. We verify that Σ is a field. Next we set $\mu(\{t_0\}) \triangleq 0$, $\mu((t', t'']) \triangleq t'' - t'$, and $\mu(\bigcup_{j=1}^{k} A_j) \triangleq \sum_{j=1}^{k} \mu(A_j)$ if $k \in \mathbb{N}$ and A_j are disjoint intervals of the form just considered. Then $\mu : \Sigma \to \mathbb{R}$ is bounded, additive, and regular and has, by I.4.14, a unique extension $\tilde{\mu}$ to a bounded regular measure on the smallest σ-field $\tilde{\Sigma}$ containing Σ. By I.2.8(2), each open subset of S is a finite or denumerable collection of open (relative to S) intervals. It is now easy to verify that each element of Σ is a Borel set and that $\tilde{\Sigma}$ contains the open subsets of S; hence $\tilde{\Sigma} = \Sigma_{\text{Borel}}(S)$. We refer to $\tilde{\mu}$ as *the Borel measure in* $[t_0, t_1]$.

Let $(S, \tilde{\Sigma}^*, \tilde{\mu}^*)$ be the Lebesgue extension of $(S, \tilde{\Sigma}, \tilde{\mu})$. We refer to the elements of $\tilde{\Sigma}^*$ as *Lebesgue measurable* subsets of $[t_0, t_1]$ and to $\tilde{\mu}^*$ as the *Lebesgue measure* in $[t_0, t_1]$.

If $A \subset \mathbb{R}$ and $A \cap [j, j+1)$ is a Borel set (as a subset of $[j, j+1]$) for each $j \in \{0, 1, -1, 2, -2, \ldots\}$, then A is a Borel set in $\mathbb{R}$. We set $\mu_{\text{B}}(A) \triangleq \sum_{j=-\infty}^{\infty} \mu_j(A \cap [j, j+1)) \in \bar{\mathbb{R}}$, where μ_j is the Borel measure in $[j, j+1]$. It is then easy to see that μ_{B} is a measure, referred to as the *Borel measure* in $\mathbb{R}$. Furthermore, if $\tilde{\mu}$ is the Borel measure in $[t_0, t_1]$ and A is a Borel subset of $[t_0, t_1]$ then, clearly, $\tilde{\mu}(A) = \sum_{t_0 \leqslant j \leqslant t_1} \mu_j([j, j+1] \cap A) = \mu_{\text{B}}(A)$.

If $A \subset \mathbb{R}$ and $A \cap [j, j+1]$ is a Lebesgue measurable subset of $[j, j+1]$ $(j = 0, \pm 1, \pm 2, \ldots)$, then we say that A is *Lebesgue measurable* as a subset of $\mathbb{R}$. It is easy to see that the Lebesgue measurable subsets of $\mathbb{R}$ form a σ-field and those contained in some bounded interval $[t_0, t_1]$ coincide with the Lebesgue measurable subsets of $[t_0, t_1]$. We extend the Lebesgue measure to $\mathbb{R}$ in the same manner as we extended the Borel measure.

I.4.B Measurable Functions

Let (S, Σ) be a measurable space and $(X, \mathscr{T})$ a topological space. A function $f : S \to X$ is *Σ-measurable* if $f^{-1}(A) \in \Sigma$ $(A \in \mathscr{T})$; and f is *Σ-measurable on B* if $B \in \Sigma$ and $f^{-1}(A) \cap B \in \Sigma$ $(A \in \mathscr{T})$. It is easy to see that for every $g : S \to X$ the set $\{A \in \mathscr{P}(X) \mid g^{-1}(A) \in \Sigma\}$ is a σ-field in X. Thus, if $\mathscr{V} \subset \mathscr{P}(X)$ and $\mathscr{T}$ is contained in the smallest σ-field containing $\mathscr{V}$, then f is Σ-measurable provided $f^{-1}(A) \in \Sigma$ $(A \in \mathscr{V})$. In particular, if X is a separable metric space, then f is Σ-measurable whenever $f^{-1}(S(x, a)) \in \Sigma$ $(x \in X,\ a > 0)$; if $X = \mathbb{R}$, then f is Σ-measurable whenever $f^{-1}(A_a) \in \Sigma$ $(a \in \mathbb{R})$, where A_a may have one of the following forms: (a) (a, ∞), or (b) $[a, \infty)$, or (c) $(-\infty, a)$, or (d) $(-\infty, a]$. If Ω is an at most denumerable set, X_ω $(\omega \in \Omega)$ are topological spaces and $X \triangleq \prod_{\omega \in \Omega} X_\omega$ with the product topology, then $f \triangleq (f_\omega)_{\omega \in \Omega} : S \to X$ is Σ-measurable if and only if f_ω is Σ-measurable for each $\omega \in \Omega$.

If (S, Σ, μ) is a finite measure space, (S, Σ^*, μ^*) its Lebesgue extension, $(X, \mathscr{T})$ a topological space, and $f : S \to X$, then we say that f is *μ-measurable* if f is Σ^*-measurable. Thus f is μ-measurable if and only if it is μ^*-measurable.

For every $A \subset S$, the *characteristic function* of A, denoted by χ_A, is a function on S to $\{0, 1\}$ defined by

$$\chi_A(s) \triangleq \begin{cases} 1 & \text{if } s \in A, \\ 0 & \text{if } s \in S \sim A. \end{cases}$$

It is clear that χ_A is μ-measurable if and only if A is μ-measurable.

It follows from the definition of a Σ-measurable function that $g \circ f$ is Σ-measurable whenever $(X, \mathscr{T})$ and $(Y, \mathscr{V})$ are topological spaces, (S, Σ) is a measurable space, $g : X \to Y$ is continuous and $f : S \to X$ is Σ-measurable. If, in particular, $\mathscr{X}$ is a topological vector space, $k \in \mathbb{N}$, and $f_j : S \to \mathscr{X}$ and $g_j : S \to \mathbb{R}$ are Σ-measurable ($j = 1, 2,\ldots k$), then $s \to \sum_{j=1}^{k} g_j(s) f_j(s)$ is Σ-measurable [since $(\alpha_1 ,\ldots, \alpha_k , x_1 ,\ldots, x_k) \to \sum_{j=1}^{k} \alpha_j x_j : \mathbb{R}^k \times \mathscr{X}^k \to \mathscr{X}$ is continuous and $(g_1 ,\ldots, g_k)$ and $(f_1 ,\ldots, f_k)$ Σ-measurable].

If (S, Σ) is a measurable space and $(S, \mathscr{T})$ and $(X, \mathscr{V})$ topological spaces, then a continuous function $f : S \to X$ is Σ-measurable whenever $\mathscr{T} \subset \Sigma$ or, equivalently, whenever $\Sigma_{\text{Borel}}(S) \subset \Sigma$.

I.4.16 ***The Metric Space*** $\mathscr{F}(S, \Sigma, \mu, X)$ Let (S, Σ, μ) be a positive finite measure space and X a metric space. We shall henceforth denote the Lebesgue extension of a measure by the same symbol used to denote the measure. We call two μ-measurable functions $f_1 : S \to X$ and $f_2 : S \to X$ *μ-equivalent*, and identify them, if $f_1(s) = f_2(s)$ μ-a.e. It is clear that f_1 and f_3 are μ-equivalent if f_1 is μ-equivalent to some f_2 and f_2 is μ-equivalent to f_3. The set of all μ-measurable functions on S to X is thus partitioned into equivalence classes, two functions belonging to the same equivalence class if they coincide μ-a.e. We identify all elements of the same equivalence class and denote by $\mathscr{F}(S, \Sigma, \mu, X)$ or, if no confusion arises, by $\mathscr{F}$ or $\mathscr{F}(\mu)$, the set of all (equivalence classes of) μ-measurable functions on S to X.

If Z is a μ-null set and $f : S \sim Z \to X$, then every extension of f to S is μ-measurable provided any one such extension is μ-measurable, and then all these extensions belong to the same equivalence class of $\mathscr{F}$. We therefore identify any such function $f : S \sim Z \to X$ with the corresponding equivalence class and with all μ-measurable functions on S to X that belong to this equivalence class.

If $f : S \to X$ and $g : S \to X$ are μ-measurable, we let

$$d_{\mathscr{F}}(f, g) \triangleq \inf\{\alpha + \mu(\{s \in S \mid d(f(s), g(s)) > \alpha\}) \mid \alpha > 0\}.$$

Then $d_{\mathscr{F}}(f, g)$ is finite and nonnegative and $d_{\mathscr{F}}(f, g) = d_{\mathscr{F}}(f_1 , g_1)$ whenever f is μ-equivalent to f_1 and g to g_1. Thus $d_{\mathscr{F}}(\cdot, \cdot)$ is a real-valued nonnegative function on $\mathscr{F}(S, \Sigma, \mu, X)$. We verify that $d_{\mathscr{F}}(f, g) = 0$ if and only if f is μ-equivalent to g and that $d_{\mathscr{F}}(f, g) = d_{\mathscr{F}}(g, f)$. If we set $A(f, g, \alpha) \triangleq \{s \in S \mid d(f(s), g(s)) > \alpha\}$, then

$$A(f_1, f_3, \alpha_1 + \alpha_3) \subset A(f_1, f_2, \alpha_1) \cup A(f_2, f_3, \alpha_3);$$

hence

$$\begin{aligned} d_{\mathscr{F}}(f_1, f_3) & \\ &= \inf\{\alpha_1 + \alpha_3 + \mu(A(f_1, f_3, \alpha_1 + \alpha_3)) \mid \alpha_1 > 0, \alpha_3 > 0\} \\ &\leqslant \inf\{\alpha_1 + \alpha_3 + \mu(A(f_1, f_2, \alpha_1)) + \mu(A(f_2, f_3, \alpha_3)) \mid \alpha_1 > 0, \alpha_3 > 0\} \\ &= d_{\mathscr{F}}(f_1, f_2) + d_{\mathscr{F}}(f_2, f_3). \end{aligned}$$

Thus $d_{\mathscr{F}}(\cdot, \cdot)$ is a metric function on $\mathscr{F}(S, \Sigma, \mu, X)$ and we shall henceforth consider $\mathscr{F}$ to be a metric space with the metric function $d_{\mathscr{F}}$.

We observe that if $f_j, g \in \mathscr{F}$ $(j \in \mathbb{N})$ and $\epsilon > 0$, then $d_{\mathscr{F}}(f_j, g) \leqslant \epsilon + \mu(\{s \in S \mid d(f_j(s), g(s)) > \epsilon\})$ for all $j \in \mathbb{N}$, and

$$\lim_j \mu(\{s \in S \mid d(f_j(s), g(s)) > \epsilon\}) = 0$$

whenever $\lim_j d_{\mathscr{F}}(f_j, g) = 0$. Thus $\lim_j d_{\mathscr{F}}(f_j, g) = 0$ if and only if $\lim_j \mu(\{s \in S \mid d(f_j(s), g(s)) > \epsilon\}) = 0$ for every $\epsilon > 0$.

If (f_j) is a Cauchy sequence in $\mathscr{F}(S, \Sigma, \mu, X)$ we also say that it is a Cauchy sequence *in μ-measure*. If f_j $(j \in \mathbb{N})$ and g are functions on S to X (not necessarily μ-measurable), we say that $\lim_j f_j = g$ μ-a.e. if $\lim_j f_j(s) = g(s)$ μ-a.e., and $\lim_j f_j = g$ *μ-uniformly* or $\lim_j f_j(s) = g(s)$ *μ-uniformly* if for every $\epsilon > 0$ there exists a set $A_\epsilon \subset S$ such that $\mu(S \sim A_\epsilon) \leqslant \epsilon$ and $\lim_j f_j(s) = g(s)$ uniformly for all $s \in A_\epsilon$.

THE VECTOR SPACE $\mathscr{F}(S, \Sigma, \mu, \mathscr{X})$

Let $\mathscr{X}$ be a Banach space. If $k \in \mathbb{N}$, f_j and g_j are μ-measurable functions on S to $\mathscr{X}$, and $f_j(s) = g_j(s)$ μ-a.e. $(j \in \mathbb{N})$, then

$$\sum_{j=1}^{k} \alpha_j f_j(s) = \sum_{j=1}^{k} \alpha_j g_j(s) \quad \mu\text{-a.e.}$$

for every choice of $\alpha_1, \ldots, \alpha_k \in \mathbb{R}$. Thus the set $\mathscr{F}(S, \Sigma, \mu, \mathscr{X})$ of equivalence classes of μ-equivalent functions on S to $\mathscr{X}$ is both a metric space and a vector space. We write $|f|_{\mathscr{F}}$ or $|f|_{\mathscr{F}(\mu)}$ for $d_{\mathscr{F}}(f, 0)$ (however, $|\cdot|_{\mathscr{F}}$ is neither a norm nor a seminorm) and observe that $d_{\mathscr{F}}(f, g) = |f - g|_{\mathscr{F}}$. Our previous remarks imply that $\lim_j |f_j|_{\mathscr{F}} = 0$ if and only if $\lim_j \mu(\{s \in S \mid |f_j(s)| > \epsilon\}) = 0$ for every $\epsilon > 0$.

I.4.17 ***Theorem*** Let (S, Σ) be a measurable space (respectively (S, Σ, μ) a finite measure space), X a separable metric space, $f_j : S \to X$ $(j \in \mathbb{N})$ Σ-measurable (respectively μ-measurable), and $\lim_j f_j(s) = f(s)$ for all $s \in S$ (respectively for μ-a.a. $s \in S$). Then f is Σ-measurable (respectively μ-measurable). If f is μ-measurable, then there exists a Σ-measurable function $\tilde{f}$ such that $f(s) = \tilde{f}(s)$ μ-a.e.

▌PROOF We recall that the metric topology of X has an at most countable base (I.2.17). Therefore f is Σ-measurable (respectively μ-measurable) if and only if $s \to d(f(s), x)$ is Σ-measurable (respectively μ-measurable) for every $x \in X$.

We first assume that f_j are Σ-measurable and $\lim_j f_j(s) = f(s)$ $(s \in S)$. If $g_j : S \to \mathbb{R}$ are Σ-measurable and $\lim_j g_j(s) = g(s)$ $(s \in S)$, then, for all $a \in \mathbb{R}$ and $J \subset (1, 2, \ldots)$, we have

$$\{s \in S \mid \sup_{j \in J} g_j(s) \leqslant a\} = \bigcap_{j \in J} \{s \in S \mid g_j(s) \leqslant a\} \in \Sigma$$

and

$$\{s \in S \mid \inf_{j \in J} g_j(s) < a\} = \bigcup_{j \in J} \{s \in S \mid g_j(s) < a\} \in \Sigma.$$

Thus $s \to \sup_{j \in J} g_j(s)$ and $s \to \inf_{j \in J} g_j(s)$ are Σ-measurable and so is, therefore, $s \to \inf_j \sup_{k \geqslant j} g_k(s) = \lim_j g_j(s) = g(s)$. It follows that, for each $x \in X$, $s \to d(f(s), x) = \lim_j d(f_j(s), x)$ is Σ-measurable.

If f_j are μ-measurable, $\mu(S \sim A) = 0$, and $\lim_j f_j(s) = f(s)$ $(s \in A)$, then our previous argument shows that f is μ-measurable on A; hence on S.

Now assume that f is μ-measurable, and let $\{x_1, x_2, \ldots\}$ be a dense subset of X. We set

$$A_k^j \triangleq f^{-1}(S(x_k, 1/j)) \qquad \text{and} \qquad H_k^j \triangleq A_k^j \sim \bigcup_{i=1}^{k-1} A_i^j \qquad (k, j \in \mathbb{N}).$$

Then $H_1^j, H_2^j, \ldots$ is a partition of S for each $j \in \mathbb{N}$ and there exist Σ-measurable $\tilde{H}_k^j \subset H_k^j$ such that $\mu(H_k^j \sim \tilde{H}_k^j) = 0$. We set

$$B \triangleq \bigcap_{j=1}^{\infty} \bigcup_{k=1}^{\infty} \tilde{H}_k^j, \qquad \tilde{f}(s) \triangleq f(s) \quad (s \in B), \qquad \tilde{f}(s) \triangleq x_1 \quad (s \notin B)$$

and, for each $j \in \mathbb{N}$,

$$f_j(s) \triangleq x_k \quad (s \in \tilde{H}_k^j \cap B, \quad k \in \mathbb{N}) \qquad \text{and} \qquad f_j(s) \triangleq x_1 \quad (s \notin B).$$

Then $B \in \Sigma$, f_j are Σ-measurable, $\lim_j f_j(s) = \tilde{f}(s)$ $(s \in S)$, and $\mu(S \sim B) = 0$. It follows that $f(s) = \tilde{f}(s)$ μ-a.e. and, by our previous argument, that $\tilde{f}$ is Σ-measurable. QED

I.4.18 ***Theorem*** (Egoroff) Let (S, Σ, μ) be a finite positive measure space, X a complete separable metric space, and $f_j : S \to X$ $(j \in \mathbb{N})$ μ-measurable. Then

(1) $\lim_j f_j = f$ μ-a.e. if and only if $\lim_j f_j = f$ μ-uniformly, and then f is μ-measurable;
(2) $\lim_j f_j = f$ in μ-measure if $\lim_j f_j = f$ μ-a.e.; and
(3) if (f_j) is a Cauchy sequence in μ-measure, then there exist a sequence $J \subset (1, 2, \ldots)$ and a μ-measurable $f: S \to X$ such that $\lim_j d_{\mathcal{F}}(f, f_j) = 0$ and $\lim_{j \in J} f_j = f$ μ-a.e. In particular, if $\lim_j f_j = f_0$ in μ-measure, then there exists a sequence $J_0 \subset (1, 2, \ldots)$ such that $\lim_{j \in J_0} f_j = f_0$ μ-a.e.

▌PROOF *Step* 1 If $\lim_j f_j = f$ μ-uniformly, then for every $i \in \mathbb{N}$ there exists a set A_i such that $\mu(S \sim A_i) \leqslant 1/i$ and $\lim_j f_j(s) = f(s)$ uniformly for all $s \in A_i$; hence $\lim_j f_j(s) = f(s)$ for all $s \in \bigcup_{i=1}^{\infty} A_i$, that is, for μ-a.a. $s \in S$. It follows then, by I.4.17, that f is μ-measurable.

Now assume that $\lim_j f_j(s) = f(s)$ for μ-a.a. $s \in S$, say for $s \in S'$. Then, by I.4.17, f is μ-measurable. We set

$$E_{k,0} \triangleq \varnothing \qquad \text{and} \qquad E_{k,j} \triangleq \bigcap_{i=j}^{\infty} \{s \in S \mid d(f_i(s), f(s)) < 1/k\} \quad (k, j \in \mathbb{N}).$$

Since the function $(x, y) \to d(x, y) : X \times X \to \mathbb{R}$ is continuous, it follows that each $E_{k,j}$ is μ-measurable. We have, for fixed k,

$$E_{k,j} \subset E_{k,j+1} \quad (j = 0, 1, \ldots) \qquad \text{and} \qquad S' \subset \bigcup_{j=1}^{\infty} E_{k,j} = \bigcup_{j=1}^{\infty} (E_{k,j} \sim E_{k,j-1});$$

hence

$$\mu(S) = \mu(S') = \lim_i \mu\left(\bigcup_{j=1}^{i} (E_{k,j} \sim E_{k,j-1})\right) = \lim_i \mu(E_{k,i}).$$

Now let $\epsilon > 0$. Then for each $k \in \mathbb{N}$ there exists $j(k) \in \mathbb{N}$ such that $\mu(S \sim E_{k,j(k)}) \leqslant \epsilon 2^{-k}$. We set $A_\epsilon \triangleq \bigcap_{k=1}^{\infty} E_{k,j(k)}$ and observe that $\mu(S \sim A_\epsilon) \leqslant \epsilon$ and $d(f_i(s), f(s)) < 1/k$ for all $k \in \mathbb{N}$, $i \geqslant j(k)$, and $s \in A_\epsilon$. This proves statement (1).

Step 2. If $\lim_j f_j = f$ μ-a.e., then, by (1), $\lim_j f_j = f$ μ-uniformly; therefore, for each $\alpha > 0$, $\lim_j \mu(\{s \in S \mid d(f(s), f_j(s)) > \alpha\}) = 0$. This shows that $\lim_j f_j = f$ in μ-measure and proves (2).

Now assume that $\lim_{i,j} d_{\mathcal{F}}(f_i, f_j) = 0$; hence $\lim_{i,j} \mu(\{s \in S \mid d(f_i(s), f_j(s)) > \alpha\}) = 0$ for each $\alpha > 0$. Then for each $k \in \mathbb{N}$ there exists i_k such that $\mu(\{s \in S \mid d(f_{i_k}(s), f_j(s)) > 2^{-k}\}) < 2^{-k}$ for all $j > i_k$. We may clearly choose $i_1, i_2, \ldots$ so that $i_{k+1} > i_k$. We set $E_k \triangleq \{s \in S \mid d(f_{i_k}(s), f_{i_{k+1}}(s)) > 2^{-k}\}$ and

$A_k \triangleq \bigcup_{j=k}^{\infty} E_j$, and observe that $\mu(A_k) < 2^{-k+1}$ and, for $s \in S \sim A_k$ and $j > l \geqslant k$,

$$(4) \qquad d(f_{i_l}(s), f_{i_j}(s)) \leqslant \sum_{n=l}^{j-1} d(f_{i_n}(s), f_{i_{n+1}}(s)) \leqslant \sum_{n=l}^{j-1} 2^{-n} < 2^{-k+1}.$$

This shows that $(f_{i_j}(s))_j$ is a Cauchy sequence for all $s \in B \triangleq S \sim \bigcap_{k=1}^{\infty} A_k$, that is, for μ-a.a. $s \in S$. Therefore, for each $s \in B$ there exists $f(s) \in X$ such that $\lim_j f_{i_j}(s) = f(s)$; hence $\lim_j f_{i_j}(s) = f(s)$ μ-a.e. By (1) and (2), f is μ-measurable and $\lim_j d_{\mathscr{F}}(f_{i_j}, f) = 0$. Since, by assumption, $\lim_j d_{\mathscr{F}}(f_{i_j}, f_j) = 0$,it follows that

$$\lim_j d_{\mathscr{F}}(f, f_j) \leqslant \lim_j [d_{\mathscr{F}}(f, f_{i_j}) + d_{\mathscr{F}}(f_{i_j}, f_j)] = 0.$$

If $\lim_j f_j = f_0$ in μ-measure, then, by the preceding argument, there exist $J_0 \subset (1, 2,...)$ and a μ-measurable g_0 such that $\lim_{j \in J_0} f_j = g_0$ μ-a.e.; hence, by (2), $\lim_{j \in J_0} d_{\mathscr{F}}(f_j, g_0) = \lim_{j \in J_0} d_{\mathscr{F}}(f_j, f_0) = 0$ and, therefore, $g_0 = f_0$ μ-a.e. QED

I.4.19 ***Theorem*** (Lusin) Let S be a topological space, (S, Σ, μ) a positive, finite, and regular measure space, $\Sigma_{\text{Borel}}(S) \subset \Sigma$, X a separable metric space, and $f: S \to X$ μ-measurable. Then for every $\epsilon > 0$ there exists a closed subset F_ϵ of S such that $\mu(S \sim F_\epsilon) \leqslant \epsilon$ and the restriction of f to F_ϵ is continuous.

▌PROOF Let $\epsilon > 0$, $j \in \mathbb{N}$, and $X_\infty \triangleq \{x_1, x_2, ...\}$ be a dense subset of X. We assume that X_∞ is infinite, the required modifications in the arguments being obvious when X_∞ is finite. The collection $\{S(x_i, 1/j) \mid i \in \mathbb{N}\}$ is an open covering of X. We let $A_i \triangleq f^{-1}(S(x_i, 1/j))$ and observe that each A_i is μ-measurable and $\bigcup_{i=1}^{\infty} A_i = S$. Thus the sets $S_i \triangleq A_i \sim \bigcup_{l=1}^{i-1} A_l$ $(i \in \mathbb{N})$ are μ-measurable and disjoint and $\mu(\bigcup_{i=1}^{\infty} S_i) = \sum_{i=1}^{\infty} \mu(S_i) = \mu(S) < \infty$; hence there exists $m \in \mathbb{N}$ such that $\mu(\bigcup_{i=m+1}^{\infty} S_i) \leqslant \epsilon/2$. Since μ is regular, there exist closed subsets B_i $(i = 1,..., m)$ of S such that $B_i \subset S_i$ and $\mu(S_i \sim B_i) \leqslant \epsilon/(2m)$ $(i = 1,... m)$. We set $F_\epsilon^j \triangleq \bigcup_{i=1}^{m} B_i$ and $h_j(s) \triangleq x_i$ for $s \in S_i$ $(i \in \mathbb{N})$, and observe that F_ϵ^j is closed, $\mu(S \sim F_\epsilon^j) \leqslant \epsilon$ and $d(f(s), h_j(s)) \leqslant 1/j$ for $s \in F_\epsilon^j$. Furthermore, since the sets B_i $(i = 1,..., m)$ are closed and disjoint, we conclude that $h_j \mid F_\epsilon^j$ is continuous. Now we set

$$F_\epsilon \triangleq \bigcap_{j=1}^{\infty} F_{2^{-j} \cdot \epsilon}^j .$$

Then F_ϵ is closed,

$$\mu(S \sim F_\epsilon) \leqslant \sum_{i=1}^{\infty} \mu(S \sim F_{2^{-i} \cdot \epsilon}^i) \leqslant \epsilon \sum_{i=1}^{\infty} 2^{-i} = \epsilon,$$

$d(f(s), h_j(s)) \leqslant 1/j$ $(s \in F_\epsilon)$, and $h_j \mid F_\epsilon$ is continuous for all $j \in \mathbb{N}$. Thus, for any $\bar{s} \in F_\epsilon$ and $\alpha > 0$, we can choose an integer $k > 3/\alpha$, and observe that $A \triangleq h_k^{-1}(S(h_k(\bar{s}), \alpha/3)) \cap F_\epsilon$ is relatively open in F_ϵ and

$$d(f(\bar{s}), f(s)) \leqslant d(f(\bar{s}), h_k(\bar{s})) + d(h_k(\bar{s}), h_k(s)) + d(h_k(s), f(s)) \leqslant \alpha$$

for every $s \in A$. We conclude that $f \mid F_\epsilon$ is continuous. QED

I.4.20 *Theorem* Let X be a separable metric space, (S, Σ) a measurable space, and f a function on S to X. Then the following statements are equivalent:

(1) f is Σ-measurable;
(2) $g \circ f$ is Σ-measurable for every continuous $g : X \to \mathbb{R}$; and
(3) the function $s \to d(f(s), x)$ is Σ-measurable for every $x \in X$.

▌PROOF If f is Σ-measurable and $g : X \to \mathbb{R}$ continuous, then, for every open $G \subset \mathbb{R}$, $g^{-1}(G)$ is open and $f^{-1}(g^{-1}(G)) = (g \circ f)^{-1}(G) \in \Sigma$. Thus (1) implies (2). Now let (2) be satisfied and $F \subset X$ be closed. Then, by I.2.10 and I.2.11, the function $x \to g_F(x) \triangleq d[x, F]$ is continuous and $g_F^{-1}(\{0\}) = F$; hence $(g_F \circ f)^{-1}(\{0\}) = f^{-1}(F) \in \Sigma$. It follows that $f^{-1}(G) = S \sim f^{-1}(X \sim G)$ is Σ-measurable for every open $G \subset X$, thus implying statement (1).

Since the function $y \to d(y, x) : X \to \mathbb{R}$ is continuous for every $x \in X$, statement (2) implies statement (3). Finally, if statement (3) is satisfied then, for every $x \in X$ and $\epsilon > 0$, the set $\{s \in S \mid d(f(s), x) < \epsilon\}$ belongs to Σ. Thus f is Σ-measurable because X has an at most countable base (I.2.17). QED

I.4.21 *Theorem* Let (S, Σ) be a measurable space, X a separable topological space, $h : S \times X \to \mathbb{R}$, $h(\cdot, x)$ Σ-measurable for each $x \in X$, and $h(s, \cdot)$ continuous and bounded for each $s \in S$. Then the function $s \to k(s) \triangleq \sup_{x \in X} h(s, x) : S \to \mathbb{R}$ is Σ-measurable.

▌PROOF Let $\{x_1, x_2, \ldots\}$ be a dense subset of X and $\beta \in \mathbb{R}$. Then $k(s) \leqslant \beta$ if and only if $h(s, x_j) \leqslant \beta$ for all $j \in \mathbb{N}$. Thus

$$k^{-1}((-\infty, \beta]) = \{s \in S \mid k(s) \leqslant \beta\} = \bigcap_{j=1}^{\infty} \{s \in S \mid h(s, x_j) \leqslant \beta\}$$

and we conclude that $k^{-1}((-\infty, \beta]) \in \Sigma$ for each $\beta \in \mathbb{R}$. Thus k is Σ-measurable. QED

I.4.22 *Theorem* Let (S, Σ) be a measurable space, X and Y separable metric spaces, $\xi : S \to X$ Σ-measurable, and $h : S \times X \to Y$ such that $h(\cdot, x)$ is Σ-measurable for each $x \in X$ and $h(s, \cdot)$ continuous for each $s \in S$. Then the function $s \to h(s, \xi(s))$ is Σ-measurable.

▌PROOF Let $\{x_1, x_2, ...\}$ be a dense subset of X and, for all $i, j \in \mathbb{N}$,

$$D_j^i \triangleq \xi^{-1}(S(x_j, 1/i)) \qquad \text{and} \qquad E_j^i \triangleq D_j^i \sim \bigcup_{k=1}^{j-1} D_k^i.$$

Then $E_1^i, E_2^i, ...$ belong to Σ and form a partition of S for each $i \in \mathbb{N}$, and we can define a function $\xi_i : S \to X$ by $\xi_i(s) \triangleq x_j$ for $s \in E_j^i$ $(j \in \mathbb{N})$. We observe that $\lim_i \xi_i(s) = \xi(s)$ for each $s \in S$ and, for each $i \in \mathbb{N}$, $s \to \xi_i(s)$ and $s \to h(s, \xi_i(s))$ are Σ-measurable. Since $h(s, \xi(s)) = \lim_i h(s, \xi_i(s))$ for each $s \in S$, our conclusion follows from I.4.17. QED

I.4.C Integrals of Simple and Nonnegative Functions

Let (S, Σ, μ) be a finite positive measure space and $\mathscr{X}$ a separable Banach space. A function $f: S \to \mathscr{X}$ is *μ-simple* if there exist $k \in \mathbb{N}$, $x_i \in \mathscr{X}$, and disjoint μ-measurable sets A_i $(i = 1, 2, ..., k)$ such that

$$f(s) = \sum_{i=1}^{k} \chi_{A_i}(s)\, x_i \qquad (s \in S).$$

We then define the *integral of f with respect to μ*, written $\int f(s)\, \mu(ds)$, by

$$\int f(s)\, \mu(ds) = \sum_{i=1}^{k} \mu(A_i)\, x_i .$$

We verify that if f is written in another form, say

$$f(s) = \sum_{j=1}^{m} \chi_{B_j}(s)\, y_j \qquad (s \in S),$$

then $x_i = y_j$ if $A_i \cap B_j \neq \varnothing$, the sets $A_i \cap B_j$ $(i = 1, ..., k;\ j = 1, ..., m)$ are disjoint, and

$$f(s) = \sum_{i=1}^{k} \sum_{j=1}^{m} \chi_{A_i}(s)\, \chi_{B_j}(s)\, x_i = \sum_{i=1}^{k} \sum_{j=1}^{m} \chi_{A_i}(s)\, \chi_{B_j}(s)\, y_j \qquad (s \in S);$$

hence

$$\sum_{i=1}^{k} \mu(A_i)\, x_i = \sum_{i=1}^{k} \sum_{j=1}^{m} \mu(A_i \cap B_j)\, x_i$$

$$= \sum_{i=1}^{k} \sum_{j=1}^{m} \mu(A_i \cap B_j)\, y_j = \sum_{j=1}^{m} \mu(B_j)\, y_j .$$

Thus $\int f(s)\, \mu(ds)$ is independent of the representation of the μ-simple function f.

If E is a μ-measurable set, we set

$$\int_E f(s)\,\mu(ds) \triangleq \int \chi_E(s) f(s)\,\mu(ds)$$

for every μ-simple f, and observe that

$$\int_S f(s)\,\mu(ds) = \int f(s)\,\mu(ds).$$

I.4.23 ***Lemma*** Let f and g be μ-simple, $\alpha, \beta \in \mathbb{R}$, the sets E and F μ-measurable, and $E_1, E_2, \ldots$ a partition of E into μ-measurable sets. Then $\alpha f + \beta g$ is μ-simple,

(1) $\int (\alpha f + \beta g)(s)\,\mu(ds) = \alpha \int f(s)\,\mu(ds) + \beta \int g(s)\,\mu(ds)$,
(2) $\int_E f(s)\,\mu(ds) \leqslant \int_F f(s)\,\mu(ds)$ if $f(s) \geqslant 0$ $(s \in S)$ and $E \subset F$,
(3) $\int_E f(s)\,\mu(ds) \leqslant \int_E g(s)\,\mu(ds)$ if $f(s) \leqslant g(s)$ $\quad (s \in E)$,
(4) $\lim_{\mu(A)\to 0} \int_A f(s)\,\mu(ds) = 0$,
(5) $\int_E f(s)\,\mu(ds) = \sum_{j=1}^{\infty} \int_{E_j} f(s)\,\mu(ds)$,
and
(6) $|\int_E f(s)\,\mu(ds)| \leqslant \int_E |f(s)|\,\mu(ds)$.

▌PROOF Direct verification.

I.4.24 ***Lemma*** Let (f_i) and (g_i) be two sequences of nonnegative μ-simple functions, $(f_i(s))_i$ and $(g_i(s))_i$ nondecreasing for each $s \in S$, and

$$\lim_i f_i(s) = \lim_i g_i(s) < \infty \qquad (s \in S).$$

Then $\lim_i \int f_i(s)\,\mu(ds)$ and $\lim_i \int g_i(s)\,\mu(ds)$ exist in $\bar{\mathbb{R}}$ and are equal.

▌PROOF Let $k \in \mathbb{N}$ and $f(s) \triangleq \lim_i f_i(s)$ $(s \in S)$. Then

$$f(s) = \lim_i f_i(s) \geqslant g_k(s) \qquad (s \in S)$$

since $(f_i(s))_i$ and $(g_i(s))_i$ are nondecreasing for each s. By Egoroff's theorem I.4.18,

$$f(s) = \lim_i f_i(s) \quad \mu\text{-uniformly;}$$

therefore, for each $\epsilon > 0$ there exist a set S_ϵ and an integer $i(\epsilon)$ such that $\mu(S \sim S_\epsilon) \leqslant \epsilon$ and $|f(s) - f_i(s)| \leqslant \epsilon$ for all $s \in S_\epsilon$ and $i \geqslant i(\epsilon)$. Thus

$$f_i(s) \geqslant g_k(s) - \epsilon \qquad \text{for all} \quad i \geqslant i(\epsilon) \quad \text{and} \quad s \in S_\epsilon\,;$$

hence, by I.4.23,

$$(1) \qquad \int_{S_\epsilon} f_i(s)\,\mu(ds) \geqslant \int_{S_\epsilon} (g_k(s) - \epsilon)\,\mu(ds) = \int_{S_\epsilon} g_k(s)\,\mu(ds) - \epsilon\mu(S_\epsilon).$$

Furthermore, for each $\eta > 0$ we can choose a positive ϵ, and therefore also $\mu(S \sim S_\epsilon)$, sufficiently small so that, by I.4.23(4),

$$\int_{S \sim S_\epsilon} g_k(s)\, \mu(ds) \leqslant \eta. \tag{2}$$

It follows now from (1), (2), and I.4.23 that

$$\begin{aligned} \int f_i(s)\, \mu(ds) &\geqslant \int_{S_\epsilon} f_i(s)\, \mu(ds) \\ &\geqslant \int_{S_\epsilon} g_k(s)\, \mu(ds) - \epsilon\mu(S_\epsilon) \\ &\geqslant \int_{S_\epsilon} g_k(s)\, \mu(ds) + \int_{S \sim S_\epsilon} g_k(s)\, \mu(ds) - \epsilon\mu(S_\epsilon) - \eta \\ &\geqslant \int g_k(s)\, \mu(ds) - \epsilon\mu(S) - \eta \qquad \text{for} \quad i \geqslant i(\epsilon). \end{aligned} \tag{3}$$

Since $f_j(s) \geqslant f_i(s)$ $(s \in S)$ if $j \geqslant i$, it follows from I.4.23(3) that the sequence $(\int f_i(s)\, \mu(ds))_i$ is nondecreasing and has a limit in $\overline{\mathbb{R}}$. Since η and the corresponding ϵ can be chosen arbitrarily small, it follows from (3) that

$$\lim_i \int f_i(s)\, \mu(ds) \geqslant \int g_k(s)\, \mu(ds) \qquad \text{for each} \quad k \in \mathbb{N};$$

hence

$$\lim_i \int f_i(s)\, \mu(ds) \geqslant \lim_k \int g_k(s)\, \mu(ds) \qquad \text{in } \overline{\mathbb{R}}.$$

The same argument, with (f_i) and (g_i) interchanged, shows that

$$\lim_i \int g_i(s)\, \mu(ds) \geqslant \lim_k \int f_k(s)\, \mu(ds) \qquad \text{in } \overline{\mathbb{R}};$$

hence

$$\lim_i \int f_i(s)\, \mu(ds) = \lim_i \int g_i(s)\, \mu(ds) \qquad \text{in } \overline{\mathbb{R}}. \quad \text{QED}$$

I.4.25 ***Lemma*** Let $f : S \to \mathbb{R}$ be nonnegative and μ-measurable. Then there exists a sequence (f_i) of nonnegative μ-simple functions such that $(f_i(s))_i$ is nondecreasing for each s and $\lim_i f_i(s) = f(s)$ $(s \in S)$.

▌PROOF Let $A_i^k(f) \triangleq \{s \in S \mid k2^{-i} \leqslant f(s) < (k+1)\, 2^{-i}\}$ $(k, i = 0, 1,...)$ and, for each $i \in \mathbb{N}$, $f_i(s) \triangleq k2^{-i}$ if $s \in A_i^k(f)$ and $k \in \{0, 1,..., i2^i - 1\}$ and $f_i(s) \triangleq 0$, otherwise. Then f_i is μ-simple and nonnegative and, whenever $f(s) < i$, $f_i(s)$ is the largest multiple of 2^{-i} not exceeding $f(s)$. Thus $(f_i(s))_i$ is nondecreasing and $\lim_i f_i(s) = f(s)$ $(s \in S)$. QED

I.4.26 ***Definition*** Let $f : S \to \mathbb{R}$ be nonnegative and μ-measurable. Then, by I.4.25, there exists a sequence (f_i) of μ-simple functions such that $(f_i(s))$ is nondecreasing and converges to $f(s)$ for all $s \in S$; furthermore, by I.4.24, $\lim_i \int f_i(s)\, \mu(ds)$ exists in $\overline{\mathbb{R}}$ and is the same for all such sequences (f_i).

We set

$$\int f(s)\, \mu(ds) \triangleq \lim_i \int f_i(s)\, \mu(ds) \quad \text{in } \overline{\mathbb{R}}$$

and refer to $\int f(s)\, \mu(ds)$ as the *integral of f with respect to μ*. We observe that this definition is consistent with the one previously given for μ-simple functions and that $\int f(s)\, \mu(ds) \geqslant 0$.

For any μ-measurable set E we write

$$\int_E f(s)\, \mu(ds) \triangleq \int \chi_E(s) f(s)\, \mu(ds).$$

We say that a nonnegative $f : S \to \mathbb{R}$ is *μ-integrable* if f is μ-measurable and $\int f(s)\, \mu(ds) < \infty$.

I.4.27 ***Lemma*** Let $f : S \to \mathbb{R}$ and $g : S \to \mathbb{R}$ be nonnegative and μ-measurable, $\alpha, \beta \geqslant 0$, E and F μ-measurable sets, and $E_1, E_2, \ldots$ a partition of E into μ-measurable sets. Then

(1) $\int (\alpha f + \beta g)(s)\, \mu(ds) = \alpha \int f(s)\, \mu(ds) + \beta \int g(s)\, \mu(ds)$;

(2) $\int_E f(s)\, \mu(ds) \leqslant \int_F f(s)\, \mu(ds)$ if $E \subset F$;

(3) $\int_E f(s)\, \mu(ds) \leqslant \int_E g(s)\, \mu(ds)$ if $f(s) \leqslant g(s) \quad (s \in E)$;

(4) $\int_E f(s)\, \mu(ds) = \sum_{j=1}^{\infty} \int_{E_j} f(s)\, \mu(ds)$;

(5) $\lim_i \int_{f^{-1}([i,\infty))} f(s)\, \mu(ds) = 0$ if f is μ-integrable;

(6) $\int_E f(s)\, \mu(ds) = 0$ only if $f(s) = 0$ μ-a.e. in E;

and

(7) $\int f(s)\, \mu(ds) = \int g(s)\, \mu(ds)$ if $f(s) = g(s)$ μ-a.e.

▌PROOF Let (f_i) be the sequence defined in I.4.25 for f and (g_i) a similar sequence for g. Then $\alpha f_i + \beta g_i$ is nonnegative and μ-simple, its value at any s is nondecreasing with i and

$$\lim_i (\alpha f_i + \beta g_i)(s) = \lim_i \big(\alpha f_i(s) + \beta g_i(s)\big) = \alpha f(s) + \beta g(s) \qquad (s \in S).$$

Thus

$$\lim_i \int \big(\alpha f_i(s) + \beta g_i(s)\big)\, \mu(ds) = \int (\alpha f + \beta g)(s)\, \mu(ds).$$

On the other hand, by I.4.23(1),

$$\int (\alpha f_i + \beta g_i)(s)\, \mu(ds) = \alpha \int f_i(s)\, \mu(ds) + \beta \int g_i(s)\, \mu(ds) \qquad (i \in \mathbb{N});$$

hence

$$\int (\alpha f + \beta g)(s)\, \mu(ds) = \alpha \int f(s)\, \mu(ds) + \beta \int g(s)\, \mu(ds),$$

proving (1).

By I.4.23(2), $\int_E f_i(s)\, \mu(ds) \leqslant \int_F f_i(s)\, \mu(ds)$ for each i if $E \subset F$. We deduce relation (2) by comparing the limits of both sides as $i \to \infty$. If $f(s) \leqslant g(s)$ for $s \in E$, then, by (1),

$$\begin{aligned}\int \chi_E(s)\, g(s)\, \mu(ds) &= \int \chi_E f(s)\, \mu(ds) + \int \chi_E(s)[g(s) - f(s)]\, \mu(ds)\\ &\geqslant \int \chi_E(s) f(s)\, \mu(ds),\end{aligned}$$

thus proving (3). By I.4.23(5), we have

$$\int_E f_i(s)\, \mu(ds) = \sum_{j=1}^{\infty} \int_{E_j} f_i(s)\, \mu(ds) = \lim_n \sum_{j=1}^{n} \int_{E_j} f_i(s)\, \mu(ds) \qquad (i \in \mathbb{N})$$

and, by I.4.23(3), the sequence $\left(\int_{E_j} f_i(s)\, \mu(ds)\right)_i$ is nondecreasing for each j. Thus, if we set $x_{n,i} \triangleq \sum_{j=1}^{n} \int_{E_j} f_i(s)\, \mu(ds)$ $(n, i \in \mathbb{N})$, then $(x_{n,1}, x_{n,2}, \ldots)$ and $(x_{1,i}, x_{2,i}, \ldots)$ are nondecreasing for each n and i and, by I.3.14(2),

$$\begin{aligned}\lim_i \int_E f_i(s)\, \mu(ds) &= \lim_i \lim_n \sum_{j=1}^{n} \int_{E_j} f_i(s)\, \mu(ds)\\ &= \lim_n \lim_i \sum_{j=1}^{n} \int_{E_j} f_i(s)\, \mu(ds)\\ &= \lim_n \sum_{j=1}^{n} \int_{E_j} f(s)\, \mu(ds) = \sum_{j=1}^{\infty} \int_{E_j} f(s)\, \mu(ds),\end{aligned}$$

thus proving (4).

Now let $A_i \triangleq \{s \in S \mid f(s) \geqslant i\}$ for $i \in \mathbb{N}$. If f is μ-integrable, then it follows from (3) and (4) that, for each $i \in \mathbb{N}$,

$$\begin{aligned}\infty > \int f(s)\,\mu(ds) &= \int_S f(s)\,\mu(ds)\\ &= \int_{A_i} f(s)\,\mu(ds) + \int_{S\sim A_i} f(s)\,\mu(ds)\\ &\geqslant \int_{A_i} f(s)\,\mu(ds) + \int_{S\sim A_i} f_i(s)\,\mu(ds)\\ &= \int_{A_i} f(s)\,\mu(ds) + \int f_i(s)\,\mu(ds);\end{aligned}$$

hence

$$\int f(s)\,\mu(ds) - \int f_i(s)\,\mu(ds) \geqslant \int_{A_i} f(s)\,\mu(ds) \geqslant 0,$$

showing that $\lim_i \int_{A_i} f(s)\,\mu(ds) = 0$ and thus proving (5).

Next let $\int_E f(s)\,\mu(ds) = 0$. We set $A_\alpha \triangleq E \cap f^{-1}([\alpha, \infty))$ for $\alpha > 0$ and observe that, by (2) and (3), $\alpha\mu(A_\alpha) \leqslant \int_{A_\alpha} f(s)\,\mu(ds) \leqslant \int_E f(s)\,\mu(ds) = 0$; hence $\mu(A_\alpha) = 0$ for all $\alpha > 0$ and $\mu(E \cap f^{-1}((0, \infty))) \leqslant \sum_{j=1}^{\infty} \mu(A_{1/j}) = 0$. This proves (6).

Finally, let $\mu(A) = 0$. Then $\int \chi_A(s) f_i(s)\,\mu(ds) = 0$ $(i \in \mathbb{N})$; hence $\int_A f(s)\,\mu(ds) = \lim_i \int \chi_A(s) f_i(s)\,\mu(ds) = 0$. If

$$f(s) = g(s) \qquad \text{for} \quad s \notin A,$$

then

$$\begin{aligned}\int f(s)\,\mu(ds) &= \int_{S\sim A} f(s)\,\mu(ds) + \int_A f(s)\,\mu(ds)\\ &= \int_{S\sim A} g(s)\,\mu(ds) = \int_{S\sim A} g(s)\,\mu(ds) + \int_A g(s)\,\mu(ds)\\ &= \int g(s)\,\mu(ds),\end{aligned}$$

thus proving (7). QED

I.4.28 ***Theorem*** Let (S, Σ, μ) be a finite positive measure space and $f: S \to \mathbb{R}$ nonnegative and μ-integrable. Then

$$\lim_{\mu(E)\to 0} \int_E f(s)\,\mu(ds) = 0.$$

▌PROOF Let $\epsilon > 0$ and $A_i \triangleq f^{-1}([i, \infty))$ $(i \in \mathbb{N})$. By I.4.27(5), there exists $j \in \mathbb{N}$ such that $\int_{A_j} f(s)\,\mu(ds) \leqslant \epsilon/2$. If $\mu(E) \leqslant \epsilon/(2j)$, then, by relations (2)–(4) of I.4.27, we have

$$0 \leqslant \int_E f(s)\,\mu(ds) = \int_{E\cap A_j} f(s)\,\mu(ds) + \int_{E\sim A_j} f(s)\,\mu(ds)$$

$$\leqslant \int_{A_j} f(s)\,\mu(ds) + \int_E j\mu(ds) \leqslant \epsilon/2 + j\mu(E) \leqslant \epsilon.$$

Thus $\lim_{\mu(E)\to 0} \int f(s)\,\mu(ds) = 0$. QED

I.4.D Bochner Integrals

I.4.29 ***Definition of*** $L^p(S, \Sigma, \mu, \mathscr{X})$ Let $\mathscr{X}$ be a separable Banach space, $1 \leqslant p < \infty$, and (S, Σ, μ) a finite positive measure space. As usual, we consider two μ-measurable functions $f: S \to \mathscr{X}$ and $g: S \to \mathscr{X}$ equivalent, and identify them, if $f(s) = g(s)$ μ-a.e. By I.3.15 and I.4.20, the function $s \to |f(s)|^p$ is μ-measurable if f is μ-measurable and, by I.4.27(7), $\int |f(s)|^p\,\mu(ds) = \int |g(s)|^p\,\mu(ds)$ if $f(s) = g(s)$ μ-a.e. We denote by $L^p(S, \Sigma, \mu, \mathscr{X})$ [or by $L^p(S, \mathscr{X})$ if Σ and μ are specified] the set of all (equivalence classes of) μ-measurable functions $f: S \to \mathscr{X}$ such that $\int |f(s)|^p\,\mu(ds) < \infty$. For each $f \in L^p(S, \Sigma, \mu, \mathscr{X})$ we write $|f|_p$ for $\{\int |f(s)|^p\,\mu(ds)\}^{1/p}$. We also write $L^p(S, \Sigma, \mu)$ [or $L^p(S)$] for $L^p(S, \Sigma, \mu, \mathbb{R})$. If $f: S \to \mathscr{X}$ and $p \in [1, \infty)$, we write $|f|_p < \infty$ for $f \in L^p(S, \Sigma, \mu, \mathscr{X})$. We say that (f_i) converges to f in $L^p(S, \Sigma, \mu, \mathscr{X})$ if $\lim_i |f - f_i|_p = 0$.

I.4.30 ***Theorem*** Let (S, Σ, μ) be a finite positive measure space, $\mathscr{X}$ a separable Banach space, and $1 \leqslant p < \infty$. If $f \in L^p(S, \Sigma, \mu, \mathscr{X})$, then there exists a sequence (f_i) of μ-simple functions on S to $\mathscr{X}$ such that $\lim_j |f - f_j|_p = 0$ and $|f_j(s)| \leqslant |f(s)| + 1$ $(j \in \mathbb{N}, s \in S)$.

▌PROOF Let $\{x_1, x_2, \ldots\}$ be a dense subset of $\mathscr{X}$,

$$A_i^j \triangleq \{s \in S \mid |f(s) - x_i|^p \leqslant 1/j\} \qquad (i, j \in \mathbb{N})$$

and

$$B_i^j \triangleq A_i^j \sim \bigcup_{k=1}^{i-1} A_k^j \qquad (i, j \in \mathbb{N}).$$

Then, for each $j \in \mathbb{N}$, $B_1^j, B_2^j, \ldots$ form a partition of S into μ-measurable sets and $\sum_{i=1}^{\infty} \mu(B_i^j) = \mu(S) < \infty$. There exists, therefore, $k(j) \in \mathbb{N}$ such that $\sum_{i=k(j)+1}^{\infty} \mu(B_i^j) \leqslant 1/j$. We set

$$f_j(s) \triangleq \begin{cases} x_i & \text{if} \quad i \in \{1, 2, \ldots, k(j)\} \quad \text{and} \quad s \in B_i^j, \\ 0 & \text{if} \quad s \in C^j \triangleq \bigcup_{i=k(j)+1}^{\infty} B_i^j. \end{cases}$$

We observe that $|f_j(s)| \leqslant |f(s)| + 1$ and $|f(s) - f_j(s)|^p \leqslant 1/j + |f(s)|^p \leqslant 1 + |f(s)|^p$ for all $s \in S$; hence, by I.4.27(3), the function $s \to |f(s) - f_j(s)|^p$ is μ-integrable for each j and, by I.4.27(2) and I.4.27(4),

$$\begin{aligned} (1) \qquad & \int |f(s) - f_j(s)|^p \, \mu(ds) \\ &= \sum_{i=1}^{k(j)} \int_{B_i^j} |f(s) - f_j(s)|^p \, \mu(ds) + \int_{C^j} |f(s) - f_j(s)|^p \, \mu(ds) \\ &\leqslant (1/j) \sum_{i=1}^{k(j)} \int_{B_i^j} \mu(ds) + \int_{C^j} |f(s)|^p \, \mu(ds) \\ &\leqslant (1/j)\, \mu(S) + \int_{C^j} |f(s)|^p \, \mu(ds). \end{aligned}$$

Since $\mu(C^j) = \sum_{i=k(j)+1}^{\infty} \mu(B_i^j) \leqslant 1/j$ and $s \to |f(s)|^p$ is μ-integrable, it follows from I.4.28 that $\lim_j \int_{C^j} |f(s)|^p \, \mu(ds) = 0$. Thus relation (1) implies that

$$\lim_j \int |f(s) - f_j(s)|^p \, \mu(ds) = \lim_j |f - f_j|_p^p = 0. \quad \text{QED}$$

I.4.31 ***Theorem*** Let (S, Σ, μ) be a finite positive measure space, $\mathscr{X}$ a separable Banach space, $1 \leqslant p < \infty$, and $f, f_j \in L^p(S, \Sigma, \mu, \mathscr{X})$ $(j \in \mathbb{N})$. Then

$$\lim_j |f - f_j|_{\mathscr{F}(\mu)} = 0 \qquad \text{whenever} \qquad \lim_j |f - f_j|_p = 0.$$

▌PROOF Assume that $\lim_j |f - f_j|_p = 0$ and let $\epsilon > 0$ and

$$A_j^\epsilon \triangleq \{s \in S \mid |f(s) - f_j(s)|^p > \epsilon\} \qquad (j \in \mathbb{N}).$$

Then, by I.4.27(3) and I.4.27(4),

$$\epsilon\mu(A_j^\epsilon) \leqslant \int_{A_j^\epsilon} |f(s) - f_j(s)|^p \, \mu(ds) \leqslant \int |f(s) - f_j(s)|^p \, \mu(ds) = |f - f_j|_p^p;$$

hence, for each fixed $\epsilon > 0$, $\lim_j \mu(A_j^\epsilon) = 0$ and therefore $\lim_j |f - f_j|_{\mathscr{F}(\mu)} = 0$.
QED

I.4.32 ***Theorem*** Let (S, Σ, μ) be a finite positive measure space, $\mathscr{X}$ a separable Banach space, $f, f_j, g_j \in L^1(S, \Sigma, \mu, \mathscr{X})$ and f_j, g_j μ-simple ($j \in \mathbb{N}$), $\lim_j |f - f_j|_1 = \lim_j |f - g_j|_1 = 0$, and E a μ-measurable set. Then

$$\lim_j \int_E f_j(s)\,\mu(ds) \qquad \text{and} \qquad \lim_j \int_E g_j(s)\,\mu(ds)$$

both exist in $\mathscr{X}$ and are equal. If, furthermore, $\mathscr{X} = \mathbb{R}$ and f is nonnegative, then

$$\int_E f(s)\,\mu(ds) = \lim_j \int_E f_j(s)\,\mu(ds) = \lim_j \int_E g_j(s)\,\mu(ds).$$

▮ PROOF For all $i, j \in \mathbb{N}$, we have, by I.4.23(6), I.4.27(2), and I.4.27(3),

$$\begin{aligned}
0 &\leqslant \left|\int_E f_i(s)\,\mu(ds) - \int_E f_j(s)\,\mu(ds)\right| \\
&= \left|\int_E (f_i(s) - f_j(s))\,\mu(ds)\right| \leqslant \int_E |f_i(s) - f_j(s)|\,\mu(ds) \\
&\leqslant \int_E |f(s) - f_i(s)|\,\mu(ds) + \int_E |f(s) - f_j(s)|\,\mu(ds) \\
&\leqslant \int |f(s) - f_i(s)|\,\mu(ds) + \int |f(s) - f_j(s)|\,\mu(ds) \\
&= |f - f_i|_1 + |f - f_j|_1 ;
\end{aligned}$$

hence

$$\lim_{i,j} \left|\int_E f_i(s)\,\mu(ds) - \int_E f_j(s)\,\mu(ds)\right| = 0.$$

This shows that the sequence $\left(\int_E f_i(s)\,\mu(ds)\right)_i$ in the complete normed vector space $\mathscr{X}$ is a Cauchy sequence and has a limit. The same argument also shows that $\lim_i \int_E g_i(s)\,\mu(ds)$ exists in $\mathscr{X}$.

For each $i \in \mathbb{N}$, we have

$$\begin{aligned}
&\left|\int_E f_i(s)\,\mu(ds) - \int_E g_i(s)\,\mu(ds)\right| \\
&\quad = \left|\int_E (f_i(s) - g_i(s))\,\mu(ds)\right| \leqslant \int_E |f_i(s) - g_i(s)|\,\mu(ds) \\
&\quad \leqslant \int_E |f(s) - f_i(s)|\,\mu(ds) + \int_E |f(s) - g_i(s)|\,\mu(ds) \\
&\quad \leqslant |f - f_i|_1 + |g - g_i|_1 ;
\end{aligned}$$

hence

$$\lim_i \int_E f_i(s)\,\mu(ds) = \lim_i \int_E g_i(s)\,\mu(ds).$$

Finally, if $\mathscr{X} = \mathbb{R}$ and $f(s) \geqslant 0$ $(s \in S)$, let $(\tilde{f}_i)$ be the sequence of non-negative μ-simple functions defined for f in I.4.25. Then $\tilde{f}_i(s) \leqslant f(s)$ $(s \in S)$ and $\int f(s)\,\mu(ds) = \lim_i \int \tilde{f}_i(s)\,\mu(ds)$. By I.4.27(1),

$$\int f(s)\,\mu(ds) = \int (f(s) - \tilde{f}_i(s))\,\mu(ds) + \int \tilde{f}_i(s)\,\mu(ds) \qquad (i \in \mathbb{N})$$

and therefore

$$\lim_i \int (f(s) - \tilde{f}_i(s))\,\mu(ds) = 0.$$

Thus, if $\varphi : S \to \mathbb{R}$ is μ-simple, then, by I.4.23 and I.4.27,

$$\begin{aligned}
&\left|\int f(s)\,\mu(ds) - \int \varphi(s)\,\mu(ds)\right| \\
&\quad = \lim_i \left|\int \tilde{f}_i(s)\,\mu(ds) - \int \varphi(s)\,\mu(ds)\right| \\
&\quad \leqslant \lim_i \int |\tilde{f}_i(s) - \varphi(s)|\,\mu(ds) \\
&\quad \leqslant \lim_i \int [|\varphi(s) - f(s)| + f(s) - \tilde{f}_i(s)]\,\mu(ds) \\
&\quad = \int |\varphi(s) - f(s)|\,\mu(ds) + \lim_i \int (f(s) - \tilde{f}_i(s)\,\mu(ds) \\
&\quad = \int |f(s) - \varphi(s)|\,\mu(ds).
\end{aligned}$$

Replacing φ by f_i for any fixed i yields

$$\left|\int f(s)\,\mu(ds) - \int f_i(s)\,\mu(ds)\right| \leqslant \int |f(s) - f_i(s)|\,\mu(ds) = |f - f_i|_1\,;$$

hence

$$\int f(s)\,\mu(ds) = \lim_i \int f_i(s)\,\mu(ds). \quad \text{QED}$$

I.4.33 ***Definition*** Let (S, Σ, μ) be a finite positive measure space, $\mathscr{X}$ a separable Banach space, E a μ-measurable set, and $f \in L^1(S, \Sigma, \mu, \mathscr{X})$. By I.4.30, there exists a sequence (f_i) of μ-simple functions on S to $\mathscr{X}$ such that $\lim_i |f - f_i|_1 = 0$ and, by I.4.32, $\lim_i \int_E f_i(s)\,\mu(ds)$ exists in $\mathscr{X}$ and is the same for all choices of such a sequence (f_i). We define the *integral of f over E with respect to* μ, written $\int_E f(s)\,\mu(ds)$, by

$$\int_E f(s)\,\mu(ds) \triangleq \lim_i \int_E f_i(s)\,\mu(ds),$$

and write $\int f(s)\,\mu(ds)$ for $\int_S f(s)\,\mu(ds)$. We observe that, by I.4.32, this definition is consistent with the definition I.4.26 of the integral for nonnegative functions. Furthermore, if $f(s) = g(s)$ μ-a.e., then, as we prove in Theorem I.4.34 below, $\int_E f(s)\,\mu(ds) = \int_E g(s)\,\mu(ds)$. Thus the integral is the same for all elements of the same equivalence class in $L^1(S, \Sigma, \mu, \mathscr{X})$. We refer to f as the *integrand* of $\int_E f(s)\,\mu(ds)$.

We call an element f of $L^1(S, \Sigma, \mu, \mathscr{X})$ a *μ-integrable function* (on S to $\mathscr{X}$), and this use of the term "μ-integrable" is consistent with its use in Definition I.4.26.

If (S, Σ, ν) is a finite measure space, then, by I.4.3, $\nu = \nu^+ - \nu^-$, where ν^+ and ν^- are finite positive measures defined on Σ. We then say that a ν-measurable function $f: S \to \mathscr{X}$ is ν-integrable or $f \in L^1(S, \Sigma, \nu, \mathscr{X})$ if $f \in L^1(S, \Sigma, |\nu|, \mathscr{X})$ and set, for each ν-measurable set E,

$$\int_E f(s)\,\nu(ds) \triangleq \int_E f(s)\,\nu^+(ds) - \int_E f(s)\,\nu^-(ds).$$

This is permissible because f is both ν^+-integrable and ν^-integrable if it is $|\nu|$-integrable, and conversely.

I.4.34 ***Theorem*** Let (S, Σ, μ) and (S, Σ, ν) be finite measure spaces, $\mathscr{X}$ and $\mathscr{Y}$ separable Banach spaces, $T \in B(\mathscr{X}, \mathscr{Y})$, $f, g \in L^1(S, \Sigma, \mu, \mathscr{X}) \cap L^1(S, \Sigma, \nu, \mathscr{X})$, $\alpha, \beta \in \mathbb{R}$, and E a μ-measurable set.

Then

(1) $$\int (\alpha f + \beta g)(s)\,\mu(ds) = \alpha \int f(s)\,\mu(ds) + \beta \int g(s)\,\mu(ds),$$

$$\int f(s)(\alpha\mu + \beta\nu)(ds) = \alpha \int f(s)\,\mu(ds) + \beta \int f(s)\,\nu(ds);$$

(2)

$$\int |f(s)|\,|\mu|\,(ds) = \lim_i \int |f_i(s)|\,|\mu|\,(ds) \text{ if } \lim_i \int |f(s) - f_i(s)|\,|\mu|(ds) = 0;$$

(3) $|\int_E f(s)\,\mu(ds)| \leqslant \int_E |f(s)|\,|\mu|\,(ds)$;
(4) $\lim_{|\mu|(A)\to 0} \int_A f(s)\,\mu(ds) = 0$;
(5) $\int_A f(s)\,\mu(ds) = \sum_{j=1}^{\infty} \int_{A_j} f(s)\,\mu(ds)$ if A is μ-measurable and $A_1, A_2, \ldots$ is a partition of A into μ-measurable sets;
(6) $\int_E f(s)\,\mu(ds) = \int_E g(s)\,\mu(ds)$ if $f(s) = g(s)$ for μ-a.a. $s \in E$;
(7) $\int_A f(s)|\,\mu\,|\,(ds) \geqslant 0$ $(A \in \Sigma)$ (respectively $= 0$, $\leqslant 0$) if and only if $f(s) \geqslant 0$ (respectively $= 0$, $\leqslant 0$) μ-a.e.; and
(8) $T \int_E f(s)\,\mu(ds) = \int_E Tf(s)\,\mu(ds)$.

▌PROOF *Proof of* (1) Let μ be positive and (f_i) and (g_i) be sequences of μ-simple functions on S to $\mathscr{X}$ such that

$$\lim_i \int |f(s) - f_i(s)|\,\mu(ds) = \lim_i \int |g(s) - g_i(s)|\,\mu(ds) = 0.$$

Then, by I.4.27(1) and I.4.27(3),

$$\int |\alpha f(s) + \beta g(s) - \alpha f_i(s) - \beta g_i(s)| \, \mu(ds)$$

$$\leqslant \int (|\alpha| \, |f(s) - f_i(s)| + |\beta| \, |g(s) - g_i(s)|) \, \mu(ds)$$

$$= |\alpha| \int |f(s) - f_i(s)| \, \mu(ds) + |\beta| \int |g(s) - g_i(s)| \, \mu(ds) \underset{i}{\to} 0.$$

Since $\int |\alpha f(s) + \beta g(s)| \, \mu(ds) \leqslant |\alpha| \int |f(s)| \, \mu(ds) + |\beta| \int |g(s)| \, \mu(ds) < \infty$, it follows, by I.4.23(1) and I.4.33, that

$$\int (\alpha f + \beta g)(s) \, \mu(ds) = \lim_i \int (\alpha f_i(s) + \beta g_i(s)) \, \mu(ds)$$

$$= \lim_i \alpha \int f_i(s) \, \mu(ds) + \lim_i \beta \int g_i(s) \, \mu(ds)$$

$$= \alpha \int f(s) \, \mu(ds) + \beta \int g(s) \, \mu(ds).$$

If μ is not positive, then we derive the above relation by applying it separately to μ^+ and μ^-.

We verify that f is $(\alpha\mu + \beta\nu)$-measurable. Since

$$\int f(s)(\alpha\mu + \beta\nu)(ds) = \alpha \int f(s) \, \mu(ds) + \beta \int f(s) \, \nu(ds)$$

if f is both μ-simple and ν-simple, it follows from Definition I.4.33 that this relation remains valid in the general case. This proves (1).

Proof of (2) It clearly suffices to prove (2) for the case of a positive measure μ. Since

$$\big| |f(s)| - |f_i(s)| \big| \leqslant |f(s) - f_i(s)| \qquad \text{for all } i \text{ and } s,$$

it follows from I.4.27 that

$$\left| \int [|f(s)| - |f_i(s)|] \, \mu(ds) \right| \leqslant \int |f(s) - f_i(s)| \, \mu(ds) \underset{i}{\to} 0,$$

thus proving (2).

Proof of (3) *and* (4) If μ is positive, then, by I.4.23(7) and I.4.33,

$$\left| \int f(s) \, \mu(ds) \right| = \lim_i \left| \int f_i(s) \, \mu(ds) \right| \leqslant \lim_i \int |f_i(s)| \, \mu(ds)$$

for every sequence (f_i) of μ-simple functions with $\lim_i \int |f(s) - f_i(s)| \, \mu(ds) = 0$, and the term on the right converges, by (2), to $\int |f(s)| \mu(ds)$. If μ is not positive, then

$$\left| \int f(s) \mu^{\#}(ds) \right| \leqslant \int |f(s)| \mu^{\#}(ds) \qquad \text{for} \quad \# = +, -$$

and

$$\left| \int f(s) \mu(ds) \right| \leqslant \left| \int f(s) \mu^{+}(ds) \right| + \left| \int f(s) \mu^{-}(ds) \right| \leqslant \int |f(s)| (\mu^{+} + \mu^{-})(ds).$$

Finally, we deduce relation (3) for $E \neq S$ by replacing f by $\chi_E f$. Relation (4) then follows from (3) and I.4.28.

Proof of (5) Next we proceed to prove relation (5) and, since $\mu = \mu^{+} - \mu^{-}$, we may clearly limit ourselves to the case where μ is positive. We first observe that, if $A \cap B = \varnothing$ and A and B are μ-measurable, then, by I.4.23,

$$\begin{aligned} \int_{A \cup B} f(s) \mu(ds) &= \lim_i \int_{A \cup B} f_i(s) \mu(ds) \\ &= \lim_i \int_A f_i(s) \mu(ds) + \lim_i \int_B f_i(s) \mu(ds) \\ &= \int_A f(s) \mu(ds) + \int_B f(s) \mu(ds); \end{aligned}$$

hence, by induction, for every finite partition of a μ-measurable set A into sets $A_1, \ldots, A_k$ in Σ, we have

$$\int_A f(s) \mu(ds) = \sum_{j=1}^{k} \int_{A_j} f(s) \mu(ds).$$

If $A_1, A_2, \ldots$ is a denumerable partition of A into μ-measurable sets, then $\sum_{j=1}^{\infty} \mu(A_j) = \mu(A) < \infty$; hence $\lim_n \sum_{j=n+1}^{\infty} \mu(A_j) = 0$. If we set $E_n \triangleq \bigcup_{j=n+1}^{\infty} A_j$, then $\lim_n \mu(E_n) = 0$ and, by (4), $\lim_n \int_{E_n} f(s) \mu(ds) = 0$. It follows that

$$\begin{aligned} \int_A f(s) \mu(ds) &= \sum_{j=1}^{n} \int_{A_j} f(s) \mu(ds) + \int_{E_n} f(s) \mu(ds) \\ &\underset{n}{\to} \lim_n \sum_{j=1}^{n} \int_{A_j} f(s) \mu(ds) \\ &= \sum_{j=1}^{\infty} \int_{A_j} f(s) \mu(ds). \end{aligned}$$

This proves (5).

Proof of (6) *and* (7) We prove (6) by observing that if $f(s) = g(s)$ for $s \in A \subset E$ and $|\mu|(E \sim A) = 0$, then, by (1), (4), and (5),

$$\int_E f(s)\,\mu(ds) - \int_E g(s)\,\mu(ds) = \int_{E\sim A} (f(s) - g(s))\,\mu(ds) = 0.$$

If $\int_A f(s)\,\mu(ds) \geqslant 0$ for all $A \in \Sigma$ and μ is positive, we set

$$A_j \triangleq f^{-1}((-\infty, -1/j]) \qquad \text{for} \quad j \in \mathbb{N}$$

and observe that $A_j \triangleq B_j \cup Z_j$, where $B_j \in \Sigma$ and $\mu(Z_j) = 0$. Thus, by I.4.27(3), $0 \geqslant \int_{B_j} [-f(s)]\,\mu(ds) \geqslant (1/j)\,\mu(B_j)$, implying that $\mu(B_j) = 0$ and

$$0 \leqslant \mu(f^{-1}((-\infty, 0)) = \mu\left(\bigcup_{j=1}^{\infty} A_j\right) = \mu\left(\bigcup_{j=1}^{\infty} B_j\right) = 0.$$

Thus $f(s) \geqslant 0$ μ-a.e. It $\int_A f(s)\,\mu(ds) \leqslant 0$ $(A \in \Sigma)$, then $\int_A (-f(s))\,\mu(ds) \geqslant 0$ and therefore $f(s) \leqslant 0$ μ-a.e. If $\int_A f(s)\,\mu(ds) = 0$ $(A \in \Sigma)$, then $0 \leqslant f(s) \leqslant 0$ μ-a.e. The converse of these statements follows from I.4.2.27(3).

Proof of (8) We assume first that μ is positive and consider a sequence (f_i) of μ-simple functions on S to $\mathscr{X}$ such that $\lim_i \int |f(s) - f_i(s)|\,\mu(ds) = 0$. Since $T : \mathscr{X} \to \mathscr{Y}$ is continuous, the function $T \circ f$ is μ-measurable and since $|Tf(s)| \leqslant |T|\,|f(s)|$ $(s \in S)$, we have $T \circ f \in L^1(S, \Sigma, \mu, \mathscr{Y})$. Furthermore,

$$|Tf(s) - Tf_i(s)| \leqslant |T|\,|f(s) - f_i(s)| \qquad (i \in \mathbb{N}, \quad s \in S)$$

and it is clear that

$$T \int_E f(s)\,\mu(ds) = \lim_i T \int_E f_i(s)\,\mu(ds) = \lim_i \int_E Tf_i(s)\,\mu(ds). \tag{9}$$

By I.4.27(3),

$$\int |Tf(s) - Tf_i(s)|\,\mu(ds) \leqslant |T| \int |f(s) - f_i(s)|\,\mu(ds) \underset{i}{\to} 0;$$

hence, by (9) and Definition I.4.33,

$$\int_E T \circ f(s)\,\mu(ds) = \lim_i \int_E Tf_i(s)\,\mu(ds) = T \int_E f(s)\,\mu(ds).$$

If μ is not a positive measure, then we apply the above relation to μ^+ and μ^- and combine the results. QED

I.4.E The Lebesgue and Radon–Nikodym Theorems

I.4.35 ***Theorem*** (Lebesgue dominated convergence theorem) Let (S, Σ, μ) be a finite positive measure space, $\mathscr{X}$ a separable Banach space, $1 \leqslant p < \infty$, $f_i : S \to \mathscr{X}$ μ-measurable, and either $\lim_i f_i = f$ μ-a.e. or $\lim_i f_i = f$ in μ-measure. Then

(1) $f, f_i \in L^p(S, \Sigma, \mu, \mathscr{X}) \subset L^1(S, \Sigma, \mu, \mathscr{X})$ and $\lim_i |f - f_i|_p = 0$ if $g \in L^p(S, \Sigma, \mu)$ and $|f_i(s)| \leqslant g(s)$ μ-a.e. ($i \in \mathbb{N}$);
(2) $\int_E f(s)\, \nu(ds) = \lim_i \int_E f_i(s)\, \nu(ds)$ if $g \in L^1(S, \Sigma, \mu)$, $|f_i(s)| \leqslant g(s)$ μ-a.e. ($i \in \mathbb{N}$), (S, Σ, ν) is a finite measure space, $\mu = |\nu|$ and E is ν-measurable; and
(3) $\int f(s)\, \mu(ds) = \lim_i \int f_i(s)\, \mu(ds)$ if f, f_i and $f - f_i$ are nonnegative.

▌ PROOF *Proof of* (1) *and* (2) Let $g \in L^p(S, \Sigma, \mu)$ and $|f_i(s)| \leqslant g(s)$ μ-a.e. ($i \in \mathbb{N}$). We first assume that $\lim_i f_i = f$ μ-a.e. By I.4.17 and I.4.18, f is μ-measurable and $\lim_i f_i = f$ in μ-measure, and it is clear that $|f(s)| = \lim_i |f_i(s)| \leqslant g(s)$ μ-a.e. Thus by I.4.27(3), $f, f_i \in L^p(S, \Sigma, \mu, \mathscr{X})$. If $\epsilon > 0$ and we set

$$A(i) \triangleq \{s \in S \mid |f(s) - f_i(s)|^p > \epsilon/[2\mu(S)]\} \qquad (i \in \mathbb{N}),$$

then, by I.4.28, there exists $i_0 \triangleq i_0(\epsilon) \in \mathbb{N}$ such that $\int_{A(i)} |g(s)|^p\, \mu(ds) \leqslant 2^{-p-1}\epsilon$ for $i \geqslant i_0$. Thus, for all $i \geqslant i_0$, we have by I.4.27,

$$\int |f(s) - f_i(s)|^p\, \mu(ds) = \int_{S \sim A(i)} |f(s) - f_i(s)|^p\, \mu(ds) + \int_{A(i)} |f(s) - f_i(s)|^p\, \mu(ds),$$

$$\int_{S \sim A(i)} |f(s) - f_i(s)|^p\, \mu(ds) \leqslant \mu(S \sim A(i)) \cdot \epsilon/(2\mu(S)) \leqslant \epsilon/2$$

and

$$\int_{A(i)} |f(s) - f_i(s)|^p\, \mu(ds) \leqslant 2^p \int_{A(i)} |g(s)|^p\, \mu(ds) \leqslant \epsilon/2;$$

hence

$$|f - f_i|_p^p = \int |f(s) - f_i(s)|^p\, \mu(ds) \leqslant \epsilon \qquad \text{for all } i \geqslant i_0(\epsilon).$$

Since $\epsilon > 0$ is arbitrary, we have $\lim_i |f - f_i|_p = 0$. Since $1 \in L^1(S, \Sigma, \mu)$ and $a \leqslant 1 + a^p$ ($a \geqslant 0$), it follows that $h \in L^1(S, \Sigma, \mu, \mathscr{X})$ whenever $h \in L^p(S, \Sigma, \mu, \mathscr{X})$.

If $p = 1$ and E is μ-measurable set, then $|\chi_E(f - f_i)|_1 \leqslant |f - f_i|_1 \underset{i}{\to} 0$; hence, by I.4.34,

$$\lim_i \left| \int_E f(s)\, \mu(ds) - \int_E f_i(s)\, \mu(ds) \right| \leqslant \lim_i \int_E |f(s) - f_i(s)|\, \mu(ds) = 0.$$

If, furthermore, (S, Σ, ν) is a finite measure space and $\mu = |\nu|$, then we can apply the above relation with μ replaced by ν^+ and ν^-, to obtain

$$\int_E f(s)\nu(ds) = \lim_i \int_E f_i(s)\, \nu(ds).$$

Now let $\lim_i f_i = f$ in μ-measure and assume that there exist $\epsilon > 0$ and $J \subset (1, 2,...)$ such that $|f - f_i|_p > \epsilon$ for $i \in J$. Then, by Egoroff's theorem I.4.18, f is μ-measurable and there exists $J_1 \subset J$ with $\lim_{i \in J_1} f_i = f$ μ-a.e. Our previous argument shows that $\lim_{i \in J_1} |f - f_i|_p = 0$, contrary to assumption. Thus $\lim_i |f - f_i|_p = 0$. The same argument applies to (2).

Proof of (3) We first observe that if ϕ is nonnegative and μ-measurable, $A(\phi, n) \triangleq \phi^{-1}([0, n])$, and $\chi_n^{\phi} \triangleq \chi_{A(\phi,n)}$, then

$$\int \phi(s)\, \mu(ds) = \lim_n \int \chi_n^{\phi}(s)\, \phi(s)\, \mu(ds). \tag{4}$$

Indeed, if (ϕ_i) is the sequence of nonnegative μ-simple functions constructed from ϕ according to the procedure of Lemma I.4.25, then $\phi_n = \chi_n^{\phi} \phi_n$ for all $n \in \mathbb{N}$ and $\int \phi(s)\, \mu(ds) = \lim_n \int \phi_n(s)\, \mu(ds)$. Since

$$\phi_n(s) \leqslant \chi_n^{\phi}(s)\, \phi(s) \leqslant \phi(s) \qquad (n \in \mathbb{N}, \quad s \in S),$$

relation (4) follows directly.

Next we observe that the conclusion of (3) follows from (2) if f is μ-integrable. We assume therefore that $\int f(s)\, \mu(ds) = \infty$. Since $\chi_n^f f$ is bounded, hence integrable, it follows from (2) that

$$\lim_i \int \chi_n^f(s) f_i(s)\, \mu(ds) = \int \chi_n^f(s) f(s)\, \mu(ds) \qquad (n \in \mathbb{N});$$

hence, by (4) and I.3.14(2),

$$\infty = \int f(s)\, \mu(ds) = \lim_n \lim_i \int \chi_n^f(s) f_i(s)\, \mu(ds) = \lim_i \lim_n \int \chi_n^f(s) f_i(s)\, \mu(ds).$$

We can now conclude that $\lim_i \int f_i(s)\, \mu(ds) = \int f(s)\, \mu(ds) = \infty$ because $\int f_i(s)\, \mu(ds) \geqslant \int \chi_n^f(s) f_i(s)\, \mu(ds)$ for all i and n in $\mathbb{N}$. QED

I.4.36 ***Theorem*** Let (S, Σ, μ) be a finite positive measure space, $1 \leqslant p < \infty$, Z an arbitrary set, $f_j : S \times Z \to \mathbb{R}$ $(j \in \mathbb{N})$, and $f : S \to \mathbb{R}$, and assume that, for $z \in Z$, $j \in \mathbb{N}$, and μ-a.a. $s \in S$, $f_j(\cdot, z)$ is μ-measurable, $f(\cdot)$ μ-integrable, $|f_j(s, z)| \leqslant f(s)$, and either $\lim_j f_j(\cdot, z) = 0$ μ-a.e., uniformly for $z \in Z$, or $\lim_j |f_j(\cdot, z)|_{\mathscr{F}(\mu)} = 0$ uniformly for $z \in Z$. Then

$$\lim_j \int f_j(s, z)\, \mu(ds) = 0 \quad \text{uniformly} \qquad \text{for} \quad z \in Z.$$

▌PROOF By I.4.35, $f_j(\cdot, z) \in L^1(S, \Sigma, \mu)$ for all j and z. Now assume, by way of contradiction, that there exist $\epsilon > 0$ and sequences $(j_i) \subset (1, 2,...)$ and (z_i) in Z such that

$$\left| \int f_{j_i}(s, z_i)\, \mu(ds) \right| > \epsilon \qquad \text{for} \quad \text{all } i \in \mathbb{N}. \tag{1}$$

Then $\lim_i f_{j_i}(\cdot, z_i) = 0$ either μ-a.e. or in μ-measure because $\lim_j f_j(\cdot, z) = 0$ either μ-a.e. or in μ-measure, uniformly for $z \in Z$; hence, by I.4.35, $\lim_i \int f_{j_i}(s, z_i)\,\mu(ds) = 0$, contradicting (1). QED

I.4.37 ***Theorem*** (Radon–Nikodym) Let (S, Σ, μ) and (S, Σ, λ) be finite measure spaces, μ positive and λ μ-continuous. Then there exists a unique $f \in L^1(S, \Sigma, \mu)$ such that

$$\lambda(E) = \int_E f(s)\,\mu(ds) \quad \text{and} \quad |\lambda|(E) = \int_E |f(s)|\,\mu(ds) \qquad (E \in \Sigma).$$

▌PROOF We first assume that λ is positive. For $i, j \in \{0, 1, 2,...\}$, let A_j^i be a negative set for $\lambda - i2^{-j}\mu$. For each j, let $E_j^i \triangleq A_j^i \sim \bigcup_{k=0}^{i-1} A_j^k$ $(i = 0, 1, 2,...)$. Then $E_j^0, E_j^1,...$ are disjoint and, if $i \in \{0, 1, 2,...\}$, $B \subset E_j^i$ and $B \in \Sigma$, we have

$$(\lambda - i2^{-j}\mu)(B) \leqslant 0 \qquad \text{and} \qquad (\lambda - (i-1)\,2^{-j}\mu)(B) \geqslant 0;$$

hence

$$(1) \quad (i-1)\,2^{-j}\mu(B) \leqslant \lambda(B) \leqslant i2^{-j}\mu(B) \qquad (B \in \Sigma,\ B \subset E_j^i,\ i = 0, 1, 2,...).$$

We now set

$$f_j(s) \triangleq \begin{cases} 0 & \text{if } s \in E_j^0, \\ (i-1)\,2^{-j} & \text{if } s \in E_j^i, \quad i = 1, 2,..., \\ 0 & \text{if } s \in Z_j \triangleq S \sim \bigcup_{i=0}^{\infty} E_j^i. \end{cases}$$

We observe that Z_j is contained in a positive set for $\lambda - i2^{-j}\mu$ for each $i \in \mathbb{N}$; hence $\lambda(Z_j) \geqslant i2^{-j}\mu(Z_j)$ for all $i \in \mathbb{N}$ and $0 \leqslant \mu(Z_j) \leqslant \lim_i i^{-1}2^j\lambda(Z_j) \leqslant \lim_i i^{-1}2^j\lambda(S) = 0$, showing that $\mu(Z_j) = 0$; hence $\lambda(Z_j) = 0$. Furthermore, f_j is μ-measurable and nonnegative. Thus, by I.4.27 and (1),

$$\begin{aligned} \int_E f_j(s)\,\mu(ds) &= \sum_{i=0}^{\infty} \int_{E_j^i \cap E} f_j(s)\,\mu(ds) \\ &\leqslant \sum_{i=0}^{\infty} \lambda(E \cap E_j^i) \\ &= \lambda(E) \leqslant \sum_{i=0}^{\infty} i2^{-j}\mu(E \cap E_j^i) \\ &\leqslant \int_E (f_j(s) + 2^{-j})\,\mu(ds) \quad (E \in \Sigma). \end{aligned}$$

It follows that

$$\lim_j \int_E f_j(s)\,\mu(ds) = \lambda(E) \qquad (E \in \Sigma). \tag{2}$$

We next observe that the set $A_j{}^i$ (a negative set for $\lambda - i2^{-j}\mu$) can be chosen for all $i, j = 0, 1, 2,...$ in such a manner that $A_j{}^i \subset A_j^{i+1}$ and $A_j{}^i = A_{j+1}^{2i}$. It follows then that $E_{j+1}^{2i} \cup E_{j+1}^{2i-1} \subset E_j{}^i$; hence

$$f_j(s) \leqslant f_{j+1}(s) \qquad \text{for all } s \in S' \triangleq S \sim \bigcup_{j=0}^{\infty} Z_j .$$

We have $\mu(S \sim S') = 0$ and

$$| f_j(s) - f_{j+1}(s)| \leqslant 2^{-j} \qquad (s \in S', \quad j \in \mathbb{N});$$

hence

$$| f_j(s) - f_m(s)| \leqslant \sum_{k=j}^{m-1} | f_k(s) - f_{k+1}(s)| < 2^{-j+1} \qquad (s \in S', \quad m > j).$$

Thus, for each $s \in S'$, $(f_j(s))$ is a nondecreasing Cauchy sequence in $\mathbb{R}$ with a limit $f(s)$, and we set $f(s) \triangleq 0$ for all $s \in S \sim S'$. By I.4.17, $f : S \to \mathbb{R}$ is μ-measurable. Thus, in view of (2), I.4.26 and I.4.28, we have

$$\lambda(E) = \lambda(E \cap S') = \int_{E\cap S'} f(s)\,\mu(ds) = \int_E f(s)\,\mu(ds) \qquad (E \in \Sigma).$$

Since $f_j(s) \geqslant 0$ $(s \in S)$, we have $f(s) \geqslant 0$ $(s \in S)$.

If λ is not necessarily positive, then, by I.4.2 and I.4.3, λ^+ and λ^- are positive and bounded measures and $\lambda = \lambda^+ - \lambda^-$. Therefore, there exist f^+, $f^- \in L^1(S, \Sigma, \mu)$ such that $f^+(s) \geqslant 0$ and $f^-(s) \geqslant 0$ μ-a.e., $\lambda^+(E) = \int_E f^+(s)\,\mu(ds)$ and $\lambda^-(E) = \int_E f^-(s)\,\mu(ds)$ $(E \in \Sigma)$ and $\lambda(E) = \int_E f(s)\mu(ds)$ for $f = f^+ - f^-$. We have

$$| \lambda |(E) = \lambda^+(E) + \lambda^-(E) = \int_E f^+(s)\,\mu(ds) + \int_E f^-(s)\,\mu(ds).$$

By I.4.5 and I.4.27(6), we have $f^+(s) \cdot f^-(s) = 0$ μ-a.e. Thus

$$| \lambda |(E) = \int_E | f(s)|\,\mu(ds).$$

Finally, if $g \in L^1(S, \Sigma, \mu)$ and $\lambda(E) = \int_E g(s)\,\mu(ds)$ $(E \in \Sigma)$, then $\int_E \big(f(s) - g(s)\big)\,\mu(ds) = 0$ for all $E \in \Sigma$ and, by I.4.34(7), $f(s) = g(s)$ μ-a.e., showing that f is equivalent to g in $L^1(S, \Sigma, \mu)$. QED

We recall that, by I.4.34(5), $E \to \int_E f(s)\,\mu(ds)$ is a finite measure on Σ if $f \in L^1(S, \Sigma, \mu)$.

I.4.38 ***Theorem*** Let (S, Σ, μ) be a positive finite measure space, $f \in L^1(S, \Sigma, \mu)$, $\lambda(E) \triangleq \int_E f(s)\,\mu(ds)$ $(E \in \Sigma)$, and $g \in L^1(S, \Sigma, \lambda)$. Then fg is μ-integrable and

$$(1) \qquad \int_E f(s)\,g(s)\,\mu(ds) = \int_E g(s)\,\lambda(ds) \qquad (E \in \Sigma).$$

▌PROOF We first assume that $f(s) \geqslant 0$ and $g(s) \geqslant 0$ $(s \in S)$. Then λ is positive. By I.4.28, $\lambda(E) = 0$ if $\mu(E) = 0$; hence E is λ-measurable if it is μ-measurable, and the functions f, g and fg are λ-measurable. On the other hand, if a set E is λ-measurable, then $E = A \cup Z$, with $A, B \in \Sigma$, $Z \subset B$ and $\lambda(B) = 0$; hence, by I.4.27(6), $f(s) = 0$ μ-a.e. in Z. It follows that every λ-measurable function is μ-measurable on $S \sim f^{-1}(\{0\})$. Furthermore, for all real α, the λ-measurable set $E_\alpha \triangleq \{s \in S \mid f(s)\,g(s) \geqslant \alpha\}$ can be represented as $A_\alpha \cup Z_\alpha$, with $A_\alpha \in \Sigma$ and $f(s) = 0$ μ-a.e. in Z_α; hence $\mu(Z_\alpha) = 0$ for $\alpha > 0$. Since $f^{-1}(\{0\}) \subset E_0$ and $E_0 \sim f^{-1}(\{0\})$ is μ-measurable, it follows that E_α is μ-measurable for all $\alpha \geqslant 0$, showing that fg is both λ-measurable and μ-measurable.

If g is λ-simple, say $g \triangleq \sum_{j=1}^k a_j \chi_{E_j}$, then

$$(2) \qquad \int_E g(s)\,\lambda(ds) = \sum_{j=1}^k a_j \lambda(E_j \cap E) = \sum_{j=1}^k a_j \int_{E_j \cap E} f(s)\,\mu(ds)$$

$$= \int_E f(s)\,g(s)\,\mu(ds) \qquad (E \in \Sigma).$$

If g is not λ-simple, then, by I.4.25, there exist a sequence (g_j) of nonnegative λ-simple functions and a sequence (f_j) of nonnegative μ-simple functions such that, for all $s \in S$, $(f_j(s))$ and $(g_j(s))$ are nondecreasing and converge to $f(s)$ and $g(s)$, respectively. Thus if we set $E' \triangleq E \sim f^{-1}(\{0\})$ $(E \in \Sigma)$, then $\chi_{E'}\,g_j$ is μ-measurable for all $j \in \mathbb{N}$ and $E \in \Sigma$. Hence, by (2) and Definition I.4.26,

$$\int_E g(s)\,\lambda(s) = \lim_j \int_E g_j(s)\,\lambda(ds)$$

$$= \lim_j \int_E f(s)\,g_j(s)\,\mu(ds)$$

$$= \lim_j \int_{E'} f(s)\,g_j(s)\,\mu(ds)$$

$$= \lim_j \lim_i \int_{E'} f_i(s)\,g_j(s)\,\mu(ds) \qquad (E \in \Sigma).$$

The double limit on the right equals $\lim_j \int_{E'} f_{i_j}(s)\, g_j(s)\, \mu(ds)$ for an appropriate sequence $(i_j)_j$. Therefore, by Definition I.4.26, we have

$$\int_E g(s)\, \lambda(ds) = \int_{E'} f(s)\, g(s)\, \mu(ds) = \int_E f(s)\, g(s)\, \mu(ds).$$

If we drop the assumption that g is nonnegative, then we set $g^+(s) \triangleq \max(g(s), 0)$ and $g^-(s) \triangleq -\min(g(s), 0)$ for all $s \in S$. Then g^+ and g^- are nonnegative, and $g = g^+ - g^-$. Thus relation (1) for a real-valued g is obtained by applying it to g^+ and g^- and combining both relations. Finally, if f is not assumed nonnegative, we set

$$f^+(s) \triangleq \max((f(s), 0), \qquad f^-(s) \triangleq -\min(f(s), 0) \qquad (s \in S)$$

and

$$\lambda_+(E) \triangleq \int_E f^+(s)\, \mu(ds), \qquad \lambda_-(E) \triangleq \int_E f^-(s)\, \mu(ds) \qquad (E \in \Sigma),$$

apply (1) separately to f^+, λ_+ and f^-, λ_- and combine the results. QED

I.4.39 ***Theorem*** Let (S, Σ, μ) be a finite measure space, T an arbitrary set, and h a function on T onto S. We set $\Sigma_h \triangleq \{h^{-1}(E) \mid E \in \Sigma\}$ and $\nu(h^{-1}(E)) \triangleq \mu(E)$ $(E \in \Sigma)$. Then (T, Σ_h, ν) is a finite measure space. If $\mathscr{X}$ is a separable Banach space and $f: S \to \mathscr{X}$ is μ-integrable, then $f \circ h$ is ν-integrable and

$$\int_E f(s)\, \mu(ds) = \int_{h^{-1}(E)} f(h(t))\, \nu(dt) \qquad (E \in \Sigma). \tag{1}$$

▌PROOF It is clear that $h^{-1}(\{s_1\}) \cap h^{-1}(\{s_2\}) = \varnothing$ if $s_1 \neq s_2$. Thus the sets $h^{-1}(\{s\})$ $(s \in S)$ form a partition of T, and we may identify all points in the same equivalence class $h^{-1}(\{s\})$ with each other and with s. It is then clear that Σ_h is identified with Σ and $\nu(A) = \mu(E)$ if A and E are identified. Thus (T, Σ_h, ν) is a finite measure space.

If $t \in h^{-1}(\{s\})$, then $f(s) = f(h(t))$ and thus $f \circ h$ is ν-measurable whenever f is μ-measurable. To prove (1), we observe that $(f_j \circ h)_j$ is a sequence of ν^+-simple respectively ν^--simple functions converging in $L^1(T, \Sigma_h, \nu^+)$ respectively $L^1(T, \Sigma_h, \nu^-)$ to $f \circ h$ whenever (f_j) is a sequence of μ^+-simple respectively μ^--simple functions converging in $L^1(S, \Sigma, \mu^+)$ respectively $L^1(S, \Sigma, \mu^-)$ to f. QED

I.4.F Absolutely Continuous Functions on an Interval

Let $T \triangleq [t_0, t_1] \subset \mathbb{R}$, $\mathscr{M}$ be the σ-field of Lebesgue measurable subsets of $\mathbb{R}$ and m the Lebesgue measure in $\mathbb{R}$. A function $f: T \to \mathbb{R}$ is *absolutely continuous on* T if for every $\epsilon > 0$ there exists $\eta(\epsilon) > 0$ such that $\sum_{j=1}^{n} |f(b_j) - f(a_j)| \leqslant \epsilon$ whenever $n \in \mathbb{N}$, $[a_j, b_j)$ are disjoint subsets of T, and $\sum_{j=1}^{n} (b_j - a_j) \leqslant \eta(\epsilon)$. A function $f: T \to \mathbb{R}$ has a *derivative* $\dot{f}(t)$ at $t \in T$ if

$$\lim_{\substack{\tau \to t \\ \tau \neq t}} (\tau - t)^{-1} \big(f(\tau) - f(t)\big) \triangleq \dot{f}(t) \in \mathbb{R}.$$

It is clear that an absolutely continuous function on T is continuous and a function with a derivative at t is continuous at t (but the converse is not true).

We shall denote by $AC(T)$ or AC the class of all absolutely continuous functions on T and by $AC_0(T)$ the set of all $f \in AC(T)$ that vanish at t_0. We shall also, as customary, write $\int_a^b f(t)\, dt$ for $\int_{[a,b]} f(t)\, m(dt)$ if $a \leqslant b$ and for $-\int_{[b,a]} f(t)\, m(dt)$ if $b < a$, use the term "measurable" instead of "m-measurable" or "Lebesgue measurable" when referring to subsets of $\mathbb{R}$ and functions on $\mathbb{R}$ and write a.e. and a.a. for m-a.e. and m-a.a. Finally, we shall write $\mathscr{M}_T$ to denote the collection of measurable subsets of T and $(T, \mathscr{M}, m)$ or $(T, \mathscr{M}_T, m)$ for $(T, \mathscr{M}_T, m_T)$, where $m_T \triangleq m \mid \mathscr{M}_T$. Similarly, $L^1(T, \mathscr{M}_T, m)$ means $L^1(T, \mathscr{M}_T, m_T)$.

For $n \in \mathbb{N}$, we shall say that a function $f = (f^1,..., f^n): T \to \mathbb{R}^n$ is *absolutely continuous* if f^j is absolutely continuous for each $j \in \{1, 2,..., n\}$; we shall say that f has a derivative at $t \in T$ if $\dot{f}^i(t)$ exists for $i = 1, 2,..., n$ and then we write $\dot{f}(t) = \big(\dot{f}^1(t),..., \dot{f}^n(t)\big)$ and verify that

$$\dot{f}(t) = \lim_{\substack{\tau \to t \\ \tau \neq t}} (\tau - t)^{-1} \big(f(\tau) - f(t)\big).$$

If $\dot{f}(t)$ exists a.e. in T, then we denote by $\dot{f}$ the function $t \to \dot{f}(t)$ defined for a.a. $t \in T$.

I.4.40 ***Theorem*** There exists a one-to-one correspondence between $AC_0(T)$ and the class of all finite, regular, and m-continuous measures on $\mathscr{M}_T$. If $f \in AC_0(T)$ corresponds to a measure φ, then

$$f(t) = \varphi([t_0, t)) = \varphi([t_0, t]) \qquad (t \in T).$$

▌PROOF Let $\varphi: \mathscr{M}_T \to \mathbb{R}$ be a finite, regular, and m-continuous measure and $f(t) \triangleq \varphi\big([t_0, t)\big)$ $(t \in T)$. Then $f(t_0) = \varphi(\varnothing) = 0$, $\varphi(\{t\}) = 0$ $(t \in T)$ and, by I.4.7, $f \in AC_0(T)$. If $\varphi_1: \mathscr{M}_T \to \mathbb{R}$ is also a finite, regular, and m-continuous measure and $\varphi([t_0, t)) = \varphi_1([t_0, t))$ for all $t \in T$, then $\varphi\big((a, b)\big) = \varphi\big([a, b]\big) = \varphi\big([t_0, b)\big) - \varphi\big([t_0, a)\big) = \varphi_1\big((a, b)\big)$ and $\varphi\big((a, t_1]\big) = \varphi_1\big((a, t_1]\big)$. Since every relatively open subset of T is, by I.2.8(2), an at most denumerable union of disjoint relatively open intervals, it follows that φ and φ_1 coincide on relatively open sets and on their complements in T, the relatively closed sets. Since both φ and φ_1 are regular, it follows that they must coincide on $\mathscr{M}_T$. Thus only a unique φ can correspond to any $f \in AC_0(T)$.

Now it must be shown that for every $f \in AC_0(T)$ we can construct an appropriate measure φ. This construction will resemble that of the Lebesgue measure in I.4.15. We consider the field Σ whose elements are the intervals $[t_0, t_0] = \{t_0\}$ and $(t', t'']$ for $t_0 \leqslant t' < t'' \leqslant t_1$ (which we shall refer to as *basic intervals*) and finite unions of basic intervals. We set $\varphi(\{t_0\}) = 0$, $\varphi((t', t'']) = f(t'') - f(t')$, and $\varphi(\bigcup_{j=1}^k A_j) = \sum_{j=1}^k \varphi(A_j)$ whenever $A_1, ..., A_k$ are disjoint basic intervals. It is clear that, as just defined, φ is an additive set function on Σ. To show that φ is bounded, we observe that, by the definition of an absolutely continuous function, there exists η_1 such that

$$\left|\varphi\left(\bigcup_{j=1}^k A_j\right)\right| \leqslant \sum_{j=1}^k |\varphi(A_j)| \leqslant 1$$

whenever $A_1, ..., A_k$ are basic intervals whose combined length does not exceed η_1. Thus, if $B_1, ..., B_m$ are disjoint basic intervals, then we can partition them into at most $k_1 \leqslant (t_1 - t_0)/\eta_1 + 1$ sets of basic intervals, the combined length in each set not exceeding η_1; hence

$$\left|\varphi\left(\bigcup_{i=1}^m B_i\right)\right| \leqslant (t_1 - t_0)/\eta_1 + 1,$$

showing that φ is bounded. Furthermore, φ is regular. Indeed, each element of Σ can be represented as a union of basic intervals with disjoint closures, and the regularity of φ on Σ follows from the continuity of f.

As shown in I.4.15, the smallest σ-field containing Σ is $\Sigma_{\text{Borel}}(T)$. It follows now from I.4.14 that φ can be uniquely extended to a bounded regular measure on $\Sigma_{\text{Borel}}(T)$ and, by the regularity of φ and the continuity of f, $\varphi((t', t'')) = \varphi((t', t'']) = f(t'') - f(t')$ $(t_0 \leqslant t' < t'' \leqslant t_1)$ and $\varphi(\{t_1\}) = 0$. The Lesbesgue extension of $(T, \Sigma_{\text{Borel}}(T), \varphi)$ yields a bounded measure φ which clearly remains regular. It remains therefore to show that $\varphi : \mathscr{M}_T \to \mathbb{R}$ is m-continuous. Since m and φ are regular, for every $A \in \mathscr{M}_T$ with $m(A) = 0$ there exist a sequence (G_j) of open sets containing A and such that $\lim_j \varphi(G_j) = \varphi(A)$ and $\lim_j m(G_j) = 0$. Each G_j is a disjoint at most denumerable union of open intervals $(a_{j,i}, b_{j,i})$ $(i \in \mathbb{N})$ and of, possibly, $\{t_0\}$ and $\{t_1\}$, and f is absolutely continuous on T. Thus, for all $\epsilon > 0$ there exists $\eta(\epsilon) > 0$ such that

$$\sum_{i \leqslant n} |f(b_{j,i}) - f(a_{j,i})| \leqslant \epsilon \quad \text{if} \quad \sum_{i \leqslant n} (b_{j,i} - a_{j,i}) \leqslant m(G_j) \leqslant \eta(\epsilon) \quad (j, n \in \mathbb{N});$$

hence

$$\lim_j \sum_i |f(b_{j,i}) - f(a_{j,i})| = 0,$$

showing that

$$|\varphi(A)| = \lim_j |\varphi(G_j)| = \lim_j \sum_i |f(b_{j,i}) - f(a_{j,i})| = 0. \quad \text{QED}$$

I.4.41 ***Lemma*** Let $\varphi : \mathscr{M}_T \to \mathbb{R}$ be a bounded, regular, and m-continuous measure. Then there exist $h \in L^1(T, \mathscr{M}, m)$ and a set $T' \subset T$ such that $m(T \sim T') = 0$, $\varphi(E) = \int_E h(t)\, dt$ $(E \in \mathscr{M}_T)$ and

$$\lim_j \frac{\varphi((a_j, b_j))}{b_j - a_j} = h(t)$$

if $t \in T'$, $a_j < t < b_j$, and $\lim_j a_j = \lim_j b_j = t$.

▌PROOF *Step* 1 Let

$$\bar{\varphi}(\alpha, t) \triangleq \sup\{\varphi((a, b))/(b - a) \mid t_0 < a < t < b < t_1, 0 < b - a < \alpha\} \in \bar{\mathbb{R}}$$
$$\text{for} \quad t \in (t_0, t_1) \quad \text{and} \quad \alpha > 0.$$

Let $\underline{\varphi}(\alpha, t)$ be similarly defined, with inf replacing sup. Since $\alpha \to \bar{\varphi}(\alpha, t)$ is nondecreasing and $\alpha \to \underline{\varphi}(\alpha, t)$ nonincreasing for each t, it follows that $\bar{\varphi}(t) \triangleq \lim_{\alpha \to 0} \bar{\varphi}(\alpha, t)$ and $\underline{\varphi}(t) \triangleq \lim_{\alpha \to 0} \underline{\varphi}(\alpha, t)$ exist in $\bar{\mathbb{R}}$ for each t and, clearly, $\underline{\varphi}(t) \leqslant \bar{\varphi}(t)$.

For all $\alpha > 0$ and $\beta \in \mathbb{R}$, the set $\{t \in T \mid \bar{\varphi}(\alpha, t) > \beta\}$ is easily seen to be open; hence (Lebesgue) measurable. Thus $t \to \bar{\varphi}(\alpha, t) : T \to \bar{\mathbb{R}}$ is measurable and so is, by I.4.17, $t \to \bar{\varphi}(t) = \lim_j \bar{\varphi}(j^{-1}, t) : T \to \bar{\mathbb{R}}$.

Step 2 We shall now show that if $\nu : \mathscr{M}_T \to \mathbb{R}$ is a positive, bounded, regular, and m-continuous measure and $\nu(A) = 0$, then $\bar{\nu}(t) = \underline{\nu}(t) = 0$ a.e. in A. Indeed, assume the contrary, and let $\eta > 0$. Since ν is positive and therefore $\bar{\nu}(t) \geqslant \underline{\nu}(t) \geqslant 0$ $(t \in T)$ and, by Step 1, $\bar{\nu}$ is measurable, there exists $\beta > 0$ such that $E \triangleq \{t \in A \mid \bar{\nu}(t) > \beta\} \in \mathscr{M}$, and $m(E) > 0$. Since m is regular, there exists a closed set $F \subset E \sim \{t_0, t_1\}$ such that $m(F) > 0$; furthermore, every point in F is covered by some open interval $I \triangleq (a, b) \subset (t_0, t_1)$, with $\nu(I) > \beta m(I)$ and $m(I) < \eta$. Since F is compact, a finite collection of such intervals, say $I_1, \ldots, I_k$, covers F and we may order them so that $m(I_1) \geqslant m(I_2) \geqslant \cdots \geqslant m(I_k)$. If we select integers $i_1, i_2, \ldots, i_n$ so that $i_1 = 1$ and i_{j+1} is the smallest integer for which $I_{i_{j+1}}$ does not intersect $I_{i_1}, I_{i_2}, \ldots, I_{i_j}$, then $I_{i_1}, \ldots, I_{i_n}$ are disjoint and $\sum_{j=1}^n m(I_{i_j}) \geqslant \frac{1}{3} m(F)$. (Indeed, if J_{i_j} is the open interval with the same center as I_{i_j} and of triple length then each I_i with $i_j \leqslant i < i_{j+1}$ intersects I_{i_l} for some $l \in \{1, 2, \ldots, j\}$ and is therefore contained in J_{i_l}).

We observe that every point in each I_{i_j} is within η of the compact set F. Thus, if G is open and $F \subset G$, then $G \supset \bigcup_{j=1}^{n} I_{i_j}$ for all sufficiently small η and

$$\nu(G) \geqslant \sum_{j=1}^{n} \nu(I_{i_j}) > \beta \sum_{j=1}^{n} m(I_{i_j}) \geqslant \tfrac{1}{3}\beta m(F) > 0.$$

This relation, valid for all open $G \supset F$, and the regularity of ν imply that $0 < \nu(F) \leqslant \nu(A)$, contradicting the assumption that $\nu(A) = 0$. Thus $\bar{\nu}(t) = \underline{\nu}(t) = 0$ a.e. in A.

Step 3 By the Radon–Nikodym theorem I.4.37, there exists

$$h \in L^1(T, \mathscr{M}_T, m) \text{ such that } \varphi(E) = \int_E h(t)\, dt$$

for all $E \in \mathscr{M}_T$. We shall prove that $\bar{\varphi}(t) = \underline{\varphi}(t) = h(t)$ a.e. in T.

Let $\alpha \in \mathbb{R}$, $A_\alpha \triangleq h^{-1}((-\infty, \alpha))$, $B_\alpha \triangleq h^{-1}([\alpha, \infty))$, and

$$\nu_\alpha(E) \triangleq \int_E \chi_{B_\alpha}(t) \cdot \big(h(t) - \alpha\big)\, dt \qquad (E \in \mathscr{M}_T).$$

Then, by I.4.34 and I.4.37, ν_α is a bounded, regular, and m-continuous measure, and it is clear that ν_α is positive and $\nu_\alpha(A_\alpha) = 0$. Thus, by Step 2, $\bar{\nu}_\alpha(t) = \underline{\nu}_\alpha(t) = 0$ a.e. in A_α. On the other hand, since $h(t) - \alpha < 0$ for $t \in A_\alpha$, we have

$$\nu_\alpha(E) \geqslant \int_E \big(h(t) - \alpha\big)\, dt = \varphi(E) - \alpha m(E) \qquad (E \in \mathscr{M}_T);$$

hence $0 = \bar{\nu}_\alpha(t) \geqslant \bar{\varphi}(t) - \alpha$ a.e. in A_α. In particular, for every rational α and $E_\alpha \triangleq \{t \in T \mid h(t) < \alpha < \bar{\varphi}(t)\}$, we have $E_\alpha \subset A_\alpha$ and $m(E_\alpha) = 0$; hence $m\big(\{t \in T \mid h(t) < \bar{\varphi}(t)\}\big) = 0$ and thus $\bar{\varphi}(t) \leqslant h(t)$ a.e.

If we replace φ by $-\varphi$ and h by $-h$, then $\overline{(-\varphi)}(t) = -\underline{\varphi}(t) \leqslant -h(t)$ a.e. and we conclude that $\underline{\varphi}(t) = \bar{\varphi}(t) = h(t)$ a.e. which completes the proof of the lemma. QED

I.4.42 ***Theorem***

(1) Let $T \triangleq [t_0, t_1] \subset \mathbb{R}$ and $f : T \to \mathbb{R}$ be absolutely continuous. Then $\dot{f}(t)$ exists a.e., $\dot{f} \in L^1(T, \mathscr{M}_T, m)$ and

$$f(t) = f(t_0) + \int_{t_0}^{t} \dot{f}(\tau)\, d\tau \qquad (t \in T).$$

(2) If $h \in L^1(T, \mathscr{M}_T, m)$, then the function g, defined by

$$g(t) \triangleq \int_{t_0}^{t} h(\tau)\, d\tau \qquad (t \in T),$$

is absolutely continuous on T and $h(t) = \dot{g}(t)$ a.e.

(3) If $[a_0, a_1] \subset \mathbb{R}$, $f_1 : T \to \mathbb{R}$, $f_2 : T \to \mathbb{R}$, and $f : T \to [a_0, a_1]$ are absolutely continuous on T, $h : [a_0, a_1] \to \mathbb{R}$ is integrable, h is bounded or f increasing, and $g(x) \triangleq \int_{a_0}^{x} h(s)\, ds$ $(x \in [a_0, a_1])$, then $f_1 + f_2$, $f_1 f_2$, and $g \circ f$ are absolutely continuous on T.

▌PROOF *Proof of* (1) Let f be absolutely continuous on T and $\varphi : \mathscr{M}_T \to \mathbb{R}$ be the measure corresponding to $f - f(t_0) \in AC_0(T)$ as defined in I.4.40. Then, by I.4.41, there exist $h \in L^1(T, \mathscr{M}_T, m)$ and $T' \subset T$ such that $m(T \sim T') = 0$, $\varphi(E) = \int_E h(t)\, dt$ $(E \in \mathscr{M}_T)$ and

$$\lim_j \frac{\varphi((a_j, b_j))}{b_j - a_j} = \lim_j \frac{f(b_j) - f(a_j)}{b_j - a_j} = h(t)$$

if $a_j < t < b_j$ and $\lim_j a_j = \lim_j b_j = t \in T'$. Since f is continuous, for every choice of $t \in T'$ and $b_j \in T$ $(j \in \mathbb{N})$ with $b_j > t$ and $\lim_j b_j = t$, we can determine $a_j < t$ $(j \in \mathbb{N})$ such that $|f(t) - f(a_j)| \leqslant (b_j - t)^2$ and $t - a_j \leqslant (b_j - t)^2$. Then

$$\begin{aligned}\lim_j \frac{f(b_j) - f(t)}{b_j - t} &= \lim_j \left(\frac{f(b_j) - f(a_j)}{b_j - t} - \frac{f(t) - f(a_j)}{b_j - t}\right) \\ &= \lim_j \frac{f(b_j) - f(a_j)}{b_j - a_j}\left(1 + \frac{t - a_j}{b_j - t}\right) = h(t).\end{aligned}$$

Similarly, by first choosing a sequence (a_j) converging to t with $a_j < t$, and then an appropriate sequence (b_j), we show that

$$\lim_j \frac{f(t) - f(a_j)}{t - a_j} = h(t) \qquad \text{for all } t \in T'.$$

Thus $\dot{f}(t) = h(t)$ a.e. in T and

$$f(t) - f(t_0) = \varphi([t_0, t]) = \int_{t_0}^{t} h(\tau)\, d\tau = \int_{t_0}^{t} \dot{f}(\tau)\, d\tau \qquad (t \in T).$$

Proof of (2) Now assume that $h \in L^1(T, \mathscr{M}_T, m)$ and let $\varphi(h)(E) \triangleq \int_E h(\tau)\, d\tau$ $(E \in \mathscr{M}_T)$. Then $\varphi(h)$ is finite and, by I.4.34(4), it is m-continuous. By the Radon–Nikodym theorem I.4.37, $|\varphi(h)|(E) = \int_E |h(\tau)|\, d\tau$ $(E \in \mathscr{M}_T)$ and thus $|\varphi(h)|$ is m-continuous. Since m is regular, it follows that $|\varphi(h)|$, and therefore also $\varphi(h)$, are regular and I.4.40 now implies that the function g, defined by $g(t) \triangleq \varphi(h)([t_0, t)) = \int_{t_0}^{t} h(\tau)\, d\tau$ $(t \in T)$, is absolutely continuous. Thus, as we have shown in the first part of the theorem, $\dot{g}(t)$ exists a.e., $\dot{g} \in L^1(T, \mathscr{M}_T, m)$, and

$$g(t) = \varphi(h)([t_0, t)) = \int_{t_0}^{t} h(\tau)\, d\tau = \int_{t_0}^{t} \dot{g}(\tau)\, d\tau = \varphi(\dot{g})([t_0, t)) \qquad (t \in T).$$

It follows that $\varphi(h)$ and $\varphi(\dot{g})$ coincide on all open intervals, hence on all open sets and, being regular and m-continuous, on all of $\Sigma_{\text{Borel}}(T)$ and of $\mathscr{M}_T$. Thus $\int_E h(\tau)\, d\tau = \int_E \dot{g}(\tau)\, d\tau$ $(E \in \mathscr{M}_T)$ and, by I.4.34(7), $h = \dot{g}$ a.e.

Proof of (3) Let $f_1 : T \to \mathbb{R}$ and $f_2 : T \to \mathbb{R}$ be absolutely continuous on $T \triangleq [t_0, t_1]$. Then we verify directly that $f_1 + f_2$ is also absolutely continuous on T. Furthermore, for every $\epsilon > 0$ there exists $\eta(\epsilon) > 0$ such that

$$\sum_{j=1}^{n} |f_1(b_j) - f_1(a_j)| \leqslant \epsilon \qquad \text{and} \qquad \sum_{j=1}^{n} |f_2(b_j) - f_2(a_j)| \leqslant \epsilon$$

if $[a_j, b_j)$ $(j = 1,..., n)$ are disjoint subsets of T and $\sum_{j=1}^{n} (b_j - a_j) \leqslant \eta(\epsilon)$. By I.2.9, $\alpha \triangleq 1 + \sup_{t\in T} |f_1(t)| + \sup_{t \in T} |f_2(t)| < \infty$. Now let $\epsilon > 0$ be fixed and $\delta \triangleq \eta(\epsilon/(2\alpha))$. Then, for $\sum_{j=1}^{n} (b_j - a_j) \leqslant \delta$, we have

$$\begin{aligned} &\sum_{j=1}^{n} |f_1(b_j) f_2(b_j) - f_1(a_j) f_2(a_j)| \\ &\qquad \leqslant \sum_{j=1}^{n} \big(|f_1(b_j)[f_2(b_j) - f_2(a_j)]| + |f_2(a_j)[f_1(b_j) - f_1(a_j)]|\big) \\ &\qquad \leqslant \alpha \sum_{j=1}^{n} \big(|f_1(b_j) - f_1(a_j)| + |f_2(b_j) - f_2(a_j)|\big) \leqslant \epsilon. \end{aligned}$$

This shows that $f_1 f_2$ is absolutely continuous.

Finally, let $f : T \to [a_0, a_1]$ be absolutely continuous on T, $h : [a_0, a_1] \to \mathbb{R}$ integrable, either h bounded or f increasing, and $g(x) \triangleq \int_{a_0}^{x} h(s)\, ds$ $(x \in [a_0, a_1])$. If $[a_j, b_j)$ $(j = 1, 2,..., n)$ are disjoint subsets of T, $A \triangleq \bigcup_{j=1}^{n} [a_j, b_j)$, and $V_\psi(A) \triangleq \sum_{j=1}^{n} |\psi(b_j) - \psi(a_j)|$ for all $\psi : T \to \mathbb{R}$, then

$$V_{g\circ f}(A) = \sum_{j=1}^{n} |g(f(b_j)) - g(f(a_j))| = \sum_{j=1}^{n} \left| \int_{f(a_j)}^{f(b_j)} h(s)\, ds \right|.$$

If $|h(s)| \leqslant c < \infty$ for all s, then $V_{g\circ f}(A) \leqslant c V_f(A)$. If f is increasing, then $V_{g\circ f}(A) \leqslant \int_{f(A)} |h(s)|\, ds$ and $V_f(A) = m(f(A))$. In either case $\lim V_{g\circ f}(A) = 0$ as $m(A) \to 0$ which shows that $g \circ f$ is absolutely continuous on T. QED

I.4.43 Theorem Let h be absolutely continuous on $T \triangleq [t_0, t_1]$, $\dot{h}(t) > 0$ a.e. in T, T' be the closed interval with endpoints $h(t_0)$ and $h(t_1)$, and $f \in L^1(T', \mathscr{M}_{T'}, m)$. Then the function $t \to f(h(t))\,\dot{h}(t)$ belongs to $L^1(T, \mathscr{M}_T, m)$,

$$\int_{h(t_0)}^{h(t_1)} f(s)\,ds = \int_{t_0}^{t_1} f(h(t))\,\dot{h}(t)\,dt,$$

h is a bijection of T onto T', h^{-1} is absolutely continuous, and $h(E) \in \mathscr{M}_{T'}$ $(E \in \mathscr{M}_T)$.

▌PROOF *Step* 1 By I.4.42, $\dot{h} \in L^1(T, \mathscr{M}_T, m)$ and $h(t'') - h(t') = \int_{t'}^{t''} \dot{h}(t)\,dt \geqslant 0$ for $t_0 \leqslant t' \leqslant t'' \leqslant t_1$. By I.4.27(6), if $h(t'') - h(t') = 0$, then $\dot{h}(t) = 0$ a.e. in $[t', t'']$; hence $t' = t''$. Thus $h : T \to h(T)$ is an increasing continuous bijection and, by I.2.9(4), has a continuous inverse. Since, by I.4.39, h and h^{-1} each map elements of a σ-field onto a σ-field and are both continuous, we conclude that $E \to h(E)$ is a bijection of $\Sigma_{\text{Borel}}(T)$ onto $\Sigma_{\text{Borel}}(T')$.

Step 2 We shall next assume that $E \in \Sigma_{\text{Borel}}(T)$ is such that $m(E) = 0$, and will show that $m(h(E)) = 0$. Since m is regular, for each j we can cover E by a relatively open set $G_j \subset T$ such that $m(G_j) \leqslant m(E) + 1/j \to_j 0$. Thus, by I.2.8(2), there exists an at most denumerable collection of disjoint open intervals $(a_{j,i}, b_{j,i})$ $[i \in J(j) \subset (1, 2,...)]$ such that

$$G_j \sim \{t_0, t_1\} = \textstyle\bigcup_{i \in J(j)} (a_{j,i}, b_{j,i})$$

and, by I.4.34(4) and I.4.42(1),

$$m(h(E)) \leqslant \sum_{i \in J(j)} (h(b_{j,i}) - h(a_{j,i})) = \int_{G_j} \dot{h}(t)\,dt \underset{j}{\to} 0.$$

This also shows, when combined with Step 1, that $h(E) \in \mathscr{M}_{T'}$ $(E \in \mathscr{M}_T)$ and $\{h^{-1}(A) \mid A \in \mathscr{M}_{T'}\}$ contains all of $\mathscr{M}_T$.

Step 3 Let

$$\nu(h^{-1}(A)) \triangleq m(A) \quad (A \in \mathscr{M}_{T'}) \qquad \text{and} \qquad \varphi(E) \triangleq \int_E \dot{h}(t)\,dt \quad (E \in \mathscr{M}_T).$$

By I.4. 39 and I.4.42, ν and φ are finite measures and, if $\alpha, \beta \in T'$ and $A \triangleq (\alpha, \beta)$, then $\varphi(h^{-1}(A)) = h(h^{-1}(\beta)) - h(h^{-1}(\alpha)) = m(A) = \nu(h^{-1}(A))$ and $\varphi(h^{-1}(\{\alpha\})) = 0 = \nu(h^{-1}(\{\alpha\}))$. Thus φ and ν coincide on open intervals and single points and therefore also on their finite or denumerable unions; hence, by I.2.8(2), on all relatively open subsets of T.

It follows from Step 2 that ν is defined on $\mathscr{M}_T$ and is m-continuous; hence regular. Since, clearly, φ is also m-continuous (and therefore regular) and φ and ν coincide on relatively open sets, we conclude that they must coincide on all of $\mathscr{M}_T$.

Step 4 We now apply I.4.39 which shows that $t \to f(h(t))$ is ν-integrable (that is, φ-integrable) and

$$\int_{h(t_0)}^{h(t_1)} f(s)\, ds = \int_{t_0}^{t_1} f(h(t))\, \nu(dt) = \int_{t_0}^{t_1} f(h(t))\, \varphi(dt).$$

It follows, by I.4.38, that $t \to f(h(t))\, \dot{h}(t)$ is m-integrable and

$$\int_{h(t_0)}^{h(t_1)} f(s)\, ds = \int_{t_0}^{t_1} f(h(t))\, \dot{h}(t)\, dt.$$

Finally, we prove that $h^{-1} : T' \to T$ is absolutely continuous. If $\varphi(E) \triangleq \int_E \dot{h}(t)\, dt = 0$,then, by I.4.27(6), $\dot{h}(t) = 0$ a.e. in E; hence $m(E) = 0$. Thus m is φ-continuous. If $[a_j , b_j)$ $(j = 1, 2,..., n)$ are disjoint subintervals of T' and $A \triangleq \bigcup_{j=1}^{n} [a_j , b_j)$, then

$$m(A) = \sum_{j=1}^{n} \int_{h^{-1}(a_j)}^{h^{-1}(b_j)} \dot{h}(t)\, dt = \varphi(h^{-1}(A)).$$

Since m is φ-continuous, this relation and I.4.7 imply that

$$\sum_{j=1}^{n} | h^{-1}(b_j) - h^{-1}(a_j)| = m(h^{-1}(A)) \to 0 \qquad \text{as} \quad m(A) \to 0.$$

Thus h^{-1} is absolutely continuous. QED

I.4.G Product Measures

Let $n \in \mathbb{N}$ and (S_i , Σ_i , μ_i) $(i = 1, 2,..., n)$ be finite measure spaces. We set $S \triangleq S_1 \times \cdots \times S_n$, $\mathscr{E} \triangleq \{E_1 \times E_2 \times \cdots \times E_n \mid E_i \in \Sigma_i \ (i = 1,..., n)\}$, and denote by $\Sigma_{\mathscr{E}}$ the smallest σ-field containing $\mathscr{E}$. It is clear that if a function g on S_1 to some topological space is Σ_1-measurable and $f(s_1 ,..., s_n) \triangleq g(s_1)$ for all $(s_1 ,..., s_n) \in S_1 \times \cdots \times S_n$, then f is $\Sigma_{\mathscr{E}}$-measurable.

I.4.44 ***Theorem*** Let (S_i , Σ_i , μ_i) $(i = 1, 2,..., n)$ be finite measure spaces. Then there exists a unique finite measure μ on $\Sigma_{\mathscr{E}}$ with the property that $\mu(E_1 \times \cdots \times E_n) = \mu_1(E_1)\, \mu_2(E_2) \cdots \mu_n(E_n)$ $(E_1 \times \cdots \times E_n \in \mathscr{E})$. If $E \in \Sigma_{\mathscr{E}}$, then, for all $(s_2 ,..., s_n) \in S_2 \times \cdots \times S_n$, the function $s_1 \to \chi_E(s_1 , s_2 ,..., s_n)$ is μ_1-integrable, the function $s_2 \to \int \chi_E(s_1 , s_2 ,..., s_n)\mu_1(ds_1)$ is defined and

μ_2-integrable, $s_3 \to \int [\int \chi_E(s_1, s_2, ..., s_n) \mu_1(ds_1)] \mu_2(ds_2)$ is defined and μ_3-integrable, etc., and

$$\mu(E) = \int \left[\cdots \left[\int \chi_E(s_1, ..., s_n) \mu_1(ds_1) \right] \cdots \right] \mu_n(ds_n).$$

▮ PROOF Let $\mathscr{A}$ be the collection of all $\Sigma_{\mathscr{E}}$-measurable functions on $S_1 \times \cdots \times S_n$ to $\mathbb{R}$ such that $s_1 \to f(s_1, s_2, ..., s_n)$ is μ_1-integrable for all $(s_2, ..., s_n)$, $s_2 \to \int_{E_1} f(s_1, s_2, ..., s_n) \mu_1(ds_1)$ is μ_2-integrable for all $(s_3, ..., s_n)$ and all $E_1 \in \Sigma_1$, $s_3 \to \int_{E_2} [\int_{E_1} f(s_1, s_2, ..., s_n) \mu_1(ds_1)] \mu_2(ds_2)$ is μ_3-integrable for all $(s_4, ..., s_n)$ and all $E_2 \in \Sigma_2$, etc. Let $\mathscr{B} \triangleq \{f \in \mathscr{A} \mid fg \in \mathscr{A} \text{ for all } g \in \mathscr{A}\}$. Then $\mathscr{B}$ is a vector space, $1 \in \mathscr{B}$ and $f_1 f_2 \in \mathscr{B}$ whenever $f_1 \in \mathscr{B}$ and $f_2 \in \mathscr{B}$. If we set $\mathscr{V} \triangleq \{A \subset S \mid \chi_A \in \mathscr{B}\}$, then we observe that $A \cap B \in \mathscr{V}$ and $S \sim A \in \mathscr{V}$ whenever $A, B \in \mathscr{V}$ (since $\chi_{A \cap B} = \chi_A \chi_B$ and $\chi_{S \sim A} = 1 - \chi_A$). Thus $A \cup B = S \sim (S \sim A) \cap (S \sim B) \in \mathscr{V}$ if $A, B \in \mathscr{V}$, showing that $\mathscr{V}$ is a field in S. If $A_1, A_2, ...$ are disjoint elements of $\mathscr{V}$, $B_k \triangleq \bigcup_{j=1}^k A_j$, $B \triangleq \bigcup_{j=1}^\infty A_j$, $E \triangleq E_1 \times \cdots \times E_n \in \mathscr{E}$ and $g \in \mathscr{A}$, then

$$\chi_B(s) g(s) = \lim_k \chi_{B_k}(s) g(s) \qquad (s \in S)$$

and, by I.4.17 and I.4.35, $s_1 \to \chi_B(s_1, s_2, ..., s_n) g(s_1, s_2, ..., s_n)$ is μ_1-integrable and

$$g_{E_1}(s_2, ..., s_n)$$

$$\triangleq \int_{E_1} \chi_B(s_1, ..., s_n) g(s_1, ..., s_n) \mu_1(ds_1)$$

$$= \lim_k \int_{E_1} \chi_{B_k}(s_1, ..., s_n) g(s_1, ..., s_n) \mu_1(ds_1) \qquad \text{for all } (s_2, ..., s_n).$$

Similarly, $s_2 \to g_{E_1}(s_2, ..., s_n)$ is μ_2-integrable for all $(s_3, ..., s_n)$ and

$$\int_{E_2} g_{E_1}(s_2, ..., s_n) \mu_2(ds_2)$$

$$= \lim_k \int_{E_2} \left[\int_{E_1} \chi_{B_k}(s_1, ..., s_n) g(s_1, ..., s_n) \mu_1(ds_1) \right] \mu_2(ds_2)$$

for all $(s_3, ..., s_n)$, etc. This shows that $\chi_B \in \mathscr{B}$ and therefore $\mathscr{V}$ is a σ-field. If we set

$$\mu(A) \triangleq \int \left[\cdots \left[\int \left[\int \chi_A(s_1, ..., s_n) \mu_1(ds_1) \right] \mu_2(ds_2) \right] \cdots \right] \mu_n(ds_n) \qquad (A \in \mathscr{V}),$$

then our previous argument, with $g = 1$, shows that $\mu : \mathscr{V} \to \mathbb{R}$ is a measure and $\mu(E_1 \times \cdots \times E_n) = \mu_1(E_1) \cdots \mu_n(E_n)$ for all $E_1 \times \cdots \times E_n \in \mathscr{E}$. Furthermore, $|\mu(A)| \leqslant |\mu_1|(S_1) \cdots |\mu_n|(S_n) < \infty$ for all $A \in \mathscr{V}$.

It is clear that $\mathscr{E} \subset \mathscr{V}$ and therefore $\Sigma_{\mathscr{E}} \subset \mathscr{V}$. Thus the restriction of μ to $\Sigma_{\mathscr{E}}$ is a finite measure and we have shown that the function χ_E has the stated properties for all $E \in \Sigma_{\mathscr{E}}$.

Now let $\nu : \Sigma_{\mathscr{E}} \to \mathbb{R}$ be a finite measure such that $\nu(E_1 \times \cdots \times E_n) = \mu_1(E_1) \cdots \mu_n(E_n)$ for all $E_1 \times \cdots \times E_n \in \mathscr{E}$. We denote by $\mathscr{A}_\nu$ the set of all real-valued ν-integrable functions f such that, for all $E \triangleq E_1 \times \cdots \times E_n \in \mathscr{E}$,

$$\int_E f(s)\, \nu(ds) = \int_{E_n} \left[\cdots \left[\int_{E_1} f(s_1, s_2, \ldots, s_n)\, \mu_1(ds_1) \right] \cdots \right] \mu_n(ds_n).$$

We let $\mathscr{B}_\nu \triangleq \{f \in \mathscr{A}_\nu \mid fg \in \mathscr{A}_\nu \text{ for all } g \in \mathscr{A}_\nu\}$ and $\mathscr{V}_\nu \triangleq \{A \subset S \mid \chi_A \in \mathscr{B}_\nu\}$. Then our previous argument, applied to $\mathscr{B}_\nu$ and $\mathscr{V}_\nu$ in place of $\mathscr{B}$ and $\mathscr{V}$, shows that $\mathscr{V}_\nu \supset \Sigma_{\mathscr{E}}$ and

$$\nu(E) = \int \left[\cdots \left[\int \chi_E(s_1, \ldots, s_n)\, \mu_1(ds_1) \right] \cdots \right] \mu_n(ds_n) = \mu(E)$$

for all $E \in \Sigma_{\mathscr{E}}$. Thus μ is the unique finite measure on $E_{\mathscr{E}}$ such that $\mu(E_1 \times \cdots \times E_n) = \mu_1(E_1) \cdots \mu_n(E_n)$ for all $E_1 \times \cdots \times E_n \in \mathscr{E}$. QED

We denote by $\mu_1 \times \cdots \times \mu_n$ the measure $\mu : \Sigma_{\mathscr{E}} \to \mathbb{R}$ defined in I.4.44 and by $\Sigma_1 \otimes \Sigma_2 \otimes \cdots \otimes \Sigma_n$ the σ-field $\Sigma_{\mathscr{E}}$. [We write $\Sigma_1 \otimes \Sigma_2 \otimes \cdots \otimes \Sigma_n$ instead of the commonly used notation $\Sigma_1 \times \Sigma_2 \times \cdots \times \Sigma_n$ since the latter represents the cartesian product $\{(E_1, \ldots, E_n) \mid E_1 \in \Sigma_1, \ldots, E_n \in \Sigma_n\}$.] The measure space

$$(S_1 \times S_2 \times \cdots \times S_n, \Sigma_1 \otimes \Sigma_2 \otimes \cdots \otimes \Sigma_n, \mu_1 \times \mu_2 \times \cdots \times \mu_n)$$

is the *product measure space of* (S_i, Σ_i, μ_i) $(i = 1, 2, \ldots, n)$, $\Sigma_1 \otimes \cdots \otimes \Sigma_n$ the *product σ-field* of $\Sigma_1, \ldots, \Sigma_n$, and $\mu_1 \times \cdots \times \mu_n$ the *product measure* of $\mu_1, \ldots, \mu_n$. We write $\int f(s_1, \ldots, s_n)\, \mu_1(ds_1) \times \cdots \times \mu_n(ds_n)$ for

$$\int f(s)\, \mu_1 \times \cdots \times \mu_n(ds)$$

and $\int_{E_n} \mu_n(ds_n) \int_{E_{n-1}} \cdots \int_{E_1} f(s_1, \ldots, s_n)\, \mu_1(ds_1)$ for

$$\int_{E_n} \left[\cdots \left[\int_{E_2} \left[\int_{E_1} f(s_1, \ldots, s_n)\, \mu_1(ds_1) \right] \mu_2(ds_2) \right] \cdots \right] \mu_n(ds_n).$$

I.4.45 ***Theorem*** (Fubini) Let (S_1, Σ_1, μ_1) and (S_2, Σ_2, μ_2) be positive finite measure spaces, (S, Σ, μ) their product measure space, $\mathscr{X}$ a separable Banach space, and $f : S \to \mathscr{X}$ μ-measurable. Then

(1) for μ_2-a.a. $s_2 \in S_2$ the function $f(\cdot, s_2)$ is μ_1-measurable and for μ_1-a.a. $s_1 \in S_1$ the function $f(s_1, \cdot)$ is μ_2-measurable;
(2) the function $s_2 \to \int f(s_1, s_2)\, \mu_1(ds_1)$ is μ_2-measurable if $s_1 \to f(s_1, s_2)$ is μ_1-integrable for μ_2-a.a. $s_2 \in S_2$; and

$$\text{(3)} \qquad \int f(s)\, \mu(ds) = \int \mu_2(ds_2) \int f(s_1, s_2)\, \mu_1(ds_1)$$
$$= \int \mu_1(ds_1) \int f(s_1, s_2)\, \mu_2(ds_2)$$

if either $f \in L^1(S, \Sigma, \mu, \mathscr{X})$, or $\int \mu_2(ds_2) \int |f(s_1, s_2)|\, \mu_1(ds_1) < \infty$, or $\int \mu_1(ds_1) \int |f(s_1, s_2)|\, \mu_2(ds_2) < \infty$, or $\mathscr{X} = \mathbb{R}$ and $f(s) \geqslant 0$ $(s \in S)$.

▌PROOF *Step* 1 Let $A \in \Sigma$ and $\mu(A) = 0$. Then, by the second part of I.4.44, $\mu(A) = \int \mu_2(ds_2) \int \chi_A(s_1, s_2)\, \mu_1(ds_1) = 0$ and, by I.4.27(6), for μ_2-a.a. $s_2 \in S_2$ we have $\chi_A(\cdot, s_2) = 0$ μ_1-a.e. If $Z \subset A$, then $\chi_Z(s) \leqslant \chi_A(s)$ $(s \in S)$ and therefore

(4) $\chi_Z(\cdot, s_2) = 0$ μ_1-a.e. for μ_2-a.a. $s_2 \in S_2$ if Z is μ-null.

Thus, by (4) and I.4.44,

$$\mu(E) = \int \mu_2(ds_2) \int \chi_E(s_1, s_2)\, \mu_1(ds_1)$$

for all μ-measurable sets E; therefore, if $g : S \to \mathscr{X}$ is μ-simple, then

(5) $g(\cdot, s_2)$ is μ_1-measurable for μ_2-a.a. $s_2 \in S_2$,
(6) the function $s_2 \to \int g(s_1, s_2)\, \mu_1(ds_1)$ is μ_2-measurable, and
(7) $\int g(s)\, \mu(ds) = \int \mu_2(ds_2) \int g(s_1, s_2)\, \mu_1(ds_1)$.

Now let $A(j) \triangleq \{s \in S \mid |f(s)| \leqslant j\}$ $(j \in \mathbb{N})$. Then $\chi_{A(j)} f$ is μ-integrable and $\lim_j \chi_{A(j)} f = f$ in μ-measure. By I.4.30 and I.4.31, there exists a sequence (f_j) of μ-simple functions such that $|f_j(s)| \leqslant |(\chi_{A(j)} f)(s)| + 1 \leqslant |f(s)| + 1$ $(s \in S)$ and $|\chi_{A(j)} f - f_j|_{\mathscr{F}(\mu)} \leqslant 1/j$; hence $\lim_j f_j = f$ in μ-measure. In view of Egoroff's theorem I.4.18, we may assume that $\lim_j f_j = f$ μ-a.e. It follows then from (4) that

$$\lim_j f_j(\cdot, s_2) = f(\cdot, s_2) \quad \mu_1\text{-a.e.} \qquad \text{for} \quad \mu_2\text{-a.a. } s_2 \in S_2$$

and therefore, by (5), (6), I.4.17, and I.4.35, that $f(\cdot, s_2)$ is μ_1-measurable for μ_2-a.a. $s_2 \in S_2$ and that $s_2 \to \int f(s_1, s_2)\, \mu_1(ds_1)$ is μ_2-measurable if $f(\cdot, s_2)$ is μ_1-integrable for μ_2-a.a. $s_2 \in S_2$. This proves (2) and the first part of (1). The second part of (1) is also valid because the product measure space (S, Σ, μ) is independent of the order of the "factor" spaces (S_1, Σ_1, μ_1) and (S_2, Σ_2, μ_2).

Step 2 Now assume that $\mathscr{X} = \mathbb{R}$ and f is nonnegative. Then, by I.4.25, there exists a sequence (g_i) of μ-simple nonnegative functions such that

$(g_i(s))_i$ is nondecreasing and $\lim_i g_i(s) = f(s)$ $(s \in S)$. By (1) and I.4.26, we have

$$\int f(s_1, s_2)\, \mu_1(ds_1) = \lim_i \int g_i(s_1, s_2)\, \mu_1(ds_1) \qquad \text{for} \quad \mu_2\text{-a.a. } s_2 \in S_2$$

and therefore, by (6), (7), and I.4.35(3),

$$\begin{aligned} \int f(s)\, \mu(ds) &= \lim_i \int g_i(s)\, \mu(ds) \\ &= \lim_i \int \mu_2(ds_2) \int g_i(s_1, s_2)\, \mu_1(ds_1) \\ &= \int \Big[\lim_i \int g_i(s_1, s_2)\, \mu_1(ds_1)\Big]\, \mu_2(ds_2) \\ &= \int \mu_2(ds_2) \int f(s_1, s_2)\, \mu_1(ds_1). \end{aligned}$$

Because the product measure space is independent of the order of the "factor" spaces, this also shows that

$$\int f(s)\, \mu(ds) = \int \mu_1(ds_1) \int f(s_1, s_2)\, \mu_2(ds_2),$$

thus proving relation (3) for nonnegative f. If f is real-valued, then (3) is valid for the functions $s \to f^+(s) \triangleq \max(f(s), 0)$ and $s \to f^-(s) \triangleq \max(-f(s), 0)$, and therefore remains valid for $f = f^+ - f^-$.

Step 3 We finally consider the case where $\mathscr{X}$ is an arbitrary separable Banach space. It follows from relations (1) and (2) and from relation (3), as applied to $s \to |f(s)|$, that $\int f(s)\, \mu(ds)$ and $\int \mu_1(ds_1) \int f(s_1, s_2)\, \mu_2(ds_2)$ are defined if and only if

$$\int \mu_2(ds_2) \int |f(s_1, s_2)|\, \mu_1(ds_1) < \infty \quad \text{or} \quad \int \mu_1(ds_1) \int |f(s_1, s_2)|\, \mu_2(ds_2) < \infty.$$

If this is the case then, by (3) (for real-valued functions) and I.4.34(8), we have for every $l \in \mathscr{X}^*$,

$$\begin{aligned} l \int f(s)\, \mu(ds) &= \int l \circ f(s)\, \mu(ds) \\ &= \int \mu_1(ds_1) \int l \circ f(s_1, s_2)\, \mu_2(ds_2) \\ &= \int [l \textstyle\int f(s_1, s_2)\, \mu_2(ds_2)]\, \mu_1(ds_1) \\ &= l \textstyle\int \mu_1(ds_1) \int f(s_1, s_2)\, \mu_2(ds_2). \end{aligned} \tag{8}$$

We now observe that for every $x \in \mathscr{X}$, $x \neq 0$, we can define a nonvanishing bounded linear functional l_x on $\{\alpha x \mid \alpha \in \mathbb{R}\}$ by $l_x(\alpha x) = \alpha$ $(\alpha \in \mathbb{R})$ and, by the Hahn–Banach theorem I.3.8, l_x can be extended to an element of $\mathscr{X}^*$; hence $lx' = lx''$ ($l \in \mathscr{X}^*$) implies that $x' = x''$. Thus it follows from (8) that

$$\int f(s)\,\mu(ds) = \int \mu_1(ds_1) \int f(s_1, s_2)\,\mu_2(ds_2),$$

and the relation

$$\int f(s)\,\mu(ds) = \int \mu_2(ds_2) \int f(s_1, s_2)\,\mu_1(ds_1)$$

follows by interchanging the order of the factor measure spaces. QED

I.4.46 ***Theorem*** Let (S_1, Σ_1) and (S_2, Σ_2) be measurable spaces, $\Sigma \triangleq \Sigma_1 \otimes \Sigma_2$ and X a topological space. If $f: S_1 \times S_2 \to X$ is Σ-measurable, then $f(s_1, \cdot)$ is Σ_2-measurable for each $s_1 \in S_1$ and $f(\cdot, s_2)$ is Σ_1-measurable for each $s_2 \in S_2$.

▌PROOF For each $E \in \Sigma$ and $s_1 \in S_1$, let $E(s_1) \triangleq \{s_2 \in S_2 \mid (s_1, s_2) \in E\}$. If A is an open subset of X, $s_1 \in S_1$, and $E \triangleq f^{-1}(A)$, then $E(s_1) = \{s_2 \in S_2 \mid f(s_1, s_2) \in A\}$. Thus $f(s_1, \cdot)$ is Σ_2-measurable if $E(s_1) \in \Sigma_2$ for each $E \in \Sigma$.

Let $s_1 \in S_1$, and let $\mathscr{A}$ be the subcollection of Σ such that $E(s_1) \in \Sigma_2$ for all $E \in \mathscr{A}$. Then

(1) $E_1 \times E_2 \in \mathscr{A}$ if $E_1 \in \Sigma_1$ and $E_2 \in \Sigma$; in particular, $\varnothing \times \varnothing = \varnothing \in \mathscr{A}$;
(2) $S_1 \times S_2 \sim E \in \mathscr{A}$ if $E \in \mathscr{A}$ since $(S_1 \times S_2 \sim E)(s_1) = S_2 \sim E(s_1)$; and
(3) $\bigcup_{j=1}^{\infty} E^j \in \mathscr{A}$ if $E^j \in \mathscr{A}$ ($j \in \mathbb{N}$) since

$$\left(\bigcup_{j=1}^{\infty} E^j\right)(s_1) = \bigcup_{j=1}^{\infty} E^j(s_1).$$

Thus $\mathscr{A} = \Sigma$, which shows that $f(s_1, \cdot)$ is Σ_2-measurable for each $s_1 \in S_1$. We similarly show that $f(\cdot, s_2)$ is Σ_1-measurable for each $s_2 \in S_2$. QED

I.4.47 ***Theorem*** Let S_1 and S_2 be separable metric spaces and $\Sigma_i \triangleq \Sigma_{\text{Borel}}(S_i)$ ($i = 1, 2$). Then $\Sigma_1 \otimes \Sigma_2 = \Sigma_{\text{Borel}}(S_1 \times S_2)$.

▌PROOF By I.2.17, each S_i has an at most countable base of topology $\{H_1^i, H_2^i, \ldots\}$, and it is easy to see that $\mathscr{H} \triangleq \{H_j^1 \times H_k^2 \mid j, k \in \mathbb{N}\}$ is a base of the topology of $S_1 \times S_2$. Thus any open G in $S_1 \times S_2$ is the union of a subcollection of $\mathscr{H}$ and, since $\mathscr{H} \subset \Sigma_1 \otimes \Sigma_2$ and $\mathscr{H}$ is at most countable, it follows that $G \in \Sigma_1 \otimes \Sigma_2$; hence $\Sigma_{\text{Borel}}(S_1 \times S_2) \subset \Sigma_1 \otimes \Sigma_2$.

On the other hand, for each open $G_2 \subset S_2$, the set

$$\{E_1 \mid E_1 \times G_2 \in \Sigma_{\text{Borel}}(S_1 \times S_2)\}$$

is a σ-field in S_1. Thus $E_1 \times G_2 \in \Sigma_{\text{Borel}}(S_1 \times S_2)$ for every $E_1 \in \Sigma_1$ and open $G_2 \subset S_2$. Since $\{E_2 \mid E_1 \times E_2 \in \Sigma_{\text{Borel}}(S_1 \times S_2)\}$ is a σ-field for each $E_1 \in \Sigma_1$, it follows that $E_1 \times E_2 \in \Sigma_{\text{Borel}}(S_1 \times S_2)$ for all $E_1 \in \Sigma_1$ and $E_2 \in \Sigma_2$. Thus $\Sigma_1 \otimes \Sigma_2 \subset \Sigma_{\text{Borel}}(S_1 \times S_2)$. QED

I.4.48 ***Theorem*** Let (S_1, Σ_1, μ_1) and (S_2, Σ_2, μ_2) be positive finite measure spaces, (S, Σ, μ) their product measure space, X and Y separable metric spaces, and $f : S_1 \times S_2 \times X \to Y$ such that $f(\cdot, \cdot, x)$ is $\mu_1 \times \mu_2$-measurable for each $x \in X$ and $f(s_1, s_2, \cdot)$ is continuous for all $(s_1, s_2) \in S_1 \times S_2$. Then there exists a subset $\tilde{S}_1$ of S_1 such that $\mu_1(S_1 \sim \tilde{S}_1) = 0$ and $f(s_1, \cdot, x)$ is μ_2-measurable for all $s_1 \in \tilde{S}_1$ and $x \in X$.

▌PROOF Let $\{x_1, x_2, ...\}$ be dense in X and $\{y_1, y_2, ...\}$ dense in Y. Then, by Fubini's theorem I.4.45(1), for each $i, j \in \mathbb{N}$ there exists a set $S_1^{i,j}$ such that $\mu_1(S_1 \sim S_1^{i,j}) = 0$ and $s_2 \to d(f(s_1, s_2, x_j), y_i)$ is μ_2-measurable for all $s_1 \in S_1^{i,j}$. We set $\tilde{S}_1 \triangleq \bigcap_{i=1}^{\infty} \bigcap_{j=1}^{\infty} S_1^{i,j}$ and observe that for each $x \in X$ there exists $J \subset (1, 2, ...)$ such that $\lim_{j \in J} f(s_1, s_2, x_j) = f(s_1, s_2, x)$ $(s_1 \in S_1, s_2 \in S_2)$. It follows, by I.4.17, that $f(s_1, \cdot, x)$ is μ_2-measurable for all $s_1 \in \tilde{S}_1$. QED

I.4.49 ***Theorem*** Let S_1 and S_2 be compact metric spaces, $\mu_1 \in \text{frm}^+(S_1)$ and $\mu_2 \in \text{frm}^+(S_2)$. Then $\mu_1 \times \mu_2 \in \text{frm}^+(S_1 \times S_2)$. If μ_1 and μ_2 are nonatomic, then $\mu_1 \times \mu_2$ is also nonatomic.

▌PROOF Let $\Sigma_1 \triangleq \Sigma_{\text{Borel}}(S_1)$ and $\Sigma_2 \triangleq \Sigma_{\text{Borel}}(S_2)$. By I.4.44, $\mu_1 \times \mu_2$ is a finite positive measure and by I.4.47, $\Sigma_1 \otimes \Sigma_2 = \Sigma_{\text{Borel}}(S_1 \times S_2)$. We shall now show that $\mu_1 \times \mu_2$ is regular.

Let $\mathscr{A}$ be the subset of $\Sigma_1 \otimes \Sigma_2$ with the property that for every $E \in \mathscr{A}$ there exist sequences (C_j^E) and (G_j^E) of, respectively, closed and open sets such that $C_j^E \subset E \subset G_j^E$ and

$$\lim_j \mu_1 \times \mu_2(G_j^E \sim E) = \lim_j \mu_1 \times \mu_2(E \sim C_j^E) = 0.$$

If $A \in \Sigma_1$ and $B \in \Sigma_2$, there exist sequences (C_j^A), (G_j^A), (C_j^B), and (G_j^B), with C_j^A closed, G_j^A open, $C_j^A \subset A \subset G_j^A$, and

$$\lim_j \mu_1(G_j^A \sim A) = \lim_j \mu_1(A \sim C_j^A) = 0;$$

and similarly for B, C_j^B, and G_j^B. Then we can easily verify that

$$C_j^A \times C_j^B \subset A \times B \subset G_j^A \times G_j^B \qquad (j \in \mathbb{N})$$

and

$$\lim_j \mu_1 \times \mu_2(G_j^A \times G_j^B \sim A \times B) = \lim_j \mu_1 \times \mu_2(A \times B \sim C_j^A \times C_j^B) = 0.$$

Thus $\{A \times B \mid A \in \Sigma_1, B \in \Sigma_2\} \subset \mathscr{A}$.

If $E, F \in \mathscr{A}$ and $E' \triangleq S_1 \times S_2 \sim E$, then the expressions

$$G_j^{E'} \triangleq S_1 \times S_2 \sim C_j^E, \qquad C_j^{E'} \triangleq S_1 \times S_2 \sim G_j^E,$$

$$G_j^{E \cup F} \triangleq G_j^E \cup G_j^F, \qquad \text{and} \qquad C_j^{E \cup F} \triangleq C_j^E \cup C_j^F \qquad (j \in \mathbb{N})$$

define appropriate sequences of closed and open sets for E' and $E \cup F$. Thus $\mathscr{A}$ is a field.

Next let (E^i) be a sequence of disjoint elements of $\mathscr{A}$. Then for each $i, j \in \mathbb{N}$ there exist a closed C_j^i and an open G_j^i such that $C_j^i \subset E^i \subset G_j^i$,

$$\mu_1 \times \mu_2(G_j^i \sim E^i) \leqslant 2^{-i-j} \qquad \text{and} \qquad \mu_1 \times \mu_2(E^i \sim C_j^i) \leqslant 2^{-i-j}.$$

We set

$$E \triangleq \bigcup_{i=1}^{\infty} E^i, \qquad G_j \triangleq \bigcup_{i=1}^{\infty} G_j^i, \qquad \text{and} \qquad C_j \triangleq \bigcup_{i=1}^{j} C_j^i \qquad (j \in \mathbb{N}).$$

Then, for each $j \in \mathbb{N}$, G_j is open, C_j closed, and $C_j \subset E \subset G_j$. Furthermore,

$$\mu_1 \times \mu_2(G_j \sim E) \leqslant \mu_1 \times \mu_2 \left(\bigcup_{i=1}^{\infty} (G_j^i \sim E^i) \right) \leqslant \sum_{i=1}^{\infty} \mu_1 \times \mu_2(G_j^i \sim E^i)$$

$$\leqslant \sum_{i=1}^{\infty} 2^{-i-j} = 2^{-j}$$

and

$$\mu_1 \times \mu_2(E \sim C_j) = \mu_1 \times \mu_2 \left(\bigcup_{i=1}^{j} (E^i \sim C_j^i) \right) + \mu_1 \times \mu_2 \left(\bigcup_{i=j+1}^{\infty} E^i \right)$$

$$\leqslant 2^{-j} + \sum_{i=j+1}^{\infty} \mu_1 \times \mu_2(E^i).$$

Since $\mu_1 \times \mu_2$ is a finite positive measure and $E_1, E_2, \ldots$ are disjoint, the right-hand side of the last inequality converges to 0 as $j \to \infty$. This shows that $E \in \mathscr{A}$ and that $\mathscr{A}$ is a σ-field containing $\{A \times B \mid A \in \Sigma_1, B \in \Sigma_2\}$. Thus $\mathscr{A} = \Sigma_1 \otimes \Sigma_2$ and $\mu_1 \times \mu_2$ is regular.

Finally, assume that μ_1 and μ_2 are nonatomic. We let

$$\mathscr{B}_1 \triangleq \{A \times B \mid A \in \Sigma_1, B \in \Sigma_2\}$$

and denote by $\mathscr{B}$ the collection of all finite unions of disjoint elements of $\mathscr{B}_1$. Then we can easily verify that $\mathscr{B}$ is a field. Since the compact metric spaces S_1 and S_2 have countable bases of topology (I.2.17), say $\{A_1, A_2, \ldots\}$ and $\{B_1, B_2, \ldots\}$, it follows that every open subset G of $S_1 \times S_2$ is a denumerable union of sets of the form $A_i \times B_j$. Therefore each open G is a denumerable union of disjoint elements of $\mathscr{B}_1$. It follows, by I.4.10, that for every $\alpha \in [0, \mu_1 \times \mu_2(G)]$ there exists $D_\alpha \subset G$ with $\mu_1 \times \mu_2(D_\alpha) = \alpha$. If $E \in \Sigma_1 \otimes \Sigma_2$ and $c \triangleq \mu_1 \times \mu_2(E) > 0$, then, by the regularity of $\mu_1 \times \mu_2$, there exists an open G_E with $E \subset G_E$ and $c \leqslant \mu_1 \times \mu_2(G_E) \leqslant 5c/4$. By our previous argument, there exists $D \subset G_E$ with $\mu_1 \times \mu_2(D) = 3c/4$. Then

$$0 < \mu_1 \times \mu_2(D \cap E) \leqslant \mu_1 \times \mu_2(D) < \mu_1 \times \mu_2(E),$$

since $\mu_1 \times \mu_2(D \cap E) = 0$ implies

$$\mu_1 \times \mu_2(D \cup E) = \mu_1 \times \mu_2(D) + \mu_1 \times \mu_2(E) = 7c/4,$$

contradicting $D \cup E \subset G_E$ and $\mu_1 \times \mu_2(G_E) = 5c/4$. This shows that $\mu_1 \times \mu_2$ is nonatomic. QED

I.4.50 ***The Lebesgue Measure in*** $\mathbb{R}^k$ Let $k \in \mathbb{N}$, $\mathscr{M}_1$ be the collection of Lebesgue measurable subsets of $[0, 1]$, and m_1 the restriction of the Lebesgue measure in $\mathbb{R}$ to $\mathscr{M}_1$. If we denote the Lebesgue extension of the product measure space $([0, 1]^k, \mathscr{M}_1 \otimes \cdots \otimes \mathscr{M}_1, m_1 \times \cdots \times m_1)$ by $([0, 1]^k, \mathscr{M}_1{}^k, m_1{}^k)$, then $\mathscr{M}_1{}^k$ is the class of *Lebesgue measurable* subsets of $[0, 1]^k$ and $m_1{}^k$ the *Lebesgue measure* in $[0, 1]^k$. A set $A \subset \mathbb{R}^k$ is *Lebesgue measurable* if $(A - (i_1, ..., i_k)) \cap [0, 1]^k$ is Lebesgue measurable for all integers $i_1, ..., i_k$. The Lebesgue measure of a Lebesgue measurable set A is

$$m^k(A) \triangleq \sum_{i_1=-\infty}^{\infty} \sum_{i_2=-\infty}^{\infty} \cdots \sum_{i_k=-\infty}^{\infty} m_1{}^k((A - (i_1, ..., i_k)) \cap [0, 1]^k).$$

I.5 The Banach Spaces $C(S, \mathscr{X})$ and $L^p(S, \Sigma, \mu, \mathscr{X})$

I.5.A The Metric Space $C(S, X)$ and the Banach Space $C(S, \mathscr{X})$

If S is a topological space and X a complete metric space, we denote by $BF(S, X)$ the collection of all bounded functions on S to X and by $C(S, X)$ the collection of all bounded continuous functions on S to X. We verify that the function

$$(f, g) \to d_B(f, g) \triangleq \sup_{s \in S} d(f(s), g(s))$$

is a metric function on $BF(S, X)$. By I.2.19, the metric space $BF(S, X)$ and its subset $C(S, X)$ are complete.

If $\mathscr{X}$ is a Banach space, then it is easily seen that $BF(S, \mathscr{X})$ and $C(S, \mathscr{X})$ are vector spaces and $f \to |f|_{\sup} \triangleq \sup_{s \in S} |f(s)| = d_B(f, 0)$ is a norm on

$BF(S, \mathscr{X})$. Thus $(BF(S, \mathscr{X}), |\cdot|_{\sup})$ and $(C(S, \mathscr{X}), |\cdot|_{\sup})$ are Banach spaces. We shall often write $|f|$ for $|f|_{\sup}$ when dealing with $BF(S, \mathscr{X})$ and $C(S, \mathscr{X})$ if there is no danger of confusing $|\cdot|_{\sup}$ with another norm. We also write $C(S)$ for $C(S, \mathbb{R})$.

If S is a compact topological space, then, by I.2.9(1), each continuous function on S to the metric space X is bounded. Thus $C(S, X)$ contains all the continuous functions whenever S is compact.

I.5.1 ***Theorem*** Let S be a compact metric space and X a complete separable metric space. Then $C(S, X)$ is separable.

▌PROOF By I.2.5, for every $j \in \mathbb{N}$ there exists a finite set $\{s_1^j, ..., s_{k(j)}^j\} \subset S$ such that $S \subset \bigcup_{k=1}^{k(j)} S(s_k^j, 1/j)$. We set

$$A(j, k) \triangleq S(s_k^j, 1/j) \sim \bigcup_{i=1}^{k-1} S(s_i^j, 1/j) \qquad [k = 1, 2, ..., k(j)]$$

and renumber the indices, if necessary, so as to leave out all empty sets $A(j, k)$. Then $A(j, 1), ..., A(j, k(j))$ is a partition of S into nonempty subsets of diameters at most $2/j$, and we can choose points $a(j, k) \in A(j, k)$ for $k = 1, 2, ..., k(j)$.

Let X_∞ be a dense at most denumerable subset of X, $Q_j \triangleq X_\infty^{k(j)}$ $(j \in \mathbb{N})$, and $Q \triangleq \bigcup_{j=1}^{\infty} Q_j$. If $q \in Q$, then, for some $j \in \mathbb{N}$, we have $q = (q_1, ..., q_{k(j)})$ and we set

$$\varphi_q(s) \triangleq q_k \qquad [k = 1, 2, ..., k(j), \quad s \in A(j, k)].$$

We now consider the denumerable subset $\Phi \triangleq \{\varphi_q \mid q \in Q\}$ of the metric space $BF(S, X)$, and its closure $\bar{\Phi}$. We shall show that the set $C(S, X)$ is contained in $\bar{\Phi}$ which, in view of I.2.17, will prove that $C(S, X)$ is separable.

Let $\varphi \in C(S, X)$. Then, by I.2.15, φ is uniformly continuous and for every $\epsilon > 0$ there exists $j \in \mathbb{N}$ such that $d(\varphi(s'), \varphi(s'')) \leqslant \epsilon/2$ whenever $d(s', s'') \leqslant 2/j$. We choose a point $q \triangleq (q_1, ..., q_{k(j)}) \in Q_j$ in such a manner that $d(q_k, \varphi(a(j, k))) \leqslant \epsilon/2$ for $k = 1, 2, ..., k(j)$. Then

$$d(\varphi(s), \varphi_q(s)) \leqslant d(\varphi(s), \varphi(a(j, k))) + d(\varphi(a(j, k)), q_k) \leqslant \epsilon$$
$$[k = 1, 2, ..., k(j), \quad s \in A(j, k)];$$

hence $d_B(\varphi, \varphi_q) \leqslant \epsilon$. Thus φ can be approximated in $BF(S, X)$ by elements of Φ, showing that $C(S, X) \subset \bar{\Phi}$. QED

I.5.2 ***Theorem*** Let S be a topological space and $n \in \mathbb{N}$. Then the Banach space $C(S, \mathbb{R}^n)$ is homeomorphic to the space $(C(S))^n$ under the identity mapping.

▌PROOF Since the sets $A_1 \times \cdots \times A_n$, where A_i is an open subset of $\mathbb{R}$ for each i, form a base of the topology in $\mathbb{R}^n$, we conclude that

$\varphi \triangleq (\varphi^1,..., \varphi^n) \in C(S, \mathbb{R}^n)$ if and only if $\varphi^i \in C(S)$ for each i. Thus the set $C(S, \mathbb{R}^n)$ coincides with $(C(S))^n$.

Furthermore, for each $\varphi \triangleq (\varphi^1,..., \varphi^n) \in C(S, \mathbb{R}^n)$,

$$|\varphi| \triangleq \sup_{s\in S} |\varphi(s)| = \sup_{s\in S} \sum_{j=1}^{n} |\varphi^j(s)| \leqslant \sum_{j=1}^{n} \sup_{s\in S} |\varphi^j(s)| = \sum_{j=1}^{n} |\varphi^j|$$

and

$$\sum_{j=1}^{n} |\varphi^j| = \sum_{j=1}^{n} \sup_{s\in S} |\varphi^j(s)| \leqslant \sum_{j=1}^{n} \sup_{s\in S} \sum_{k=1}^{n} |\varphi^k(s)| = n |\varphi| .$$

Thus the topologies of $C(S, \mathbb{R}^n)$ and $C(S)^n$ coincide. QED

I.5.3 ***Theorem*** Let S be a compact metric space, $\mathscr{X}$ a Banach space, $f_j : S \to \mathscr{X}$ $(j \in \mathbb{N})$ have a modulus of continuity Ω, and $f : S \to \mathscr{X}$. If $\lim_j f_j(s) = f(s)$ $(s \in S)$, then $\lim_j f_j(s) = f(s)$ uniformly for $s \in S$ and Ω is a modulus of continuity for f.

▌PROOF By I.2.16, Ω is a modulus of continuity for f. Now assume that $(f_j(s))$ does not converge to $f(s)$ uniformly for $s \in S$. Then there exist $\epsilon > 0$, $\bar{s} \in S$ and sequences $J \subset (1, 2,...)$ and $(s_j)_{j\in J}$ in S such that $\lim_{j\in J} s_j = \bar{s}$ and

$$(1) \qquad |f_j(s_j) - f(s_j)| > \epsilon \qquad (j \in J).$$

Let η be such that $\Omega(h) \leqslant \epsilon/3$ $(h \leqslant \eta)$, and let $j_0 \in \mathbb{N}$ be such that $|f_j(\bar{s}) - f(\bar{s})| \leqslant \epsilon/3$ and $d(s_j, \bar{s}) \leqslant \eta$ $(j \geqslant j_0)$. Then, for all $j \in J$, $j \geqslant j_0$, we have

$$\begin{aligned} |f_j(s_j) - f(s_j)| &\leqslant |f_j(s_j) - f_j(\bar{s})| + |f_j(\bar{s}) - f(\bar{s})| + |f(\bar{s}) - f(s_j)| \\ &\leqslant 2\Omega(\eta) + \epsilon/3 \leqslant \epsilon, \end{aligned}$$

contrary to (1). Thus $\lim_j f_j(s) = f(s)$ uniformly for $s \in S$. QED

I.5.4 ***Theorem*** (Ascoli) Let S be a compact metric space, $\mathscr{X}$ a Banach space, and $A \subset C(S, \mathscr{X})$. Then A is conditionally compact if and only if A is equicontinuous and the set $F \triangleq \{f(s) \mid s \in S, f \in A\}$ is conditionally compact. In particular, if $n \in \mathbb{N}$ and $\mathscr{X} \triangleq \mathbb{R}^n$, then A is conditionally compact if and only if it is equicontinuous and bounded.

▌PROOF We first assume that A is equicontinuous and $\bar{F}$ compact. Let $\epsilon > 0$ and

$$\delta(\epsilon, \bar{s}) \triangleq \sup\{h > 0 \mid |f(s) - f(\bar{s})| \leqslant \epsilon \text{ for all } f \in A \text{ if } d(s, \bar{s}) \leqslant h\} \qquad (\bar{s} \in S).$$

Then, by equicontinuity, $\delta(\epsilon, \bar{s}) > 0$ for each $\bar{s} \in S$. If $\inf_{\bar{s}\in S} \delta(\epsilon, \bar{s}) = 0$, then there exists a sequence (s_j) such that $\lim_j \delta(\epsilon, s_j) = 0$. Since S is metric and compact, it is sequentially compact (I.2.5) and we may assume (extracting a subsequence) that (s_j) converges to some s_0. Then

$$|f(s)-f(s_j)| \leqslant |f(s)-f(s_0)| + |f(s_0)-f(s_j)| \leqslant \epsilon$$

for all $f \in A$ if $d(s, s_0) < \delta(\epsilon/2, s_0)$ and $d(s_0, s_j) < \delta(\epsilon/2, s_0)$; in particular, if $d(s_0, s_j) < \frac{1}{2}\delta(\epsilon/2, s_0)$ and $d(s, s_j) < \frac{1}{2}\delta(\epsilon/2, s_0)$. Thus $\delta(\epsilon, s_j) \geqslant \frac{1}{2}\delta(\epsilon/2, s_0) > 0$ for all suffciently large j, contradicting the assumption that $\lim_j \delta(\epsilon, s_j) = 0$. We conclude that, for every $\epsilon > 0$, $\delta(\epsilon) \triangleq \inf_{\bar{s}\in S} \delta(\epsilon, \bar{s}) > 0$. It follows now from I.2.12 that all the functions $f \in A$ have a common modulus of continuity and therefore, by I.2.18 and I.5.3, the set $\bar{A}$ is sequentially compact. Thus, by I.2.5, $\bar{A}$ is compact.

Now we consider the converse, and assume that $\bar{A}$ is compact. If A is not equicontinuous, then there exist $\bar{s} \in S$, $\epsilon > 0$, and sequences (f_j) in A and (s_j) in S such that $\lim_j s_j = \bar{s}$ and

$$(1) \qquad |f_j(s_j) - f_j(\bar{s})| > \epsilon \qquad (j \in \mathbb{N})$$

By I.2.5, $\bar{A}$ is sequentially compact and we may assume, therefore, (extracting a subsequence, if necessary) that $\lim_j |f_j - \bar{f}| = 0$ for some $\bar{f} \in C(S, \mathscr{X})$. Then there exists $\delta > 0$ such that $|\bar{f}(s) - \bar{f}(\bar{s})| \leqslant \epsilon/3$ if $d(s, \bar{s}) \leqslant \delta$; and there exists $j_0 \in \mathbb{N}$ such that $|\bar{f}(s) - f_j(s)| \leqslant \epsilon/3$ $(s \in S)$ and $d(s_j, \bar{s}) \leqslant \delta$ if $j \geqslant j_0$. Thus

$$|f_j(s_j) - f_j(\bar{s})| \leqslant |f_j(s_j) - \bar{f}(s_j)| + |\bar{f}(s_j) - \bar{f}(\bar{s})| + |\bar{f}(\bar{s}) - f_j(\bar{s})| \leqslant \epsilon \qquad (j \geqslant j_0),$$

contradicting (1). Therefore A is equicontinuous.

If $\bar{F}$ is not compact, then there exist sequences (s_j) in S and (f_j) in A such that $(f_j(s_j))$ has no convergent subsequence. Since S and $\bar{A}$ are compact, there exist $J \subset (1, 2, ...)$, $\bar{s} \in S$, and $\bar{f} \in \bar{A}$ such that $\lim_{j\in J} s_j = \bar{s}$ and

$$\lim_{j\in J} f_j(s) = \bar{f}(s) \quad \text{uniformly} \qquad \text{for} \quad s \in S;$$

hence

$$\lim_{j\in J} f_j(s_j) = \lim_{j\in J} \bar{f}(s_j) = \bar{f}(\bar{s}),$$

contradicting the assertion that $(f_j(s_j))$ has no convergent subsequence. Thus $\bar{F}$ is compact.

Finally, we assume that $n \in \mathbb{N}$ and $\mathscr{X} = \mathbb{R}^n$. It is clear that F is bounded if and only if A is bounded. If this is the case, say $F \subset S(0, \alpha)$, then $F \subset [-\alpha, \alpha]^n$ and, by I.2.6 and I.2.8, $[-\alpha, \alpha]^n$ and its closed subset $\bar{F}$ are compact. Conversely, if $\bar{F}$ is compact then F, hence also A, are bounded. QED

I.5.5 ***Theorem*** Let (S, Σ, μ) be a measure space, S a topological space, $\Sigma \supset \Sigma_{\text{Borel}}(S)$, μ finite and regular, $g : S \to \mathbb{R}$ μ-integrable, and $A(g) \triangleq \{\int g(s) f(s) \mu(ds) \mid f \in C(S), |f|_{\text{sup}} \leqslant 1\}$. Then

(1) μ is positive if $\int f(s)\,\mu(ds) \geqslant 0$ for all continuous $f: S \to [0, 1]$;
(2) $g(s) \geqslant 0$ μ-a.e. if μ is positive and $\int f(s)\,g(s)\,\mu(ds) \geqslant 0$ for all continuous $f: S \to [0, 1]$; and
(3) $\int_E g(s)\,\mu(ds) \in \overline{A(g)}$ and $\int_E |g(s)|\,|\mu|(ds) \in \overline{A(g)}$ for every μ-measurable set E, and

$$\sup A(g) = \int |g(s)|\,|\mu|(ds).$$

▮ PROOF Assume first that $\int g(s) f(s)\,\mu(ds) \geqslant 0$ for all continuous $f: S \to [0, 1]$. Since μ is regular, for any $E \in \Sigma$ and $\epsilon > 0$ there exist a closed F_ϵ and an open G_ϵ such that $F_\epsilon \subset E \subset G_\epsilon$ and $|\mu|(G_\epsilon \sim F_\epsilon) \leqslant \epsilon$. We set $f(s) \triangleq 1$ if $G_\epsilon = S$ and otherwise let

$$f(s) \triangleq d[s, S \sim G_\epsilon]/(d[s, F_\epsilon] + d[s, S \sim G_\epsilon]);$$

then $f(s) = 1$ for $s \in F_\epsilon$, $f(s) = 0$ for $s \in S \sim G_\epsilon$, $0 \leqslant f(s) \leqslant 1$, and f is continuous. It follows that

$$\int_E g(s)\,\mu(ds) + 2 \int_{G_\epsilon \sim F_\epsilon} |g(s)|\,|\mu|(ds) \geqslant \int g(s) f(s)\,\mu(ds) \geqslant 0;$$

hence, by I.4.28,

$$\text{(4)} \quad \int_E g(s)\,\mu(ds) \geqslant -2 \lim_{\epsilon \to 0} \int_{G_\epsilon \sim F_\epsilon} |g(s)|\,|\mu|(ds) = 0 \qquad \text{for all } E \in \Sigma.$$

Statement (1) now follows from (4) by setting $g(s) = 1$ for all $s \in S$ and statement (2) follows from (4) and I.4.34(7).

Now let E^1 and E^2 be disjoint μ-measurable sets. Since μ is regular there exist, for $i = 1, 2$, sequences $(F_j^i)_j$ of closed sets and $(G_j^i)_j$ of open sets such that $F_j^i \subset E^i \subset G_j^i$ and $\lim_j |\mu|(G_j^i \sim F_j^i) = 0$. We set $\varphi_j^i(s) \triangleq 1$ if $G_j^i = S$ and otherwise

$$\varphi_j^i(s) \triangleq d[s, S \sim G_j^i]/(d[s, S \sim G_j^i] + d[s, F_j^i]) \qquad (j \in \mathbb{N}, \quad s \in S),$$

and observe that $\varphi_j^i(F_j^i) = \{1\}$, $\varphi_j^i(S \sim G_j^i) = \{0\}$, and $0 \leqslant \varphi_j^i(s) \leqslant 1$. Then, by I.4.34(4),

$$\begin{aligned} \lim_j \int g(s)\,\varphi_j^i(s)\,\mu(ds) &= \lim_j \Big[\int_{E^i} g(s)\,\mu(ds) - \int_{E^i \sim F_j^i} g(s)\,\mu(ds) \\ &\qquad + \int_{G_j^i \sim F_j^i} g(s)\,\varphi_j^i(s)\,\mu(ds)\Big] \\ &= \int_{E^i} g(s)\,\mu(ds) \qquad (i = 1, 2); \end{aligned}$$

hence

$$\int_{E^1} g(s)\,\mu(ds) - \int_{E^2} g(s)\,\mu(ds) \tag{5}$$

$$= \lim_j \int g(s)[\varphi_j^1(s) - \varphi_j^2(s)]\,\mu(ds) \in \overline{A(g)}.$$

In particular, for $E^2 = \varnothing$, $\int_{E^1} g(s)\,\mu(ds) \in \overline{A(g)}$.

Now let E_0 be a positive set for μ, E any μ-measurable set,

$$E^1 \triangleq \{s \in E \cap E_0 \mid g(s) \geqslant 0\} \cup \{s \in E \sim E_0 \mid g(s) \leqslant 0\}$$

and $E^2 \triangleq E \sim E^1$. Then, by (5),

$$\int_E |\,g(s)|\,|\,\mu\,|\,(ds) = \int_{E^1} g(s)\,\mu(ds) - \int_{E^2} g(s)\,\mu(ds) \in \overline{A(g)}. \tag{6}$$

Finally, since

$$\int g(s) f(s)\,\mu(ds) \leqslant \int |\,g(s)|\,|\,\mu\,|(ds) \cdot |\,f\,|_{\sup} \qquad [f \in C(S)],$$

it follows that $\int |\,g(s)|\,|\,\mu\,|\,(ds) \geqslant \sup A(g)$ and, by (6), that

$$\int |\,g(s)|\,|\,\mu(ds)| = \sup A(g). \quad \text{QED}$$

I.5.6 ***Lemma*** Let S be a compact metric space, $\mathscr{P}(S)$ the field of all the subsets of S, and $\omega : \mathscr{P}(S) \to \mathbb{R}$ bounded, nonnegative, and additive. Then there exists some $\nu \in \text{frm}^+(S)$ such that $\nu(S) = \omega(S)$ and $\nu(G) \leqslant \omega(G)$ for all open $G \subset S$.

▌PROOF We shall denote by $\mathscr{F}$ (respectively $\mathscr{G}$) the collection of all the closed (respectively open) subsets of S, and by F and G, with or without subscripts, elements of $\mathscr{F}$ and $\mathscr{G}$. We set $\nu_1(F) \triangleq \inf_{G \supset F} \omega(G)$ $(F \in \mathscr{F})$ and $\nu_2(A) \triangleq \sup_{F \subset A} \nu_1(F)$ $[A \in \mathscr{P}(S)]$. Then $\nu_2(\varnothing) = 0$, $\nu_2(S) = \omega(S)$, ν_2 is nonnegative and bounded, and, by Caratheodory's theorem I.4.13,

$$\mathscr{P}_{\nu_2} \triangleq \{E \subset S \mid \nu_2(A) = \nu_2(A \cap E) + \nu_2(A \sim E)\ (A \subset S)\}$$

is a field and $\nu_2 \mid \mathscr{P}_{\nu_2}$ is additive.

We shall show that $\mathscr{F} \subset \mathscr{P}_{\nu_2}$. We first observe that $0 \leqslant \nu_1(F_1) \leqslant \nu_1(F_2)$ if $F_1 \subset F_2$, $0 \leqslant \nu_2(A) \leqslant \nu_2(B)$ if $A \subset B$, and $\nu_2(F) = \nu_1(F)$ for all F. For every choice of disjoint sets F_1 and F_2 there exist disjoint $G_1 \supset F_1$ and $G_2 \supset F_2$. Since for every $G \supset F_1 \cup F_2$ we have $\omega(G) \geqslant \omega(G \cap G_1) + \omega(G \cap G_2)$, it follows that

$$\nu_1(F_1 \cup F_2) \geqslant \nu_1(F_1) + \nu_1(F_2).$$

Thus, for all F and all $A \subset S$, we have

$$\sup\{\nu_1(F_1 \cup F_2) | F_1 \subset A \cap F, F_2 \subset A \sim F\} \geqslant \sup_{F_1 \subset A \cap F} \nu_1(F_1) + \sup_{F_2 \subset A \sim F} \nu_1(F_2);$$

hence

$$\nu_2(A) \geqslant \nu_2(A \cap F) + \nu_2(A \sim F). \tag{1}$$

On the other hand, for arbitrary F_1 and G_1 and all $G \supset F_1 \sim G_1$, we have

$$\nu_1(F_1) \leqslant \omega(G_1 \cup G) \leqslant \omega(G_1) + \omega(G)$$

and therefore

$$\nu_1(F_1) \leqslant \omega(G_1) + \inf_{G \supset F_1 \sim G_1} \omega(G) = \omega(G_1) + \nu_1(F_1 \sim G_1)$$
$$= \omega(G_1) + \nu_2(F_1 \sim G_1).$$

Thus, for each F,

$$\nu_1(F_1) \leqslant \inf_{G_1 \supset F \cap F_1} [\omega(G_1) + \nu_2(F_1 \sim G_1)] \leqslant \nu_1(F \cap F_1) + \nu_2(F_1 \sim F)$$

and, for each $A \subset S$,

$$\nu_2(A) = \sup_{F_1 \subset A} \nu_1(F_1) \leqslant \sup_{F_1 \subset A} \nu_1(F \cap F_1) + \sup_{F_1 \subset A} \nu_2(F_1 \sim F)$$
$$\leqslant \nu_2(F \cap A) + \nu_2(A \sim F).$$

When combined with (1), this relation shows that $\mathscr{F} \in \mathscr{P}_{\nu_2}$.

The nonnegative, bounded, and additive function $\nu_2 | \mathscr{P}_{\nu_2}$ must be regular. Indeed, $\nu_2(F) = \nu_1(F)$ for all F and thus $\nu_2(E) = \sup_{F \subset E} \nu_2(F)$ $(E \in \mathscr{P}_{\nu_2})$; hence, setting $E = S \sim A$, we obtain

$$\nu_2(S) - \nu_2(A) = \nu_2(S \sim A) = \sup_{F \subset S \sim A} \nu_2(F) = \sup_{G \supset A} \nu_2(S \sim G)$$
$$= \nu_2(S) - \inf_{G \supset A} \nu_2(G)$$

and, therefore,

$$\nu_2(A) = \inf_{G \supset A} \nu_2(G) = \sup_{F \subset A} \nu_2(F) \qquad (A \in \mathscr{P}_{\nu_2})$$

Thus $\nu_2 | \mathscr{P}_{\nu_2}$ is regular and it follows from I.4.14 that there exists a unique bounded and regular measure ν on the smallest σ-field $\tilde{\mathscr{P}}_{\nu_2}$ containing $\mathscr{P}_{\nu_2}$

that is an extension of $\nu_2 \mid \mathscr{P}_{\nu_2}$ to $\tilde{\mathscr{P}}_{\nu_2}$. Since $\mathscr{F} \subset \mathscr{P}_{\nu_2}$, we have $\Sigma_{\text{Borel}}(S) \subset \tilde{\mathscr{P}}_{\nu_2}$. Furthermore, $\nu_2(F) = \nu_1(F) = \inf_{G \supset F} \omega(G)$ for all F; hence $\nu_2(F) \leqslant \omega(G)$ if $F \subset G$ and $\nu(G) = \nu_2(G) = \sup_{F \subset G} \nu_2(F) \leqslant \omega(G)$. Finally, $\nu(S) = \nu_2(S) = \nu_1(S) = \omega(S)$. QED

I.5.7 ***Lemma*** Let the conditions of Lemma I.5.6 be satisfied, $i \in \mathbb{N}$, $f : S \to [0, 1]$ be continuous,

$$C_k \triangleq f^{-1}\left(\left(\frac{k-1}{i}, \frac{k}{i}\right]\right) \quad (k = 1, 2, \ldots, i) \qquad \text{and} \qquad c_i(f) \triangleq \sum_{k=1}^{i} \frac{k}{i}\, \omega(C_k).$$

Then

$$\left| c_i(f) - \int f(s)\, \nu(ds) \right| \leqslant 2\nu(S)/i.$$

▌PROOF Let

$$A_k \triangleq f^{-1}\left(\left(\frac{k-1}{i}, \frac{k}{i}\right)\right) \qquad \text{and} \qquad B_k \triangleq f^{-1}\left(\left\{\frac{k}{i}\right\}\right) \qquad (k = 1, 2, \ldots, i).$$

We observe that $C_k = A_k \cup B_k$, A_k is open and B_k closed. We can determine closed $F_k \subset A_k$ $(k = 1, 2, \ldots, i)$ and $\epsilon > 0$ such that the open sets

$$G_k \triangleq \{s \in S \mid d[s, F_k] < \epsilon\} \quad \text{and} \quad G_k' \triangleq \{s \in S \mid d[s, B_k] < \epsilon\} \quad (k = 1, 2, \ldots, i)$$

are disjoint, $G_k \subset A_k$, $G_k' \subset A_k \cup B_k \cup A_{k+1}$ (where $A_{i+1} \triangleq \varnothing$), and $\nu(A_k \sim F_k) \leqslant \nu(S)/i^2$. Then

$$\nu(A_k) \leqslant \nu(F_k) + \nu(S)/i^2 \leqslant \nu(G_k) + \nu(S)/i^2 \leqslant \omega(G_k) + \nu(S)/i^2$$

and

$$\nu(B_k) \leqslant \nu(G_k') \leqslant \omega(G_k');$$

hence

$$\text{(1)} \qquad \int f(s)\, \nu(ds) = \sum_{k=1}^{i} \left[\int_{A_k} f(s)\, \nu(ds) + \frac{k}{i}\, \nu(B_k)\right]$$

$$\leqslant \sum_{k=1}^{i} \frac{k}{i} [\nu(A_k) + \nu(B_k)]$$

$$\leqslant \sum_{k=1}^{i} \frac{k}{i} [\omega(G_k) + \omega(G_k')] + \frac{\nu(S)}{i}.$$

Now let $G_0'' = C_{i+1} \triangleq \varnothing$ and $G_k \cup G_k' \triangleq G_k''$. Since $G_k'' \subset C_k \cup C_{k+1}$, $G_k \cap G_k' = \varnothing$, $G_1'',..., G_i''$ are disjoint and $C_1,..., C_i$ are disjoint, we have

$$\sum_{k=1}^{i} \frac{k}{i} [\omega(G_k) + \omega(G_k')] = \sum_{k=1}^{i} \frac{k}{i} \sum_{j=k}^{k+1} \omega(G_k'' \cap C_j)$$

$$= \sum_{j=1}^{i+1} \sum_{k=j-1}^{j} \frac{k}{i} \omega(G_k'' \cap C_j)$$

$$\leqslant \sum_{j=1}^{i+1} \frac{j}{i} \sum_{k=j-1}^{j} \omega(G_k'' \cap C_j) \leqslant \sum_{j=1}^{i} \frac{j}{i} \omega(C_j) \triangleq c_i(f).$$

This relation combines with (1) to yield

$$\int f(s)\, \nu(ds) \leqslant c_i(f) + \nu(S)/i. \tag{2}$$

Next we observe that, if we write A_k^f and B_k^f for the sets previously denoted by A_k and B_k and set $A_0^f \triangleq \varnothing$ and $B_0^f \triangleq f^{-1}(\{0\})$, then

$$A_k^{1-f} = A_{i-k+1}^f \qquad \text{and} \qquad B_k^{1-f} = B_{i-k}^f \qquad (k = 0, 1, 2,..., i).$$

Thus

$$c_i(1-f) = \sum_{k=1}^{i} \frac{k}{i} \omega(A_{i-k+1}^f) + \sum_{k=1}^{i} \frac{k}{i} \omega(B_{i-k}^f) \tag{3}$$

$$= \sum_{j=1}^{i} \left(1 - \frac{j-1}{i}\right) \omega(A_j^f) + \sum_{j=0}^{i} \left(1 - \frac{j}{i}\right) \omega(B_j^f)$$

$$\leqslant \left(1 + \frac{1}{i}\right) \omega(S) - c_i(f) = \left(1 + \frac{1}{i}\right) \nu(S) - c_i(f).$$

We replace f by $1-f$ in relation (2) and combine it with relation (3) to yield

$$\nu(S) - \int f(s)\, \nu(ds) \leqslant (1 + 2/i)\, \nu(S) - c_i(f);$$

hence

$$\int f(s)\, \nu(ds) \geqslant c_i(f) - 2\nu(S)/i.$$

This relation and (2) yield our assertion. QED

I.5.8 ***Theorem*** (Riesz representation theorem) Let S be a compact metric space. Then there exists an isomorphism $\mathscr{I}$ on frm(S) to $C(S)^*$ defined by

$$\mathscr{I}(\mu)(\varphi) \triangleq \int \varphi(s)\,\mu(ds) \qquad [\mu \in \mathrm{frm}(S), \quad \varphi \in C(S)],$$

and we have

$$|\mathscr{I}(\mu)| = |\mu|(S).$$

▌PROOF Let $\mu \in \mathrm{frm}(S)$, $\varphi_j, \varphi \in C(S)$ $(j \in \mathbb{N})$, and $\lim_j |\varphi - \varphi_j|_{\sup} = 0$. Then φ and φ_j are μ-measurable and bounded and, by the dominated convergence theorem I.4.35, $\lim_j \int \varphi_j(s)\,\mu(ds) = \int \varphi(s)\,\mu(ds)$. Thus $\mathscr{I}(\mu)(\cdot)$ is continuous and, since the mapping $\varphi \to \mathscr{I}(\mu)(\varphi) \triangleq \int \varphi(s)\,\mu(ds) : C(S) \to \mathbb{R}$ is clearly linear, it belongs to $C(S)^*$. It is also clear [I.4.34(1)] that $\mu \to \mathscr{I}(\mu) : \mathrm{frm}(S) \to C(S)^*$ is linear.

If $\int \varphi(s)\,\mu_1(ds) = \int \varphi(s)\,\mu_2(ds)$ $[\varphi \in C(S)]$ for $\mu_1, \mu_2 \in \mathrm{frm}(S)$, then, by I.5.5(1), $\mu_1 = \mu_2$. Thus $\mathscr{I}$ is injective. It remains to show that, for every $l \in C(S)^*$, there exists some $\mu \in \mathrm{frm}(S)$ such that $l = \mathscr{I}(\mu)$ and $|l| = |\mu|(S)$.

Since $C(S)$ is a subspace of $BF(S, \mathbb{R})$, every element l of $C(S)^*$ has, by the Hahn–Banach theorem I.3.8, a continuous linear extension l_1 to $BF(S, \mathbb{R})^*$. For each $A \subset S$, let $\sigma(A) \triangleq l_1(\chi_A)$. Then it is clear that σ is a bounded additive set function on $\mathscr{P}(S)$ and $|\sigma(A)| \leqslant |l_1|$. By I.4.1 and I.5.6, there exist

$$\mu_+ \in \mathrm{frm}^+(S) \qquad \text{and} \qquad \mu_- \in \mathrm{frm}^+(S)$$

such that $\mu_+(G) \leqslant \sigma^+(G)$ and $\mu_-(G) \leqslant \sigma^-(G)$ for all open $G \subset S$, $\mu_+(S) = \sigma^+(S)$ and $\mu_-(S) = \sigma^-(S)$. We set $\mu \triangleq \mu_+ - \mu_-$ and observe that $\mu \in \mathrm{frm}(S)$.

Let $f : S \to [0, 1]$ be continuous and, for $i \in \mathbb{N}$ and $k \in \{0, 1, 2,...\}$,

$$C(i, k) \triangleq f^{-1}\left(\left(\frac{k-1}{i}, \frac{k}{i}\right]\right),$$

$$c_i^+(f) \triangleq \sum_{k=1}^{i} \frac{k}{i}\,\sigma^+(C(i, k)), \qquad c_i^-(f) \triangleq \sum_{k=1}^{i} \frac{k}{i}\,\sigma^-(C(i, k)),$$

and

$$f_i(s) \triangleq \sum_{k=1}^{i} \frac{k}{i}\,\chi_{C(i,k)}(s) \qquad (s \in S).$$

Then it follows from I.5.7 that

$$\text{(1)} \quad \lim_i c_i^+(f) = \int f(s)\,\mu_+(ds) \qquad \text{and} \qquad \lim_i c_i^-(f) = \int f(s)\,\mu_-(ds).$$

Since $\lim_i |f_i - f|_{\sup} = 0$, we have

$$l(f) = \lim_i l_1(f_i) = \lim_i \sum_{k=1}^{i} \frac{k}{i} \sigma(C(i, k)) = \lim_i (c_i^+(f) - c_i^-(f));$$

hence, by (1),

$$l(f) = \int f(s)\, \mu(ds). \tag{2}$$

If $\varphi \in C(S)$ and $|\varphi|_{\sup} \neq 0$, we set $\varphi_1(s) = |\varphi|_{\sup}^{-1} \max(\varphi(s), 0)$ and $\varphi_2(s) = |\varphi|_{\sup}^{-1} \max(-\varphi(s), 0)$. Then φ_1 and φ_2 are continuous functions on S to $[0, 1]$ and $\varphi = |\varphi|_{\sup} \varphi_1 - |\varphi|_{\sup} \varphi_2$. Relation (2) yields therefore

$$l(\varphi) = |\varphi|_{\sup} l(\varphi_1) - |\varphi|_{\sup} l(\varphi_2) = \int \varphi(s) \mu(ds). \tag{3}$$

Finally, we observe that (3) and I.5.5(3) imply that $|l| = |\mathscr{I}(\mu)| = \int |\mu|(ds) = |\mu|(S)$. QED

I.5.9 ***Theorem*** Let S be a compact metric space and $n \in \mathbb{N}$. Then

(1) there exists an isomorphism $\mathscr{I}$ on $\mathrm{frm}(S)^n$ to $C(S, \mathbb{R}^n)^*$ defined by the relation

$$\mathscr{I}(\mu)(\varphi) \triangleq \int \varphi(s) \cdot \mu(ds) \triangleq \sum_{j=1}^{n} \int \varphi^j(s)\, \mu^j(ds)$$

for $\mu \triangleq (\mu^1, ..., \mu^n) \in \mathrm{frm}(S)^n$ and $\varphi \triangleq (\varphi^1, ..., \varphi^n) \in C(S, \mathbb{R}^n)$; and
(2) for each $l \triangleq \mathscr{I}(\mu) \in C(S, \mathbb{R}^n)^*$ there exist $\lambda \in \mathrm{frm}^+(S)$ and $\tilde{\lambda} \in L^1(S, \Sigma_{\mathrm{Borel}}(S), \lambda, \mathbb{R}^n)$ such that $|\tilde{\lambda}(s)| = 1$ $(s \in S)$ and

$$l(\varphi) = \int \tilde{\lambda}(s) \cdot \varphi(s)\, \lambda(ds) \qquad [\varphi \in C(S, \mathbb{R}^n)].$$

PROOF By I.5.2, the topologies of $C(S, \mathbb{R}^n)$ and $C(S)^n$ coincide and therefore the sets $C(S, \mathbb{R}^n)^*$ and $(C(S)^n)^*$ coincide. Every $l \in (C(S)^n)^*$ corresponds uniquely to some $(l^1, ..., l^n) \in (C(S)^*)^n$ in such a manner that $l(\varphi) = \sum_{j=1}^n l^j(\varphi^j)$ for each $\varphi \triangleq (\varphi^1, ..., \varphi^n) \in C(S)^n$. Our first conclusion now follows from I.5.8.

Now let $l \in C(S, \mathbb{R}^n)^*$ and $l \triangleq \mathscr{I}(\mu)$. We set $\zeta \triangleq \sum_{i=1}^n |\mu^i|$ and observe that μ^i is ζ-continuous for $i = 1, 2, ..., n$. By the Radon–Nikodym theorem I.4.37 there exists $\tilde{\zeta} \triangleq (\tilde{\zeta}^1, ..., \tilde{\zeta}^n) \in L^1(S, \Sigma_{\mathrm{Borel}}(S), \zeta, \mathbb{R}^n)$ such that

$$\mu^i(A) = \int_A \tilde{\zeta}^i(s)\, \zeta(ds) \qquad [A \in \Sigma_{\mathrm{Borel}}(S)].$$

We set

$$\lambda(A) \triangleq \int_A |\tilde{\zeta}(s)|\, \zeta(ds) \qquad [A \in \Sigma_{\mathrm{Borel}}(S)],$$

and observe that $|\tilde{\zeta}(s)| \neq 0$ for λ-a.a. $s \in S$. If we let

$$\tilde{\lambda}(s) \triangleq |\tilde{\zeta}(s)|^{-1}\, \zeta(s) \qquad \text{if} \quad |\tilde{\zeta}(s)| \neq 0$$

and

$$\tilde{\lambda}(s) \triangleq (1, 0, \ldots, 0) \in \mathbb{R}^n \qquad \text{if} \quad |\tilde{\zeta}(s)| = 0$$

then $|\tilde{\lambda}(s)| = 1$ $(s \in S)$ and, by I.4.38,

$$l(\varphi) = \int \varphi(s) \cdot \mu(ds) = \int \tilde{\zeta}(s) \cdot \varphi(s)\, \zeta(ds) = \int \tilde{\lambda}(s) \cdot \varphi(s)\, \lambda(ds)$$

$$[\varphi \in C(S, \mathbb{R}^n)] \quad \text{QED}$$

I.5.10 ***Theorem*** Let S be a compact metric space, $\Sigma \triangleq \Sigma_{\mathrm{Borel}}(S)$, and K a compact linear operator in $C(S)$. Then there exist a finite measure $\mu : \Sigma \to \mathbb{R}$ and a function $k : S \times S \to \mathbb{R}$ such that

(1) $$\mu \in \mathrm{frm}^+(S) \quad \text{and} \quad k(s, \cdot) \in L^1(S, \Sigma, \mu) \qquad (s \in S)$$

and

(2) $$K(\phi)(s) = \int k(s, t)\, \phi(t)\, \mu(dt) \qquad [\phi \in C(S),\ s \in S].$$

Furthermore, if k and μ satisfy (1) and (2), then

(3) $$k \in C\big(S, L^1(S, \Sigma, \mu)\big)$$

and

(4) $$|k| \triangleq \sup_{s \in S} \int |k(s, t)|\, \mu(dt) = |K|.$$

▌PROOF By I.2.17, S has an at most denumerable dense subset $\{s_1, s_2, \ldots\}$ which we shall assume infinite, the modifications being clear when it is finite. For each $\bar{s} \in S$, the function $\phi \to K(\phi)(\bar{s}) : C(S) \to \mathbb{R}$ is linear and continuous and, by the Riesz representation theorem I.5.8, there exists some $\nu(\bar{s}) \in \mathrm{frm}(S)$ such that

$$K(\phi)(\bar{s}) = \int \phi(s)\, \nu(\bar{s})(ds) \qquad [\phi \in C(S)]$$

and

$$|\nu(\bar{s})|(S) = \sup\{|K(\phi)(\bar{s})| \mid \phi \in C(S),\ |\phi| \leqslant 1\} \leqslant |K|.$$

Let $\mu_j \triangleq |\nu(s_j)|$ $(j \in \mathbb{N})$. By I.4.6, μ_j are regular and thus, by I.4.9, $\mu \triangleq \sum_{j=1}^{\infty} 2^{-j} \mu_j \in \mathrm{frm}^+(S)$.

We shall now show that, for each $\bar{s} \in S$, $\nu(\bar{s})$ is μ-continuous. Let $\mu(A)=0$. Then $\mu_j(A) = 0$ $(j \in \mathbb{N})$ and therefore $\nu(s_j)(A) = 0$. Since K is a compact operator, it follows from I.5.4 that $\{K(\phi) \mid |\phi| \leqslant 1\}$ is equicontinuous and for every $\bar{s} \in S$ and $\epsilon > 0$ there exists $j \in \mathbb{N}$ such that

$$|K(\phi)(\bar{s}) - K(\phi)(s_j)| \leqslant \epsilon/2 \qquad \text{if} \quad |\phi| \leqslant 1;$$

hence

$$\left| \int \phi(s)[\nu(\bar{s}) - \nu(s_j)](ds) \right| \leqslant \epsilon/2 \qquad [\phi \in C(S), \quad |\phi| \leqslant 1].$$

This relation and I.5.5.(3) imply that

$$\left| \int_A [\nu(\bar{s}) - \nu(s_j)](ds) \right| = |\nu(\bar{s})(A) - \nu(s_j)(A)| = |\nu(\bar{s})(A)| \leqslant \epsilon/2,$$

thus showing that $\nu(\bar{s})(A) = 0$ and $\nu(\bar{s})$ is μ-continuous for each $\bar{s} \in S$.

By the Radon–Nikodym theorem I.4.37, for each $\bar{s} \in S$ there exists a μ-integrable $k(\bar{s}, \cdot)$ such that

$$\nu(\bar{s})(E) = \int_E k(\bar{s}, t)\, \mu(dt) \qquad (E \in \Sigma)$$

and, by I.4.38,

$$K(\phi)(\bar{s}) = \int \phi(t)\, \nu(\bar{s})(dt) = \int k(\bar{s}, t)\, \phi(t)\, \mu(dt) \qquad [\phi \in C(S)].$$

This proves (1) and (2).

Now let k and μ satisfy (1) and (2). We first prove (3). Indeed, assume the contrary. Then there exist $\epsilon > 0$, $\bar{s} \in S$ and a sequence (s_j) in S converging to $\bar{s}$ such that

$$\int |k(\bar{s}, t) - k(s_j, t)|\, \mu(dt) > \epsilon \qquad (j \in \mathbb{N}).$$

This relation, as well as (2) and I.5.5(3), imply that

$$\sup\{|K(\phi)(\bar{s}) - K(\phi)(s_j)| \,\big|\, |\phi| \leqslant 1\} = \int |k(\bar{s}, t) - k(s_j, t)|\, \mu(dt) > \epsilon \qquad (j \in \mathbb{N}).$$

Thus, for each $j \in \mathbb{N}$ there exists some $\phi_j \in C(S)$ such that

$$|\phi_j| \leqslant 1 \qquad \text{and} \qquad |K(\phi_j)(\bar{s}) - K(\phi_j)(s_j)| > \epsilon,$$

showing that $K(\phi_j)(\cdot)$ are not equicontinuous and, by I.5.4, K is not a compact operator, contrary to assumption. Thus $k \in C(S, L^1(S, \Sigma, \mu))$.

Finally, we observe that by (1) and I.5.5(3),

$$\int |k(s,t)|\,\mu(dt) = \sup\{K(\phi)(s) \,\big|\, |\phi| \leqslant 1\} \leqslant |K| \qquad (s \in S);$$

hence

$$|k| \triangleq \sup_{s\in S} \int |k(s,t)|\,\mu(dt) \leqslant |K|.$$

On the other hand,

$$|K| = \sup\{|K(\phi)(s)| \,\big|\, s\in S, |\phi| \leqslant 1\} \leqslant \sup_{s\in S} \int |k(s,t)|\,\mu(dt) = |k|. \text{ QED}$$

I.5.11 ***Theorem*** Let S be a compact metric space, $\Sigma \triangleq \Sigma_{\text{Borel}}(S)$, $n \in \mathbb{N}$, and $K: C(S, \mathbb{R}^n) \to C(S, \mathbb{R}^n)$. Then K is a compact linear operator in $C(S, \mathbb{R}^n)$ if and only if there exist

$$\mu \in \text{frm}^+(S) \qquad \text{and} \qquad k \in C\big(S, L^1(S, \Sigma, \mu, B(\mathbb{R}^n, \mathbb{R}^n))\big)$$

such that

$$K(\phi)(s) = \int k(s,t)\,\phi(t)\,\mu(dt) \qquad [s\in S, \phi \in C(S, \mathbb{R}^n)] \tag{1}$$

and

$$|K| \leqslant |k| \leqslant n\,|K|. \tag{2}$$

Furthermore, if K is a compact operator, $\mu \in \text{frm}^+(S)$, $k(s,\cdot)$ μ-integrable for all $s \in S$, and (1) is satisfied, then $k \in C\big(S, L^1(S, \Sigma, \mu, B(\mathbb{R}^n, \mathbb{R}^n))\big)$.

▌PROOF Assume first that there exist

$$\mu \in \text{frm}^+(S) \qquad \text{and} \qquad k \in C\big(S, L^1(S, \Sigma, \mu, B(\mathbb{R}^n, \mathbb{R}^n))\big)$$

satisfying (1). Then, for all $\phi \in C(S, \mathbb{R}^n)$ and $\bar{s} \in S$, we have

$$\begin{aligned} &\lim_{s\to\bar{s}} \left| \int k(\bar{s},t)\,\phi(t)\,\mu(dt) - \int k(s,t)\,\phi(t)\,\mu(dt) \right| \\ &\qquad \leqslant |\phi| \lim_{s\to\bar{s}} \int |k(\bar{s},t) - k(s,t)|\,\mu(dt) = 0. \end{aligned} \tag{3}$$

Thus relation (1) defines a mapping $K: C(S, \mathbb{R}^n) \to C(S, \mathbb{R}^n)$ which is clearly linear. We have, for each $\phi \in C(S, \mathbb{R}^n)$,

$$\sup_{s\in S} \left| \int k(s,t)\,\phi(t)\,\mu(dt) \right| \leqslant |k|\,|\phi|$$

and relation (3) shows that the set $\{K(\phi) \mid |\phi| \leqslant 1\}$ is equicontinuous. It follows therefore from I.5.4 that K is compact.

Now assume that K is a compact linear operator in $C(S, \mathbb{R}^n)$. By I.5.2, $C(S, \mathbb{R}^n)$ is homeomorphic to $C(S)^n$ under the identity mapping. It follows that $K \in B(C(S)^n, C(S)^n)$; hence

$$K(\phi) = \sum_{i=1}^{n} K_i(\phi^i) \qquad \text{and} \qquad K_j(\phi^j) = (K_j^1(\phi^j), \ldots, K_j^n(\phi^j))$$

$$[\phi \triangleq (\phi^1, \ldots, \phi^n) \in C(S)^n, \quad j = 1, 2, \ldots, n],$$

where $K_j^i \in B(C(S), C(S))$. For each $j \in \{1, 2, \ldots, n\}$, the set

$$\{K_j(\phi^j) \mid | \phi^j | \leqslant 1\} = \{K(\phi) \mid \phi \triangleq (\phi^1, \ldots, \phi^n), \phi^m = 0 \ (m \neq j), | \phi^j | \leqslant 1\}$$

is conditionally compact in $C(S, \mathbb{R}^n)$ (because K is a compact operator) and it follows that $\{K_j^i(\phi^j) \mid | \phi^j | \leqslant 1\}$ $(i, j = 1, 2, \ldots, n)$ is conditionally compact in $C(S)$. Thus K_j^i is a compact linear operator in $C(S)$.

By I.5.10, for each $i, j = 1, 2, \ldots, n$, there exist $\mu_j^i \in \text{frm}^+(S)$ and $\tilde{k}_j^i \in C(S, L^1(S, \Sigma, \mu_j^i))$ such that

$$K_j^i(\phi)(s) = \int \tilde{k}_j^i(s, t)\, \phi(t)\, \mu_j^i(dt) \qquad [\phi \in C(S), s \in S] \tag{4}$$

and

$$| K_j^i | = \sup_{s \in S} \int | \tilde{k}_j^i(s, t)|\, \mu_j^i(dt). \tag{5}$$

We set $\mu \triangleq \sum_{i=1}^{n} \sum_{j=1}^{n} \mu_j^i$. Then $\mu \in \text{frm}^+(S)$ and μ_j^i is μ-continuous for each i and j. It follows, by I.4.37 and I.4.38, that there exist $m_j^i : S \to \mathbb{R}$ $(i, j = 1, 2, \ldots, n)$ such that

$$\mu_j^i(E) = \int_E m_j^i(s)\, \mu(ds) \qquad (E \in \Sigma), \tag{6}$$

$$\int f(s)\, \mu_j^i(ds) = \int f(s)\, m_j^i(s)\, \mu(ds) \qquad [f \in L^1(S, \Sigma, \mu)], \tag{7}$$

and $t \to k_j^i(s, t) \triangleq \tilde{k}_j^i(s, t)\, m_j^i(t)$ is μ-integrable for each i, j and s. We denote by $k(s, t)$ the element of $B(\mathbb{R}^n, \mathbb{R}^n)$ with the matrix $(k_j^i(s, t))$ $(i, j = 1, \ldots, n)$. Then relation (1) follows from (4) and (7). Furthermore, by (1), (5), and (7),

$$| K | \leqslant \sup_{s \in S} \int | k(s, t)|\, \mu(dt) \leqslant \sup_{s \in S} \sum_{i=1}^{n} \sum_{j=1}^{n} \int | k_j^i(s, t)|\, \mu(dt)$$

$$\leqslant \sum_{i=1}^{n} \sum_{j=1}^{n} | K_j^i | \leqslant n \, | K | ,$$

which proves (2).

We shall now prove the last statement of the theorem which will complete the proof of the entire theorem. Let $i, j \in \{1, \ldots, n\}$ and K_j^i be the previously defined compact operator. Then, by (1),

$$K_j^i(\phi) = \int k_j^i(s, t)\, \phi(t)\, \mu(dt) \qquad [s \in S, \phi \in C(S)]$$

and it follows now from I.5.10(3) that $k_j^i \in C(S, L^1(S, \Sigma, \mu))$. This implies that $k \in C(S, \Sigma, \mu, B(\mathbb{R}^n, \mathbb{R}^n)))$. QED

I.5.B The Space $L^p(S, \Sigma, \mu, \mathscr{X})$ $(1 \leqslant p \leqslant \infty)$

I.5.12 ***The Banach Space*** $L^\infty(S, \Sigma, \mu, \mathscr{X})$ Let (S, Σ, μ) be a positive finite measure space and $\mathscr{X}$ a separable Banach space. A function $f: S \to \mathscr{X}$ is *μ-essentially bounded* if $|f(s)| \leqslant c$ μ-a.e. for some $c \in \mathbb{R}$. We set

$$\mu\text{-ess}\sup_{s\in A} |f(s)| \triangleq \inf\{c > 0 \mid |f(s)| \leqslant c \ \mu\text{-a.e. in } A\} \qquad (A \subset S),$$

and write μ-ess sup $|f(s)|$ for μ-ess $\sup_{s\in S} |f(s)|$.

We denote by $L^\infty(S, \Sigma, \mu, \mathscr{X})$ [or by $L^\infty(S, \mathscr{X})$ if Σ and μ are specified] the set of all (equivalence classes of) μ-measurable and μ-essentially bounded functions on S to $\mathscr{X}$. We observe that μ-ess sup $|f_1(s)| = \mu$-ess sup $|f_2(s)|$ if $f_1 = f_2$ μ-a.e. and thus $|f|_\infty \triangleq \mu$-ess sup $|f(s)|$ is defined on $L^\infty(S, \Sigma, \mu, \mathscr{X})$. It is clear that $L^\infty(S, \Sigma, \mu, \mathscr{X})$ is a vector space and the function $f \to |f|_\infty$ is a norm on $L^\infty(S, \Sigma, \mu, \mathscr{X})$. If (f_i) is a Cauchy sequence in $L^\infty(S, \Sigma, \mu, \mathscr{X})$, then $\lim\sup_i |f_i|_\infty < \infty$ and for each $i, j \in \mathbb{N}$ there exists a μ-null set $A_{i,j}$ such that $|f_i - f_j|_\infty = \sup_{s\notin A_{i,j}} |f_i(s) - f_j(s)|$. Then $(f_i(s))$ is a Cauchy sequence in $\mathscr{X}$ for all $s \notin A' \triangleq \bigcup_{i,j} A_{i,j}$ and has a limit $f(s)$ such that μ-ess sup $|f(s)| \leqslant \lim\sup_i |f_i|_\infty < \infty$. By I.4.17, f is μ-measurable and therefore an element of $L^\infty(S, \Sigma, \mu, \mathscr{X})$. If $\epsilon > 0$ and $|f_i - f_j|_\infty \leqslant \epsilon$ for $i \geqslant j \geqslant k$, then

$$|f(s)-f_j(s)| \leqslant \lim_i |f(s)-f_i(s)| + \lim_i |f_i(s)-f_j(s)| \leqslant \epsilon \qquad (s \notin A', \ j \geqslant k);$$

hence $\lim_j |f - f_j|_\infty = 0$, showing that $L^\infty(S, \Sigma, \mu, \mathscr{X})$ is complete, and therefore a Banach space.

We write $L^\infty(S, \Sigma, \mu)$ [or $L^\infty(S)$] for $L^\infty(S, \Sigma, \mu, \mathbb{R})$.

Definition *Conjugate Exponents* We say that p and p' are *conjugate exponents* if $p, p' \in [1, \infty]$ and $1/p + 1/p' = 1$. It is clear that for every $p \in [1, \infty)$ there exists a unique conjugate exponent p' and that $(p')' = p$.

I.5.13 ***Theorem*** (Hölder's inequality) Let (S, Σ, μ) be a finite positive measure space, p and p' conjugate exponents, $f \in L^p(S, \Sigma, \mu)$ and $g \in L^{p'}(S, \Sigma, \mu)$. Then fg is μ-integrable and

$$\left| \int f(s)\, g(s)\, \mu(ds) \right| \leqslant |f|_p\, |g|_{p'} \, .$$

▌PROOF If $p = 1$ (hence $p' = \infty$), then $| g(s)| \leqslant | g |_\infty$ for μ-a.a. $s \in S$; hence $| f(s) g(s)| \leqslant | g |_\infty |f(s)| \mu$-a.e. and

$$\left| \int f(s) g(s) \mu(ds) \right| \leqslant \int | g |_\infty | f(s)| \mu(ds) = | g_\infty | | f |_1 .$$

If $p = \infty$, then the same argument applies, with f and g interchanged. It remains therefore to consider the case where $1 < p < \infty$.

We first assume that $| f |_p \neq 0$ and $| g |_{p'} \neq 0$. We recall that, by I.3.15,

$$(1) \qquad x^\alpha y^{1-\alpha} \leqslant \alpha x + (1 - \alpha) y \qquad [\alpha \in (0, 1), \quad x, y > 0],$$

and this relation remains obviously valid if $x = 0$ or $y = 0$. We set, for an arbitrary $s \in S$, $x = | f(s)|^p / | f |_p^p$, $y = | g(s)|^{p'} / | g |_{p'}^{p'}$ and $\alpha = 1/p$. Then (1) yields

$$f(s) g(s) \leqslant | f(s)| \, | g(s)| \leqslant \frac{1}{p} | f |_p^{1-p} | g |_{p'} | f(s)|^p + \frac{1}{p'} | g |_{p'}^{1-p'} | f |_p | g(s)|^{p'} ;$$

hence

$$\int f(s) g(s) \mu(ds) \leqslant \frac{1}{p} | f |_p | g |_{p'} + \frac{1}{p'} | f |_p | g |_{p'} = | f |_p | g |_{p'} .$$

Finally, if either $| f |_p = 0$ or $| g |_{p'} = 0$, then, by I.4.27(6), $f(s) = 0$ μ-a.e. or $g(s) = 0$ μ-a.e. and therefore $\int f(s) g(s) \mu(ds) = 0$. QED

I.5.14 ***Theorem*** Let (S, Σ, μ) be a finite positive measure space, $k \in \mathbb{N}$, $r, p_1, ..., p_k \in [1, \infty]$, $\sum_{j=1}^k 1/p_j = 1/r$, and $f_i \in L^{p_i}(S, \Sigma, \mu)$ $(i = 1, ..., k)$. Then $\phi \triangleq f_1 f_2 \cdots f_k \in L^r(S, \Sigma, \mu)$ and

$$| \phi |_r \leqslant | f_1 |_{p_1} \cdots | f_k |_{p_k} .$$

▌PROOF Let $p, q, r \in [1, \infty]$, $1/p + 1/q = 1/r$, $f \in L^p(S)$, and $g \in L^q(S)$. Then p/r, $q/r \in [1, \infty]$, and the functions $s \to \tilde{f}(s) \triangleq | f(s)|^r$ and $s \to \tilde{g}(s) \triangleq | g(s)|^r$ belong to $L^{p/r}(S)$ and $L^{q/r}(S)$, respectively. Then, by Hölder's inequality I.5.13,

$$(1) \quad \int | f(s) g(s)|^r \mu(ds) = \int | \tilde{f}(s) \tilde{g}(s)| \mu(ds) \leqslant | \tilde{f} |_{p/r} | \tilde{g} |_{q/r} = | f |_p^r | g |_q^r .$$

We now observe that, if $f_j \in L^\infty(S)$ $(k' + 1 \leqslant j \leqslant k)$ for some $k' \in \{1, ..., k\}$, then

$$| \phi |_r \leqslant | f_1 f_2 \cdots f_{k'} |_r | f_{k'+1} |_\infty \cdots | f_k |_\infty .$$

We may therefore assume that $p_j < \infty$ for all j; hence $r < \infty$.

Now let $2 \leqslant l < k$, $\sum_{j=1}^{l} 1/p_j \triangleq 1/r_l$ and assume that

$$\int |f_1(s) \cdots f_l(s)|^{r_l} \mu(ds) \leqslant (|f_1|_{p_1} \cdots |f_l|_{p_l})^{r_l}. \tag{2}$$

Then

$$\frac{1}{r_l} + \frac{1}{p_{l+1}} = \frac{1}{r_{l+1}}$$

and $s \to \phi_l(s) \triangleq |f_1(s) \cdots f_l(s)|$ belongs to $L^{r_l}(s)$. It follows, by (1) and (2), that

$$\begin{aligned}\int |f_1(s) \cdots f_l(s) f_{l+1}(s)|^{r_{l+1}} \mu(ds) &\leqslant (|\phi_l|_{r_l} |f_{l+1}|_{p_{l+1}})^{r_{l+1}} \\ &\leqslant (|f_1|_{p_1} \cdots |f_l|_{p_l} |f_{l+1}|_{p_{l+1}})^{r_{l+1}}.\end{aligned}$$

Thus (2) remains valid with l replaced by $l + 1$. Since (2) is valid for $l = 2$ as a consequence of (1), we conclude that (2) is valid for $2 \leqslant l \leqslant k$. QED

We shall show that $(L^p(S, \Sigma, \mu, \mathscr{X}), |\cdot|_p)$ is a normed vector space and a fundamental prerequisite for establishing this is *Minkowski's inequality*

$$|f + g|_p \leqslant |f|_p + |g|_p .$$

I.5.15 ***Theorem*** (Minkowski's inequality) Let (S, Σ, μ) be a finite positive measure space, $\mathscr{X}$ a separable Banach space, and $1 \leqslant p < \infty$. Then $L^p(S, \Sigma, \mu, \mathscr{X})$ is a vector space and $|\cdot|_p$ is a norm on it.

▌PROOF Let $f, g \in L^p(S, \Sigma, \mu, \mathscr{X})$ and $\alpha \in \mathbb{R}$. Then $|\alpha f|_p = |\alpha| |f|_p < \infty$. Next we observe that, for all $a, b \geqslant 0$, we have

$$|a + b|^p \leqslant [2 \max(a, b)]^p \leqslant 2^p(a^p + b^p);$$

hence

$$\begin{aligned}\int |f(s) + g(s)|^p \mu(ds) &\leqslant \int [|f(s)| + |g(s)|]^p \mu(ds) \\ &\leqslant 2^p |f|_p^p + 2^p |g|_p^p < \infty.\end{aligned}$$

Thus $f + g \in L^p(S, \Sigma, \mu, \mathscr{X})$, showing that $L^p(S, \Sigma, \mu, \mathscr{X})$ is a vector space. If $|f|_p = 0$, then, by I.4.27(6), $f(s) = 0$ μ-a.e. Since $|f|_p \geqslant 0$, it remains to prove that $|f + g|_p \leqslant |f|_p + |g|_p$.

When $p = 1$ this last inequality follows from I.4.34(3), and the inequality is obvious if $p > 1$ and $|f + g|_p = 0$. When $p > 1$ and $|f+g|_p \neq 0$, we have, by I.4.27(3),

$$(1)\qquad \int |f(s) + g(s)|^p \mu(ds) = \int |f(s) + g(s)|\,|f(s) + g(s)|^{p-1} \mu(ds)$$
$$\leqslant \int |f(s)|\,|f(s) + g(s)|^{p-1} \mu(ds)$$
$$+ \int |g(s)|\,|f(s) + g(s)|^{p-1} \mu(ds).$$

Since $p' = p/(p-1)$, we observe that $s \to |f(s) + g(s)|^{p-1}$ belongs to $L^{p'}(S, \Sigma, \mu)$; hence, applying Holder's inequality I.5.13 to each integral on the right of (1), we find that

$$|f + g|_p^p \leqslant |f|_p \left\{\int |f(s) + g(s)|^p \mu(ds)\right\}^{(p-1)/p}$$
$$+ |g|_p \left\{\int |f(s) + g(s)|^p \mu(ds)\right\}^{(p-1)/p}$$
$$= |f|_p\,|f+g|_p^{p-1} + |g|_p\,|f+g|_p^{p-1};$$

hence

$$|f + g|_p \leqslant |f|_p + |g|_p. \quad \text{QED}$$

I.5.16 ***Lemma*** Let (S, Σ, μ) be a finite positive measure space, $\mathscr{X}$ a separable Banach space, $1 \leqslant p < \infty$, $c \in \mathbb{R}$, $f : S \to \mathscr{X}$ μ-measurable, $f_i \in L^p(S, \Sigma, \mu, \mathscr{X})$, $|f_i|_p \leqslant c$ $(i \in \mathbb{N})$, and $\lim_i |f - f_i|_{\mathscr{F}(\mu)} = 0$. Then $f \in L^p(S, \Sigma, \mu, \mathscr{X})$.

▌PROOF Let

$$A_n \triangleq \{s \in S \mid |f(s)|^p \leqslant n\} \qquad \text{and} \qquad B_i \triangleq \{s \in S \mid |f_i(s) - f(s)| \geqslant 1\}$$
$$(i, n \in \mathbb{N}).$$

Since $\lim_i |f - f_i|_{\mathscr{F}(\mu)} = 0$, there exists $k \triangleq k(n) \in \mathbb{N}$ such that $\mu(B_k) \leqslant 1/n$. Then

$$(1)\qquad \int_{A_n} |f(s)|^p \mu(ds) = \int_{A_n \sim B_k} |f(s)|^p \mu(ds) + \int_{A_n \cap B_k} |f(s)|^p \mu(ds)$$
$$\leqslant \int_{S \sim B_k} |f(s)|^p \mu(ds) + 1.$$

Now, for $s \in S \sim B_k$ we have $||f(s)| - |f_k(s)|| \leqslant |f(s) - f_k(s)| < 1$; hence

$|f(s)| \leqslant |f_k(s)| + 1$. Furthermore, since $(a+1)^p \leqslant 2^p(a^p+1)$ for all $a \geqslant 0$, we have $|f(s)|^p \leqslant 2^p[|f_k(s)|^p + 1]$. It follows that

$$\int_{S\sim B_k} |f(s)|^p \, \mu(ds) \leqslant 2^p[c^p + \mu(S)]$$

and, by (1),

$$\text{(2)} \qquad \int \chi_{A_n}(s)|f(s)|^p \, \mu(ds) \leqslant 2^p[c^p + \mu(S)] + 1.$$

By I.4.25, there exists a sequence (g_n) of μ-simple functions such that $(g_n(s))$ is nondecreasing and $\lim_n g_n(s) = |f(s)|^p$ $(s \in S)$. Since the functions $\chi_{A_n} g_n$ have the properties listed for g_n, it follows from (2) and Definition I.4.26 that

$$\int |f(s)|^p \, \mu(ds) = \lim_n \int \chi_{A_n}(s)\, g_n(s)\, \mu(ds) \leqslant 2^p[c^p + \mu(S)] + 1 < \infty. \quad \text{QED}$$

I.5.17 ***Theorem*** Let (S, Σ, μ) be a finite positive measure space, $\mathscr{X}$ a separable Banach space, and $1 \leqslant p < \infty$. Then $(L^p(S, \Sigma, \mu, \mathscr{X}), |\cdot|_p)$ is a Banach space.

▌PROOF By I.5.15, $(L^p(S, \Sigma, \mu, \mathscr{X}), |\cdot|_p)$ is a normed vector space. Now let (f_i) be a Cauchy sequence in $L^p(S, \Sigma, \mu, \mathscr{X})$. Then $(|f_i|_p)$ is a Cauchy sequence in $\mathbb{R}$ and there exists, therefore, some $c \in \mathbb{R}$ such that $|f_i|_p \leqslant c$ for all $i \in \mathbb{N}$. If we set $A(\epsilon, i, j) \triangleq \{s \in S \mid |f_i(s) - f_j(s)| \geqslant \epsilon\}$ $(i, j \in \mathbb{N}, \epsilon > 0)$, then

$$\epsilon^p \mu(A(\epsilon, i, j)) \leqslant \int_{A(\epsilon,i,j)} |f_i(s) - f_j(s)|^p \, \mu(ds) \leqslant |f_i - f_j|_p^p \underset{i,j}{\rightarrow} 0$$

for every $\epsilon > 0$; hence $\lim_{i,j} |f_i - f_j|_{\mathscr{F}(\mu)} = 0$ and, by Egoroff's theorem I.4.18, there exists a μ-measurable $f: S \to \mathscr{X}$ such that $\lim_i |f_i - f|_{\mathscr{F}(\mu)} = 0$. By I.5.16, $f \in L^p(S, \Sigma, \mu, \mathscr{X})$.

We next show that for every $\epsilon > 0$ there exists $\eta(\epsilon) > 0$ such that $|\chi_E f|_p \leqslant \epsilon$ and $|\chi_E f_i|_p \leqslant \epsilon$ $(i \in \mathbb{N})$ if $\mu(E) \leqslant \eta(\epsilon)$. Indeed, let $k \triangleq k(\epsilon) \in \mathbb{N}$ be such that $|f_i - f_j|_p \leqslant \epsilon/2$ if $i, j \geqslant k$. By I.4.28, there exist $\eta_i(\epsilon)$ $(i = 0, 1, 2, \ldots)$ such that $\int_E |f(s)|^p \mu(ds) \leqslant \epsilon^p$ if $\mu(E) \leqslant \eta_0(\epsilon)$ and $\int_E |f_i(s)|^p \mu(ds) \leqslant 2^{-p}\epsilon^p$ if $\mu(E) \leqslant \eta_i(\epsilon)$ $(i \in \mathbb{N})$. We set $\eta(\epsilon) \triangleq \operatorname{Min}_{0 \leqslant i \leqslant k} \eta_i(\epsilon)$. For $\mu(E) \leqslant \eta(\epsilon)$, we have $|\chi_E f|_p \leqslant \epsilon$ and $|\chi_E f_i|_p \leqslant \epsilon/2$ for $i = 1, 2, \ldots, k$. It follows that

$$|\chi_E f_i|_p \leqslant |\chi_E(f_i - f_k)|_p + |\chi_E f_k|_p \leqslant |f_i - f_k|_p + \epsilon/2 \leqslant \epsilon \quad \text{for } i \geqslant k,$$

showing that

$$\text{(1)} \quad |\chi_E f|_p \leqslant \epsilon \quad \text{and} \quad |\chi_E f_i|_p \leqslant \epsilon \ \ (i \in \mathbb{N}) \quad \text{if} \ \ \mu(E) \leqslant \eta(\epsilon).$$

We now set, for any fixed $\epsilon > 0$,

$$B(\epsilon, i) \triangleq \{s \in S \mid |f(s) - f_i(s)|^p > \epsilon^p/\mu(S)\},$$

and choose $m(\epsilon) \in \mathbb{N}$ such that $\mu(B(\epsilon, i)) \leqslant \eta(\epsilon)$ for $i \geqslant m(\epsilon)$. For all $i \geqslant m(\epsilon)$, we have

$$(2) \quad |f - f_i|_p^p = \int_{B(\epsilon,i)} |f(s) - f_i(s)|^p \mu(ds) + \int_{S \sim B(\epsilon,i)} |f(s) - f_i(s)|^p \mu(ds)$$
$$\leqslant \int_{B(\epsilon,i)} |f(s) - f_i(s)|^p \mu(ds) + \epsilon^p.$$

Since $|f(s) - f_i(s)|^p \leqslant [|f(s)| + |f_i(s)|]^p \leqslant 2^p[|f(s)|^p + |f_i(s)|^p]$ for all i and s, it follows, by (1) and (2), that

$$|f - f_i|_p^p \leqslant 2^p |\chi_{B(\epsilon,i)} f|_p^p + 2^p |\chi_{B(\epsilon,i)} f_i|^p + \epsilon^p \leqslant (2^{p+1} + 1)\epsilon^p$$

for all $i \geqslant m(\epsilon)$. This shows that $\lim_i |f - f_i|_p = 0$. Thus every Cauchy sequence (f_i) in $L^p(S, \Sigma, \mu, \mathscr{X})$ has a limit f. QED

I.5.18 ***Theorem*** Let S be a compact metric space, $\mu \in \text{frm}^+(S)$, $\Sigma \triangleq \Sigma_{\text{Borel}}(S)$, $\mathscr{X}$ a separable Banach space, $\mathscr{X}_\infty \triangleq \{x_1, x_2, ...\}$ dense in $\mathscr{X}$, and $1 \leqslant p < \infty$. Then $L^p(S, \Sigma, \mu, \mathscr{X})$ is separable and the following subsets are dense in $L^p(S, \Sigma, \mu, \mathscr{X})$:

(a) the set of μ-simple functions on S to $\mathscr{X}_\infty$,
(b) the set $\{\sum_{j=1}^k c_j(\cdot)\, x_{i_j} \mid k, i_1, i_2, ..., i_k \in \mathbb{N},\ c_j \in C(S)\}$, and
(c) $C(S, \mathscr{X})$.

▌PROOF Let $E \in \Sigma$ and $\epsilon > 0$. Then there exist a closed set $F \subset E$ and an open $G \supset E$ such that $\mu(G \sim F) \leqslant \epsilon$. We set $h \triangleq 1$ if $G = S$; otherwise

$$h(s) \triangleq d[s, S \sim G]/(d[s, F] + d(s, S \sim G]).$$

Then h is continuous, $0 \leqslant h(s) \leqslant 1$, $h(s) = 1\ (s \in F)$, and $h(s) = 0\ (s \in S \sim G)$; hence

$$\int |\chi_E(s) - h(s)|^p \mu(ds) = \int_{G \sim F} |\chi_E(s) - h(s)|^p \mu(ds) \leqslant \epsilon.$$

Thus the characteristic function of any $E \in \Sigma$, and therefore also of any μ-measurable set, can be approximated in $L^p(S, \Sigma, \mu)$ by continuous functions.

Now let $f \in L^p(S, \mathscr{X})$ and $\epsilon > 0$. By I.4.30, there exists a μ-simple function

$f_\epsilon : S \to \mathscr{X}$ such that $|f - f_\epsilon|_p \leqslant \epsilon/2$. If $f_\epsilon \triangleq \sum_{j=1}^k \chi_{E_j} y_i$, we choose for each $j = 1, 2, \ldots, k$ a point $x_{i_j} \in \mathscr{X}_\infty$ such that $|y_j - x_{i_j}| \leqslant \epsilon/[2\mu(S)^{1/p}]$. Then

$$\bar{f}_\epsilon \triangleq \sum_{j=1}^{k} \chi_{E_j} x_{i_j} : S \to \mathscr{X}_\infty$$

is μ-simple and $|f - \bar{f}_\epsilon|_p \leqslant |f - f_\epsilon|_p + |f_\epsilon - \bar{f}_\epsilon|_p \leqslant \epsilon$. This shows that the set of μ-simple functions on S to $\mathscr{X}_\infty$ is dense in $L^p(S, \mathscr{X})$.

As we have shown before, the characteristic function of each μ-measurable set E can be approximated in $L^p(S)$ by a sequence of continuous functions. We choose, therefore, for each E_j in the expression for $\bar{f}_\epsilon$ a continuous c_j such that $|c_j - \mathscr{X}_{E_j}|_p \leqslant \epsilon/(k\,|x_{i_j}|)$. Then

$$\left| \sum_{j=1}^{k} c_j(\cdot)\, x_{i_j} - \bar{f}_\epsilon \right|_p \leqslant \sum_{j=1}^{k} \left\{ \int | c_j(s) - \chi_{E_j}(s)|^p \, |x_{i_j}|^p \, \mu(ds) \right\}^{1/p} \leqslant \epsilon$$

and thus the set defined in (*b*) is dense in $L^p(S, \mathscr{X})$. If we denote this set by C', then we observe that $C' \subset C(S, \mathscr{X}) \subset L^p(S, \mathscr{X})$ and thus $C(S, \mathscr{X})$ is dense in $L^p(S, \mathscr{X})$.

Finally, we recall that, by I.5.1, the Banach space $(C(S), |\cdot|_{\sup})$ is separable. Since the convergence in the $|\cdot|_{\sup}$-topology clearly implies convergence in the $|\cdot|_p$-topology, the functions c_j in the expression in (b) can be chosen from an at most denumerable set. This shows that $L^p(S, \Sigma, \mu, \mathscr{X})$ is separable.

QED

I.5.19 ***Theorem*** Let (S, Σ, μ) be a positive finite measure space, $1 \leqslant p < \infty$, and p' the conjugate exponent to p. Then there exists an isomorphism $\mathscr{I}$ on $L^{p'}(S, \Sigma, \mu)$ to $L^p(S, \Sigma, \mu)^*$, defined by

$$\mathscr{I}(g)(f) \triangleq \int g(s) f(s)\, \mu(ds) \qquad [g \in L^{p'}(S), \quad f \in L^p(S)],$$

and we have

$$|\mathscr{I}(g)| = |g|_{p'}.$$

▌PROOF We shall write $L^q(S)$ for $L^q(S, \Sigma, \mu)$ and $|\cdot|_q$ for the norm in $L^q(S, \Sigma, \mu)$.

Step 1 Let $g \in L^{p'}(S)$. It follows from I.4.34(3) and Hölder's inequality I.5.13 that $|\int g(s) f(s)\, \mu(ds)| < \infty$ for all $f \in L^p(S)$, and it is clear that $f \to \int g(s) f(s)\, \mu(ds)$ is linear on $L^p(S)$. If $\lim_j |f - f_j|_p = 0$, then

$$\left| \int g(s) f(s)\, \mu(ds) - \int g(s) f_j(s)\, \mu(ds) \right| \leqslant \int |g(s)|\, |f(s) - f_j(s)|\, \mu(ds)$$
$$\leqslant |g|_{p'}\, |f - f_j|_p \underset{j}{\to} 0.$$

Thus $\mathscr{I}(g)$ is continuous.

If $\mathscr{I}(g_1)(f) = \mathscr{I}(g_2)(f)$ $[f \in L^p(S)]$, then we set $f_1(s) = 1$ if $g_1(s) - g_2(s) \geqslant 0$ and $f_1(s) = -1$ if $g_1(s) - g_2(s) < 0$. It is clear that $f_1 \in L^p(S)$ and therefore

$$\left| \int g_1(s) f_1(s) \mu(ds) - \int g_2(s) f_1(s) \mu(ds) \right| = \left| \int (g_1(s) - g_2(s)) f_1(s) \mu(ds) \right|$$

$$= \int | g_1(s) - g_2(s) | \mu(ds) = 0;$$

hence, by I.4.27(6), $g_1(s) = g_2(s)$ μ-a.e. Thus the mapping

$$g \to \mathscr{I}(g) : L^{p'}(S) \to L^p(S)^*$$

is an injection.

Step 2 We shall next show that for every $l \in L^p(S)^*$ there exists $g \in L^1(S)$ such that $l(f) = \int g(s) f(s) \mu(ds)$ for all $f \in L^p(S)$.

Let $\nu(E) \triangleq l(\chi_E)$ $(E \in \Sigma)$. It is clear that ν is a bounded additive set function. If $E_1, E_2, \ldots$ are disjoint elements of Σ and $E \triangleq \bigcup_{j=1}^{\infty} E_j$, then

$$\lim_i \left| \chi_E - \sum_{j=1}^{i} \chi_{E_j} \right|_p = \lim_i \mu \left(E \sim \bigcup_{j=1}^{i} E_j \right)^{1/p} = 0;$$

hence $\lim_i \nu(\bigcup_{j=1}^{i} E_j) = \lim_i l(\sum_{j=1}^{i} \chi_{E_j}) = l(\chi_E) = \nu(E)$ and thus $\nu : \Sigma \to \mathbb{R}$ is a finite measure. Since $| \chi_E |_p = 0$ if $\mu(E) = 0$, ν is μ-continuous. It follows, by the Radon–Nikodym theorem I.4.37, that there exists some $g \in L^1(S)$ such that

$$l(\chi_E) = \nu(E) = \int g(s) \chi_E(s) \mu(ds) \qquad (E \in \Sigma).$$

Since both l and $\mathscr{I}(g)$ are linear, it follows that

$$l(h) = \int g(s) h(s) \mu(ds)$$

for all μ-simple $h : S \to \mathbb{R}$.

Let, for all $u : S \to \mathbb{R}$, $s_u(s) \triangleq 1$ if $u(s) \geqslant 0$ and $s_u(s) \triangleq -1$ if $u(s) < 0$. Since $| f\alpha |_p = | f |_p$ for all $f \in L^p(S)$ if $| \alpha(s) | = 1$ $(s \in S)$, we have, for all μ-simple h,

$$\text{(1)} \qquad \int | g(s) | \, | h(s) | \mu(ds) = \int g(s) s_g(s) h(s) s_h(s) \mu(ds)$$

$$= l(s_g s_h h) \leqslant | l | \, | h |_p .$$

If $f \in L^p(S)$, then, by I.4.30, there exists a sequence (f_j) of μ-simple functions such that $\lim_j f_j = f$ in $L^p(S)$ and $| f_j(s) | \leqslant | f(s) | + 1$ $(s \in S)$, from which it follows that $| f_j |_p \leqslant | f |_p + \mu(S)^{1/p}$ for all $j \in \mathbb{N}$. In view of I.4.31 and

Egoroff's theorem I.4.18, we may choose (f_j) in such a manner that $\lim_j f_j = f$ μ-a.e. and $\lim_j |gf_j - gf|_{\mathscr{F}(\mu)} = 0$ and, by (1), we have $\int |g(s)f_j(s)| \mu(ds) \leqslant |l|[|f|_p + \mu(S)^{1/p}]$. It follows now from I.5.16 that $gf \in L^1(S)$. Since $|f_j(s)g(s)| \leqslant [|f(s)| + 1]|g(s)|$ $(s \in S)$, we conclude, applying the dominated convergence theorem I.4.35, that

$$(2) \qquad l(f) = \lim_j l(f_j) = \lim_j \int g(s) f_j(s) \mu(ds) = \int g(s) f(s) \mu(ds).$$

Step 3 To show that $g \in L^{p'}(S)$, we first assume that $p > 1$. Let s_g be defined as in Step 2,

$$A_n \triangleq \{s \in S \mid |g(s)| \leqslant n\} \quad (n \in \mathbb{N}) \qquad \text{and} \qquad \chi_n \triangleq \chi_{A_n}.$$

Then $s \to \gamma_n(s) \triangleq |g(s)|^{p'-1} \chi_n(s) s_g(s)$ is bounded and therefore in $L^p(S)$; hence, by (2),

$$(3) \quad \int \chi_n(s)|g(s)|^{p'} \mu(ds) = \int g(s) \gamma_n(s) \mu(ds) = l(\gamma_n) \leqslant |l| \, |\gamma_n|_p \quad (n \in \mathbb{N}).$$

Since $p(p'-1) = p'$, this relation shows that $|\chi_n g|_{p'}^{p'} \leqslant |l| |\chi_n g|_{p'}^{p'/p}$, implying that $|\chi_n g|_{p'} \leqslant |l|$ and, by I.5.16, $g \in L^{p'}(S)$.

If $p = 1$, we have

$$\int_E |g(s)| \mu(ds) = \int g(s) s_g(s) \chi_E(s) \mu(ds) \leqslant |l| \, |s_g \chi_E|_1 = |l| \mu(E) \quad (E \in \Sigma).$$

By I.4.34(7), this relation is only possible if $|g(s)| \leqslant |l|$ μ-a.e.; hence

$$(4) \qquad |g|_\infty \leqslant |l| \qquad \text{and} \qquad g \in L^\infty(S).$$

Step 4 Finally, we prove that $|l| = |g|_{p'}$. Relation

$$(5) \qquad |l| \leqslant |g|_{p'}$$

follows from (2) and Hölder's inequality I.5.13. If $p = 1$, relations (4) and (5) yield $|l| = |g|_\infty$. If $p > 1$ and we set $f(s) \triangleq |g(s)|^{p'-1} s_g(s)$ $(s \in S)$, then $f \in L^p(S)$ and $|f|_p = \{\int |g(s)|^{p'} \mu(ds)\}^{1/p} = |g|_{p'}^{1/(p-1)}$; from which, by (2),

$$|l| \, |g|_{p'}^{1/(p-1)} \geqslant l(f) = \int |g(s)|^{p'} \mu(ds) = |g|_{p'}^{p'}.$$

Thus $|l| \geqslant |g|_{p'}$, and, by (5), $|l| = |g|_{p'}$. QED

I.5.20 ***Theorem*** Let (S, Σ, μ) be a positive finite measure space, $1 \leqslant p < \infty$, p' the conjugate exponent of p, and $n \in \mathbb{N}$. Then there exists an isomorphism $\mathscr{I}$ on $L^{p'}(S, \Sigma, \mu, \mathbb{R}^n)$ to $L^p(S, \Sigma, \mu, \mathbb{R}^n)^*$, defined by

$$\mathscr{I}(g)(f) \triangleq \int g(s) \cdot f(s)\, \mu(ds) \qquad [g \in L^{p'}(S, \Sigma, \mu, \mathbb{R}^n), \quad f \in L^p(S, \Sigma, \mu, \mathbb{R}^n)]$$

▌PROOF We shall write $L^q(S, \mathbb{R}^n)$ for $L^q(S, \Sigma, \mu, \mathbb{R}^n)$. It can be easily verified that the sets $L^p(S, \mathbb{R}^n)$ and $L^p(S)^n$ and their respective norm topologies coincide. Thus $L^p(S, \mathbb{R}^n)^* = (L^p(S)^n)^*$ and for each $l \in L^p(S, \mathbb{R}^n)^*$ there exist unique $l^j \in L^p(S)^*$ $(j = 1,..., n)$ such that $l(f) = \sum_{j=1}^n l^j(f^j)$ for all $f \triangleq (f^1,..., f^n) \in L^p(S, \mathbb{R}^n)$. Our conclusion now follows from I.5.19. QED

I.5.C Special Spaces

Let S_1, S_2, and X be given sets, $F(S_2, X)$ a collection of functions on S_2 to X, and $\mathscr{H}$ a collection of functions $h: S_1 \times S_2 \to X$ with the property that $h(s_1, \cdot) \in F(S_2, X)$ $(s_1 \in S_1)$. Then there exists a natural mapping of $\mathscr{H}$ into the class of functions on S_1 to $F(S_2, X)$ that assigns to each $h \in \mathscr{H}$ the function $s_1 \to h(s_1, \cdot)$. We shall often identify in such cases the function $h : S_1 \times S_2 \to X$ with the function $s_1 \to h(s_1, \cdot) : S_1 \to F(S_2, X)$.

If $\mathscr{H}_1$ is a collection of functions on S_1 to $\mathbb{R}$, $\mathscr{X}$ a Banach space, and $\mathscr{H}_2$ a collection of functions on S_2 to $\mathscr{X}$, we denote by $\mathscr{H}_1 \otimes \mathscr{H}_2$ the collection of functions $h : S_1 \times S_2 \to \mathscr{X}$ such that $h(s_1, s_2) = \sum_{j=1}^k h_j^1(s_1)\, h_j^2(s_2)$, where $k \in \mathbb{N}$, $h_j^1 \in \mathscr{H}_1$, and $h_j^2 \in \mathscr{H}_2$. We also write $h_j^1 \otimes h_j^2$ for the function $(s_1, s_2) \to h_j^1(s_1)\, h_j^2(s_2) : S_1 \times S_2 \to \mathscr{X}$.

We shall often apply, without further reference, the results of I.5.1 and I.5.18 where it was established that the Banach spaces $C(S, \mathscr{X})$ and $L^p(S, \Sigma, \mu, \mathscr{X})$ are separable if S is a compact metric space, $\mu \in \mathrm{frm}^+(S)$, $\Sigma \triangleq \Sigma_{\mathrm{Borel}}(S)$, and $\mathscr{X}$ is a separable Banach space.

I.5.21 ***Theorem*** Let $\mathscr{X}$ be a separable Banach space, $q \in [1, \infty)$, and, for $i = 1, 2$, S_i a compact metric space, $\mu_i \in \mathrm{frm}^+(S_i)$, and $\Sigma_i \triangleq \Sigma_{\mathrm{Borel}}(S_i)$. If $H : S_1 \to L^q(S_2, \Sigma_2, \mu_2, \mathscr{X})$ is μ_1-measurable, then there exists a $\mu_1 \times \mu_2$-measurable $h : S_1 \times S_2 \to \mathscr{X}$ such that, for all $s_1 \in S_1$, $h(s_1, \cdot) = H(s_1)(\cdot)$ μ_2-a.e. If, furthermore, H is μ_1-integrable and $k \triangleq \int H(s_1)\, \mu_1(ds_1)$, then

$$k(s_2) = \int h(s_1, s_2)\, \mu_1(ds_1) \qquad \text{for} \quad \mu_2\text{-a.a. } s_2 \in S_2 .$$

▌PROOF We first assume that H is μ_1-integrable. Then, by Theorem I.4.30, there exists a sequence (H_i) of μ_1-simple functions such that

$$\lim_i \int | H(s_1) - H_i(s_1)|_q\, \mu_1(ds_1) = 0 \tag{1}$$

and

$$H_i(s_1) \triangleq \sum_{j=1}^{k(i)} \chi_j^i(s_1)\, y_j^i \qquad (s_1 \in S_1, \quad i \in \mathbb{N}), \tag{2}$$

where χ_j^i is the characteristic function of a μ_1-measurable set A_j^i. Now let $i \in \mathbb{N}$ be fixed. We choose, for each $j \in \{1, 2, ..., k(i)\}$, a Σ_2-measurable function $\hat{y}_j^i$ in the equivalence class of y_j^i (I.4.17), set

$$h_i(s_1, s_2) \triangleq \sum_{j=1}^{k(i)} \chi_j^i(s_1)\, \hat{y}_j^i(s_2) \qquad (s_1 \in S_1, \quad s_2 \in S_2), \tag{3}$$

and observe that, for every $\beta \in \mathbb{R}$ and $x \in \mathscr{X}$, we have

$$\begin{aligned} B(\beta, x) &\triangleq \{(s_1, s_2) \in S_1 \times S_2 \mid \, |h_i(s_1, s_2) - x| < \beta\} \\ &= \bigcup_{j=1}^{k(i)} A_j^i \times \{s_2 \in S_2 \mid \, |\hat{y}_j^i(s_2) - x| < \beta\}. \end{aligned}$$

Since $\hat{y}_j^i$ are Σ_2-measurable, it follows that $B(\beta, x)$ differs by a $\mu_1 \times \mu_2$-null set from an element of $\Sigma_1 \otimes \Sigma_2$; hence h_i is $\mu_1 \times \mu_2$-measurable. By I.4.35, $L^q(S_2, \Sigma_2, \mu_2, \mathscr{X}) \subset L^1(S_2, \Sigma_2, \mu_2, \mathscr{X})$ and therefore $y_j^i \in L^1(S_2, \Sigma_2, \mu_2, \mathscr{X})$; hence it follows from Fubini's theorem I.4.45 that h_i is $\mu_1 \times \mu_2$-integrable.

We next observe that, by (2), (3), and Hölder's inequality I.5.13,

$$\int |h_i(s_1, s_2) - H(s_1)(s_2)|\, \mu_2(ds_2) \leqslant |H_i(s_1) - H(s_1)|_q\, \mu(S_2)^{(q-1)/q} \qquad (i \in \mathbb{N}, \quad s_1 \in S);$$

hence, by (1),

$$\lim_i \int \mu_1(ds_1) \int |h_i(s_1, s_2) - H(s_1)(s_2)|\, \mu_2(ds_2) = 0. \tag{4}$$

Thus, by Fubini's theorem I.4.45,

$$\begin{aligned} &\lim_{i,k} \int |h_i(s_1, s_2) - h_k(s_1, s_2)|\, \mu_1(ds_1) \times \mu_2(ds_2) \\ &\quad = \lim_{i,k} \int \mu_1(ds_1) \int |h_i(s_1, s_2) - h_k(s_1, s_2)|\, \mu_2(ds_2) \\ &\quad \leqslant \lim_i \int \mu_1(ds_1) \int |h_i(s_1, s_2) - H(s_1)(s_2)|\, \mu_2(ds_2) \\ &\qquad + \lim_k \int \mu_1(ds_1) \int |h_k(s_1, s_2) - H(s_1)(s_2)|\, \mu_2(ds_2) \\ &\quad = 0. \end{aligned}$$

Thus (h_i) is a Cauchy sequence in $L^1(S_1 \times S_2, \Sigma_1 \otimes \Sigma_2, \mu_1 \times \mu_2, \mathscr{X})$ and, by I.5.17, there exists a $\mu_1 \times \mu_2$-integrable $\tilde{h} : S_1 \times S_2 \to \mathscr{X}$ such that $\lim_i \int | h_i(s) - \tilde{h}(s)| \, \mu_1 \times \mu_2(ds) = 0$. It follows, in view of (4), that

$$\begin{aligned}
&\int \mu_1(ds_1) \int | \tilde{h}(s_1, s_2) - H(s_1)(s_2)| \, \mu_2(ds_2) \\
&\quad \leqslant \lim_i \int \mu_1(ds_1) \int | \tilde{h}(s_1, s_2) - h_i(s_1, s_2)| \, \mu_2(ds_2) \\
&\qquad + \lim_i \int \mu_1(ds_1) \int | h_i(s_1, s_2) - H(s_1)(s_2)| \, \mu_2(ds_2) \\
&\quad = 0.
\end{aligned}$$

We conclude, applying I.4.27(6), that there exists a set $S_1' \subset S_1$ such that $\mu_1(S_1 \sim S_1') = 0$ and, for each $s_1 \in S_1'$, $\tilde{h}(s_1, s_2) = H(s_1)(s_2)$ for μ_2-a.a. $s_2 \in S_2$. We set $h(s_1, s_2) \triangleq \tilde{h}(s_1, s_2)$ for all $(s_1, s_2) \in S_1' \times S_2$ and $h(s_1, s_2) \triangleq H(s_1)(s_2)$ for all $(s_1, s_2) \in (S_1 \sim S_1') \times S_2$. Then h is $\mu_1 \times \mu_2$-integrable since $\tilde{h}$ is $\mu_1 \times \mu_2$-integrable and h differs from $\tilde{h}$ on the set $(S_1 \sim S_1') \times S_2$ of $\mu_1 \times \mu_2$-measure 0.

Now let

$$T_E \phi \triangleq \int_E \phi(s_2) \, \mu_2(ds_2) \qquad [E \in \Sigma_2, \quad \phi \in L^q(S_2, \mathscr{X})].$$

Then, by Hölder's inequality I.5.13,

$$| T_E \phi | \leqslant \int \chi_E(s_2) | \, \phi(s_2)| \, \mu_2(ds_2) \leqslant \mu_2(S_2)^{1-1/q} \, | \, \phi \, |_q ,$$

showing that T_E is a bounded linear operator on $L^q(S_2, \Sigma_2, \mu_2, \mathscr{X})$ to $\mathscr{X}$. It follows, by I.4.34(8) and Fubini's theorem I.4.45(3), that, setting $k \triangleq \int H(s_1) \, \mu_1(ds_1)$, we have

$$\begin{aligned}
\int_E k(s_2) \, \mu_2(ds_2) &= T_E \int h(s_1, \cdot) \, \mu_1(ds_1) \\
&= \int T_E h(s_1, \cdot) \, \mu_1(ds_1) \\
&= \int \mu_1(ds_1) \int \chi_E(s_2) \, h(s_1, s_2) \, \mu_2(ds_2) \\
&= \int_E \mu_2(ds_2) \int h(s_1, s_2) \, \mu_1(ds_1).
\end{aligned}$$

Thus, by I.4.34(7), $k(s_2) = \int h(s_1, s_2) \, \mu_1(ds_1)$ for μ_2-a.a. $s_2 \in S_2$. This completes the proof in the case where H is μ_1-integrable.

If H is not μ_1-integrable, we set

$$D_n \triangleq \{s_1 \in S_1 \mid n \leqslant |H(s_1)|_q < n+1\} \qquad (n = 0, 1, 2,...).$$

Then $\chi_{D_n} H$ is μ_1-integrable for each $n = 0, 1, 2,...$ and, by our previous argument, there exists a $\mu_1 \times \mu_2$-measurable $h^n : S_1 \times S_2 \to \mathscr{X}$ such that, for all $s_1 \in D_n$, $h^n(s_1, \cdot) = H(s_1)(\cdot)$ μ_2-a.e. We then set $h(s_1, s_2) \triangleq h^n(s_1, s_2)$ $(n = 0, 1, 2,..., s_1 \in D_n, s_2 \in S_2)$. QED

I.5.22 ***Theorem*** Let $\mathscr{X}$ be a separable Banach space, $q \in [1, \infty)$, and, for $i = 1, 2$, S_i a compact metric space, $\mu_i \in \text{frm}^+(S_i)$, and $\Sigma_i \triangleq \Sigma_{\text{Borel}}(S_i)$. If $H : S_1 \to L^q(S_2, \Sigma_2, \mu_2, \mathscr{X})$ is continuous, then there exists a $\mu_1 \times \mu_2$-measurable $h : S_1 \times S_2 \to \mathscr{X}$ such that, for each $s_1 \in S_1$, $h(s_1, \cdot) = H(s_1)(\cdot)$ μ_2-a.e.

▌PROOF Since $\Sigma_1 = \Sigma_{\text{Borel}}(S_1)$, H is μ_1-measurable. Our conclusion now follows from I.5.21. QED

I.5.23 ***Theorem*** Under the conditions of Theorem I.5.11, for every fixed choice of $\lambda \in \text{frm}(S)$, the function $k : S \times S \to B(\mathbb{R}^n, \mathbb{R}^n)$ may be assumed $\lambda \times \mu$-measurable.

▌PROOF Let $(k_j^i)(i, j = 1,..., n)$ be the matrix of k. By I.5.22, we can modify each $k_j^i(\bar{s}, \cdot)$ on a set of μ-measure 0 in such a manner that the modified function $k_j^i(\cdot, \cdot)$ is $\lambda \times \mu$-measurable. Clearly, such a modification does not invalidate the conclusions of I.5.11. QED

I.5.24 ***Theorem*** For $i = 1, 2$, let S_i be a compact metric space, $\mu_i \in \text{frm}^+(S_i)$, and $\Sigma_i \triangleq \Sigma_{\text{Borel}}(S_i)$. Furthermore, let $\mathscr{X}$ be a separable Banach space, $p, q \in [1, \infty)$, $L^p(S_1)$ denote $L^p(S_1, \Sigma_1, \mu_1)$, $L^q(S_2, \mathscr{X})$ denote $L^q(S_2, \Sigma_2, \mu_2, \mathscr{X})$, $|\cdot|_p$ be the norm in $L^p(S_1)$, $|\cdot|_q$ the norm in $L^q(S_2, \mathscr{X})$, and $\mathscr{H}$ the set of (equivalence classes of) $\mu_1 \times \mu_2$-measurable functions $h : S_1 \times S_2 \to \mathscr{X}$ such that the function $s_1 \to |h(s_1, \cdot)|_q$ belongs to $L^p(S_1)$. Then $\mathscr{H}$ is isomorphic to $L^{p,q} \triangleq L^p(S_1, \Sigma_1, \mu_1, L^q(S_2, \Sigma_2, \mu_2, \mathscr{X}))$ under the mapping $\mathscr{I}$ that assigns to each $h \in \mathscr{H}$ the function $s_1 \to h(s_1, \cdot)$, and the set $C(S_1) \otimes C(S_2, \mathscr{X})$ is dense in $L^{p,q}$.

▌PROOF Let $h \in \mathscr{H}$ and $H(s_1) \triangleq h(s_1, \cdot)$ $(s_1 \in S_1)$. If $f \in L^q(S_2, \mathscr{X})$, then the function $(s_1, s_2) \to h(s_1, s_2) - f(s_2) : S_1 \times S_2 \to \mathscr{X}$ belongs to $\mathscr{H}$ and, by Fubini's theorem I.4.45(2), the function $s_1 \to |H(s_1) - f|_q$ is μ_1-measurable, implying that $s_1 \to H(s_1)$ is μ_1-measurable. Since, by assumption, $s_1 \to |H(s_1)|_q$ belongs to $L^p(S_1)$, we conclude that $H \in L^{p,q}$.

Now let $H \in L^{p,q}$. Then, by I.5.21, there exists a $\mu_1 \times \mu_2$-measurable function $h : S_1 \times S_2 \to \mathscr{X}$ such that, for all $s_1 \in S_1$, $h(s_1, \cdot) = H(s_1)(\cdot)$ μ_2-a.e., and it is clear that $h \in \mathscr{H}$. If h_1 is another such function, then h_1 differs from h on a set of $\mu_1 \times \mu_2$-measure 0 and therefore h is the unique element of $\mathscr{H}$ corresponding to H. Since the mapping $\mathscr{I}$ is clearly linear, we have shown that $\mathscr{I}$ is an isomorphism of $\mathscr{H}$ to $L^{p,q}$.

Now we observe that, by I.5.18, $L^q(S_2, \mathscr{X})$ is separable, the set $C(S_2, \mathscr{X})$ is dense in $L^q(S_2, \mathscr{X})$, and the set $C(S_1) \otimes \{f_1, f_2, ...\}$ is dense in $L^{p,q}$ for every choice of a dense subset $\{f_1, f_2, ...\}$ of $L^q(S_2, \mathscr{X})$. We may choose $\{f_1, f_2, ...\}$ out of $C(S_2, \mathscr{X})$ which implies that $C(S_1) \otimes C(S_2, \mathscr{X})$ is dense in $L^{p,q}$. QED

I.5.25 ***Theorem*** Let (S_1, Σ, μ) be a positive finite measure space, S_2 a separable metric space, $\mathscr{X}$ a separable Banach space, and

$$\mathscr{B} \triangleq \mathscr{B}(S_1, \Sigma, \mu, S_2; \mathscr{X})$$

the vector space of (equivalence classes of) functions $\phi : S_1 \times S_2 \to \mathscr{X}$ such that $\phi(s_1, \cdot) \in C(S_2, \mathscr{X})$ $(s_1 \in S_1)$, $\phi(\cdot, s_2)$ is μ-measurable for each $s_2 \in S_2$, for each $\phi \in \mathscr{B}$ there exists a μ-integrable ψ_ϕ with $|\phi(s_1, \cdot)|_{\sup} \leqslant \psi_\phi(s_1)$ $(s_1 \in S_1)$, and two elements ϕ_1 and ϕ_2 are identified if $\phi_1(s_1, \cdot) = \phi_2(s_1, \cdot)$ for μ-a.a. $s_1 \in S_1$. Then

(1) the function $s_1 \to |\phi(s_1, \cdot)|_{\sup}$ is μ-integrable, the function $\phi \to |\phi|_{\mathscr{B}} \triangleq \int |\phi(s_1, \cdot|_{\sup} \mu(ds_1) : \mathscr{B} \to \mathbb{R}$ is a norm on $\mathscr{B}$, and the function $s_1 \to \phi(s_1, \xi(s_1)) : S_1 \to \mathscr{X}$ is μ-integrable for every $\phi \in \mathscr{B}$ and μ-measurable $\xi : S_1 \to S_2$;
(2) if S_2 is compact, then the mapping $\mathscr{I}$ that assigns to each $\phi \in \mathscr{B}$ the function $s_1 \to \phi(s_1, \cdot) : S_1 \to C(S_2, \mathscr{X})$ is an isometric isomorphism of $(\mathscr{B}, |\cdot|_{\mathscr{B}})$ to $L^1(S_1, \Sigma, \mu, C(S_2, \mathscr{X}))$; and
(3) if both S_1 and S_2 are compact metric spaces, μ is regular, and $\Sigma \triangleq \Sigma_{\text{Borel}}(S_1)$, then $(\mathscr{B}, |\cdot|_{\mathscr{B}})$ is separable and $C(S_1) \otimes C(S_2, \mathscr{X})$ dense in $\mathscr{B}$.

▌PROOF *Proof of* (1) Since the function $x \to |x| : \mathscr{X} \to \mathbb{R}$ is continuous, it follows from I.4.21 that $s_1 \to |\phi(s_1, \cdot)|_{\sup}$ is μ-measurable for each $\phi \in \mathscr{B}$. Since $|\phi(s_1, \cdot)|_{\sup} \leqslant \psi_\phi(s_1)$ $(s_1 \in S_1)$, the function $s_1 \to |\phi(s_1, \cdot)|_{\sup}$ is μ-integrable and $|\phi|_{\mathscr{B}}$ is defined in $\mathbb{R}$ for all $\phi \in \mathscr{B}$. It is easily verified that $|\phi|_{\mathscr{B}} \geqslant 0$, $|\phi_1 + \phi_2|_{\mathscr{B}} \leqslant |\phi_1|_{\mathscr{B}} + |\phi_2|_{\mathscr{B}}$, and $|0|_{\mathscr{B}} = 0$. Now assume that $|\phi|_{\mathscr{B}} = 0$. Then, by I.4.27(6), $|\phi(s_1, \cdot)|_{\sup} = 0$ for μ-a.a. $s_1 \in S_1$ and thus ϕ is equivalent to 0. We conclude that $|\cdot|_{\mathscr{B}}$ is a norm on $\mathscr{B}$. If $\phi \in \mathscr{B}$ and $\xi : S_1 \to S_2$ is μ-measurable, then, by I.4.22, $s_1 \to \phi(s_1, \xi(s_1))$ is μ-measurable and, since $|\phi(s_1, \xi(s_1))| \leqslant |\phi(s_1, \cdot)|_{\sup} \leqslant \psi_\phi(s_1)$ $(s_1 \in S_1)$, it follows that $s_1 \to \phi(s_1, \xi(s_1))$ is μ-integrable.

Proof of (2) By I.5.1, $C(S_2, \mathscr{X})$ is separable. Furthermore, the function $(s_1, s_2) \to \phi(s_1, s_2) - c(s_2)$ belongs to $\mathscr{B}$ for every $\phi \in \mathscr{B}$ and $c \in C(S_2, \mathscr{X})$. It follows that the set $\{s_1 \in S_1 \mid |\phi(s_1, \cdot) - c(\cdot)|_{\sup} < \beta\}$ is μ-measurable for every $\beta \in \mathbb{R}$, showing that the function $s_1 \to \phi(s_1, \cdot) : S_1 \to C(S_2, \mathscr{X})$ is μ-measurable. Since $|\phi(s_1, \cdot)|_{\sup} \leqslant \psi_\phi(s_1)$ $(s_1 \in S_1)$ and $C(S_2, \mathscr{X})$ is separable, this shows that $\mathscr{I}(\phi) \in L^1(S_1, \Sigma, \mu, C(S_2, \mathscr{X}))$. Conversely, if $f \in L^1(S_1, \Sigma, \mu, C(S_2, \mathscr{X}))$, then we observe that $f(s_1) \in C(S_2, \mathscr{X})$ $(s_1 \in S_1)$, $\psi_f(s_1) \triangleq |f(s_1)(\cdot)|_{\sup}$ is μ-integrable, and $s_1 \to f(s_1)(\bar{s}_2)$ is μ-measurable for each $\bar{s}_2 \in S_2$ [since the function $c \to c(\bar{s}_2) : C(S_2, \mathscr{X}) \to \mathscr{X}$ is continuous]. Thus $(s_1, s_2) \to f(s_1)(s_2)$ belongs to $\mathscr{B}$. Since $\mathscr{I}$ is clearly linear and injective, this shows that $\mathscr{I}$ is an isomorphism of $\mathscr{B}$ onto $L^1(S_1, \Sigma, \mu, C(S_2, \mathscr{X}))$. We also observe that

$$|\phi|_{\mathscr{B}} = \int |\phi(s_1, \cdot)|_{\sup} \mu(ds_1) = |\mathscr{I}(\phi)| .$$

Proof of (3) By I.5.1, $(C(S_1), |\cdot|_{\sup})$ and $(C(S_2, \mathscr{X}), |\cdot|_{\sup})$ contain dense at most denumerable subsets $\{c_1^1, c_2^1, ...\}$ and $\{c_1^2, c_2^2, ...\}$, and it is easily verified that $\{c_1^1, c_2^1, ...\}$ is also dense in $C(S_1)$ in the $L^1(S_1, \Sigma, \mu)$-topology. Hence, by I.5.18, $C(S_1) \otimes \{c_1^2, c_2^2, ...\}$ is dense in $\mathscr{B}$. It follows that the at most denumerable set $\{c_1^1, c_2^1, ...\} \otimes \{c_1^2, c_2^2, ...\}$ is dense in $\mathscr{B}$ and, as a consequence, $C(S_1) \otimes C(S_2, \mathscr{X})$ is dense in $\mathscr{B}$. QED

I.5.26 ***Theorem*** Let X be a complete separable metric space, (S_1, Σ_1) a measurable space, S_2 a compact metric space, and $h : S_1 \times S_2 \to X$ such that $h(\cdot, s_2)$ is Σ_1-measurable and $h(s_1, \cdot)$ continuous for all $(s_1, s_2) \in S_1 \times S_2$. Then

(1) $s_1 \to h(s_1, \cdot) : S_1 \to C(S_2, X)$ is Σ_1-measurable.
If, furthermore, S_1 is a compact metric space, $\mu_1 \in \text{frm}^+(S_1)$, and (S_1, Σ_1, μ_1^*) the Lebesgue extension of $(S_1, \Sigma_{\text{Borel}}(S_1), \mu_1)$, then
(2) for every $\epsilon > 0$ there exists a closed subset F_ϵ of S_1 such that $\mu_1(S_1 \sim F_\epsilon) \leqslant \epsilon$ and $h \mid F_\epsilon \times S_2$ is continuous; and
(3) for every $\mu_2 \in \text{frm}(S_2)$, h is $\mu_1 \times \mu_2$-measurable.

▌PROOF By I.5.1, $C(S_2, X)$ is separable and contains an at most denumerable dense subset $\{c_1, c_2, ...\}$. If we denote by $d_B(\cdot, \cdot)$ the distance function in $C(S_2, X)$ [that is, $d_B(c', c'') \triangleq \sup_{s_2 \in S_2} d(c'(s_2), c''(s_2))$], then, by I.4.20 and I.4.21, the function $s_1 \to d_B(h(s_1, \cdot), c_i)$ is Σ_1-measurable for each $i \in \mathbb{N}$. It follows that the sets $\{s_1 \in S_1 \mid d_B(h(s_1, \cdot), c_i) \leqslant \beta\}$ are Σ_1-measurable for each $\beta \in \mathbb{R}$ and $i \in \mathbb{N}$, and therefore $s_1 \to h(s_1, \cdot) : S_1 \to C(S_2, X)$ is Σ_1-measurable.

Now assume that S_1 is a compact metric space, $\mu_1 \in \text{frm}^+(S_1)$, and (S_1, Σ_1, μ_1^*) is the Lebesgue extension of $(S_1, \Sigma_{\text{Borel}}(S_1), \mu_1)$, and let $H(s_1) \triangleq h(s_1, \cdot) \in C(S_2, X)$ $(s_1 \in S_1)$. Then H is μ_1-measurable and, by Lusin's theorem I.4.19, for each $\epsilon > 0$ there exists a closed subset F_ϵ of S_1 such that

$\mu_1(S_1 \sim F_\epsilon) \leqslant \epsilon$ and $H \mid F_\epsilon$ is continuous. If $\lim_j t_j = \bar{t}$ in F_ϵ and $\lim_j r_j = \bar{r}$ in S_2, then

$$\lim_j d(h(\bar{t}, \bar{r}), h(t_j, r_j)) \leqslant \lim_j [d(h(\bar{t}, \bar{r}), h(\bar{t}, r_j)) + d(h(\bar{t}, r_j), h(t_j, r_j))]$$
$$\leqslant \lim_j [d(h(\bar{t}, \bar{r}), h(\bar{t}, r_j)) + d_B(H(\bar{t}), H(t_j))] = 0,$$

and this shows that $h \mid F_\epsilon \times S_2$ is continuous.

Finally, if $\mu_2 \in \text{frm}(S_2)$, then, by I.4.47, the continuous function $h \mid F_\epsilon \times S_2$ is $\Sigma_1 \otimes \Sigma_{\text{Borel}}(S_2)$-measurable and therefore also $\mu_1 \times \mu_2$-measurable. Since $\mu_1 \times \mu_2(S_1 \times S_2 \sim \bigcup_{j=1}^\infty F_{1/j} \times S_2) = 0$, it follows that h is $\mu_1 \times \mu_2$-measurable. QED

I.5.27 ***Theorem*** Let $\mathscr{X}$ be a separable Banach space, (S_1, Σ_1) a measurable space, S_2 a compact metric space, $\nu \in \text{frm}(S_2)$, and $\phi : S_1 \to C(S_2, \mathscr{X})$ Σ_1-measurable. Then the function $s_1 \to \int \phi(s_1)(s_2)\, \nu(ds_2)$ is defined on S_1 to $\mathscr{X}$ and it is Σ_1-measurable.

▌PROOF Each $c(\cdot) \in C(S_2, \mathscr{X})$ is bounded and continuous, hence ν-integrable. By I.4.34(3),

$$\left| \int c_1(s_2)\, \nu(ds_2) - \int c_2(s_2)\, \nu(ds_2) \right| \leqslant | c_1 - c_2 |_{\sup} | \nu |(S_2);$$

therefore the function

$$c \to T_\nu c \triangleq \int c(s_2)\, \nu(ds_2) : C(S_2, \mathscr{X}) \to \mathscr{X}$$

is continuous. It follows that the function $T_\nu \circ \phi : S_1 \to \mathscr{X}$ is Σ_1-measurable.
QED

I.6 Convex Sets

Let $\mathscr{X}$ be a vector space. A set $A \subset \mathscr{X}$ is *symmetric* if $A = -A$. A set $A \subset \mathscr{X}$ is *convex* if $\{\alpha a + (1 - \alpha) b \mid 0 \leqslant \alpha \leqslant 1\} \subset A$ whenever $a, b \in A$. A real-valued function ψ defined on a convex subset A of $\mathscr{X}$ is *convex* if

$$\psi(\alpha x + (1 - \alpha) y) \leqslant \alpha\psi(x) + (1 - \alpha) \psi(y) \qquad (\alpha \in [0, 1], \quad x, y \in A).$$

It is easy to verify that $\alpha A + \beta B$, $A \cap B$, $A \times C$, and $T(A)$ are convex whenever $A, B \subset \mathscr{X}$ and $C \subset \mathscr{Y}$ are convex, $\alpha, \beta \in \mathbb{R}$, and $T \in \mathscr{L}(\mathscr{X}, \mathscr{Z})$.

We say that $l \in \mathscr{L}(\mathscr{X}, \mathbb{R})$ *separates* subsets A and B of $\mathscr{X}$ if $l \neq 0$ and either

$$l(x) \leqslant \alpha \leqslant l(y) \qquad \text{or} \qquad l(x) \geqslant \alpha \geqslant l(y) \qquad (x \in A, y \in B)$$

for some $\alpha \in \mathbb{R}$. If $\bar{x} \in A \subset \mathscr{X}$, $0 \neq l \in \mathscr{L}(\mathscr{X}, \mathbb{R})$, and $l(x) \geqslant l(\bar{x})$ $(x \in A)$, we say that l is an *inward normal* to A at $\bar{x}$. If $A \subset \mathscr{X}$, we denote by $\text{co}(A)$, and call the *convex hull* of A, the intersection of all the convex subsets of $\mathscr{X}$ containing A.

We set

$$\mathscr{T}_n \triangleq \left\{(\theta^1, \ldots, \theta^n) \in \mathbb{R}^n \mid \theta^j \geqslant 0, \sum_{j=1}^n \theta^j \leqslant 1\right\} \qquad (n \in \mathbb{N})$$

and

$$\mathscr{T}_n' \triangleq \left\{(\theta^0, \theta^1, \ldots, \theta^n) \in \mathbb{R}^{n+1} \mid \theta^j \geqslant 0, \sum_{j=0}^n \theta^j = 1\right\} \qquad (n = 0, 1, 2, \ldots).$$

We observe that $(\theta^1, \ldots, \theta^n) \in \mathscr{T}_n$ if $(\theta^0, \theta^1, \ldots, \theta^n) \in \mathscr{T}_n'$, and that

$$\left(1 - \sum_{j=1}^n \theta^j, \theta^1, \ldots, \theta^n\right) \in \mathscr{T}_n' \qquad \text{if} \quad (\theta^1, \ldots, \theta^n) \in \mathscr{T}_n .$$

We refer to a linear combination $\sum_{j=0}^n \theta^j x_j$ in a vector space as a *convex combination* if $n \in \{0, 1, 2, \ldots\}$ and $(\theta^0, \ldots, \theta^n) \in \mathscr{T}_n'$.

If $\mathscr{X}$ is a topological vector space, we denote by $\mathscr{T}_0(\mathscr{X})$ the collection of the open neighborhoods of 0 in $\mathscr{X}$. Since the function

$$(x, y) \to x + y : \mathscr{X} \times \mathscr{X} \to \mathscr{X}$$

is continuous, for every $U \in \mathscr{T}_0(\mathscr{X})$ there exists some $V \in \mathscr{T}_0(\mathscr{X})$ such that $V + V \subset U$, and we shall often apply this result without further justification. A topological vector space is *locally convex* if every open neighborhood of 0 in $\mathscr{X}$ contains an open convex neighborhood U of 0; we may assume that such U is symmetric since otherwise $U' \triangleq U \cap (-U) = -U' \subset U$, $U' \in \mathscr{T}_0(\mathscr{X})$, and U' is convex. It is easy to verify that every normed vector space is a locally convex topological vector space. A closed convex set A in a topological vector space is a *convex body* if $A^\circ \neq \varnothing$. If $A \subset \mathscr{X}$, we denote by $\overline{\text{co}}(A)$, and call the *convex closure* or the *closed convex hull* of A, the intersection of all the closed convex subsets of $\mathscr{X}$ containing A.

If K is a convex subset of a topological vector space $\mathscr{X}$ and $0 \in K^\circ$, then for every $x \in \mathscr{X}$ there exists $a > 0$ such that $(1/a)\, x \in K$; thus

$$\phi(x) \triangleq \inf \left\{a > 0 \mid \frac{1}{a} x \in K\right\} < \infty \qquad (x \in \mathscr{X}).$$

We refer to the function $x \to \phi(x) : \mathscr{X} \to [0, \infty)$ as the *gauge* (or *support*) function of K.

I.6.1 ***Theorem*** Let $\mathscr{X}$ be a vector space and $A \subset \mathscr{X}$. Then

$$\text{co}(A) = \left\{ \sum_{j=0}^{k} \theta^j x_j \mid k \in \{0, 1,...\}, (\theta^0,..., \theta^k) \in \mathscr{T}_k', x_0 ,..., x_k \in A \right\}.$$

▌PROOF Let C be the set on the right side of the relation. Then C is convex and contains A. Now let K be convex and contain A. Then we shall show that $C \subset K$ which will complete the proof.

Assume that $\sum_{j=0}^{k} \theta^j x_j \in K$ if $(\theta^0,..., \theta^k) \in \mathscr{T}_k'$, $x_0 ,..., x_k \in A$, and $1 \leqslant k \leqslant n$. This is clearly true for $n = 1$. Now let $(\theta^0,..., \theta^{n+1}) \in \mathscr{T}'_{n+1}$, $x_j \in A$ $(j = 0,..., n + 1)$, and $\alpha \triangleq \sum_{i=0}^{n} \theta^i$. If some $\theta^j = 0$, then, by our assumption, $\sum_{j=0}^{n+1} \theta^j x_j \in K$. If $\theta^j > 0$ $(j = 0,..., n + 1)$, then

$$\sum_{j=0}^{n+1} \theta^j x_j = \alpha \sum_{j=0}^{n} \frac{1}{\alpha} \theta^j x_j + \theta^{n+1} x_{n+1} .$$

By our assumption, $\sum_{j=0}^{n} (1/\alpha) \theta^j x_j \in K$ and we observe that $x_{n+1} \in A \subset K$ and $\alpha + \theta^{n+1} = 1$. Since K is convex, it follows that $\sum_{j=0}^{n+1} \theta^j x_j \in K$. This completes the induction and shows that $C \subset K$. QED

I.6.2 ***Theorem*** (Caratheodory) Let $n \in \mathbb{N}$ and $A \subset \mathbb{R}^n$. Then

$$\text{co}(A) = \left\{ \sum_{j=0}^{n} \theta^j x_j \mid (\theta^0,..., \theta^n) \in \mathscr{T}_n', x_0 ,..., x_n \in A \right\}.$$

▌PROOF Let $x \in \text{co}(A)$. Then, by I.6.1,

$$x = \sum_{j=0}^{k} \theta^j x_j \quad \text{for some } k \in \{0, 1,...\},\ (\theta^0,..., \theta^k) \in \mathscr{T}_k' \ \text{ and } \ x_0 ,..., x_k \in A,$$

and we may clearly assume that $\theta^j > 0$ for $j = 0,..., k$ and that k is the smallest such integer. If $k > n$, then, by I.3.1, the set $\{x_0 - x_k ,..., x_{k-1} - x_k\}$ is dependent and there exists $(\alpha^0,..., \alpha^{k-1}) \neq 0$ in $\mathbb{R}^k$ such that $\sum_{j=0}^{k-1} \alpha^j x_j = (\sum_{j=0}^{k-1} \alpha^j) x_k$. We set $\alpha^k \triangleq -\sum_{j=1}^{k-1} \alpha^j$ and conclude that $\sum_{j=0}^{k} \alpha^j x_j = 0$, $\sum_{j=0}^{k} \alpha^j = 0$, and $\alpha \triangleq (\alpha^0,..., \alpha^k) \neq 0$. Thus, for all $t \in \mathbb{R}$, we have

$$x = \sum_{j=0}^{k} \theta^j x_j = \sum_{j=0}^{k} (\theta^j + t\alpha^j) x_j$$

and $\sum_{j=0}^{k} (\theta^j + t\alpha^j) = 1$. Since $\theta^j > 0$ $(j = 0, 1,..., k)$ and $\alpha^i < 0$ for some $i \in \{0, 1,..., k\}$, we conclude that

$$\beta \triangleq \sup\{t \geqslant 0 \mid \theta^j + t\alpha^j \geqslant 0 \text{ for } j = 0,..., k\} < \infty$$

and that $\theta^j + \beta\alpha^j \geqslant 0$ for $j = 0,\ldots, k$ and $\theta^i + \beta\alpha^i = 0$ for some $i \in \{0,\ldots, k\}$. Thus $x = \sum_{j=0}^{k} (\theta^j + \beta\alpha^j)\, x_j$ is a convex combination of k points of A, contradicting our assumption about k. We conclude that $k \leqslant n$. QED

I.6.3 ***Theorem*** Let $\mathscr{X}$ be a topological vector space and K a convex subset of $\mathscr{X}$. Then $\overline{K}$ and K° are convex, $\overline{K} = \overline{K^\circ}$ if $K^\circ \neq \varnothing$, and $\alpha x + (1 - \alpha)\, y \in K^\circ$ if $x \in K^\circ$, $y \in \overline{K}$ and $0 < \alpha \leqslant 1$. If $A \subset \mathscr{X}$, then $\overline{\mathrm{co}}(A) = \overline{\mathrm{co}(A)}$.

▌PROOF Let $a, b \in \overline{K}$ and $U \in \mathscr{T}_0(\mathscr{X})$. Then there exists $V \in \mathscr{T}_0(\mathscr{X})$ such that $V + V \subset U$. For any $\alpha \in (0, 1)$, we have $(1/\alpha)\, V, (1 - \alpha)^{-1}\, V \in \mathscr{T}_0(\mathscr{X})$, and there exist points $a_1 \in K \cap [a + (1/\alpha)\, V]$ and $b_1 \in K \cap [b + (1 - \alpha)^{-1}\, V]$. We have

$$\begin{aligned}\alpha a_1 + (1 - \alpha)\, b_1 &\in K \cap [\alpha a + (1 - \alpha)\, b + V + V]\\ &\subset K \cap [\alpha a + (1 - \alpha)\, b + U].\end{aligned}$$

Thus $\alpha a + (1 - \alpha)\, b \in \overline{K}$ and we conclude that $\overline{K}$ is convex.

Now let $x \in K^\circ$, $y \in \overline{K}$, and $0 < \alpha < 1$. Then $x + U \subset K$ for some symmetric $U \in \mathscr{T}_0(\mathscr{X})$. Since $y \in \overline{K}$, there exists some $y' \in K \cap [\, y + \alpha(1 - \alpha)^{-1}\, U]$. It follows that

$$y \in y' + \frac{\alpha}{1 - \alpha}\, U \quad \text{and} \quad \alpha(x + U) + (1 - \alpha)\, y' \subset \alpha K + (1 - \alpha)\, K \subset K;$$

hence

$$\alpha x + (1 - \alpha)\, y \in \alpha x + (1 - \alpha)\, y' + \alpha U \subset K.$$

Thus $\alpha x + (1 - \alpha)\, y$ belongs to an open subset of K.

Now assume that $K^\circ \neq \varnothing$. Then our last argument shows that K° is convex and that $y + \beta(x - y) \in K^\circ$ $(\beta \in (0, 1])$. If $U \in \mathscr{T}_0(\mathscr{X})$, then $\beta(x - y) \in U$ for all sufficiently small positive β. Thus

$$y + \beta(x - y) \in K^\circ \cap (\, y + U),$$

showing that $K^\circ \cap (\, y + U) \neq \varnothing$; hence $y \in \overline{K^\circ}$.

Now let $A \subset \mathscr{X}$. Then $\overline{\mathrm{co}(A)}$, being the closure of a convex set, is itself convex and therefore contains $\overline{\mathrm{co}}(A)$. On the other hand, $\overline{\mathrm{co}(A)}$ is contained in every closed set containing $\mathrm{co}(A)$; hence $\overline{\mathrm{co}(A)} \subset \overline{\mathrm{co}}(A)$. QED

I.6.4 ***Theorem*** Let $\mathscr{X}$ be a topological vector space. If $K \subset \mathscr{X}$ is convex, $0 \in K^\circ$, and ϕ is the gauge function of K, then, for all $x, y \in \mathscr{X}$, we have

(1) $\phi(x) \geqslant 0$,
(2) $\phi(x + y) \leqslant \phi(x) + \phi(\, y)$,

(3) $\phi(\alpha x) = \alpha\phi(x) \quad (\alpha \geqslant 0)$,
(4) $\lim_{x\to 0} \phi(x) = 0$, and
(5) $\phi(x) < 1$, $\phi(x) = 1$, and $\phi(x) > 1$ if and only if $x \in K^\circ$, $x \in \partial K$, and $x \in \mathscr{X} \sim \overline{K}$, respectively.

Conversely, if $\phi : \mathscr{X} \to \mathbb{R}$ satisfies relations (1)–(4), then ϕ is continuous, the set $K \triangleq \{x \in \mathscr{X} \mid \phi(x) < 1\}$ is an open convex set containing 0, and the gauge function of K is ϕ. In particular, every gauge function is continuous.

▌PROOF Let $0 \in K^\circ$, K be convex, and ϕ the gauge function of K. Then properties (1) and (3) are obvious. We next prove (5). Let $x \in K^\circ$. Then $x + U \subset K$ for some $U \in \mathscr{T}_0(\mathscr{X})$ and there exists $\epsilon > 0$ such that $\epsilon x \in U$. We have $(1 + \epsilon)\, x \in x + U \subset K$ implying $\phi(x) \leqslant 1/(1 + \epsilon) < 1$. Conversely, if $\phi(x) < 1$, then $(1/a)\, x \in K$ for some $a \in (0, 1)$. Then, by I.6.3, $(1 - a)\, 0 + a\big((1/a)\, x\big) = x \in K^\circ$. Thus $\phi(x) < 1$ if and only if $x \in K^\circ$.

Now let $x \in \partial K$. Then, by I.6.3, $\alpha x + (1 - \alpha)\, 0 = \alpha x \in K^\circ$ for $0 \leqslant \alpha < 1$; hence $\phi(\alpha x) = \alpha\phi(x) \leqslant 1$ for all $\alpha \in [0, 1)$, implying $\phi(x) \leqslant 1$. Since $x \notin K^\circ$, we have $\phi(x) \geqslant 1$; hence $\phi(x) = 1$. Next let $x \in \mathscr{X} \sim \overline{K}$, and let $U \in \mathscr{T}_0(\mathscr{X})$ be such that $(x + U) \cap \overline{K} = \varnothing$. Then $-\epsilon x \in U$ for some $\epsilon \in (0, 1)$ and $(1 - \epsilon)\, x \in x + U$. Thus $(1 - \epsilon)\, x \notin K^\circ$, hence

$$\phi((1 - \epsilon)\, x) = (1 - \epsilon)\, \phi(x) \geqslant 1 \qquad \text{and} \qquad \phi(x) \geqslant 1/(1 - \epsilon) > 1.$$

This completes the proof of (5).

Now let $\phi(x) \leqslant \alpha_1$, $\phi(y) \leqslant \alpha_2$. Then

$$x \in \alpha_1 \overline{K}, \qquad y \in \alpha_2 \overline{K}, \qquad \text{and} \qquad x + y \in (\alpha_1 + \alpha_2)\, \overline{K},$$

implying $\phi(x + y) \leqslant \alpha_1 + \alpha_2$. Thus

$$\phi(x + y) \leqslant \inf\{\alpha_1 + \alpha_2 \mid \phi(x) \leqslant \alpha_1,\ \phi(y) \leqslant \alpha_2\} = \phi(x) + \phi(y)$$

and relation (2) is satisfied. Finally, let $\epsilon > 0$ and $U = \epsilon K^\circ$. Then $U \in \mathscr{T}_0(\mathscr{X})$ and $(1/\epsilon)\, x \in K^\circ$ for all $x \in U$. Thus, by (3) and (5), $\phi\big((1/\epsilon)\, x\big) = (1/\epsilon)\, \phi(x) < 1$; hence $\phi(x) < \epsilon$ for all $x \in U$, proving (4).

Conversely, let ϕ satisfy relation (1)–(4). Then, by (4), ϕ is continuous at 0. By (2), $\phi(z) - \phi(x) \leqslant \phi(z - x)$ and $\phi(x) - \phi(z) \leqslant \phi(x - z)$. Thus $|\phi(z) - \phi(x)| \leqslant \text{Max}\big(\phi(z - x), \phi(x - z)\big)$. By (4),

$$\lim_{z\to x} \phi(x - z) = \lim_{z\to x} \phi(z - x) = 0,$$

showing that ϕ is continuous at x for every $x \in \mathscr{X}$. Thus the set $K \triangleq \phi^{-1}\big((-\infty, 1)\big) \triangleq \{x \in \mathscr{X} \mid \phi(x) < 1\}$ is open. By (3), $\phi(0) = 0 < 1$;

hence $0 \in K$. By (2) and (3), ϕ is convex; hence K is convex. Now let ϕ_1 be the gauge function of K. Then

$$\phi_1(y) = \inf \left\{ a > 0 \mid \frac{1}{a} y \in K \right\} = \inf \left\{ a > 0 \mid \phi\left(\frac{1}{a} y\right) = \frac{1}{a} \phi(y) < 1 \right\} = \phi(y)$$

$(y \in \mathscr{X})$. QED

I.6.5 ***Theorem*** Let $\mathscr{X}$ be a topological vector space. A function $\psi : \mathscr{X} \to \mathbb{R}$ is continuous and a seminorm if and only if ψ is the gauge function of a symmetric convex set K containing 0 in its interior.

▌PROOF A continuous seminorm ψ satisfies conditions (1)–(4) of I.6.4, showing that it is the gauge function of the convex set $K \triangleq \{x \in \mathscr{X} \mid \psi(x) < 1\}$. Since $\psi(x) = \psi(-x)$, K is symmetric.

Conversely, let $K \in \mathscr{T}_0(\mathscr{X})$ be convex and symmetric. Then its gauge function ψ satisfies conditions (1)–(4) of I.6.4 and $\psi(x) = \psi(-x)$. Thus ψ is a seminorm and, by I.6.4, it is continuous. QED

I.6.6 ***Theorem*** Let $\mathscr{X}$ be a topological vector space, $A, B \subset \mathscr{X}$, and $A^\circ \neq \varnothing$. If a linear functional l separates A and B, then l is continuous.

▌PROOF We may assume that $l(x) \leqslant \alpha \leqslant l(y)$ for some $\alpha \in \mathbb{R}$ and all $x \in A$ and $y \in B$, otherwise replacing l by $-l$. Now let $U \in \mathscr{T}_0(\mathscr{X})$ and $a \in A^\circ$ be such that $U = -U$ and $a + U \subset A$. Then $l(a) + l(U) \subset (-\infty, \alpha]$; hence $l(U) = -l(U) \subset (-\infty, \alpha - l(a)]$ or $l(U) \subset [-\beta, \beta]$, where $\beta \triangleq \alpha - l(a) \geqslant 0$. Now let $\epsilon > 0$. Then $|l(x)| \leqslant \epsilon$ if $x \in \epsilon(\beta + 1)^{-1} U$, showing that l is continuous at 0, hence continuous. QED

I.6.7 ***Theorem*** Let K and M be nonempty disjoint convex subsets of a topological vector space $\mathscr{X}$ and $K^\circ \neq \varnothing$. Then there exists $l \in \mathscr{X}^*$ that separates K and M.

▌PROOF Let $k \in K^\circ$ and $m \in M$. Then $l \in \mathscr{X}^*$ separates K and M if and only if it separates $K - M - k + m$ and $\{m - k\}$. We may therefore assume that $0 \in K^\circ$ and $M = \{y\}$.

Let ϕ be the gauge function of K. Since $y \notin K$, I.6.4(5) implies that $\phi(y) \geqslant 1$ and $\phi(x) \leqslant 1$ $(x \in K)$. Now let l be the linear functional on $\mathscr{X}_y \triangleq \{\alpha y \mid \alpha \in \mathbb{R}\}$ defined by $l(\alpha y) \triangleq \alpha\phi(y)$. Then $l(\alpha y) = \phi(\alpha y)$ for $\alpha \geqslant 0$ and $l(\alpha y) < 0 \leqslant \phi(\alpha y)$ for $\alpha < 0$, implying $l(x) \leqslant \phi(x)$ for all $x \in \mathscr{X}_y$. It follows now, by the Hahn–Banach theorem I.3.8, that l can be extended to $\mathscr{X}$ as a linear functional in such a manner that $l(x) \leqslant \phi(x)$ for all $x \in \mathscr{X}$. Thus $l(x) \leqslant 1$ for all $x \in K$ and $l(y) = \phi(y) \geqslant 1$; hence l separates K and $\{y\}$.

It remains to show that l is continuous. This follows directly from I.6.6.

QED

I.6.8 ***Theorem*** Let K be a convex subset of a topological vector space $\mathscr{X}$ and $K^\circ \neq \varnothing$. Then for every point $\bar{x} \in \partial K$ there exists $l \in \mathscr{X}^*$ such that l is an inward normal to $\bar{K}$ at $\bar{x}$.

▌PROOF By I.6.3, K° is convex and, by I.6.7, there exists $l \in \mathscr{X}^*$ that separates K° and $\{\bar{x}\}$. We may assume that $l(x) \geqslant l(\bar{x})$ for all $x \in K^\circ$, otherwise replacing l by $-l$. Since l is continuous, the set $l^{-1}([l(\bar{x}), \infty))$ is closed, and we have just shown that it contains K°. Thus, by I.6.3, $l(x) \geqslant l(\bar{x})$ for all $x \in \overline{K^\circ} = \bar{K}$. QED

I.6.9 ***Theorem*** Let $n \in \mathbb{N}$, K be a nonempty convex subset of $\mathbb{R}^n$, and $0 \in K$. Then K has a nonempty interior in the relative topology of $\text{sp}(K)$.

▌PROOF If $K = \{0\}$, then our statement is trivially satisfied. Otherwise, let m be the largest integer for which there exists an independent subset $\{y_1, ..., y_m\}$ of K. Then $m \geqslant 1$ and, by I.3.1, $m \leqslant n$. It is easy to see that

$$\text{sp}(K) = \left\{ \sum_{j=1}^{k} \alpha^j x_j \mid k \in \mathbb{N},\ \alpha^j \in \mathbb{R},\ x_j \in K \right\}$$

and therefore $\{y_1, ..., y_m\}$ is a Hamel basis for $\text{sp}(K)$. Thus, by I.3.4, the mapping

$$\alpha \triangleq (\alpha^1, ..., \alpha^m) \to \mathscr{H}(\alpha) \triangleq \sum_{j=1}^{m} \alpha^j y_j : \mathbb{R}^m \to \text{sp}(K)$$

is a homeomorphism. It follows that the set $\mathscr{H}(\mathscr{T}_m)$ has a nonempty interior in the relative topology of $\text{sp}(K)$. By I.6.1, $\mathscr{H}(\mathscr{T}_m) = \text{co}(\{0, y_1, ..., y_m\}) \subset K$, showing that K has a nonempty interior in the relative topology of $\text{sp}(K)$.

QED

I.6.10 ***Theorem*** Let $n \in \mathbb{N}$ and let K and M be nonempty disjoint convex subsets of $\mathbb{R}^n$. Then there exists a continuous linear functional on $\mathbb{R}^n$ that separates K and M.

▌PROOF Let $k \in K$ and $m \in M$. A linear functional l separates K and M if and only if it separates $K - M - (k - m)$ and $\{m - k\}$. We may therefore assume that $0 \in K$ and $M = \{y\}$.

If $y \notin \text{sp}(K)$, let $V \triangleq \text{sp}(K \cup \{y\})$. Then every $v \in V$ can be represented as $k + \beta y$, with $k \in \text{sp}(K)$, and this representation is unique since $k + \beta y = 0$ implies $\beta = 0$. We can therefore define a linear functional l_1 on V by the relation $l_1(k + \beta y) \triangleq \beta$, and we have $l_1(k) = 0$ $(k \in K)$ and $l_1(y) = 1$. By I.3.6, l_1 is continuous and therefore, by the Hahn–Banach theorem I.3.8, we can extend l_1 to a continuous linear functional l on $\mathbb{R}^n$. Since $l(k) = l_1(k) = 0$ $(k \in K)$ and $l(y) = l_1(y) = 1$, we conclude that l separates K and $\{y\}$.

If $y \in \text{sp}(K)$, then we observe that, by I.6.9, K has a nonempty interior in the relative topology of $\text{sp}(K)$ and, by I.6.7, there exists a linear functional l on $\text{sp}(K)$ that separates K and $\{y\}$. As before, we can extend l to $\mathbb{R}^n$. QED

I.6.11 ***Theorem*** Let $n \in \mathbb{N}$ and K be a convex subset of $\mathbb{R}^n$. Then for every $\bar{x} \in \partial K$ there exists an inward normal to $\bar{K}$ at $\bar{x}$.

▌PROOF If $K = \{\bar{x}\}$, then any nonzero linear functional on $\mathbb{R}^n$ is an inward normal to $\bar{K}$ at $\bar{x}$. We assume, therefore, that $K \neq \{\bar{x}\}$. By I.6.9, $\bar{K}$ has a nonempty interior in the relative topology of $\text{sp}(\bar{K})$. By I.6.8, there exists a linear functional l_1 on $\text{sp}(\bar{K})$ that is an inward normal to $\bar{K}$ at $\bar{x}$. As in I.6.10, we apply the Hahn–Banach theorem I.3.8 to extend l_1 to all of $\mathbb{R}^n$. QED

I.6.12 ***Theorem*** Let $n \in \mathbb{N}$. Any two bounded convex bodies in $\mathbb{R}^n$ are homeomorphic. In particular, any bounded convex body in $\mathbb{R}^n$ is homeomorphic to $\mathscr{T}_n$.

▌PROOF Since $\mathscr{T}_n$ is a convex body in $\mathbb{R}^n$, it suffices to prove the first part of our statement.

Let K_1 and K_2 be bounded convex bodies in $\mathbb{R}^n$. We may assume that $0 \in K_1^\circ \cap K_2^\circ$ (otherwise replacing K_1 and K_2 by the homeomorphic sets $K_1 - k_1$, $K_2 - k_2$, where $k_1 \in K_1^\circ$, $k_2 \in K_2^\circ$). Let ϕ_i be the gauge function of K_i $(i = 1, 2)$. Since K_1 and K_2 are bounded, we can verify that $\phi_1(x) \neq 0$ and $\phi_2(x) \neq 0$ if $x \neq 0$. We can therefore define a mapping $f: K_2 \to \mathbb{R}^n$ by

$$f(x) \triangleq \frac{\phi_2(x)}{\phi_1(x)}\, x \quad (x \in K_2 \sim \{0\}), \qquad f(0) \triangleq 0.$$

We shall now apply the properties of ϕ_1 and ϕ_2 described in I.6.4. We observe that

$$\phi_1(f(x)) = \frac{\phi_2(x)}{\phi_1(x)}\, \phi_1(x) = \phi_2(x) \qquad \text{for} \quad x \neq 0$$

and

$$\phi_1(f(0)) = \phi_1(0) = \phi_2(0) = 0.$$

Thus $f(K_2) \subset K_1$. Conversely, if $0 \neq y \in K_1$ and $c \triangleq \phi_1(y)/\phi_2(y)$, then $f(cy) = y$ and $\phi_2(cy) = \phi_1(y) \leqslant 1$; hence $cy \in K_2$ and $y \in f(K_2)$. If $f(x) = f(y)$, then either $x = y = 0$, or $x \neq 0$, $y \neq 0$ and $y = \alpha x$ for

$$\alpha = \frac{\phi_2(x)}{\phi_1(x)} \frac{\phi_1(y)}{\phi_2(y)} = \frac{\phi_2(x)}{\phi_1(x)} \frac{\phi_1(\alpha x)}{\phi_2(\alpha x)} = 1.$$

Thus $f: K_2 \to K_1$ is bijective.

We next show that f is continuous. Since ϕ_1 and ϕ_2 are both continuous, it is clear that f is continuous at every $x \neq 0$. Now let $\lim_j x_j = 0$, $x_j \neq 0$, and assume that $\lim_j[\phi_1(x_j)/\phi_2(x_j)] = 0$. The sequence $(y_j) \triangleq ((1/|x_j|)\, x_j)$ in $\mathbb{R}^n$ is bounded and, by I.3.5 and I.2.5, has a subsequence $(y_j)_{j\in J}$ converging to some $\bar{y}$, with $|\bar{y}| = 1$. Thus

$$\lim_{j\in J} \frac{\phi_1(x_j)}{\phi_2(x_j)} = \lim_{j\in J} \frac{\phi_1(y_j)}{\phi_2(y_j)} = \frac{\phi_1(\bar{y})}{\phi_2(\bar{y})} \neq 0,$$

contrary to assumption. It follows that $\lim_{x\to 0} f(x) = 0$ and thus f is continuous at 0; hence on K_2. Finally, we observe that $f^{-1}(y) = [\phi_1(y)/\phi_2(y)]\, y$ for all $y \in K_1$ and the same argument shows that f^{-1} is continuous. QED

I.6.13 ***Theorem*** Let (S, Σ, μ) be a probability measure space, $\mathscr{X}$ a separable Banach space, C a closed convex subset of $\mathscr{X}$, and $f: S \to C$ μ-integrable. Then

$$\int f(s)\, \mu(ds) \in C.$$

▌PROOF Assume, by way of contradiction, that

$$a \triangleq \int f(s)\, \mu(ds) \notin C.$$

Then there exists $\alpha > 0$ such that $S(a, \alpha)$ and C are disjoint. It follows, by I.6.7, that there exists a (nonzero) $l \in \mathscr{X}^*$ that separates $S(a, \alpha)$ and C. We may assume that $lx \leqslant ly$ for all $x \in S(a, \alpha)$ and $y \in C$, otherwise replacing l by $-l$. It follows that

$$\sup_{|\xi|\leqslant 1} l(a + \alpha\xi) = l(a) + \alpha\, |\, l\, | \leqslant l(f(s)) \qquad (s \in S);$$

hence, integrating both sides over S with respect to μ, we obtain

$$\text{(1)} \qquad l(a) + \alpha\, |\, l\, | \leqslant \int l(f(s))\, \mu(ds).$$

By I.4.34(8), $\int l(f(s))\, \mu(ds) = l \int f(s)\, \mu(ds) = l(a)$ and (1) yields, therefore,

$$l(a) + \alpha\, |\, l\, | \leqslant l(a),$$

a contradiction. This shows that $\int f(s)\, \mu(ds) \in C$. QED

I.6.14 ***Theorem*** Let S be a compact metric space, $n \in \mathbb{N}$,

$$f \triangleq (f^1, \ldots, f^n) \in C(S, \mathbb{R}^n)$$

and rpm$(S, n + 1)$ the subset of rpm(S) whose elements are measures with supports containing at most $n + 1$ points of S. Then

$$\text{co}(f(S)) = \left\{ \int f(s)\, \mu(ds) | \, \mu \in \text{rpm}(S) \right\} = \left\{ \int f(s)\, \mu(ds) | \, \mu \in \text{rpm}(S, n + 1) \right\}.$$

▌PROOF Let $C_0 \triangleq \text{co}(f(S))$, $C_1 \triangleq \{\int f(s)\, \mu(ds) \mid \mu \in \text{rpm}(S)\}$, and $C_2 \triangleq \{\int f(s)\, \mu(ds) \mid \mu \in \text{rpm}(S, n + 1)\}$. By Caratheodory's theorem I.6.2, for every $c \in C_0$ there exists points $s_0, s_1, \ldots, s_n \in S$ and $(\theta^0, \ldots, \theta^n) \in \mathscr{T}_n'$ such that $c = \sum_{j=0}^{n} \theta^j f(s_j)$. We consider the measure $\mu : \Sigma_{\text{Borel}}(S) \to \mathbb{R}$ with support on $\{s_0, s_1, \ldots, s_n\}$ and such that

$$\mu(\{s_j\}) = \theta^j \qquad (j = 0, 1, \ldots, n),$$

and observe that $\mu \in \text{rpm}(S, n + 1)$ and $c = \int f(s)\, \mu(ds)$, Thus $C_0 \subset C_2$. It is obvious that $C_2 \subset C_1$ and it remains therefore to show that $C_1 \subset C_0$. By I.2.9, I.2.6, and I.6.2,

$$C_0 \triangleq \text{co}(f(S)) = \left\{ \sum_{j=0}^{n} \theta^j x_j \mid (\theta^0, \ldots, \theta^n) \in \mathscr{T}_n',\ x_j \in f(S) \right\}$$

is compact. It follows, by I.6.13, that $\int f(s)\, \mu(ds) \in \text{co}(f(S))$ for every $\mu \in \text{rpm}(S)$; hence $C_1 \subset C_0$. QED

I.7 Measurable Set-Valued Mappings

Let T and X be compact metric spaces, $\mu \in \text{frm}^+(T)$, $\Sigma \triangleq \Sigma_{\text{Borel}}(T)$, and $\mathscr{P}'(X) \triangleq \{A \mid A \subset X, A \neq \varnothing\}$ the class of all nonempty subsets of X. We refer to a mapping on some subset F of T to $\mathscr{P}'(X)$ as a *set-valued mapping on F into X*. If $A \in \mathscr{P}'(X)$ and $\epsilon > 0$, we write $S(A, \epsilon)$ for $\{x \in X \mid d[x, A] < \epsilon\}$ and $S^F(A, \epsilon)$ for $\{x \in X \mid d[x, A] \leqslant \epsilon\}$. We say that a mapping $\Gamma : F \to \mathscr{P}'(X)$ is *upper semicontinuous at* $\bar{t} \in F$ if for every $\epsilon > 0$ there exists $\eta > 0$ such that $\Gamma(t) \subset S(\Gamma(\bar{t}), \epsilon)$ provided $d(\bar{t}, t) < \eta$. A mapping $\Gamma : F \to \mathscr{P}'(X)$ is *upper semicontinuous* if it is upper semicontinuous at $\bar{t}$ for every $\bar{t} \in F$.

Since X is a compact metric space, it is bounded. It follows that we can define a function $\delta : \mathscr{P}'(X) \times \mathscr{P}'(X) \to \mathbb{R}$ (the *Hausdorff semimetric*) by

$$\delta(A, B) \triangleq \inf\{\alpha > 0 \mid A \subset S(B, \alpha), \qquad B \subset S(A, \alpha)\}.$$

It is easy to verify that $\delta(A, B) \geqslant 0$, $\delta(A, B) = \delta(B, A)$, and $\delta(A, C) \leqslant \delta(A, B) + \delta(B, C)$ for all $A, B, C \in \mathscr{P}'(X)$. We observe that the relation $\delta(A, B) = 0$ is equivalent to the relation $\bar{A} = \bar{B}$. Thus the Hausdorff semimetric is *not* a distance function. We define $\mathscr{P}'(X)$ as a topological space by choosing the smallest topology containing the sets $\{B \in \mathscr{P}'(X) \mid \delta(A, B) < \epsilon\}$ for $A \in \mathscr{P}'(X)$ and $\epsilon > 0$.

We denote by $\mathscr{K}(X)$ the class of all nonempty closed (and therefore compact) subsets of X. The restriction of δ to $\mathscr{K}(X) \times \mathscr{K}(X)$ is a distance function because $\delta(A, B) = 0$ implies $A = B$ for compact sets. The function $\delta \mid \mathscr{K}(X) \times \mathscr{K}(X)$ is referred to as the *Hausdorff metric* and defines $\mathscr{K}(X)$ as a metric space. The metric topology of $\mathscr{K}(X)$ clearly coincides with the relativization to $\mathscr{K}(X)$ of the topology of $\mathscr{P}'(X)$.

We observe that if $\Gamma : F \to \mathscr{K}(X)$ is continuous, then Γ is also upper semicontinuous.

If Γ is a function on $F \subset T$ to $\mathscr{P}'(X)$, we write

$$G(\Gamma) \triangleq \{(t, x) \in F \times X \mid x \in \Gamma(t)\}.$$

I.7.1 ***Theorem*** The topological space $\mathscr{P}'(X)$ and the metric space $\mathscr{K}(X)$ are separable and each has an at most countable base of topology.

▌PROOF We first observe that there exists a one-to-one correspondence between subbases of $\mathscr{P}'(X)$ and $\mathscr{K}(X)$, the set

$$G(A, \epsilon) \triangleq \{B \in \mathscr{P}'(X) \mid \delta(A, B) < \epsilon\}$$

coinciding with $G(A', \epsilon)$ if $\bar{A}' = \bar{A}$ and corresponding to the set $\{B \in \mathscr{K}(X) \mid \delta(\bar{A}, B) < \epsilon\}$; furthermore, any set dense in $\mathscr{K}(X)$ is also dense in $\mathscr{P}'(X)$. It suffices therefore to prove our contention for $\mathscr{K}(X)$ alone.

Let $X_\infty \triangleq \{x_1, x_2, \ldots\}$ be a dense subset of X (I.2.17), $A \in \mathscr{K}(X)$ and $\epsilon > 0$. By I.2.5, A is totally bounded and there exists a finite subset $\{a_1, \ldots, a_k\}$ of A such that the collection $\{S(a_i, \epsilon/2) \mid i = 1, \ldots, k\}$ is an open covering of A. We can choose $b_i \in X_\infty$ so that $d(a_i, b_i) < \epsilon/2$ for $i = 1, \ldots, k$, and conclude that $\{S(b_i, \epsilon) \mid i = 1, \ldots, k\}$ is a covering of A. It follows that $\delta(A, \{b_1, \ldots, b_k\}) \leqslant \epsilon$, showing that the at most countable collection $\mathscr{K}_\infty$ of all finite subsets of X_∞ is dense in $\mathscr{K}(X)$, and $\mathscr{K}(X)$ is separable. We also observe that the at most denumerable collection

$$\{\{A \in \mathscr{K}(X) \mid \delta(A, B) < 1/i\} \mid B \in \mathscr{K}_\infty, i \in \mathbb{N}\}$$

is a base of the topology of $\mathscr{K}(X)$. QED

I.7.2 ***Theorem*** (Berge) Let F be a closed subset of T and $\Gamma : F \to \mathscr{P}'(X)$. Then $G(\Gamma)$ is a closed subset of $F \times X$ if and only if $\Gamma(t) \in \mathscr{K}(X)$ $(t \in F)$ and Γ is upper semicontinuous.

▌PROOF Let $G(\Gamma)$ be closed, (t_j) a sequence in F converging to $\bar{t}$, and $\epsilon > 0$. Then clearly $\Gamma(t) \in \mathscr{K}(X)$ $(t \in F)$. We shall show that there exists $j_0 \in \mathbb{N}$ such that $\Gamma(t_j) \subset S(\Gamma(\bar{t}), \epsilon)$ for all $j \geqslant j_0$. Indeed, assume the contrary. Then there exists a sequence $J \subset (1, 2,...)$ and points $x_j \in \Gamma(t_j)$ $(j \in J)$ such that

$$(1) \qquad d[\Gamma(\bar{t}), x_j] \geqslant \epsilon.$$

Since X is sequentially compact (I.2.5), we may assume that J was chosen so that $\lim_{j\in J} x_j = \bar{x} \in X$. It follows that $(\bar{t}, \bar{x}) = \lim_{j\in J}(t_j, x_j) \in G(\Gamma)$; hence $\bar{x} \in \Gamma(\bar{t})$, contradicting (1).

Conversely, let $\Gamma: F \to \mathscr{K}(X)$ be upper semicontinuous and $((t_j, x_j))$ a sequence in $G(\Gamma)$ converging to some $(\bar{t}, \bar{x}) \in F \times X$. Then $x_j \in \Gamma(t_j)$ for all $j \in \mathbb{N}$ and, for every $\epsilon > 0$, $x_j \in S^F(\Gamma(\bar{t}), \epsilon)$ for all sufficiently large j. It follows, since $\Gamma(\bar{t})$ is compact, that $\bar{x} \in S^F(\Gamma(\bar{t}), \epsilon)$ for all $\epsilon > 0$; hence $\bar{x} \in \Gamma(\bar{t})$ and $(\bar{t}, \bar{x}) \in G(\Gamma)$. QED

I.7.3 ***Theorem*** (Castaing) The topology of $\mathscr{K}(X)$ has a subbase consisting of sets of the form

$$\mathscr{I}(G) \triangleq \{K \in \mathscr{K}(X) \mid K \cap G \neq \varnothing\} \qquad \text{or} \qquad \mathscr{J}(G) \triangleq \{K \in \mathscr{K}(X) \mid K \subset G\},$$

where G is any open subset of X.

▌PROOF Let $G \subset X$ be open. We first show that $\mathscr{I}(G)$ is open. Indeed, let $K \in \mathscr{I}(G)$ and $x \in K \cap G$.Then there exists $\epsilon > 0$ such that $S(x, \epsilon) \subset G$. If $\delta(K', K) < \epsilon$, then $K \subset S(K', \epsilon)$; hence $d(x, x') < \epsilon$ for some $x' \in K'$, implying that $K' \cap G \neq \varnothing$ and $K' \in \mathscr{I}(G)$. Thus every $K \in \mathscr{I}(G)$ is an interior point and $\mathscr{I}(G)$ is open.

Now let $K \subset G$. Since K is compact and G open, the sets K and $X \sim G$ are disjoint and compact and there exists $\epsilon > 0$ such that $S(K, \epsilon) \subset G$. If $\delta(K', K) < \epsilon$, then $K' \subset S(K, \epsilon) \subset G$, showing that K is an interior point of $\mathscr{J}(G)$ and $\mathscr{J}(G)$ is open.

Now let $K \in \mathscr{K}(X)$ and $\epsilon > 0$. By I.2.5, there exist $m \in \mathbb{N}$ and $x_1, x_2, ..., x_m \in K$ such that the open sets $G_i \triangleq S(x_i, \epsilon/2)$ $(i = 1,..., m)$ cover K. The open subset $\mathscr{H} \triangleq \mathscr{J}(S(K, \epsilon)) \cap \mathscr{I}(G_1) \cap \cdots \cap \mathscr{I}(G_m)$ of $\mathscr{K}(X)$ contains the point $K \in \mathscr{K}(X)$. Furthermore, if $K' \in \mathscr{H}$, then $K' \subset S(K, \epsilon)$ and $K' \cap G_i \neq \varnothing$ $(i = 1, 2,..., m)$. If we choose some points $g_i \in K' \cap G_i$, then

$$S(K', \epsilon) \supset \bigcup_{i=1}^{m} S(g_i, \epsilon) \supset \bigcup_{i=1}^{m} S(x_i, \epsilon/2) \supset K;$$

hence $\delta(K', K) \leqslant \epsilon$. This shows that the collection

$$\{\mathscr{I}(G) \mid G \text{ open}\} \cup \{\mathscr{J}(G) \mid G \text{ open}\}$$

is a subbase of the topology of $\mathscr{K}(X)$. QED

I.7.4 ***Theorem*** (Castaing) Let F be a closed subset of T, $\Gamma : F \to \mathscr{K}(X)$, and $\Gamma^{-}A \triangleq \{t \in F \mid \Gamma(t) \cap A \neq \varnothing\}$ for $A \subset X$. Then the following statements are equivalent:

(1) Γ is μ-measurable,
(2) $\Gamma^{-}A$ is μ-measurable for every closed $A \subset X$, and
(3) $\Gamma^{-}A$ is μ-measurable for every open $A \subset X$.

PROOF *Step* 1 Let Γ be μ-measurable. Then, by I.7.1 and Lusin's theorem I.4.19, for every $\epsilon > 0$ there exists a closed subset F_ϵ of F such that $\mu(F \sim F_\epsilon) \leqslant \epsilon$ and $\Gamma \mid F_\epsilon$ is continuous. Now let $A \subset X$ be open, $\epsilon > 0$, and $\bar{t} \in F_\epsilon \cap \Gamma^{-}A$. Since A is open, there exist $\bar{x} \in \Gamma(\bar{t})$ and $h > 0$ such that $S(\bar{x}, h) \subset A$. Since $\Gamma \mid F_\epsilon$ is continuous, there exists $\eta > 0$ such that, for $t \in F_\epsilon$ and $d(\bar{t}, t) < \eta$, we have $\Gamma(\bar{t}) \subset S(\Gamma(t), h)$; hence there exists some $x \in \Gamma(t)$ such that $x \in S(\bar{x}, h) \subset A$. Thus $\Gamma^{-}A \cap F_\epsilon$ is relatively open in the closed set F_ϵ and is therefore μ-measurable. Since $\mu(F \sim \bigcup_{j=1}^{\infty} F_{1/j}) = 0$, it follows that $\Gamma^{-}A$ is μ-measurable. Thus (1) implies (3).

Step 2 Now assume that statement (3) is valid. If we assume that $A \subset X$ is closed [and recall that $\Gamma(t)$ is closed], then $A = \bigcap_{j=1}^{\infty} S(A, 1/j)$, and $\Gamma(t) \cap A \neq \varnothing$ if and only if $\Gamma(t) \cap S(A, 1/j) \neq \varnothing$ for all $j \in \mathbb{N}$. Then $\Gamma^{-}A = \bigcap_{j=1}^{\infty} \Gamma^{-}S(A, 1/j)$ and, by (3), $\Gamma^{-}A$ is a denumerable intersection of μ-measurable sets. Thus (3) implies (2).

We now show that (3) implies (1). By I.7.1 and I.7.3, Γ is μ-measurable if $\Gamma^{-1}(B)$ is μ-measurable for every B of the form $\mathscr{I}(G)$ or $\mathscr{J}(G)$, with G open. We have

$$\Gamma^{-1}(\mathscr{I}(G)) = \{t \in F \mid \Gamma(t) \cap G \neq \varnothing\} \triangleq \Gamma^{-}G$$

and

$$\begin{aligned}\Gamma^{-1}(\mathscr{J}(G)) &= \{t \in F \mid \Gamma(t) \subset G\} = F \sim \{t \in F \mid \Gamma(t) \cap (X \sim G) \neq \varnothing\} \\ &= F \sim \Gamma^{-}(X \sim G).\end{aligned}$$

Thus $\Gamma^{-1}(\mathscr{I}(G))$ is μ-measurable because (3) is assumed valid, and $\Gamma^{-1}(\mathscr{J}(G))$ is μ-measurable because (3) implies (2). It follows that (3) implies (1).

Step 3 It remains to show that (2) implies (3). Let $A \subset X$ be open. Then for every $x \in A$ there exists $\epsilon_x > 0$ such that $S^F(x, \epsilon_x) \subset A$, and the collection $\{S(x, \epsilon_x) \mid x \in A\}$ is an open covering of A. By Lindelöf's theorem I.2.1 and I.7.1, there exists a subset $\{a_1, a_2, \ldots\}$ of A such that

$$A \subset \bigcup_{i=1}^{\infty} S(a_i, \epsilon_{a_i}) \subset \bigcup_{i=1}^{\infty} S^F(a_i, \epsilon_{a_i}) \subset A;$$

hence

$$\Gamma^{-}A = \bigcup_{i=1}^{\infty} \Gamma^{-}S^{F}(a_i, \epsilon_{a_i}),$$

and this last set is μ-measurable if $\Gamma^{-}S^{F}(a_i, \epsilon_{a_i})$ is μ-measurable for each i. Thus (2) implies (3). QED

I.7.5 ***Lemma*** Let F be a closed subset of T, $J \subset \mathbb{N}$, $\Gamma_j : F \to \mathscr{K}(X)$ $(j \in J)$ either upper semicontinuous or μ-measurable, and

$$\Gamma_J(t) \triangleq \bigcap_{j \in J} \Gamma_j(t) \neq \varnothing \qquad (t \in F).$$

Then $\Gamma_J : F \to \mathscr{K}(X)$ is μ-measurable. Furthermore, a function

$$t \to \Delta(t) \triangleq \prod_{j=1}^{k} \Delta_j(t) : F \to \mathscr{K}(X^k)$$

is μ-measurable if $k \in \mathbb{N}$ and each $\Delta_j : F \to \mathscr{K}(X)$ is μ-measurable. In particular, an upper semicontinuous $\Omega : F \to \mathscr{K}(X)$ is μ-measurable.

▌PROOF We shall first assume that each Γ_j is upper semicontinuous. Let $A \subset X$ be closed, $\Gamma_A(t) \triangleq A$ $(t \in F)$, and $\tilde{\Gamma}(t) \triangleq \Gamma_J(t) \cap \Gamma_A(t) (t \in F)$. Then, by I.7.2, $G(\Gamma_j)$ is closed for each $j \in J$; hence $G(\tilde{\Gamma}) = \bigcap_{j \in J} G(\Gamma_j) \cap G(\Gamma_A)$ is closed and $\Gamma_J^{-}A = \{t \in F \mid \tilde{\Gamma}(t) \neq \varnothing\} = \mathrm{pr}_F\, G(\tilde{\Gamma})$ also closed. Thus $\Gamma_J^{-}A$ is μ-measurable for every closed $A \subset X$ and our first conclusion follows from I.7.4.

If Γ_j is μ-measurable for $j \in J$, then, by I.7.1 and Lusin's theorem I.4.19, for every $\epsilon > 0$ there exists a closed set $F_\epsilon^j \subset F$ such that $\Gamma_j \mid F_\epsilon^j$ is continuous and $\mu(F \sim F_\epsilon^j) \leqslant 2^{-j}\epsilon$. Then our previous argument, applied to $F_\epsilon \triangleq \bigcap_{j \in J} F_\epsilon^j$ shows that $\Gamma_J \mid F_\epsilon$ is μ-measurable for each $\epsilon > 0$. Since $\mu(F \sim F_\epsilon) \leqslant \epsilon \sum_{j \in J} 2^{-j} \leqslant \epsilon$, it follows that $\mu(F \sim F') = 0$ for $F' \triangleq \bigcup_{i=1}^{\infty} F_{1/i}$, and Γ_J is μ-measurable on F' and therefore also on F.

Finally, let $\Delta_j : F \to \mathscr{K}(X)$ $(j = 1,..., k)$ be μ-measurable, and set $\Delta_j'(t) \triangleq \prod_{i=1}^{k} A_i^j(t)$, where $A_i^j(t) \triangleq X$ $(i \neq j)$ and $A_j^j(t) \triangleq \Delta_j(t)$. If $X_j \triangleq X$ $(j = 1,..., k)$, G is an open subset of $X^k \triangleq \prod_{j=1}^{k} X_j$, and $G_j \triangleq \mathrm{pr}_j\, G$, then, by I.7.4(3), the set

$$\{t \in F \mid \Delta_j'(t) \cap G \neq \varnothing\} = \{t \in F \mid \Delta_j(t) \cap G_j \neq \varnothing\}$$

is μ-measurable and therefore Δ_j' is μ-measurable. It follows, by our previous argument, that the function $t \to \prod_{j \in J} \Delta_j(t) = \bigcap_{j \in J} \Delta_j'(t) : F \to \mathscr{K}(X^k)$ is μ-measurable. QED

I.7.6 ***Theorem*** (Castaing) Let $\Gamma : T \to \mathscr{K}(X)$ be μ-measurable, Y a complete separable metric space, $h : T \times X \to Y$ and $g : T \to Y$ such that

$h(\cdot, x)$ and $g(\cdot)$ are μ-measurable and $h(t, \cdot)$ is continuous for all $(t, x) \in T \times X$, and $\tilde{\Gamma}(t) \triangleq \{x \in \Gamma(t) \mid h(t, x) = g(t)\} \neq \varnothing$ $(t \in T)$. Then $\tilde{\Gamma}(t) \in \mathscr{K}(X)$ $(t \in T)$ and $\tilde{\Gamma} : T \to \mathscr{K}(X)$ is μ-measurable.

▌PROOF Since $h(t, \cdot)$ is continuous for each $t \in T$, the set

$$H_t \triangleq \{x \in X \mid h(t, x) = g(t)\}$$

is closed and so is

$$\tilde{\Gamma}(t) = H_t \cap \Gamma(t).$$

By Lusin's theorem I.4.19 and I.5.26, for every $\epsilon > 0$ there exists a closed subset F_ϵ of T such that $\mu(T \sim F_\epsilon) \leqslant \epsilon$ and $\Gamma \mid F_\epsilon$, $g \mid F_\epsilon$, and $h \mid F_\epsilon \times X$ are continuous. Thus the set $D \triangleq \{(t, x) \in F_\epsilon \times X \mid h(t, x) = g(t)\}$ is closed and so is, by I.7.2, the set $G_\epsilon(\Gamma) \triangleq \{(t, x) \in F_\epsilon \times X \mid x \in \Gamma(t)\}$. It follows that the set

$$G_\epsilon(\tilde{\Gamma}) \triangleq \{(t, x) \in F_\epsilon \times X \mid x \in \Gamma(t), h(t, x) = g(t)\} = D \cap G_\epsilon(\Gamma)$$

is closed and, by I.7.2, $\tilde{\Gamma} \mid F_\epsilon$ is upper semicontinuous; hence, by I.7.5, $\tilde{\Gamma} \mid F_\epsilon$ is μ-measurable. Since ϵ is arbitrary, we conclude that $\tilde{\Gamma}$ is μ-measurable. QED

If $\Gamma : T \to \mathscr{P}'(X)$, then we say that a function $\xi : T \to X$ is a *selection* of Γ (and Γ *has a selection* ξ) if $\xi \in \prod_{t \in T} \Gamma(t)$, that is, $\xi(t) \in \Gamma(t)$ for all $t \in T$.

I.7.7 ***Theorem*** (von Neumann–Aumann–Castaing) Let $\Gamma : T \to \mathscr{K}(X)$ be μ-measurable. Then Γ has a μ-measurable selection.

▌PROOF Let $\{x_1, x_2, ...\}$ be a dense subset of X. For each $t \in T$, set $\Gamma_0(t) \triangleq \Gamma(t)$ and, recursively,

$$(1) \quad \Gamma_{i+1}(t) \triangleq \{x \in \Gamma_i(t) \mid d(x, x_{i+1}) = d[\Gamma_i(t), x_{i+1}]\} \qquad (i = 0, 1, 2, ...).$$

Since $\Gamma(t)$ is compact and nonempty, it follows that each $\Gamma_i(t)$ is also compact and nonempty.

By Lusin's theorem I.4.19, for every $\epsilon > 0$ there exists a closed $F_\epsilon^0 \subset T$ such that $\mu(T \sim F_\epsilon^0) \leqslant \epsilon/2$ and $\Gamma_0 \mid F_\epsilon^0$ is continuous. Now assume, for purposes of induction, that $i \in \{0, 1, 2, ...\}$ and there exist closed sets $F_\epsilon^i \subset F_\epsilon^{i-1} \subset \cdots \subset F_\epsilon^0$ such that $\mu(T \sim F_\epsilon^k) \leqslant \epsilon \sum_{j=0}^{k} 2^{-j-1}$ $(k = 0, 1, ..., i)$ and $\Gamma_k \mid F_\epsilon^k$ is continuous. Then the functions $t \to d[\Gamma_i(t), x_{i+1}] : F_\epsilon^i \to \mathbb{R}$ and $x \to d(x, x_{i+1}) : X \to \mathbb{R}$ are continuous, and it follows from (1) and I.7.6 that $\Gamma_{i+1} \mid F_\epsilon^i$ is μ-measurable. By Lusin's theorem I.4.19 there exists a closed $F_\epsilon^{i+1} \subset F_\epsilon^i$ such that $\mu(F_\epsilon^i \sim F_\epsilon^{i+1}) \leqslant 2^{-i-2}\epsilon$ [hence $\mu(T \sim F_\epsilon^{i+1}) \leqslant \sum_{j=0}^{i+1} 2^{-j-1}\epsilon$] and $\Gamma_{i+1} \mid F_\epsilon^{i+1}$ is continuous. This completes the induction.

We observe that the set $F_\epsilon \triangleq \bigcap_{i=0}^{\infty} F_\epsilon^i$ is closed, $\mu(T \sim F_\epsilon) \leqslant \epsilon$, and $\Gamma_i \mid F_\epsilon$ are continuous for $i = 0, 1, 2, ...$. Since ϵ can be chosen arbitrarily, this shows that Γ_i are μ-measurable on T and so is, by I.7.5, $t \to \Gamma_P(t) \triangleq \bigcap_{i=0}^{\infty} \Gamma_i(t) \neq \varnothing$. If $\xi_1(t), \xi_2(t) \in \Gamma_P(t)$ for some $t \in T$, then $d(\xi_1(t), x_i) = d(\xi_2(t), x_i)$ for $i \in \mathbb{N}$. Since $\{x_1, x_2, ...\}$ is dense in X, it follows that $\xi_1(t) = \xi_2(t)$. Thus $\Gamma_P(t)$ consists of a single point $\xi(t)$ and, since $t \to \{\xi(t)\} : T \to \mathscr{K}(X)$ is μ-measurable, so is $t \to \xi(t) : T \to X$. QED

I.7.8 ***Theorem*** (Castaing) Let $\Gamma : T \to \mathscr{K}(X)$ be μ-measurable. Then there exists a denumerable collection $\{\xi_1, \xi_2, ...\}$ of μ-measurable selections of Γ such that the set $\{\xi_1(t), \xi_2(t), ...\}$ is dense in $\Gamma(t)$ for all $t \in T$.

▌PROOF Let $\{x_1, x_2, ...\}$ be a dense subset of X and, for each $i, j \in \mathbb{N}$, let $T_{i,j} \triangleq \{t \in T \mid \Gamma(t) \cap S^F(x_i, 1/j) \neq \varnothing\}$ and let $\Gamma_{i,j} : T \to \mathscr{K}(X)$ be defined by

$$\Gamma_{i,j}(t) \triangleq \begin{cases} \Gamma(t) \cap S^F(x_i, 1/j) & \text{if} \quad t \in T_{i,j}, \\ \Gamma(t) & \text{if} \quad t \in T \sim T_{i,j}. \end{cases}$$

By I.7.4, each $T_{i,j}$ is μ-measurable and, by I.7.5, each $\Gamma_{i,j}$ is μ-measurable [since the mapping $t \to S^F(x_i, 1/j)$ for $t \in T_{i,j}$ and $t \to X$ for $t \in T \sim T_{i,j}$ is, like $T_{i,j}$, μ-measurable]. By I.7.7, each $\Gamma_{i,j}$ has a μ-measurable selection $\xi_{i,j}$. For each choice of $t \in T$, $x \in \Gamma(t)$, and $j \in \mathbb{N}$, there exists $i \in \mathbb{N}$ such that $d(x, x_i) \leqslant 1/j$; hence $\xi_{i,j}(t) \in \Gamma_{i,j}(t) \subset S^F(x_i, 1/j)$, showing that $d(x, \xi_{i,j}(t)) \leqslant 2/j$. This shows that the denumerable set $\{\xi_{i,j}(t) \mid i, j \in \mathbb{N}\}$ is dense in $\Gamma(t)$. QED

I.7.9 ***Theorem*** Let $\Gamma : T \to \mathscr{P}'(X)$ be μ-measurable and either

(1) $\Gamma(t)$ is closed for all $t \in T$, or
(2) $\Gamma(t) \subset \overline{\Gamma(t)^\circ}$ $(t \in T)$ and for every $\epsilon > 0$ there exists a closed $T_\epsilon \subset T$ such that $\mu(T \sim T_\epsilon) \leqslant \epsilon$ and the set $G_\epsilon(\Gamma^\circ) \triangleq \{(t, x) \in T_\epsilon \times X \mid x \in \Gamma(t)^\circ\}$ is an open subset of $T_\epsilon \times X$.

Then there exists a denumerable collection $\{\xi_1, \xi_2, ...\}$ of μ-measurable selections of Γ such that the set $\{\xi_1(t), \xi_2(t), ...\}$ is dense in $\Gamma(t)$ for μ-a.a. $t \in T$.

▌PROOF If (1) is valid, then the assertion is a restatement of I.7.8. We assume, therefore, that (2) is valid and let $\epsilon > 0$ be fixed. Since $G_\epsilon(\Gamma^\circ)$ is open in $T_\epsilon \times X$ and $\mathrm{pr}_{T_\epsilon} G_\epsilon(\Gamma^\circ) = T_\epsilon$, for every $\bar{t} \in T_\epsilon$ there exist $x(\bar{t}) \in \Gamma(\bar{t})^\circ$ and a neighborhood $N(\bar{t})$ of $\bar{t}$ in T_ϵ such that

$$x(\bar{t}) \in \Gamma(t)^\circ \qquad [t \in N(\bar{t})].$$

Since T_ϵ is compact, a finite subcollection $\{N(\bar{t}_1), N(\bar{t}_2), ..., N(\bar{t}_k)\}$ of $\{N(\bar{t}) \mid \bar{t} \in T_\epsilon\}$ covers T_ϵ, and we set

$$\bar{\xi}(t) \triangleq x(\bar{t}_i) \qquad \left[i = 1, 2, ..., k, \quad t \in N(\bar{t}_i) \sim \bigcup_{j=1}^{i-1} N(\bar{t}_j)\right].$$

It is clear that $\bar{\xi}$ is a μ-measurable selection of $\Gamma \mid T_\epsilon$.

We next observe that if $\{\bar{t}\} \times A \subset G_\epsilon(\Gamma^\circ)$ and A is closed, then there exists a neighborhood $U(\bar{t})$ of $\bar{t}$ in T_ϵ such that $U(\bar{t}) \times A \subset G_\epsilon(\Gamma^\circ)$. Indeed, otherwise there exist sequences (t_j) in T and (a_j) in A such that $\lim_j t_j = \bar{t}$ and

$$(t_j, a_j) \notin G_\epsilon(\Gamma^\circ) \qquad (j \in \mathbb{N}). \tag{3}$$

Since A is compact, we may assume that $\lim_j a_j = \bar{a}$ for some $\bar{a} \in A$, hence $\lim_j(t_j, a_j) \in \{\bar{t}\} \times A \subset G_\epsilon(\Gamma^\circ)$, contrary to (3).

Now let $\{x_1, x_2, \ldots\}$ be a dense subset of X and, for all $i, j \in \mathbb{N}$,

$$T_{i,j}^\epsilon \triangleq \{t \in T_\epsilon \mid S^F(x_i, 1/j) \subset \Gamma(t)^\circ\}$$

and

$$\xi_{i,j}^\epsilon(t) \triangleq \begin{cases} x_i & \text{if } t \in T_{i,j}^\epsilon, \\ \bar{\xi}(t) & \text{elsewhere in } T_\epsilon. \end{cases}$$

It follows from our previous remark that $T_{i,j}^\epsilon$ is open relative to T_ϵ, hence μ-measurable on T_ϵ, and therefore $\xi_{i,j}^\epsilon$ is μ-measurable on T_ϵ. For every choice of $t \in T_\epsilon$ and $x \in \Gamma(t)^\circ$ there exists $j_0 \in \mathbb{N}$ such that $S^F(x, 2/j_0) \subset \Gamma(t)^\circ$. For each $j \geqslant j_0$ there exists $i \in \mathbb{N}$ such that $d(x_i, x) \leqslant 1/j$; hence $S^F(x_i, 1/j) \subset S^F(x, 2/j) \subset \Gamma(t)^\circ$, implying $t \in T_{i,j}^\epsilon$ and

$$d(x, \xi_{i,j}^\epsilon(t)) = d(x, x_i) \leqslant 1/j.$$

Since j can be arbitrarily large, this shows that $\{\xi_{i,j}^\epsilon(t) \mid i, j \in \mathbb{N}\}$ is dense in $\Gamma(t)^\circ$ and, by (2), in $\Gamma(t)$.

Now we let $\epsilon = 1/1, 1/2, 1/3, \ldots$ and set, for all $i, j \in \mathbb{N}$,

$$\xi_{i,j}(t) \triangleq \xi_{i,j}^{1/m}(t) \qquad \left(t \in T_{1/m} \sim \bigcup_{k=1}^{m-1} T_{1/k}, \quad m \in \mathbb{N}\right)$$

and

$$\xi_{i,j}(t) \triangleq \bar{\xi}(t) \qquad \left(t \in T \sim \bigcup_{k=1}^{\infty} T_{1/k}\right).$$

Since $\mu(\bigcup_{k=1}^\infty T_{1/k}) = \mu(T)$, it follows that each $\xi_{i,j}$ is a μ-measurable selection of Γ and $\{\xi_{i,j}(t) \mid i, j \in \mathbb{N}\}$ is dense in $\Gamma(t)$ for μ-a.a. $t \in T$. QED

I.7.10 ***Theorem*** (Filippov–Castaing) Let $\Gamma : T \to \mathscr{K}(X)$ be μ-measurable, Y a complete separable metric space, and $h : T \times X \to Y$ and $g : T \to Y$ such that $h(\cdot, x)$ and $g(\cdot)$ are μ-measurable, $h(t, \cdot)$ is continuous, and $g(t) \in h(t, \Gamma(t))$ for all $(t, x) \in T \times X$. Then there exists a μ-measurable selection ξ of Γ such that $g(t) = h(t, \xi(t))$ for all $t \in T$.

▌PROOF Since $g(t) \in h(t, \Gamma(t))$ for all $t \in T$, it follows that

$$\tilde{\Gamma}(t) \triangleq \{x \in \Gamma(t) \mid h(t, x) = g(t)\} \neq \varnothing$$

and, by I.7.6, $\tilde{\Gamma}(t) \in \mathscr{K}(X)$ for all $t \in T$ and $\tilde{\Gamma}$ is μ-measurable. It follows, by I.7.7, that $\tilde{\Gamma}$ has a μ-measurable selection ξ which clearly satisfies the relation $g(t) = h(t, \xi(t))$ for all $t \in T$. QED

Notes

In preparing this chapter I have consulted several well-known texts and, in particular, those of Dunford and Schwartz [1], Dieudonné [1], Munroe [1], and Rudin [1], as well as mimeographed notes of Fan [1] and a paper of Castaing [1]. I have attempted to use standard definitions, notation, and terminology. When a "standard" was lacking (which was very often the case), I have chosen what appeared most consistent with other conventions or most appropriate for our purposes. Thus I use the notation of Dunford and Schwartz [1] for the integral because we often deal with functions and measures depending on parameters and an expression such as $\int f(r, a)\, \sigma(t)\, (dr)$ has a clear meaning whereas other notations for this integral would either cause confusion or be inconvenient. On the other hand, I write $\Sigma_1 \otimes \Sigma_2$ for the product σ-field because the usual notation $\Sigma_1 \times \Sigma_2$ is inconsistent with the standard notation for the cartesian product of the sets Σ_1 and Σ_2. For the same reason, the symbol $d[A, x]$ is used to denote the distance from x to A in order to avoid the customary notation $d(A, x)$ which means $\{d(y, x) \mid y \in A\}$. The expression "A is a sequentially compact subset of X" is used to mean "the relative topology of A in X is sequentially compact," contrary to its use in Dunford and Schwartz [1] (but consistently with its use by Köthe [1]), because this convention establishes a certain parallelism in the use of the terms "compact," "closed," and "sequentially compact," "sequentially closed." Finally, the term "measurable" is defined in the sense most appropriate for a function on a measure space to a topological space, which is the case we are most interested in.

Practically all of the material of Chapter I is well known, and many parts of it are developed along the lines suggested in the previously mentioned sources. On the other hand, other parts of the exposition are developed independently because a different approach appeared simpler or more natural within the context in which we have placed ourselves. Thus, for example, the construction of regular Borel measures (I.4.11–I.4.15) and the derivation of the Riesz representation theorem (I.5.6–I.5.8) largely follow the lines of Dunford and Schwartz [1]. On the other hand, integration of scalar and vector-valued functions (I.4.23–I.4.36) is defined and developed without any particular exposition in mind.

I have followed to varying extents the following sources as indicated: Dunford and Schwartz [1] on I.2.1, I.2.5, I.3.8, much of I.4.A, I.4.44, I.4.45, I.5.6–I.5.8, I.5.21, and I.6.7; Munroe [1] on I.4.37, I.5.13, and I.5.19; Dieudonné [1] on I.3.13; Rudin [1] on I.4.40–I.4.42; Fan [1] on I.3.4, I.6.4, and I.6.5; R. Palais (lectures) on I.3.5; and Castaing [1] on I.7.1–I.7.4, I.7.6–I.7.8, and I.7.10. I should also mention that many results of Section I.7, while based on the approach of Castaing, have been obtained in various forms by other authors (see Castaing [1] for references), while alternative (2) of I.7.9 may represent a new result. Furthermore, the equivalence of the weak star topology of the (strong norm) unit ball of $\mathscr{X}^*$ with a topology derived from a weak norm on $\mathscr{X}^*$ (I.3.11) has apparently been first observed by Bishop [1].

CHAPTER II

Functional Equations

II.1 Definitions and Background

If sets Y and Z and functions $h_1 : Y \to Z$ and $h_2 : Y \to Z$ are given, then the statement

$$h_1(y) = h_2(y) \tag{1}$$

is defined for each $y \in Y$ whether or not it is valid for a particular choice of y. We refer to statement (1) as an *equation in* Y or an *equation* (if Y has been specified). If this statement is valid for $y = \bar{y}$, then we say that $\bar{y}$ is a *solution* of Eq. (1) and $\bar{y}$ *satisfies* Eq. (1). If statement (1) is satisfied by only one point $\bar{y}$ of Y, we say that Eq. (1) has a *unique solution* and $\bar{y}$ is *the unique solution* of (1). If Y_1 is a given set, $\bar{y} \in Y_1 \cap Y$, and Eq. (1) is satisfied by $\bar{y}$ but is not satisfied by any other point of $Y_1 \cap Y$, then we say that Eq. (1) has *the unique solution* $\bar{y}$ *in* Y_1. We refer to the argument y in Eq. (1) as the *unknown* (point of Y). Equation (1) is a *functional equation* if Y is a class of functions with a common domain and a common range.

If Y_1, Y_2, Z_1, and Z_2 are given sets and $h_j^i : Y_i \to Z_i$ $(i, j = 1, 2)$ given functions, then we say that the equations

$$h_1^1(y_1) = h_2^1(y_1) \quad \text{in } Y_1 \tag{2}$$

and

$$h_1^2(y_2) = h_2^2(y_2) \quad \text{in } Y_2 \tag{3}$$

are *equivalent* if there exists a one-to-one correspondence between the *solution sets*

$$\{y_1 \in Y_1 \mid h_1^1(y_1) = h_2^1(y_1)\} \qquad \text{and} \qquad \{y_2 \in Y_2 \mid h_1^2(y_2) = h_2^2(y_2)\}.$$

We *transform* an equation by determining an equivalent equation; thus we say that (2) is *transformed* into (3).

If Y, Z, and Ω are given sets and $h_i^\omega : Y \to Z$ $(i = 1, 2, \omega \in \Omega)$ are given functions, then, for each $\omega \in \Omega$, the statement

$$h_1^\omega(y) = h_2^\omega(y)$$

is an equation. The collection of statements

$$h_1^\omega(y) = h_2^\omega(y) \qquad (\omega \in \Omega) \tag{4}$$

is referred to as a *system of simultaneous equations*, and $\bar{y}$ is a *solution* of this system if each of these statements is valid for $y = \bar{y}$. If we define, for $y \in Y$ and $i \in \{1, 2\}$, $h_i(y)$ as the function $\omega \to h_i^\omega(y) : \Omega \to Z$, then system (4) is equivalent to Eq. (1) in the sense that $\bar{y}$ is a solution of (4) if and only if it is a solution of (1). For this reason, we often refer to systems of simultaneous equations as "equations."

If Y, Z, and P are given sets and $h_i : Y \times P \to Z$ $(i = 1, 2)$ are given functions, then the statement

$$h_1(y, p) = h_2(y, p)$$

is an equation in Y for each choice of p in P, and we refer to p as a *parameter* of this equation.

We shall be particularly interested in functional equations referred to as *ordinary differential equations*, *integral equations*, and *functional-integral equations*. The study of ordinary differential equations and of certain integral equations, in particular, had inspired many of the original investigations into functional analysis and the latter, in turn, had provided powerful tools for a simpler and more general analysis of these equations and many of their generalizations. Arguments quite analogous to those that had been used to study the existence of solutions of ordinary differential equations also apply when difference-differential, delay-differential, and certain functional-integral equations are considered. Similarly, the solutions of these equations exhibit many similarities in their dependence on parameters that may appear in these equations.

Many functional equations of the types we have just mentioned were first studied in various branches of mathematical physics and more recently arose in engineering applications and in biological sciences. The methods most frequently used in their initial study consisted in what Volterra [1, p. 3] called "the passage from the discrete to the continuous"; specifically, in the study of discrete "approximations" to these equations and of the limits of their solutions as the approximations become more refined. Such approximations are obtained, for example, when the unknown function is represented by a finite number of real unknowns which are the values of the function at preassigned points, the value at other points being computed by linear or other interpolation. These arguments often led to rigorous proofs as in the case of Peano's existence theorem for ordinary differential equations or Fredholm's study of linear integral equations. When many of the basic methods and concepts thus developed were defined in a more abstract setting, it became possible to devise simpler and more general proofs in the framework of function spaces, eliminating the need for the more laborious and highly technical procedures involved in the "passage from the discrete to the continuous."

Among the main topics that concern us are the existence of solutions of the various functional equations that we consider and their dependence on parameters that may appear in these equations. The existence proofs are based on Schauder's fixed point theorem II.2.8 which is topological in nature. The dependence on parameters is mostly studied in terms of certain differentiability properties of functional transformations, and a basic tool will be the implicit function theorem II.3.8 for Banach spaces. The study of the special case of linear functional-integral equations will be based on the theory of compact operators (I.3.13) due to F. Riesz who had generalized Fredholm's theory of certain integral equations.

We shall now consider some special equations, in each of which the unknown function y has its range in $\mathbb{R}^n$. As in I.4.F, we shall use the terms a.e., a.a., measurable, and integrable as referring to the Lebesgue measure on a subset of $\mathbb{R}$.

First let $T \triangleq [t_0, t_1] \subset \mathbb{R}$, $n \in \mathbb{N}$, $V \subset \mathbb{R}^n$, $f : T \times V \to \mathbb{R}^n$, Z be the set of (equivalence classes of) measurable functions on T to $\mathbb{R}^n$, and Y the set of absolutely continuous functions $y : T \to V$ such that $t \to f(t, y(t))$ is measurable. Then, by I.4.42, the function $y \to h_1(y) \triangleq \dot{y}$ is defined on Y to Z and, by assumption, so is the function h_2 defined by $h_2(y)(t) \triangleq f(t, y(t))$ a.e. in T. The functional equation (1) is then written in the form

$$\dot{y}(t) = f(t, y(t)) \quad \text{a.e. in } T, \tag{5}$$

and called an *ordinary differential equation.* (The term *ordinary* is used to distinguish this equation from a *partial differential equation* that involves an unknown function y defined on a subset of $\mathbb{R}^k$ ($k > 1$) and its partial derivatives). If the set Y is further restricted to contain only functions y such that, for given $\bar{t} \in T$ and $\eta \in \mathbb{R}^n$, we have $y(\bar{t}) = \eta$, then Eq. (5) represents an *initial value problem* (in ordinary differential equations) and this problem is described by the (simultaneous) equations

$$\dot{y}(t) = f(t, y(t)) \quad \text{a.e. in } T, \qquad y(\bar{t}) = \eta. \tag{6}$$

According to our previous definition, an absolutely continuous function $y : T \to \mathbb{R}^n$ is a solution of the initial value problem if it satisfies Eq. (6). If $\tilde{T}$ is a subinterval of T, $\bar{t} \in \tilde{T}$, $y : \tilde{T} \to \mathbb{R}^n$ is absolutely continuous, $y(\bar{t}) = \eta$, and $\dot{y}(t) = f(t, y(t))$ a.e. in $\tilde{T}$, then we say that y is a *local solution* of (6).

The reason it is customary to include only absolutely continuous functions in the set Y when dealing with ordinary differential equations is to avoid certain "pathological" solutions. If, for example, the solutions y were only restricted to have derivatives a.e. in $[0, 1]$ instead of being absolutely continuous, then the equations $\dot{y}(t) = 0$ a.e. in $[0, 1]$, $y(0) = 0$ would be satisfied by a "reasonable" solution, namely $y(t) = 0$ ($t \in [0, 1]$); these equations would also be satisfied, however, by the "pathological" Cantor's function (see Munroe [1, p. 193]) that vanishes at 0 and has a 0 derivative a.e. but is not constant.

We can combine Eqs. (6) with the condition of absolute continuity to derive the equivalent equation

$$y(t) = \eta + \int_{\bar{t}}^{t} f(\tau, y(\tau))\, d\tau \qquad (t \in T), \tag{7}$$

where

$$Y = Z \triangleq C(T, \mathbb{R}^n), \quad h_1(y) \triangleq y, \quad \text{and} \quad h_2(y)(t) \triangleq \int_{\bar{t}}^{t} f(\tau, y(\tau))\, d\tau \quad (t \in T);$$

its solutions, if any, are (by I.4.42) absolutely continuous and satisfy Eq. (5).

The ordinary differential equation in the form of Eq. (7) is a special case of an *integral equation of Uryson type*

$$y(t) = \int f(t, \tau, y(\tau))\, \mu(d\tau) \qquad (t \in T), \tag{8}$$

where T may now be some closed bounded domain of $\mathbb{R}^m$ or, more generally, a compact metric space and (T, Σ, μ) a finite measure space. Equation (8) is often investigated in two cases: where the solutions are restricted to $C(T, \mathbb{R}^n)$ and where they are restricted to $L^p(T, \Sigma, \mu, \mathbb{R}^n)$ for $1 \leqslant p < \infty$.

The ordinary differential equation can also be generalized in a different direction. One such generalization is the difference-differential system of equations

$$\dot{y}(t) = f(t, y(t - h_1), \ldots, y(t - h_k)) \quad \text{a.e. in } T \triangleq [t_0, t_1],$$
$$y(t) = \tilde{y}(t) \qquad (t \in [t_0 - h_k, t_0]),$$

where $0 \leqslant h_1 < h_2 < \cdots < h_k$, $\tilde{y} : [t_0 - h_k, t_0] \to \mathbb{R}^n$ is preassigned, and a solution y is required to be absolutely continuous on T. This yields, as in the case of Eqs. (6), the equations

$$\text{(9)} \qquad \begin{aligned} y(t) &= \tilde{y}(t_0) + \int_{t_0}^{t} f(\tau, y(\tau - h_1), \ldots, y(\tau - h_k))\, d\tau \qquad (t \in T), \\ y(t) &= \tilde{y}(t) \qquad (t \in [t_0 - h_k, t_0]) \end{aligned}$$

in $C(T, \mathbb{R}^n)$. Instead of the *constant delays* $h_1, \ldots, h_k$, we may consider nonnegative variable delays $h_1(\tau), \ldots, h_k(\tau)$, yielding the *delay-differential* (system of) *equations*

$$\text{(10)} \qquad \begin{aligned} y(t) &= y(t_0) + \int_{t_0}^{t} f(\tau, y(\tau - h_1(\tau)), \ldots, y(\tau - h_k(\tau)))\, d\tau \qquad (t \in T), \\ y(t) &= \tilde{y}(t) \qquad (t \in [t_0 - \sup_{\tau \in T} h_k(\tau), t_0]). \end{aligned}$$

System (10) [and its special case (9)] can be transformed into another form, more suitable for our present purposes. If we set

$$\hat{y}(t) = y(t) - \tilde{y}(t_0) \qquad (t \in T),$$

$$\hat{h}_j(\tau) = h_j(\tau) \quad \text{and} \quad w_j(v, \tau) = v + \tilde{y}(t_0)$$
$$(\tau \in T, \quad v \in \mathbb{R}^n, \quad j = 1, \ldots, k, \quad \tau - h_j(\tau) \in T),$$

$$\hat{h}_j(\tau) = 0 \quad \text{and} \quad w_j(v, \tau) = \tilde{y}(\tau - h_j(\tau)) + \tilde{y}(t_0)$$
$$(\tau \in T, \quad v \in \mathbb{R}^n, \quad j = 1, \ldots, k, \tau - h_j(\tau) \notin T),$$

and

$$\hat{f}(\tau, v_1, \ldots, v_k) = f(\tau, w_1(v_1, \tau), \ldots, w_k(v_k, \tau)) \qquad (\tau \in T, \quad v_1, \ldots, v_k \in \mathbb{R}^n),$$

then system (10) is equivalent to the equation

$$\text{(11)} \qquad \hat{y}(t) = \int_{t_0}^{t} \hat{f}(\tau, \hat{y}(\tau - \hat{h}_1(\tau)), \ldots, \hat{y}(\tau - \hat{h}_k(\tau)))\, d\tau \qquad (t \in T),$$

where now $\tau \to \tau - \hat{h}_i(\tau)$ are functions on T to T.

The equations (7), (8), and (11) are all special cases of the functional-integral equation

$$y(t) = \int f(t, \tau, \xi(y)(\tau))\, \mu(d\tau) \qquad (t \in T), \tag{12}$$

where (T, Σ, μ) is a finite measure space and ξ a given functional transformation (that is, a function on X to Y, where X and Y are given classes of functions).

Our general approach will be to treat all these equations as of the form

$$y = F(y),$$

where F is a function on some topological vector space $\mathscr{Y}$ into itself. In fact, for the problems we consider, $\mathscr{Y}$ will always be either the Banach space $C(T, \mathbb{R}^n)$ or the Banach space $L^p\,(T, \Sigma, \mu, \mathbb{R}^n)$ for $1 \leqslant p < \infty$.

II.2 Brouwer's, Schauder's, and Tychonoff's Fixed Point Theorems

If $n \in \mathbb{N}$, $x_0 ,..., x_n \in \mathbb{R}^n$, and the set $\{x_1 - x_0 ,..., x_n - x_0\}$ is independent, we say that the points $x_0 ,..., x_n$ are in a *general position.* By I.6.1, the set $\{\sum_{j=0}^n \theta^j x_j \mid (\theta^0, \theta^1,..., \theta^n) \in \mathscr{T}_n'\}$ coincides with $\text{co}(\{x_0 ,..., x_n\})$. If $x_0 ,..., x_n$ are in a general position, we call $\text{co}(\{x_0 ,..., x_n\})$ an *n-simplex* with *vertices* $x_0 ,..., x_n$. [We observe that, for $x_0 = (0,..., 0)$, $x_1 = (1, 0,..., 0)$, $x_2 = (0, 1, 0,... 0)$,..., $x_n = (0, 0,..., 0, 1)$, $\mathscr{T}_n$ is the n-simplex $\text{co}(\{x_0 ,..., x_n\})$]. If $k \in \{0, 1,..., n\}$ and $j_0 ,..., j_k$ are $k + 1$ distinct elements of $\{0, 1,..., n\}$, we refer to $\text{co}(\{x_{j_0},..., x_{j_k}\})$ as a *k-dimensional face* of the n-simplex $\text{co}(\{x_0 ,..., x_n\})$ with vertices $x_{j_0} ,..., x_{j_k}$. If $S_1 ,..., S_l$ are n-simplices such that $\bigcup_{j=1}^l S_j = \text{co}(\{x_0 ,..., x_n\})$ and, for $i \neq j$, $S_i \cap S_j$ is either empty or an m-dimensional face of both S_i and S_j for some $m \in \{0, 1,..., n\}$, then we say that $S \triangleq \{S_1 ,..., S_l\}$ is a *simplicial subdivision* of $\text{co}(\{x_0 ,..., x_n\})$ into subsimplices $S_1 ,..., S_l$. We call a vertex of any of these subsimplices a *vertex of* (*the subdivision*) S. We denote by $V(S)$ the set of the vertices of S.

We observe that if $x_0 ,..., x_n$ are in a general position, then the mapping

$$(\theta^0,..., \theta^n) \to \sum_{j=0}^n \theta^j x_j : \mathscr{T}_n' \to \text{co}(\{x_0 ,..., x_n\})$$

is bijective since $\sum_{j=0}^n \theta^j x_j = \sum_{j=0}^n \alpha^j x_j$ implies $\sum_{j=1}^n (\theta^j - \alpha^j)(x_j - x_0) = 0$; hence $\theta^j = \alpha^j$ for $j = 1,..., n$ and $\theta^0 = 1 - \sum_{j=1}^n \theta^j = 1 - \sum_{j=1}^n \alpha^j = \alpha^0$.

If $x \triangleq \sum_{j=0}^{n} \theta^j x_j \in \mathrm{co}(\{x_0, \ldots, x_j\})$, then $(\theta^0, \ldots, \theta^n) \in \mathscr{T}_n'$ and we call $\theta^0, \ldots, \theta^n$ the *barycentric coordinates* of x [and we have just shown that, for each x, $(\theta^0, \ldots, \theta^n)$ is unique].

Let S be a simplicial subdivision of an n-simplex $\mathrm{co}(\{x_0, \ldots, x_n\})$ and $m(\cdot)$ a function on $V(S)$ to $\{0, 1, \ldots, n\}$. We say that $m(\cdot)$ is a *proper label* of S if $m(p) \in \{j_0, \ldots, j_k\}$ whenever $p \in \mathrm{co}\{x_{j_0}, \ldots, x_{j_k}\}$. Such a function $m(\cdot)$ labels any vertex of S with an index from $\{0, 1, \ldots, n\}$ in such a manner that a vertex lying on a k-dimensional face $\mathrm{co}(\{x_{j_0}, \ldots, x_{j_k}\})$ of $\mathrm{co}(\{x_0, \ldots, x_n\})$ is labeled by the index of one of the vertices of that face. We say that a subsimplex $\mathrm{co}(\{p_0', \ldots, p_n'\})$ of S is *m-special* if its $n + 1$ vertices are labeled with the indices $\{0, \ldots, n\}$, that is, if $\{m(p_0'), \ldots, m(p_n')\} = \{0, 1, \ldots, n\}$.

A topological space A has the *fixed point property* if for every continuous $f: A \to A$ there exists a point $x \in A$ such that $x = f(x)$. We then refer to x as a *fixed point* of f.

II.2.1 ***Lemma*** For every $n \in \mathbb{N}$ and $\epsilon > 0$ there exists a simplicial subdivision $S \triangleq \{S_1, \ldots, S_l\}$ of $\mathscr{T}_n$ such that $\mathrm{diam}(S_i) \leqslant \epsilon$ for $i = 1, \ldots, l$.

▌PROOF Let N be a positive integer larger than n/ϵ and $P \triangleq \{(j_1/N, \ldots, j_n/N) \mid j_i \in \{0, 1, 2, \ldots\}$ and $\sum_{i=1}^{n} j_i \leqslant N\}$. We consider all the n-simplices of the form $p \pm (1/N)\mathscr{T}_n$ that are contained in $\mathscr{T}_n$ and where $p \in P$. We can verify that they form a simplicial subdivision S of $\mathscr{T}_n$ and that each of them has a diameter not exceeding ϵ. QED

II.2.2 ***Theorem*** (Sperner) Let $n \in \mathbb{N}$, $A \triangleq \mathrm{co}(\{x_0, \ldots, x_n\})$ be an n-simplex, S a simplicial subdivision of A, and $m(\cdot)$ a proper label of S. Then the number of m-special subsimplices of S is odd.

▌PROOF When $n = 1$, the theorem is easily verified (for example, by induction on the total number of subintervals in a simplicial subdivision). Now assume, for purposes of induction, that $n \geqslant 2$ and the theorem is true for all k-simplices $(1 \leqslant k \leqslant n - 1)$. We shall call an $(n - 1)$ – dimensional face of a subsimplex of S m-special if its vertices are labeled by $m(\cdot)$ with the indices $0, 1, \ldots, n - 1$. Let a be the number of m-special subsimplices, b the number of m-special faces that belong to the boundary of A, and $c(T)$ the number of m-special faces of a subsimplex T of S. If T is m-special, then $c(T) = 1$; otherwise $c(T)$ is either 0 or 2. It follows that a and $\sum_{T \in S} c(T)$ are both even or both odd. On the other hand, if an m-special face belongs to the boundary of A, then it belongs to exactly one subsimplex of S; otherwise it belongs to two such subsimplices exactly. Thus $\sum_{T \in S} c(T)$ is even or odd according to whether b is even or odd; hence a is odd if b is odd. However, b is odd by our inductive assumption because an m-special face belongs to the boundary of A if and only if it is an m-special subsimplex of the simplicial subdivision of $\mathrm{co}(\{x_0, \ldots, x_{n-1}\})$ induced by S. This completes the induction. QED

II.2.3 ***Theorem*** (Brouwer) Let a topological space A be homeomorphic to a bounded convex body in $\mathbb{R}^n$. Then A has the fixed point property.

▌PROOF Let A and B be topological spaces and ϕ a homeomorphism on A onto B. If A has the fixed point property and $f : B \to B$ is continuous, then $\phi^{-1} \circ f \circ \phi$ is a continuous mapping on A to A and has a fixed point a. Then $\phi^{-1} \circ f(\phi(a)) = a$; hence $\phi(a)$ is a fixed point of f. Thus the fixed point property is preserved under a homeomorphism.

By I.6.12, any bounded convex body in $\mathbb{R}^n$ is homeomorphic to $\mathscr{T}_n$. It suffices therefore to show that $\mathscr{T}_n$ has the fixed point property. Let $f : \mathscr{T}_n \to \mathscr{T}_n$ be continuous, $x_0, ..., x_n$ be the vertices of $\mathscr{T}_n$, and $\theta^0(x), ..., \theta^n(x)$ the barycentric coordinates of $x \in \mathscr{T}_n$. Since $x - x_0 = \sum_{j=1}^n \theta^j(x)(x_j - x_0)$ and $\{x_1 - x_0, ..., x_n - x_0\}$ is independent, it follows from I.3.4 that the mapping $x \to (\theta^1(x), ..., \theta^n(x)) : \mathscr{T}_n \to \mathscr{T}_n$ is a homeomorphism. Thus

$$x \to (\theta^0(x), ..., \theta^n(x)) : \mathscr{T}_n \to \mathscr{T}_n'$$

and

$$x \to (\theta^0(f(x)), ..., \theta^n(f(x))) : \mathscr{T}_n \to \mathscr{T}_n'$$

are continuous.

Now let S be a simplicial subdivision of $\mathscr{T}_n$ and assume that f has no fixed point. Then, for every vertex y of S, the relation

$$\sum_{j=0}^{n} \theta^j(y) = \sum_{j=0}^{n} \theta^j(f(y)) = 1$$

implies that the set $M_y \triangleq \{j \in \{0, 1, ..., n\} \mid \theta^j(f(y)) < \theta^j(y)\}$ is nonempty, and we can define $m(y) \triangleq \min M_y$. We observe that if $k \in \{0, 1, ..., n\}$ and $y \in \text{co}(\{x_{j_0}, ..., x_{j_k}\})$, then $\theta^j(y) = 0$ for $j \notin \{j_0, ..., j_k\}$; hence $m(y) \in \{j_0, ..., j_k\}$. Thus $m(\cdot)$ is a proper label of S and, by II.2.2, there exists at least one m-special subsimplex $T \triangleq \text{co}(\{p_0, ..., p_n\})$ of S. Therefore

$$\theta^{m(p)}(f(p)) < \theta^{m(p)}(p) \qquad \text{for} \quad p = p_0, ..., p_n,$$

and $\{m(p_0), ... m(p_n)\} = \{0, ..., n\}$; hence

$$\theta^i(f(p_{j_i})) < \theta^i(p_{j_i}) \qquad (i = 0, 1, ..., n), \tag{1}$$

where $m(p_{j_i}) = i$.

By II.2.1, for every $l \in \mathbb{N}$ there exists a simplicial subdivision S^l of $\mathscr{T}_n$ such that $\text{diam}(T) \leqslant 1/l$ for $T \in S^l$. Let $T^l \triangleq \text{co}(\{p_0{}^l, ..., p_n{}^l\})$ be an m-special subsimplex of S^l. Since $\mathscr{T}_n^{n+1}$ is closed and bounded in $(\mathbb{R}^n)^{n+1}$, it follows from I.2.5, I.2.6, and I.2.8 that $\mathscr{T}_n^{n+1}$ is sequentially compact. Thus there exist a sequence $J \subset (1, 2, ...)$ and $(\bar{p}_0, ..., \bar{p}_n) \in \mathscr{T}_n^{n+1}$ such that $\lim_{l \in J} p_i{}^l =$

$\bar{p}_i$ for $i = 0, 1, \ldots, n$. Since $\operatorname{diam}(T^l) \leqslant 1/l$, we conclude that $\bar{p}_0 = \bar{p}_1 = \cdots = \bar{p}_n \triangleq \bar{p}$ and $\lim_{l \in J} p(l) = \bar{p}$ whenever $p(l) \in T^l$ $(l \in \mathbb{N})$.

We may now apply relation (1) for $S = S^l$, $T = T^l$ $(l \in \mathbb{N})$. Since $p_{j_i} \in T^l$ for each l and $x \to \theta^i(f(x))$ is continuous for each i, it follows that

$$\theta^i(f(\bar{p})) \leqslant \theta^i(\bar{p}) \qquad \text{for} \quad i = 0, 1, \ldots, n.$$

Since $\sum_{j=0}^{n} \theta^j(f(\bar{p})) = \sum_{j=0}^{n} \theta^j(\bar{p}) = 1$, we conclude that

$$\theta^i(f(\bar{p})) = \theta^i(\bar{p}) \qquad (i = 0, \ldots, n);$$

hence $f(\bar{p}) = \bar{p}$, contradicting the assumption that f has no fixed point.
QED

II.2.4 ***Theorem*** (Mazur) If $\mathscr{X}$ is a Banach space, $A \subset \mathscr{X}$, and A is compact, then $\overline{\text{co}}(A)$ is compact.

▌PROOF We first observe that if $n \in \{0, 1, 2, \ldots\}$ and $a_0, \ldots, a_n \in \mathscr{X}$, then $\text{co}(\{a_0, \ldots, a_n\})$ is the image of the compact set $\mathscr{T}_n'$ under the continuous mapping $(\theta^0, \ldots, \theta^n) \to \sum_{j=0}^{n} \theta^j a_j$. Thus, by I.2.9, $\text{co}(\{a_0, \ldots, a_n\})$ is compact and therefore also closed; hence $\overline{\text{co}}(\{a_0, \ldots, a_n\}) = \text{co}(\{a_0, \ldots, a_n\})$.

Now let A be a compact subset of $\mathscr{X}$. In view of I.2.5, it suffices to prove that $\overline{\text{co}}(A)$ is totally bounded. For any $\epsilon > 0$ there exists a finite subset $A_\epsilon \triangleq \{a_0, \ldots, a_n\}$ of the compact set A such that

$$A \subset \bigcup_{j=0}^{n} S^F(a_j, \epsilon/2) \subset \text{co}(A_\epsilon) + S^F(0, \epsilon/2).$$

Since $\text{co}(A_\epsilon)$ and $S^F(0, \epsilon/2)$ are closed and convex, so is $\text{co}(A_\epsilon) + S^F(0, \epsilon/2)$; hence $\overline{\text{co}}(A) \subset \text{co}(A_\epsilon) + S^F(0, \epsilon/2)$. As we have shown before, $\text{co}(A_\epsilon)$ is compact; hence

$$\text{co}(A_\epsilon) \subset \bigcup_{j=1}^{m} S(b_j, \epsilon/2) = \{b_1, \ldots, b_m\} + S^F(0, \epsilon/2)$$

for some points $b_1, \ldots, b_m$ in $\text{co}(A_\epsilon)$, and

$$\overline{\text{co}}(A) \subset \{b_1, \ldots, b_m\} + S^F(0, \epsilon).$$

Since $\{b_1, \ldots, b_m\} \subset \text{co}(A_\epsilon) \subset \overline{\text{co}}(A)$, this shows that $\overline{\text{co}}(A)$ is totally bounded.
QED

▌*Remark* If $k \in \mathbb{N}$, $\mathscr{X} = \mathbb{R}^k$, and A is a compact subset of $\mathbb{R}^k$, then, by Caratheodory's theorem I.6.2, $\text{co}(A) = \{\sum_{j=0}^{k} \theta^j x_j \mid \theta \triangleq (\theta^0, \ldots, \theta^k) \in \mathscr{T}_k',$ $x_j \in A\}$ and therefore $\text{co}(A)$ is the image of the compact set $\mathscr{T}_k' \times A^{k+1}$ under the continuous mapping $(\theta, x_0, \ldots, x_k) \to \sum_{j=0}^{k} \theta^j x_j$. Thus, by I.2.9, $\text{co}(A)$ is compact and the question arises whether this may remain true when $\mathscr{X}$ is an infinite-dimensional Banach space. The following counterexample shows that $\overline{\text{co}}(A)$ cannot be replaced by $\text{co}(A)$ in Mazur's theorem II.2.4.

Let $\mathscr{X} \triangleq C([0, \frac{1}{2}])$,

$$\bar{f}(t) \triangleq 0 \quad \text{and} \quad f_j(t) \triangleq (t)^j \quad (t \text{ to the power } j)$$
$$(j = 0, 1, 2, \ldots; \quad t \in [0, \tfrac{1}{2}]).$$

Then $A \triangleq \{\bar{f}, f_0, f_1, \ldots\}$ is compact [since $\lim_{j\in J} f_j(t) = \bar{f}(t)$ uniformly for all $t \in [0, \frac{1}{2}]$ if $J \subset (1, 2, \ldots)$]. If we set

$$\alpha_i \triangleq \sum_{j=0}^{i} 1/j! \quad \text{and} \quad h_i(t) \triangleq \alpha_i^{-1} \sum_{j=0}^{i} f_j(t)/j! \quad (i \in \mathbb{N}, \quad t \in [0, \tfrac{1}{2}]),$$

then, by I.6.1, $h_i \in \text{co}(A)$ for all $i \in \mathbb{N}$ and, by I.3.15, $\lim_i h_i(t) = e^{t-1}$ for each $t \in [0, \frac{1}{2}]$. In fact, we can verify that the functions $h_1, h_2, \ldots$ are equicontinuous and therefore, by I.5.3, $\lim_i h_i(t) = e^{t-1}$ uniformly for $t \in [0, \frac{1}{2}]$. Since every element of $\text{co}(A)$ is a polynomial and $t \to e^t$ is not a polynomial, this shows that $\text{co}(A)$ is not closed and, a fortiori, not compact.

II.2.5 ***Theorem*** Let $\mathscr{X}$ be a locally convex topological vector space and S the family of all continous seminorms on $\mathscr{X}$. Then $\psi(x) = 0$ $(\psi \in S)$ if and only if $x = 0$.

▌PROOF We have $\psi(0) = 0$ for every seminorm ψ. Now let x be such that $\psi(x) = 0$ $(\psi \in S)$. If $x \neq 0$, then there exists a convex symmetric neighborhood U of 0 not containing x. The gauge function $\bar{\psi}$ of U is, by I.6.5, a continuous seminorm and, by I.6.4, $\bar{\psi}(x) \geqslant 1$. This contradicts the assumption that $\psi(x) = 0$ for all $\psi \in S$. QED

II.2.6 ***Lemma*** Let K be a compact convex set in a topological vector space $\mathscr{X}$ and $f(\cdot, \cdot) : K \times K \to \mathbb{R}$ continuous. If, for every fixed $y \in K$, $f(\cdot, y)$ is convex on K, then there exists $\bar{y} \in K$ such that

$$f(x, \bar{y}) \geqslant f(\bar{y}, \bar{y}) \quad (x \in K). \tag{1}$$

▌PROOF For each $x \in K$, let $F(x) \triangleq \{y \in K \mid f(x, y) \geqslant f(y, y)\}$. For each x, $F(x)$ is closed and nonempty. The assertion we wish to prove is that $\bigcap_{x\in K} F(x) \neq \varnothing$. Since K is compact, this is true if $\bigcap_{j=0}^{n} F(x_j) \neq \varnothing$ for every finite subset $\{x_0, \ldots, x_n\}$ of K.

Let $n \in \{0, 1, 2, \ldots\}$, $x_0, \ldots, x_n \in K$, $C \triangleq \text{co}(\{x_0, \ldots, x_n\})$,

$$g_j(y) \triangleq \max(f(y, y) - f(x_j, y), 0) \quad (y \in C, \quad j = 0, \ldots, n)$$

and

$$y(\theta) \triangleq \sum_{j=0}^{n} \theta^j x_j \qquad [\theta \triangleq (\theta^0,..., \theta^n) \in \mathscr{T}_n'].$$

Then each g_j is continuous and nonnegative. Therefore the function $h \triangleq (h^0,..., h^n) : \mathscr{T}_n' \to \mathbb{R}^{n+1}$, defined by

$$h^i(\theta) \triangleq \frac{\theta^i + g_i(y(\theta))}{1 + \sum_{k=0}^{n} g_k(y(\theta))} \qquad [i = 0, 1,..., n, \quad \theta \triangleq (\theta^0,..., \theta^n) \in \mathscr{T}_n'],$$

is a continuous mapping of $\mathscr{T}_n'$ into itself. Since $\mathscr{T}_n'$ is homeomorphic to $\mathscr{T}_n$, it follows from Brouwer's fixed point theorem II.2.3 that there exists $\bar{\theta} \in \mathscr{T}_n'$ such that $\bar{\theta} = h(\bar{\theta})$, yielding

$$\text{(2)} \qquad \bar{\theta}^i + \bar{\theta}^i \sum_{k=0}^{n} g_k(y(\bar{\theta})) = \bar{\theta}^i + g_i(y(\bar{\theta})) \qquad (i = 0, 1,..., n).$$

We now recall that, by assumption, $f(\cdot, y)$ is convex for each $y \in K$; hence

$$f(y(\bar{\theta}), y(\bar{\theta})) = f\left(\sum_{j=0}^{n} \bar{\theta}^j x_j, y(\bar{\theta})\right) \leqslant \sum_{j=0}^{n} \bar{\theta}^j f(x_j, y(\bar{\theta})).$$

This implies that there exists $m \in \{0, 1,..., n\}$ such that

$$f(x_m, y(\bar{\theta})) \geqslant f(y(\bar{\theta}), y(\bar{\theta})) \qquad \text{and} \qquad \bar{\theta}^m > 0;$$

or

$$g_m(y(\bar{\theta})) = 0 \qquad \text{and} \qquad \bar{\theta}^m > 0.$$

This relation, when combined with (2), implies that

$$\sum_{k=0}^{n} g_k(y(\bar{\theta})) = g_i(y(\bar{\theta})) = 0 \qquad (i = 0, 1,..., n);$$

hence

$$f(x_i, y(\bar{\theta})) \geqslant f(y(\bar{\theta}), y(\bar{\theta})) \qquad (i = 0, 1,..., n).$$

This shows that $y(\bar{\theta}) \in \bigcap_{j=0}^{n} F(x_j)$, thus proving our assertion. QED

II.2.7 ***Theorem*** (Tychonoff) Let K be a compact and convex subset of a locally convex topological vector space $\mathscr{X}$. Then every continuous $f: K \to K$ has a fixed point.

▌PROOF Let S be the family of all continuous seminorms on $\mathscr{X}$ and

$$F_\psi \triangleq \{y \in K \mid \psi(y - f(y)) = 0\} \qquad (\psi \in S).$$

The set F_ψ is closed because the function $y \to \psi(y - f(y))$ is continuous. Therefore, by II.2.5, there exists a fixed point of f if $\bigcap_{\psi \in S} F_\psi \neq \varnothing$. Since K is compact and F_ψ closed, it suffices to show that $\bigcap_{i=1}^n F_{\psi_i} \neq \varnothing$ for every finite subset $\{\psi_1, ..., \psi_n\}$ of S.

Let $h(x, y) \triangleq \sum_{i=1}^n \psi_i(x - f(y))$. Then $h(\cdot, \cdot)$ satisfies the conditions of II.2.6 and there exists $\bar{y} \in K$ such that $h(x, \bar{y}) \geqslant h(\bar{y}, \bar{y})$ for all $x \in K$, in particular, for $x = f(\bar{y})$. Thus

$$0 = \sum_{i=1}^n \psi_i(0) \geqslant \sum_{i=1}^n \psi_i(\bar{y} - f(\bar{y})),$$

implying that $\psi_i(\bar{y} - f(\bar{y})) = 0$ $(i = 1, ..., n)$ and $\bigcap_{i=1}^n F_{\psi_i} \neq \varnothing$. QED

II.2.8 ***Theorem*** (Schauder) Let K be a closed convex subset of a Banach space $\mathscr{X}$, $f: K \to K$ continuous, and $\overline{f(K)}$ compact. Then f has a fixed point.

PROOF Let $M \triangleq \overline{\text{co}}(f(K)) = \overline{\text{co}(\overline{f(K)})}$. By Mazur's theorem II.2.4, M is compact, and clearly $M \subset K$. Thus the restriction of f to M satisfies the conditions of Tychonoff's theorem II.2.7 and has a fixed point. QED

II.3 Derivatives and the Implicit Function Theorem

Derivatives

Let $\mathscr{X}$ be a Banach space, $\mathscr{Y}$ a topological vector space, $A \subset \mathscr{X}$, and $f: A \to \mathscr{Y}$. We say that F is the *derivative of f at $\bar{a} \in A$* and that f is *differentiable at $\bar{a}$* if $\bar{a}$ is contained in a convex subset of A with a nonempty interior, $F \in B(\mathscr{X}, \mathscr{Y})$ and

$$\lim |a - \bar{a}|^{-1}[f(a) - f(\bar{a}) - F(a - \bar{a})] = 0 \qquad \text{as} \quad a \to \bar{a}, \quad a \in A \sim \{\bar{a}\}.$$

It is not obvious from this definition that the derivative of f at $\bar{a}$, if it exists, is unique. However, we shall prove in II.3.1 below that this is the case and it is for this reason that we refer to "the derivative" of f at $\bar{a}$ and not to "a derivative" of f at $\bar{a}$. We shall use the symbols $f'(\bar{a})$ or $\mathscr{D}f(\bar{a})$ to denote the derivative of f at $\bar{a}$.

It is clear from the definition that f is continuous at $\bar{a}$ if it is differentiable at $\bar{a}$ (but the converse is not true as shown by the example: $A = \mathscr{X} = \mathscr{Y} = \mathbb{R}$, $f(a) = |a|$, $\bar{a} = 0$). It is also easy to verify that differentiation at $\bar{a}$ is a linear operation, that is, if $f: A \to \mathscr{Y}$ and $g: A \to \mathscr{Y}$ are differentiable at $\bar{a}$, then

$$(f + g)'(\bar{a}) = f'(\bar{a}) + g'(\bar{a}) \qquad \text{and} \qquad (\alpha f)'(\bar{a}) = \alpha f'(\bar{a}) \quad (\alpha \in \mathbb{R}).$$

If $\mathscr{Y}$ is a Banach space and $f : A \to \mathscr{Y}$ is differentiable at $\bar{a}$, we denote by $|f'(\bar{a})|$ the usual norm of $f'(\bar{a})$ in $B(\mathscr{X}, \mathscr{Y})$, that is, $|f'(\bar{a})| \triangleq \sup_{|x| \leqslant 1} |f'(\bar{a})x|$.

If $\mathscr{X} = \mathbb{R}$ and $A = [a_0, a_1] \subset \mathbb{R}$, then the derivative $f'(\bar{a})$ at $\bar{a} \in [a_0, a_1]$ corresponds uniquely to the element $\bar{y} \triangleq f'(\bar{a}) \cdot 1$ of $\mathscr{Y}$, and we have $f'(\bar{a})x = x\bar{y}$ $(x \in \mathbb{R})$. We then choose to identify $f'(\bar{a})$ with $f'(\bar{a}) \cdot 1$, it being clear from the context whether we consider $f'(\bar{a})$ as an element of $\mathscr{Y}$ or of $B(\mathbb{R}, \mathscr{Y})$. Furthermore, we verify that

$$f'(\bar{a}) \cdot 1 = \lim h^{-1}[f(\bar{a} + h) - f(\bar{a})] \qquad \text{as} \quad h \to 0, \qquad h \neq 0.$$

Thus, if $n \in \mathbb{N}$, $\mathscr{X} = \mathbb{R}$, $\mathscr{Y} = \mathbb{R}^n$, and $A = [a_0, a_1] \subset \mathbb{R}$, our present definition of the derivative $f'(\bar{a})$ coincides with the definition of the derivative $\dot{f}(\bar{a})$ of I.4.F.

If $\mathscr{X}$ is a Banach space, $\mathscr{Y}$ a topological vector space, $A \subset \mathscr{X}$, $f : A \to \mathscr{Y}$, and f is differentiable at a for all $a \in A$, then we say that f is *differentiable*. If f is differentiable, then we denote by f' or $\mathscr{D}f$ the function

$$a \to f'(a) : A \to B(\mathscr{X}, \mathscr{Y})$$

and, for each $\Delta a \in \mathscr{X}$, we denote by $f' \, \Delta a$ or $\mathscr{D}f \, \Delta a$ the function $a \to f'(a) \, \Delta a : A \to \mathscr{Y}$. If $\mathscr{Y}$ is a Banach space and unless otherwise specified, we shall assume $B(\mathscr{X}, \mathscr{Y})$ to be a Banach space with the norm

$$|T| \triangleq \sup_{|x| \leqslant 1} |Tx|;$$

in this sense we say that f' is continuous at $\bar{a}$ whenever

$$\lim_{a \to \bar{a}} |f'(a) - f'(\bar{a})| = 0.$$

If f is differentiable at some point and we denote the arguments of f by some letter, with or without bars, subscripts, etc. (say $\bar{a}$ or x or θ_1), then we shall usually denote the arguments of $f'(\bar{a})$ by Δa [or of $f'(x)$ by Δx, etc.).

Partial Derivatives

If $\mathscr{X}_1, \mathscr{X}_2, ..., \mathscr{X}_n$ are Banach spaces, $\mathscr{Y}$ is a topological vector space, $\bar{a}_i \in A_i \subset \mathscr{X}_i$ $(i = 1,..., n)$, $f : A_1 \times \cdots \times A_n \to \mathscr{Y}$, and the function

$$a_i \to f_i(a_i) \triangleq f(\bar{a}_1, ..., \bar{a}_{i-1}, a_i, \bar{a}_{i+1}, ..., \bar{a}_n) : A_i \to \mathscr{Y}$$

has a derivative at $\bar{a}_i$, we call this derivative the *partial derivative* of f with respect to a_i at $(\bar{a}_1, ..., \bar{a}_n)$ and denote it by $f_{a_i}(\bar{a}_1, ..., \bar{a}_n)$ or $\mathscr{D}_i f(\bar{a}_1, ..., \bar{a}_n)$. We refer on occasions to the derivative $f'((\bar{a}_1, ..., \bar{a}_n))$ as the *total derivative* of f at $(\bar{a}_1, ..., \bar{a}_n)$.

If $\mathscr{X} \triangleq \mathscr{X}_1 \times \cdots \times \mathscr{X}_n$, $\bar{a} \triangleq (\bar{a}_1, ..., \bar{a}_n) \in A \subset \mathscr{X}$, and $f : A \to \mathscr{Y}$ has a derivative at $\bar{a}$, then we may consider the functions

$$x_i \to F_i x_i \triangleq f'(\bar{a})\, \tilde{x}^i : \mathscr{X}_i \to \mathscr{Y} \qquad (i = 1, 2, ..., n),$$

where $\tilde{x}^i \triangleq (0, ..., 0, x_i, 0, ..., 0)$ [that is, $\tilde{x}^i \triangleq (\tilde{x}_1{}^i, ..., \tilde{x}_n{}^i)$ with $\tilde{x}_i{}^i \triangleq x_i$ and $\tilde{x}_j{}^i \triangleq 0$ for $j \neq i$]. Then $F_i \in B(\mathscr{X}_i, \mathscr{Y})$ and

$$f'(\bar{a})(x_1, ..., x_n) = \sum_{i=1}^{n} F_i x_i,$$

and we refer to F_i as the *partial derivative of* f with *respect to* a_i at $\bar{a} \triangleq (\bar{a}_1, ..., \bar{a}_n)$ and denote F_i by $\mathscr{D}_i f(\bar{a})$ or $f_{a_i}(\bar{a})$. This definition of the partial derivative is consistent with the one given previously since it follows directly from the definitions of $f'(\bar{a})$ and F_i that F_i is the derivative of the function

$$f(\bar{a}_1, ..., \bar{a}_{i-1}, \cdot, \bar{a}_{i+1}, ..., \bar{a}_n) : A_i \to \mathscr{Y}$$

in the special case where $A \triangleq A_1 \times \cdots \times A_n$, $\bar{a}_i \in A_i \subset \mathscr{X}_i$, and $\bar{a}_i$ belongs to a convex subset of A_i with a nonempty interior.

We have thus defined $\mathscr{D}_i f(\bar{a})$ for $f : A \to \mathscr{Y}$ in two cases: (a) when $\bar{a} \triangleq (\bar{a}_1, ..., \bar{a}_n) \in A_1 \times \cdots \times A_n$, $A_i \subset \mathscr{X}_i$ and $f(\bar{a}_1, ..., a_{i-1}, \cdot, \bar{a}_{i+1}, ..., \bar{a}_n)$ is differentiable at $\bar{a}_i$, and (b) when f is differentiable at $\bar{a}$.

If $\bar{a} \triangleq (\bar{a}_1, ..., \bar{a}_n) \in A \triangleq A_1 \times \cdots \times A_n$ and $A_i \subset \mathscr{X}_i$ $(i = 1, 2, ..., n)$, then, as we have just observed, $\mathscr{D}_i f(\bar{a})$ exist for $i = 1, 2, ..., n$ whenever $f'(\bar{a})$ exists; but the converse is not true. For example, the function $f : \mathbb{R}^2 \to \mathbb{R}$, defined by

$$f(x_1, x_2) \triangleq \frac{x_1 x_2}{(x_1)^2 + (x_2)^2} \qquad \text{for} \quad (x_1, x_2) \neq (0, 0) \quad \text{and} \quad f(0, 0) \triangleq 0,$$

has partial derivatives $\mathscr{D}_1 f(0, 0) = \mathscr{D}_2 f(0, 0) = 0$, but f is not differentiable at (0, 0).

If $\mathscr{D}_i f(a)$ exists for all $a \in A$, we write f_{a_i} or $\mathscr{D}_i f$ for the function

$$a \to f_{a_i}(a) : A \to B(\mathscr{X}_i, \mathscr{Y})$$

and, for each $\Delta a_i \in \mathscr{X}_i$, $f_{a_i} \Delta a_i$ or $\mathscr{D}_i f \Delta a_i$ for the function

$$a \to f_{a_i}(a)\, \Delta a_i : A \to \mathscr{Y}.$$

Second Derivatives

Let $\mathscr{X}$ and $\mathscr{Y}$ be Banach spaces, $A \subset \mathscr{X}$, $f : A \to \mathscr{Y}$, and assume that f has a derivative f'. Then f' is a function on A to the Banach space $B(\mathscr{X}, \mathscr{Y})$. If $f' : A \to B(\mathscr{X}, \mathscr{Y})$ has a derivative at $\bar{a}$, which we denote by $f''(\bar{a})$ or $\mathscr{D}^2 f(\bar{a})$, then we say that f is *twice differentiable at* $\bar{a}$ and refer to $f''(\bar{a})$ as the *second derivative of* f *at* $\bar{a}$. If $f''(\bar{a})$ exists for all $\bar{a} \in A$, then we denote the function $a \to f''(a) : A \to B(\mathscr{X}, B(\mathscr{X}, \mathscr{Y}))$ by f'' or $\mathscr{D}^2 f$. We say that f has a continuous second derivative if f'' is continuous on A when the usual norm topology is chosen for $B(\mathscr{X}, \mathscr{Y})$ and $B(\mathscr{X}, B(\mathscr{X}, \mathscr{Y}))$.

Directional Derivatives

Let $\mathscr{X}$ be a vector space, $\mathscr{Y}$ a topological vector space, A a subset of $\mathscr{X}$, $a, \bar{a} \in A$, and $f : A \to \mathscr{Y}$. If, for some $\bar{\alpha} > 0$, we have $\bar{a} + \alpha(a - \bar{a}) \in A$ $(\alpha \in [0, \bar{\alpha}])$ and

$$Df(\bar{a}; a - \bar{a}) \triangleq \lim_{\alpha \to +0} \alpha^{-1}[f(\bar{a} + \alpha(a - \bar{a})) - f(\bar{a})]$$

exists, then we refer to $Df(\bar{a}; a - \bar{a})$ as the *directional derivative of* f *at* $\bar{a}$ *in the direction of* a. Since, by definition, a topological vector space is Hausdorff, $Df(\bar{a}; a - \bar{a})$ is unique and we observe that it coincides with the derivative at 0 of the function $\alpha \to f(\bar{a} + \alpha(a - \bar{a})) : [0, \bar{\alpha}] \to \mathscr{Y}$. We also verify that if $a_1 \triangleq \bar{a} + \beta(a - \bar{a})$ for some $\beta \in [0, 1]$, then $Df(\bar{a}; a_1 - \bar{a}) = \beta Df(\bar{a}; a - \bar{a})$. If $f : A_1 \times A_2 \times \cdots \times A_k \to \mathscr{Y}$, $j \in \{1, ..., k\}$, A_j is a subset of some vector space, $a_j, \bar{a}_j \in A_j$, and $\mathscr{Y}$ is a topological vector space, we write $D_j f(\bar{a}_1, ..., \bar{a}_k ; a_j - \bar{a}_j)$ for the directional derivative at $\bar{a}_j$ in the direction of a_j of the function $a \to f(\bar{a}_1, ..., \bar{a}_{j-1}, a, \bar{a}_{j+1}, ..., \bar{a}_k) : A_j \to \mathscr{Y}$.

n-Differentiable Functions

Let $\mathscr{X}$ be a vector space, $\mathscr{Y}$ a topological vector space, $a_0 \in A \subset \mathscr{X}$, and $f : A \to \mathscr{Y}$. If $n \in \mathbb{N}$ and for every choice of $a \triangleq (a_1, ..., a_n) \in A^n$ there exists $\alpha_a > 0$ such that $a_0 + \sum_{j=1}^n \theta^j(a_j - a_0) \in A$ $(\theta^j \in [0, \alpha_a])$ and the function

$$\theta \triangleq (\theta^1, ..., \theta^n) \to f\Big(a_0 + \sum_{j=1}^n \theta^j(a_j - a_0)\Big) : [0, \alpha_a]^n \to \mathscr{Y}$$

has a derivative at 0, we say that f is *n-differentiable at* a_0. It is clear that if $\mathscr{X}$ is a Banach space, A is convex, and f has a derivative at a_0, then f is n-differentiable at a_0 for all $n \in \mathbb{N}$. Furthermore, f is m-differentiable at a_0 whenever f is n-differentiable at a_0 and $1 \leqslant m \leqslant n$.

II.3.1 ***Theorem*** If $\mathscr{X}$ is a Banach space, $\mathscr{Y}$ a topological vector space, $\bar{a} \in A \subset \mathscr{X}$, $f : A \to \mathscr{Y}$, and F is a derivative of f at $\bar{a}$, then F is the unique derivative of f at $\bar{a}$.

▌ PROOF Let F_1 and F_2 be derivatives of f at $\bar{a}$. Then there exists a convex subset A_0 of A containing $\bar{a}$, with $A_0^\circ \neq \varnothing$. Now let $b \in A_0^\circ, b \neq \bar{a}$. Then there exists $\alpha > 0$ such that the closed ball $S^F(b, \alpha)$ does not contain $\bar{a}$ and belongs to A_0. Let (η_j) be a sequence in $(0, 1]$ converging to 0, $c \in S^F(b, \alpha)$, and $a_j \triangleq \bar{a} + \eta_j(c - \bar{a})$. Then $a_j \in A_0$ and

$$\lim_j |a_j - \bar{a}|^{-1} [f(a_j) - f(\bar{a}) - F_i(a_j - \bar{a})] = 0 \qquad (i = 1, 2);$$

hence

$$|c - \bar{a}|^{-1} \lim_j (\eta_j)^{-1} [f(a_j) - f(\bar{a})] = |c - \bar{a}|^{-1} F_i(c - \bar{a}) \qquad (i = 1, 2).$$

Thus $F_1(c - \bar{a}) = F_2(c - \bar{a})$ for all $c \in S^F(b, \alpha)$. Now let $x \in \mathscr{X}$, $|x| \leqslant 1$. Then $b + \alpha x \in S^F(b, \alpha)$; hence

$$F_1(b - \bar{a} + \alpha x) = F_2(b - \bar{a} + \alpha x) \qquad (x \in \mathscr{X}, \quad |x| \leqslant 1)$$

and therefore

$$F_1(x) = F_2(x) \qquad (x \in \mathscr{X}). \quad \text{QED}$$

II.3.2 ***Theorem*** Let $k \in \mathbb{N}$, $\bar{\alpha} > 0$, $\mathscr{X}$ be a vector space,

$$A \triangleq \left\{a_0 + \sum_{j=1}^{k} \theta^j(a_j - a_0) \,\middle|\, \theta^j \in [0, \bar{\alpha}]\right\} \subset \mathscr{X},$$

$\mathscr{Y}$ a topological vector space, $f: A \to \mathscr{Y}$, and

$$\theta \to f_A(\theta) \triangleq f\left(a_0 + \sum_{j=1}^{k} \theta^j(a_j - a_0)\right) : [0, \bar{\alpha}]^k \to \mathscr{Y}$$

differentiable at 0. Then

$$f_A'(0)\, \Delta\theta = \sum_{j=1}^{k} \Delta\theta^j\, Df(a_0\,; a_j - a_0) \qquad [\Delta\theta \triangleq (\Delta\theta^1, \ldots, \Delta\theta^k) \in \mathbb{R}^k].$$

▌PROOF Let $\delta_i \triangleq (\delta_i^1, \ldots, \delta_i^k)$ $(i = 1, \ldots, k)$, with $\delta_i^j \triangleq 0$ for $j \neq i$ and $\delta_i^i \triangleq 1$. For each $i \in \{1, 2, \ldots, k\}$, we have

$$\begin{aligned} 0 &= \lim_{\alpha \to +0} \alpha^{-1}[f_A(\alpha\delta_i) - f_A(0) - \alpha f_A'(0)\, \delta_i] \\ &= \lim_{\alpha \to +0} \alpha^{-1}[f(a_0 + \alpha(a_i - a_0)) - f(a_0) - \alpha f_A'(0)\, \delta_i]; \end{aligned}$$

hence $f_A'(0)\, \delta_i = Df(a_0\,; a_i - a_0)$. Our conclusion follows since $f_A'(0)$ is linear and $\Delta\theta = \sum_{j=1}^{k} \Delta\theta^j\, \delta_j$. QED

II.3.3 ***Theorem*** Let A be a subset of a Banach space $\mathscr{X}$, $\mathscr{Y}$ a topological vector space, and $f : A \to \mathscr{Y}$ 2-differentiable at $a_0 \in A$. Then, for all a_1, $a_2 \in A$ and $\beta \in [0, 1]$,

$$(1)\qquad \begin{aligned} &Df(a_0\,;\beta a_1 + (1-\beta)\,a_2 - a_0) \\ &\quad = \beta Df(a_0\,; a_1 - a_0) + (1-\beta)\,Df(a_0\,; a_2 - a_0). \end{aligned}$$

If A is convex, then $\{Df(a_0\,; a - a_0) | \, a \in A\}$ is also convex.

▌PROOF Let a_1, $a_2 \in A$ and $\beta \in [0, 1]$, and let $\bar{\alpha} > 0$ be such that

$$a_0 + \theta^1(a_1 - a_0) + \theta^2(a_2 - a_0) \in A \qquad \text{for all } \theta \triangleq (\theta^1, \theta^2) \in [0, \bar{\alpha}]^2.$$

The function

$$\theta \triangleq (\theta^1, \theta^2) \to h(\theta) \triangleq f(a_0 + \theta^1(a_1 - a_0) + \theta^2(a_2 - a_0)) : [0, \bar{\alpha}]^2 \to \mathscr{Y}$$

has a derivative $h'(0)$; hence, for $b \triangleq (\beta, 1-\beta)$,

$$\lim_{\alpha \to +0} \alpha^{-1}[h(\alpha b) - h(0) - h'(0)(\alpha b)] = 0.$$

Thus

$$\begin{aligned} h'(0)b &= \lim_{\alpha \to +0} \alpha^{-1}[f(a_0 + \alpha(\beta a_1 + (1-\beta)\,a_2 - a_0)) - f(a_0)] \\ &= Df(a_0\,; \beta a_1 + (1-\beta)\,a_2 - a_0). \end{aligned}$$

On the other hand, by II.3.2,

$$h'(0)b = \beta Df(a_0\,; a_1 - a_0) + (1-\beta)\,Df(a_0\,; a_2 - a_0),$$

thus proving (1). If A is convex, then (1) implies that $\{Df(a_0\,; a - a_0) | \, a \in A\}$ is also convex. QED

II.3.4 ***Theorem*** (The chain rule) Let A and B be subsets of Banach spaces, C a subset of a topological vector space, $f : A \to B$, $g : B \to C$, $a_0 \in A$, and $b_0 \triangleq f(a_0)$. If there exist derivatives $f'(a_0)$ and $g'(b_0)$, then $g'(b_0) \circ f'(a_0)$ is the derivative of $g \circ f$ at a_0. If f' is continuous at a_0 and g' at $f(a_0)$, then $(g \circ f)'$ is continuous at a_0.

▌PROOF Let $h \triangleq g \circ f$ and $T \triangleq g'(b_0) \circ f'(a_0)$. Then, for all $a \in A \sim \{a_0\}$,

$$\begin{aligned} &|a - a_0|^{-1}[h(a) - h(a_0) - T(a - a_0)] \\ &\quad = |a - a_0|^{-1}[g \circ f(a) - g(b_0) - g'(b_0)(f(a) - b_0)] \\ &\qquad + g'(b_0) \cdot |a - a_0|^{-1}[f(a) - f(a_0) - f'(a_0)(a - a_0)]. \end{aligned}$$

Let the first and the second terms on the right be denoted by $\alpha_1(a)$ and $\alpha_2(a)$, respectively. Since $g'(b_0)$ is a continuous operator and f is differentiable at

a_0, we conclude that $\lim \alpha_2(a) = 0$ as $a \to a_0$. Now $\alpha_1(a) = 0$ if $f(a) = f(a_0)$. If $\lim_j a_j = a_0$, $a_j \in A \sim \{a_0\}$, and $f(a_j) \neq f(a_0)$, then

$$(1) \qquad \lim_j \alpha_1(a_j) = \lim_j \frac{|f(a_j) - f(a_0)|}{|a_j - a_0|} \cdot \frac{1}{|f(a_j) - f(a_0)|} \cdot [g(f(a_j)) - g(b_0) - g'(b_0)(f(a_j) - b_0)].$$

We observe that, since

$$\lim_j |a_j - a_0|^{-1} |f(a_j) - f(a_0) - f'(a_0)(a_j - a_0)| = 0,$$

it follows that

$$\lim_j \sup \frac{|f(a_j) - f(a_0)|}{|a_j - a_0|} = \lim_j \sup |f'(a_0)|a_j - a_0|^{-1}(a_j - a_0)| \leqslant |f'(a_0)|.$$

We can now conclude from (1) that $\lim \alpha_1(a_j) = 0$, which completes the proof of the first part of our assertion.

If f' is continuous at a_0 and g' is continuous at $f(a_0)$, then f is continuous at a_0 and $a \to (g \circ f)'(a) = g'(f(a)) \circ f'(a)$ is continuous at a_0. QED

II.3.5 ***Theorem*** Let $\mathscr{Y}$ be a Banach space, $f: [0, 1] \to \mathscr{Y}$, $c \in \mathbb{R}$, and assume that f has a derivative f' and $|f'(\xi)| \leqslant c$ for all $\xi \in [0, 1]$. Then $|f(1) - f(0)| \leqslant c$.

▌PROOF Let $\epsilon > 0$. For every $\xi \in [0, 1]$ there exists $\delta(\xi) > 0$ such that

$$(1) \qquad |f(\eta) - f(\xi) - f'(\xi)(\eta - \xi)| \leqslant \epsilon |\eta - \xi|$$

if $|\eta - \xi| \leqslant \delta(\xi)$. Let $A_\xi \triangleq (\xi - \delta(\xi), \xi + \delta(\xi)) \cap [0, 1]$, η, $\eta' \in A_\xi$ and $\eta' < \xi < \eta''$. Then, by (1),

$$\begin{aligned} |f(\eta'') - f(\eta') - f'(\xi)(\eta'' - \eta')| &\leqslant |f(\eta'') - f(\xi) - f'(\xi)(\eta'' - \xi)| \\ &\quad + |f(\xi) - f(\eta') - f'(\xi)(\xi - \eta')| \\ &\leqslant \epsilon(|\eta'' - \xi| + |\xi - \eta'|) = \epsilon |\eta'' - \eta'|; \end{aligned}$$

hence

$$(2) \qquad |f(\eta'') - f(\eta')| \leqslant (|f'(\xi)| + \epsilon)|\eta'' - \eta'| \leqslant (c + \epsilon)|\eta'' - \eta'|.$$

Since the sets A_ξ ($\xi \in [0, 1]$) are an open covering of $[0, 1]$ (relative to $[0, 1]$) and $[0, 1]$ is compact, there exists a finite subcovering of $[0, 1]$ by $A_{\xi_1}, \ldots, A_{\xi_k}$. We may assume that $\xi_1 < \xi_2 < \cdots < \xi_k$. Thus there exist $\eta_0 = 0 \in A_{\xi_1}$, $\eta_i \in A_{\xi_i} \cap A_{\xi_{i+1}}$ ($i = 1, 2, \ldots, k - 1$), and $\eta_k = 1 \in A_{\xi_k}$ such that

$$\eta_0 \leqslant \xi_1 \leqslant \eta_1 \leqslant \cdots \leqslant \xi_k \leqslant \eta_k.$$

Then, applying (2) for $\xi = \xi_i$, $\eta' = \eta_{i-1}$, $\eta'' = \eta_i$ ($i = 1, 2, \ldots, k$), we find that

$$|f(1) - f(0)| \leqslant \sum_{j=1}^{k} |f(\eta_j) - f(\eta_{j-1})| \leqslant (c + \epsilon) \sum_{j=1}^{k} (\eta_j - \eta_{j-1}) = c + \epsilon.$$

Since ϵ is arbitrary, this proves the theorem. QED

II.3.6 ***Theorem*** (Mean value theorem) Let $\mathscr{X}$ and $\mathscr{Y}$ be Banach spaces, $\text{co}(\{a_1, a_1 + \Delta a\}) \subset A \subset \mathscr{X}$, $f : A \to \mathscr{Y}$ have a derivative $f'(a)$ for all $a \in \text{co}(\{a_1, a_1 + \Delta a\})$, and $F \in B(\mathscr{X}, \mathscr{Y})$. Then

$$\begin{aligned} |f(a_1 + \Delta a) - f(a_1) - F\,\Delta a| &\leqslant \sup_{0 \leqslant \alpha \leqslant 1} |(f'(a_1 + \alpha\,\Delta a) - F)\,\Delta a| \\ &\leqslant \sup_{0 \leqslant \alpha \leqslant 1} |f'(a_1 + \alpha\,\Delta a) - F|\,|\Delta a|. \end{aligned}$$

In particular, if f is also differentiable at $a_0 \in A$, then

$$\begin{aligned} |f(a_1 + \Delta a) - f(a_1) - f'(a_0)\,\Delta a| &\leqslant \sup_{0 \leqslant \alpha \leqslant 1} |[f'(a_1 + \alpha\,\Delta a) - f'(a_0)]\,\Delta a| \\ &\leqslant \sup_{0 \leqslant \alpha \leqslant 1} |f'(a_1 + \alpha\,\Delta a) - f'(a_0)|\,|\Delta a|. \end{aligned}$$

▌PROOF Let $g : [0, 1] \to \mathscr{Y}$ be defined by

$$g(\alpha) \triangleq f(a_1 + \alpha\,\Delta a) - F(a_1 + \alpha\,\Delta a) \qquad (\alpha \in [0, 1]).$$

Then, by the chain rule I.3.4,

$$g'(\alpha) \cdot 1 = f'(a_1 + \alpha\,\Delta a)\,\Delta a - F\,\Delta a \qquad (\alpha \in [0, 1])$$

and, by II.3.5,

$$\begin{aligned} |g(1) - g(0)| &= |f(a_1 + \Delta a) - f(a_1) - F\,\Delta a| \\ &\leqslant \sup_{0 \leqslant \alpha \leqslant 1} |g'(\alpha)| \\ &= \sup_{0 \leqslant \alpha \leqslant 1} |[f'(a_1 + \alpha\,\Delta a) - F]\,\Delta a| \\ &\leqslant \sup_{0 \leqslant \alpha \leqslant 1} |f'(a_1 + \alpha\,\Delta a) - F|\,|\Delta a|. \quad \text{QED} \end{aligned}$$

II.3.7 ***Theorem*** Let $\mathscr{Y}$ be a Banach space, Z a topological space, $0 < k < 1$, $\beta > 0$, $S^F(y_0, \beta) \subset Y \subset \mathscr{Y}$, and $v : Y \times Z \to \mathscr{Y}$ be continuous. If

(1) $$|v(y', z) - v(y'', z)| \leqslant k\,|y' - y''| \qquad (z \in Z, \quad y', y'' \in Y)$$

and

(2) $$|v(y_0, z) - y_0| \leqslant \beta(1-k) \qquad (z \in Z),$$

then there exists a continuous $u : Z \to Y$ such that

$$u(z) = v(u(z), z) \qquad (z \in Z).$$

PROOF Let $z \in Z$ be fixed. We shall show that the relation

$$y_{j+1} \triangleq v(y_j, z) \qquad (j = 0, 1, 2, \ldots)$$

defines a sequence (y_j) in $S^F(y_0, \beta)$. Indeed, by (2), $y_1 \triangleq v(y_0, z) \in S^F(y_0, \beta)$. Now assume inductively that $p \in \mathbb{N}$ and $y_j \in S^F(y_0, \beta)$ for $j = 0, 1, \ldots, p$. Then, by (1),

$$|y_{j+1} - y_j| = |v(y_j, z) - v(y_{j-1}, z)| \leqslant k\,|y_j - y_{j-1}| \qquad (j = 1, \ldots, p),$$

implying that

(3) $$|y_{j+1} - y_j| \leqslant k^j |y_1 - y_0| \qquad \text{for} \quad j = 0, 1, \ldots, p;$$

hence, by (2),

$$\begin{aligned} |y_{p+1} - y_0| &\leqslant \sum_{j=0}^{p} |y_{j+1} - y_j| \\ &\leqslant (1 + k + k^2 + \cdots + k^p)\,|v(y_0, z) - y_0| \leqslant \beta. \end{aligned}$$

Thus $y_{p+1} \in S^F(y_0, \beta)$ and this completes the induction.

If we write $u_j(z)$ for y_j $(j \in \mathbb{N})$, then we can verify by induction that $z \to u_j(z) : Z \to S^F(y_0, \beta)$ is continuous for all $j \in \mathbb{N}$. We have, by (2) and (3),

$$|u_{j+1}(z) - u_j(z)| \leqslant k^j |v(y_0, z) - y_0| \leqslant k^j(1-k)\beta \qquad (j = 0, 1, 2, \ldots, z \in Z).$$

Thus, for $p < q$, $p, q \in \mathbb{N}$,

$$|u_q(z) - u_p(z)| \leqslant \sum_{j=p}^{q-1} |u_{j+1}(z) - u_j(z)| \leqslant \beta(1-k)\,k^p \sum_{j=0}^{\infty} k^j = \beta k^p \qquad (z \in Z)$$

and $\lim_p k^p = 0$. It follows, by I.2.19, that there exists a continuous $u : Z \to \mathscr{Y}$ such that

$$\lim_j u_j(z) = u(z) \quad \text{uniformly for} \quad z \in Z.$$

Since $u_j(z) \in S^F(y_0, \beta) \subset Y$ ($j \in \mathbb{N}$, $z \in Z$), we have $u(z) \in Y$ ($z \in Z$). Furthermore,

$$u(z) = \lim_j u_j(z) = \lim_j v(u_{j-1}(z), z) = v(u(z), z) \qquad (z \in Z). \quad \text{QED}$$

II.3.8 ***Theorem*** (Implicit function theorem) Let Y be an open subset of a Banach space $\mathscr{Y}$, Z a topological space, $f: Y \times Z \to \mathscr{Y}$ and $f_y : Y \times Z \to B(\mathscr{Y}, \mathscr{Y})$ continuous, $(y_0, z_0) \in Y \times Z$, $f(y_0, z_0) = 0$, and $f_y(y_0, z_0)$ a homeomorphism of $\mathscr{Y}$ onto $\mathscr{Y}$. Then

(1) there exist open subsets Y_1 of Y and Z_1 of Z and a unique continuous $u : Z_1 \to Y_1$ such that $z_0 \in Z_1$ and

$$f(u(z), z) = 0 \qquad (z \in Z_1);$$

(2) if, furthermore, Z is a subset of a Banach space and f has a partial derivative $f_z(y_0, z_0)$, then u has a derivative $u'(z_0)$ and

$$u'(z_0) = -f_y(y_0, z_0)^{-1} \circ f_z(y_0, z_0);$$

and

(3) if W is a topological space, A a subset of a Banach space $\mathscr{X}_A$, $Z \triangleq A \times W$, $z \triangleq (a, w)$, and f_a exists and is continuous on $Y \times Z$, then there exists a derivative $u_a(a, w)$ for all (a, w) in some relatively open subset $\tilde{A} \times \tilde{W}$ of Z_1, defined by

$$u_a(a, w) = -f_y(a, w, u(a, w))^{-1} \circ f_a(a, w, u(a, w)),$$

and the function $(a, w) \to u_a(a, w) : \tilde{A} \times \tilde{W} \to B(\mathscr{X}_A, \mathscr{Y})$ is continuous.

▌PROOF Let $T_0 \triangleq f_y(y_0, z_0)$ and

$$v(y, z) \triangleq y - T_0^{-1} f(y, z) \qquad [(y, z) \in Y \times Z].$$

We have, for $y_1, y_2 \in Y$ and $z \in Z$,

$$(4) \quad v(y_1, z) - v(y_2, z) = -T_0^{-1}[f(y_1, z) - f(y_2, z) - T_0(y_1 - y_2)].$$

Let $\beta > 0$ be such that $S^F(y_0, \beta) \subset Y$. Then, for all $y_1, y_2 \in S^F(y_0, \beta)$ and $z \in Z$ we have, by the mean-value theorem II.3.6,

$$(5) \quad |f(y_1, z) - f(y_2, z) - f_y(y_0, z)(y_1 - y_2)| \leqslant |y_1 - y_2| \sup_{0 \leqslant \alpha \leqslant 1} |f_y(y_1 + \alpha(y_2 - y_1), z) - f_y(y_0, z)|.$$

Since f_y is continuous, we can determine open neighborhoods Z_1 of z_0 and $Y_1 \triangleq S(y_0, \beta_1) \subset Y$ such that $\beta_1 \leqslant \beta$,

$$|f_y(y, z) - f_y(y_0, z)| \leqslant \tfrac{1}{4} |T_0^{-1}|^{-1} \qquad (y \in Y_1, \quad z \in Z_1)$$

and

$$|f_y(y_0, z) - f_y(y_0, z_0)| \leqslant \tfrac{1}{4} | T_0^{-1} |^{-1} \qquad (z \in Z_1).$$

Thus, by (5),

$$\begin{aligned} |f(y_1, z) - f(y_2, z) - T_0(y_1 - y_2)| &\leqslant |f(y_1, z) - f(y_2, z) \\ &\qquad - f_y(y_0, z)(y_1 - y_2)| \\ &\qquad + |f_y(y_0, z) - f_y(y_0, z_0)| \, | y_1 - y_2 | \\ &\leqslant \tfrac{1}{2} | T_0^{-1} |^{-1} | y_1 - y_2 |; \end{aligned}$$

hence, by (4),

(6) $$|v(y_1, z) - v(y_2, z)| \leqslant \tfrac{1}{2} | y_1 - y_2| \qquad (y_1, y_2 \in Y_1, z \in Z_1).$$

Since $f(y_0, z_0) = 0$, we have $v(y_0, z_0) - y_0 = 0$. Since f and v are continuous, we may assume that Z_1 was chosen so that

$$|v(y_0, z) - y_0| \leqslant \beta_1/2 \qquad (z \in Z_1).$$

Thus the conditions of II.3.7 are satisfied with $k = \frac{1}{2}$ and Y, Z, and β replaced by Y_1, Z_1, and β_1, respectively. It follows that there exists a continuous $u : Z_1 \to Y_1$ such that $f(u(z), z) = 0$ $(z \in Z_1)$. This function u must be unique since $u_1 : Z_1 \to Y_1$, $u_2 : Z_1 \to Y_1$, and $f(u_i(z), z) = 0$ $(i = 1, 2; z \in Z_1)$ imply $u_i(z) = v(u_i(z), z)$ and, by (6),

$$|u_1(z) - u_2(z)| \leqslant \tfrac{1}{2} | u_1(z) - u_2(z)|.$$

Now assume that the conditions of (2) are satisfied, and set

$$\tilde{u}(z) \triangleq u(z) - u(z_0) = u(z) - y_0 \qquad (z \in Z_1).$$

Then, by the mean value theorem II.3.6,

(7) $$\begin{aligned} &|f(y_0 + \tilde{u}(z), z) - f(y_0, z) - f_y(y_0, z)\, \tilde{u}(z)| \\ &\quad \leqslant |\tilde{u}(z)| \sup_{0 \leqslant \alpha \leqslant 1} | f_y(y_0 + \alpha\tilde{u}(z), z) - f_y(y_0, z)| \quad (z \in Z_1). \end{aligned}$$

Furthermore, for every $\epsilon > 0$ there exists $\delta(\epsilon) > 0$ such that

(8) $$\begin{aligned} &|f(y_0, z) - f(y_0, z_0) - f_z(y_0, z_0)(z - z_0)| \\ &\quad \leqslant \epsilon | z - z_0| \qquad \text{if} \quad | z - z_0| \leqslant \delta(\epsilon). \end{aligned}$$

Let $0 < \epsilon < \frac{1}{4} | T_0^{-1}|^{-1}$. We may choose $\delta(\epsilon)$ [and consequently $| \tilde{u}(z)|$ for $| z - z_0| \leqslant \delta(\epsilon)$] sufficiently small so that

$$\sup_{0 \leqslant \alpha \leqslant 1} | f_y(y_0 + \alpha\tilde{u}(z), z) - f_y(y_0, z)| \leqslant \epsilon$$

and

$$|f_y(y_0, z) - f_y(y_0, z_0)| \leqslant \epsilon$$

if $|z - z_0| \leqslant \delta(\epsilon)$. Then (7) and (8) yield

$$\begin{aligned}|f(y_0 + \tilde{u}(z), z) - f(y_0, z_0) - f_z(y_0, z_0)(z - z_0) - f_y(y_0, z_0)\tilde{u}(z)| \\ \leqslant \text{left side of (7)} + \text{left side of (8)} + |f_y(y_0, z)\tilde{u}(z) - f_y(y_0, z_0)\tilde{u}(z)| \\ \leqslant 2\epsilon\,|\tilde{u}(z)| + \epsilon\,|z - z_0|.\end{aligned}$$

Since $f(y_0 + \tilde{u}(z), z) = f(u(z), z) = f(y_0, z_0) = 0$, the last inequality implies

$$|T_0\tilde{u}(z) + f_z(y_0, z_0)(z - z_0)| \leqslant 2\epsilon\,|\tilde{u}(z)| + \epsilon\,|z - z_0|;$$

hence

$$\text{(9)} \quad \begin{aligned}|\tilde{u}(z) + T_0^{-1} \circ f_z(y_0, z_0)(z - z_0)| &\leqslant |T_0^{-1}|\,\epsilon(2\,|\tilde{u}(z)| + |z - z_0|) \\ &\leqslant \tfrac{1}{2}\,|\tilde{u}(z)| + \tfrac{1}{4}\,|z - z_0|\end{aligned}$$

and

$$\text{(10)} \quad |\tilde{u}(z)| \leqslant \alpha\,|z - z_0| \qquad \text{for} \quad \alpha \triangleq 2(\tfrac{1}{4} + |T_0^{-1}|\,|f_z(y_0, z_0)|).$$

The first inequality of (9) and (10) now yield

$$|\tilde{u}(z) + T_0^{-1} \circ f_z(y_0, z_0)(z - z_0)| \leqslant |T_0^{-1}|\,(2\alpha + 1)\,\epsilon|\,z - z_0|$$

for all $z \in Z_1$ with $|z - z_0| \leqslant \delta(\epsilon)$. This shows that $\tilde{u}$ and u have a derivative at z_0 defined by the relation

$$\tilde{u}'(z_0) = u'(z_0) = -f_y(y_0, z_0)^{-1} \circ f_z(y_0, z_0).$$

Finally, assume that the conditions of (3) are satisfied, and let $z_0 \triangleq (a_0, w_0)$. By I.3.7, there exist relatively open sets $\tilde{Y} \subset Y$, $\tilde{A} \subset A$, and $\tilde{W} \subset W$ such that

$$(y_0, a_0, w_0) \in \tilde{Y} \times \tilde{A} \times \tilde{W}$$

and

$$f_y(y, a, w)^{-1} \in B(\mathscr{Y}, \mathscr{Y}) \qquad \text{for all} \quad (y, a, w) \in \tilde{Y} \times \tilde{A} \times \tilde{W}.$$

We may clearly assume that $\tilde{A}$ is convex. Then our previous argument holds if we fix $w \in \tilde{W}$, with Y_1 replaced by $\tilde{Y} \cap Y_1$ and Z_1 by $(\tilde{A} \times \{w\}) \cap Z_1$, and this argument shows [replacing $z_0 \triangleq (a_0, w_0)$ by $(a, w) \in (\tilde{A} \times \tilde{W}) \cap Z_1$] that

$$u_a(a, w) \triangleq -f_y(u(a, w), a, w)^{-1} \circ f_a(u(a, w), a, w)$$

is the derivative of the function $a \to u(a, w)$ for all a sufficiently close to a_0.

Since u, f_y, and f_a are continuous, it follows that $(a, w) \to u_a(a, w)$ is also continuous. QED

II.3.9 ***Theorem*** Let $\mathscr{X}$ and $\mathscr{Y}$ be separable Banach spaces, (S, Σ, μ) a finite measure space, $\bar{a} \in A \subset \mathscr{X}$, and $h : S \times A \to \mathscr{Y}$. If the function $h(s, \cdot)$ has a derivative $h_a(s, \bar{a})$ for all $s \in S$ and $h(\cdot, a)$ is μ-measurable for all $a \in A$, then, for every $\Delta a \in \mathscr{X}$, the functions $s \to |h_a(s, \bar{a})|$ and $s \to h_a(s, \bar{a})\, \Delta a$ are μ-measurable. If, furthermore, $n \in \mathbb{N}$ and $\mathscr{X} = \mathbb{R}^n$, then $s \to h_a(s, \bar{a})$ is μ-measurable.

▌PROOF By the definition of a derivative, $\bar{a}$ belongs to a convex subset A_0 of A with a nonempty interior. Now let $\Delta a \in A_0 - \bar{a}$. We have

$$\lim_{\alpha \to +0} \alpha^{-1} | h(s, \bar{a} + \alpha\, \Delta a) - h(s, \bar{a}) - \alpha h_a(s, \bar{a})\, \Delta a | = 0 \qquad \text{for each} \quad s \in S;$$

hence

$$h_a(s, \bar{a})\, \Delta a = \lim_{\alpha \to +0} \alpha^{-1}[h(s, \bar{a} + \alpha\, \Delta a) - h(s, \bar{a})] \qquad \text{for all} \quad s \in S.$$

Since, by assumption, $h(\cdot, a)$ is μ-measurable for all $a \in A$, it follows from I.4.17 that the function $s \to h_a(s, \bar{a})\, \Delta a$ is μ-measurable.

Now let $\Delta a \in \mathscr{X}$. Then there exist a_1, $a_2 \in A_0^\circ$ and $\beta \in \mathbb{R}$ such that

$$\Delta a = \beta(a_2 - a_1) = \beta(a_2 - \bar{a}) - \beta(a_1 - \bar{a}).$$

It follows that

$$s \to h_a(s, \bar{a})\, \Delta a = \beta h_a(s, \bar{a})(a_2 - \bar{a}) - \beta h_a(s, \bar{a})(a_1 - \bar{a})$$

is μ-measurable and so is, by I.4.20 and I.4.21,

$$s \to | h_a(s, \bar{a})| = \sup_{|\Delta a| \leqslant 1} | h_a(s, \bar{a})\, \Delta a |.$$

Now let $\{x_1, x_2, \ldots\}$ be a dense subset of the closed unit ball $S^F(0, 1)$ in $\mathscr{X}$, $\beta \in \mathbb{R}$, and $T \in B(\mathscr{X}, \mathscr{Y})$. Then

$$\begin{aligned} \{s \in S \mid | h_a(s, \bar{a}) - T | \leqslant \beta\} &= \{s \in S \mid \sup_{|x| \leqslant 1} |(h_a(s, \bar{a}) - T)x | \leqslant \beta\} \\ &= \{s \in S \mid \sup_i |(h_a(s, \bar{a})\, x_i - Tx_i | \leqslant \beta\} \\ &= \bigcap_{i=1}^{\infty} \{s \in S \mid | h_a(s, \bar{a})\, x_i - Tx_i | \leqslant \beta\}. \end{aligned}$$

This last set is μ-measurable, showing that $s \to h_a(s, \bar{a})$ is μ-measurable if $B(\mathscr{X}, \mathscr{Y})$ is separable. This is the case, in particular, if $\mathscr{X} = \mathbb{R}^n$ because $B(\mathbb{R}^n, \mathscr{Y})$ is homeomorphic to $\mathscr{Y}^n$ and therefore separable. QED

II.3.10 *Theorem* Let $\mathscr{X}$ and $\mathscr{Y}$ be Banach spaces, $\mathscr{Y}$ separable, S a compact metric space, $\lambda \in \text{frm}(S)$, $A \subset \mathscr{X}$, $h : S \times A \to \mathscr{Y}$ continuous, and $f(a) \triangleq \int h(s, a)\lambda(ds)$ $(a \in A)$. Then $f : A \to \mathscr{Y}$ is continuous and

$$|f(a)| \leqslant |\lambda|(S) \cdot \sup_{s \in S} |h(s, a)| \qquad (a \in A).$$

If, furthermore, $h(s, \cdot)$ has a derivative $h_a(s, a)$ for all $(s, a) \in S \times A$, A is convex, and $h_a : S \times A \to B(\mathscr{X}, \mathscr{Y})$ is continuous, then f has a derivative $f'(a)$ for all $a \in A$,

$$f'(a)\, \Delta a = \int h_a(s, a)\, \Delta a\, \lambda(ds) \qquad (a \in A, \quad \Delta a \in \mathscr{X}),$$

$$f' : A \to B(\mathscr{X}, \mathscr{Y}) \qquad \text{is continuous,}$$

and

$$|f'(a)| \leqslant |\lambda|(S) \cdot \sup_{s \in S} |h_a(s, a)| \qquad (a \in A).$$

▌PROOF For each $a \in A$, $h(\cdot, a)$ is continuous on the compact set S and therefore λ-integrable. If $\lim_j a_j = a$ in A, then, by I.2.15(2), $\lim_j h(s, a_j) = h(s, a)$ uniformly for $s \in S$ and therefore $\lim_j f(a_j) = f(a)$. Thus f is continuous on A and we verify directly that

$$|f(a)| \leqslant |\lambda|(S) \cdot \sup_{s \in S} |h(s, a)| \qquad (a \in A).$$

Now assume that h_a is continuous on $S \times A$. Then the functions $s \to h_a(s, a)\, \Delta a : S \to \mathscr{Y}$ $(a \in A,\ \Delta a \in \mathscr{X})$, being continuous on the compact set S, are λ-integrable. Let $a \in A$, A be convex, and $0 \neq \Delta a \in A - a$. Then, by the mean value theorem II.3.6,

$$\begin{aligned}
&|\Delta a|^{-1} \left| f(a + \Delta a) - f(a) - \int h_a(s, a)\, \Delta a\, \lambda(ds) \right| \\
&\quad = |\Delta a|^{-1} \left| \int (h(s, a + \Delta a) - h(s, a) - h_a(s, a)\, \Delta a)\, \lambda(ds) \right| \\
&\quad \leqslant \int \sup_{0 \leqslant \alpha \leqslant 1} |h_a(s, a + \alpha\, \Delta a) - h_a(s, a)|\, |\lambda|(ds) \\
&\quad \leqslant |\lambda|(S) \cdot \sup_{(s, \beta) \in S \times [0,1]} |h_a(s, a + \beta\, \Delta a) - h_a(s, a)|.
\end{aligned}$$

By I.2.15, the right-hand side converges to 0 with $|\Delta a|$. Furthermore, the function $\Delta a \to \int h_a(s, a)\, \Delta a\, \lambda(ds)$ is clearly linear and

$$\left| \int h_a(s, a)\, \Delta a\, \lambda(ds) \right| \leqslant |\lambda|(S)\, |\Delta a| \cdot \sup_{s \in S} |h_a(s, a)|.$$

This shows that the function $\Delta a \to \int h_a(s, a)\, \Delta a\, \lambda(s)$ is a derivative of f at a and $|f'(a)| \leqslant |\lambda|(S) \cdot \sup_{s\in S} |h_a(s, a)|$ for all $a \in A$. If (a_j) is a sequence in A converging to a, then [I.2.15(2)] $\lim_j h_a(s, a_j) = h_a(s, a)$ uniformly for $s \in S$. This implies that for $|\Delta a| \leqslant 1$

$$|[f'(a_j) - f'(a)]\, \Delta a| \leqslant \int |h_a(s, a_j) - h_a(s, a)|\, |\lambda|\, (ds) \underset{j}{\to} 0,$$

and thus $\lim_j f'(a_j) = f'(a)$; hence f' is continuous. QED

II.3.11 ***Theorem*** Let $\mathscr{X}$ and $\mathscr{Y}$ be Banach spaces, $A \subset \mathscr{X}$, $B \subset \mathscr{Y}$, $\phi : A \to B$ a homeomorphism with a derivative $\phi'(\bar{a})$ at $\bar{a} \in A$, and $\phi'(\bar{a})$ a homeomorphism of $\mathscr{X}$ onto $\mathscr{Y}$. If $\phi(\bar{a})$ is contained in a convex subset of B with a nonempty interior, then $\phi'(\bar{a})^{-1}$ is the derivative of ϕ^{-1} at $\phi(\bar{a})$.

▌PROOF Let $\bar{b} \triangleq \phi(\bar{a})$. For every $\epsilon > 0$ there exist $\eta(\epsilon) > 0$ and $\delta(\epsilon) > 0$ such that

$$|\phi^{-1}(b) - \phi^{-1}(\bar{b})| \leqslant \epsilon \qquad \text{if} \quad |b - \bar{b}| \leqslant \eta(\epsilon)$$

and

$$|\phi(a) - \phi(\bar{a}) - \phi'(\bar{a})(a - \bar{a})| \leqslant \epsilon\, |a - \bar{a}| \qquad \text{if} \quad |a - \bar{a}| \leqslant \delta(\epsilon).$$

Now let $\epsilon > 0$ be fixed and $|b - \bar{b}| \leqslant \eta(\delta(\epsilon))$. Then

$$|\phi^{-1}(b) - \phi^{-1}(\bar{b})| \leqslant \delta(\epsilon)$$

and

$$|\phi(\phi^{-1}(b)) - \phi(\phi^{-1}(\bar{b})) - \phi'(\bar{a})(\phi^{-1}(b) - \phi^{-1}(\bar{b}))| \leqslant \epsilon\, |\phi^{-1}(b) - \phi^{-1}(\bar{b})|;$$

hence

$$\text{(1)} \qquad |b - \bar{b} - \phi'(\bar{a})(\phi^{-1}(b) - \phi^{-1}(\bar{b}))| \leqslant \epsilon\, |\phi^{-1}(b) - \phi^{-1}(\bar{b})|$$

and therefore

$$|\phi^{-1}(b) - \phi^{-1}(\bar{b})| \leqslant |\phi'(\bar{a})^{-1}|\big(|b - \bar{b}| + \epsilon\, |\phi^{-1}(b) - \phi^{-1}(\bar{b})|\big).$$

It follows that for $\epsilon \leqslant \frac{1}{2} |\phi'(\bar{a})^{-1}|^{-1}$ we have

$$\text{(2)} \qquad |\phi^{-1}(b) - \phi^{-1}(\bar{b})| \leqslant 2\, |\phi'(\bar{a})^{-1}|\, |b - \bar{b}|.$$

We now observe that, by (1) and (2),

$$\begin{aligned} |\phi^{-1}(b) - \phi^{-1}(\bar{b}) - \phi'(\bar{a})^{-1}(b - \bar{b})| &\leqslant \epsilon\, |\phi'(\bar{a})^{-1}|\, |\phi^{-1}(b) - \phi^{-1}(\bar{b})| \\ &\leqslant 2\epsilon\, |\phi'(\bar{a})^{-1}|^2\, |b - \bar{b}|. \end{aligned}$$

This shows that $\phi'(\bar{a})^{-1}$ is the derivative of ϕ^{-1} at $\bar{b} \triangleq \phi(\bar{a})$. QED

II.3.12 ***Derivatives of* exp(·), log(·) *and of Products*** By I.3.15, we have $\exp(x + y) = \exp(x)\exp(y)$ $(x, y \in \mathbb{R})$. Furthermore, for $h \neq 0$, $|h| \leqslant \frac{1}{2}$, we have

$$|h^{-1}[\exp(h) - 1] - 1| = \left|\sum_{j=2}^{\infty} h^{j-1}/j!\right| \leqslant |h| \sum_{j=2}^{\infty} |h|^{j-2} \leqslant 2|h|;$$

hence $\lim_{h\to 0} h^{-1}[\exp(h) - 1] = 1$. Thus, for any $x \in \mathbb{R}$,

$$\lim_{h\to 0} h^{-1}[\exp(x + h) - \exp(x)] = \exp(x),$$

yielding

$$\exp'(x) = \exp(x) \qquad (x \in \mathbb{R}). \tag{1}$$

By I.3.15, $\exp : \mathbb{R} \to (0, \infty)$ is a bijection and, since $\exp'(x)$ exists for each $x \in \mathbb{R}$, $\exp(\cdot)$ is continuous. It follows, by I.2.9(4) that, for every closed interval $[a_0, a_1] \subset \mathbb{R}$, $\exp(\cdot) : [a_0, a_1] \to \exp([a_0, a_1])$ is a homeomorphism, from which we conclude that $\exp(\cdot) : \mathbb{R} \to (0, \infty)$ is a homeomorphism. Thus $\log(\cdot) : (0, \infty) \to \mathbb{R}$ is continuous.

Since $\exp(x) \neq 0$ $(x \in \mathbb{R})$, we may apply II.3.11 and (1) to obtain

$$\log'(\exp(x)) = \exp(x)^{-1} \qquad (x \in \mathbb{R})$$

and, setting $t = \exp(x)$,

$$\log'(t) = 1/t \qquad [t \in (0, \infty)]. \tag{2}$$

Now let $[t_0, t_1] \subset \mathbb{R}$, $k \in \mathbb{N}$, and

$$t_0 \leqslant a_1 < b_1 \leqslant a_2 < b_2 < \cdots \leqslant a_k < b_k \leqslant t_1.$$

Since $\exp(\cdot)$ is increasing on $[t_0, t_1]$, we have $0 < \exp(t) \leqslant \exp(t_1)$ $(t \in [t_0, t_1])$. We apply (1) and the mean value theorem II.3.5 to obtain

$$|\exp(t'') - \exp(t')| \leqslant \exp(t_1)|t'' - t'| \qquad (t', t'' \in [t_0, t_1]);$$

hence

$$\sum_{j=1}^{n} |\exp(b_j) - \exp(a_j)| \leqslant \exp(t_1) \sum_{j=1}^{n} (b_j - a_j).$$

This shows that $\exp(\cdot) : [t_0, t_1] \to [\exp(t_0), \exp(t_1)]$ is absolutely continuous. It follows, by I.4.42 and (1), that

$$\exp(t) = \exp(t_0) + \int_{t_0}^{t} \exp(\tau)\, d\tau \qquad (t \in [t_0, t_1]). \tag{3}$$

A similar argument, employing (2) instead of (1), shows that if $0 < t_0 < t_1 < \infty$, then

$$\log(t) = \log t_0 + \int_{t_0}^{t} d\tau/\tau \qquad (t \in [t_0, t_1]). \tag{4}$$

Finally, we observe that if $\mathscr{Y}$, $\mathscr{Z}$, and $\mathscr{V}$ are Banach spaces and $p(T, S) \triangleq TS$ $[T \in B(\mathscr{Z}, \mathscr{V}), S \in B(\mathscr{Y}, \mathscr{Z})]$, then

$$\begin{aligned}(|\Delta T| + |\Delta S|)^{-1} &| p(T + \Delta T, S + \Delta S) - T \cdot S - T \cdot \Delta S - \Delta T \cdot S | \\ &= (|\Delta T| + |\Delta S|)^{-1} | \Delta T \, \Delta S | \leqslant (|\Delta T| + |\Delta S|)^{-1} | \Delta T | \, | \Delta S | \\ &\leqslant |\Delta T| + |\Delta S| \qquad [\Delta T \in B(\mathscr{Z}, \mathscr{V}), \ \ \Delta S \in B(\mathscr{Y}, \mathscr{Z})];\end{aligned}$$

hence

$$p'(T, S)(\Delta T, \Delta S) = \Delta T \cdot S + T \cdot \Delta S.$$

Now let $\mathscr{X}$ be a Banach space, $\bar{a} \in A \subset \mathscr{X}$, the functions $f : A \to B(\mathscr{Z}, \mathscr{V})$ and $g : A \to B(\mathscr{Y}, \mathscr{Z})$ differentiable at $\bar{a}$, and $(f \cdot g)(a) \triangleq f(a) \cdot g(a)$ $(a \in A)$. Then, by the chain rule II.3.4,

$$\begin{aligned}(f \cdot g)'(\bar{a}) \, \Delta x &= \mathscr{D}[p \circ (f, g)](\bar{a}) \, \Delta x \\ &= f'(\bar{a}) \, \Delta x \cdot g(\bar{a}) + f(\bar{a}) \cdot g'(a) \, \Delta x \qquad (\Delta x \in \mathscr{X})\end{aligned}$$

which we write in the form

$$(f \cdot g)'(\bar{a}) = f'(\bar{a}) \cdot g(\bar{a}) + f(\bar{a}) \cdot g'(\bar{a}), \tag{5}$$

where $f'(\bar{a}) \cdot g(\bar{a})$ denotes the function

$$\Delta x \to (f'(a) \, \Delta x) \cdot g(\bar{a}) : \mathscr{X} \to B(\mathscr{Y}, \mathscr{V}).$$

II.4 Ordinary Differential Equations

Caratheodory Functions

We shall find it convenient, when discussing particular types of functional equations (such as ordinary differential or functional-integral equations), to refer to certain spaces whose elements we call *Caratheodory functions*. Some of these spaces have been introduced in I.5.25. Specifically, let (T, Σ, μ) be a positive finite measure space, S a topological space, and $\mathscr{X}$ a Banach space. We shall identify two functions ϕ_1 and ϕ_2 on $T \times S$ to $\mathscr{X}$ if $\phi_1(t, \cdot) = \phi_2(t, \cdot)$ μ-a.e., and will denote by $\mathscr{B}(T, \Sigma, \mu, S; \mathscr{X})$ [or $\mathscr{B}(T, S; \mathscr{X})$ or $\mathscr{B}$] the vector space of (equivalence classes of) functions $\phi : T \times S \to \mathscr{X}$ such that, for all $(t, s) \in T \times S$, $\phi(t, \cdot) \in C(S, \mathscr{X})$, $\phi(\cdot, s)$ is μ-measurable and there exists a μ-integrable $\psi_\phi : T \to \mathbb{R}$ with $|\phi(t, \cdot)|_{\sup} \leqslant \psi_\phi(t)$. We refer to elements of $\mathscr{B}$ as *Caratheodory functions* because the properties of the elements of $\mathscr{B}$ were apparently first systematically exploited by Caratheodory in his existence theorem for differential equations (see II.4.1 below).

We have shown in I.5.25 that if S is a separable metric space and $\mathscr{X}$ a separable Banach space, then the function

$$\phi \to |\phi|_{\mathscr{B}} \triangleq \int |\phi(t, \cdot)|_{\sup}\, \mu(dt) : \mathscr{B} \to \mathbb{R}$$

is a norm on $\mathscr{B}$. We shall consider $\mathscr{B}$ in such cases to be a normed vector space, and shall also use, without further justification, another result of I.5.25, stating that the function $t \to \phi(t, \xi(t)) : T \to \mathscr{X}$ is μ-integrable for every choice of $\phi \in \mathscr{B}$ and a μ-measurable $\xi : T \to S$.

We have also shown, in I.5.25, that $\mathscr{B}$ is isometrically isomorphic to the Banach space $L^1(T, \Sigma, \mu, C(S, \mathscr{X}))$ whenever S is a compact metric space and $\mathscr{X}$ is separable.

THE SPACE $B(\mathbb{R}^m, \mathbb{R}^n)$

For every Banach space $\mathscr{X}$ and $m \in \mathbb{N}$, the space $B(\mathbb{R}^m, \mathscr{X})$ is homeomorphic to $\mathscr{X}^m$. Thus $B(\mathbb{R}^m, \mathbb{R}^n)$ is separable for all $m, n \in \mathbb{N}$, and we shall use this result without specific mention.

As remarked at the beginning of I.5.C, we identify a function $h : A \times B \to C$ with the function $a \to h(a, \cdot)$.

II.4.A Existence of Local Solutions

Let $T \triangleq [t_0, t_1] \subset \mathbb{R}$, $n \in \mathbb{N}$, $V \subset \mathbb{R}^n$, and f be a function on $T \times V$ to $\mathbb{R}^n$. For a given $\eta \in V$, we consider the initial value problem defined by the equations

$$(*) \qquad \begin{aligned} \dot{y}(t) &= f(t, y(t)) \quad \text{a.e. in } T, \\ y(t_0) &= \eta, \end{aligned}$$

where $y : T \to V$ is restricted to be absolutely continuous. We recall that, by I.4.42, a function $y(\cdot)$ on T is absolutely continuous if and only if

$$y(t'') - y(t') = \int_{t'}^{t''} \dot{y}(\tau)\, d\tau$$

for all $t', t'' \in T$. Thus the problem defined by (*) is equivalent to the problem of finding a continuous function $y(\cdot) : T \to \mathbb{R}^n$ satisfying

$$(**) \qquad y(t) = \eta + \int_{t_0}^{t} f(\tau, y(\tau))\, d\tau \qquad (t \in T),$$

since any solution of (**) is absolutely continuous and satisfies (*), and any absolutely continuous solution of (*) must satisfy (**).

Equations (*) and (**) are defined in terms of a function f on $T \times V$. If the function f is independent of its argument in T [that is, if $f(t, v) = g(v)$ $(t \in T,\ v \in V)$ for some $g : V \to \mathbb{R}^n$], then we write $f(v)$ for $f(t, v)$ and refer to Eqs. (*) and (**) as *autonomous*. We can transform every equation of the form (*) or (**) into an autonomous one. This we do by transforming our equation into another equation with the unknown y replaced by $t \to (y(t), t)$. We set

$$\hat{V} \triangleq V \times T, \qquad \hat{\eta} \triangleq (\eta, t_0),$$
$$\hat{f}^i(\hat{v}) \triangleq f^i(v^{n+1}, v) \qquad [i = 1, 2, \ldots, n, \quad \hat{v} \triangleq (v, v^{n+1}) \triangleq (v^1, v^2, \ldots, v^{n+1}) \in \hat{V}],$$
$$\hat{f}^{n+1}(\hat{v}) \triangleq 1 \qquad (\hat{v} \in \hat{V}),$$

and consider the equation

$$(***) \qquad \hat{y}(t) = \hat{\eta} + \int_{t_0}^{t} \hat{f}(\hat{y}(\tau))\, d\tau \qquad (t \in T).$$

It is clear that for any solution $\hat{y} \triangleq (y^1, \ldots, y^{n+1})$ of this equation we have $y^{n+1}(t) = t$ $(t \in T)$ and that $y \triangleq (y^1, \ldots, y^n)$ satisfies (**). Conversely, if $y \triangleq (y^1, \ldots, y^n)$ satisfies (**), then the function $t \to \hat{y}(t) \triangleq (y^1(t), \ldots, y^n(t), t)$ satisfies (***). Thus Eqs. (*) and (**) are equivalent to the autonomous equation (***).

We begin by investigating the existence of local solutions. We write, throughout II.4, $\mathscr{B}(T, S; \mathscr{X})$ for $\mathscr{B}(T, \mathscr{M}, m, S; \mathscr{X})$, where $(T, \mathscr{M}, m)$ refers to the Lebesgue measure in T.

II.4.1 ***Theorem*** (Caratheodory) Let $T \triangleq [t_0, t_1] \subset \mathbb{R}$, $n \in \mathbb{N}$, $V \subset \mathbb{R}^n$, and $f \in \mathscr{B}(T, V; \mathbb{R}^n)$. If $\eta \in V$, $b > 0$, $S^F(\eta, b) \subset V$,

$$\tilde{t}_1 \triangleq \sup \left\{ t \in T \,\middle|\, \int_{t_0}^{t} |f(\tau, \cdot)|_{\sup}\, d\tau \leqslant b \right\}$$

and $\tilde{T} \triangleq [t_0, \tilde{t}_1]$, then there exists a function $\bar{y}(\cdot) : \tilde{T} \to V$ such that

$$\bar{y}(t) = \eta + \int_{t_0}^{t} f(\tau, \bar{y}(\tau))\, d\tau \qquad (t \in \tilde{T}).$$

▌PROOF Let

$$K \triangleq \{y \in C(\tilde{T}, \mathbb{R}^n) \mid |y(t) - \eta| \leqslant b \ (t \in \tilde{T})\}.$$

Since $f \mid \tilde{T} \times V$ belongs to $\mathscr{B}(\tilde{T}, V; \mathbb{R}^n)$, the function

$$\tau \to f(\tau, y(\tau)) : \tilde{T} \to \mathbb{R}^n$$

is integrable for each $y \in K$. Thus

$$F(y)(t) \triangleq \eta + \int_{t_0}^{t} f(\tau, y(\tau))\, d\tau$$

is defined for all $t \in \tilde{T}$ and $y \in K$, and $F(y)(\cdot) \in C(\tilde{T}, \mathbb{R}^n)$. We have

$$|F(y)(t) - \eta| \leqslant \int_{t_0}^{t} |f(t, \cdot)|_{\sup}\, d\tau \leqslant b \qquad (t \in \tilde{T}),$$

which shows that $F(K) \subset K$. Furthermore,

$$|F(y)(t) - F(y)(t')| \leqslant \int_{t'} |f(\tau, \cdot)|_{\sup}\, d\tau \qquad (t_0 \leqslant t' \leqslant t \leqslant \tilde{t}_1, \quad y \in K),$$

and thus $F(K)$ is a bounded and equicontinuous subset of $C(T, \mathbb{R}^n)$. It follows, by I.5.4, that $\overline{F(K)}$ is compact.

Now let $\lim_j y_j = y$ in K. Then

$$\lim_j f(\tau, y_j(\tau)) = f(\tau, y(\tau)) \qquad (\tau \in \tilde{T})$$

and

$$|f(\tau, y_j(\tau))| \leqslant |f(\tau, \cdot)|_{\sup} \qquad (\tau \in \tilde{T}, \quad j \in \mathbb{N}).$$

Therefore, by the dominated convergence theorem I.4.35,

$$\lim_j F(y_j)(t) = F(y)(t) \qquad (t \in \tilde{T})$$

and, since $F(K)$ is equicontinuous and $\tilde{T}$ compact, it follows from I.5.3 that $\lim_j F(y_j) = F(y)$ in $C(\tilde{T}, \mathbb{R}^n)$. Thus $F : K \to K$ is continuous. We verify that K is closed and convex. It follows now, by Schauder's fixed point theorem II.2.8, that $F : K \to K$ has a fixed point $\bar{y}$, thus completing the proof. QED

II.4.B Extension of Local Solutions and Uniqueness

The existence theorem that we have obtained applies to solutions defined in some subinterval $\tilde{T}$ of the original domain T. The question arises as to whether these *local solutions* can be extended to the entire interval T.

II.4.2 ***Theorem*** Let $n \in \mathbb{N}$, V be an open subset of $\mathbb{R}^n$, T a closed interval $[t_0, t_1] \subset \mathbb{R}$, $\eta \in V$, and $f \in \mathscr{B}(T, V; \mathbb{R}^n)$. Then either there exists a function $\bar{y} : T \to V$ such that, for all $t \in T$,

$$\bar{y}(t) = \eta + \int_{t_0}^{t} f(\tau, \bar{y}(\tau))\, d\tau, \tag{1}$$

or there exists a point $t_1' \in (t_0, t_1]$ and a function $\bar{y} : [t_0, t_1') \to V$ satisfying relation (1) for all $t \in [t_0, t_1')$, and such that $\lim_{t \to t_1'} \bar{y}(t)$ exists and

$$\lim_{t \to t_1'} \bar{y}(t) \in \partial V.$$

▌PROOF Let

$$h(\beta, \tau) \triangleq \sup \left\{ \alpha \in [0, t_1 - \tau) \;\middle|\; \int_{\tau}^{\tau+\alpha} |f(s, \cdot)|_{\sup}\, ds \leqslant \beta \right\}$$
$$(\beta \in (0, \infty], \quad \tau \in T).$$

Then, for each fixed $\tau \in T$, $\beta \to h(\beta, \tau)$ is nonnegative and nondecreasing and, by I.4.28, for all sequences (τ_j) in T and (β_j) in $(0, \infty)$ with $\tau_{j+1} \geqslant \tau_j$,

$$\lim_j h(\beta_j, \tau_j) = 0 \quad \text{implies that either} \quad \lim_j \beta_j = 0 \quad \text{or} \quad \lim_j \tau_j = t_1. \tag{2}$$

Now let

$$d(\xi) \triangleq d[\xi, \mathbb{R}^n \sim V] \qquad (\xi \in V)$$

[which implies, in particular, that $d(\xi) \triangleq \infty$ if $V = \mathbb{R}^n$] and

$$\tilde{T}(\xi, \tau) \triangleq [\tau, \tau + h(d(\xi)/2, \tau)].$$

Then, by II.4.1, for every $(\tau, \xi) \in T \times V$ there exists a function

$$y(\tau, \xi)(\cdot) \in C(\tilde{T}(\tau, \xi), V)$$

satisfying the equation

$$y(t) = \xi + \int_{\tau}^{t} f(s, y(s))\, ds \qquad [t \in \tilde{T}(\tau, \xi)].$$

We set

$$\tau_0 \triangleq t_0, \qquad \xi_0 \triangleq \eta, \qquad \tau_1 \triangleq \tau_0 + h(d(\xi_0)/2, \tau_0),$$

and

$$y_0 \triangleq y(\tau_0, \xi_0) : [\tau_0, \tau_1] \to V.$$

If $\tau_1 < t_1$, we set

$$\xi_1 \triangleq y_0(\tau_1), \quad \tau_2 \triangleq \tau_1 + h(d(\xi_1)/2, \tau_1), \quad \text{and} \quad y_1 \triangleq y(\tau_1, \xi_1) : [\tau_1, \tau_2] \to V,$$

and so on. This process will either stop if $\tau_l = t_1$ for some $l \in \mathbb{N}$ or will continue indefinitely. In the first case, we set

$$\tau_j \triangleq \tau_l, \qquad \xi_j \triangleq \xi_l, \qquad \text{and} \qquad y_j(\tau_j) \triangleq \xi_l \qquad (j = l+1, l+2, \ldots).$$

In either case, we let

$$\bar{y}(t) \triangleq y_j(t) \qquad \text{for} \quad t \in [\tau_j, \tau_{j+1}] \qquad (j = 0, 1, 2, \ldots).$$

Then (τ_j) is a nondecreasing sequence in T and therefore convergent, and

$$\lim_j h(d(y_{j-1}(\tau_j))/2, \tau_j) = \lim_j (\tau_{j+1} - \tau_j) = 0.$$

Thus, by (2), either

$$\lim_j d(y_{j-1}(\tau_j)) = \lim_j d(\bar{y}(\tau_j)) = 0 \qquad \text{or} \qquad \lim_j \tau_j = t_1.$$

Furthermore, if we set $t_1' \triangleq \lim_j \tau_j$, then $\bar{y}$ satisfies (1) for all $t \in [t, t_1')$. Since

$$|\bar{y}(t) - \bar{y}(\tau_j)| \leqslant \int_{\tau_j}^{\tau_{j+1}} |f(\tau, \cdot)|_{\sup} \, d\tau \qquad (j \in \mathbb{N}, \quad t \in [\tau_j, \tau_{j+1}])$$

and

$$\sum_{j=0}^{\infty} \int_{\tau_j}^{\tau_{j+1}} |f(\tau, \cdot)|_{\sup} \, d\tau \leqslant \int_{t_0}^{t_1} |f(\tau, \cdot)|_{\sup} \, d\tau < \infty,$$

it follows that $\lim_{t \to t_1'} \bar{y}(t) = \lim_j \bar{y}(\tau_j)$ and both limits exist. If $t_1' = t_1$ and $\lim_{t \to t_1'} \bar{y}(t) \in V$, then we set $\bar{y}(t_1) = \lim_{t \to t_1} \bar{y}(t)$ and $\bar{y} : [t_0, t_1] \to V$ satisfies (1). Otherwise, we have $\lim_{t \to t_1'} \mathrm{d}(\bar{y}(t)) = 0$ whence $\lim_{t \to t_1'} \bar{y}(t) \in \partial V$.

QED

II.4.3 ***Theorem*** Let $T \triangleq [t_0, t_1] \subset \mathbb{R}$, $\eta \in \mathbb{R}^n$, $f : T \times \mathbb{R}^n \to \mathbb{R}^n$, and assume that $f(t, \cdot)$ is continuous for all $t \in T$, $f(\cdot, v)$ is measurable for all $v \in \mathbb{R}^n$, and there exist an integrable $\psi : T \to \mathbb{R}$ and a positive, increasing, and continuous $\phi : (0, \infty) \to (0, \infty)$ such that

(1) $$|f(t, v)| \leqslant \phi(|v|)\, \psi(t) \qquad (t \in T, \quad v \in \mathbb{R}^n)$$

and

(2) $$\lim_{r \to \infty} \int_0^r ds/\phi(s) = \infty.$$

Then there exists a function $\bar{y} : T \to \mathbb{R}^n$ such that, for all $t \in T$,

(3) $$\bar{y}(t) = \eta + \int_{t_0}^{t} f(\tau, \bar{y}(\tau))\, d\tau.$$

▌PROOF Let $\Phi(r) \triangleq \int_0^r d\alpha/\phi(\alpha)$ for $r \geqslant 0$. Then Φ is a continuous function increasing from 0 to ∞ as $r \to \infty$ and there exists, therefore, $\beta \geqslant 0$ such that

$$\Phi(\beta) = \Phi(|\,\eta\,|) + \int_{t_0}^{t_1} \psi(\tau)\, d\tau. \tag{4}$$

Now let $V \triangleq S(0, \beta + 1)$. Then $0 < \phi(|\,v\,|) \leqslant \phi(\beta + 1)$ $(v \in V)$. By II.4.2, there exist a point $t_1' \in [t_0, t_1]$ and a function $\bar{y} : [t_0, t_1') \to V$ satisfying (3) for all $t \in T' \triangleq [t_0, t_1')$ and such that either $\lim_{t \to t_1'} \bar{y}(t) \in \partial V$ or $\bar{y}$ can be extended to T and satisfies (3) for all $t \in T$.

Let

$$x(t) \triangleq |\,\eta\,| + \int_{t_0}^{t} \psi(\tau)\, \phi(\bar{y}(\tau))\, d\tau \qquad (t \in T').$$

Then, by I.4.42, x is absolutely continuous on every closed subinterval of T' and $\dot{x}(t) = \psi(t)\, \phi(\bar{y}(t))$ a.e. in T'. Since $\bar{y}$ satisfies (3) for $t \in T'$, we have

$$|\,\bar{y}(t)| \leqslant |\,\eta\,| + \int_{t_0}^{t} |\,f(\tau, \bar{y}(\tau))|\, d\tau \leqslant x(t) \qquad (t \in T').$$

Since ϕ is increasing, it follows that

$$\dot{x}(\tau) = \psi(\tau)\, \phi(\bar{y}(\tau)) \leqslant \psi(\tau)\, \phi(x(\tau)) \quad \text{a.e. in } T';$$

hence

$$\dot{x}(\tau)/\phi(x(\tau)) \leqslant \psi(\tau) \quad \text{a.e. in } T' \tag{5}$$

By I.4.43, the function $t \to \dot{x}(t)/\phi(x(t))$ is integrable on every closed subinterval of T' and

$$\int_{t_0}^{t} \frac{\dot{x}(\tau)}{\phi(x(\tau))}\, d\tau = \int_{x(t_0)}^{x(t)} \frac{ds}{\phi(s)} = \Phi(x(t)) - \Phi(x(t_0)) \qquad (t \in T');$$

hence, by (5),

$$\Phi(x(t)) - \Phi(|\,\eta\,|) \leqslant \int_{t_0}^{t} \psi(\tau)\, d\tau \leqslant \int_{t_0}^{t_1} \psi(\tau)\, d\tau \qquad (t \in T').$$

Thus, by (4), $\Phi(x(t)) \leqslant \Phi(\beta)$ $(t \in T')$, implying that $|\,\bar{y}(t)| \leqslant x(t) \leqslant \beta$ $(t \in T')$ and therefore $\lim_{t \to t_1'} \bar{y}(t) \notin \partial V$. It follows that $t_1' = t_1$ and $\bar{y} : T \to \mathbb{R}^n$ satisfies relation (3) for all $t \in T$. QED

II.4.4 ***Theorem*** (Gronwall's inequality) Let $T \triangleq [t_0, t_1] \subset \mathbb{R}$, $\psi : T \to \mathbb{R}$ be nonnegative and integrable, $h, u \in C(T)$, h nonnegative, and

$$u(t) \leqslant \int_{t_0}^{t} \psi(\tau)\, u(\tau)\, d\tau + h(t) \qquad (t \in T). \tag{1}$$

Then

$$u(t) \leqslant h(t) + c\int_{t_0}^{t} \psi(\tau)\, h(\tau)\, d\tau \qquad (t \in T),$$

where

$$c \triangleq \exp\left(\int_{t_0}^{t_1} \psi(\tau)\, d\tau\right).$$

▌PROOF Let

$$w(t) \triangleq \int_{t_0}^{t} \psi(\tau)\, u(\tau)\, d\tau \qquad \text{and} \qquad p(t) \triangleq \int_{t_0}^{t} \psi(\tau)\, d\tau \qquad (t \in T).$$

By I.4.42, II.3.4, and II.3.12, the functions $w(\cdot)$, $p(\cdot)$, and $t \to e^{-p(t)}w(t)$ are absolutely continuous on T and

$$\int_{t_0}^{t} e^{-p(\tau)}[\dot{w}(\tau) - \psi(\tau)\, w(\tau)]\, d\tau = e^{-p(t)}w(t) \qquad (t \in T);$$

furthermore $\dot{w}(\tau) = \psi(\tau)\, u(\tau)$ a.e. It follows, by (1), that

$$\dot{w}(t) \leqslant \psi(t)\, w(t) + \psi(t)\, h(t) \quad \text{a.e. in } T,$$
$$[e^{-p(t)}w(t)]^{\cdot} \leqslant e^{-p(t)}h(t)\, \psi(t) \qquad \text{a.e. in } T$$

and

$$e^{-p(t)}w(t) \leqslant \int_{t_0}^{t} e^{-p(\tau)}h(\tau)\, \psi(\tau)\, d\tau \qquad (t \in T).$$

Thus

$$w(t) \leqslant \exp[p(t_1)] \int_{t_0}^{t} \psi(\tau)\, h(\tau)\, d\tau$$

and, by (1),

$$u(t) \leqslant \exp[p(t_1)] \int_{t_0}^{t} \psi(\tau)\, h(\tau)\, d\tau + h(t),$$

from which our conclusion follows. QED

II.4.5 ***Theorem*** Let $T \triangleq [t_0, t_1] \subset \mathbb{R}$, $n \in \mathbb{N}$, $V \subset \mathbb{R}^n$, $f : T \times V \to \mathbb{R}^n$, and let $\bar{y} : T \to V$ satisfy the equation

$$\bar{y}(t) = \eta + \int_{t_0}^{t} f(\tau, y(\tau))\, d\tau \qquad (t \in T). \tag{1}$$

If there exists an integrable ψ_1 on T such that

$$|f(t, v_1) - f(t, v_2)| \leqslant \psi_1(t)|\, v_1 - v_2\,| \qquad (t \in T, \quad v_1, v_2 \in V), \tag{2}$$

then $\bar{y}$ is the unique solution of Eq. (1).

▌PROOF Let $\bar{y}_1, \bar{y}_2$ be two solutions. Then

$$(3) \qquad |\bar{y}_1(t) - \bar{y}_2(t)| = \left| \int_{t_0}^{t} [f(\tau, \bar{y}_1(\tau)) - f(\tau, \bar{y}_2(\tau))] \, d\tau \right|$$

$$\leqslant \int_{t_0}^{t} \psi_1(\tau) \, | \bar{y}_1(\tau) - \bar{y}_2(\tau)| \, d\tau \qquad (t \in T).$$

Set $u(t) \triangleq |\bar{y}_1(t) - \bar{y}_2(t)|$. Then, by (3) and Gronwall's inequality II.4.4, $u(t) \leqslant 0$; hence $u(\cdot) = 0$ and $\bar{y}_1 = \bar{y}_2$. QED

II.4.C Linear Ordinary Differential Equations

Let $T \triangleq [t_0, t_1] \subset \mathbb{R}$, $n \in \mathbb{N}$, $A(\cdot) : T \to B(\mathbb{R}^n, \mathbb{R}^n)$, and $b(\cdot) : T \to \mathbb{R}^n$. If $(\alpha_{i,j}(t))$ $(i, j = 1,..., n)$ is the matrix of $A(t)$ for $t \in T$, we observe that $A(\cdot)$ is continuous (measurable, differentiable) if and only if $\alpha_{ij}(\cdot)$ is continuous (measurable, differentiable) for each i and j. We identify the function $A(\cdot)$ with the matrix-valued function $t \to [A(t)] \triangleq (\alpha_{i,j}(t))$ $(i, j = 1,..., n)$. If $\alpha_{i,j}(\cdot)$ is differentiable at $\bar{t}$, respectively integrable on T, for all $i, j \in \{1, 2,..., n\}$, then so is $A(\cdot)$, and we have

$$\dot{A}(\bar{t}) = (\dot{\alpha}_{i,j}(\bar{t}))(i, j = 1, 2,..., n)$$

respectively

$$\int_{t'}^{t''} A(\tau) \, d\tau = \left(\int_{t}^{t''} \alpha_{i,j}(\tau) \, d\tau \right) (i, j = 1, 2,..., n) \quad (t', t'' \in T).$$

We refer to the initial value problem

$$\dot{y}(t) = A(t) \, y(t) + b(t) \quad \text{a.e. in } T, \qquad y(t_0) = \eta$$

which is equivalent to the equation

$$y(t) = \eta + \int_{t_0}^{t} [A(\tau) \, y(\tau) + b(\tau)] \, d\tau \qquad (t \in T)$$

as a *linear ordinary differential equation*. Thus a linear ordinary differential equation is an ordinary differential equation of the type that we have studied, with

$$f(t, v) \triangleq A(t) \, v + b(t) \qquad (t \in T, \quad v \in \mathbb{R}^n).$$

For a given function $A(\cdot) : T \to B(\mathbb{R}^n, \mathbb{R}^n)$, we can consider the *linear matrix differential equations*

$$(*) \qquad \dot{Y}(t) = A(t) \, Y(t) \quad \text{a.e. in } T$$

and

(**) $$\dot{Y}(t) = Y(t)\,A(t) \quad \text{a.e. in } T.$$

A solution of (*) or (**) is a matrix-valued absolutely continuous function on T satisfying the respective equation. (By a matrix-valued absolutely continuous function we mean one each of whose coefficients is absolutely continuous.) We observe that if Y is a solution of Eq. (*), then, for $j = 1, 2,..., n$, its j th column Y_j is a solution of the linear differential equation

(*j) $$\dot{Y}_j(t) = A(t)\,Y_j \quad \text{a.e. in } T.$$

Similarly, if Y is a solution of Eq. (**), then, for $j = 1, 2,..., n$, its j th row Y^{jT} is a solution of the linear differential equation

(**j) $$\dot{Y}^j(t)^T = Y^j(t)^T A(t) \quad \text{a.e. in } T.$$

II.4.6 ***Theorem*** Let $T \triangleq [t_0, t_1] \subset \mathbb{R}$, $A(\cdot) : T \to B(\mathbb{R}^n, \mathbb{R}^n)$ and $b(\cdot) : T \to \mathbb{R}^n$ be integrable, $\bar{t} \in T$, and $\eta \in \mathbb{R}^n$. Then there exists a unique absolutely continuous function $\bar{y} : T \to \mathbb{R}^n$ such that, for a.a. $t \in T$,

(1) $$\dot{y}(t) = A(t)\,\bar{y}(t) + b(t), \qquad y(\bar{t}) = \eta.$$

PROOF We denote by $\alpha_{i,j}(t)$ the coefficients of $A(t)$, set

$$\psi(t) \triangleq \|A(t)\| + |b(t)| \qquad (t \in T),$$

and observe that ψ is integrable and

$$|A(t)v + b(t)| = \sum_{i=1}^{n} \left| \sum_{j=1}^{n} \alpha_{i,j}(t)\,v^j + b^i(t) \right|$$
$$\leqslant \psi(t)(|v| + 1) \qquad (t \in T, \quad v \in \mathbb{R}^n).$$

Since, by II.3.12,

$$\lim_{r\to\infty} \int_0^r d\alpha/(\alpha + 1) = \lim_{r\to\infty} \log(r + 1) = \infty,$$

it follows that the assumptions of II.4.3 are satisfied with T replaced by $[\bar{t}, t_1]$ and there exists, therefore, $\bar{y} : [\bar{t}, t_1] \to \mathbb{R}^n$ satisfying (1) for a.a. $t \in [\bar{t}, t_1]$. By II.4.5, $\bar{y} : [\bar{t}, t_1] \to \mathbb{R}^n$ is unique.

Now let

$$\tilde{A}(s) \triangleq A(\bar{t} - s) \qquad \text{and} \qquad \tilde{b}(s) \triangleq b(\bar{t} - s) \qquad (s \in [0, \bar{t} - t_0]).$$

Then, by our previous argument, there exists a unique absolutely continuous $\tilde{y} : [0, \bar{t} - t_0] \to \mathbb{R}^n$ such that

$$\dot{\tilde{y}}(s) = -\tilde{A}(s)\, \tilde{y}(s) - \tilde{b}(s) \quad \text{a.e. in } [0, \bar{t} - t_0], \qquad \tilde{y}(0) = \eta.$$

We set $\bar{y}(t) \triangleq \tilde{y}(\bar{t} - t)$ $(t \in [t_0, \bar{t}])$ and observe that $\dot{\bar{y}}(t) = -\dot{\tilde{y}}(\bar{t} - t)$; hence

$$\dot{\bar{y}}(t) = A(t)\, \bar{y}(t) + b(t) \quad \text{a.e. in } [t_0, \bar{t}], \qquad \bar{y}(\bar{t}) = \eta. \quad \text{QED}$$

II.4.7 ***Theorem*** Let $A(\cdot) : T \to B(\mathbb{R}^n, \mathbb{R}^n)$ be integrable, $\bar{t} \in T$, and, for each $\eta \in \mathbb{R}^n$, $\bar{y}(\eta)(\cdot) : T \to \mathbb{R}^n$ be absolutely continuous on T and satisfy the equation

(1) $$\dot{y}(t) = A(t)\, y(t) \text{ a.e. in } T, \qquad y(\bar{t}) = \eta.$$

Then

(2) $\bar{y}(\eta)(\tilde{t}) = 0$ for some $\tilde{t} \in T$ if and only if $\bar{y}(\eta)(\cdot) = 0$,
(3) the set $\{\bar{y}(\eta_1)(t),\ldots, \bar{y}(\eta_n)(t)\}$ is independent for $t = \tilde{t} \in T$ if and only if it is independent for all $t \in T$, and
(4) the mapping

$$\eta \to \bar{y}(\eta)(\cdot) : \mathbb{R}^n \to C(T, \mathbb{R}^n)$$

is linear and continuous.

▌PROOF By II.4.6, there exists a unique absolutely continuous solution $\bar{y}(\eta)(\cdot)$ of (1) for all $\eta \in \mathbb{R}^n$. We verify that the mapping

$$\eta \to \bar{y}(\eta)(\cdot) : \mathbb{R}^n \to C(T, \mathbb{R}^n)$$

is linear. If $\eta \in \mathbb{R}^n$ and $\bar{y}(\eta)(\tilde{t}) = 0$ for some $\tilde{t} \in T$, then $\bar{y}(\eta)(\cdot)$ satisfies

$$\dot{y}(t) = A(t)\, y(t) \quad \text{a.e. in } T, \qquad y(\tilde{t}) = 0.$$

By II.4.6, this equation has a unique absolutely continuous solution and, since $y = 0$ is a solution, we have $\bar{y}(\eta)(t) = 0$ $(t \in T)$ and $\eta = \bar{y}(\eta)(\bar{t}) = 0$.

If $\{\bar{y}(\eta_1)(\tilde{t}),\ldots, \bar{y}(\eta_n)(\tilde{t})\}$ is dependent for some $\tilde{t} \in T$, then there exists $\alpha \triangleq (\alpha^1,\ldots, \alpha^n) \neq 0$ such that

$$\sum_{j=1}^{n} \alpha^j \bar{y}(\eta_j)(\tilde{t}) = \bar{y}\left(\sum_{j=1}^{n} \alpha^j \eta_j\right)(\tilde{t}) = 0;$$

hence, as just shown,

$$\sum_{j=1}^{n} \alpha^j \eta_j = 0 \qquad \text{and} \qquad \bar{y}\left(\sum_{j=1}^{n} \alpha^j \eta_j\right)(t) = \sum_{j=1}^{n} \alpha^j \bar{y}(\eta_j)(t) = 0 \quad (t \in T).$$

Now let $\psi(t) = \| A(t)\|$. Then ψ is integrable. We have, for each $\eta \in \mathbb{R}^n$,

$$\bar{y}(\eta)(t) = \eta + \int_{\bar{t}}^{t} A(\tau)\, \bar{y}(\eta)(\tau)\, d\tau \qquad (t \in T);$$

hence

$$| \bar{y}(\eta)(t)| \leqslant | \eta | + \int_{\bar{t}}^{t} \psi(\tau)\, | \bar{y}(\eta)(\tau)|\, d\tau \qquad (t \in [\bar{t}\,, t_1]).$$

It follows, by Gronwall's inequality II.4.4, that there exists $c \in \mathbb{R}$ such that

$$(5) \qquad | \bar{y}(\eta)(t)| \leqslant c \,| \eta | \qquad (t \in [\bar{t}, t_1],\ \eta \in \mathbb{R}^n).$$

An argument patterned on the last paragraph in the proof of II.4.6 shows that

$$(6) \qquad | \bar{y}(\eta)(t)| \leqslant c \,| \eta | \qquad (t \in [t_0\,, \bar{t}],\ \eta \in \mathbb{R}^n).$$

Relations (5) and (6) show that the linear mapping

$$\eta \to \bar{y}(\eta)(\cdot) : \mathbb{R}^n \to C(T, \mathbb{R}^n)$$

is continuous. QED

II.4.8 ***Theorem*** Let $A(\cdot) : T \to B(\mathbb{R}^n, \mathbb{R}^n)$ and $b(\cdot) : T \to \mathbb{R}^n$ be integrable, $\tilde{t}$, $\bar{t} \in T$, and $\eta \in \mathbb{R}^n$. Then there exist unique absolutely continuous functions $Y : T \to B(\mathbb{R}^n, \mathbb{R}^n)$ and $Z : T \to B(\mathbb{R}^n, \mathbb{R}^n)$ such that

(1) $\dot{Y}(t) = A(t)\, Y(t)$ a.e. in T, $Y(\tilde{t}) = I$,
(2) $\dot{Z}(t) = -Z(t)\, A(t)$ a.e. in T, $Z(\tilde{t}) = I$ and $Y(t) = Z(t)^{-1}$ $(t \in T)$.

The unique solution $\bar{y}(\eta, b)(\cdot)$ of the initial value problem

(3) $\dot{y}(t) = A(t)\, y(t) + b(t)$ a.e. in T, $y(\bar{t}) = \eta$

satisfies each of the relations

(4) $\bar{y}(\eta, b)(t) = Z(t)^{-1}\, Z(\bar{t})\, \eta + Z(t)^{-1} \int_{\bar{t}}^{t} Z(\tau)\, b(\tau)\, d\tau \quad (t \in T)$

and

(5) $\bar{y}(\eta, b)(t) = Y(t)\, Y(\bar{t})^{-1}\, \eta + Y(t) \int_{\bar{t}}^{t} Y(\tau)^{-1}\, b(\tau)\, d\tau \quad (t \in T)$.

▌PROOF Let δ_j be the jth column of the $n \times n$ unit matrix and δ^{iT} its ith row. Then, by II.4.6, the equation

$$(6) \qquad \dot{y}(t) = A(t)\, y(t) \quad \text{a.e. in } T, \qquad y(\tilde{t}) = \delta_j$$

has a unique absolutely continuous solution $y_j \triangleq (y_j^1, \ldots, y_j^n)$ for each $j \in \{1, 2, \ldots, n\}$, and we verify that $Y(\cdot) \triangleq \big(y_j^i(\cdot)\big)$ $(i, j = 1, \ldots, n)$ satisfies (1). Conversely, every solution of (1) must have its jth column satisfy (6). Thus

$Y(\cdot)$ is the unique solution of (1). By II.4.7, the set $\{y_1(t),..., y_n(t)\}$ is independent for all $t \in T$; hence, by I.3.3, $Y(t)^{-1}$ exists for all $t \in T$.

Equation (2) is equivalent to the equations

$$\dot{Z}(t)^T = -A(t)^T Z(t)^T \quad \text{a.e. in } T, \qquad Z(\bar{t})^T = I,$$

and, by our previous argument, has a unique solution $Z^T(\cdot)$. We premultiply both sides of the equation $\dot{Y}(t) = A(t)\, Y(t)$ by $Z(t)$, postmultiply those of $\dot{Z}(t) = -Z(t)\, A(t)$ by $Y(t)$, and add, obtaining

$$Z(t)\, \dot{Y}(t) + \dot{Z}(t)\, Y(t) = 0 \quad \text{a.e. in } T$$

and, by II.3.12, $\big(Z(t)\, Y(t)\big)^{\cdot} = 0$ a.e. in T. Since Z and Y are absolutely continuous and $Z(\bar{t}) = Y(\bar{t}) = I$, it follows from I.4.42 that

$$Z(t)\, Y(t) = I \qquad (t \in T);$$

hence $Y(t) = Z(t)^{-1}$ $(t \in T)$.

If $\eta \in \mathbb{R}^n$ and we write $y(t)$ for $\bar{y}(\eta, b)(t)$, then, by (2) and (3), we have $\dot{Z}(\tau)\, y(\tau) = -Z(\tau)\, A(\tau)\, y(\tau)$ and $Z(\tau)\, \dot{y}(\tau) = Z(\tau)\, A(\tau)\, y(\tau) + Z(\tau)\, b(\tau)$ a.e. in T; hence

$$Z(\tau)\, \dot{y}(\tau) + \dot{Z}(\tau)\, y(\tau) = \big(Z(\tau)\, y(\tau)\big)^{\cdot} = Z(\tau)\, b(\tau) \quad \text{a.e. in } T.$$

We conclude, integrating both sides from $\bar{t}$ to t, that

$$Z(t)\, y(t) - Z(\bar{t})\, y(\bar{t}) = Z(t)\, y(t) - Z(\bar{t})\eta = \int_{\bar{t}}^{t} Z(\tau)\, b(\tau)\, d\tau \qquad (t \in T)$$

and

$$\begin{aligned} y(t) &= Z(t)^{-1}\, Z(\bar{t})\eta + Z(t)^{-1} \int_{\bar{t}}^{t} Z(\tau)\, b(\tau)\, d\tau \\ &= Y(t)\, Y(\bar{t})^{-1}\, \eta + Y(t) \int_{\bar{t}}^{t} Y(\tau)^{-1}\, b(\tau)\, d\tau \qquad (t \in T) \quad \text{QED.} \end{aligned}$$

II.4.D Dependence on Parameters of Solutions of Ordinary Differential Equations

II.4.9 ***Theorem*** Let $n, k \in \mathbb{N}$, V be an open subset of $\mathbb{R}^n$, A a convex subset of $\mathbb{R}^k$, $T \triangleq [t_0, t_1] \subset \mathbb{R}$, and $f : T \times V \times A \to \mathbb{R}^n$. Assume that, for all $(t, v, a) \in T \times V \times A$, $f(t, \cdot, \cdot)$ is differentiable at (v, a) and that

$$f \in \mathscr{B}(T, V \times A; \mathbb{R}^n) \qquad \text{and} \qquad f_{(v,a)} \in \mathscr{B}\big(T, V \times A; B(\mathbb{R}^{n+k}, \mathbb{R}^n)\big).$$

Then the relation

$$F(y, a)(t) \triangleq \int_{t_0}^{t} f(\tau, y(\tau), a)\, d\tau \qquad [t \in T, \quad y \in C(T, V), \quad a \in A]$$

defines a continuous mapping $F: C(T, V) \times A \to C(T, \mathbb{R}^n)$ and F has a continuous derivative F' such that

$$F'(y, a)(\Delta y, \Delta a)(t) = \int_{t_0}^{t} f_{(v,a)}(\tau, y(\tau), a)(\Delta y(\tau), \Delta a)\, d\tau \tag{1}$$

$$[\Delta y \in C(T, \mathbb{R}^n), \Delta a \in \mathbb{R}^k, t \in T].$$

▌PROOF *Step* 1 Let $Y \triangleq C(T, V)$, $y \in Y$, and $d_y \triangleq d[y(T), \mathbb{R}^n \sim V]$ (with $d_y \triangleq \infty$ if $V = \mathbb{R}^n$). Since y is continuous, V open, and $y(T) \subset V$, we have $d_y > 0$. Thus Y is an open subset of $C(T, \mathbb{R}^n)$. Since

$$f(\cdot, \cdot, a) \in \mathscr{B}(T, V; \mathbb{R}^n)$$

for each $a \in A$, the function $t \to f(t, y(t), a)$ is integrable for all $(y, a) \in Y \times A$ and thus (by I.4.42) $t \to F(y, a)(t)$ is continuous. Now let $\lim_j y_j = \bar{y}$ in Y and $\lim a_j = \bar{a}$ in A. Then, by the dominated convergence theorem I.4.35, $\lim_j F(y_j, a_j)(t) = F(\bar{y}, a)(t)$ for each $t \in T$. Since the set $F(Y \times A)$ is equicontinuous [because $t \to |f(t, \cdot, \cdot)|_{\sup}$ is integrable], the convergence is uniform (I.5.3). Thus $F: Y \times A \to C(T, \mathbb{R}^n)$ is continuous.

Step 2 Let $y \in Y$, $0 \neq \Delta y \in Y - y$, $a \in A$, $0 \neq \Delta a \in A - a$, and

$$p(\alpha)(\tau) \triangleq (y(\tau) + \alpha\, \Delta y(\tau), a + \alpha\, \Delta a) \qquad (\tau \in T, \alpha \in [0, 1]).$$

Since Y is open in $C(T, \mathbb{R}^n)$, we have $p(\alpha)(\tau) \in V \times A$ provided $|\Delta y|$ is sufficiently small, and we assume that this is the case. We easily verify that the function

$$(v, \Delta v, a') \to f_{(v,a)}(t, v, a')(\Delta v, \Delta a) : V \times \mathbb{R}^n \times A \to \mathbb{R}^n$$

is continuous for each $t \in T$. Thus, by I.4.22, the function

$$\tau \to h(\tau, \alpha) \triangleq (|\Delta y| + |\Delta a)^{-1} f_{(v,a)}(\tau, p(\alpha)(\tau))(\Delta y(\tau), \Delta a)$$

is measurable for each $\alpha \in [0, 1]$. Since, furthermore, $h(\tau, \cdot)$ is continuous on $[0, 1]$ for each $\tau \in T$, it follows from I.4.21 that

$$\tau \to \sup_{0 \leqslant \alpha \leqslant 1} |h(\tau, \alpha) - h(\tau, 0)|$$

is measurable.

By the mean value theorem II.3.6, we have

$$(2)\quad (|\Delta y| + |\Delta a|)^{-1} \,|\, F(y + \Delta y, a + \Delta a)(t) - F(y, a)(t)$$
$$- \int_{t_0}^{t} f_{(v,a)}(\tau, y(\tau), a)(\Delta y(\tau), \Delta a)\, d\tau \,|$$
$$= (|\Delta y| + |\Delta a|)^{-1} \left| \int_{t_0}^{t} [f(\tau, y(\tau) + \Delta y(\tau), a + \Delta a) - f(\tau, (y\tau), a)\right.$$
$$\left. - f_{(v,a)}(\tau, y(\tau), a)(\Delta y(\tau), \Delta a)]\, d\tau \right|$$
$$\leqslant \int_{t_0}^{t_1} \sup_{0 \leqslant \alpha \leqslant 1} |\, h(\tau, \alpha) - h(\tau, 0)|\, d\tau.$$

As $|\Delta y| + |\Delta a| \to 0$, $(|\Delta y| + |\Delta a|)^{-1} |(\Delta y(\tau), \Delta a)|$ remains bounded by 1 for all $\tau \in T$ and, since $f_{(v,a)}(\tau, \cdot, \cdot)$ is continuous for each $\tau \in T$, the integrand on the right of (2) converges to 0 with $|\Delta y| + |\Delta a|$. Since this integrand is bounded by $2\,|f_{(v,a)}(\tau, \cdot, \cdot)|_{\sup}$, it follows from the dominated convergence theorem I.4.35 that the right side of (2) converges to 0. Thus the left side of (2) converges to 0 uniformly for all $t \in T$. Furthermore, the linear operator on $C(T, \mathbb{R}^n) \times \mathbb{R}^k$ defined by

$$H(\Delta y, \Delta a)(t) \triangleq \int_{t_0}^{t} f_{(v,a)}(\tau, y(\tau), a)(\Delta y(\tau), \Delta a)\, d\tau$$

satisfies the relation

$$|\, H(\Delta y, \Delta a)| \leqslant \int_{t_0}^{t_1} |f_{(v,a)}(\tau, \cdot, \cdot)|_{\sup}\, d\tau \cdot |(\Delta y, \Delta a)|$$

and is therefore continuous. We can now conclude that relation (1) is valid.

Finally, we prove that F' is continuous. Let $\lim_j(y_j, a_j) = (\bar{y}, \bar{a})$ in $C(T, V) \times A$ and $|\Delta y| + |\Delta a| \leqslant 1$. Then, by (1),

$$(3)\quad |\, F'(y_j, a_j)(\Delta y, \Delta a)(t) - F'(\bar{y}, \bar{a})(\Delta y, \Delta a)(t)|$$
$$\leqslant \int_{t_0}^{t_1} |[f_{(v,a)}(\tau, y_j(\tau), a_j) - f_{(v,a)}(\tau, \bar{y}(\tau), \bar{a})](\Delta y(\tau), \Delta a)|\, d\tau$$
$$\leqslant \int_{t_0}^{t_1} |\, f_{(v,a)}(\tau, y_j(\tau), a_j) - f_{(v,a)}(\tau, \bar{y}(\tau), \bar{a})|\, d\tau.$$

By the dominated convergence theorem I.4.35, the right side of (3) converges to 0 as $j \to \infty$. Thus $\lim_j |\, F'(y_j, a_j) - F'(\bar{y}, \bar{a})| = 0$. QED

II.4.10 ***Theorem*** Under the conditions and with the notation of II.4.9, let

$$G(y, \eta, a) \triangleq y - \eta - F(y, a) \qquad [(y, \eta, a) \in Y \times \mathbb{R}^n \times A].$$

Then, for all $(y, \eta, a) \in Y \times \mathbb{R}^n \times A$, $G_y(y, \eta, a)$ is a homeomorphism of $C(T, \mathbb{R}^n)$ onto itself.

▌PROOF Let I be the identity operator in $C(T, \mathbb{R}^n)$. Our statement is equivalent to the assertion that, for every $h \in C(T, \mathbb{R}^n)$, the equation

$$(I - F_y(y, a))\, \Delta y = h$$

has a unique solution Δy and that there exists a number c such that $|\Delta y| \leqslant c\,|h|$ for all h. By II.4.9, the above equation can be written

$$(1) \qquad \Delta y(t) = \int_{t_0}^{t} f_v(\tau, y(\tau), b)\, \Delta y(\tau)\, d\tau + h(t) \qquad (t \in T)$$

or, setting $p \triangleq \Delta y - h$,

$$p(t) = \int_{t_0}^{t} f_v(\tau, y(\tau), b)\big(p(\tau) + h(\tau)\big)\, d\tau \qquad (t \in T).$$

By II.4.3 [with $\phi(r) = r + 1$] and II.4.5, there exists a unique p in $C(T, \mathbb{R}^n)$ satisfying the last equation. Furthermore, if we set

$$\psi(\tau) \triangleq |f_v(\tau, \cdot, \cdot)|_{\sup} \qquad (\tau \in T),$$

then (1) yields

$$|\Delta y(t)| \leqslant \int_{t_0}^{t} \psi(\tau)\, |\Delta y(\tau)|\, d\tau + |h(t)| \qquad (t \in T);$$

hence, by Gronwall's inequality II.4.4, there exists $c' \in \mathbb{R}$ such that

$$|\Delta y(t)| \leqslant |h(t)| + c' \int_{t_0}^{t} \psi(\tau)\, |h(\tau)|\, d\tau \qquad (t \in T).$$

It follows that

$$|\Delta y| \leqslant \left[1 + c' \int_{t_0}^{t_1} \psi(\tau)\, d\tau\right] |h|. \quad \text{QED}$$

II.4.11 ***Theorem*** Let $k, n \in \mathbb{N}$, V be an open subset of $\mathbb{R}^n$, A a convex subset of $\mathbb{R}^k$, $T \triangleq [t_0, t_1] \subset \mathbb{R}$, and $f : T \times V \times A \to \mathbb{R}^n$. Assume that, for all $(t, v, a) \in T \times V \times A$, $f(t, \cdot, \cdot)$ is differentiable at (v, a), and that

$$f \in \mathscr{B}(T, V \times A; \mathbb{R}^n), \qquad f_{(v,a)} \in \mathscr{B}\big(T, V \times A; B(\mathbb{R}^{n+k}, \mathbb{R}^n)\big),$$

$\eta_0 \in V$, $a_0 \in A$, $y_0 \in C(T, V)$, and

$$y_0(t) = \eta_0 + \int_{t_0}^{t} f(\tau, y_0(\tau), a_0)\, d\tau \qquad (t \in T).$$

Then there exist neighborhoods $\tilde{V}$ of η_0 in V and $\tilde{A}$ of a_0 relative to A such that the equation

$$(1) \qquad y(t) = \eta + \int_{t_0}^{t} f(\tau, y(\tau), a)\, d\tau \qquad (t \in T)$$

has a unique solution $y(\cdot) = u(\eta, a)(\cdot)$ in $C(T, V)$ for all $(\eta, a) \in \tilde{V} \times \tilde{A}$, the function $u : \tilde{V} \times \tilde{A} \to C(T, \mathbb{R}^n)$ is continuous and has a continuous derivative u' on $\tilde{V} \times \tilde{A}$, and, for $\eta \in \tilde{V}$, $\Delta\eta \in \mathbb{R}^n$, $a \in A$, $\Delta a \in \mathbb{R}^k$, and

$$p \triangleq u'(\eta, a)(\Delta\eta, \Delta a),$$

we have

$$(2) \qquad p(t) = \int_{t_0}^{t} f_{(v,a)}(\tau, u(\eta, a)(\tau), a)(p(\tau), \Delta a)\, d\tau + \Delta\eta \qquad (t \in T).$$

▌PROOF Let $Y \triangleq \{y(\cdot) \in C(T, \mathbb{R}^n) |\, y(T) \subset V\}$. By II.4.9, the relation

$$G(y, \eta, a)(t) \triangleq y(t) - \eta - \int_{t_0}^{t} f(\tau, y(\tau), a)\, d\tau \qquad (t \in T)$$

defines a continuous mapping $G : Y \times \mathbb{R}^n \times A \to C(T, \mathbb{R}^n)$ with a continuous derivative G' satisfying the relation

$$(3) \quad G'(y, \eta, a)(\Delta y, \Delta\eta, \Delta a)(t)$$

$$= \Delta y(t) - \Delta\eta - \int_{t_0}^{t} f_{(v,a)}(\tau, y(\tau), a)(\Delta y(\tau), \Delta a)\, d\tau \qquad (t \in T).$$

By II.4.10, $G_y(y_0, \eta_0, a_0)$ is a homeomorphism of $C(T, \mathbb{R}^n)$ onto itself and, by assumption, $G(y_0, \eta_0, a_0) = 0$. Thus, by the implicit function theorem II.3.8, there exist neighborhoods $\tilde{V}$ of η_0 and $\tilde{A}$ of a_0 (relative to A) and a solution $u(\eta, a)$ of Eq. (1) for all $(\eta, a) \in \tilde{V} \times \tilde{A}$, $u(\cdot, \cdot) : \tilde{V} \times \tilde{A} \to C(T, V)$ is continuous and has a continuous derivative u' on $\tilde{V} \times \tilde{A}$, and

$$(4) \qquad u'(\eta, a) = -G_y(u(\eta, a), \eta, a)^{-1} \circ G_{(\eta,a)}(u(\eta, a), \eta, a).$$

By II.4.5 and II.3.6, $u(\eta, a)(\cdot)$ is the unique solution of (1).

Finally, relation (2) follows directly from (3) and (4). QED

II.5 Functional-Integral Equations in $C(T, \mathbb{R}^n)$

II.5.A Existence of Solutions of Functional-Integral Equations in $C(T, \mathbb{R}^n)$

Having examined the special case of ordinary differential equations, we shall consider the more general *functional-integral equation* of the form

$$(*) \qquad y(t) = \int f(t, \tau, \xi(y)(\tau))\, \mu(d\tau) \qquad (t \in T).$$

Here T is a compact metric space, (T, Σ, μ) a positive finite measure space, $n, m \in \mathbb{N}$, $V \subset \mathbb{R}^m$, $W \subset \mathbb{R}^n$, $(t, \tau, v) \to f(t, \tau, v) : T \times T \times V \to \mathbb{R}^n$, and $\xi : C(T, W) \to L^\infty(T, \Sigma, \mu, V)$ [where

$$L^\infty(T, \Sigma, \mu, V) \triangleq \{z \in L^\infty(T, \Sigma, \mu, \mathbb{R}^n) |\ z(\tau) \in V\ \mu\text{-a.e. in } T\}].$$

As an example of such a functional-integral equation we may consider the Uryson-type integral equation II.1(8). A more general example is provided by the *integral equation with pseudodelays* for which ξ is defined in the following manner. We assume that $k \in \mathbb{N}$, $V = W^{k+1}$, and we are given measurable functions $h_j : T \to T$ $(j = 1,..., k)$. We then set

$$\xi(y)(\tau) \triangleq (y(h_1(\tau)), y(h_2(\tau)),..., y(h_k(\tau))) \qquad [y \in C(T, W), \quad \tau \in T].$$

Then $\xi(y)(\cdot)$ is bounded and μ-measurable and belongs, therefore, to $L^\infty(T, \Sigma, \mu, \mathbb{R}^m)$. In the special case where T is the interval $[t_0, t_1] \subset \mathbb{R}$,

$$\tau \geqslant h_1(\tau) \geqslant \cdots \geqslant h_k(\tau) \qquad (\tau \in T, j = 1, 2,..., k),$$

and $f(t, \tau, v) = 0$ for all $v \in V$ and $\tau > t$, our equation becomes a delay-differential equation.

Before returning to the consideration of the general equation $(*)$ we recall our convention of I.5C concerning functions with values in a set of functions. Specifically, if S_1, S_2, and X are given sets, $F(S_2, X)$ a collection of functions on S_2 to X, and $H : S_1 \to F(S_2, X)$, then we identify H with the function $h : S_1 \times S_2 \to X$ defined by

$$h(s_1, s_2) \triangleq H(s_1)(s_2) \qquad (s_1 \in S_1, \quad s_2 \in S_2).$$

In this sense, if S is a separable metric space, $\mathscr{Y}$ a separable Banach space, (T, Σ, μ) a positive finite measure space, and $f \in C(T, \mathscr{B}(T, S; \mathscr{Y}))$, then we identify f with the function $(t, \tau, s) \to f(t)(\tau, s) : T \times T \times S \to \mathscr{Y}$ and write $f(t, \tau, s)$ for $f(t)(\tau, s)$. Thus $|f| = \sup_{t \in T} \int |f(t, \tau, \cdot)|_{\sup} \mu(d\tau)$.

II.5.1 ***Theorem*** Let T be a compact metric space, (T, Σ, μ) a positive finite measure space, $n, m \in \mathbb{N}$, $V \subset \mathbb{R}^m$, $W \subset \mathbb{R}^n$, and $\xi : C(T, W) \to L^\infty(T, V)$ be continuous. Then the equation

$$y(t) = \int f(t, \tau, \xi(y)(\tau))\, \mu(d\tau) \qquad (t \in T)$$

has a solution $\bar{y} \in C(T, W)$ if either

$$(1) \qquad f \in C(T, \mathscr{B}(T, V; \mathbb{R}^n)) \quad \text{and} \quad S^F(0, |f|) \subset W$$

or

$$(2) \qquad V = \mathbb{R}^m, \qquad W = \mathbb{R}^n,$$

$f | T \times T \times V_1$ belongs to $C(T, \mathscr{B}(T, V_1; \mathbb{R}^n))$ for every bounded $V_1 \subset \mathbb{R}^m$ and there exist $\alpha, \beta, \gamma \in \mathbb{R}$ and $\psi : T \times T \to \mathbb{R}$ such that

$$\sup_{t \in T} \int \psi(t, \tau)\, \mu(d\tau) < \infty,$$

$$|\xi(y)|_\infty \leqslant \alpha(|y|_{\sup} + 1)^\beta \qquad [y \in C(T, \mathbb{R}^n)],$$

$$|f(t, \tau, v)| \leqslant \psi(t, \tau)(|v| + 1)^\gamma \qquad (t, \tau \in T, \quad v \in \mathbb{R}^m),$$

and

$$\beta\gamma < 1.$$

▌PROOF Assume first that (1) is valid, and let $K \triangleq C(T, S^F(0, |f|))$ and $y \in K$. Since $f(t, \cdot, \cdot) \in \mathscr{B}(T, V; \mathbb{R}^n)$ for each $t \in T$ and $\xi(y)(\cdot)$ is μ-measurable, it follows that the function $\tau \to f(t, \tau, \xi(y)(\tau))$ is μ-integrable for each $t \in T$. Thus $F(y)(t) \triangleq \int f(t, \tau, \xi(y)(\tau))\, \mu(d\tau)$ is defined for all $y \in K$ and $t \in T$, and

$$|F(y)(t)| \leqslant \sup_{t \in T} \int |f(t, \tau, \cdot)|_{\sup}\, \mu(d\tau) = |f|.$$

Now, for each $t \in T$,

$$(1) \qquad \lim_{t' \to t} \left| \int f(t, \tau, \xi(y)(\tau))\, \mu(d\tau) - \int f(t', \tau, \xi(y)(\tau))\, \mu(d\tau) \right|$$

$$\leqslant \lim_{t' \to t} \int |f(t, \tau, \cdot) - f(t', \tau, \cdot)|_{\sup}\, \mu(d\tau)$$

$$= \lim_{t' \to t} |f(t, \cdot, \cdot) - f(t', \cdot, \cdot)|_{\mathscr{B}} = 0.$$

Thus $F(y)(\cdot)$ is continuous and $F(y) \in K$ $(y \in K)$. Relation (1) also shows that $F(K)$ is equicontinuous. Since K is bounded and $F(K) \subset K$, $F(K)$ is also bounded. Thus, by I.5.4, $\overline{F(K)}$ is compact. It is also clear that K is a closed and convex subset of $C(T, \mathbb{R}^n)$. Our conclusion will follow, therefore, from Schauder's fixed point theorem II.2.8 if we prove that $F(\cdot)$ is continuous.

Let (y_j) be a sequence in K converging to some $y \in K$. Then $\lim_j \xi(y_j)(\tau) = \xi(y)(\tau)$ for μ-a.a. $\tau \in T$ and, for each $t \in T$ and μ-a.a. $\tau \in T$,

$$\lim_j f(t, \tau, \xi(y_j)(\tau)) = f(t, \tau, \xi(y)(\tau))$$

and

$$| f(t, \tau, \xi(y_j)(\tau))| \leqslant | f(t, \tau, \cdot)|_{\sup} \qquad (j \in \mathbb{N});$$

hence, by the dominated convergence theorem I.4.35, $\lim_i F(y_i)(t) = F(y)(t)$ for each $t \in T$. Since the set $F(K)$ is equicontinuous and T is compact, it follows from I.5.3 that the convergence is uniform for all $t \in T$. Thus $\lim_i F(y_i) = F(y)$ in $C(T, \mathbb{R}^n)$ and F is continuous. This completes the proof if assumption (1) is satisfied.

Now assume that condition (2) is satisfied, and let

$$\delta \triangleq \sup_{t \in T} \int \psi(t, \tau)\, \mu(d\tau).$$

Since $\beta\gamma < 1$, we can easily verify that there exists some finite positive c_w such that

$$\delta[\alpha(c_w + 1)^\beta + 1]^\gamma \leqslant c_w .$$

We set $W_1 \triangleq S^F(0, c_w)$, $c_v \triangleq \alpha(c_w + 1)^\beta$, $V_1 \triangleq S^F(0, c_v)$, $f_1 \triangleq f \,|\, T \times T \times V_1$, and $\xi_1 \triangleq \xi \,|\, C(T, W_1)$. Then

$$| \xi_1(y)|_\infty \leqslant \alpha(| y |_{\sup} + 1)^\beta \leqslant \alpha(c_w + 1)^\beta = c_v \qquad [y \in C(T, W_1)];$$

hence ξ_1 is a function on $C(T, W_1)$ to $L^\infty(T, V_1)$. It is also clear that

$$| f_1 | = \sup_{t \in T} \int \sup_{v \in V_1} | f(t, \tau, v)| \, \mu(d\tau) \leqslant \sup_{t \in T} \int \psi(t, \tau)\, \mu(d\tau) \cdot (c_v + 1)^\gamma \leqslant c_w ;$$

hence $S^F(0, |f_1|) \subset W_1$. It follows, by our previous argument, that the equation

$$y(t) = \int f(t, \tau, \xi(y)(\tau))\, \mu(d\tau) \qquad (t \in T)$$

has a solution in $C(T, W_1)$. QED

II.5.B Local Solutions and Unique Solutions of Hereditary Functional-Integral Equations

We shall now consider functional-integral equations of the type discussed in II.5.1 but with certain additional properties that enable us to define "local" solutions and to derive a uniqueness theorem. (The main reason for introducing local solutions is to eliminate the bounds on f assumed in II.5.1.) Perhaps the best way to introduce these properties is to first consider the special case of ordinary differential equations. In the notation of II.5.1, let $T \triangleq [t_0, t_1] \subset \mathbb{R}$, Σ and μ refer to the Lebesgue measure on T, $V = W$, ξ be the identity operator in $C(T, V)$, and f have the property that $f(t, \tau, v) = 0$ if $\tau > t$ and $f(t, \tau, v) = \tilde{f}(\tau, v)$ if $\tau \leqslant t$. Then the functional-integral equation of II.5.1 is equivalent to an ordinary differential equation written in the form

$$y(t) = \int_{t_0}^{t} \tilde{f}(\tau, y(\tau))\, d\tau \qquad (t \in T),$$

and the condition that

$$f \in C(T, \mathscr{B}(T, V; \mathbb{R}^n))$$

becomes

$$\tilde{f} \in \mathscr{B}(T, V; \mathbb{R}^n).$$

If $S^F(0, \beta) \subset V$, $\tilde{T} \triangleq \{t \in T \mid \int_{t_0}^{t} |\tilde{f}(\tau, \cdot)|_{\sup}\, d\tau \leqslant \beta\}$, and f_1 and $\tilde{f}_1$ are obtained by restricting f and $\tilde{f}$, respectively, to $\tilde{T}$ [so that $f_1 \in C(\tilde{T}, \mathscr{B}(\tilde{T}, V; \mathbb{R}^n))$ and $\tilde{f}_1 \in \mathscr{B}(T, V; \mathbb{R}^n)$], then we observe that

$$|f_1| = \sup_{t \in \tilde{T}} \int_{\tilde{T}} |f_1(t, \tau, \cdot)|_{\sup}\, d\tau = \sup_{t \in \tilde{T}} \int_{t_0}^{t} |\tilde{f}_1(\tau, \cdot)|_{\sup}\, d\tau \leqslant \beta,$$

and therefore $S^F(0, |f_1|) \subset V = W$. Thus Theorem II.5.1 yields Caratheodory's theorem II.4.1 about the existence of local solutions (in the special case $\eta = 0$).

We now consider once more the general functional-integral equation of II.5.1 but under less restrictive conditions than those imposed by ordinary differential equations. We first introduce the concept of a p-hereditary transformation.

p-Hereditary Transformation

Let T and W be topological spaces, $F(T, V)$ a collection of functions on T to some set V, and $p : T \to [0, 1]$. We say that a function $\xi : C(T, W) \to F(T, V)$ is *p-hereditary* if $\xi(y_1)(\tau) = \xi(y_2)(\tau)$ $[p(\tau) \leqslant \alpha]$ whenever $\alpha \in [0, 1]$ and $y_1(\tau) = y_2(\tau)$ $[p(\tau) \leqslant \alpha]$. If ξ is p-hereditary,

$\alpha \in [0, 1]$, and $\tilde{y}$ is the restriction to $p^{-1}([0, \alpha])$ of some $y \in C(T, W)$, then we write $\xi(\tilde{y})(\tau)$ $[\tau \in p^{-1}([0, \alpha])]$ for $\xi(y)(\tau)$ $[\tau \in p^{-1}([0, \alpha])]$, it being clear from the definition of a p-hereditary function that $\xi(\tilde{y})(\tau)$ $[\tau \in p^{-1}([0, \alpha])]$ is the same for all choices of a continuous extension of $\tilde{y}$ to T.

Local Solutions

In the setting of Theorem II.5.1, whenever ξ is p-hereditary, $\beta \in (0, 1]$, and $f(t, \tau, v) = 0$ $[p(t) < p(\tau)]$, we say that a function $\bar{y} \in C(p^{-1}([0, \beta]), W)$ is a *local solution* of the equation

$$y(t) = \int f(t, \tau, \xi(y)(\tau))\, \mu(d\tau) \qquad (t \in T)$$

if $\bar{y}$ is the restriction to $p^{-1}([0, \beta])$ of some continuous function on T to W and

$$\bar{y}(t) = \int f(t, \tau, \xi(\bar{y})(\tau))\, \mu(d\tau) \qquad [t \in p^{-1}([0, \beta])].$$

It is clear that this definition of a local solution applies to the solution of an ordinary differential equation introduced in Caratheodory's existence theorem II.4.1 [and corresponding to the case where $T = [t_0, t_1] \subset \mathbb{R}$, $m = n$, $V = W$, ξ is the identity operator in $C(T, V)$, and $p(t) = t$ $(t \in T)$.]

II.5.2 ***Lemma*** Let T be a compact metric space, (T, Σ, μ) a positive finite measure space, $m, n \in \mathbb{N}$, $V \subset \mathbb{R}^m$, and $f \in C(T, \mathscr{B}(T, V; \mathbb{R}^n))$. Then

$$\lim_{\mu(E)\to 0} \int_E |f(t, \tau, \cdot)|_{\sup}\, \mu(d\tau) = 0 \quad \text{uniformly for } t \in T.$$

▌PROOF Assume the contrary. Then there exist sequences (t_j) in T and (E_j) in Σ and $\epsilon > 0$ such that $\lim_j \mu(E_j) = 0$ and

$$(1) \qquad \int_{E_j} |f(t_j, \tau, \cdot)|_{\sup}\, \mu(d\tau) > \epsilon \qquad (j \in \mathbb{N}).$$

Since T is a compact metric space, we may assume that (t_j) converges to some $\bar{t}$. Since $f \in C(T, \mathscr{B}(T, V; \mathbb{R}^n))$, we may choose $j_0 \in \mathbb{N}$ such that

$$a_j \triangleq \int_{E_j} |f(\bar{t}, \tau, \cdot)|_{\sup}\, \mu(d\tau) \leqslant \epsilon/2 \qquad (j \geqslant j_0)$$

and

$$b_j \triangleq \int |f(\bar{t}, \tau, \cdot) - f(t_j, \tau, \cdot)|_{\sup}\, \mu(d\tau) \leqslant \epsilon/2 \qquad (j \geqslant j_0);$$

hence

$$\int_{E_j} |f(t_j, \tau, \cdot)|_{\sup} \mu(d\tau) \leqslant a_j + b_j \leqslant \epsilon \qquad (j \geqslant j_0),$$

contrary to (1). QED

II.5.3 ***Theorem*** Let T be a compact metric space, (T, Σ, μ) a positive finite measure space, $n, m \in \mathbb{N}$, $V \subset \mathbb{R}^m$ and $W \subset \mathbb{R}^n$ open, $0 \in W$,

$$f \in C(T, \mathscr{B}(T, V; \mathbb{R}^n)),$$

and

$$\xi : C(T, W) \to L^\infty(T, \Sigma, \mu, V) \quad \text{continuous.}$$

Assume, furthermore, that there exists a μ-measurable function $p : T \to [0, 1]$ such that

(1) $$\lim_{\alpha \to +0} \mu(p^{-1}([0, \alpha])) = 0,$$

(2) $$\xi \text{ is } p\text{-hereditary},$$

and

(3) $$f(t, \tau, v) = 0 \qquad [t, \tau \in T, v \in V, p(\tau) > p(t)].$$

Then there exist $\beta \in (0, 1]$ and a local solution $\bar{y} \in C(p^{-1}([0, \beta]), W)$ of the equation

$$y(t) = \int f(t, \tau, \xi(y)(\tau))\, \mu(d\tau) \qquad (t \in T).$$

▌PROOF Let $T(\alpha) \triangleq p^{-1}([0, \alpha])$ and $\mu_\alpha(A) \triangleq \mu(T(\alpha) \cap A)$ $(\alpha \in [0, 1], A \in \Sigma)$. By (1) and II.5.2, we have

(4) $$\lim_{\alpha \to 0} \int_{T(\alpha)} |f(t, \tau, \cdot)|_{\sup} \mu(d\tau) = 0 \quad \text{uniformly for} \quad t \in T.$$

Since W is open and $0 \in W$, there exists some $a > 0$ such that $S^F(0, a) \subset W$. In view of (4), we can determine $\beta \in (0, 1]$ such that

(5) $$\int_{T(\beta)} |f(t, \tau, \cdot)|_{\sup} \mu(d\tau) = \int |f(t, \tau, \cdot)|_{\sup} \mu_\beta(d\tau) \leqslant a \qquad (t \in T).$$

It is clear that

$$f \in C(T, \mathscr{B}(T, \Sigma, \mu_\beta, V; \mathbb{R}^n))$$

and

$$\xi : C(T, W) \to L^\infty(T, \Sigma, \mu_\beta, V) \quad \text{is continuous.}$$

Furthermore, by (5), $\sup_{t \in T} \int |f(t, \tau, \cdot)|_{\sup} \mu_\beta(d\tau) \leqslant a$. It follows thus from II.5.1 that there exists $\bar{y} \in C(T, W)$ such that

$$\bar{y}(t) = \int f(t, \tau, \xi(\bar{y})(\tau))\, \mu_\beta(d\tau) \qquad (t \in T).$$

Since $\mu(A) = \mu_\beta(A)$ for all $A \in \Sigma$, $A \subset T(\beta)$, and $f(t, \tau, v) = 0$ for $t \in T(\beta)$ and $\tau \notin T(\beta)$, we conclude that

$$\bar{y}(t) = \int f(t, \tau, \xi(\bar{y})(\tau))\, \mu(d\tau) \qquad [t \in T(\beta)]. \quad \text{QED}$$

II.5.4 ***Theorem*** Let T be a compact metric space, (T, Σ, μ) a positive finite measure space, $n, m \in \mathbb{N}$, $V \subset \mathbb{R}^m$, $W \subset \mathbb{R}^n$, $f : T \times T \times V \to \mathbb{R}^n$, $\xi : C(T, W) \to L^\infty(T, V)$, and assume that there exist $c \in \mathbb{R}$, $\psi_1 \in C(T, L^1(T))$, and a μ-measurable $p : T \to [0, 1]$ such that

(1) $$\alpha \to \mu(p^{-1}([0, \alpha])) : [0, 1] \to \mathbb{R} \quad \text{is continuous,}$$

(2) $$f(t, \tau, v) = 0 \qquad [p(\tau) > p(t)],$$

(3) $$|f(t, \tau, v_1) - f(t, \tau, v_2)| \leqslant \psi_1(t, \tau)|\, v_1 - v_2 \,| \qquad (t, \tau \in T,\ v_1, v_2 \in V),$$

and

(4) $$|\xi(y_1)(t) - \xi(y_2)(t)| \leqslant c \sup\nolimits_{p(\tau) \leqslant p(t)} |\, y_1(\tau) - y_2(\tau)| \qquad [y_1, y_2 \in C(T, W),\ \mu\text{-a.a. } t \in T].$$

Then the equation

(5) $$y(t) = \int f(t, \tau, \xi(y)(\tau))\, \mu(d\tau) \qquad (t \in T)$$

can have at most one solution in $C(T, W)$.

▌ PROOF Assume that $\bar{y}_1$ and $\bar{y}_2$ are both solutions of (5) in $C(T, W)$, and set

$$u(t) \triangleq |\, \bar{y}_1(t) - \bar{y}_2(t)|, \qquad s(t) \triangleq \sup\{u(\tau) \mid p(\tau) \leqslant p(t)\} \qquad (t \in T)$$

and

$$\beta \triangleq \sup\{\alpha \in [0, 1] \mid s(t) = 0 \quad \text{if} \quad p(t) \leqslant \alpha\}.$$

Then

$$u(t) = \left| \int_{p^{-1}([\beta, p(t)])} [f(t, \tau, \xi(\bar{y}_1)(\tau)) - f(t, \tau, \xi(\bar{y}_2)(\tau))]\, \mu(d\tau) \right|$$

$$\leqslant c \int_{p^{-1}([\beta, p(t)])} \psi_1(t, \tau)\, s(\tau)\, \mu(d\tau) \qquad [p(t) \geqslant \beta];$$

hence

(6) $$s(t) \leqslant cs(t) \int_{p^{-1}([\beta, p(t)])} \psi_1(t, \tau)\, \mu(d\tau) \qquad [p(t) \geqslant \beta].$$

Now assume that $\beta < 1$. Since $\psi_1 \in C(T, L^1(T))$, it follows from (1) and II.5.2 that there exists $\epsilon > 0$ such that $\beta + \epsilon \leqslant 1$ and

$$\int_{p^{-1}([\beta, \beta+\epsilon])} \psi_1(t, \tau)\, \mu(d\tau) < 1/c \qquad (t \in T).$$

Thus (6) implies that $s(t) = 0$ for $t \in p^{-1}([\beta, \beta + \epsilon])$, contrary to the definition of β. Thus $\beta = 1$; hence $\bar{y}_1(t) = \bar{y}_2(t)$ if $t \in p^{-1}([0, 1))$ and, by (4), $\xi(\bar{y}_1)(t) = \xi(\bar{y}_2)(t)$ for μ-a.a. $t \in p^{-1}([0, 1))$. In view of (1), this implies that $\xi(\bar{y}_1)(t) = \xi(\bar{y}_2)(t)$ μ-a.e. and, since both $\bar{y}_1$ and $\bar{y}_2$ are solutions of (5), that

$$\bar{y}_1(t) = \int f(t, \tau, \xi(\bar{y}_1)(\tau))\, \mu(d\tau) = \int f(t, \tau, \xi(\bar{y}_2)(\tau))\, \mu(d\tau) = \bar{y}_2(t) \qquad (t \in T)$$

QED.

II.5.C Linear Functional-Integral Equations in $C(T, \mathbb{R}^n)$

RESOLVENT KERNEL

Let T be a compact metric space, (T, Σ, μ) a positive finite measure space, $n \in \mathbb{N}$, and A a linear operator in $C(T, \mathbb{R}^n)$. We say that (k^*, μ^*) is a *resolvent kernel* of A if $I - A$ is a homeomorphism of $C(T, \mathbb{R}^n)$ onto itself, $\mu^* \in \text{frm}^+(T)$, $k^* : T \times T \to B(\mathbb{R}^n, \mathbb{R}^n)$, and

$$([(I - A)^{-1} - I]\, \Delta y)(t) = \int k^*(t, \tau)\, \Delta y(\tau)\, \mu^*(d\tau) \qquad [t \in T,\ \ \Delta y \in C(T, \mathbb{R}^n)].$$

II.5.5 ***Theorem*** Let T be a compact metric space, (T, Σ, μ) a positive finite measure space, $m, n \in \mathbb{N}$,

$$\eta \in B(C(T, \mathbb{R}^n), L^\infty(T, \Sigma, \mu, \mathbb{R}^m)),$$

and

$$a \in C(T, L^1(T, \Sigma, \mu, B(\mathbb{R}^m, \mathbb{R}^n))).$$

Then

(1) the expression

$$A(y)(t) \triangleq \int a(t, \tau)\, \eta(y)(\tau)\, \mu(d\tau) \qquad (t \in T)$$

defines a compact operator A in $C(T, \mathbb{R}^n)$.

(2) If $I - A$ is injective, then $I - A$ is a homeomorphism of $C(T, \mathbb{R}^n)$ onto itself, $A^* \triangleq (I - A)^{-1} - I$ is a compact operator in $C(T, \mathbb{R}^n)$, and A has a resolvent kernel (k^*, μ^*) such that

$$k^* \in C(T, L^1(T, \Sigma_{\text{Borel}}(T), \mu^*, B(\mathbb{R}^n, \mathbb{R}^n))).$$

For each choice of $\lambda \in \text{frm}^+(T)$, k^* can be chosen as a $\lambda \times \mu^*$-measurable function.

(3) If, furthermore, $\Sigma \triangleq \Sigma_{\text{Borel}}(T)$, $\mu \in \text{frm}^+(T)$, and

$$\lim_i \int a(t, \tau)\, \eta(\Delta y_i)(\tau)\, \mu(d\tau) = 0$$

whenever $t \in T$, $\Delta y_i \in C(T, \mathbb{R}^n)$, $|\Delta y_i|_{\sup} \leqslant 1$, and $\lim_i \mu(\{\tau \in T \mid \Delta y_i(\tau) \neq 0\}) = 0$, then we may assume that $\mu^* = \mu$. [This is the case, in particular,if $\eta = I$ and $\mu \in \text{frm}^+(T)$].

▌ *Remark.* It is clear that Theorem II.5.5 implies the following statement: *Fredholm alternative* Under the conditions of Theorem II.5.5, either the equation $y = A(y)$ has a solution $y = y_1 \neq 0$ in $C(T, \mathbb{R}^n)$, or the equation

$$y = A(y) + c$$

has a unique solution y in $C(T, \mathbb{R}^n)$ for every choice of $c \in C(T, \mathbb{R}^n)$.

▌ PROOF For every $y \in C(T, \mathbb{R}^n)$ and $t, t' \in T$, we have

$$|A(y)(t) - A(y)(t')| \leqslant |\eta|\, |y| \int |a(t, \tau) - a(t', \tau)|\, \mu(d\tau).$$

Thus $A(y)(\cdot)$ is continuous and $A(\cdot)$ is a mapping on $C(T, \mathbb{R}^n)$ into itself which is clearly linear. We have

$$|A(y)| \leqslant \sup_{t \in T} \int |a(t, \tau)|\, \mu(d\tau) \cdot |\eta|\, |y| = |a|\, |\eta|\, |y|;$$

hence A is a bounded, and therefore continuous, linear operator in $C(T, \mathbb{R}^n)$. These relations also show that the set $\{A(y) \mid |y| \leqslant 1\}$ is bounded and equicontinuous; hence, by I.5.4, conditionally compact. Thus A is a compact operator in $C(T, \mathbb{R}^n)$ and statement (2) follows from I.3.13, I.5.11, and I.5.23.

Now let the conditions of (2) and (3) be satisfied and

$$\left(k^*_{l,j}(t, \tau)\right) \; (l, j = 1, \ldots, n)$$

be the matrix of $k^*(t, \tau)$ for $t, \tau \in T$. We shall show that, for each fixed $t \in T$ and $l, j \in \{1, 2, \ldots, n\}$, the measure $\omega_{l,j}(t)$, defined by

$$\omega_{l,j}(t)(E) \triangleq \int_E k^*_{l,j}(t, \tau)\, \mu^*(d\tau) \qquad (E \in \Sigma), \tag{4}$$

is μ-continuous. We first observe that $(I + A^*)(I - A) = I$, hence $A^* = (I + A^*)A$, therefore $\lim_i A^* \Delta y_i = 0$ if $\lim_i A \Delta y_i = 0$. Now let $Z \in \Sigma$ and $\mu(Z) = 0$. Since both μ and μ^* belong to $\text{frm}^+(T)$, we can determine a sequence (G_i) of open sets and (F_i) of closed sets such that $F_i \subset Z \subset G_i$ and

$$\lim_i [\mu(G_i) + \mu^*(G_i \sim F_i)] = 0.$$

Let $\phi_i(t) \triangleq 1$ if $G_i = T$ and otherwise let

$$\phi_i(t) \triangleq d[t, T \sim G_i]/(d[t, T \sim G_i] + d[t, F_i]) \qquad (i \in \mathbb{N}, \quad t \in T).$$

Then each ϕ_i is continuous, $\phi_i(T) \subset [0, 1]$, and $\phi_i(T \sim G_i) = \{0\}$. We set $\Delta y_i \triangleq (\Delta y_i^1, ..., \Delta y_i^n)$, $\Delta y_i^\alpha \triangleq 0$ $(\alpha \neq l)$, $\Delta y_i^l \triangleq \phi_i$, and observe that, by the assumption of (3),

$$\lim_i \int a(t, \tau)(\eta\, \Delta y_i)(\tau)\, \mu(d\tau) = 0 \qquad \text{for each} \quad t \in T,$$

and therefore $\lim_i (A\, \Delta y_i)(t) = 0$ $(t \in T)$. Since A is compact, the last limit is uniform for $t \in T$ (I.5.3 and I.5.4); hence $\lim_i A\, \Delta y_i = 0$ implying $\lim_i A^*\, \Delta y_i = 0$. On the other hand, if we denote by k_l^* the lth column of k*, then

$$\begin{aligned}(A^*\, \Delta y_i)(t) &= \int k_l^*(t, \tau)\, \phi_i(\tau)\, \mu^*(d\tau) \\ &= \int_Z k_l^*(t, \tau)\, \mu^*(d\tau) \\ &\quad + \int_{G_i \sim Z} k_l^*(t, \tau)\, \phi_i(\tau)\, \mu^*(d\tau).\end{aligned}$$

Since $\lim_i \mu^*(G_i \sim Z) \leqslant \lim_i \mu^*(G_i \sim F_i) = 0$, we conclude that

$$\int_Z k_l^*(t, \tau)\, \mu^*(d\tau) = \lim_i (A^*\, \Delta y_i)(t) = 0$$

for all $l \in \{1, 2, ..., n\}$ and $t \in T$, showing that $\omega_{l,j}(t)$ is μ-continuous for $j, l \in \{1, 2, ...\}$ and $t \in T$.

By (4), the Radon–Nikodym theorem I.4.37 and I.4.38, there exists, for each j, l, and t, a function $k_{l,j}^{\#}(t, \cdot) \in L^1(T, \Sigma, \mu)$ such that

(5) $$\int_E k_{l,j}^{\#}(t, \tau)\, \mu(d\tau) = \omega_{l,j}(t)(E) = \int_E k_{l,j}^*(t, \tau)\, \mu^*(d\tau) \qquad (E \in \Sigma)$$

and

(6) $$\int k_{l,j}^{\#}(t, \tau)\, \phi(\tau)\, \mu(d\tau) = \int \phi(\tau)\, \omega_{l,j}(t)\, (d\tau) = \int k_{l,j}^*(t, \tau)\, \phi(\tau)\, \mu^*(d\tau) \qquad [\phi \in C(T)].$$

If we denote by $k^{\#}(t, \tau)$ the element of $B(\mathbb{R}^n, \mathbb{R}^n)$ with the matrix

$$(k_{l,j}^{\#}(t, \tau))\ (l, j = 1, ..., n),$$

then (2) and (6) imply that

$$A^*(\Delta y)(t) = \int k^{\#}(t, \tau)\, \Delta y(\tau)\, \mu(d\tau) \qquad [t \in T, \quad \Delta y \in C(T, \mathbb{R}^n)]$$

and thus $(k^{\#}, \mu)$ is a resolvent kernel of A. It follows now from the second part of I.5.11 that $k^{\#} \in C(T, L^1(T, \Sigma, \mu, B(\mathbb{R}^n, \mathbb{R}^n)))$ and from I.5.23 that, for every choice of $\lambda \in \text{frm}^+(T)$, $k^{\#}$ may be assumed $\lambda \times \mu$-measurable. Thus statement (3) is valid with $k^{\#}$ replacing k^*. QED

II.5.6 ***Theorem*** Under the conditions of II.5.5, assume that there exists a μ-measurable $p : T \to [0, 1]$ such that

$$\eta \quad \text{is } p\text{-hereditary,} \tag{1}$$

$$a(t, \tau) = 0 \qquad [p(\tau) > p(t)], \tag{2}$$

and

$$\alpha \to \mu(p^{-1}([0, \alpha])) : [0, 1] \to \mathbb{R} \quad \text{is continuous.} \tag{3}$$

Then $I - A$ is injective.

▌PROOF We shall first show that

$$\operatorname*{ess\,sup}_{p(\tau) \leqslant \alpha} |\eta(y)(\tau)| \leqslant n \,|\eta|\, \sup_{p(\tau)\leqslant\alpha} |y(\tau)| \qquad (\alpha \in [0, 1],\ y \in C(T, \mathbb{R}^n)). \tag{4}$$

Let $y \triangleq (y^1,..., y^n) \in C(T, \mathbb{R}^n)$, $\alpha \in [0, 1]$, and $B \triangleq p^{-1}([0, \alpha])$. Then y^i is continuous on the compact set T and therefore

$$s^i \triangleq \sup_{\tau\in B} |y^i(\tau)| < \infty \qquad (i = 1,..., n).$$

It is easy to verify that

$$t \to \max(h_1(t), h_2(t)) \qquad \text{and} \qquad t \to \min(h_1(t), h_2(t))$$

are continuous if $h_1, h_2 \in C(T)$. Thus, if we set

$$\zeta^i(t) \triangleq \max(-s^i, \min(y^i(t), s^i)) \qquad (t \in T, \quad i = 1, 2,..., n)$$

and $\zeta \triangleq (\zeta^1,..., \zeta^n)$, then

$$\zeta \in C(T, \mathbb{R}^n), \qquad \zeta \mid B = y \mid B \qquad \text{and} \qquad |\eta(\zeta)(\tau)| \leqslant |\eta| \textstyle\sum_{i=1}^n s^i \ \mu\text{-a.e.} \tag{5}$$

We have $\sum_{i=1}^n s^i \leqslant n \sup_{\tau\in B} |y(\tau)|$ and, since η is p-hereditary, $\eta(\zeta)(\tau) = \eta(y)(\tau)$ $(\tau \in B)$. Thus (4) follows from (5). Our conclusion now follows from II.5.4. QED

II.5.D Dependence on Parameters of Solutions of Functional-Integral Equations in $C(T, \mathbb{R}^n)$

We shall now consider a functional-integral equation in $C(T, \mathbb{R}^n)$ involving an independent parameter, namely

$$y(t) = \int f(t, \tau, \xi(y)(\tau), a)\, \mu(d\tau) \qquad (t \in T).$$

II.5.7 ***Lemma*** Let T be a compact metric space, (T, Σ, μ) a positive finite measure space, $\mathscr{Y}$ a separable Banach space, S a separable metric space, $x_j : T \times [0, 1] \to S$ $(j \in \mathbb{N})$, $\bar{x} : T \to S$, $\bar{x}(\cdot)$ and $x_j(\cdot, \alpha)$ μ-measurable for each $\alpha \in [0, 1]$, $x_j(t, \cdot)$ continuous for each $t \in T$,

$$\lim_j x_j(\tau, \alpha) = \bar{x}(\tau) \quad \text{for } \mu\text{-a.a. } \tau \in T, \quad \text{uniformly for } \alpha \in [0, 1]$$

and

$$\phi \in C(T, \mathscr{B}(T, S; \mathscr{Y})).$$

Then

$$\lim_j \int \sup_{0 \leqslant \alpha \leqslant 1} | \phi(t, \tau, x_j(\tau, \alpha)) - \phi(t, \tau, \bar{x}(\tau))| \, \mu(d\tau) = 0$$

uniformly for $t \in T$.

▌PROOF We first observe that, by I.4.21 and I.4.22, the functions

$$\tau \to \phi(t, \tau, x_j(\tau, \alpha)), \qquad \tau \to \phi(t, \tau, \bar{x}(\tau)),$$

and

$$\tau \to \sup_{0 \leqslant \alpha \leqslant 1} | \phi(t, \tau, x_j(\tau, \alpha)) - \phi(t, \tau, \bar{x}(\tau))|$$

are μ-measurable for all $t \in T, j \in \mathbb{N}$, and $\alpha \in [0, 1]$. Now assume that the conclusion of the lemma is not valid. Then there exist $\epsilon > 0$, $J \subset (1, 2,...)$ and a sequence (t_j) in T such that

$$(1) \qquad \int \sup_{0 \leqslant \alpha \leqslant 1} | \phi(t_j, \tau, x_j(\tau, \alpha)) - \phi(t_j, \tau, \bar{x}(\tau))| \, \mu(d\tau) > \epsilon \qquad (j \in J).$$

Since T is sequentially compact, we may assume that $\lim_j t_j = \bar{t}$. By assumption, there exists $j_0 \in \mathbb{N}$ such that

$$(2) \qquad \int | \phi(\bar{t}, \tau, \cdot) - \phi(t_j, \tau, \cdot)|_{\sup} \, \mu(d\tau) \leqslant \epsilon/2 \qquad (j \geqslant j_0)$$

and

$$(3) \qquad \lim_j \sup_{0 \leqslant \alpha \leqslant 1} | \phi(\bar{t}, \tau, x_j(\tau, \alpha)) - \phi(\bar{t}, \tau, \bar{x}(\tau))| = 0 \quad \mu\text{-a.e.}$$

Since $\sup_{0 \leqslant \alpha \leqslant 1} | \phi(\bar{t}, \tau, x_j(\tau, \alpha)) - \phi(\bar{t}, \tau, \bar{x}(\tau))| \leqslant 2 \, | \phi(\bar{t}, \tau, \cdot)|_{\sup}$ $(\tau \in T)$, it follows from (3) and the dominated convergence theorem I.4.35 that

$$(4) \qquad \lim_j \int \sup_{0 \leqslant \alpha \leqslant 1} | \phi(\bar{t}, \tau, x_j(\tau, \alpha)) - \phi(\bar{t}, \tau, \bar{x}(\tau))| \, \mu(d\tau) = 0.$$

Finally, by assumption,

$$\text{(5)} \qquad \lim_{j\in J} \int \sup_{0\leqslant\alpha\leqslant 1} |\phi(t_j, \tau, x_j(\tau, \alpha)) - \phi(\bar{t}, \tau, x_j(\tau, \alpha))| \, \mu(d\tau)$$

$$\leqslant \lim_{j\in J} \int |\phi(t_j, \tau, \cdot) - \phi(\bar{t}, \tau, \cdot)|_{\sup} \, \mu(d\tau) = 0.$$

Thus, by (4) and (5),

$$\lim_{j\in J} \int \sup_{0\leqslant\alpha\leqslant 1} |\phi(t_j, \tau, x_j(\tau, \alpha)) - \phi(t_j, \tau, \bar{x}(\tau))| \, \mu(d\tau)$$

$$\leqslant \lim_{j\in J} \int \sup_{0\leqslant\alpha\leqslant 1} |\phi(t_j, \tau, x_j(\tau, \alpha)) - \phi(\bar{t}, \tau, x_j(\tau, \alpha))| \, \mu(d\tau)$$

$$+ \lim_{j\in J} \int \sup_{0\leqslant\alpha\leqslant 1} |\phi(\bar{t}, \tau, x_j(\tau, \alpha)) - \phi(\bar{t}, \tau, \bar{x}(\tau))| \, \mu(d\tau) = 0,$$

contrary to (1). QED

II.5.8 ***Theorem*** Let T be a compact metric space, (T, Σ, μ) a positive finite measure space, $m, k, n \in \mathbb{N}$, $V \subset \mathbb{R}^m$ and $W \subset \mathbb{R}^n$ open, A a convex subset of $\mathbb{R}^k$,

$$f \in C(T, \mathscr{B}(T, V \times A; \mathbb{R}^n)), \qquad f_{(v,a)} \in C(T, \mathscr{B}(T, V \times A; B(\mathbb{R}^m \times \mathbb{R}^k, \mathbb{R}^n))),$$

and let $\xi : C(T, W) \to L^\infty(T, V)$ have a continuous derivative. Then the relation

$$F(y, a)(t) \triangleq \int f(t, \tau, \xi(y)(\tau), a) \, \mu(d\tau) \qquad [y \in C(T, W), \quad a \in A, \quad t \in T]$$

defines a continuous function $F : C(T, W) \times A \to C(T, \mathbb{R}^n)$ with a continuous derivative F' such that

$$\text{(1)} \quad F'(y, a)(\Delta y, \Delta a)(t) = \int f_{(v,a)}(t, \tau, \xi(y)(\tau), a)((\xi'(y)\,\Delta y)(\tau), \Delta a) \, \mu(d\tau)$$
$$[\Delta y \in C(T, \mathbb{R}^n), \Delta a \in \mathbb{R}^k, t \in T].$$

PROOF We define two transformations

$$G : L^\infty(T, V) \times A \to C(T, L^1(T, \mathbb{R}^n))$$

and

$$H : C(T, L^1(T, \mathbb{R}^n)) \to C(T, \mathbb{R}^n)$$

by the following relations:

$$\text{(2)} \quad G(z, a)(t, \tau) \triangleq f(t, \tau, z(\tau), a) \qquad [t, \tau \in T, \quad z \in L^\infty(T, V), \quad a \in A]$$

and

$$H(\phi)(t) \triangleq \int \phi(t, \tau)\, \mu(d\tau) \qquad [\phi \in C(T, L^1(T, \mathbb{R}^n)), \quad t \in T]. \tag{3}$$

We shall prove that both G and H have domains and ranges as indicated, are continuous, and have continuous derivatives.

Step 1 We first consider G. For each t, z, and a, the function

$$\tau \to f(t, \tau, z(\tau), a)$$

is μ-integrable. Furthermore,

$$\begin{aligned} &\lim_{t' \to t} \int |f(t, \tau, z(\tau), a) - f(t', \tau, z(\tau), a)|\, \mu(d\tau) \\ &\qquad \leqslant \lim_{t' \to t} \int |f(t, \tau, \cdot, \cdot) - f(t', \tau, \cdot, \cdot)|_{\sup}\, \mu(d\tau) = 0, \end{aligned}$$

thus showing that $G(z, a) \in C(T, L^1(T, \mathbb{R}^n))$.

Now we observe that, for each $z \in L^\infty(T, V)$ and $a \in A$, the operator $\tilde{G}(z, a)$, defined by

$$\begin{aligned} \tilde{G}(z, a)(\Delta z, \Delta a)(t, \tau) &\triangleq f_{(v,a)}(t, \tau, z(\tau), a)(\Delta z(\tau), \Delta a) \\ &\qquad [\Delta z \in L^\infty(T, \mathbb{R}^m),\ \Delta a \in \mathbb{R}^k,\ t, \tau \in T], \end{aligned} \tag{4}$$

belongs to

$$B(L^\infty(T, \mathbb{R}^m) \times \mathbb{R}^k, C(T, L^1(T, \mathbb{R}^n))).$$

Indeed, for fixed z, a, Δz, and Δa, $\tau \to \tilde{G}(z, a)(\Delta z, \Delta a)(t, \tau)$ is μ-integrable for each $t \in T$ (I.5.25), and we have

$$\begin{aligned} &\lim_{t' \to t} \int |[f_{(v,a)}(t, \tau, z(\tau), a) - f_{(v,a)}(t', \tau, z(\tau), a)](\Delta z(\tau), \Delta a)|\, \mu(d\tau) \\ &\quad \leqslant (|\Delta z| + |\Delta a|) \lim_{t' \to t} \int |f_{(v,a)}(t, \tau, \cdot, \cdot) - f_{(v,a)}(t', \tau, \cdot, \cdot)|_{\sup}\, \mu(d\tau) = 0, \end{aligned}$$

showing that $\tilde{G}(z, a)(\Delta z, \Delta a) \in C(T, L^1(T, \mathbb{R}^n))$. It is clear that

$$(\Delta z, \Delta a) \to \tilde{G}(z, a)(\Delta z, \Delta a)$$

is linear and, for each z and a,

$$\begin{aligned} |\tilde{G}(z, a)(\Delta z, \Delta a)| &= \sup_{t \in T} \int |f_{(v,a)}(t, \tau, z(\tau), a)(\Delta z(\tau), \Delta a)|\, \mu(d\tau) \\ &\leqslant (|\Delta z| + |\Delta a|) \sup_{t \in T} \int |f_{(v,a)}(t, \tau, \cdot, \cdot)|_{\sup}\, \mu(d\tau); \end{aligned}$$

hence $\tilde{G}(z, a)$ is a bounded linear operator.

We now show that $G'(z, a) = \tilde{G}(z, a)$. Indeed, for fixed z and a and for all $\Delta z \in L^\infty(T, V) - z$ and $\Delta a \in A - a$ with $|\Delta z| + |\Delta a \neq 0$ and

$$S^F(z, |\Delta z|) \subset L^\infty(T, V),$$

the mean value theorem II.3.6 implies that

$$\begin{aligned}(|\Delta z| + |\Delta a|)^{-1} |f(t, \tau, z(\tau) + \Delta z(\tau), a + \Delta a) - f(t, \tau, z(\tau), a)\\ - f_{(v,a)}(t, \tau, z(\tau), a)(\Delta z(\tau), \Delta a)|\\ \leqslant \sup_{0 \leqslant \alpha \leqslant 1} |f_{(v,a)}(t, \tau, z(\tau) + \alpha\,\Delta z(\tau), a + \alpha\,\Delta a)\\ - f_{(v,a)}(t, \tau, z(\tau), a)| \qquad (t, \tau \in T);\end{aligned}$$

hence, by II.5.7,

$$\lim(|\Delta z| + |\Delta a|)^{-1} |G(z + \Delta z, a + \Delta a) - G(z, a) - \tilde{G}(z, a)(\Delta z, \Delta a)| = 0 \quad \text{as} \quad |\Delta z| + |\Delta a| \to 0.$$

This shows that G is differentiable at (z, a) and $G'(z, a) = \tilde{G}(z, a)$, which implies, in turn, that G is continuous at (z, a).

To prove that G' is continuous, we consider a point $(z, a) \in L^\infty(T, V) \times A$ and a sequence $((z_j, a_j))$ in $L^\infty(T, V) \times A$ converging to (z, a). We have

$$\begin{aligned}|G'(z_j, a_j) - G'(z, a)|\\ = \sup \Big\{ \Big| \int |[f_{(v,a)}(t, \tau, z_j(\tau), a_j) - f_{(v,a)}(t, \tau, z(\tau), a)](\Delta z(\tau), \Delta a)|\, \mu(d\tau) \Big|\\ t \in T, \Delta z \in L^\infty(T, \mathbb{R}^m), \Delta a \in \mathbb{R}^k, |\Delta z| + |\Delta a| \leqslant 1 \Big\}\\ \leqslant \sup_{t \in T} \int |f_{(v,a)}(t, \tau, z_j(\tau), a_j) - f_{(v,a)}(t, \tau, z(\tau), a)|\, \mu(d\tau) \qquad (j \in \mathbb{N})\end{aligned}$$

and, by II.5.7, the right-hand side of the above relation converges to 0. Thus G' is continuous.

Step 2 We next consider the much simpler transformation H. It is clear that H is linear, and we have

$$\lim_{t' \to t} |H(\phi)(t) - H(\phi)(t')| \leqslant \lim_{t' \to t} \int |\phi(t, \tau) - \phi(t', \tau)|\, \mu(d\tau) = 0$$

and

$$|H(\phi)| = \sup_{t \in T} \left| \int \phi(t, \tau)\, \mu(d\tau) \right| \leqslant \sup_{t \in T} \int |\phi(t, \tau)|\, \mu(d\tau) = |\phi|;$$

hence $H(\phi) \in C(T, \mathbb{R}^n)$ and $|H| \leqslant 1$, implying that

$$H \in B(C(T, L^1(T, \mathbb{R}^n)), C(T, \mathbb{R}^n)).$$

Thus $H'(\phi) = H$ for all ϕ and H' is continuous (and in fact constant).

Step 3 We can now complete the proof. We observe that, for all $z \in L^\infty(T, V)$, $a \in A$, and $t \in T$,

$$H \circ G(z, a)(t) = \int f(t, \tau, z(\tau), a)\, \mu(d\tau);$$

hence for all $y \in C(T, W)$ and $a \in A$,

$$F(y, a) = H \circ G(\xi(y), a).$$

Since H, G, and ξ are continuous and have continuous derivatives, it follows that F is continuous and, by the chain rule II.3.4, F has a continuous derivative F' such that

$$F'(y, a)(\Delta y, \Delta a) = H'(G(\xi(y), a)) \circ G'(\xi(y), a)(\xi'(y)\, \Delta y, \Delta a)$$
$$[\Delta y \in C(T, \mathbb{R}^n),\ \Delta a \in \mathbb{R}^k].$$

Since $H'(\phi) = H$ and $G' = \tilde{G}$, relations (3) and (4) yield (1). QED

II.5.9 ***Theorem*** Let the conditions of II.5.8 be satisfied, $y_0 \in C(T, W)$, $a_0 \in A$, and assume that

$$y_0(t) = \int f(t, \tau, \xi(y_0)(\tau), a_0)\, \mu(d\tau) \qquad (t \in T) \tag{1}$$

and the equation

$$\Delta y(t) = \int f_v(t, \tau, \xi(y_0)(\tau), a_0)(\xi'(y_0)\, \Delta y)(\tau)\, \mu(d\tau) \qquad (t \in T) \tag{2}$$

has only the solution $\Delta y = 0$ in $C(T, \mathbb{R}^n)$. Then there exist a neighborhood $\tilde{Y}$ of y_0 in $C(T, W)$ and a relative neighborhood $\tilde{A}$ of a_0 in A such that the equation

$$y(t) = \int f(t, \tau, \xi(y)(\tau), a)\, \mu(d\tau) \qquad (t \in T) \tag{3}$$

has a unique solution $y(\cdot) = u(a)(\cdot)$ in $\tilde{Y}$ for all $a \in \tilde{A}$; furthermore, the function $u : \tilde{A} \to \tilde{Y}$ is continuous and has a continuous derivative u' on $\tilde{A}$ satisfying the relation

$$\begin{aligned} w(t) &\triangleq (u'(a)\, \Delta a)(t) \\ &= \int f_{(v,a)}(t, \tau, \xi(u(a))(\tau), a)((\xi'(u(a))w)(\tau), \Delta a)\, \mu(d\tau) \\ &\qquad (t \in T,\quad \Delta a \in \mathbb{R}^k). \end{aligned} \tag{4}$$

If there exists a μ-measurable $p : T \to [0, 1]$ such that $\xi'(y_0)$ is p-hereditary, $\alpha \to \mu(p^{-1}([0, \alpha])) : [0, 1] \to \mathbb{R}$ continuous and $f_{(v,a)}(t, \tau, \xi(y_0)(\tau), a_0) = 0$ $[p(\tau) > p(t)]$, then Eq. (2) has only the solution 0.

▌PROOF Let $G(y, a) \triangleq y - F(y, a)$ $[(y, a) \in C(T, W) \times A]$. By II.5.8, G and G' are continuous on $C(T, W) \times A$ and, furthermore, $C(T, W)$ is open, A is convex, and $G(y_0, a_0) = 0$. By II.5.8(1), if we set $\xi_0 \triangleq \xi(y_0)$, then

$$|(F_y(y_0, a_0)\,\Delta y)(t) - (F_y(y_0, a_0)\,\Delta y)(t')|$$

$$\leqslant \int |[f_v(t, \tau, \xi_0(\tau), a_0) - f_v(t', \tau, \xi_0(\tau), a_0)](\xi'(y_0)\,\Delta y)(\tau)|\,\mu(d\tau)$$

$$\leqslant |\xi'(y_0)|\,|\Delta y| \int |f_v(t, \tau, \xi_0(\tau), a_0) - f(t', \tau, \xi_0(\tau), a_0)|\,\mu(d\tau),$$

showing that $\{F_y(y_0, a_0)\,\Delta y \mid |\Delta y| \leqslant 1\}$ is equicontinuous and bounded in $C(T, \mathbb{R}^n)$; hence, by I.5.4, $F_y(y_0, a_0)$ is a compact operator and, by Riesz's theorem I.3.13, $G_y(y_0, a_0) = I - F_y(y_0, a_0)$ is a homeomorphism. Thus it follows from the implicit function theorem II.3.8 that there exist neighborhoods $\tilde{Y}$ of y_0 in $C(T, W)$ and $\tilde{A}$ of a_0 (relative to A) such that Eq. (3) has a unique solution $u(a)$ in $\tilde{Y}$ for all $a \in \tilde{A}$, u is continuous and has a continuous derivative u' on $\tilde{A}$, and

$$u'(a) = -G_y(u(a), a)^{-1} \circ G_a(u(a), a) \qquad (a \in \tilde{A}).$$

We conclude that

$$G_y(u(a), a)\,u'(a)\,\Delta a + G_a(u(a), a)\,\Delta a = 0 \qquad (a \in \tilde{A})$$

and, by II.5.8(1),

$$w(t) - \int f_{(v,a)}(t, \tau, \xi(u(a))(\tau), a)((\xi'(u(a))w)(\tau), \Delta a)\,\mu(d\tau) = 0$$

$$(\Delta a \in \mathbb{R}^k, \quad t \in T).$$

Finally, if there exists a function p as described in the last part of the theorem then, by II.5.6, Eq. (2) has at most one solution which obviously must be 0. QED

II.5.E Integral Equations with Pseudodelays in $C(T, \mathbb{R}^n)$

We next consider solutions in $C(T, \mathbb{R}^n)$ of the integral equation with pseudodelays II.1(6).

II.5.10 ***Theorem*** Let T be a compact metric space, (T, Σ, μ) a positive finite measure space,

$$\Sigma \triangleq \Sigma_{\text{Borel}}(T), \quad n, l \in \mathbb{N}, \quad W \subset \mathbb{R}^n, \quad \text{and} \quad h_i : T \to T \quad (i = 1, 2..., l)$$

μ-measurable. If we set

$$\xi(y)(t) = \big(y \circ h_1(t), ..., y \circ h_l(t)\big) \qquad [y \in C(T, W), \quad t \in T],$$

then $\xi \in B\big(C(T, W), L^\infty(T, \Sigma, \mu, W^l)\big)$ and ξ satisfies the conditions of Theorems II.5.1, II.5.5, and II.5.8. If, furthermore, $p : T \to [0, 1]$ is μ-measurable and

$$p\big(h_i(\tau)\big) \leqslant p(\tau) \qquad (\tau \in T, \quad i = 1, 2,..., l),$$

then ξ is p-hereditary and satisfies the conditions of Theorems II.5.3, II.5.4, and II.5.6.

▌PROOF Direct verification.

II.6 Functional-Integral Equations in $L^p(T, \mathbb{R}^n)$

II.6.A Existence of Solutions

All of the specific functional equations that we have examined so far were equations in $C(T, \mathbb{R}^n)$. We shall now consider an equation in $L^p(T, \mathbb{R}^n)$. Let (T, Σ, μ) be a finite positive measure space, $n, m \in \mathbb{N}$, $f : T \times T \times \mathbb{R}^m \to \mathbb{R}^n$, $1 \leqslant p < \infty$, $1 \leqslant p_1 \leqslant \infty$, and

$$\xi : L^p(T, \Sigma, \mu, \mathbb{R}^n) \to L^{p_1}(T, \Sigma, \mu, \mathbb{R}^m).$$

We shall consider the equation

$$y(t) = \int f\big(t, \tau, \xi(y)(\tau)\big)\, \mu(d\tau) \quad \mu\text{-a.e.}$$

which formally resembles the functional-integral equation discussed in II.5. When convenient, we shall write $| t \to h(t)|_p$ for $| h |_p$ when

$$h \in L^p(T, \Sigma, \mu, \mathscr{X})$$

for some separable Banach space $\mathscr{X}$ and we shall agree, in all of II.6, that $c/0 \triangleq \infty$ $(c > 0)$ and $\infty/\infty \triangleq 1$.

II.6.1 ***Assumption***

(1) T is a compact metric space, $\mu \in \text{frm}^+(T)$, and $\Sigma \triangleq \Sigma_{\text{Borel}}(T)$;
(2) $n, m \in \mathbb{N}$, $1 \leqslant p < \infty$, $1 \leqslant p_1 \leqslant \infty$, and $\xi : L^p(T, \Sigma, \mu, \mathbb{R}^n) \to L^{p_1}(T, \Sigma, \mu, \mathbb{R}^m)$ is continuous;

(3) $\beta \in \mathbb{R} \cap [0, p_1]$, d is the conjugate exponent to p_1/β, ψ is $\mu \times \mu$-measurable, and $|\, t \to |\, \psi(t, \cdot)|_d\, |_p < \infty$;
(4) $\chi \in L^{p_1}(T, \Sigma, \mu)$ and $c \in \mathbb{R}$ are such that $|\, \xi(y)(t)| \leqslant \chi(t)$ μ-a.e. whenever $y \in L^p(T, \Sigma, \mu, \mathbb{R}^n)$ and $|\, y(t)| \leqslant c\, |\, \psi(t, \cdot)|_d$ μ-a.e.

The next two lemmas, II.6.2 and II.6.3, are somewhat more general than needed for our present purposes. They will be used in VIII.1.

II.6.2 ***Lemma*** Let Assumption II.6.1 be satisfied, S be a compact metric space, Q a collection of μ-measurable functions on T to S, and $\tilde{f} : T \times T \times \mathbb{R}^m \times S \to \mathbb{R}^n$ such that, for all $(t, \tau, v, s) \in T \times T \times \mathbb{R}^m \times S$, $\tilde{f}(\cdot, \cdot, v, s)$ is $\mu \times \mu$-measurable, $\tilde{f}(t, \tau, \cdot, \cdot)$ is continuous, and

$$|\, \tilde{f}(t, \tau, v, s)| \leqslant \psi(t, \tau)(1 + |\, v\, |^\beta).$$

Furthermore, let

$$K \triangleq \{ y \in L^p(T, \mathbb{R}^n) \,\big|\, |y(t)| \leqslant c\, |\, \psi(t, \cdot)|_d \quad \mu\text{-a.e.}\}$$

and

$$\psi^{\#}(t, \tau) \triangleq \psi(t, \tau)[1 + \chi(\tau)^\beta].$$

Then

(1) the expression

$$\tilde{F}(y, q)(t) \triangleq \int \tilde{f}(t, \tau, \xi(y)(\tau), q(\tau))\, \mu(d\tau) \quad \mu\text{-a.e.} \qquad (y \in K, \quad q \in Q)$$

defines a mapping $\tilde{F} : K \times Q \to L^p(T, \mathbb{R}^n)$,

(2) $$|\, \psi^{\#}(t, \cdot)|_1 \leqslant [\mu(T)^{\beta/p_1} + |\, \chi\, |_{p_1}^\beta]\, |\, \psi(t, \cdot)|_d \qquad (t \in T)$$

and

(3) $|\, \tilde{f}(t, \tau, \xi(y)(\tau), q(\tau))| \leqslant \psi^{\#}(t, \tau) \qquad (y \in K,\ q \in Q,\ t \in T,\ \mu\text{-a.a. } \tau \in T)$.

▌PROOF Let $y \in K$ and $q \in Q$. By II.6.1(2), the function

$$(t, \tau) \to (\xi(y)(\tau), q(\tau))$$

is $\mu \times \mu$-measurable and it follows from I.4.22 and I.4.45(1) that

$$(t, \tau) \to \tilde{f}(t, \tau, \xi(y)(\tau), q(\tau))$$

is $\mu \times \mu$-measurable and $\tau \to \tilde{f}(t, \tau, \xi(y)(\tau), q(\tau))$ is μ-measurable for μ-a.a. $t \in T$. By II.6.1(4),

$$\begin{aligned} |\, \tilde{f}(t, \tau, \xi(y)(\tau), q(\tau))| &\leqslant \psi(t, \tau)(1 + |\, \xi(y)(\tau)|^\beta) \\ &\leqslant \psi(t, \tau)(1 + \chi(\tau)^\beta) \\ &\triangleq \psi^{\#}(t, \tau) \qquad (t \in T, \quad \mu\text{-a.a. } \tau \in T), \end{aligned}$$

thus proving (3); hence, by Hölder's inequality I.5.13 and Minkowski's inequality I.5.15,

$$\int |\tilde{f}(t, \tau, \xi(y)(\tau), q(\tau))| \mu(d\tau) \leqslant \int \psi^{\#}(t, \tau) \mu(d\tau)$$

$$\leqslant |\psi(t, \cdot)|_d [\mu(T)^{\beta/p_1} + |\chi|_{p_1}^{\beta}] \qquad (t \in T).$$

This shows that (2) is valid and, by I.4.35 and I.4.45(2), that

$$\tilde{F}(y, q) \in L^p(T, \mathbb{R}^n). \qquad \text{QED}$$

II.6.3 ***Lemma*** Under the conditions of Lemma II.6.2, $\tilde{F}(K \times Q)$ is a conditionally compact subset of $L^p(T, \mathbb{R}^n)$.

PROOF Let $B_N \triangleq S^F(0, N) \subset \mathbb{R}^m$,

$$T_N \triangleq \{t \in T \mid \chi(t) \leqslant N, [\mu(T)^{\beta/p_1} + |\chi|_{p_1}^{\beta}] \, |\psi(t, \cdot)|_d \leqslant N\} \qquad (N > 0)$$

and $\epsilon > 0$. By II.6.2(2), II.6.1(3), I.4.35, and I.4.45, ψ and $\psi^{\#}$ are $\mu \times \mu$-integrable and, by I.4.27(5), I.4.28, and the regularity of μ, there exist $N \triangleq N(\epsilon)$ and a closed $\tilde{T} \subset T_N$ such that

$$\text{(1)} \qquad \int \mu(dt) \int_{T \sim \tilde{T}} \psi^{\#}(t, \tau) \mu(d\tau) \leqslant \epsilon/8.$$

We denote by $\tilde{\Sigma}$ the collection of Borel subsets of $\tilde{T}$ and set $\tilde{\mu} = \mu \mid \tilde{\Sigma}$. Since $\tilde{f}(\cdot, \cdot, v, s) | T \times \tilde{T}$ is $\mu \times \tilde{\mu}$-measurable for each $(v, s) \in B_N \times S$, $\tilde{f}(t, \tau, \cdot, \cdot)$ is continuous on $B_N \times S$ for each $(t, \tau) \in T \times \tilde{T}$ and

$$|\tilde{f}(t, \tau, v, s)| \leqslant \psi(t, \tau)(1 + N^{\beta})$$

for $(t, \tau, v, s) \in T \times \tilde{T} \times B_N \times S$, it follows that $\tilde{f} | T \times \tilde{T} \times B_N \times S$ is an element of the space $\mathscr{B}(T \times \tilde{T}, \Sigma \otimes \tilde{\Sigma}, \mu \times \tilde{\mu}, B_N \times S; \mathbb{R}^n)$. Thus, by I.5.25(2) and I.4.30, $\tilde{f}$ can be approximated in

$$L^1(T \times \tilde{T}, \Sigma \otimes \tilde{\Sigma}, \mu \times \tilde{\mu}, C(B_N \times S, \mathbb{R}^n))$$

by a $\mu \times \tilde{\mu}$-simple function; there exist, therefore, $k \triangleq k(\epsilon) \in \mathbb{N}$, $\mu \times \tilde{\mu}$-measurable $\alpha_j : T \times \tilde{T} \to \{0, 1\}$ and $\beta_j \in C(B_N \times S, \mathbb{R}^n)$ $(j = 1, 2, \ldots, k)$ such that

$$\text{(2)} \qquad \int \mu(dt) \int_{\tilde{T}} \left| \tilde{f}(t, \tau, \cdot, \cdot) - \sum_{j=1}^{k} \alpha_j(t, \tau) \beta_j(\cdot, \cdot) \right|_{\sup} \mu(d\tau) \leqslant \epsilon/8.$$

Furthermore, it follows from I.5.24 that for each $j \in \{1, 2,..., k\}$ there exist $k_j \in \mathbb{N}$ and continuous $b_l^j : \tilde{T} \to \mathbb{R}$ and $a_l^j : T \to \mathbb{R}$ $(l = 1, 2,..., k_j)$ such that

$$\int \mu(dt) \int_{\tilde{T}} \left| \alpha_j(t, \tau) - \sum_{l=1}^{k_j} b_l^j(\tau)\, a_l^j(t) \right| \mu(d\tau) \leqslant \epsilon/[8k \sup_{(v,s)\in B_N \times S} |\beta_j(v, s)|].$$

We may therefore rewrite relation (2), after suitably redefining k and β_j, in the form

$$(3) \qquad \int \mu(dt) \int_{\tilde{T}} \left| \tilde{f}(t, \tau, \cdot, \cdot) - \sum_{j=1}^{k} a_j(t)\, b_j(\tau)\, \beta_j(\cdot, \cdot) \right|_{\sup} \mu(d\tau) \leqslant \epsilon/4,$$

where $a_j : T \to \mathbb{R}$, $b_j : \tilde{T} \to \mathbb{R}$, and $\beta_j : B_N \times S \to \mathbb{R}^n$ are continuous. We may also assume that

$$|\beta_j(v, s)| \leqslant 1 \text{ and } |b_j(\tau)| \leqslant 1 \qquad [j = 1, 2,..., k,\ (v, s) \in B_N \times S,\ \tau \in \tilde{T}]$$

(after multiplying a_j and dividing b_j and β_j by suitable constants).

Now let $((y_i, q_i))$ be a sequence in $K \times Q$. By II.6.1(4), we have

$$|\xi(y_i)(\tau)| \leqslant \chi(\tau) \quad \mu\text{-a.e.};$$

hence

$$|\xi(y_i)(\tau)| \leqslant N \quad \tilde{\mu}\text{-a.e. in } \tilde{T}.$$

We may therefore define the points

$$\gamma_{j,i}(\epsilon) \triangleq \gamma_{j,i} \triangleq \int_{\tilde{T}} b_j(\tau)\, \beta_j(\xi(y_i)(\tau), q_i(\tau))\, \mu(d\tau) \qquad (j = 1, 2,... k; \quad i \in \mathbb{N}).$$

Since $y_i \in K$, it follows from II.6.2 that, for all $\rho, \sigma \in \mathbb{N}$ and μ-a.a. $t \in T$, we have

$$\begin{aligned}
&|\tilde{F}(y_\rho, q_\rho)(t) - \tilde{F}(y_\sigma, q_\sigma)(t)| \\
&\quad = \left| \int [\tilde{f}(t, \tau, \xi(y_\rho)(\tau), q_\rho(\tau)) - \tilde{f}(t, \tau, \xi(y_\sigma)(\tau), q_\sigma(\tau))]\, \mu(d\tau) \right| \\
&\quad \leqslant 2 \int_{T \sim \tilde{T}} \psi^{\#}(t, \tau)\, \mu(d\tau) \\
&\qquad + \sum_{i \in \{\rho, \sigma\}} \int_{\tilde{T}} \Big| \tilde{f}(t, \tau, \xi(y_i)(\tau), q_i(\tau)) \\
&\qquad - \sum_{j=1}^{k} a_j(t)\, b_j(\tau)\, \beta_j(\xi(y_i)(\tau), q_i(\tau)) \Big| \mu(d\tau) \\
&\qquad + \sum_{j=1}^{k} |\gamma_{j,\rho} - \gamma_{j,\sigma}|\, |a_j(t)|;
\end{aligned}$$

hence, by (1) and (3),

$$\int |\tilde{F}(y_\rho, q_\rho)(t) - \tilde{F}(y_\sigma, q_\sigma)(t)| \, \mu(dt) \leqslant 3\epsilon/4 + \sum_{j=1}^{k} |\gamma_{j,\rho} - \gamma_{j,\sigma}| \int |a_j(t)| \, \mu(dt). \tag{4}$$

Since $|\gamma_{j,i}| \leqslant \mu(\tilde{T}) \leqslant \mu(T)$ $(j = 1, 2,..., k; i \in \mathbb{N})$, for any sequence $\tilde{J} \subset (1, 2,...)$ and $\epsilon > 0$ we can determine a subsequence $J' \triangleq J'(\tilde{J}, \epsilon)$ of $\tilde{J}$ such that each of the sequences $(\gamma_{j,i}(\epsilon))_{i \in J'}$ in $\mathbb{R}^n$ has a limit for $j = 1, 2,..., k(\epsilon)$. Now let $J_1 \triangleq (1, 2,...)$ and $J_{l+1} \triangleq J'(J_l, 1/l)$ $(l \in \mathbb{N})$, and let J be the diagonal subsequence of $J_1, J_2, ...$. Then $(\gamma_{j,i}(1/m))_{i \in J}$ converges for each $m \in \mathbb{N}$ and $j = 1, 2,..., k(1/m)$ and there exists $i_0(m) \in \mathbb{N}$ such that, by (4),

$$\int |\tilde{F}(y_\rho, q_\rho)(t) - \tilde{F}(y_\sigma, q_\sigma)(t)| \, \mu(dt) \leqslant 1/m$$

for $\rho \geqslant \sigma \geqslant i_0(m)$ and $\rho, \sigma \in J$. We conclude that $(\tilde{F}(y_i, q_i))_{i \in J}$ is a Cauchy sequence in $L^1(T, \mathbb{R}^n)$ which, by I.5.17 and I.4.31, converges to some $\bar{y} \in L^1(T, \mathbb{R}^n)$ in μ-measure. By II.6.2, we have

$$|\tilde{F}(y, q)(t)| \leqslant [\mu(T)^{\beta/p_1} + |\chi|_{p_1}^{\beta}] \, |\psi(t, \cdot)|_d \, ;$$

hence, by I.4.35, $\bar{y} \in L^p(T, \mathbb{R}^n)$ and $\lim_{i \in J} \tilde{F}(y_i, q_i) = \bar{y}$ in $L^p(T, \mathbb{R}^n)$. Thus the closure of $\tilde{F}(K \times Q)$ is a sequentially compact subset of $L^p(T, \mathbb{R}^n)$ and, by I.2.5, $\tilde{F}(K \times Q)$ is conditionally compact. QED

II.6.4 ***Lemma*** Under the conditions of Lemma II.6.2, let

$$\tilde{f}(t, \tau, v, s) \triangleq f(t, \tau, v) \qquad (t, \tau \in T, v \in \mathbb{R}^m, s \in S)$$

and $\tilde{F}(y, q) \triangleq F(y)$ $(y \in K, q \in Q)$. Then K is closed and $F : K \to L^p(T, \mathbb{R}^n)$ continuous.

▌PROOF Let (y_i) be a sequence in K converging to some $\bar{y} \in L^p(T, \mathbb{R}^n)$. Then, by I.4.31 and I.4.18(3), some subsequence $(y_i)_{i \in J}$ converges to $\bar{y}$ μ-a.e.; hence $|\bar{y}(t)| \leqslant c \, |\psi(t, \cdot)|_d$ μ-a.e., showing that K is closed.

Now let $\lim_i y_i = \bar{y}$ in K, and assume that, for some $J \subset (1, 2,...)$ and $\epsilon > 0$, we have

$$|F(y_i) - F(\bar{y})|_p > \epsilon \qquad \text{for all} \quad i \in J. \tag{1}$$

By II.6.1(2), $\lim_{i \in J} \xi(y_i) = \xi(\bar{y})$ in $L^{p_1}(T, \mathbb{R}^m)$ and, by I.4.31 and I.4.18(3), some subsequence $(\xi(y_i))_{i \in J_1}$ converges to $\xi(\bar{y})$ μ-a.e. Therefore

$$\lim_{i \in J_1} f(t, \tau, \xi(y_i)(\tau)) = f(t, \tau, \xi(\bar{y})(\tau)) \qquad \text{for all } t \in T \text{ and } \mu\text{-a.a. } \tau \in T.$$

It follows, applying the dominated convergence theorem I.4.35 and taking account of II.6.2(2) and II.6.2(3) that

$$\lim_{i \in J_1} | F(y_i) - F(\bar{y})|_p^p$$
$$\leqslant \lim_{i \in J_1} \int \left\{ \int \left| f(t, \tau, \xi(y_i)(\tau)) - f(t, \tau, \xi(\bar{y})(\tau)) \right| \mu(d\tau) \right\}^p \mu(dt) = 0,$$

contradicting (1). Thus $\lim_i F(y_i) = F(\bar{y})$ if $\lim_i y_i = \bar{y}$ in K; hence F is continuous. QED

II.6.5 ***Theorem*** Let Assumption II.6.1 be satisfied and

$$\mu(T)^{\beta/p_1} + | \chi |_{p_1}^{\beta} \leqslant c.$$

Let, furthermore, $f : T \times T \times \mathbb{R}^m \to \mathbb{R}^n$ be such that, for all

$$(t, \tau, v) \in T \times T \times \mathbb{R}^m,$$

$f(\cdot, \cdot, v)$ is $\mu \times \mu$-measurable, $f(t, \tau, \cdot)$ continuous, and $| f(t, \tau, v)| \leqslant \psi(t, \tau)(1 + | v |^{\beta})$. Then the equation

$$y(t) = \int f(t, \tau, \xi(y)(\tau))\, \mu(d\tau) \quad \mu\text{-a.e.} \tag{1}$$

has a solution $\bar{y} \in L^p(T, \mathbb{R}^n)$.

▌PROOF Let F and K be defined as in II.6.2 and II.6.4. Then, clearly, K is a convex subset of $L^p(T, \mathbb{R}^n)$. By II.6.2–II.6.4, K is closed, $F : K \to L^p(T, \mathbb{R}^n)$ is continuous, and $\overline{F(K)}$ is compact. By II.6.2,

$$| F(y)(t)| \leqslant \int | f(t, \tau, \xi(y)(\tau))| \, \mu(d\tau) \leqslant c \, | \psi(t, \cdot)|_d$$
$$\text{for all} \quad y \in K \quad \text{and} \quad t \in T;$$

hence $F(K) \subset K$. Our conclusion now follows from Schauder's fixed point theorem II.2.8. QED

II.6.B Linear Functional-Integral Equations in $L^p(T, \mathbb{R}^n)$

As before, we assume that T is a compact metric space, (T, Σ, μ) a measure space, $\Sigma_{\text{Borel}}(T) = \Sigma$, and μ positive, finite, and regular. We also identify a function $(t, \tau) \to a(t, \tau)$ with the function $t \to a(t, \cdot)$.

II.6.6 ***Theorem*** Let $m, n \in \mathbb{N}$, $p \in [1, \infty)$, $p_1 \in (1, \infty)$, p_1' be the conjugate exponent to p_1, $\eta \in B(L^p(T, \mathbb{R}^n), L^{p_1}(T, \mathbb{R}^m))$, and

$$a \in L^p(T, L^{p_1'}(T, B(\mathbb{R}^m, \mathbb{R}^n))).$$

Then

(1) the expression

$$A(y)(t) \triangleq \int a(t, \tau)(\eta y)(\tau)\, \mu(d\tau) \qquad [\mu\text{-a.a. } t \in T, \quad y \in L^p(T, \mathbb{R}^n)]$$

defines a compact operator A in $L^p(T, \mathbb{R}^n)$;
(2) if $I - A$ is an injection, then $I - A$ is a homeomorphism of $L^p(T, \mathbb{R}^n)$ onto itself; and
(3) if $I - A$ is an injection and $\tau \to a(\cdot, \tau)$ belongs to

$$L^{p_1'}(T, L^p(T, B(\mathbb{R}^m, \mathbb{R}^n))),$$

then there exists a function $h : T \times T \to B(\mathbb{R}^m, \mathbb{R}^n)$ such that $\tau \to h(\cdot, \tau)$ belongs to $L^{p_1'}(T, L^p(T, B(\mathbb{R}^m, \mathbb{R}^n)))$ and

$$((I - A)^{-1} y)(t) = y(t) + \int h(t, \tau)(\eta y)(\tau)\, \mu(d\tau)$$
$$[\mu\text{-a.a. } t \in T, \quad y \in L^p(T, \mathbb{R}^n)].$$

▌PROOF *Step* 1 By Hölder's inequality I.5.13, for each $y \in L^p(T, \mathbb{R}^n)$ we have

$$| A(y)(t)| \leqslant \int | a(t, \tau)| \, | \eta(y)(\tau)| \, \mu(d\tau)$$
$$\leqslant | a(t, \cdot)|_{p_1'} \, | \eta(y)|_{p_1} \leqslant | a(t, \cdot)|_{p_1'} \, | \eta | \, | y |_p \quad \mu\text{-a.e.};$$

hence, by I.4.35, I.5.21, and I.4.45,

$$A(y) \in L^p(T, \mathbb{R}^n) \qquad \text{and} \qquad | Ay |_p \leqslant | a | \, | \eta | \, | y |_p . \tag{4}$$

Thus $A : L^p(T, \mathbb{R}^n) \to L^p(T, \mathbb{R}^n)$ is defined and it is clearly linear. The second relation in (4) shows that A is continuous.

Now let $\epsilon > 0$ and (y_i) be a sequence in the closed unit ball of $L^p(T, \mathbb{R}^n)$. By I.5.24, there exist $j(\epsilon) \in \mathbb{N}$,

$$\alpha_j^\epsilon \triangleq \alpha_j \in C(T) \qquad \text{and} \qquad \beta_j^\epsilon \triangleq \beta_j \in C(T, B(\mathbb{R}^m, \mathbb{R}^n)) \qquad [j = 1, 2, \ldots, j(\epsilon)]$$

such that

$$| t \to | a(t, \cdot) - \textstyle\sum_{j=1}^{j(\epsilon)} \alpha_j(t)\, \beta_j(\cdot)|_{p_1'}|_p \leqslant \epsilon/(4 \, | \eta |). \tag{5}$$

We set

$$e^\epsilon(t, \tau) \triangleq e(t, \tau) \triangleq | a(t, \tau) - \sum_{j=1}^{j(\epsilon)} \alpha_j(t)\, \beta_j(\tau)| \qquad (t, \tau \in T).$$

We may assume that $|\beta_j(\tau)| \leqslant 1$ for all $j \in \mathbb{N}$ and $\tau \in T$ (otherwise multiply α_j and divide β_j by an appropriate constant). We have, for all $q, r \in \mathbb{N}$,

$$(6) \quad |A(y_q)(t) - A(y_r)(t)| \leqslant \left| \sum_{j=1}^{j(\epsilon)} \alpha_j(t) \int \beta_j(\tau)\, \eta(y_q - y_r)(\tau)\, \mu(d\tau) \right| + \int e(t, \tau)\, |\eta(y_q - y_r)(\tau)|\, \mu(d\tau) \quad \mu\text{-a.e.}$$

By Hölder's inequality I.5.13,

$$\begin{aligned} \int e(t, \tau)\, |\eta(y_q - y_r)(\tau)|\, \mu(d\tau) &\leqslant |e(t, \cdot)|_{p_1'}\, |\eta(y_q - y_r)|_{p_1} \\ &\leqslant |e(t, \cdot)|_{p_1'}\, |\eta|\, |y_q - y_r|_p \\ &\leqslant 2\, |e(t, \cdot)|_{p_1'}\, |\eta|; \end{aligned}$$

hence, by (5),

$$(7) \qquad |t \to \int e(t, \tau)|\, \eta(y_q - y_r)(\tau)|\, \mu(d\tau)|_p \leqslant \epsilon/2.$$

Let

$$\gamma_{j,i}^{\epsilon} \triangleq \gamma_{j,i} \triangleq \int \beta_j(\tau)\, \eta(y_i)(\tau)\, \mu(d\tau) \qquad [j = 1, 2, \ldots, j(\epsilon); \quad i \in \mathbb{N}].$$

Then, by (6) and (7),

$$(8) \qquad |A(y_q) - A(y_r)|_p \leqslant \textstyle\sum_{j=1}^{j(\epsilon)} |\gamma_{j,q} - \gamma_{j,r}|\, |\alpha_j|_p + \epsilon/2.$$

Furthermore, by Hölder's inequality I.5.13,

$$\begin{aligned} |\gamma_{j,i}| &\leqslant \int |\beta_j(\tau)|\, |\eta(y_i)(\tau)|\, \mu(d\tau) \\ &\leqslant \mu(T)^{1/p_1'}\, |\eta(y_i)|_{p_1} \leqslant \mu(T)^{1/p_1'}\, |\eta|. \end{aligned}$$

We can now deduce from (8) that $(A(y_i))$ has a subsequence converging in $L^p(T, \mathbb{R}^n)$ to some limit $\bar{y}$, using exactly the same argument as was used in II.6.3 starting with II.6.3(4). This shows that A is a compact operator in $L^p(T, \mathbb{R}^n)$, thus proving (1). Statement (2) now follows from I.3.13.

Step 2 Now assume that the equation $y = A(y)$ has only the solution 0 in $L^p(T, \mathbb{R}^n)$ and that the function $\tau \to a(\cdot, \tau)$ belongs to

$$L^{p_1'}(T, L^p(T, B(\mathbb{R}^m, \mathbb{R}^n))).$$

By (2), $I - A$ has a bounded inverse $B \triangleq (I - A)^{-1}$. By I.4.30, there exists a Cauchy sequence $(\tau \to a_i(\cdot, \tau))_i$ of μ-simple functions on T to $L^p(T, B(\mathbb{R}^m, \mathbb{R}^m))$ such that

$$a_i(t, \tau) = \sum_{j=1}^{\alpha(i)} \chi_{i,j}(\tau)\, \phi_{i,j}(t) \qquad (i \in \mathbb{N})$$

and

$$\lim_i |\tau \to |a(\cdot, \tau) - a_i(\cdot, \tau)|_p|_{p_1'} = 0,$$

where, for each i, $\chi_{i,1}, \chi_{i,2}, \ldots$ are characteristic functions of disjoint μ-measurable sets and $\phi_{i,j} \in L^p(T, B(\mathbb{R}^m, \mathbb{R}^n))$. We set

$$\psi_{i,j} \triangleq B\phi_{i,j}$$

and

$$h_i(t, \tau) \triangleq \sum_{j=1}^{\alpha(i)} \chi_{i,j}(\tau)\, \psi_{i,j}(t) \qquad [t, \tau \in T, \quad i \in \mathbb{N}, \quad j = 1, 2, \ldots, \alpha(i)].$$

Then, for μ-a.a. $\tau \in T$ and all $i \in \mathbb{N}$, $h_i(\cdot, \tau) \in L^p(T, B(\mathbb{R}^m, \mathbb{R}^n))$ and $h_i(\cdot, \tau) = Ba_i(\cdot, \tau)$. Furthermore, for all $i, l \in \mathbb{N}$,

$$|h_i(\cdot, \tau) - h_l(\cdot, \tau)|_p = |B(a_i(\cdot, \tau) - a_l(\cdot, \tau))|_p \leqslant |B|\, |a_i(\cdot, \tau) - a_l(\cdot, \tau)|_p\,;$$

hence

$$|\tau \to |h_i(\cdot, \tau) - h_l(\cdot, \tau)|_p|_{p_1'} \leqslant |B|\, |\tau \to |a_i(\cdot, \tau) - a_l(\cdot, \tau)|_p|_{p_1'} \xrightarrow[i,l]{} 0.$$

This shows that $(\tau \to h_i(\cdot, \tau))_i$ is a Cauchy sequence in the Banach space $L^{p_1'}(T, L^p(T, B(\mathbb{R}^m, \mathbb{R}^n)))$ (I.5.17) and converges, therefore, to an element $\tau \to h(\cdot, \tau)$ of that space.

We now observe that, by Hölder's inequality I.5.13, we have

$$\int |h(\cdot, \tau)|_p\, |\eta(y)(\tau)|\, \mu(d\tau) < \infty \qquad [y \in L^p(T, \mathbb{R}^n)],$$

showing that the function $\tau \to h(\cdot, \tau)\, \eta(y)(\tau)$ belongs to $L^1(T, L^p(T, \mathbb{R}^n))$ for each $y \in L^p(T, \mathbb{R}^n)$. We may therefore define

$$H(y) \triangleq \int h(\cdot, \tau)(\eta y)(\tau)\, \mu(d\tau) \qquad [y \in L^p(T, \mathbb{R}^n)]$$

and observe that, by Hölder's inequality I.5.13, we have

$$|H(y)|_p \leqslant \int |h(\cdot, \tau)|_p\, |(\eta y)(\tau)|\, \mu(d\tau) \leqslant |\tau \to |h(\cdot, \tau)|_p|_{p_1'}\, |\eta|\, |y|_p\,.$$

Thus H is a bounded linear operator in $L^p(T, \mathbb{R}^n)$ and the same remark applies to the operators

$$y \to H_i(y) \triangleq \int h_i(\cdot, \tau)(\eta y)(\tau)\, \mu(d\tau) : L^p(T, \mathbb{R}^n) \to L^p(T, \mathbb{R}^n).$$

Furthermore,

$$|(H - H_i)y|_p = \left| \int [h(\cdot, \tau) - h_i(\cdot, \tau)](\eta y)(\tau)\, \mu(d\tau) \right|_p$$

$$\leqslant |\tau \to |h(\cdot, \tau) - h_i(\cdot, \tau)|_p|_{p_1'}\, |\eta|\, |y|_p$$

$$[i \in \mathbb{N}, \quad y \in L^p(T, \mathbb{R}^n)];$$

hence

$$\lim_i |H - H_i| = 0.$$

It follows from I.4.34(8) and the definitions of h_i and H_i that

$$H_i(y) = B \int a_i(\cdot, \tau)(\eta y)(\tau)\, \mu(d\tau)$$

and

$$(H_i - BA)(y) = B \int [a_i(\cdot, \tau) - a(\cdot, \tau)](\eta y)(\tau)\, \mu(d\tau) \qquad [y \in L^p(T, \mathbb{R}^n)].$$

Therefore, by Hölder's inequality I.5.13, we have

$$|H_i - BA| \leqslant |B|\, |\tau \to |a_i(\cdot, \tau) - a(\cdot, \tau)|_p|_{p_1'}\, |\eta| \underset{i}{\to} 0,$$

showing that $H = \lim_i H_i = BA$.

Since $B = (I - A)^{-1}$, we have $B = I + BA = I + H$ and therefore

$$[(I - A)^{-1} y](t) = y(t) + \int h(t, \tau)(\eta y)(\tau)\, \mu(d\tau)$$

$$(\mu\text{-a.a. } t \in T, \quad y \in L^p(T, \mathbb{R}^n)]. \qquad \text{QED}$$

II.6.C Dependence on Parameters of Solutions of Functional-Integral Equations in $L^p(T, \mathbb{R}^n)$

We shall now consider a functional-integral equation in $L^p(T, \mathbb{R}^n)$ that contains a parameter, and will discuss the dependence of solutions on this parameter.

II.6.7 ***Lemma*** Let T and Γ be compact metric spaces, $\mu \in \text{frm}^+(T)$, $\Sigma \triangleq \Sigma_{\text{Borel}}(T)$, $\mathscr{X}$ and $\mathscr{X}_A$ separable Banach spaces, A a convex subset of $\mathscr{X}_A$, $m \in \mathbb{N}$, $p \in [1, \infty)$, $p_1 \in (1, \infty)$, $0 \leqslant r_1 \leqslant r_2 \leqslant p_1$,

$$r \triangleq p_1/(p_1 - r_2 + r_1), \qquad |t \to |\psi(t, \cdot)|_{p_1/(p_1 - r_2)}|_p < \infty$$

and

$$\phi : T \times T \times \mathbb{R}^m \times A \times \Gamma \to \mathscr{X}.$$

Assume that, for all $(t, \tau, v, a, \gamma) \in T \times T \times \mathbb{R}^m \times A \times \Gamma$, $\phi(\cdot, \cdot, v, a, \gamma)$ is $\mu \times \mu$-measurable, $\phi(t, \tau, \cdot, \cdot, \cdot)$ continuous, and

$$(1) \qquad | \phi(t, \tau, v, a, \gamma)| \leqslant \psi(t, \tau)(1 + | v |^{r_1}).$$

Then, for every choice of a continuous $h : \Gamma \to [0, 1]$, $z \in L^{p_1}(T, \mathbb{R}^m)$, and $a \in A$, we have

$$(2) \qquad \lim | t \to | \tau \to \sup_{\gamma \in \Gamma} | \phi(t, \tau, z(\tau) + h(\gamma) \Delta z(\tau), a + h(\gamma) \Delta a, \gamma) - \phi(t, \tau, z(\tau), a, \gamma)| |_r |_p = 0$$

$$\text{as } | \Delta z |_{p_1} + | \Delta a | \to 0, \quad \Delta a \in A - a.$$

PROOF We first observe that, by I.4.48, there exists a set T^ϕ such that $\mu(T \sim T^\phi) = 0$ and $\phi(t_1, \cdot, v, a, \gamma)$ is μ-measurable for all $t_1 \in T^\phi$ and $(v, a, \gamma) \in \mathbb{R}^m \times A \times \Gamma$. By I.4.22, for each

$$(t_1, z, a, \gamma) \in T^\phi \times L^{p_1}(T, \mathbb{R}^m) \times A \times \Gamma,$$

the function $\tau \to \phi(t_1, \tau, z(\tau), a, \gamma)$ is μ-measurable and the function $(t, \tau) \to \phi(t, \tau, z(\tau), a, \gamma)$ is $\mu \times \mu$-measurable. Now let

$$(h, z, a) \in C(\Gamma, [0, 1]) \times L^{p_1}(T, \mathbb{R}^m) \times A$$

be fixed and

$$e(t, \tau, \Delta z, \Delta a, \gamma) \triangleq | \phi(t, \tau, z(\tau) + h(\gamma) \Delta z(\tau), a + h(\gamma) \Delta a, \gamma) - \phi(t, \tau, z(\tau), a, \gamma)|$$

$$[t, \tau \in T, \Delta z \in L^{p_1}(T, \mathbb{R}^m), \Delta a \in A - a, \gamma \in \Gamma].$$

Then, for each $t, \tau, \Delta z$, and Δa, the function $\gamma \to e(t, \tau, \Delta z, \Delta a, \gamma)$ is continuous on Γ and therefore bounded; hence

$$E(t, \tau, \Delta z, \Delta a) \triangleq \sup_{\gamma \in \Gamma} e(t, \tau, \Delta z, \Delta a, \gamma) \in \mathbb{R}.$$

By I.4.21, $E(t_1, \cdot, \Delta z, \Delta a)$ is μ-measurable for each $t_1 \in T^\phi$ and $E(\cdot, \cdot, \Delta z, \Delta a)$ is $\mu \times \mu$-measurable. We shall write henceforth $\lim_\Delta$ for

$$\lim \quad \text{as } | \Delta z |_{p_1} + | \Delta a | \to 0, \qquad \Delta a \in A - a,$$

and will denote by T_ψ the set

$$\{t \in T^\phi \mid \psi(t, \cdot) \in L^{p_1/(p_1 - r_2)}(T)\}.$$

We shall next show that

(3) $$\lim_{\Delta} | E(t, \cdot, \Delta z, \Delta a)|_r = 0 \qquad (t \in T_\psi).$$

Indeed, assume the contrary. Then there exist $\bar{t} \in T_\psi$, $\epsilon > 0$, and a sequence $((\Delta z_j, \Delta a_j))$ in $L^{p_1}(T, \mathbb{R}^m) \times (A - a)$ converging to 0 such that

(4) $$| E(\bar{t}, \cdot, \Delta z_j, \Delta a_j)|_r > \epsilon \qquad (j \in \mathbb{N})$$

and, in view of I.4.31 and Egoroff's theorem I.4.18, we may assume that $\lim_j \Delta z_j = 0$ μ-a.e. Now let, for $j \in \mathbb{N}$,

$$B_j \triangleq \{\tau \in T \mid | \Delta z_j(\tau)| > | z(\tau)| + 1\},$$

$$E_j(\tau) \triangleq E(\bar{t}, \tau, \Delta z_j, \Delta a_j)\, \chi_{T \sim B_j}(\tau) \qquad (\tau \in T),$$

and

$$\tilde{E}_j(\tau) = E(\bar{t}, \tau, \Delta z_j, \Delta a_j)\, \chi_{B_j}(\tau) \qquad (\tau \in T).$$

Assumption (1) implies that

(5) $$E(t, \tau, \Delta z, \Delta a) \leqslant \psi(t, \tau)(2 + [| z(\tau)| + | \Delta z(\tau)|]^{r_1} + | z(\tau)|^{r_1})$$

for all $t, \tau, \Delta z, \Delta a$; hence

(6) $$\begin{aligned} E_j(\tau) &\leqslant \psi(\bar{t}, \tau)(2 + [2 \mid z(\tau)| + 1]^{r_1} + | z(\tau)|^{r_1}) \\ &\triangleq \psi(\bar{t}, \tau)\, w(\tau) \qquad (j \in \mathbb{N}, \quad \tau \in T), \end{aligned}$$

where $w \in L^{p_1/r_1}(T)$. By I.5.14,

(7) $$| \psi(\bar{t}, \cdot)\, w(\cdot)|_r \leqslant | \psi(\bar{t}, \cdot)|_{p_1/(p_1 - r_2)}\, | w |_{p_1/r_1} < \infty.$$

Since $\phi(t, \tau, \cdot, \cdot, \cdot)$ is continuous and $\lim \Delta z_j = 0$ μ-a.e., it follows, by I.2.15(2), that $\lim_j E_j(\cdot) = 0$ μ-a.e.; hence, by (6), (7), and the dominated convergence theorem I.4.35, we have

(8) $$\lim_j | E_j |_r = 0.$$

We next consider $| \tilde{E}_j|_r$. We first observe that

(9) $$\mu(B_j) \leqslant \int_{B_j} (| z(\tau)| + 1)^{p_1}\, \mu(d\tau) \leqslant | \Delta z_j |_{p_1}^{p_1} \underset{j}{\rightarrow} 0.$$

By (5), we have

$$\tilde{E}_j(\tau) \leqslant \psi(\bar{t}, \tau)[2 + (2^{r_1} + 1)\, | \Delta z_j(\tau)|^{r_1}]\, \chi_{B_j}(\tau);$$

hence, by I.5.14, I.5.15, and (9),

$$(10) \qquad |\tilde{E}_j|_r \leqslant |\psi(\bar{t}, \cdot)|_{p_1/(p_1-r_2)} \{2\mu(B_j)^{r_1/p_1} + (2^{r_1}+1)|\Delta z_j|_{p_1}^{r_1}\} \underset{j}{\to} 0.$$

Since

$$E(\bar{t}, \tau, \Delta z_j, \Delta a_j) = E_j(\tau) + \tilde{E}_j(\tau) \qquad (j \in \mathbb{N}, \quad \tau \in T),$$

it follows from (8), (10), and Minkowski's inequality I.5.15 that

$$\lim_j |E(\bar{t}, \cdot, \Delta z_j, \Delta_j)|_r = 0,$$

contrary to (4). This shows that relation (3) is valid.

Since $\mu(T \sim T_\psi) = 0$, relation (2) will follow from (3) and the dominated convergence theorem I.4.35 once we show that there exists $g \in L^p(T)$ such that $|E(t, \cdot, \Delta z, \Delta a)|_r \leqslant g(t)$ for all Δz and Δa, with $|\Delta z|_{p_1} + |\Delta a| \leqslant 1$. This follows immediately from (5) and I.5.14 since

$$|\tau \to 2 + (|z(\tau)| + |\Delta z(\tau)|)^{r_1} + |z(\tau)|^{r_1})|_{p_1/r_1}$$

is easily seen to be bounded for all Δz with $|\Delta z|_{p_1} \leqslant 1$. QED

II.6.8 ***Theorem*** Let T be a compact metric space, $\mu \in \text{frm}^+(T)$, $\Sigma \triangleq \Sigma_{\text{Borel}}(T)$, $k, m, n \in \mathbb{N}$, $p \in [1, \infty)$, $p_1 \in (1, \infty)$, A a convex subset of $\mathbb{R}^k$, $\xi : L^p(T, \mathbb{R}^n) \to L^{p_1}(T, \mathbb{R}^m)$, and $f : T \times T \times \mathbb{R}^m \times A \to \mathbb{R}^n$. Assume that, for all choices of points $(t, \tau, v, a) \in T \times T \times \mathbb{R}^m \times A$, ξ and $f(t, \tau, \cdot, \cdot)$ have continuous derivatives, $f(\cdot, \cdot, v, a)$ is $\mu \times \mu$-measurable, and there exist $\alpha, \beta \in \mathbb{R}$, $\psi_0 : T \times T \to \mathbb{R}$ and $\psi_1 : T \times T \to \mathbb{R}$ such that

$$|t \to |\psi_0(t, \cdot)|_{p_1/(p_1-\beta)}|_p < \infty, \qquad |t \to |\psi_1(t, \cdot)|_{p_1/(p_1-\alpha-1)}|_p < \infty,$$

$$0 \leqslant \alpha \leqslant p_1 - 1, \qquad 0 \leqslant \beta \leqslant p_1,$$

$$|f(t, \tau, v, a)| \leqslant (1 + |v|^\beta)\psi_0(t, \tau),$$

$$|f_a(t, \tau, v, a)| \leqslant (1 + |v|^\beta)\psi_0(t, \tau),$$

and

$$|f_v(t, \tau, v, a)| \leqslant (1 + |v|^\alpha)\psi_1(t, \tau).$$

Then the expression

$$F(y, a)(t) \triangleq \int f(t, \tau, \xi(y)(\tau), a)\, \mu(d\tau) \quad \mu\text{-a.e.} \qquad [y \in L^p(T, \mathbb{R}^n), \quad a \in A]$$

defines a continuous mapping $F : L^p(T, \mathbb{R}^n) \times A \to L^p(T, \mathbb{R}^n)$, and F has a continuous derivative F' on $L^p(T, \mathbb{R}^n) \times A$ such that

$$(1) \qquad F'(y, a)(\Delta y, \Delta a)(t)$$

$$= \int f_{(v,a)}(t, \tau, \xi(y)(\tau), a)((\xi'(y)\, \Delta y)(\tau), \Delta a)\, \mu(d\tau)$$

$$[\mu\text{-a.a. } t \in T, \Delta y \in L^p(T, \mathbb{R}^n), \Delta a \in \mathbb{R}^k]$$

PROOF We proceed in a manner similar to that of II.5.8. We define the transformations

$$G : L^{p_1}(T, \mathbb{R}^m) \times A \to L^p(T, L^1(T, \mathbb{R}^n))$$

and

$$H : L^p(T, L^1(T, \mathbb{R}^n)) \to L^p(T, \mathbb{R}^n)$$

by the following expressions:

$$\text{(2)} \quad G(z, a)(t, \tau) \triangleq f(t, \tau, z(\tau), a) \qquad [t, \tau \in T, \quad z \in L^{p_1}(T, \mathbb{R}^m), \quad a \in A]$$

and

$$\text{(3)} \qquad H(\phi)(t) \triangleq \int \phi(t, \tau)\, \mu(d\tau) \qquad [\phi \in L^p(T, L^1(T, \mathbb{R}^n)), \quad t \in T].$$

We shall prove that both G and H have domains and ranges as indicated, are continuous, and have continuous derivatives.

Step 1 Let $z \in L^{p_1}(T, \mathbb{R}^m)$ and $a \in A$. Since the function $(t, \tau) \to z(\tau)$ is clearly $\mu \times \mu$-measurable, it follows from I.4.22 that

$$(t, \tau) \to G(z, a)(t, \tau) = f(t, \tau, z(\tau), a)$$

is $\mu \times \mu$-measurable. By Hölder's inequality I.5.13 and Minkowski's inequality I.5.15, we have

$$\begin{aligned}\text{(4)} \qquad \int |f(t, \tau, z(\tau), a)|\, \mu(d\tau) &\leqslant \int \psi_0(t, \tau)(1 + |z(\tau)|^\beta)\, \mu(d\tau) \\ &\leqslant |\psi_0(t, \cdot)|_{p_1/(p_1-\beta)}\, (\mu(T)^{\beta/p_1} + |z|_{p_1}^\beta) \\ &\qquad \text{for} \quad \mu\text{-a.a. } t \in T\end{aligned}$$

and, by Fubini's theorem I.4.45(2), $t \to \int |f(t, \tau, z(\tau), a)|\, \mu(d\tau)$ is μ-measurable. Thus, by (2), (4), and I.5.24, $G(z, a) \in L^p(T, L^1(T, \mathbb{R}^n))$ which shows that G is a mapping on $L^{p_1}(T, \mathbb{R}^m) \times A$ into $L^p(T, L^1(T, \mathbb{R}^n))$.

Step 2 Next we consider, for each fixed $(z, a) \in L^{p_1}(T, \mathbb{R}^m) \times A$, the linear operator $\tilde{G}(z, a)$ defined by

$$\begin{aligned}\text{(5)} \quad \tilde{G}(z, a)(\Delta z, \Delta a)(t, \tau) &\triangleq f_{(v,a)}(t, \tau, z(\tau), a)(\Delta z(\tau), \Delta a) \\ &= f_v(t, \tau, z(\tau), a)\, \Delta z(\tau) + f_a(t, \tau, z(\tau), a)\, \Delta a \\ &\qquad [\Delta z \in L^{p_1}(T, \mathbb{R}^m), \Delta a \in \mathbb{R}^k, t, \tau \in T].\end{aligned}$$

For each fixed Δz and Δa, each term in (5) is $\mu \times \mu$-measurable (II.3.9 and I.4.22) and, by I.5.14 and I.5.15,

$$\begin{aligned}(6)\qquad &\int |f_v(t, \tau, z(\tau), a)\, \Delta z(\tau)|\, \mu(d\tau)\\ &\quad\leqslant \int \psi_1(t, \tau)(1 + |z(\tau)|^\alpha)\, |\Delta z(\tau)|\, \mu(d\tau)\\ &\quad\leqslant |\psi_1(t, \cdot)|_{p_1/(p_1-\alpha-1)}\, (\mu(T)^{\alpha/p_1} + |z|_{p_1}^\alpha)\, |\Delta z|_{p_1}\\ &\qquad\qquad \text{for } \mu\text{-a.a. } t \in T.\end{aligned}$$

Thus, by (6) and Fubini's theorem I.4.45 (2), $t \to \int |f_v(t, \tau, z(\tau), a)\, \Delta z(\tau)|\, \mu(d\tau)$ belongs to $L^p(T, \mathbb{R}^n)$. We similarly derive the relation

$$\begin{aligned}(7)\quad \int |f_a(t, \tau, z(\tau), a)\, \Delta a\,|\, \mu(d\tau) &\leqslant |\psi_0(t, \cdot)|_{p_1/(p_1-\beta)}\, (\mu(T)^{\beta/p_1}\\ &\quad + |z|_{p_1}^\beta)\, |\Delta a| \qquad \text{for } \mu\text{-a.a. } t \in T,\end{aligned}$$

which, together with (6) and I.5.24, shows that $\tilde{G}(z, a)$ is a mapping on $L^{p_1}(T, \mathbb{R}^m) \times \mathbb{R}^k$ into $L^p(T, L^1(T, \mathbb{R}^n))$. Furthermore, (5)–(7) imply that there exists $\tilde{\psi} \in L^p(T)$ such that

$$\int |\tilde{G}(z, a)(\Delta z, \Delta a)(t, \tau)|\, \mu(d\tau) \leqslant \tilde{\psi}(t)(|\Delta z|_{p_1} + |\Delta a|) \quad \mu\text{-a.e.}$$

for all Δz and Δa, thus showing that $\tilde{G}(z, a)$ is a bounded linear operator.

Step 3 We can now show that $\tilde{G}(z, a) = G'(z, a)$. Let $\Delta z \in L^{p_1}(T, \mathbb{R}^m)$, $\Delta a \in A - a$, and $0 \neq |\Delta z|_{p_1} + |\Delta a| \leqslant 1$. Then, by the mean value theorem II.3.6,

(8)

$$\begin{aligned}&|G(z + \Delta z, a + \Delta a)(t, \tau) - G(z, a)(t, \tau) - \tilde{G}(z, a)(\Delta z, \Delta a)(t, \tau)|\\ &\quad = |f(t, \tau, z(\tau) + \Delta z(\tau), a + \Delta a) - f(t, \tau, z(\tau), a)\\ &\qquad - f_{(v,a)}(t, \tau, z(\tau), a)(\Delta z(\tau), \Delta a)|\\ &\quad \leqslant \sup_{0\leqslant\gamma\leqslant 1} |f_v(t, \tau, z(\tau) + \gamma\, \Delta z(\tau), a + \gamma\, \Delta a) - f_v(t, \tau, z(\tau), a)|\, |\Delta z(\tau)|\\ &\qquad + \sup_{0\leqslant\gamma\leqslant 1} |f_a(t, \tau, z(\tau) + \gamma\, \Delta z(\tau), a + \gamma\, \Delta a) - f_a(t, \tau, z(\tau), a)|\, |\Delta a|\\ &\qquad\qquad (t, \tau \in T).\end{aligned}$$

Let $E_1(t, \tau, \Delta z, \Delta a)$ and $E_2(t, \tau, \Delta z, \Delta a)$ denote, respectively, the coefficients of $|\Delta z(\tau)|$ and $|\Delta a|$ on the right of (8). Then II.6.7 yields [setting $r_1 = \alpha$, $r_2 = \alpha + 1$, and $h(\gamma) = \gamma$]

$$\lim |t \to |E_1(t, \cdot, \Delta z, \Delta a)|_{p_1/(p_1-1)}|_p = 0 \qquad \text{as} \quad |\Delta z|_{p_1} + |\Delta a| \to 0$$

and II.6.7 yields [setting $r_1 = r_2 = \beta$ and $h(\gamma) = \gamma$]

$$\lim |\, t \to |\, E_2(t, \cdot, \Delta z, \Delta a)|_1 \,|_p = 0 \qquad \text{as} \quad |\, \Delta z \,|_{p_1} + |\, \Delta a \,| \to 0.$$

It follows, integrating both sides of (8) as functions of τ with respect to μ, applying Hölder's inequality I.5.13 to the first term on the right, dividing by $(|\, \Delta z \,|_{p_1} + |\, \Delta a \,|)$, and finally taking the L^p-norms of the resulting functions of t, that

$$\lim(|\, \Delta z \,|_{p_1} + |\, \Delta a \,|)^{-1} \,|\, G(z + \Delta z, a + \Delta a) - G(z, a) - \tilde{G}(z, a)(\Delta z, \Delta a)| = 0$$

as $0 \neq |\, \Delta z \,|_{p_1} + |\, \Delta a \,| \to 0$, $\Delta a \in A - a$. Thus $\tilde{G}(z, a)$ is the derivative of G at (z, a), which also shows that G is continuous at (z, a).

Step 4 We can now show that G' (that is, $\tilde{G}$) is continuous. Let

$$\lim_j (z_j, a_j) = (\bar{z}, \bar{a}) \text{ in } L^{p_1}(T, \mathbb{R}^m) \times A.$$

Then, by (5), for all $\Delta z \in L^{p_1}(T, \mathbb{R}^m)$ and $\Delta a \in A - a$,

$$\begin{aligned} &|[G'(z_j, a_j) - G'(\bar{z}, \bar{a})](\Delta z, \Delta a)(t, \tau)| \\ &\quad \leqslant |\, f_v(t, \tau, z_j(\tau), a_j) - f_v(t, \tau, \bar{z}(\tau), \bar{a})|\, |\, \Delta z(\tau)| \\ &\qquad + |\, f_a(t, \tau, z_j(\tau), a_j) - |\, f_a(t, \tau, \bar{z}(\tau), \bar{a})|\, |\, \Delta a \,|, \end{aligned}$$

and the same argument as was applied to (8) [but with $h(\gamma) = 1$ and $(\Delta z, \Delta a)$ replaced by $(z_j - z, a_j - a)$ "inside" $f_{(v,a)}$] shows that

$$\lim_j |[G'(z_j, a_j) - G'(\bar{z}, \bar{a})](\Delta z, \Delta a)| = 0 \quad \text{uniformly for all } (\Delta z, \Delta a) \text{ with } |\, \Delta z \,|_{p_1} + |\, \Delta a \,| \leqslant 1.$$

Thus G' is continuous.

Step 5 We finally consider the transformation H defined in (3). It is clear that H is defined on $L^p(T, L^1(T, \mathbb{R}^n))$ into $L^p(T, \mathbb{R}^n)$ and that it is linear and bounded. Thus $H'(\phi)$ exists for all $\phi \in L^p(T, L^1(T, \mathbb{R}^n))$ and $H'(\phi) = H$; hence H' is constant and therefore continuous.

Step 6 We now complete the proof exactly as in II.5.8. We observe that

$$F(y, a) = H \circ G(\xi(y), a) \in L^p(T, \mathbb{R}^n) \qquad [y \in L^p(T, \mathbb{R}^n), \quad a \in A],$$

and therefore F is continuous and has a continuous derivative F' such that

$$F'(y, a)(\Delta y, \Delta a) = H'(G(\xi(y), a)) \circ G'(\xi(y), a)(\xi'(y)\, \Delta y, \Delta a).$$

Since $H'(\phi) = H$ and $G' = \tilde{G}$, relation (1) now follows from (3) and (5).

QED

II.6.9 *Theorem* Let the conditions of II.6.8 be satisfied, $y_0 \in L^p(T, \mathbb{R}^n)$,

$$y_0(t) = \int f(t, \tau, \xi(y_0)(\tau), a_0)\, \mu(d\tau) \quad \mu\text{-a.e.},$$

and the equation

$$\Delta y(t) = \int f_v(t, \tau, \xi(y_0)(\tau), a_0)(\xi'(y_0)\, \Delta y)(\tau)\, \mu(d\tau) \quad \mu\text{-a.e.}$$

have only the solution $\Delta y = 0$ in $L^p(T, \mathbb{R}^n)$. Then there exist neighborhoods $\tilde{Y}$ of y_0 in $L^p(T, \mathbb{R}^n)$ and $\tilde{A}$ of a_0 relative to A such that the equation

$$y(t) = \int f(t, \tau, \xi(y)(\tau), a)\, \mu(d\tau) \quad \mu\text{-a.e.}$$

has a unique solution $y(\cdot) = u(a)(\cdot) \in \tilde{Y}$ for all $a \in \tilde{A}$, and the function $a \to u(a) : \tilde{A} \to \tilde{Y}$ is continuous and has a continuous derivative u' on $\tilde{A}$ satisfying the relation

$$\begin{aligned} w(t) &\triangleq (u'(a)\, \Delta a)(t) \\ &= \int f_{(v,a)}(t, \tau, \xi(u(a))(\tau), a)((\xi'(u(a))w)(\tau), \Delta a)\, \mu(d\tau) \quad \mu\text{-a.e.} \qquad (\Delta a \in \mathbb{R}^k). \end{aligned}$$

▌PROOF Let

$$\eta \triangleq \xi'(y_0),$$

$$k(t, \tau) \triangleq f_v(t, \tau, \xi(y_0)(\tau), a_0) \qquad (t, \tau \in T),$$

and

$$K(\Delta y)(t) \triangleq \int k(t, \tau)(\eta\, \Delta y)(\tau)\, \mu(d\tau) \qquad [\mu\text{-a.a. } t \in T, \quad \Delta y \in L^p(T, \mathbb{R}^n)].$$

By I.4.22, $k(\cdot, \cdot)$ is μ-measurable; furthermore

$$| k(t, \tau)| \leqslant \psi_1(t, \tau)(1 + | \xi(y_0)(\tau)|^\alpha).$$

Hence, by I.5.14,

$$| k(t, \cdot)|_{p_1/(p_1-1)} \leqslant | \psi_1(t, \cdot)|_{p_1/(p_1-\alpha-1)}\, | \tau \to 1 + | \xi(y_0)(\tau)|^\alpha |_{p_1/\alpha} \,.$$

It follows, by I.5.24, that $k \in L^p(T, L^{p_1/(p_1-1)}(T, B(\mathbb{R}^m, \mathbb{R}^n)))$. Thus, by II.6.6, $I - K$ is a homeomorphism of $L^p(T, \mathbb{R}^n)$ onto itself and our conclusion follows from II.6.8 and the implicit function theorem II.2.8. QED

II.6.D Integral Equations with Pseudodelays in $L^p(T, \mathbb{R}^n)$

We shall now show that, with suitable assumptions about the pseudodelays $h_1, \ldots, h_k$, we can apply Theorems II.6.5, II.6.6, II.6.8, and II.6.9 to study solutions in $L^p(T, \mathbb{R}^n)$ of the integral equation with pseudodelays II.1(6).

II.6.10 ***Theorem*** Let T be a compact metric space,

$$\mu \in \mathrm{frm}^+(T), \qquad \Sigma \triangleq \Sigma_{\mathrm{Borel}}(T),$$

(T, Σ^*, μ) the Lebesgue extension of (T, Σ, μ), $l \in \mathbb{N}$, $1 < p < \infty$, $c \in \mathbb{R}$, and $h_i : T \to T$ $(i = 1, 2,..., l)$ such that

$$h_i^{-1}(E) \in \Sigma^* \qquad \text{and} \qquad \mu(h_i^{-1}(E)) \leqslant c\mu(E) \qquad (E \in \Sigma^*).$$

If we set

$$\xi(y)(t) \triangleq \big(y(h_1(t)),..., y(h_l(t))\big) \qquad \text{for all } y \in L^p(T, \Sigma, \mu, \mathbb{R}^n) \text{ and } t \in T,$$

then $\xi \in B\big(L^p(T, \Sigma, \mu, \mathbb{R}^n), L^p(T, \Sigma, \mu, \mathbb{R}^{ln})\big)$, ξ satisfies the pertinent conditions of Assumption II.6.1 and of II.6.8, and Theorems II.6.5, II.6.6, II.6.8, and II.6.9 remain valid with $m = ln$, $p_1 = p$, and $\eta = \xi$.

▌ PROOF Let $i \in \{1, 2,..., l\}$, $\nu_i(h_i^{-1}(E)) \triangleq \mu(E)$ $(E \in \Sigma^*)$, and

$$\Sigma_i \triangleq \{h_i^{-1}(E) \mid E \in \Sigma^*\}.$$

Then $\Sigma_i \subset \Sigma^*$, (T, Σ_i, ν_i) is a finite positive measure space (see I.4.39), $y \circ h_i$ is both Σ^*-measurable and Σ_i-measurable if $y \in L^p(T, \Sigma, \mu, \mathbb{R}^n)$, and

$$\int |y(h_i(\tau))|^p \, \nu_i(d\tau) = \int |y(s)|^p \, \mu(ds) < \infty. \tag{1}$$

By our assumption, $c\nu_i(h^{-1}(E)) = c\mu(E) \geqslant \mu(h^{-1}(E))$ $(E \in \Sigma^*)$; hence $c\nu_i(A) \geqslant \mu(A)$ for all $A \in \Sigma_i$. Thus it follows from (1) that

$$\int |y(h_i(\tau))|^p \, \mu(d\tau) \leqslant c \int |y(h_i(\tau))|^p \, \nu_i(d\tau) < \infty, \tag{2}$$

showing that $y \circ h_i \in L^p(T, \Sigma, \mu, \mathbb{R}^n)$ $(i = 1, 2,..., l)$ and, consequently, $\xi(y) \in L^p(T, \Sigma, \mu, \mathbb{R}^{ln})$.

It is clear that ξ is linear. By (1) and (2), we have

$$\begin{aligned} \int |\xi(y)(\tau)|^p \, \mu(d\tau) &= \int \Big\{\sum_{i=1}^{l} |y(h_i(\tau))|\Big\}^p \mu(d\tau) \\ &\leqslant l^p \sum_{i=1}^{l} \int |y(h_i(\tau))|^p \, \mu(d\tau) \\ &\leqslant l^{p+1} c \int |y(s)|^p \, \mu(ds); \end{aligned}$$

hence

$$|\xi(y)|_p \leqslant l(cl)^{1/p}|y|_p .$$

This shows that $\xi \in B(L^p(T, \Sigma, \mu, \mathbb{R}^n), L^p(T, \Sigma, \mu, \mathbb{R}^{ln}))$. It follows that $\xi'(y) = \xi$ for all y. Thus ξ and ξ' are continuous.

Since $|y(\tau)| \leqslant \phi(\tau)$ implies $|y(h_i(\tau))| \leqslant \phi(h_i(\tau))$ for all i and ϕ, and relation (2) remains valid with y replaced by any $\phi \in L^p(T, \Sigma, \mu)$, we conclude that Condition II.6.1(4) is valid with

$$\chi(t) \triangleq c \sum_{i=1}^{l} |\psi(h_i(t), \cdot)|_d \qquad (t \in T)$$

and an arbitrary positive c. QED

Notes

The proof of Sperner's lemma II.2.2 is to some extent patterned on the exposition of Graves [1], the proofs of Tychonoff's and Schauder's fixed point theorems (II.2.6–II.2.8) on the notes of Fan [1], and the proofs of the mean value and implicit function theorems (II.3.6–II.3.8) on the text of Dieudonné [1]. The statement of the implicit function theorem and the definition of the (Fréchet) derivative are somewhat more general than is customary because of our requirements in Part Two. In particular, our definition of $f'(\bar{a})$ includes the (customary) case where $\bar{a}$ is an interior point of the domain of definition A of f as well as the case where $\bar{a}$ is an arbitrary point of A and A is convex with a nonempty interior. A definition of a derivative for the case where A is a convex cone was used by Krasnoselskii [1].

While the material in II.4 on ordinary differential equations is standard, I believe that Sections II.5 and II.6 on functional-integral equations contain a number of essentially new results. However, it must be borne in mind that many of these results are related to known theorems or techniques (see Krasnoselskii [1, p. 368] for some of the pertinent references).

The concepts of local solutions and p-hereditary transformations are motivated by Volterra integral equations and by certain of their multidimensional analogs such as the equation

$$y(t) = \int_{D(t)} f(\tau, y(\tau))\, d\tau_1 \times d\tau_2 \qquad (t \in T),$$

where $T \triangleq [0, a_1] \times [0, a_2]$, the integral is defined with respect to the Lebesgue measure on T, and

$$D(t) \triangleq D(t_1, t_2) \triangleq [0, t_1] \times [0, t_2].$$

[Under certain conditions, this equation is equivalent to a "Cauchy problem" defined by the partial differential equation

$$\mathscr{D}_1\mathscr{D}_2 y(t_1, t_2) = f(t_1, t_2, y(t_1, t_2)) \qquad ((t_1, t_2) \in T).]$$

The p-hereditary transformation is related to several concepts of "causal transformations," the most common of which correspond to the case where $T \triangleq [t_0, t_1] \subset \mathbb{R}$ and $p(t) \triangleq t$ (see, e.g., Saeks [1] for additional references).

Theorem II.5.5 [on linear functional-integral equations in $C(T, \mathbb{R}^n)$] had been originally stated in Warga [11] in the special case where η is the identity operator. Theorem II.6.6 [on linear functional-integral equations in $L^p(T, \mathbb{R}^n)$] is a generalization of a theorem of F. Riesz (see Riesz and Sz.-Nagy [1, p. 177ff]) to which it reduces itself in the special case where $p = p_1 = 2$ and η is the identity operator, and my proof is patterned in part on the argument of Riesz and Sz.-Nagy [1].

PART TWO

Optimal Control

CHAPTER III

Basic Problems and Concepts, and Heuristic Considerations

III.1 The Subject of the Optimal Control Theory

A function $g : S \to \mathbb{R}$ has a *minimum* m on $S_1 \subset S$ and s_1 *minimizes* g on S_1 if $g(s_1) = \inf g(S_1) = m$. The point s_1 is also referred to as a *minimizing point* of g on S_1. The basic problem of the optimal control theory is one of optimization, that is, the search for a minimum and a minimizing point of a given function on a subset of its domain defined by certain relations (both in the form of equations and of more general constraints). If $g_0 : \mathbb{R}^n \to \mathbb{R}$, $g_1 : \mathbb{R}^n \to \mathbb{R}^m$, and $a \in \mathbb{R}^m$ are given and we wish to minimize g_0 on the set $\{w \in \mathbb{R}^n \mid g_1(w) = a\}$, then our search is for a *restricted minimum* in $\mathbb{R}^n$, a problem long considered in calculus. If g_0 and g_1 are defined on (or restricted to) a particular subset of $\mathbb{R}^n$, say the positive octant $\{w \triangleq (w^1,..., w^n) \mid w^j \geqslant 0\}$, or if the constraint $g_1(w) = a$ is replaced by $g_1(w) \in A$ for a given set $A \subset \mathbb{R}^m$, then our problem is referred to as a *mathematical programming* problem (or a *linear* respectively *nonlinear programming* problem depending on whether both g_0 and g_1 are linear or not). A general problem of optimization is one in which some general sets W, $\mathscr{X}_1$, and $C_1 \subset \mathscr{X}_1$ and functions $g_0 : W \to \mathbb{R}$ and $g_1 : W \to \mathscr{X}_1$ are given, and it is desired to minimize g_0 on the set $g^{-1}(C_1) \triangleq \{w \in W \mid g_1(w) \in C_1\}$. Clearly, such a problem is too general for study unless some (say, topological or algebraic) structure is provided for $\mathscr{X}_1$ (and possibly W) and the functions g_0 and g_1 are endowed with certain properties (such as continuity or differentiability or special characteristics of the image). The function g_0 to be minimized is variously referred to as the *cost functional*, the *objective function*, or the *criterion* (*of optimization*).

What primarily distinguishes the optimal control problems from other optimization problems of the general form just described is the particular nature of the set W. In optimal control problems, stated in the form that we find most convenient, W is usually a subset of the cartesian product of three sets: the set $\mathscr{Y}$ of *states*, whose elements are usually functions with values in some vector space (*state functions*), the set $\mathscr{U}$ of *control functions* on some measure space T into some space R, and the set B of *control parameters* (endowed with some algebraic or topological structure). The optimal control theory lays a special emphasis on the nature of the elements of $\mathscr{U}$ as functions on a measure space and seeks to exploit this property to better characterize the conditions for a minimum.

The following example may help to illustrate the nature of the sets $\mathscr{Y}$, $\mathscr{U}$, B, and W and their relationship. Suppose that a space vehicle orbits the Earth and it is desired to transfer the vehicle to the vicinity of the Moon in a preassigned period of time in such a manner as to burn up the least amount of fuel. We choose, as the origin of time, the instant when the rockets are activated, and designate by $y(t) \triangleq (y^1(t),..., y^n(t))$ the element of $\mathbb{R}^n$ that represents the "state" of the vehicle at the time t. Thus, among the components y^j of y, we include the coordinates (with respect to a previously chosen coordinate system) of the vehicle, the components of its velocity vector, the mass of the remaining fuel, and, in a more complex analysis, data representing the inclination of the vehicle relative to its velocity. Let t_1 represent the preassigned duration of the flight, B_0 the known collection of states while in Earth orbit [that is, $\{y(t) \mid t \leqslant 0\}$], B_1 the set of permitted states at time t_1 (essentially, the combinations of location and velocity components on "arrival" that permit a transfer to an acceptable Moon orbit), and $u(t)$ the control configuration at time t (for example, data representing the orientation of the engines, the magnitude of thrust, etc.). The control function u can be chosen as an arbitrary measurable function on $[0, t_1]$ with values in a set R of *feasible control configurations*. In the present case, a feasible control configuration may be one for which the thrust does not exceed the capacity of the engine and the angle that the direction of the thrust makes with the axis of the vehicle does not exceed a preassigned value. While in flight, the vehicle obeys an equation of motion (derived from the basic laws of dynamics) of the form

$$\dot{y}(t) = f(t, y(t), u(t)) \quad \text{a.e. in } T \triangleq [0, t_1], \tag{1}$$

where $y(\cdot)$ is an absolutely continuous function on $[0, t_1]$. This equation may contain terms representing the thrust of the engines, the attraction of the Earth, the Moon, the Sun, and possibly other celestial bodies, solar pressure, etc.

If we choose to activate the engines at a time when the state of the vehicle in Earth orbit is represented by a point $b_0 \in B_0$ and if we choose to set the controls to fit the configuration $u(t)$ for all $t \in [0, t_1]$, then the state function $y(\cdot)$ of the vehicle will satisfy Eq. (1) and the relation $y(0) = b_0$. Both of these relations, together with the condition that y be absolutely continuous, can be combined into the equivalent relation

$$y(t) = b_0 + \int_0^t f(\tau, y(\tau), u(\tau))\, d\tau \qquad (t \in T). \tag{2}$$

In many problems of the optimal control theory, Eq. (2) has a unique continuous solution y for each choice of $b_0 \in B_0$ and of a measurable $u : T \triangleq [0, t_1] \to R$. Whether or not this is the case, we shall consider triplets (y, u, b_0) (with a continuous $y : T \to \mathbb{R}^n$, a measurable $u : T \to R$ and $b_0 \in B_0$) that are solutions of Eq. (2). Such a triplet (y, u, b_0) will be *admissible* if $y(t_1) \in B_1$, that is, if it yields a successful trajectory. If $\tilde{g}_0(y(t))$ represents the expenditure of fuel during the interval $[0, t]$, then an admissible triplet $(\bar{y}, \bar{u}, \bar{b}_0)$ is *optimal* if $\tilde{g}_0(\bar{y}(t_1)) \leq \tilde{g}_0(y(t_1))$ for all admissible triplets (y, u, b_0).

We can express the problem just described in the form of a general optimization problem by setting $\mathscr{Y} \triangleq C(T, \mathbb{R}^n)$, representing by $\mathscr{U}$ the class of measurable functions on T to R, letting

$$B \triangleq B_0, \qquad W \triangleq \mathscr{Y} \times \mathscr{U} \times B,$$
$$\mathscr{X}_1 \triangleq \mathbb{R}^n \times C(T, \mathbb{R}^n), \qquad C_1 \triangleq B_1 \times \{0\} \subset \mathscr{X}_1,$$
$$g_0(y, u, b) \triangleq \tilde{g}_0(y(t_1)), \qquad g^1(y, u, b) \triangleq y(t_1),$$

and

$$g^2(y, u, b)(t) \triangleq y(t) - b - \int_0^t f(\tau, y(\tau), u(\tau))\, d\tau \qquad (t \in T).$$

Then our problem consists in minimizing g_0 on the set

$$\{(y, u, b) \in \mathscr{Y} \times \mathscr{U} \times B \mid g^1(y, u, b) \in B_1,\ g^2(y, u, b)(t) = 0\ (t \in T)\}$$
$$= \{(y, u, b) \in \mathscr{Y} \times \mathscr{U} \times B \mid (g^1, g^2)(y, u, b) \in C_1\}.$$

In this formulation of the optimal control problem, the two restrictions, $g^1(y, u, b) \in B_1$ and $g^2(y, u, b) = 0 \in C(T, \mathbb{R}^n)$, have been consolidated into one, namely $(g^1, g^2)(y, u, b) \in C_1$. However, these restrictions are of a fundamentally different nature. The second one, $g^2(y, u, b) = 0$, represents the equation of motion, that is, a physical law that holds absolutely; any triplet (y, u, b) violating the equation of motion (2) [equivalent to $g^2(y, u, b) = 0$] is literally "out of this world." On the other hand, the condition $g^1(y, u, b) \triangleq y(t_1) \in B_1$ is a desideratum but it is not absolute, so that a triplet (y, u, b) violating this condition represents an abortive or an undesirable mission to the Moon, but one consistent with the laws of nature. Indeed, if $y(t_1)$ is not "far" from B_1, then such a mission might even achieve partial success.

Because of these considerations, that also hold true for other control problems, we shall treat the equation of motion differently from other constraints of the problem. Accordingly, we shall restate the "mission to the Moon" problem as a special case of a general optimization problem in the following terms: We define $\mathscr{Y}, \mathscr{U}, B, g_0$, and g^1 as before, set $C_1 \triangleq B_1$, $g_1 \triangleq g^1$,

$$F(y, u, b)(t) \triangleq b + \int_0^t f(\tau, y(\tau), u(\tau))\, d\tau \qquad (t \in T),$$

and let

$$W \triangleq \mathscr{H}(\mathscr{U}) \triangleq \{(y, u, b) \in \mathscr{Y} \times \mathscr{U} \times B \mid y = F(y, u, b)\}.$$

(The notation $\mathscr{H}(\mathscr{U})$, emphasizing the role of $\mathscr{U}$, is employed because, at a later stage, we shall embed $\mathscr{U}$ in a larger set $\mathscr{S}^\#$, extend F to $\mathscr{Y} \times \mathscr{S}^\# \times B$, and consider the set $\mathscr{H}(\mathscr{S}^\#)$ defined similarly to $\mathscr{H}(\mathscr{U})$.) Our problem, in its new form, has as its objective the determination of a point $(\bar{y}, \bar{u}, \bar{b})$ that minimizes g_0 on the set

$$\mathscr{A}(\mathscr{U}) \triangleq \{(y, u, b) \in \mathscr{H}(\mathscr{U}) \mid g_1(y, u, b) \in C_1\}.$$

We define in an analogous manner a general *optimal control problem.* Let $T, R, B, \mathscr{Y}$, and $\mathscr{X}_1$ be given sets, $C_1 \subset \mathscr{X}_1$, $\mathscr{U}$ a class of functions on T to R, and $F : \mathscr{Y} \times \mathscr{U} \times B \to \mathscr{Y}$, $g_0 : \mathscr{Y} \times \mathscr{U} \times B \to \mathbb{R}$, and $g_1 : \mathscr{Y} \times \mathscr{U} \times B \to \mathscr{X}_1$ given functions. We let

$$\mathscr{H}(\mathscr{U}) \triangleq \{(y, u, b) \in \mathscr{Y} \times \mathscr{U} \times B \mid y = F(y, u, b)\}$$

and consider the problem of minimizing g_0 on the set

$$\mathscr{A}(\mathscr{U}) \triangleq \{(y, u, b) \in \mathscr{H}(\mathscr{U}) \mid g_1(y, u, b) \in C_1\}.$$

We refer to the equation $y = F(y, u, b)$ in $\mathscr{Y} \times \mathscr{U} \times B$ as *the equation of motion*, to elements y of $\mathscr{Y}$ as *states*, to elements u of $\mathscr{U}$ as *control functions*, and to points $b \in B$ as *control parameters*. If $\mathscr{Y}$ is a collection of functions, then we use the terms "states" and "state functions" interchangingly. We refer to a couple $(u, b) \in \mathscr{U} \times B$ as a *control*. The *optimal control problem* has a *minimizing $\mathscr{U}$-solution* if there exists a point $(\bar{y}, \bar{u}, \bar{b})$ that minimizes g_0 on $\mathscr{A}(\mathscr{U})$. Since the study of the equations of motion plays a central role in the investigation of optimal control problems, we refer to the latter as the *optimal control* of *differential*, or *integral*, or *functional equations*, according to the nature of the equations of motion.

It is perfectly clear that a problem may be of the type that we have just described even if it is defined without specifying each of the objects $\mathscr{Y}$, $\mathscr{U}$, B, $\mathscr{X}_1$, C_1, F, g_0, and g_1. If F, g_0, and g_1 are functions defined on $\mathscr{Y} \times \mathscr{U}$ and we are searching for a minimum of g_0 on $\{(y, u) \in \mathscr{Y} \times \mathscr{U} \mid y = F(y, u), g_1(y, u) \in C_1\}$, then we might choose some arbitrary set as B and think of F, g_0, and g_1 as defined on $\mathscr{Y} \times \mathscr{U} \times B$ but with values independent of the choice of the argument in B. Similar devices apply to problems where g_0 and g_1 are independent of y and the equation of motion $y = F(y, u, b)$ is therefore immaterial, or where the restriction $g_1(y, u, b) \in C_1$ is absent.

The preceding description of an optimal control problem is one that we find convenient and that would probably apply to most problems referred to as such in the mathematical literature. While other formulations have been suggested and studied (see Notes at the end of Chapter V), this diversity of definitions is perhaps natural in a rapidly evolving field.

Until rather recently, a number of authors (including the present one) used the terms *variational* and *calculus of variations* when referring to problems of the type just described. The present trend is to reserve the term "variational" for control problems in which T is a subset of some $\mathbb{R}^k$, $B = \mathbb{R}^l$, both $\mathscr{Y}$ and $\mathscr{U}$ are collections of all functions on T to $\mathbb{R}^n$ that satisfy certain measurability, continuity or differentiability properties, and the restriction $g_1(y, u, b) \in C_1$ corresponds to a finite set of equalities. Problems with inequalities, termed *unilateral*, have been studied throughout the history of the calculus of variations (one of them by Newton, see Goursat [1, p. 658]) but no satisfactory theory has evolved within the variational framework and these problems seem much more easily tractable by the methods of the optimal control theory; it appears therefore justified to classify them alongside other optimal control problems. If we accept this classification, then optimal control problems are primarily distinguished from variational problems by the presence of control functions whose range is not a vector space and/or by the presence of restrictions other than equalities.

The term "control" has been coined by Pontryagin and his collaborators [1] within the context of problems defined by ordinary differential equations. Such problems have accounted, at least until very recently, for most of the research in the optimal control theory, and the term "optimal control" is frequently used in the narrow sense of "optimal control theory of ordinary differential equations." However, optimal control problems defined by other functional equations are being increasingly investigated. They include optimal control problems involving delay-differential and functional-differential equations, integral equations, partial differential equations, and stochastic differential equations.

The expansion of the optimal control theory to encompass problems defined by functional equations more general than differential equations is also accompanied by a more systematic use of the concepts and the tools of functional analysis. Not only the fundamental concepts and the foundations but also the methods and the techniques of the optimal control theory increasingly emphasize the more general "soft" and abstract approach of functional analysis at the expense of the more detailed and specialized approach of "hard" analysis. In this respect, the evolution of the optimal control theory follows the path paved by the theory of functional equations with which it remains closely associated.

III.2 Original, Approximate, and Relaxed Solutions

We now return to our model of an optimal control problem defined by sets $\mathscr{Y}$, $\mathscr{U}$, B, $\mathscr{X}_1$, and C_1 and functions

$$F : \mathscr{Y} \times \mathscr{U} \times B \to \mathscr{Y} \qquad \text{and} \qquad (g_0, g_1) : \mathscr{Y} \times \mathscr{U} \times B \to \mathbb{R} \times \mathscr{X}_1,$$

where $\mathscr{U}$ is a collection of functions on a measure space T into a set R (which we shall assume to be a topological space). In defining this problem, the role of the scientist or of the engineer is to provide a realistic translation of the physical situation. Such a translation often leaves certain loose ends that have to be tied by the mathematician. For example, certain equations may be provided by the scientist, and the mathematician may be free to specify, within certain limits, the subset of the domain of definition of the equation to which the solutions must be restricted.

In a large number of cases, illustrated by the example of III.1, there appears to be little room for the mathematician to maneuver except in the specification of the set $\mathscr{U}$. In discussing the Moon shot, we have defined $\mathscr{U}$ to be the class of all measurable functions on $[0, t_1]$ to R. However, one might as well have considered the cases where $\mathscr{U}$ is the set of all continuous functions, or piecewise linear functions, or piecewise constant functions, etc. In other problems, the control function may be "physically" restricted by conditions of the type

$$u(t) \in R^{\#}(t) \qquad (t \in T),$$

where the mapping $R^{\#} : T \to \mathscr{P}'(R)$ is specified, but there is still room to decide whether to consider only continuous functions, or measurable ones, etc. In all these cases, it is presumed that any choice of a function u out of $\mathscr{U}$ can be physically implemented with sufficient accuracy.

We would certainly be treading on dangerous ground if we tried to decide for the engineering community whether continuous, or piecewise continuous, or constant, or measurable functions can be implemented in physical systems. In any event, all these terms imply a certain idealization of the physical reality and must be interpreted according to the physical context. For this reason, it seems quite reasonable to desist from any attempt to apply purely physical criteria in specifying $\mathscr{U}$ and to give a larger weight to mathematical considerations. This is, indeed, what mathematicians have been doing since the beginnings of the calculus of variations (and it is quite consistent with the general practice of theoretical science). In fact, the mathematical criteria were usually very simple and their selection seemed to obey the following directive: "Choose the class of functions that you are most familiar with and that provides a convenient tool, and make assumptions that allow you to prove interesting theorems!" Accordingly, the classical calculus of variations usually dealt with continuous, and even differentiable, counterparts of control functions. More recently, when certain phenomena were discovered that could not be accommodated by continuous functions (see, e.g., Goursat [1, p. 647]), piecewise continuous functions came to the fore. Nowadays the dominant fashion is to define $\mathscr{U}$ as the largest class of measurable functions on T to R for which the various restrictions are defined.

If we follow this pragmatic approach, we would select for consideration as $\mathscr{U}$ those sets of "realistic" control functions that best lend themselves to the achievement of our mathematical objectives and to the use of available techniques. For the model that we are considering, these objectives would include (a) proving the existence, under reasonable conditions, of a minimizing $\mathscr{U}$-solution, and (b) inquiring into the properties of such a solution, if one exists. As we shall demonstrate in what follows, objective (a) cannot be achieved in general, except for quite restricted classes of optimal control problems. We shall show, furthermore, that even if a minimizing $\mathscr{U}$-solution exists for a particular problem, it may have to be discarded because it turns out, after all, not to be "optimal."

Because of these considerations, we shall not only study the original problem that we have previously defined but also a related extended problem. For both of these problems we shall consider sets $\mathscr{U}$ whose elements are measurable selections of an appropriate set-valued mapping $R^{\#} : T \to \mathscr{P}'(R)$, and we shall require that $\mathscr{U}$ satisfy two basic "technical" properties (IV.3.3). We shall refer to such a set of control functions as an *abundant set*, and will show that, in particular, the sets of all measurable or simple selections of

$R^{\#}$ and, in special cases, the sets of all piecewise continuous or continuous functions on T to R, are abundant sets. Many of our results will be equally valid for any choice of an abundant set $\mathscr{U}$.

We shall illustrate some of our contentions and motivate the need for new concepts with two examples.

Problem I Let $\mathscr{U}$ be the class of all measurable functions on $T \triangleq [0, 1]$ to $R \triangleq [-1, 1]$, $\mathscr{Y} \triangleq C(T)$, and F and g_0 be functions defined by

$$F(y, u)(t) \triangleq \int_0^t u(\tau)\, d\tau \qquad (y \in \mathscr{Y}, \quad u \in \mathscr{U}, \quad t \in T)$$

and

$$g_0(y, u) \triangleq \int_0^1 ([y(t)]^2 - [u(t)]^2)\, dt \qquad (y \in \mathscr{Y}, \quad u \in \mathscr{U}).$$

The objective of the problem is to minimize g_0 on the set

$$\mathscr{H}(\mathscr{U}) \triangleq \{(y, u) \in \mathscr{Y} \times \mathscr{U} \mid y = F(y, u)\}.$$

We shall verify that Problem I has no minimizing $\mathscr{U}$-solution. Indeed, we have $u(t) \in [-1, 1]$; hence $(u(t))^2 \leqslant 1$ $(t \in T,\ u \in \mathscr{U})$ and

$$g_0(y, u) \triangleq \int_0^1 [(y(t))^2 - (u(t))^2]\, dt \geqslant -1 \qquad (y \in \mathscr{Y}, \quad u \in \mathscr{U}).$$

If $j \in \mathbb{N}$ and we choose $\bar{u}_j(t)$ to equal alternately $+1$ and -1 on successive subintervals of length $(2j)^{-1}$ of $[0, 1]$, and if we set $\bar{y}_j(t) \triangleq \int_0^t \bar{u}_j(\tau)\, d\tau$ $(t \in T)$, then

$$(\bar{y}_j, \bar{u}_j) \in \mathscr{H}(\mathscr{U}), \qquad 0 \leqslant \int_0^t \bar{u}_j(\tau)\, d\tau \leqslant (2j)^{-1},$$

and

$$-1 \leqslant g_0(\bar{y}_j, \bar{u}_j) \leqslant (2j)^{-2} - 1.$$

Thus

$$\lim_j g_0(\bar{y}_j, \bar{u}_j) = -1 = \inf g_0(\mathscr{H}(\mathscr{U})).$$

On the other hand, if there exists $(\bar{y}, \bar{u}) \in \mathscr{H}(\mathscr{U})$ such that $g_0(\bar{y}, \bar{u}) = -1$, then

$$|\bar{u}(t)| = 1 \quad \text{a.e. in } [0, 1] \qquad \text{and} \qquad \bar{y}(t) = \int_0^t \bar{u}(\tau)\, d\tau = 0 \quad (t \in [0, 1]).$$

The second of these relations yields $\bar{u}(t) = 0$ a.e. in $[0, 1]$, contradicting the first relation. Thus $g_0(\mathscr{H}(\mathscr{U}))$ does not contain its infimum. The same argument also applies if we define $\mathscr{U}$ to be the set of all the piecewise continuous functions.

We shall also find it instructive to consider a variant of Problem I.

Problem II Let $\mathscr{Y}$, $\mathscr{U}$, g_0, $\bar{y}_j$, and $\bar{u}_j$ be defined as in Problem I, and let

$$g_1(y, u) \triangleq \int_0^1 [y(t)]^2 \, dt \qquad [(y, u) \in \mathscr{Y} \times \mathscr{U}].$$

We now consider how to minimize g_0 on the set $\mathscr{A}(\mathscr{U}) \triangleq \{(y, u) \in \mathscr{Y} \times \mathscr{U} \mid y = F(y, u),\ g_1(y, u) = 0\}$. Clearly $y = F(y, u)$ and $g_1(y, u) = 0$ imply $y(t) = \int_0^t u(\tau)\, d\tau = 0$ for all $t \in [0, 1]$; hence $u(t) = 0$ a.e. in $[0, 1]$. Thus $(y, u) \in \mathscr{A}(\mathscr{U})$ implies $y(t) = 0$ $(t \in T)$ and $u(t) = 0$ a.e. in T; it follows that $\bar{y} = 0$ and $\bar{u} = 0$ yield the minimum $g_0(\bar{y}, \bar{u}) = 0$ of g_0 on $\mathscr{A}(\mathscr{U})$. On the other hand, we observe that

$$g_0(\bar{y}_j, \bar{u}_j) \leqslant -1 + (2j)^{-2}$$

and

$$0 \leqslant \int_0^t \bar{u}_j(\tau)\, d\tau = \bar{y}_j(t) \leqslant (2j)^{-1} \qquad (j \in \mathbb{N}, \quad t \in [0, 1)]);$$

hence $0 \leqslant g_1(\bar{y}_j, \bar{u}_j) \leqslant (2j)^{-2}$ $(j \in \mathbb{N})$. Thus, while the restricted minimum of g_0 exists and equals 0, we can lower the value of g_0 to nearly -1 by choosing $y = \bar{y}_j$ and $u = \bar{u}_j$ for large j and, in so doing, we violate the restriction $g_1(y, u) = 0$ by an arbitrarily small amount [bounded by $(2j)^{-2}$].

The two examples that we have just discussed suggest that, in addition to points of $\mathscr{H}(\mathscr{U}) \triangleq \{(y, u, b) \in \mathscr{Y} \times \mathscr{U} \times B \mid y = F(y, u, b)\}$ that may yield a solution to the (restricted or unrestricted) minimization problem, we ought to consider "approximate" solutions of such a problem. We shall now proceed to define such solutions in the general case. To do so, we assume that the set $\mathscr{X}_1$ is a topological vector space. We let $g \triangleq (g_0, g_1) : \mathscr{Y} \times \mathscr{U} \times B \to \mathbb{R} \times \mathscr{X}_1$ and refer to a sequence $((y_j, u_j, b_j))$ in $\mathscr{H}(\mathscr{U})$ as an *approximate $\mathscr{U}$-solution* if for every neighborhood G of 0 in $\mathscr{X}_1$ there exists $j(G) \in \mathbb{N}$ such that $g_1(y_j, u_j, b_j) \in C_1 + G$ for $j \geqslant j(G)$. We say that an approximate $\mathscr{U}$-solution $((\bar{y}_j, \bar{u}_j, \bar{b}_j))$ is a *minimizing approximate $\mathscr{U}$-solution* if

$$\lim_j g_0(\bar{y}_j, \bar{u}_j, \bar{b}_j) \leqslant \liminf_j g_0(y_j, u_j, b_j)$$

for every approximate $\mathscr{U}$-solution $((y_j, u_j, b_j))$. In this sense, the sequence $((\bar{y}_j, \bar{u}_j))$ defined above is a minimizing approximate $\mathscr{U}$-solution for both Problems I and II (neither of which involves the control parameter b).

It is rather easy to see that, in a large number of cases, there exists a minimizing approximate $\mathscr{U}$-solution if there exists any approximate $\mathscr{U}$-solution at all (in particular, if g_0 is bounded below, $\mathscr{A}(\mathscr{U}) \neq \varnothing$, and the topology of $\mathscr{X}_1$ is metric). Thus, the introduction of approximate $\mathscr{U}$-solutions ensures the existence of a minimizing (approximate) solution to the optimization problem in a large number of cases of interest. No less important is a second consideration; indeed, as Problem II demonstrates, a minimizing approximate $\mathscr{U}$-solution may yield better (that is, lower) values of g_0 than do minimizing $\mathscr{U}$-solutions, and it clearly never yields a worse value. Since the relation $g_1(y, u, b) \in C_1$ presumably represents human or engineering restrictions that are always measured with some error, one is inclined to accept points (y, u, b) that violate this restriction to "an arbitrarily small extent" only.

If we admit a minimizing approximate $\mathscr{U}$-solution as the desired answer (or, at least, one answer) to the optimal control problem, and after we show that a minimizing approximate $\mathscr{U}$-solution exists, we are still left with the task of determining such a solution. We shall often accomplish this task in two steps; first, by embedding the set $\mathscr{U}$ in a topological space $\mathscr{S}^{\#}$ of which $\mathscr{U}$ is a dense subset, and extending the definition of F, g_0, and g_1 to $\mathscr{Y} \times \mathscr{S}^{\#} \times B$ in such a manner that the set

$$\mathscr{H}(\mathscr{S}^{\#}) \triangleq \{(y, \sigma, b) \in \mathscr{Y} \times \mathscr{S}^{\#} \times B \mid y = F(y, \sigma, b)\}$$

is sequentially compact, and F, g_0, and g_1 are sequentially continuous when restricted to $\mathscr{H}(\mathscr{S}^{\#})$. We shall show then that, under suitable (and realistic) conditions, there exists a point $(\bar{y}, \bar{\sigma}, \bar{b})$ in $\mathscr{H}(\mathscr{S}^{\#})$ that minimizes g_0 on the set

$$\mathscr{A}(\mathscr{S}^{\#}) \triangleq \{(y, \sigma, b) \in \mathscr{H}(\mathscr{S}^{\#}) \mid g_1(y, \sigma, b) \in C_1\},$$

and we shall determine certain relations that must be satisfied by such a *minimizing relaxed solution* $(\bar{y}, \bar{\sigma}, \bar{b})$ (*necessary conditions for a relaxed minimum*). These necessary conditions often admit only a finite number of solutions, in which case the solution yielding a lowest value of g_0 is a minimizing relaxed solution. The next step consists in devising a procedure (see IV.2.6 respectively IV.3.9 and V.1.2) for determining a sequence (u_j) in $\mathscr{U}$ corresponding to $\bar{\sigma}$ and a sequence (y_j) in $\mathscr{Y}$ corresponding to (u_j) and $\bar{b}$ such that $((y_j, u_j, \bar{b}))$ is a minimizing approximate $\mathscr{U}$-solution and

$$\lim_j g_i(y_j, u_j, \bar{b}) = g_i(\bar{y}, \bar{\sigma}, \bar{b}) \qquad (i = 0, 1).$$

The program that we have just outlined is carried out for a large class of optimal control problems. The set $\mathscr{S}^{\#}$ is the set of *relaxed control functions*. The topology of $\mathscr{S}^{\#}$ (suggested by Young's work on *generalized curves* [1, 2]) is metric and compact. In problems that we study, $\mathscr{Y}$ is a Banach space and B a convex subset of a vector space, and we assume, when necessary, that B has a sequentially compact topology. Under appropriate conditions, the functions F, g_0, and g_1 are sequentially continuous when restricted to $\mathscr{H}(\mathscr{S}^{\#})$, and the set $\mathscr{H}(\mathscr{S}^{\#})$ is a sequentially compact subset of $\mathscr{Y} \times \mathscr{S}^{\#} \times B$.

This program is modified for two classes of problems. In "ordinary differential problems with unbounded contingent sets" which we discuss in VI.4, we are able to dispense with certain boundedness and compactness assumptions by embedding the state functions in a larger set of "curve parametrizations" and transforming the control functions prior to "relaxing" them. This procedure enables us to define a related "compactified parametric problem" which can be handled in the previously described manner and whose solutions yield an answer to the original problem with unbounded contingent sets.

In "conflicting control problems" (which we study in Chapters IX and X) the situation is more complex. If the problem is characterized by "additively coupled conflicting controls," then our previous remarks remain valid and the problem exhibits the "usual" pattern encountered in Chapters VI–VIII. If the conflicting controls are not additively coupled, then new phenomena appear and the function corresponding to $g_1 \mid \mathscr{H}(\mathscr{S}^{\#})$ is not, in general, sequentially continuous. Nevertheless, by resorting to "hyperrelaxed adverse controls" (Chapter X), we are able to determine a minimizing relaxed solution and to construct a corresponding minimizing approximate $\mathscr{U}$-solution.

III.3 Measure-Valued Control Functions

III.3.A "Limits" of Rapidly Oscillating Functions

We shall now endeavor to demonstrate, using special examples and heuristic arguments, how relaxed control functions arise from the consideration of approximate $\mathscr{U}$-solutions and why they "work," our purpose being to motivate the rigorous definitions and arguments of Chapters IV and V. We first observe that, in the special cases of Problems I and II of III.2, the minimizing approximate $\mathscr{U}$-solution $((\bar{y}_j, \bar{u}_j))$ is characterized by the highly oscillatory nature of each $\bar{u}_j(\cdot)$ for large values of j. The function $\bar{u}_j(\cdot)$ takes on alternately the values $+1$ and -1 on successive intervals of length $(2j)^{-1}$, and it is clear that neither $(\bar{u}_j)$ nor any of its subsequences have a limit in any usual mode of convergence of functions (such as convergence for each $t \in T$, uniform convergence, convergence in measure, convergence in L^p). We shall attempt to simulate "limits" of such nonconverging sequences by embedding the set $\mathscr{U}$ of functions in some larger set whose elements will not even be functions on T to R. This type of approach is quite common (see Notes at the end of Chapter IV) and it is exemplified by the embedding of the rational numbers between 0 and 1 in the compact interval $[0, 1]$ of the real numbers. However, while we may be quite content with defining real numbers as equivalence classes of Cauchy sequences of rationals, we shall find it convenient to embed $\mathscr{U}$ in a set whose elements can be characterized more "concretely" than as sequences in $\mathscr{U}$.

We shall use as an example an optimal control problem defined by the ordinary differential equation

$$(1) \qquad y(t) = F(y, u)(t) \triangleq \int_0^t f(\tau, y(\tau), u(\tau))\, d\tau \qquad (t \in T \triangleq [0, 1]),$$

where $R \triangleq [-1, 1]$, $f: T \times \mathbb{R}^n \times R \to \mathbb{R}^n$ is continuous and bounded, and the equation has a unique solution $\tilde{y}(u)(\cdot)$ for each u in the set $\mathscr{U}$ of measurable functions on T to R. Our first purpose is to study what happens when u is replaced by highly oscillatory functions such as $\bar{u}_j$. Therefore, let j be some fixed large integer. The function $\bar{y}_j \triangleq \tilde{y}(\bar{u}_j)$ satisfies the equation

$$(1') \qquad y(t) = \int_0^t f(\tau, y(\tau), \bar{u}_j(\tau))\, d\tau \qquad (t \in T).$$

If we denote by $T_k \triangleq [t_k', t_{k+1}']$ $(k = 0, 1,..., j - 1)$ successive subintervals of $[0, 1]$ of length $1/j$ and by t_k'' the midpoint of T_k, and if we set $T_k' \triangleq [t_k', t_k'')$ and $T_k'' \triangleq [t_k'', t_{k+1}')$, then $\bar{u}_j(t) = 1$ $(t \in T_k')$ and $\bar{u}_j(t) = -1$ $(t \in T_k'')$ for each k and, since T_k is of small length, $\bar{y}_j$ is approximately constant on each T_k. Thus, using $\approx$ to denote approximate equality and setting

$$f_k(r) \triangleq f(t_k', \tilde{y}(\bar{u}_j)(t_k'), r) \qquad (r \in R),$$

we have

$$\begin{aligned}
\bar{y}_j(t_{k+1}') - \bar{y}_j(t_k') &\approx \int_{T_k} f_k(\bar{u}_j(\tau))\, d\tau \\
&= \int_{t_k'}^{t_k''} f_k(1)\, d\tau + \int_{t_k''}^{t_{k+1}'} f_k(-1)\, d\tau \\
&= (t_{k+1}' - t_k')[\tfrac{1}{2} f_k(1) + \tfrac{1}{2} f_k(-1)] \qquad (k = 0, 1,..., j - 1).
\end{aligned}$$

This shows that $\bar{y}_j$ "approximately" satisfies the differential equation

$$(2) \qquad y(t) = \int_0^t [\tfrac{1}{2} f(\tau, y(\tau), 1) + \tfrac{1}{2} f(\tau, y(\tau), -1)]\, d\tau \qquad (t \in T)$$

and, indeed, the accuracy of the approximation improves as j grows large.

The functions $\bar{u}_j(\cdot)$ have a relatively simple mode of oscillation; they spend equal amounts of time at $+1$ and -1. In more complicated situations, we ought to be prepared to deal with functions u_j that take on several, or infinitely many, values on any subinterval of T. The form of Eq. (2) suggests that an analogous equation for some general measurable u_j can be obtained as follows. We divide the interval T into equal subintervals $T_{j,k}$ $[k = 0, 1,..., m(j)]$, where $m(j)$ is chosen for each j to fit the nature of u_j but always so that $\lim_j m(j) = \infty$. In each subinterval $T_{j,k}$ we study the "distribution" of the values of u_j. This can be done by determining, for each k and each $A \in \Sigma_{\mathrm{Borel}}(R)$, the fraction $\sigma_{j,k}(A)$ of the time $T_{j,k}$ that $u_j(\cdot)$ "spends" in A; specifically $\sigma_{j,k}(A)$ is defined by

$$T_{j,k}(A) \triangleq u_j^{-1}(A) \cap T_{j,k} \qquad \text{and} \qquad \sigma_{j,k}(A) \triangleq \int_{T_{j,k}(A)} dt \Big/ \int_{T_{j,k}} dt.$$

It is easily verified that each $\sigma_{j,k} : \Sigma_{\mathrm{Borel}}(R) \to [0, 1]$ is a probability measure. When $m(j) = j - 1$ and $u_j = \bar{u}_j$, the integrand in (2) is

$$\tfrac{1}{2}f(\tau, y(\tau), 1) + \tfrac{1}{2}f(\tau, y(\tau), -1) = \int f(\tau, y(\tau), r)\, \sigma_{j,k}(dr) \qquad (\tau \in T_{j,k}),$$

and it so happens that the probability measures $\sigma_{j,0}, \sigma_{j,1}, ..., \sigma_{j,j-1}$ are all equal to $\frac{1}{2}\delta_1 + \frac{1}{2}\delta_{-1}$, where δ_r is the Dirac measure at r. In other, more complicated, situations we obtain, instead of (2), the equation

$$y(t) = \sum_{k=0}^{m(j)} \int_{T_{j,k} \cap [0,t]} d\tau \int f(\tau, y(\tau), r)\, \sigma_{j,k}(dr) \qquad (t \in T)$$

which we rewrite in the form

$$y(t) = \int_0^t d\tau \int f(\tau, y(\tau), r)\, \sigma_j(\tau)\,(dr) \qquad (t \in T), \tag{3}$$

where $\sigma_j(\tau) \triangleq \sigma_{j,k}$ for $\tau \in T_{j,k}$.

The comparison of Eq. (3) with Eq. (1′) shows that we have replaced a sequence $(\bar{u}_j)$ of functions on T to R with a sequence (σ_j) of piecewise constant functions on T whose values are probability measures on $\Sigma_{\mathrm{Borel}}(R)$. We have simultaneously "redefined" the function

$$(y, u) \to F(y, u) : C(T, \mathbb{R}^n) \times \mathscr{U} \to C(T, \mathbb{R}^n)$$

satisfying the relation $F(y, u)(t) = \int_0^t f(\tau, y(\tau), u(\tau))\, d\tau$ $(t \in T)$, as a function $(y, \sigma) \to F(y, \sigma)$ satisfying the relation

$$F(y, \sigma)(t) = \int_0^t d\tau \int f(\tau, y(\tau), r)\, \sigma(\tau)\,(dr) \qquad (t \in T),$$

where $y \in C(T, \mathbb{R}^n)$ and σ is a piecewise constant function on $T \triangleq [0, 1]$ whose values are probability measures. In fact, Eq. (3) remains meaningful even if the σ_j's are not piecewise constant, so long as, for each $y \in C(T, \mathbb{R}^n)$, the functions $\tau \to \int f(\tau, y(\tau), r)\, \sigma_j(\tau)\,(dr)$ are integrable on T. In particular, if $u \in \mathscr{U}$ and $\delta_u(t) \triangleq \delta_{u(t)}$ is the Dirac measure at $u(t)$, then

$$F(y, \delta_u)(t) = \int_0^t d\tau \int f(\tau, y(\tau), r)\, \delta_u(\tau)\,(dr)$$

$$= \int_0^t f(\tau, y(\tau), u(\tau))\, d\tau \qquad (t \in T),$$

suggesting that we may identify a function $u : T \to R$ with the function $t \to \delta_{u(t)}$ on T.

We shall now verify that, in the case of Problem II of III.2, this procedure yields a "limit" of the sequence $(\bar{u}_j)$. If we denote by y^2 the function previously designated as y, then Problem II can be thought of as defined by the differential equations [corresponding to Eq. (1)] of the form

$$y^1(t) = \int_0^t ([y^2(\tau)]^2 - [u(\tau)]^2)\, d\tau, \quad y^2(t) = \int_0^t u(\tau)\, d\tau, \quad y^3(t) = \int_0^t [y^2(\tau)]^2\, d\tau,$$

[where $y \triangleq (y^1, y^2, y^3) \in \mathbb{R}^3$], with $g_0(y, u) \triangleq y^1(1)$ and $g_1(y, u) \triangleq y^3(1)$. The corresponding form of Eq. (3) for $\sigma_j = \bar{\sigma} \triangleq \frac{1}{2}\delta_1 + \frac{1}{2}\delta_{-1}$ is

$$y^1(t) = \int_0^t ([y^2(\tau)]^2 - 1)\, d\tau, \qquad y^2(t) = 0, \qquad y^3(t) = \int_0^t [y^2(\tau)]^2\, d\tau,$$

yielding $y^1(t) = -t$, $y^2(t) = 0$, $y^3(t) = 0$ $(t \in T)$, $g_0(y, \bar{\sigma}) = -1$ and $g_1(y, \bar{\sigma}) = 0$. Thus the measure-valued constant function $\bar{\sigma} \triangleq \frac{1}{2}\delta_1 + \frac{1}{2}\delta_{-1}$ and the corresponding solution y of Eq. (3) yield values of g_0 and g_1 that are limits, as $j \to \infty$, of $g_0(\tilde{y}(\bar{u}_j), \bar{u}_j)$ and $g_1(\tilde{y}(\bar{u}_j), \bar{u}_j)$. A single measure-valued function $\bar{\sigma}$ achieves as much, in a sense, as the sequence $((\tilde{y}(\bar{u}_j), \bar{u}_j))_j$ which is a minimizing approximate $\mathscr{U}$-solution.

It appears appropriate to emphasize, at this stage, that our procedure was successful only because we chose partitions $T_{j,0}, \ldots, T_{j,j-1}$ of $[0, 1)$ that "fitted" the behavior of $\bar{u}_j$ for each $j \in \mathbb{N}$. If we had chosen $m(j) = 2j - 1$ instead of $m(j) = j - 1$, then, on the new smaller subintervals, σ_j would have alternately equaled δ_1 and δ_{-1}, and then Eq. (3) would have been identical with the original Eq. (1′). We were successful because, for each j, we had chosen subintervals that were sufficiently large to "smooth out" the oscillations of $\bar{u}_j$ and, at the same time, our choice caused each subinterval of the jth partition to shrink in size to 0 as j increased to ∞. In this respect, our procedure bears a similarity to the arguments of physicists who define the local temperature of a gas at some point of space by an averaging process taken over volumes that are "microscopically" large and "macroscopically" small; that is, such that each volume contains a sufficiently large number of molecules for statistical averages to be "almost certainly" achieved and each volume is sufficiently small so that the corresponding average represents the temperature at the given point and not elsewhere.

What about more complicated problems than Problem II that are defined by Eq. (1) and whose minimizing approximate $\mathscr{U}$-solutions $((y_j, u_j))$ do not yield σ_j's that are the same for each j? The answer is that, for each choice of a sequence $((y_j, u_j))$ in $C(T, \mathbb{R}^n) \times \mathscr{U}$ satisfying Eq. (1), we can produce a sequence $J \subset (1, 2, ...)$, a measure-valued $\bar{\sigma}$ (a *relaxed control function*), and a solution $\bar{y}$ of the equation

$$y(t) = \int_0^t d\tau \int f(\tau, y(\tau), r)\, \bar{\sigma}(\tau)\,(dr) \qquad (t \in T)$$

such that $\lim_{j\in J} y_j(t) = \bar{y}(t)$ uniformly for $t \in T$. In all such cases we can think of $\bar{\sigma}$ as being a "limit" of $(u_j)_{j\in J}$ in the sense that the effect of u_j on the differential equation for large j simulates the effect of $\bar{\sigma}$.

III.3.B Relaxed Controls as Linear Functionals

The "smoothing" operations that we have described above and that transform an original control function into a piecewise constant relaxed (measure-valued) control function may provide an intuitive explanation of why the set of relaxed control functions has a sequentially compact topology that "fits" the problem. However, when it comes to defining relaxed control functions in a rigorous manner, we approach this problem from a somewhat different angle. As we have mentioned before, when discussing the optimal control problem defined by Eq. (1), we view original control functions u_j as "converging" to a relaxed control function $\bar{\sigma}$ if the solutions of the differential equation corresponding to u_j converge to the solution for $\bar{\sigma}$ in the topology of $C(T, \mathbb{R}^n)$. It can be shown that this will be the case if

$$\lim_j \int_T f(\tau, y(\tau), u_j(\tau))\, d\tau = \int_T d\tau \int f(\tau, y(\tau), r)\, \bar{\sigma}(\tau)\,(dr)$$

for every $y \in C(T, \mathbb{R}^n)$. In fact, rather then considering the functions $(\tau, r) \to f(\tau, y(\tau), r)$ for various choices of $y \in C(T, \mathbb{R}^n)$, it is simpler to compare the effects of u_j and $\bar{\sigma}$ on a larger set of functions, namely on $L^1(T, C(R))$ [which includes the functions $(\tau, r) \to f(\tau, y(\tau), r)$ for all continuous y when we identify, as we did in I.5.C and Chapter II, the function

$\phi : T \to C(R)$ with the function $(t, r) \to \phi(t)(r) \triangleq \phi(t, r) : T \times R \to \mathbb{R}$.] Thus we agree to say that "(u_j) converges to $\bar{\sigma}$" if

$$\lim_j \int_0^1 \phi(\tau, u_j(\tau))\, d\tau = \int_0^1 d\tau \int \phi(\tau, r)\, \bar{\sigma}(\tau)\,(dr) \qquad [\phi \in L^1(T, C(R))].$$

Indeed, in the same spirit, we use the expression "a sequence (σ_j) of relaxed control functions converges to $\bar{\sigma}$" to mean

$$\lim_j \int \mu(d\tau) \int \phi(\tau, r)\, \sigma_j(\tau)\,(dr) = \int \mu(d\tau) \int \phi(\tau, r)\, \bar{\sigma}(\tau)\,(dr)$$
$$[\phi \in L^1(T, C(R))]$$

in a situation where the equation of motion $y = F(y, u, b)$ is no longer an ordinary differential equation but some more general functional equation involving a control function $u : T \to R$ between compact metric spaces and with an appropriate measure space (T, Σ, μ) defined for the problem. When the σ_j's have values that are Dirac measures, this last definition yields the concept of the convergence of original control functions to a relaxed control function.

This definition of convergence is patterned after the concept of "generalized curves" introduced in 1937 by Young for certain variational problems. In line with this definition, we view both original and relaxed control functions as continuous linear functionals on $L^1(T, C(R))$ in the sense that we identify each σ with the functional

$$\phi \to \int \mu(d\tau) \int \phi(\tau, r)\, \sigma(\tau)\,(dr) : L^1(T, C(R)) \to \mathbb{R}.$$

It follows then that the set $\mathscr{S}$ of relaxed control functions is embedded in $L^1(T, C(R))^*$, and the topology consistent with this definition of convergence is precisely the weak star topology of $L^1(T, C(R))^*$. In this topology, the set $\mathscr{S}$ of relaxed control functions is the closure of the set $\mathscr{R}$ of all μ-measurable functions on T to R (with $\rho \in \mathscr{R}$ identified with the relaxed control function $t \to \delta_{\rho(t)}$).

In the most general case that we consider, the set $\mathscr{U}$ of original control functions may be contained in the set

$$\mathscr{R}^{\#} \triangleq \{\rho \in \mathscr{R} \mid \rho(t) \in R^{\#}(t)\ \mu\text{-a.e.}\},$$

where $R^{\#}(t)$ is a given nonempty subset of R for each $t \in T$. In such a case, we define relaxed control functions as the elements of the set

$$\mathscr{S}^{\#} \triangleq \{\sigma \in \mathscr{S} \mid \sigma(\overline{R^{\#}(t)}) = 1\ \mu\text{-a.e.}\};$$

that is, elements of $\mathscr{S}^{\#}$ are those elements σ of $\mathscr{S}$ for which $\sigma(t)$ is supported, for μ-a.a. $t \in T$, on the closure of $R^{\#}(t)$. If $R^{\#}$ satisfies appropriate conditions, then we can show that $\mathscr{S}^{\#}$ is the closure of $\mathscr{R}^{\#}$. Indeed, if $\mathscr{U}$ is an "abundant" subset of $\mathscr{R}^{\#}$, then $\mathscr{S}^{\#}$ is the closure of $\mathscr{U}$. In the special case where $R^{\#}(t) = R$ for all $t \in T$, we have $\mathscr{R}^{\#} = \mathscr{R}$ and $\mathscr{S}^{\#} = \mathscr{S}$.

With this construction of the set $\mathscr{S}^{\#}$ of relaxed control functions accomplished, we can begin to carry out the program outlined in the preceding section. The existence of minimizing relaxed solutions and of corresponding minimizing approximate $\mathscr{U}$-solutions usually follows fairly directly from two properties of $\mathscr{S}^{\#}$; namely, that $\mathscr{S}^{\#}$ is compact and that it is the closure of $\mathscr{U}$. Our search for necessary conditions for minimum is then facilitated by a third property of $\mathscr{S}^{\#}$; it is a convex subset of a normed vector space. For appropriate convex finite-dimensional subsets B' of B the function $g_i \mid \mathscr{Y} \times \mathscr{S}^{\#} \times B'$ is a mapping on a convex subset of a normed vector space into a topological vector space, and with these ingredients we can manufacture appropriate derivatives and apply the techniques of the differential calculus.

III.4 Necessary Conditions for a Minimum

Let Q be a convex subset of a vector space and g_0 a real-valued function on Q possessing, for all $\bar{q}$, $q \in Q$, a directional derivative

$$Dg_0(\bar{q}; q - \bar{q}) \triangleq \lim_{\alpha \to +0} \alpha^{-1}[g_0(\bar{q} + \alpha(q - \bar{q})) - g_0(\bar{q})].$$

If $\bar{q}$ is a minimizing point of g_0, then, clearly, $g_0(\bar{q} + \alpha(q - \bar{q})) \geqslant g_0(\bar{q})$ $(\alpha \in [0, 1], q \in Q)$; hence

$$(1) \qquad Dg_0(\bar{q}; q - \bar{q}) \geqslant 0 \qquad (q \in Q).$$

Relation (1) must be satisfied by every minimizing point $\bar{q}$ but it may also be valid for other points. For example, if $Q \triangleq [-2, 2] \subset \mathbb{R}$ and

$$g_0(q) \triangleq 3q^4 - 8q^3 + 6q^2 \qquad (q \in Q),$$

then

$$Dg_0(\bar{q}; q - \bar{q}) = (12\bar{q}^3 - 24\bar{q}^2 + 12\bar{q})(q - \bar{q})$$

and $Dg_0(\bar{q}; q - \bar{q}) \geqslant 0$ $(q \in Q)$ for $\bar{q} = 0, 1$ but only 0 is a minimizing point of g_0. Thus (1) is a necessary condition for $\bar{q}$ to be a minimizing point but it need not be sufficient. A condition of this type is customarily referred to as a *necessary condition for minimum*. If a relation is satisfied by minimizing points only, then it is a *sufficient condition for minimum*. More desirable than either are *necessary and sufficient conditions for minimum* (that is, relations that are both necessary conditions and sufficient conditions) but these are few and far between and difficult to obtain except in rather simple situations. For certain variational problems, a necessary and sufficient condition corresponding to (1) has been derived by Young [3, (69.3), p. 182].

A simple example of a necessary and sufficient condition for minimum can be provided by assuming that the function g_0 considered above is convex. If $\bar{q}, q_1 \in Q$ and $g_0(\bar{q}) > g_0(q_1)$, then

$$\begin{aligned} g_0(\bar{q} + \alpha(q_1 - \bar{q})) &= g_0(\alpha q_1 + (1 - \alpha)\bar{q}) \\ &\leqslant \alpha g_0(q_1) + (1 - \alpha)\, g_0(\bar{q}) \qquad (\alpha \in [0, 1]); \end{aligned}$$

hence

$$\alpha^{-1}[g_0(\bar{q} + \alpha(q_1 - \bar{q})) - g_0(\bar{q})] \leqslant g_0(q_1) - g_0(\bar{q}) < 0 \qquad (\alpha \in (0, 1])$$

and

$$Dg_0(\bar{q}; q_1 - \bar{q}) < 0.$$

Thus, for a convex function g_0, relation (1) is valid if and only if $\bar{q}$ is a minimizing point and this relation is therefore a necessary and sufficient condition for minimum.

We must bear in mind that, even if we should be able to determine a necessary and sufficient condition for minimum, such a condition may turn out to be useless. The function $q \to g_0(q) \triangleq q^2 : (0, 1) \to \mathbb{R}$ is convex and has the directional derivative $Dg_0(\bar{q}; q - \bar{q}) = 2\bar{q}(q - \bar{q})$, but relation (1) is not satisfied by any point $\bar{q} \in (0, 1)$. In the absence of an existence theorem [one ensuring either the existence of a minimizing point $\bar{q}$ or the existence of a point $\bar{q}$ satisfying relation (1)], the precious necessary and sufficient condition may first whet our appetite and then leave us hungry.

We next consider an example of necessary conditions for minimum in the presence of restrictions. Let Q be a convex subset of a vector space and $g_i : Q \to \mathbb{R}$ convex functions for $i = 0, 1,..., m$, each with a directional derivative $Dg_i(\bar{q}; q - \bar{q})$ for all $\bar{q}, q \in Q$. If $\bar{q}$ yields the minimum of g_0 on the set $\{q \in Q \mid g_i(q) \leqslant 0\, (i = 1, 2,..., m)\}$, then we can derive a relation analogous to (1). We first observe that

$$g_0(\bar{q}) \leqslant g_0(q) + \alpha_0$$

for all $q \in Q$ and $\alpha_0, \alpha_1,..., \alpha_m \geqslant 0$ such that

$$g_i(q) + \alpha_i = 0 \qquad (i = 1, 2,..., m).$$

Since the functions $g_0,..., g_m$ are convex, it follows that the set

$$G \triangleq \{(g_0(q) + \alpha_0, g_1(q) + \alpha_1,..., q_m(q) + \alpha_m) \mid q \in Q, \alpha_0,..., \alpha_m \geqslant 0\}$$

is a convex subset of $\mathbb{R}^{m+1}$ and, setting $q = \bar{q}$, $\alpha_0 = 0$, and $\alpha_i = -g_i(\bar{q})$ $(i = 1,..., m)$, we find that

$$\bar{x} \triangleq (g_0(\bar{q}), 0,..., 0) \in G.$$

Since $x_0 \geqslant g_0(\bar{q})$ for all $(x_0, 0,..., 0) \in G$, we have $(g_0(\bar{q}) - \epsilon, 0,..., 0) \notin G$ for all $\epsilon > 0$ and therefore $\bar{x} \in \partial G$. It follows, by I.6.11, that there exists $l \triangleq (l_0, l_1,..., l_m) \in (\mathbb{R}^{m+1})^*$ such that $l \neq 0$ and

$$l(\bar{x}) = l_0 g_0(\bar{q}) \leqslant l(x) \qquad (x \in G);$$

hence

$$l_0 g_0(\bar{q}) \leqslant \sum_{i=0}^{m} l_i g_i(q) + \sum_{i=0}^{m} l_i \alpha_i \qquad (q \in Q, \quad \alpha_1,..., \alpha_m \geqslant 0).$$

If we set $q = \bar{q}$, then this relation implies that $l_i \geqslant 0$ $(i = 0, 1,..., m)$ and that $l_i = 0$ whenever $g_i(\bar{q}) < 0$ and $i \in \{1, 2,..., m\}$. If we set $\alpha_0 = 0$ and $\alpha_i = -g_i(\bar{q})$ $(i = 1, 2,..., m)$, then we obtain

$$\sum_{i=0}^{m} l_i g_i(\bar{q}) \leqslant \sum_{i=0}^{m} l_i g_i(q) \qquad (q \in Q).$$

Thus $\bar{q}$ minimizes the function $l \circ g$ [where $g \triangleq (g_0,..., g_m)$] and, by (1),

$$\begin{aligned} &Dl \circ g(\bar{q}; q - \bar{q}) = l(Dg(\bar{q}; q - \bar{q})) \geqslant 0 \qquad (q \in Q), \\ &l \neq 0, \quad l_i \geqslant 0 \quad (i = 0, 1,..., m), \quad l_i g_i(\bar{q}) = 0 \quad (i = 1, 2,..., m). \end{aligned} \tag{2}$$

Relation (2) is the generalization of (1) to the case of restricted minimization problems, and while we have derived (2) under the assumption that the functions g_i are convex, similar necessary conditions for minimum remain valid in much more general situations. Indeed, a large part of our efforts will be directed toward the derivation of analogous conditions for optimal control problems.

The numbers $l_0, l_1,..., l_m$ that define the linear functional l of relation (2) are referred to as *Lagrange coefficients* or *Lagrange multipliers*, and the function $l \circ g$ is often called the *Hamiltonian* of the optimization problem. In more general situations, when g_0 is a real-valued function on Q, $\mathscr{X}_1$ a topological vector space, C_1 a convex subset of $\mathscr{X}_1$, and $g_1 : Q \to \mathscr{X}_1$, we can derive necessary conditions for $\bar{q}$ to minimize g_0 on the set $\{q \in Q \mid g_1(q) \in C_1\}$, and these conditions are of the form

$$\begin{aligned} &l_0 g_0(\bar{q}; q - \bar{q}) + l_1(Dg_1(\bar{q}; q - \bar{q})) \geqslant 0 \qquad (q \in Q), \\ &(l_0, l_1) \neq 0, \quad l_0 \geqslant 0, \quad l_1(c) \leqslant l_1(g_1(\bar{q})) \quad (c \in C_1), \end{aligned} \tag{3}$$

where $l_1 \in \mathscr{X}_1{}^*$. The second line in (3) yields the second line in (2) when $C_1 \triangleq \{(x_1, x_2,..., x_m) \in \mathbb{R}^m \mid x_i \leqslant 0\}$ and $g_1 : Q \to \mathscr{X}_1$ is replaced by

$(g_1, \ldots, g_m) : Q \to \mathbb{R}^m$. We still refer to l_0 and l_1 as "Lagrange multipliers of the optimization problem" even though the term "multiplier" refers to a functional rather than a number.

In the "relaxed" optimal control problems that we consider, the set Q corresponds to $\mathscr{S}^{\#} \times B$, and we derive conditions analogous to (3) (*necessary conditions for a relaxed minimum*). For "original" optimal control problems, the set Q is replaced by $\mathscr{U} \times B$. If we view $\mathscr{U}$ as embedded in $\mathscr{S}^{\#}$, then the set $\mathscr{U}$ of original control functions need not contain any convex subsets. Still, in the case of problems discussed in Chapters VI–VIII, we are able to derive necessary conditions for an original minimum of exactly the same form as those that apply to minimizing relaxed solutions but under slightly more restrictive conditions.

On the other hand, there are reasonably simple optimal control problems (like certain conflicting control problems of Chapters IX and X) with minimizing $\mathscr{U}$-solutions that do not satisfy the necessary conditions applying to minimizing relaxed solutions. We can derive for many such problems alternate necessary conditions (*weak necessary conditions for an original minimum*) that are based on a different concept of convexity. This second type of convexity applies when we deal with optimal control problems (or appropriate equivalent problems) for which the space R is a subset of some vector space (usually $\mathbb{R}^k$) and $\mathscr{R}^{\#}$ is a convex subset of the space of μ-measurable functions on T to R. It follows then that the set $\mathscr{R}^{\#} \times B$ (with the corresponding algebraic structure) is convex. We replace B by $\mathscr{R}^{\#} \times B$, thus treating the original control functions as control parameters, and derive the weak necessary conditions for an original minimum as the corresponding form of necessary conditions for a relaxed minimum of the new problem. We use the term "weak" when referring to these conditions because in those problems of Chapters VI–VIII where both types of conditions are valid, the necessary conditions of the form applying to relaxed problems imply the validity of the weak conditions, but not conversely.

In the classical calculus of variations there are two types of necessary conditions for an original minimum, the *Euler–Lagrange equations* and the *Weierstrass E-condition.* The Euler–Lagrange equations are analogous to the conditions that we call "weak" while the "strong" Weierstrass E-condition is analogous to the condition that applies to minimizing relaxed solutions.

III.5 Minimizing Original Solutions

We have attempted to show in III.2 that a minimizing approximate $\mathscr{U}$-solution yields the most satisfactory answer to the optimal control problem from the point of view of engineering applications. Since a minimizing relaxed solution can serve as a means for constructing a minimizing approximate $\mathscr{U}$-solution, this relaxed solution plays a more important "practical" role than any *strict $\mathscr{U}$-solution* (that is, a minimizing $\mathscr{U}$-solution that is not at the same time a minimizing relaxed solution). This more "practical" aspect of a minimizing relaxed solution is reinforced by the existence theorems of Section 1 in each of Chapters V–X that ensure the existence of such a solution in very general situations whereas many examples show that minimizing $\mathscr{U}$-solutions are often absent in such situations.

Under rather special conditions (that we discuss in Theorems VI.3.3, VII.1.4, and VIII.1.3), it can be shown that to every minimizing relaxed solution $(\bar{y}, \bar{\sigma}, \bar{b})$ there corresponds a point $\bar{u} \in \mathscr{U}$ such that $(\bar{y}, \bar{u}, \bar{b})$ satisfies the equation of motion $y = F(y, \sigma, b)$. If, furthermore, the function $g \triangleq (g_0, g_1) : \mathscr{Y} \times \mathscr{S}^{\#} \times B \to \mathbb{R} \times \mathscr{X}_1$ is independent of the argument in $\mathscr{S}^{\#}$ [i.e., if $g(y, \sigma, b) \triangleq \tilde{g}(y, b)$], then $(\bar{y}, \bar{u}, \bar{b})$ is a minimizing relaxed solution whenever $(\bar{y}, \bar{\sigma}, \bar{b})$ is such a solution. Since $\bar{u}$ belongs to $\mathscr{U}$, it follows that $(\bar{y}, \bar{u}, \bar{b})$ is also a minimizing $\mathscr{U}$-solution. (A theorem of this type was established in 1940 by McShane [4] for the Bolza problem of the calculus of variations). It may also happen with some particular problems not subject to Theorems VI.3.3, VII.1.4, and VIII.1.3 that a minimizing relaxed solution $(\bar{y}, \bar{\sigma}, \bar{b})$ has the property that $\bar{\sigma} \in \mathscr{U}$. (This is the case, in particular, for certain problems investigated by Neustadt [1]). In all such cases a point $(\bar{y}, \bar{u}, \bar{b}) \in \mathscr{Y} \times \mathscr{U} \times B$ is a minimizing $\mathscr{U}$-solution by virtue of its being a minimizing relaxed solution.

It appears therefore that, from the point of view of applications, minimizing relaxed solutions are much more useful than minimizing $\mathscr{U}$-solutions. In very general situations they (a) exist and (b) provide a means for computing minimizing approximate $\mathscr{U}$-solutions (which the engineer ought to be concerned with even if he does not always realize it), and (c) whenever a class of problems is shown to possess minimizing $\mathscr{U}$-solutions, it is because these minimizing $\mathscr{U}$-solutions are also minimizing relaxed solutions.

A partisan of minimizing $\mathscr{U}$-solutions might retort to these arguments by saying that man does not live by bread alone nor does a mathematician abandon all research whose value for scientific or engineering applications is in doubt. A mathematician is often motivated by aesthetic considerations, and if he is able to throw some light on an unknown subject and to discover new laws or regularities, then he feels that such activities ought to be pursued whether or not their usefulness can be demonstrated. A philosophically inclined reader, or one mindful of the history of mathematics, will also adduce examples of originally "pure" mathematical subjects that later acquired a great significance for science or for other branches of mathematics when the time was ripe. In the present case, minimizing approximate $\mathscr{U}$-solutions appear to us to provide the desired answer in the physical situations to which mathematical models of optimal control problems are today applied. It may be, however, that these, or similar, models will also be found applicable to entirely unrelated physical situations, and that minimizing $\mathscr{U}$-solutions will hold the answer in such cases.

Let us, therefore, accept these arguments in favor of a continued investigation of necessary conditions for an original minimum (which today accounts for a very large proportion of all published research in the optimal control theory). Our previous remarks suggest that, if such a study is to be continued independently of the theory of minimizing relaxed solutions, then this study will prove fruitful to the extent that it will shed a light on the properties of *strict $\mathscr{U}$-solutions*, that is, those minimizing $\mathscr{U}$-solutions that are not also minimizing relaxed solutions. What information do we have that refers specifically to strict $\mathscr{U}$-solutions?

We have mentioned so far two items of this nature. Problem II of III.2 demonstrates that strict $\mathscr{U}$-solutions actually exist under certain circumstances. On the other hand, Theorems VI.3.3, VII.1.4, and VIII.1.3 define certain quite restricted, but infinite, classes of problems that admit no strict $\mathscr{U}$-solutions. We shall now consider a third item which appears to us of prime importance. We shall prove in Theorem V.3.4 that a general optimal control problem cannot admit a strict $\mathscr{U}$-solution unless it has certain "exceptional" properties that are referred to in the classical calculus of variations as *abnormal*.

In order to describe the nature of these "abnormal" conditions we must digress and consider once more the form of our necessary conditions for a relaxed minimum. As was mentioned in III.4, these conditions bear a strong resemblance to relation III.4(3), and we shall illustrate them for slightly simplified forms of the optimal control problem. Assume, therefore, that the function $g \triangleq (g_0, g_1) : \mathscr{Y} \times \mathscr{S}^{\#} \times B \to \mathbb{R} \times \mathscr{X}_1$ that defines this problem is independent of the argument in $\mathscr{Y}$ [i.e., $g(y, \sigma, b)$ is independent of y and written $g(\sigma, b)$]. Then the equation of motion is immaterial, and the necessary conditions for a relaxed minimum are precisely those in relation III.4(3), with $\bar{q} \triangleq (\bar{\sigma}, \bar{b})$ and $q \triangleq (\sigma, b) \in \mathscr{S}^{\#} \times B \triangleq Q$.

We refer to a triplet $(\bar{q}, l_0, l_1)$ that satisfies III.4(3) as an *extremal*; the extremal $(\bar{q}, l_0, l_1)$ is *normal* if $l_0 \neq 0$, *abnormal* if $l_0 = 0$, and *admissible* if $g_1(\bar{q}) \in C_1$. The optimal control problem is *normal* if there exist no abnormal admissible extremals, and the problem is *abnormal* otherwise. A problem can be abnormal only if the function g_1 and the set C_1 are related in a special manner. This relation is most transparent when g_1 has certain convexity properties; for example, when $\mathscr{X}_1 \triangleq \mathbb{R}^m$, $C_1 \triangleq (-\infty, 0]^m$, and g_1 is replaced by $(g_1, g_2, \ldots, g_m) : Q \to \mathbb{R}^m$, where each g_i is a convex function. Then relation III.4(3) reduces to III.4(2) and, for the abnormal case ($l_0 = 0$), we have

$$\sum_{i=1}^{m} l_i Dg_i(\bar{q}; q - \bar{q}) \geqslant 0 \qquad (q \in Q), \quad l_i \geqslant 0,$$
$$l_i g_i(\bar{q}) = 0 \qquad (i = 1, 2, \ldots, m). \tag{1}$$

Since $l_i \geqslant 0$, it follows that the function $\sum_{i=1}^{m} l_i g_i$ is also convex and, as shown in III.4, the relation

$$\sum_{i=1}^{m} l_i Dg_i(\bar{q}; q - \bar{q}) = D\left(\sum_{i=1}^{m} l_i g_i\right)(\bar{q}; q - \bar{q}) \geqslant 0 \qquad (q \in Q)$$

implies that $\bar{q}$ minimizes $\sum_{i=1}^{m} l_i g_i$; hence

$$\sum_{i=1}^{m} l_i g_i(q) \geqslant \sum_{i=1}^{m} l_i g_i(\bar{q}) = 0 \qquad (q \in Q).$$

This shows that the linear functional $(l_1, l_2, \ldots, l_m)$ on $\mathbb{R}^m$ separates the sets C_1 ($\triangleq (-\infty, 0]^m$) and $(g_1, g_2, \ldots, g_m)(Q)$. If we were to slightly disturb the restriction $g_i(q) \leqslant 0$ $(i = 1, \ldots, m)$ by requiring instead that $g_i(q) \leqslant -\epsilon$ $(i = 1, 2, \ldots, m)$ for some positive ϵ, then this new restriction cannot be satisfied and our restricted minimization problem reduces to a search for a minimum of g_0 on the empty set $\{q \in Q \mid g_i(q) \leqslant -\epsilon \ (i = 1, \ldots, m)\}$.

The extreme situation that we have just described does not occur for every abnormal problem, and the relation between g_1 and C_1 in many abnormal problems may be much subtler and less threatening to the "stability" of the optimization problem. Still, the example we have just given appears to justify the common attitude that abnormal problems are rather special and exceptional.

We can now return to the topic from which we have digressed, and consider once more the third item that sheds a light on strict $\mathscr{U}$-solutions. Theorem V.3.4 asserts that, under very general conditions (of the same type as are required to derive "strong" necessary conditions for an original minimum), only abnormal optimal control problems possessing very special properties can have strict $\mathscr{U}$-solutions. It appears, therefore, that the only refuge of an adherent of an independent theory of minimizing $\mathscr{U}$-solutions is to be found in the class of abnormal problems, and within such a limited framework the prospects for a meaningful theory are not very promising.

A more fruitful approach to a theory of minimizing $\mathscr{U}$-solutions might be patterned after McShane's [4] study of the Bolza problem in the related field of the calculus of variations. This approach would be to investigate conditions under which a minimizing relaxed solution $(\bar{y}, \bar{\sigma}, \bar{b})$ is "equivalent" to a minimizing relaxed solution $(\bar{y}, \bar{u}, \bar{b})$, with $\bar{u} \in \mathscr{U}$. As mentioned before, this is the path that leads to Theorems VI.3.3, VII.1.4, and VIII.1.3 (the first of which represents results derived, independently, by Filippov [1] in 1959, Roxin [1], Warga [1], and Ważewski [1, 2] in 1962, and Ghouila–Houri [1] in 1965).

Lest our use of the term "optimal control theory" lead to a misunderstanding, let us emphasize that the above remarks are primarily meant to apply to the type of problems investigated in this book. There are some indications, however, that these remarks may also be applicable to some other areas of optimal control, in particular to many problems defined by partial differential equations that can be transformed into the form discussed in Chapter VII.

CHAPTER IV

Original and Relaxed Control Functions

IV.0 Summary

Let T and R be compact metric spaces, $\mu \in \text{frm}^+(T)$, μ nonatomic, $\Sigma \triangleq \Sigma_{\text{Borel}}(T)$, and $\mathscr{R}$ the collection of μ-measurable functions on T to R. We embed $\mathscr{R}$ in the topological dual of $L^1(T, C(R))$ by identifying each $\rho \in \mathscr{R}$ with the continuous linear functional l_ρ on $L^1(T, C(R))$ defined by

$$l_\rho(\phi) \triangleq \int \phi(t, \rho(t))\, \mu(dt) \qquad [\phi \in L^1(T, C(R))].$$

A version of the Dunford–Pettis theorem (IV.1.8) states that $L^1(T, C(R))^*$ is isomorphic to the set $\mathscr{N}$ of μ-measurable functions $\nu : T \to (\text{frm}(R), |\cdot|_w)$ such that μ-ess sup $|\nu(t)|(R) < \infty$. We topologize $C(R)^*$ [$=\text{frm}(R)$] and $L^1(T, C(R))^*$ by weak norms (I.3.11) and show that the closure of $\mathscr{R}$ in $\mathscr{N}$ coincides with the set

$$\mathscr{S} \triangleq \{\nu \in \mathscr{N} \mid \nu(t) \in \text{rpm}(R)\ \mu\text{-a.e.}\}.$$

The "original control functions" that we consider are among the elements ρ of $\mathscr{R}$ such that

$$\rho(t) \in R^{\#}(t) \qquad \mu\text{-a.e.},$$

where $R^{\#}$ is a given mapping on T to the class $\mathscr{P}'(R)$ of nonempty subsets of R. If $R^{\#}$ satisfies Condition IV.3.1 below, then the set $\mathscr{R}^{\#}$ of μ-measurable selections of $R^{\#}$ has as its closure in $(\mathscr{N}, |\cdot|_w)$ the set

$$\mathscr{S}^{\#} \triangleq \{\sigma \in \mathscr{S} \mid \sigma\left(\overline{R^{\#}(t)}\right) = 1 \ \mu\text{-a.e.}\}$$

of "relaxed control functions," and we study certain properties of $\mathscr{S}^{\#}$. We also indicate certain criteria, based on the results of I.7, for verifying Condition IV.3.1.

The "abundant subsets of $\mathscr{R}^{\#}$" (Definition IV.3.3) that we shall consider as acceptable sets $\mathscr{U}$ of original control functions are characterized by two rather "technical" properties. We establish certain criteria (such as Condition IV.3.4, Lemmas IV.3.5 and IV.3.6, and Remark following Theorem IV.3.9) that enable us to classify certain simply defined sets, in particular $\mathscr{R}^{\#}$, among the abundant sets.

We also establish certain technical results in preparation for future use.

IV.1 The Spaces $C(R)$ and $L^1(T, C(R))$ and Their Conjugate Spaces

IV.1.1 ***Strong and Weak Norms on Dual Spaces*** We shall be dealing from now on with certain separable normed vector spaces and with their topological duals. If $\mathscr{X}$ is such a space and $\mathscr{X}^*$ its dual, we consider two different norms on $\mathscr{X}^*$; the *strong norm* (usually referred to as the *norm*), defined by

$$|l| \triangleq \sup\{|l(x)| \mid x \in \mathscr{X}, |x| \leqslant 1\},$$

and a *weak norm*, introduced in I.3.11. To define a weak norm on $\mathscr{X}^*$ we choose a denumerable dense subset $\{x_1, x_2, ...\}$ of $\mathscr{X}$ and let

$$|l|_w \triangleq \sum_{j=1}^{\infty} 2^{-j} |lx_j|/(1 + |x_j|).$$

If we set $U(\mathscr{X}^*) \triangleq \{l \in \mathscr{X}^* \mid |l| \leqslant 1\}$, then Theorem I.3.11 shows that $|\cdot|_w$ is a norm on $\mathscr{X}^*$, that $\lim_j |l - l_j|_w = 0$ for $l, l_j \in U(\mathscr{X}^*)$ if and only if $\lim_j l_j x = lx$ for all $x \in \mathscr{X}$, and that the relative topology of $U(\mathscr{X}^*)$ in $(\mathscr{X}^*, |\cdot|_w)$ [the weak norm topology of $U(\mathscr{X}^*)$] is the same for all choices of the set $\{x_1, x_2, ...\}$ and coincides with the relative weak star topology.

IV.1.2 ***The Spaces $T, R, (T, \Sigma, \mu)$, frm(R), and $\mathscr{B}$*** We next define certain spaces that we shall constantly refer to in the chapters to follow and whose properties, stated below, will be assumed to hold without further mention.

We consider compact metric spaces T and R, a nonzero, nonatomic measure $\mu \in \text{frm}^+(T)$, and $\Sigma \triangleq \Sigma_{\text{Borel}}(T)$. We define the space $\mathscr{B}$ as the space $\mathscr{B}(T, \Sigma, \mu, R; \mathbb{R})$ of I.5.25 whose elements are functions $\phi : T \times R \to \mathbb{R}$ such that $\phi(\cdot, r)$ is μ-measurable, $\phi(t, \cdot) \in C(R)$, and $|\phi(t, \cdot)|_{\sup} \leqslant \psi_\phi(t)$, where ψ_ϕ is μ-integrable. As shown in I.5.25,

$$\phi \to |\phi|_{\mathscr{B}} \triangleq \int |\phi(\tau, \cdot)|_{\sup} \, \mu(d\tau) : \mathscr{B} \to \mathbb{R}$$

is a norm on $\mathscr{B}$, $(\mathscr{B}, |\cdot|_{\mathscr{B}})$ is isometrically isomorphic to the space $L^1(T, \Sigma, \mu, C(R))$ (and therefore, by I.5.1 and I.5.17, it is a Banach space), $(\mathscr{B}, |\cdot|_{\mathscr{B}})$ is separable and its subset

$$\mathscr{B}' \triangleq C(T) \otimes C(R) \triangleq \left\{(t, r) \to \sum_{i=1}^{k} f_i(t)\, c_i(r) \mid k \in \mathbb{N}, f_i \in C(T), c_i \in C(R)\right\}$$

is dense in $\mathscr{B}$. We henceforth identify $L^1(T, C(R))$ and $(\mathscr{B}, |\cdot|_{\mathscr{B}})$.

IV.1.3 ***The Isomorphic Spaces $C(R)^*$ and frm(R)*** By the Riesz representation theorem I.5.8, there exists an algebraic isomorphism $\mathscr{I} \triangleq \mathscr{I}(\text{frm}(R), C(R)^*)$ of frm(R) onto $C(R)^*$, defined by the expression

$$\mathscr{I}(s)(c) \triangleq \int c(r)\, s(dr),$$

and with

$$|\mathscr{I}(s)| = |s|(R) \qquad [s \in \text{frm}(R), \quad c \in C(R)].$$

We shall henceforth identify frm(R) and $C(R)^*$ and, unless otherwise specified, consider each as a normed vector space with a weak norm defined by some dense denumerable subset of $C(R)$ (which, by I.5.1, is separable). Our remarks in IV.1.1 show that the weak norm topology of

$$U(\text{frm}(R)) \triangleq \{s \in \text{frm}(R) \mid |s|(R) \leqslant 1\}$$

is the same for every weak norm.

IV.1.4 ***Theorem*** The normed vector space $(\text{frm}(R), |\cdot|_w)$ is separable and its subsets $U(\text{frm}(R))$ and rpm(R) are compact.

▌PROOF By I.3.12, the set $U(\text{frm}(R))$ is compact; hence, by I.2.17, it is separable. It is now easy to verify that if $\{\alpha_1, \alpha_2, \ldots\}$ is the set of all the rational numbers and $\{s_1, s_2, \ldots\}$ is a dense subset of $U(\mathrm{frm}(R))$, then the denumerable set $\{\alpha_i s_j \mid i, j \in \mathbb{N}\}$ is dense in $\mathrm{frm}(R)$.

Now let (s_j) be a sequence in $\mathrm{rpm}(R)$ converging to some $\bar{s} \in \mathrm{frm}(R)$. Then

$$\int c(r)\, \bar{s}(dr) = \lim_j \int c(r)\, s_j(dr) \geqslant 0 \qquad \text{if} \quad c \in C(R, [0, \infty))$$

and

$$\bar{s}(R) = \lim_j \int_R s_j(dr) = 1.$$

Thus, by I.5.5(1), we have $\bar{s} \in \mathrm{rpm}(R)$. QED

IV.1.5 ***The Embedding of R in* frm(*R*) *and the Extension of a Continuous Function from R to* frm(*R*)** We next observe that R can be embedded in $\mathrm{frm}(R)$ by identifying each $\bar{r} \in R$ with the Dirac measure $\delta_{\bar{r}}$ at $\bar{r}$ [or, equivalently, identifying $\bar{r}$ with the element $l_{\bar{r}}$ of $C(R)^*$ such that $l_{\bar{r}}(c) = c(\bar{r})$ for all $c \in C(R)$]. We can also uniquely extend any real-valued continuous function c on R to $\mathrm{frm}(R)$ by setting

$$c(s) \triangleq \int c(r)\, s(dr) \qquad [s \in \mathrm{frm}(R)]$$

which yields, in particular,

$$c(\delta_{\bar{r}}) = c(\bar{r}) \qquad (\bar{r} \in R).$$

Since $\lim_j c(s_j) = c(\bar{s})$ if $s_j, \bar{s} \in U(\mathrm{frm}(R))$ and $\lim_j |s_j - \bar{s}|_w = 0$, and since $(\mathrm{frm}(R), |\cdot|_w)$ is metric, we conclude that c is continuous on $(\mathrm{frm}(R), |\cdot|_w)$.

We observe that if $\mathscr{I}$ is the isomorphism of $\mathrm{frm}(R)$ onto $C(R)^*$ referred to in IV.1.3, then $\mathscr{I}(s)(c) = c(s)$ for all $s \in \mathrm{frm}(R)$ and $c \in C(R)$ and

$$|s|_w = |\mathscr{I}(s)|_w = \sum_{j=1}^{\infty} 2^{-j}\, |c_j(s)| / (1 + |c_j|_{\sup}),$$

where $\{c_1, c_2, \ldots\}$ is the subset of $C(R)$ that defines $|\cdot|_w$.

IV.1.6 ***Theorem*** Let $\nu : T \to (\mathrm{frm}(R), |\cdot|_w)$ be such that

$$\mu\text{-ess sup } |\nu(t)|\,(R) < \infty.$$

Then ν is μ-measurable if and only if the function $t \to c(\nu(t)) \triangleq \int c(r)\, \nu(t)\,(dr)$ is μ-measurable for every $c \in C(R)$. Furthermore, if ν is μ-measurable, then $t \to \phi(t, \nu(t))$ is μ-integrable for every $\phi \in L^1(T, C(R))$.

▌PROOF Let $m_\nu \triangleq \mu\text{-ess sup } |\nu(t)|\,(R) \neq 0$ and $\tilde{\nu}(t) \triangleq m_\nu^{-1} \nu(t)$. Then $\tilde{\nu}(t) \in U(\mathrm{frm}(R))$ μ-a.e. and $\tilde{\nu}$ is μ-measurable if and only if ν is μ-measurable. We may assume, therefore, that $\nu(t) \in U(\mathrm{frm}(R))$ μ-a.e.

If ν is μ-measurable, then $t \to c(\nu(t))$ is μ-measurable for each $c \in C(R)$ since the extension of c to $U(\text{frm}(R))$ is continuous. Now assume that $t \to c(\nu(t))$ is μ-measurable for each $c \in C(R)$, and let $\{c_1, c_2, ...\}$ be the dense subset of $C(R)$ that defines $|\cdot|_w$, $s_1 \in \text{frm}(R)$, and $\epsilon > 0$. Then the function $t \to c_j(\nu(t))$ is μ-measurable for each $j \in \mathbb{N}$; hence, by I.4.17,

$$t \to |\nu(t) - s_1|_w = \sum_{j=1}^{\infty} 2^{-j} |c_j(\nu(t)) - c_j(s_1)|/(1 + |c_j|)$$

is also μ-measurable. Thus the set $\{t \in T \mid |\nu(t) - s_1|_w < \epsilon\}$ is μ-measurable and since, by I.3.12 and I.2.17, $U(C(R)^*)$ [identified with $U(\text{frm}(R))$] is separable (in the $|\cdot|_w$-topology) and the sets

$$\{s \in U(\text{frm}(R)) \mid |s - s_1|_w < \epsilon\} \qquad [s_1 \in \text{frm}(R), \quad \epsilon > 0]$$

form a base of its relative weak norm topology, we conclude that ν is μ-measurable.

If $\phi \in L^1(T, C(R))$, then, as observed in IV.1.5,

$$s \to \phi(t, s) : (\text{frm}(R), |\cdot|_w) \to \mathbb{R}$$

is continuous for each $t \in T$ and, by I.5.27, $t \to \phi(t, s) : T \to \mathbb{R}$ is μ-measurable for each $s \in \text{frm}(R)$. It follows, by I.4.22, that $t \to \phi(t, \nu(t)) : T \to \mathbb{R}$ is μ-measurable and, since

$$|\phi(t, \nu(t))| \leqslant \mu\text{-ess sup} |\nu(t)| (R) \cdot |\phi(t, \cdot)|_{\sup} \qquad (t \in T),$$

it follows that $t \to \phi(t, \nu(t))$ is μ-integrable. QED

IV.1.7 ***The Space $\mathscr{N}$ and Its Subsets $\mathscr{S}$ and $\mathscr{R}$*** We denote by $\mathscr{N}$ the set of all (equivalence classes of) μ-measurable functions $\nu : T \to (\text{frm}(R), |\cdot|_w)$ with the property that $c_\nu \triangleq \mu\text{-ess sup} |\nu(t)| (R) < \infty$. As usual, we consider ν_1 and ν_2 as belonging to the same equivalence class, and identify them, whenever $\nu_1(t) = \nu_2(t)$ μ-a.e.

We set

$$\mathscr{S} \triangleq \{\nu \in \mathscr{N} \mid \nu(t) \in \text{rpm}(R) \ \mu\text{-a.e.}\},$$

denote by δ_r the Dirac measure at r, and set

$$\mathscr{S}_{\mathscr{R}} \triangleq \{\nu \in \mathscr{S} \mid \nu(t) = \delta_{\rho(t)} \ \mu\text{-a.e. for some } \rho : T \to R\}.$$

It follows from IV.1.6 that $\nu \in \mathscr{N}$ if $\mu\text{-ess sup} |\nu(t)| (R) < \infty$ and the function $t \to c(\nu(t))$ is μ-measurable for each $c \in C(R)$. Thus, if $\nu \in \mathscr{S}_{\mathscr{R}}$ and $\nu(t) = \delta_{\rho(t)}$ μ-a.e., then $t \to c(\nu(t)) = c(\rho(t))$ is μ-measurable for each $c \in C(R)$ and, by I.4.20, the function $t \to \rho(t)$ is μ-measurable. Conversely, if $\rho : T \to R$ is μ-measurable, then $t \to c(\rho(t)) = c(\delta_{\rho(t)})$ is μ-measurable and μ-ess sup $|\delta_{\rho(t)}|(R) = 1$; hence, by IV.1.6, the function $t \to \delta_{\rho(t)}$ belongs to $\mathscr{S}_{\mathscr{R}}$. Thus there exists a one-to-one correspondence between $\mathscr{S}_{\mathscr{R}}$ and the set $\mathscr{R}$ of μ-measurable functions $\rho : T \to R$. We identify each element ρ of $\mathscr{R}$ with the function $t \to \delta_{\rho(t)}$ in $\mathscr{S}_{\mathscr{R}}$ and thus embed $\mathscr{R}$ as a subset of $\mathscr{S}$.

We next consider the space $\mathscr{B}^*$ [that is, $L^1(T, C(R))^*$] and show that it is algebraically isomorphic to $\mathscr{N}$.

IV.1.8 ***Theorem*** (Dunford–Pettis) There exists an algebraic isomorphism $\mathscr{I}$ of $\mathscr{N}$ onto $\mathscr{B}^*$ [$=L^1(T, C(R))^*$], defined by

$$\mathscr{I}(\nu)(\phi) \triangleq \int \phi(t, \nu(t))\, \mu(dt)$$

$$\triangleq \int \mu(dt) \int \phi(t, r)\, \nu(t)\,(dr) \qquad (\nu \in \mathscr{N}, \quad \phi \in \mathscr{B}),$$

and we have

$$|\mathscr{I}(\nu)| \triangleq \sup_{|\phi|_{\mathscr{B}} \leqslant 1} |\mathscr{I}(\nu)(\phi)| = \mu\text{-ess sup}\, |\nu(t)|\,(R).$$

▌PROOF We first show that $\mathscr{I}(\nu) \in \mathscr{B}^*$ for each $\nu \in \mathscr{N}$. Indeed, let $c_\nu \triangleq \mu$-ess sup $|\nu(t)|(R)$ and $\phi \in \mathscr{B}$. Then, by IV.1.6, $t \to \phi(t, \nu(t))$ is μ-integrable and we have

$$|\phi(t, \nu(t))| = \left| \int \phi(t, r)\, \nu(t)\,(dr) \right| \leqslant |\phi(t, \cdot)|_{\sup}\, |\nu(t)|\,(R)$$

$$\leqslant c_\nu\, |\phi(t, \cdot)|_{\sup} \quad \mu\text{-a.e. in } T;$$

hence,

$$|\mathscr{I}(\nu)(\phi)| = \left| \int \phi(t, \nu(t))\, \mu(dt) \right| \leqslant c_\nu\, |\phi|_{\mathscr{B}} \qquad (\phi \in \mathscr{B}).$$

Thus $\mathscr{I}(\nu)$ exists for all $\nu \in \mathscr{N}$, $\mathscr{I}(\nu)$ is clearly linear, and $\mathscr{I}(\nu)$ is bounded; hence, by I.3.6, $\mathscr{I}(\nu) \in \mathscr{B}^*$.

If $\mathscr{I}(\nu) = 0$, then we choose a dense subset $\{c_1, c_2, \ldots\}$ of the separable space $C(R)$ and consider the elements

$$(t, r) \to \chi_E(t)\, c_j(r) \qquad (j \in \mathbb{N}, \quad E \in \Sigma)$$

of $\mathscr{B}$. For each $j \in \mathbb{N}$, we have

$$\int \chi_E(t)\, c_j(\nu(t))\, \mu(dt) = \int_E c_j(\nu(t))\, \mu(dt) = 0 \qquad (E \in \Sigma);$$

hence, by I.4.34(7),

$$c_j(\nu(t)) \triangleq \int c_j(r)\, \nu(t)\,(dr) = 0 \quad \mu\text{-a.e.},$$

say for $t \in T_j$. For each

$$t \in T' \triangleq \bigcap_{j=1}^{\infty} T_j \qquad \text{and} \qquad c \in C(R),$$

we have $\lim_{j\in J} | c_j - c | = 0$ for some $J \subset (1, 2, ...)$ and therefore $c(\nu(t)) = \lim_{j\in J} \int c_j(r)\, \nu(t)\,(dr) = 0$. It follows, by I.5.5(1), that $\nu(t) = 0$ for $t \in T'$, that is, for μ-a.a. $t \in T$. We conclude that $\mathscr{I}$ is injective.

Now we proceed to show that $\mathscr{I}$ is surjective so that, for each $l \in \mathscr{B}^*$, there exists $\nu \in \mathscr{N}$ with $l = \mathscr{I}(\nu)$. We denote by C' a dense denumerable subset of $C(R)$, and we may assume C' to be closed under addition and multiplication by rationals. (Specifically, if C'' is a dense denumerable subset of $C(R)$ and Q the set of all rationals, we let

$$C' \triangleq \left\{ \sum_{i=1}^{s} \alpha_i c_i \mid s \in \mathbb{N},\ \alpha_i \in Q,\ c_i \in C'' \right\}$$

and verify that C' is denumerable and that $c' + c''$, $\alpha c' \in C'$ whenever $c', c'' \in C'$ and $\alpha \in Q$.) We choose an arbitrary $l \in \mathscr{B}^*$ and set $| l | \triangleq \sup_{|\phi|_{\mathscr{B}} \leqslant 1} | l\phi |$.

For $f \in L^1(T, \Sigma, \mu) \triangleq L^1(T)$ and $c \in C(R)$, let $f \otimes c$ represent the function $(t, r) \to f(t)\, c(r) : T \times R \to \mathbb{R}$. For every fixed $c \in C'$, the function $f \to l(f \otimes c) : L^1(T) \to \mathbb{R}$ is an element of $L^1(T)^*$; indeed, this function is linear and $| l(f \otimes c)| \leqslant | l | | f \otimes c |_{\mathscr{B}} = | l | | c | | f |_1$, implying that this function is continuous. It follows, by I.5.19, that there exists $k(\cdot; c) \in L^\infty(T, \Sigma, \mu)$ such that

$$l(f \otimes c) = \int k(t; c) f(t)\, \mu(dt) \qquad [f \in L^1(T),\ c \in C'] \tag{1}$$

and

$$\mu\text{-ess}\sup_{t\in T} | k(t; c)| = \sup_{|f|_1 \leqslant 1} | l(f \otimes c)| \leqslant | l | | c | \qquad (c \in C'). \tag{2}$$

Thus $| k(t; c)| \leqslant | l | | c |$ for each $c \in C'$ and all t in some set T_c of μ-measure $\mu(T)$; hence, setting $T' \triangleq \bigcap_{c\in C'} T_c$, we have $\mu(T \sim T') = 0$ and

$$| k(t; c)| \leqslant | l | | c | \qquad (t \in T',\quad c \in C'). \tag{3}$$

We now observe that if $c_1, c_2 \in C'$, α_1 and α_2 are rational and $E \in \Sigma$, then

$$l(\chi_E \otimes (\alpha_1 c_1 + \alpha_2 c_2)) = \alpha_1 l(\chi_E \otimes c_1) + \alpha_2 l(\chi_E \otimes c_2);$$

hence, by (1),

$$\int_E [k(t; \alpha_1 c_1 + \alpha_2 c_2) - \alpha_1 k(t; c_1) - \alpha_2 k(t; c_2)]\, \mu(dt) = 0$$

and, by I.4.34(7),

$$k(t; \alpha_1 c_1 + \alpha_2 c_2) = \alpha_1 k(t; c_1) + \alpha_2 k(t; c_2) \quad \mu\text{-a.e.}$$

It follows that for any $m \in \mathbb{N}$, any rationals $\alpha_1, \ldots, \alpha_m$, and any $c_1, \ldots, c_m \in C'$, we have

$$k\left(t; \sum_{i=1}^{m} \alpha_i c_i\right) = \sum_{i=1}^{m} \alpha_i k(t; c_i) \quad \mu\text{-a.e.,} \tag{4}$$

say for $t \in T_\beta$, where $\beta \triangleq (m, \alpha_1, \ldots, \alpha_m, c_1, \ldots, c_m)$. Since the set of β's is denumerable, the set $T'' \triangleq \bigcap_\beta T_\beta \cap T'$ has μ-measure $\mu(T)$.

Thus $k(t; \cdot)$ satisfies relation (4) for each $t \in T''$ and, by (3), it is uniformly continuous. Therefore, by I.2.13, it has a uniformly continuous extension $k(t; \cdot)$ to $C(R)$ satisfying

$$|\, k(t; c)| \leqslant |\, l\,|\,|\, c\,| \qquad [t \in T'', \quad c \in C(R)]. \tag{5}$$

In view of (4) and the continuity of $k(t; \cdot)$, we conclude that $k(t; \cdot) \in C(R)^*$ for each $t \in T''$. Since each $c \in C(R)$ is a limit of a sequence $(c_1', c_2', \ldots)$ in C', we have

$$k(t; c) = \lim_j k(t; c_j') \qquad (t \in T'').$$

We now set $k(t; c) \triangleq 0$ $[t \in T \sim T'', c \in C(R)]$ and infer from I.4.17 that, for each $c \in C(R)$, $k(\cdot; c)$ is μ-measurable.

We easily verify that the function

$$c \to l(f \otimes c) : C(R) \to \mathbb{R}$$

is continuous for each $f \in L^1(T)$. Since $k(t; \cdot)$ is continuous for each $t \in T$, it follows from (1), (3), and the dominated convergence theorem I.4.35 that

$$l(f \otimes c) = \int k(t; c) f(t)\, \mu(dt) \qquad [f \in L^1(T), \quad c \in C(R)]. \tag{6}$$

We can also apply the Riesz representation theorem I.5.8 to each $k(t; \cdot) \in C(R)^*$ to show that there exists $\nu(t) \in \text{frm}(R)$ such that

$$k(t; c) = \int c(r)\, \nu(t)\, (dr) \triangleq c(\nu(t)) \qquad [c \in C(R), \quad t \in T] \tag{7}$$

and

(8) $$\sup_{|c|\leqslant 1} | k(t; c)| = | \nu(t)| (R) \qquad (t \in T).$$

We now combine relations (6) and (7) into

(9) $$l(f \otimes c) = \int f(t)\, c(\nu(t))\, \mu(dt) \qquad [f \in L^1(T), \quad c \in C(R)],$$

and relations (5) and (8) into

(10) $$\mu\text{-ess sup} \, | \nu(t)|(R) \leqslant | l |.$$

Since $k(\cdot; c)$ is μ-measurable for each $c \in C(R)$, it follows from (7), (10), and IV.1.6 that $\nu \in \mathscr{N}$.

By I.5.25, for each $\phi \in \mathscr{B}$ and $\epsilon > 0$ there exist $m \in \mathbb{N}$, $f_i \in C(T)$, and $c_i \in C(R)$ $(i = 1,..., m)$ such that

$$\int \Big| \phi(t, \cdot) - \sum_{i=1}^{m} f_i(t)\, c_i(\cdot) \Big|_{\sup} \mu(dt) < \epsilon.$$

It follows, by (9) and (10), that

$$\begin{aligned} &\Big| l(\phi) - \int \phi(t, \nu(t))\, \mu(dt) \Big| \\ &\quad \leqslant \Big| l \Big(\phi - \sum_{i=1}^{m} f_i \otimes c_i\Big) \Big| + \Big| \int \Big(\sum_{i=1}^{m} f_i(t)\, c_i(\nu(t)) - \phi(t, \nu(t))\Big)\, \mu(dt) \Big| \\ &\quad \leqslant | l | \,\epsilon + \int \Big| \sum_{i=1}^{m} f_i(t)\, c_i(\cdot) - \phi(t, \cdot) \Big|_{\sup} | \nu(t)| \,(R)\, \mu(dt) \\ &\quad \leqslant 2\, | l |\, \epsilon. \end{aligned}$$

Since ϵ is arbitrary, we have

$$l(\phi) = \int \phi(t, \nu(t))\, \mu(dt),$$

showing that $\mathscr{I}$ is surjective. Thus $\mathscr{I}$ is bijective and, since it is clearly linear, $\mathscr{I}$ is an algebraic isomorphism.

Finally, we observe that, whenever $| \phi |_{\mathscr{B}} \leqslant 1$, we have

$$| l(\phi)| = | \mathscr{I}(\nu)(\phi)| = \Big| \int \mu(dt) \int \phi(t, r)\, \nu(t)\, (dr) \Big| \leqslant \mu\text{-ess sup} \, | \nu(t)| \,(R);$$

hence $| l | \leqslant \mu\text{-ess sup} \, | \nu(t)|(R)$. In view of (10), this implies that $| l | = \mu\text{-ess sup} \, | \nu(t)|(R)$. QED

IV.1.9 ***The Isomorphic Spaces $\mathscr{B}^*$ and $\mathscr{N}$ and Their Subsets $\mathscr{S}$ and $\mathscr{R}$ with the Weak Norm Topology*** As we have done for $C(R)^*$ and frm(R), we now identify the isomorphic spaces $\mathscr{B}^*$, $L^1(T, C(R))^*$, and $\mathscr{N}$ and, when no confusion arises, represent by $\mathscr{I}$ the isomorphism defined in IV.1.8. Thus we identify any $\nu \in \mathscr{N}$ with the continuous linear functional $\mathscr{I}(\nu)$ on $\mathscr{B}$. We write $\langle \nu, \phi \rangle$ for $\mathscr{I}(\nu)(\phi)$. We also introduce, once and for all, the weak norm topology for $\mathscr{B}^*$ and $\mathscr{N}$. Since, by I.5.25, $\mathscr{B}$ is separable, we choose some denumerable dense subset $\{\phi_1, \phi_2, ...\}$ in $\mathscr{B}$ and set

$$
\begin{aligned}
| \mathscr{I}(\nu)|_w = | \nu |_w &\triangleq \sum_{j=1}^{\infty} 2^{-j} \, | \langle \nu, \phi_j \rangle | /(1 + \phi_j |_{\mathscr{B}} \\
&= \sum_{j=1}^{\infty} 2^{-j} \left| \int \phi_j(t, \nu(t)) \, \mu(dt) \right| /(1 + | \phi_j |_{\mathscr{B}}).
\end{aligned}
$$

The corresponding weak norm topology in $\mathscr{B}^*$ and $\mathscr{N}$ is independent of the choice of $\{\phi_1, \phi_2, ...\}$ and defines henceforth $\mathscr{B}^*$ and $\mathscr{N}$ as metrizable topological spaces. Similarly, the subsets $\mathscr{S}$ and $\mathscr{R}$ are assumed to have the corresponding relative topologies.

By IV.1.8, the strong norm of an element ν of $\mathscr{N}$ is $| \nu | = \mu\text{-ess sup} \, | \nu(t)|(R)$. We write $U(\mathscr{N}) \triangleq \{\nu \in \mathscr{N} \mid | \nu | \leqslant 1\}$ and state a theorem analogous to IV.1.4.

IV.1.10 ***Theorem*** The space $(\mathscr{N}, | \cdot |_w)$ is separable and its subset $U(\mathscr{N})$ is compact.

▌PROOF The proof of IV.1.4 is directly applicable, with frm(R) replaced by $\mathscr{N}$. QED

IV.1.11 ***Theorem*** Let $\nu, \nu_j \in U(\mathscr{N})$ $(j \in \mathbb{N})$. Then $\lim_j \nu_j = \nu$ if and only if $\lim_j \langle \nu_j, \phi \rangle = \langle \nu, \phi \rangle$ for all $\phi \in \mathscr{B}$.

▌PROOF Follows directly from I.3.11 and IV.1.8. QED

IV.2 The Sets $\mathscr{R}$ and $\mathscr{S}$

We discuss certain properties of the subsets $\mathscr{R}$ and $\mathscr{S}$ of $\mathscr{N}$ that will play a central role in our future considerations. We continue to represent by $U(\mathscr{N})$ the set $\{\nu \in \mathscr{N} \mid | \nu | = \mu\text{-ess sup} \, | \nu(t)|(R) \leqslant 1\}$.

IV.2.1 ***Theorem*** The set $\mathscr{S}$ is convex, compact, and sequentially compact.

▌PROOF If $\sigma_1, \sigma_2 \in \mathscr{S}$ and $0 \leqslant \alpha \leqslant 1$, then

$$(\alpha\sigma_1 + (1 - \alpha)\,\sigma_2)(t) \in \operatorname{frm}^+(R) \quad \mu\text{-a.e.}$$

and

$$(\alpha\sigma_1 + (1 - \alpha)\,\sigma_2)(t)(R) = 1 \quad \mu\text{-a.e.}$$

Thus $\alpha\sigma_1 + (1 - \alpha)\sigma_2 \in \mathscr{S}$, showing that $\mathscr{S}$ is convex.

If $\sigma \in \mathscr{S}$, then

$$|\langle\sigma, \phi\rangle| \triangleq \left|\int \phi(t, \sigma(t))\,\mu(dt)\right| \leqslant |\phi|_{\mathscr{B}} \qquad (\phi \in \mathscr{B});$$

hence $|\sigma| \leqslant 1$ and $\mathscr{S} \subset U(\mathscr{N})$. It follows, by I.3.12, that $\mathscr{S}$ is conditionally compact and $\bar{\mathscr{S}}$ is sequentially compact. It remains to show that $\mathscr{S}$ is closed.

Let $\sigma_j \in \mathscr{S}$, $\lim_j \sigma_j = \nu$, and $E \in \Sigma$. By I.5.1 and I.2.17, $C(R)$ contains an at most denumerable dense subset $\{c_1, c_2, ...\}$ of the collection of all nonnegative continuous functions on R. Since $\mathscr{S} \subset U(\mathscr{N})$, it follows from IV.1.10 that $\nu \in U(\mathscr{N})$. Furthermore, the function $(t, r) \to \chi_E(t)\,c_i(r)$ belongs to $\mathscr{B}$ for each $i \in \mathbb{N}$. Thus, by IV.1.11, for each $i \in \mathbb{N}$ we have

$$0 \leqslant \lim_j \int \chi_E(t)\,c_i(\sigma_j(t))\,\mu(dt) = \int \chi_E(t)\,c_i(\nu(t))\,\mu(dt). \tag{1}$$

Since this relation holds for all $E \in \Sigma$, it follows, by I.4.34(7), that $c_i(\nu(t)) \triangleq \int c_i(r)\,\nu(t)\,(dr) \geqslant 0$ μ-a.e., say for $t \in T_i$. Since every nonnegative $c \in C(R)$ is a uniform limit of some sequence $(c_i)_{i \in J}$, we conclude that $\int c(r)\,\nu(t)\,(dr) \geqslant 0$ for all $t \in \bigcap_{i=1}^{\infty} T_i$ and every nonnegative $c \in C(R)$; hence by I.5.5(1), $\nu(t)$ is a positive measure μ-a.e. For $c(r) = 1$ $(r \in R)$ and $E \in \Sigma$, we have, by IV.1.11,

$$\int \chi_E(t)\,c(\sigma_j(t))\,\mu(dt) = \int \chi_E(t)\,\sigma_j(t)(R)\,\mu(dt)$$

$$\underset{j}{\to} \int \chi_E(t)\,\nu(t)(R)\,\mu(dt);$$

hence

$$\mu(E) = \int_E \nu(t)(R)\,\mu(dt) \qquad \text{and} \qquad \int_E [\nu(t)(R) - 1]\,\mu(dt) = 0.$$

Thus, by I.4.34(7), $\nu(t)(R) = 1$ μ-a.e. Since $\nu(t)$ is a positive measure and $\nu(t)(R) = 1$ μ-a.e., we have $\nu \in \mathscr{S}$, showing that $\mathscr{S}$ is a closed subset of the compact set $U(\mathscr{N})$. Thus $\mathscr{S}$ is compact and, by I.2.5, sequentially compact.

QED

IV.2.2 ***Lemma*** Let $\{\nu_1, ..., \nu_k\}$ be an independent subset of $\mathcal{N}$ and $A \triangleq \text{sp}(\{\nu_1, ..., \nu_k\})$. Then the mapping

$$\alpha \to \sum_{j=1}^{k} \alpha_j \nu_j : \mathbb{R}^k \to A$$

is a linear homeomorphism.

▌PROOF Since $(\mathcal{N}, |\cdot|_w)$ is a normed vector space and A a finite-dimensional subspace, our conclusion follows from I.3.4. QED

IV.2.3 ***Lemma*** Let $\sigma, \sigma_j \in \mathcal{S}$ and $A_j \triangleq \{t \in T \mid \sigma_j(t) \neq \sigma(t)\}$ $(j \in \mathbb{N})$. Then $\lim_j \mu(A_j) = 0$ implies $\lim_j \sigma_j = \sigma$.

▌PROOF For each $\phi \in \mathcal{B}$ we have

$$\lim_j \langle \sigma_j - \sigma, \phi \rangle = \lim_j \left| \int \mu(dt) \int \phi(t, r)(\sigma_j(t) - \sigma(t))(dr) \right|$$

$$\leqslant 2 \lim_j \int_{A_j} |\phi(t, \cdot)|_{\sup} \mu(dt) = 0.$$

Thus, by IV.1.11, $\lim_j \sigma_j = \sigma$. QED

We continue to write

$$\mathcal{T}_n \triangleq \left\{ \theta \triangleq (\theta^1, ..., \theta^n) \in \mathbb{R}^n \mid \theta^j \geqslant 0, \sum_{j=1}^{n} \theta^j \leqslant 1 \right\}$$

and

$$\mathcal{T}_n' \triangleq \left\{ \theta \triangleq (\theta^0, \theta^1, ..., \theta^n) \in \mathbb{R}^{n+1} \mid \theta^j \geqslant 0, \sum_{j=0}^{n} \theta^j = 1 \right\}.$$

IV.2.4 ***Lemma*** Let $\theta \to \sigma_j(\theta) : \mathcal{T}_n' \to \mathcal{S}$ $(j = 0, 1, 2, ...)$. Then

(1) $\qquad \lim_j \sigma_j(\theta) = \sigma_0(\theta) \qquad$ uniformly for all $\theta \in \mathcal{T}_n'$

if, for each $f \in C(T)$ and $c \in C(R)$,

(2) $\lim_j \int f(t)\, c(\sigma_j(\theta)(t))\, \mu(dt) = \int f(t)\, c(\sigma_0(\theta)(t))\, \mu(dt)$ uniformly for all $\theta \in \mathcal{T}_n'$.

▌PROOF Since the (weak norm) topology in $\mathcal{S}$ is metric and compact and has a subbase consisting of the sets

$$\{\sigma \in \mathcal{S} \mid \langle \sigma - \sigma', \phi \rangle < \epsilon\} \qquad (\sigma' \in \mathcal{S}, \quad \phi \in \mathcal{B}, \quad \epsilon > 0),$$

it follows that statement (1) is satisfied if for every $\phi \in \mathcal{B}$ and $\epsilon > 0$ there exists $j_0 \in \mathbb{N}$ such that $\langle \sigma_0(\theta) - \sigma_j(\theta), \phi \rangle < \epsilon$ for all $\theta \in \mathcal{T}_n'$ and all $j \geqslant j_0$. It suffices, in fact, to select ϕ from a dense subset of $\mathcal{B}$.

By I.5.25, $C(T) \otimes C(R)$ is a dense subset of $\mathscr{B}$. We conclude that statement (1) follows from (2). QED

IV.2.5 ***Lemma*** For every $\sigma \in \mathscr{S}$ and all open subsets R' and R'' of R, the function $t \to \sigma(t)\,(R' \sim R'')$ is μ-measurable.

▌PROOF Let F be a closed subset of R, $F_j \triangleq \{r \in R \mid d[r, F] \geqslant 1/j\}$, $c_j(r) \triangleq 1$ if $F_j = \varnothing$ and otherwise

$$c_j(r) \triangleq d[r, F_j]/(d[r, F] + d[r, F_j]) \qquad (j \in \mathbb{N}, \quad r \in R).$$

Then the c_j are continuous, $c_j(R) \subset [0, 1]$, $\lim_j c_j(r) = 1$ for $r \in F$, and $\lim_j c_j(r) = 0$ for $r \in R \sim F$. Thus, by IV.1.6, the function

$$t \to \sigma(t)(F) = \lim_j \int c_j(r)\, \sigma(t)\,(dr)$$

is μ-measurable. It follows that

$$t \to \sigma(t)(R') = 1 - \sigma(t)(R \sim R'),$$

$$t \to \sigma(t)(R' \cap R'') = 1 - \sigma(t)(R \sim R' \cap R''),$$

and

$$t \to \sigma(t)(R' \sim R'') = \sigma(t)(R') - \sigma(t)(R' \cap R'')$$

are μ-measurable. QED

IV.2.6 ***Theorem*** The set $\mathscr{S}$ is the closure of $\mathscr{R}$ in $\mathscr{N}$.

▌PROOF By IV.2.1, $\mathscr{S}$ is metric and compact. It suffices to show, therefore, that $\mathscr{R}$ is dense in $\mathscr{S}$.

Let $\sigma \in \mathscr{S}$. Since R is metric and compact, for every $i \in \mathbb{N}$ we can cover R by a finite collection of open sets $\tilde{R}_k^{\,i}$ $[k = 1, ..., k(i)]$ of diameter at most $1/i$. Setting

$$R_k^{\,i} \triangleq \tilde{R}_k^{\,i} \sim \bigcup_{j=1}^{k-1} \tilde{R}_j^{\,i} \qquad [k = 1, ..., k(i)]$$

and eliminating all empty $R_k^{\,i}$, we can partition R into sets $R_k^{\,i}$ $[k = 1, ..., k(i)]$ that are differences of open sets. Similarly, we can partition T into nonempty disjoint Borel subsets $T_j^{\,i}$ $[j = 1, ..., j(i)]$ of diameter at most $1/i$.

Let

$$\alpha_{j,k}^{i} \triangleq \int_{T_j^{\,i}} \sigma(t)(R_k^{\,i})\, \mu(dt).$$

Since $\sum_{k=1}^{k(i)} \alpha_{j,k}^{i} = \mu(T_j^{\,i})$ and μ is nonatomic, it follows from I.4.10 that we can partition each T_j^i into subsets $T_{j,k}^i$ such that $\mu(T_{j,k}^i) = \alpha_{j,k}^i$. We choose, for each $k \in \{1,..., k(i)\}$, a point $r_k^i \in R_k^i$ and let

$$\rho_i(t) \triangleq r_k^i \qquad \text{for} \quad t \in \bigcup_{j=1}^{j(i)} T_{j,k}^i \qquad [k = 1,..., k(i)].$$

We shall show that $\lim_i \rho_i = \sigma$. Since $\mathscr{S}$ is metric, this will prove that $\mathscr{R}$ is dense in $\mathscr{S}$.

In view of IV.2.4, it suffices to prove that

$$\lim_i \int f(t)\, c(\rho_i(t))\, \mu(dt) = \int f(t)\, c(\sigma(t))\, \mu(dt)$$

for arbitrary $f \in C(T)$ and $c \in C(R)$. Let $\Omega_f(\cdot)$ respectively $\Omega_c(\cdot)$ represent a modulus of continuity of f respectively c, and let t_j^i be an arbitrary point of T_j^i $[i \in \mathbb{N}, j = 1,..., j(i)]$. We have

$$\left| \int f(t)\, \mu(dt) \int c(r)\, \sigma(t)\, (dr) - \int f(t)\, c(\rho_i(t))\, \mu(dt) \right|$$

$$\leqslant \left| \int f(t) \sum_{k=1}^{k(i)} c(r_k^i)\, \sigma(t)(R_k^i)\, \mu(dt) - \sum_{k=1}^{k(i)} c(r_k^i) \sum_{j=1}^{j(i)} \int_{T_{j,k}^i} f(t)\, \mu(dt) \right|$$

$$+ \Omega_c(1/i) \int |f(t)|\, \mu(dt)$$

$$= \left| \sum_{k=1}^{k(i)} \sum_{j=1}^{j(i)} c(r_k^i) \left[\int_{T_j^i} f(t)\, \sigma(t)(R_k^i)\, \mu(dt) - \int_{T_{j,k}^i} f(t)\, \mu(dt) \right] \right|$$

$$+ \Omega_c(1/i) \int |f(t)|\, \mu(dt)$$

$$\leqslant \sum_{k=1}^{k(i)} \sum_{j=1}^{j(i)} c(r_k^i)\, f(t_j^i) \left[\int_{T_j^i} \sigma(t)(R_k^i)\, \mu(dt) - \mu(T_{j,k}^i) \right]$$

$$+ 2\Omega_f(1/i)\, \mu(T)\, |c|_{\sup} + \Omega_c(1/i)\, |f|_1$$

$$= 2\Omega_f(1/i)\, \mu(T)\, |c|_{\sup} + \Omega_c(1/i)\, |f|_1 \xrightarrow[i]{} 0. \quad \text{QED}$$

IV.2.7 ***Theorem*** Let $k, n \in \mathbb{N}$, A be a convex subset of $\mathbb{R}^k$, $\sigma \in \mathscr{S}$,

$$g \in C(T, \mathscr{B}(T, R \times A; \mathbb{R}^n))$$

and

$$f(t, \tau, a) \triangleq g(t, \tau, \sigma(\tau), a) \triangleq \int g(t, \tau, r, a)\, \sigma(\tau)\, (dr) \qquad (t,\tau \in T, a \in A).$$

Then

$$f \in C(T, \mathscr{B}(T, A; \mathbb{R}^n)).$$

If, furthermore, $g(t, \tau, r, \cdot)$ has a derivative $g_a(t, \tau, r, a)$ for all $(t, \tau, r, a) \in T \times T \times R \times A$ and $g_a \in C(T, \mathscr{B}(T, R \times A; B(\mathbb{R}^k, \mathbb{R}^n)))$, then $f(t, \tau, \cdot)$ has a derivative $f_a(t, \tau, a)$,

$$f_a(t, \tau, a)\, \Delta a = \int g_a(t, \tau, r, a)\, \Delta a\, \sigma(\tau)\,(dr) \qquad (t, \tau \in T,\ a \in A,\ \Delta a \in \mathbb{R}^k)$$

and

$$f_a \in C(T, \mathscr{B}(T, A; B(\mathbb{R}^k, \mathbb{R}^n))).$$

▌PROOF By II.3.10,

$$f(t, \tau, \cdot) \in C(A, \mathbb{R}^n) \qquad \text{and} \qquad |f(t, \tau, \cdot)|_{\sup} \leqslant |g(t, \tau, \cdot, \cdot)|_{\sup} \qquad (t, \tau \in T).$$

If $g \triangleq (g^1, \ldots, g^n)$, then

$$g^i(t, \cdot, \cdot, a) \in L^1(T, C(R)) \qquad (i = 1, \ldots, n, \quad t \in T, \quad a \in A);$$

hence, by IV.1.6, $f^i(t, \cdot, a)$ is μ-measurable. Thus

$$f(t, \cdot, \cdot) \in \mathscr{B}(T, A; \mathbb{R}^n) \qquad (t \in T).$$

Furthermore

$$\begin{aligned} |f(t, \tau, \cdot) - f(t', \tau, \cdot)|_{\sup} &= \sup_{a \in A} \left| \int [g(t, \tau, r, a) - g(t', \tau, r, a)]\, \sigma(\tau)\,(dr) \right| \\ &\leqslant |g(t, \tau, \cdot, \cdot) - g(t', \tau, \cdot, \cdot)|_{\sup} \qquad (t, t', \tau \in T); \end{aligned}$$

hence

$$\lim_{t' \to t} \int |f(t, \tau, \cdot) - f(t', \tau, \cdot)|_{\sup}\, \mu(d\tau) = 0 \qquad (t \in T),$$

thus showing that $f \in C(T, \mathscr{B}(T, A; \mathbb{R}^n))$.

If $g_a \in C(T, \mathscr{B}(T, R \times A; B(\mathbb{R}^k, \mathbb{R}^n)))$, then, by II.3.10, $f_a(t, \tau, a)$ exists for all $(t, \tau, a) \in T \times T \times A$,

$$f_a(t, \tau, a)\, \Delta a = \int g_a(t, \tau, r, a)\, \Delta a\, \sigma(\tau)\,(dr) \qquad (\Delta a \in \mathbb{R}^k, \qquad t, \tau \in T, a \in A),$$

$$f_a(t, \tau, \cdot) \in C(A, B(\mathbb{R}^k, \mathbb{R}^n)),$$

and

$$|f_a(t, \tau, \cdot)|_{\sup} \leqslant |g_a(t, \tau, \cdot, \cdot)|_{\sup}$$

Since $f(t, \cdot, a)$ is μ-measurable, it follows from II.3.9 that $f_a(t, \cdot, a)$ is

μ-measurable for all $(t, a) \in T \times A$. Thus $f_a(t, \cdot, \cdot) \in \mathscr{B}(T, A; B(\mathbb{R}^k, \mathbb{R}^n))$. Finally,

$$\lim_{t' \to t} \int |f_a(t, \tau, \cdot) - f_a(t', \tau, \cdot)|_{\sup} \mu(d\tau) \leqslant \lim_{t' \to t} \int | g_a(t, \tau, \cdot, \cdot) - g_a(t', \tau, \cdot, \cdot)|_{\sup} \mu(dt) = 0,$$

thus showing that $f_a \in C(T, \mathscr{B}(T, A; B(R^k, \mathbb{R}^n)))$. QED

IV.2.8 ***Theorem*** Let $k, n \in \mathbb{N}$, A be a convex subset of $\mathbb{R}^k$, $\sigma \in \mathscr{S}$, $g : T \times T \times R \times A \to \mathbb{R}^n$, and

$$f(t, \tau, a) \triangleq g(t, \tau, \sigma(\tau), a) \triangleq \int g(t, \tau, r, a)\, \sigma(\tau)\, (dr) \qquad (t, \tau \in T, \quad a \in A);$$

and assume that, for all $(t, \tau, r, a) \in T \times T \times R \times A$, $g(t, \tau, \cdot, \cdot)$ is continuous and $g(\cdot, \cdot, r, a)$ $\mu \times \mu$-measurable. Then $f(t, \tau, \cdot)$ is continuous and $f(\cdot, \cdot, a)$ $\mu \times \mu$-measurable.

If, furthermore, for all $(t, \tau, r, a) \in T \times T \times R \times A$, $g(t, \tau, r, \cdot)$ has a derivative $g_a(t, \tau, r, a)$ and $g_a(t, \tau, \cdot, \cdot)$ is continuous, then

$$f_a(t, \tau, a)\, \Delta a = \int g_a(t, \tau, r, a)\, \Delta a\, \sigma(\tau)\, (dr) \qquad (t, \tau \in T, \quad a \in A, \quad \Delta a \in \mathbb{R}^k),$$

$f_a(t, \tau, \cdot)$ is continuous, and $f_a(\cdot, \cdot, a)$ $\mu \times \mu$-measurable.

▌PROOF By II.3.10, $f(t, \tau, \cdot)$ is continuous. Since

$$s \to g(t, \tau, s, a) : (\mathrm{frm}(R), |\cdot|_w) \to \mathbb{R}^n$$

is continuous for all $t, \tau \in T$ and $a \in A$, it follows from I.5.26 and I.5.27 that $g(\cdot, \cdot, s, a)$ is $\mu \times \mu$-measurable for all $s \in \mathrm{frm}(R)$ and $a \in A$. Hence, by I.4.22, $(t, \tau) \to g(t, \tau, \sigma(\tau), a) \triangleq f(t, \tau, a)$ is $\mu \times \mu$-measurable for all $a \in A$.

If g_a exists and has the specified properties, then, by II.3.10,

$$f_a(t, \tau, a)\, \Delta a = \int g_a(t, \tau, r, a)\, \Delta a\, \sigma(\tau)\, (dr) \qquad (t, \tau \in T, \quad a \in A, \quad \Delta a \in \mathbb{R}^k)$$

and $f_a(t, \tau, \cdot)$ is continuous for all $t, \tau \in T$. By IV.1.6 and II.3.9, $f_a(\cdot, \cdot, a)$ is $\mu \times \mu$-measurable for all $a \in A$. QED

IV.2.9 ***Theorem*** Let $k, n \in \mathbb{N}$, $V \subset \mathbb{R}^n$, Z be a topological space, $h : T \times V \times R \times Z \to \mathbb{R}^k$, $x_j : T \to V$ $(j \in \mathbb{N})$ μ-measurable, $\lim_j x_j = \bar{x}$ μ-a.e., $\lim_j (\sigma_j, z_j) = (\bar{\sigma}, \bar{z})$, and $\psi \in L^1(T, \Sigma, \mu)$. Assume, furthermore, that $h(\cdot, v, r, z)$ is μ-measurable and $h(\tau, \cdot, \cdot, \cdot)$ continuous for all $(\tau, v, r, z) \in T \times V \times R \times Z$ and

$$| h(\tau, x_j(\tau), \sigma_j(\tau), z_j)| \leqslant \psi(\tau) \qquad (\tau \in T, \quad j \in \mathbb{N}).$$

Then

$$\lim_j \int h(\tau, x_j(\tau), \sigma_j(\tau), z_j)\, \mu(d\tau) = \int h(\tau, \bar{x}(\tau), \bar{\sigma}(\tau), \bar{z})\, \mu(d\tau).$$

▌PROOF By IV.1.6, $h(\tau, x_j(\tau), \sigma_j(\tau), z_j)$ and $h(\tau, \bar{x}(\tau), \bar{\sigma}(\tau), z_j)$ are μ-measurable functions of τ and, since their norms are bounded by $\psi(\tau)$, they are μ-integrable. By I.2.15(2), for μ-a.a. $\tau \in T$ we have

$$\lim_j h(\tau, x_j(\tau), r, z_j) = h(\tau, \bar{x}(\tau), r, \bar{z}) \quad \text{uniformly for } r \in R;$$

hence

$$\lim_j [h(\tau, x_j(\tau), \sigma_j(\tau), z_j) - h(\tau, \bar{x}(\tau), \sigma_j(\tau), \bar{z})] = 0. \tag{1}$$

Since

$$| h(\tau, x_j(\tau), \sigma_j(\tau), z_j)| \leqslant \psi(\tau) \qquad (\tau \in T, \quad j \in \mathbb{N}),$$

it follows from (1) and the dominated convergence theorem I.4.35 that

$$\begin{aligned}(2)\qquad \lim_j \int h(\tau, x_j(\tau), \sigma_j(\tau), z_j)\, \mu(d\tau) &= \lim_j \int [h(\tau, x_j(\tau), \sigma_j(\tau), z_j) - h(\tau, \bar{x}(\tau), \sigma_j(\tau), \bar{z})]\, \mu(d\tau) \\ &\quad + \lim_j \int h(\tau, \bar{x}(\tau), \sigma_j(\tau), \bar{z})\, \mu(d\tau) \\ &= \lim_j \int \mu(d\tau) \int h(\tau, \bar{x}(\tau), r, \bar{z})\, \sigma_j(\tau)\, (dr).\end{aligned}$$

If $h \triangleq (h^1, \ldots, h^k)$, then $(\tau, r) \to h^i(\tau, \bar{x}(\tau), r, \bar{z})$ belongs to $L^1(T, C(R))$ for each $i \in \{1, 2, \ldots, k\}$ and therefore

$$\lim_j \int \mu(d\tau) \int h(\tau, \bar{x}(\tau), r, \bar{z})\, \sigma_j(\tau)\, (dr) = \int h(\tau, \bar{x}(\tau), \bar{\sigma}(\tau), \bar{z})\, \mu(d\tau).$$

Our conclusion now follows from (2). QED

IV.3 The Sets $\mathscr{R}^{\#}$ and $\mathscr{S}^{\#}$ and Abundant Sets

The original control functions ρ that we shall consider from now on may be subjected to restrictions of the form

$$\rho(t) \in R^{\#}(t) \qquad \text{for} \quad \mu\text{-a.a. } t \in T, \tag{$*$}$$

where the nonempty subset $R^{\#}(t)$ of R is preassigned for every t. For example, we may require that $h(t, \rho(t)) \in H$ μ-a.e. for given Y, $H \subset Y$, and $h : T \times R \to Y$. Our study of particular classes of optimal control problems will not be much affected by this restriction. The readers who prefer (at least at first reading) to study optimal control problems that are not restricted by condition $(*)$ will find the remainder of this section greatly simplified if they assume that $R^{\#}(t) = R$ $(t \in T)$ and replace $\mathscr{R}^{\#}$ and $\mathscr{S}^{\#}$ by $\mathscr{R}$ and $\mathscr{S}$, respectively. They will not need in that case to refer to I.7 (on measurable set-valued mappings).

THE MAPPING $R^{\#}$ AND THE SETS $\mathscr{R}^{\#}$ AND $\mathscr{S}^{\#}$

Let $R^{\#} : T \to \mathscr{P}'(R)$ be a given mapping. We write

$$\mathscr{R}^{\#} \triangleq \{\rho \in \mathscr{R} \mid \rho(t) \in R^{\#}(t) \ \mu\text{-a.e.}\}, \qquad \bar{R}^{\#}(t) \triangleq \overline{R^{\#}(t)} \quad (t \in T),$$

$$\mathscr{S}^{\#} \triangleq \{\sigma \in \mathscr{S} \mid \sigma(t)(\bar{R}^{\#}(t)) = 1 \ \mu\text{-a.e.}\},$$

and

$$\mathscr{S}_{\mathscr{R}}^{\#} \triangleq \{\sigma \in \mathscr{S} \mid \sigma(t)(\{\rho(t)\}) = 1 \ \mu\text{-a.e.} \ \text{ for some } \rho \in \mathscr{R}^{\#}\} = \mathscr{S}^{\#} \cap \mathscr{S}_{\mathscr{R}}$$

As in the case of the sets $\mathscr{S}$ and $\mathscr{R}$, we identify $\mathscr{S}_{\mathscr{R}}^{\#}$ and $\mathscr{R}^{\#}$ and thus consider $\mathscr{R}^{\#}$ to be a subset of $\mathscr{S}^{\#}$.

As in I.7, we define the class $\mathscr{P}'(R)$ of nonempty subsets of R as a topological space by means of the Hausdorff semimetric δ. Accordingly, we shall say that the restriction of $R^{\#}$ to some subset F of T is continuous if for every $\bar{t} \in F$ and $\epsilon > 0$ there exists $\eta > 0$ such that $\delta(R^{\#}(\bar{t}), R^{\#}(t)) < \epsilon$ if $t \in F$ and $d(\bar{t}, t) < \eta$.

Our future arguments require that the mapping $R^{\#} : T \to \mathscr{P}'(R)$ satisfy the following condition.

IV.3.1 ***Condition*** The mapping $R^{\#} : T \to \mathscr{P}'(R)$ is μ-measurable and there exists an at most denumerable subset $\mathscr{R}_{\infty}^{\#}$ of $\mathscr{R}^{\#}$ such that the set $\{\rho(t) \mid \rho \in \mathscr{R}_{\infty}^{\#}\}$ is dense in $R^{\#}(t)$ for μ-a.a. $t \in T$.

▌ *Remark* Since $R^{\#}$ is μ-measurable respectively continuous if and only if $\bar{R}^{\#}$ has the same property, it follows from I.7.1 and Lusin's theorem I.4.19 that for every $\epsilon > 0$ there exists a closed set $T_{\epsilon} \subset T$ such that $\mu(T \sim T_{\epsilon}) \leqslant \epsilon$ and $R^{\#} \mid T_{\epsilon}$ is continuous.

Theorem IV.3.2 below shows that Condition IV.3.1 is satisfied in many cases of interest and establishes certain criteria for verifying this condition.

IV.3.2 ***Theorem*** Let $R^{\#} : T \to \mathscr{P}'(R)$ be μ-measurable and assume that either

(1) $R^{\#}(t)$ is closed for all $t \in T$, or
(2) $R^{\#}(t) \subset \overline{R^{\#}(t)^{\circ}}$ $(t \in T)$, and for every $\epsilon > 0$ there exists a closed subset T_{ϵ} of T such that $\mu(T \sim T_{\epsilon}) \leqslant \epsilon$ and the set $\{(t, r) \in T_{\epsilon} \times R \mid r \in R^{\#}(t)^{\circ}\}$ is open relative to $T_{\epsilon} \times R$.

Then $R^{\#}$ satisfies Condition IV.3.1.

This is the case, in particular, if Y is a complete separable metric space, H a closed subset of Y, $h : T \times R \to Y$, $h(t, \cdot)$ continuous and $h(\cdot, r)$ μ-measurable for all $(t, r) \in T \times R$, and

$$R^{\#}(t) \triangleq \{r \in R \mid h(t, r) \in H\} \neq \varnothing \qquad \mu\text{-a.e.}$$

▌PROOF The first part of our assertion follows directly from I.7.9. To prove the second part we observe that since $h(t, \cdot)$ is continuous and H closed, $R^{\#}(t) \triangleq \{r \in R \mid h(t, r) \in H\}$ is closed for all $t \in T$. By I.5.26, for every $\epsilon > 0$ there exists a closed $F_{\epsilon} \subset T$ such that $\mu(T \sim F_{\epsilon}) \leqslant \epsilon$ and $h \mid F_{\epsilon} \times R$ is continuous. Then the set

$$G_{\epsilon}(R^{\#}) \triangleq \{(t, r) \in F_{\epsilon} \times R \mid h(t, r) \in H\}$$

is closed and, by I.7.2 and I.7.5, $R^{\#} \mid F_{\epsilon}$ is μ-measurable. Since ϵ is arbitrary, $R^{\#}$ is μ-measurable and the first part of the theorem applies. QED

We next consider a class of subsets $\mathscr{U}$ of $\mathscr{R}^{\#}$ that lend themselves to the study of approximate $\mathscr{U}$-solutions and of minimizing $\mathscr{U}$-solutions. As in the case of Condition IV.3.1, we shall follow up the definition of these "abundant sets" with the study of certain criteria that define a class of abundant sets.

IV.3.3 ***Definition*** *Abundant Sets* We say that $\mathscr{U}$ is an *abundant set* or an *abundant subset* of $\mathscr{R}^{\#}$ if $\mathscr{U} \subset \mathscr{R}^{\#}$,

(1) $\mathscr{U}$ contains a denumerable subset $\mathscr{U}_{\infty}$ such that the set $\{u(t) \mid u \in \mathscr{U}_{\infty}\}$ is dense in $R^{\#}(t)$ for μ-a.a. $t \in T$, and
(2) for every choice of $n \in \mathbb{N}$, $\sigma_j \in \mathscr{S}^{\#}$ $(j = 0, .,..., n)$ and $\theta \in \mathscr{T}_n'$, there exists a sequence $(u_i(\theta))$ in $\mathscr{U}$ such that $\lim_i u_i(\theta) = \sum_{j=0}^{n} \theta^j \sigma_j$ in $\mathscr{S}$ uniformly for all $\theta \in \mathscr{T}_n'$ and the function

$$\theta \to u_i(\theta) : \mathscr{T}_n' \to \mathscr{S}$$

is continuous for each $i \in \mathbb{N}$.

We shall prove, in Theorem IV.3.9 below, that $\mathscr{U}$ is an abundant set if it satisfies Condition IV.3.4 below. This will prove, in particular (in view of Condition IV.3.1), that $\mathscr{R}^{\#}$ is an abundant set.

IV.3.4 ***Condition*** The set $\mathscr{U}$ is a subset of $\mathscr{R}^{\#}$ satisfying condition IV.3.3(1) and such that, for every μ-measurable set T_1 and all choices of $u_1, u_2 \in \mathscr{U}$, the function $u : T \to R$, defined by

$$u(t) \triangleq u_1(t) \quad (t \in T_1), \qquad u(t) \triangleq u_2(t) \quad (t \in T \sim T_1)$$

belongs to $\mathscr{U}$.

Lemmas IV.3.5 and IV.3.6 below are obvious.

IV.3.5 ***Lemma*** Let $\tilde{R}$ be a dense subset of R and $R^{\#}(t) = \tilde{R}$ $(t \in T)$. Then the set of μ-simple functions on T to $\tilde{R}$ satisfies Condition IV.3.4.

We shall refer to a function $\rho : T \to R$ as *μ-piecewise continuous* if there exists a finite partition of T into μ-measurable sets $T_1, ..., T_l$ such that $\rho \mid T_j$ is continuous for $j = 1, ..., l$.

IV.3.6 ***Lemma*** Let $R^{\#}(t) = R$ $(t \in T)$. Then the class of μ-piecewise continuous functions on T to R satisfies Condition IV.3.4.

IV.3.7 ***Lemma*** Let $\mathscr{U}$ satisfy Condition IV.3.4 and $\epsilon > 0$. Then there exists a closed subset $T_\epsilon^{\mathscr{U}}$ of T such that

(1) $\mu(T \sim T_\epsilon^{\mathscr{U}}) \leqslant \epsilon$, and
(2) for every nonempty $R' \subset R$ there exists some $u \in \mathscr{U}$ such that

$$d[u(t), R'] < \epsilon \qquad \text{if} \quad t \in T_\epsilon^{\mathscr{U}} \quad \text{and} \quad d[R^{\#}(t), R'] = 0.$$

▌PROOF Let $\mathscr{U}_\infty \triangleq \{u^1, u^2, ...\}$ be as defined in IV.3.3(1) and T_ϵ as in the remark following Condition IV.3.1. Then, by Lusin's theorem I.4.19, there exist closed subsets T_ϵ^i of T $(i \in \mathbb{N})$ such that $\mu(T \sim T_\epsilon^i) \leqslant \epsilon/2^{i+1}$, $u^i \mid T_\epsilon^i$ is continuous and $\{u^1(t), u^2(t), ...\}$ is dense in $R^{\#}(t)$ for all $t \in T_\epsilon^i$. We set

$$T_\epsilon^{\mathscr{U}} \triangleq \bigcap_{i=1}^{\infty} T_\epsilon^i \cap T_{\epsilon/2}$$

and observe that $\mu(T \sim T_\epsilon^{\mathscr{U}}) \leqslant \epsilon$, $R^{\#} \mid T_\epsilon^{\mathscr{U}}$ and $u^i \mid T_\epsilon^{\mathscr{U}}$ are continuous for all i and $\{u^1(t), u^2(t), ...\}$ is dense in $R^{\#}(t)$ for all $t \in T_\epsilon^{\mathscr{U}}$.

Now let $R' \subset R$ and $R' \neq \varnothing$. Since $R^{\#} \mid T_\epsilon^{\mathscr{U}}$ is continuous, it follows easily that

$$t \to d[R', R^{\#}(t)] : T_\epsilon^{\mathscr{U}} \to \mathbb{R}$$

is continuous. Thus, for every $\tau \in T_\epsilon^{\mathscr{U}}$ there exists some $u_\tau \in \mathscr{U}_\infty$ such that

$$d[u_\tau(\tau), R'] < \epsilon/2 + d[R', R^{\#}(\tau)].$$

It follows that there exists some relative open neighborhood N_τ of τ in $T_\epsilon^{\mathscr{U}}$ such that

$$d[u_\tau(t), R'] < \epsilon + d[R', R^{\#}(t)] \qquad (t \in N_\tau).$$

Since $T_\epsilon^{\mathscr{U}}$ is closed (hence compact), it can be covered by a finite subfamily of such neighborhoods, say $N_{\tau_1},..., N_{\tau_k}$. We set

$$u(t) \triangleq \begin{cases} u_{\tau_j}(t) & \text{for } \ t \in N_{\tau_j} \sim \bigcup_{i=1}^{j-1} N_{\tau_i} \quad (j = 1, 2,..., k), \\ u_{\tau_1}(t) & \text{for } \ t \in T \sim T_\epsilon^{\mathscr{U}}, \end{cases}$$

and observe that $u \in \mathscr{U}$ and $d[u(t), R'] < \epsilon$ if $t \in T_\epsilon^{\mathscr{U}}$ and $d[R', R^{\#}(t)] = 0$.
QED

IV.3.8 ***Lemma*** Let F be a μ-measurable subset of T, $\{R_1,..., R_m\}$ a partition of R such that $R_k = R_k' \sim R_k'' \neq \varnothing$ for some open sets R_k' and R_k'' $(k = 1,..., m)$, and $\sigma_j \in \mathscr{S}^{\#}$ $(j = 0, 1,..., n)$. Then, for every $\theta \in \mathscr{T}_n'$, F can be partitioned into disjoint μ-measurable subsets $F_{j,k}(\theta)$ $(j = 0,..., n;$ $k = 1,..., m)$ such that

$$\theta^j \int_F \sigma_j(t)(R_k)\, \mu(dt) = \mu\big(F_{j,k}(\theta)\big), \tag{1}$$

$$\lim_{\theta_1 \to \theta} \sum_{j=0}^{n} \sum_{k=1}^{m} \{\mu\big(F_{j,k}(\theta) \sim F_{j,k}(\theta_1)\big) + \mu\big(F_{j,k}(\theta_1) \sim F_{j,k}(\theta)\big)\} = 0, \tag{2}$$

and

$$d[R^{\#}(t), R_k] = 0 \qquad [t \in F_{j,k}(\theta)]. \tag{3}$$

▌PROOF Let

$$A_k \triangleq \left\{ t \in F \,\middle|\, \sum_{j=0}^{n} \sigma_j(t)(R_k) \neq 0 \right\} \qquad (k = 1,..., m),$$

P_m be the class of nonempty subsets of $\{1, 2,..., m\}$, and

$$G_D \triangleq \bigcap_{k \in D} A_k \bigcap_{k \notin D} (F \sim A_k) \qquad (D \in P_m);$$

hence G_D is the set of those $t \in F$ for which $\sum_{j=0}^{n} \sigma_j(t)(R_k) \neq 0$ for all $k \in D$ and no other k. Thus the sets G_D $(D \in P_m)$ form a partition of F. Since $(n+1)^{-1} \sum_{j=0}^{n} \sigma_j \in \mathscr{S}^{\#}$, it follows from IV.2.5 that $A_1,..., A_m$ are μ-measurable, and therefore the sets G_D are μ-measurable.

Now let $|D|$ be the number of elements in D and

$$\alpha_{j,k}^{D} \triangleq \begin{cases} 1/|D| & \text{if } \mu(G_D) = 0 \text{ and } k \in D, \\ 0 & \text{if } \mu(G_D) = 0 \text{ and } k \notin D, \\ [1/\mu(G_D)] \int_{G_D} \sigma_j(t)(R_k)\, \mu(dt) & \text{if } \mu(G_D) \neq 0 \quad (j = 0, 1,..., n; \\ & \qquad k = 1,..., m; \quad D \in P_m). \end{cases}$$

Since μ is nonatomic, it follows from I.4.10 that we can define, for each $D \in P_m$, sets $G_D(\alpha)$ ($\alpha \in [0, 1]$) such that

$$G_D(\alpha) \subset G_D(\beta) \quad (\alpha \leqslant \beta), \qquad \mu(G_D(\alpha)) = \alpha\mu(G_D),$$
$$G_D(0) = \varnothing, \qquad \text{and} \qquad G_D(1) = G_D.$$

We observe that

$$\sum_{j=0}^{n} \sum_{k=1}^{m} \theta^j \alpha_{j,k}^{D} = 1 \qquad (D \in P_m, \theta \in \mathscr{T}_n').$$

Let the couples (j, k) ($j = 0,..., n, k = 1,..., m$) be ordered lexicographically, that is, $(j_1, k_1) < (j_2, k_2)$ if either $j_1 < j_2$ or $j_1 = j_2$ and $k_1 < k_2$. For each $D \in P_m$ and $\theta \in \mathscr{T}_n'$, we can partition the interval $[0, 1)$ into consecutive intervals

$$I_{j,k}^{D}(\theta) \triangleq [a_{j,k}^{D}(\theta), b_{j,k}^{D}(\theta))$$

of lengths $\theta^j \alpha_{j,k}^{D}$, arranged in the increasing lexicographic order of (j, k). We set

$$F_{j,k}(\theta) \triangleq \bigcup_{D \in P_m} [G_D(b_{j,k}^{D}(\theta)) \sim G_D(a_{j,k}^{D}(\theta))] \qquad (j = 0, 1,..., n; \quad k = 1,..., m, \quad \theta \in \mathscr{T}_n').$$

Since the sets

$$G_D(b_{j,k}^{D}(\theta)) \sim G_D(a_{j,k}^{D}(\theta)) \qquad (j = 0, 1,..., n; \quad k = 1,...,m)$$

are μ-measurable and disjoint for each $D \in P_m$ and $\theta \in \mathscr{T}_n'$, it follows that the sets $F_{j,k}(\theta)$ are μ-measurable and disjoint for each $\theta \in \mathscr{T}_n'$. Their union is $\bigcup_{D \in P_m} G_D = F$. We have

$$\begin{aligned} \mu(F_{j,k}(\theta)) &= \sum_{D \in P_m} [b_{j,k}^{D}(\theta) - a_{j,k}^{D}(\theta)]\, \mu(G_D) \\ &= \sum_{D \in P_m} \theta^j \alpha_{j,k}^{D} \mu(G_D) \\ &= \sum_{D \in P_m} \theta^j \int_{G_D} \sigma_j(t)(R_k)\, \mu(dt) \\ &= \theta^j \int_F \sigma_j(t)(R_k)\, \mu(dt) \end{aligned}$$

Thus relation (1) is satisfied. If $t \in F_{j,k}(\theta)$, then

$$t \in G_D(b_{j,k}^D(\theta)) \sim G_D(a_{j,k}^D(\theta)) \qquad \text{for some} \quad D \in P_m\,;$$

hence $b_{j,k}^D(\theta) - a_{j,k}^D(\theta) = \theta^j \alpha_{j,k}^D > 0$. Therefore $k \in D$ and $\sum_{l=0}^n \sigma_l(t)(R_k) \neq 0$, implying that $d[\bar{R}^{\#}(t), R_k] = 0$. Thus (3) holds.

We finally consider relation (2). By construction,

$$b_{j,k}^D(\theta) = \sum_{(j_1,k_1) \leqslant (j,k)} \alpha_{j_1,k_1}^D \theta^{j_1} \qquad \text{and} \qquad a_{j,k}^D = \sum_{(j_1,k_1) < (j,k)} \alpha_{j_1,k_1}^D \theta^{j_1}$$

for all j, k, D, and θ. Thus the functions $\theta \to b_{j,k}^D(\theta)$ and $\theta \to a_{j,k}^D(\theta)$ are continuous on $\mathscr{T}_n{}'$. Now let $A \mathbin{\Delta} B \triangleq (A \sim B) \cup (B \sim A)$. Then

$$\begin{aligned}
&\mu([G_D(\beta_1) \sim G_D(\alpha_1)] \mathbin{\Delta} [G_D(\beta_2) \sim G_D(\alpha_2)]) \\
&\quad \leqslant \mu([G_D(\mathrm{Max}(\beta_1, \beta_2)) \sim G_D(\mathrm{Min}(\beta_1, \beta_2))] \\
&\qquad \cup [G_D(\mathrm{Max}(\alpha_1, \alpha_2)) \sim G_D(\mathrm{Min}(\alpha_1, \alpha_2))]) \\
&\quad = (|\beta_2 - \beta_1| + |\alpha_2 - \alpha_1|)\, \mu(G_D)
\end{aligned}$$

for all $D \in P_m$ and $\alpha_1, \alpha_2, \beta_1, \beta_2 \in [0, 1]$. Since the sets G_D $(D \in P_m)$ are disjoint, it follows that

$$\begin{aligned}
\lim_{\theta_1 \to \theta} \sum_{j=0}^n \sum_{k=1}^m \mu(F_{j,k}(\theta) \mathbin{\Delta} F_{j,k}(\theta_1)) &\leqslant \lim_{\theta_1 \to \theta} \sum_{D \in P_m} \sum_{j=0}^n \sum_{k=1}^m (|b_{j,k}^D(\theta) - b_{j,k}^D(\theta_1)| \\
&\qquad + |a_{j,k}^D(\theta) - a_{j,k}^D(\theta_1)|)\, \mu(G_D) = 0,
\end{aligned}$$

proving relation (2). QED

IV.3.9 ***Theorem*** Let $\mathscr{U}$ satisfy Condition IV.3.4. Then $\mathscr{U}$ is an abundant set. In particular, $\mathscr{R}^{\#}$ is an abundant set.

▌PROOF Let $T_\epsilon^{\mathscr{U}}$ be defined for each $\epsilon > 0$ as in IV.3.7, $n \in \mathbb{N}$, $\sigma_0, \ldots, \sigma_n \in \mathscr{S}^{\#}$, and $\tilde{\sigma}(\theta) \triangleq \sum_{j=0}^n \theta^j \sigma_j$ $(\theta \in \mathscr{T}_n{}')$. By I.2.5, for each fixed $i \in \mathbb{N}$ we can cover the compact metric space R with a finite collection of open sets $R_1'^i, R_2'^i, \ldots, R_{k_i}'^i$, each of diameter at most $1/i$. We set

$$R_k''^i \triangleq \bigcup_{j=1}^{k-1} R_j'^i \qquad \text{and} \qquad R_k{}^i \triangleq R_k'^i \sim R_k''^i \qquad (k = 1, \ldots, k_i),$$

and observe that $R_1{}^i, \ldots, R_{k_i}^i$ form a partition of R and each $R_j{}^i$ can be assumed nonempty. We can similarly partition the set $T_{1/i}^{\mathscr{U}}$ (leaving out, if necessary, a set of μ-measure 0) into Borel subsets $T_1{}^i, \ldots, T_{l_1}^i$, each of diameter at most $1/i$ and of positive μ-measure.

For every $l \in \{1, 2, \ldots, l_i\}$ we can apply IV.3.8, replacing R_k, F, and m by, respectively, $R_k{}^i$, $T_l{}^i$, and k_i, to partition $T_l{}^i$ into disjoint μ-measurable sets

$T^i_{l,j,k}(\theta)$ $(j = 0,\ldots, n;\ \theta \in \mathscr{T}_n')$ satisfying relations (1)–(3) of IV.3.8 with $F_{j,k}(\theta) \triangleq T^i_{l,j,k}(\theta)$. It follows then from relation IV.3.8(3) and IV.3.7 that there exist $u_k^i \in \mathscr{U}$ such that $d[u_k^i(t), R_k^i] \leqslant 1/i$ for $t \in T^i_{l,j,k}(\theta)$ and all l, j, k and θ. We set, for each $\theta \in \mathscr{T}_n'$,

$$u_i(\theta)(t) \triangleq \begin{cases} u_k^i(t) & \text{for } \ t \in T^i_{l,j,k}(\theta) \ \ (l = 1,\ldots, l_i\,;\, j = 0,\ldots, n,\ k = 1,\ldots, k_i) \\ u_1^i(t) & \text{otherwise.} \end{cases}$$

These relations define $u_i(\theta)(t)$ for all $\theta \in \mathscr{T}_n'$ and $t \in T$.

We now observe that the function $\theta \to u_i(\theta) : \mathscr{T}_n' \to \mathscr{U}$ is continuous. Indeed, this follows from IV.2.3 and IV.3.8(2) because

$$\mu(\{t \in T \mid u_i(\theta)(t) \neq u_i(\theta')(t)\}) \leqslant \sum_{l=1}^{l_i} \cdot \sum_{j=0}^{n} \sum_{k=1}^{k_i} \mu\big(T^i_{l,j,k}(\theta)\, \Delta\, T^i_{l,j,k}(\theta')\big) \xrightarrow[\theta' \to \theta]{} 0.$$

In view of IV.2.4, it remains to prove that for each $f \in C(T)$ and $c \in C(R)$ we have

$$\text{(1)} \qquad \lim_i \int f(t)\, c\big(u_i(\theta)(t)\big)\, \mu(dt) = \int f(t)\, \mu(dt) \int c(r)\, \tilde{\sigma}(\theta)(t)\, (dr)$$

uniformly for all $\theta \in \mathscr{T}_n'$.

For each i, l, and k, we choose points $t_l^i \in T_l^i$ and $r_k^i \in R_k^i$. The symbol $O(\epsilon)$ will represent a quantity whose absolute value does not exceed ϵ. Let $\epsilon > 0$, and let $i_0 \in \mathbb{N}$ be sufficiently large so that

$$i_0 \geqslant 6 |f|_{\sup} |c|_{\sup}/\epsilon,$$

$$|c(r) - c(r')| \leqslant \epsilon/(6 |f|_{\sup}\, \mu(T)) \quad \text{and} \quad |f(t) - f(t')| \leqslant \epsilon/(6 |c|_{\sup}\, \mu(T))$$

if $d(r, r') \leqslant 2/i_0$ and $d(t, t') \leqslant 1/i_0$. Then for every $\theta \in \mathscr{T}_n'$ and $i \geqslant i_0$, and with summations taken for $j = 0,\ldots, n$, $l = 1,\ldots, l_i$, and $k = 1,\ldots, k_i$, we have

$$\begin{aligned}
&\int f(t)\, \mu(dt) \int c(r)\, \tilde{\sigma}(\theta)(t)\, (dr) \\
&\quad = \sum_l \int_{T_l^i} f(t)\, \mu(dt) \int_R c(r)\, \tilde{\sigma}(\theta)(t)\, (dr) + O(\epsilon/6) \\
&\quad = \sum_l f(t_l^i) \int_{T_l^i} \mu(dt) \int_R c(r)\, \tilde{\sigma}(\theta)(t)\, (dr) + O(2\epsilon/6) \\
&\quad = \sum_{l,k} f(t_l^i) \int_{T_l^i} \mu(dt) \int_{R_k^i} c(r)\, \tilde{\sigma}(\theta)(t)\, (dr) + O(2\epsilon/6) \\
&\quad = \sum_{l,j,k} f(t_l^i)\, c(r_k^i)\, \theta^j \int_{T_l^i} \mu(dt) \int_{R_k^i} \sigma_j(t)\, (dr) + O(3\epsilon/6)
\end{aligned}$$

$$= \sum_{l,j,k} f(t_l^{\,i})\, c(r_k^{\,i})\, \mu\big(T^i_{l,j,k}(\theta)\big) + O(3\epsilon/6)$$

$$= \sum_{l,j,k} f(t_l^{\,i}) \int_{T^i_{l,j,k}(\theta)} c(r_k^{\,i})\, \mu(dt) + O(3\epsilon/6)$$

$$= \sum_{l,j,k} f(t_l^{\,i}) \int_{T^i_{l,j,k}(\theta)} c(u_i(\theta)(t))\, \mu(dt) + O(4\epsilon/6)$$

$$= \sum_{l,j,k} \int_{T^i_{l,j,k}(\theta)} f(t)\, c((u_i(\theta)(t))\, \mu(dt) + O(5\epsilon/6)$$

$$= \sum_{l} \int_{T_l^i} f(t)\, c\big(u_i(\theta)(t)\big)\, \mu(dt) + O(5\epsilon/6)$$

$$= \int_{T} f(t)\, c\big(u_i(\theta)(t)\big)\, \mu(dt) + O(\epsilon).$$

Since i_0 was chosen independently of θ, statement (1) is thus verified. QED

▌ *Remark* In the special case where $T \triangleq [t_0, t_1] \subset \mathbb{R}$, μ is the Borel measure in T, and $R^{\#}(t) \triangleq R$ $(t \in T)$, we can specify certain simply defined subsets of $\mathscr{R}^{\#}$ that are abundant sets without satisfying Condition IV.3.4. It can be shown that each of the sets $\mathscr{R}'_{\text{pc}}$ of piecewise continuous and $\mathscr{R}''_{\text{pc}}$ of piecewise constant functions is an abundant set. Condition IV.3.3(1) is verified by fairly simple constructions. Furthermore, the argument of IV.3.9 can be modified to show that $\mathscr{R}''_{\text{pc}}$ (and, a fortiori, $\mathscr{R}'_{\text{pc}}$) also satisfies condition IV.3.3(2). This modification consists in choosing the sets $T_l^{\,i}$ and $T^i_{l,j,k}$ as intervals; the results of IV.3.7 with $T_\epsilon^{\mathscr{U}} = T$ and of IV.3.9 with F, $F_{jk}(\theta)$ as intervals being almost obvious.

If we also assume that R is a convex subset of some vector space, then we can show that $C(T, R)$ is an abundant set. Indeed, $C(T, R)$ is easily seen to satisfy IV.3.3(1). We can also show that it satisfies condition IV.3.3(2), the construction consisting in "smoothing out" into appropriate polygons the functions $u_i(\theta)(\cdot)$ defined for $\mathscr{R}''_{\text{pc}}$.

IV.3.10 ***Theorem*** If $\mathscr{U}$ is an abundant subset of $\mathscr{R}^{\#}$, then $\mathscr{U}$ is dense in $\mathscr{S}^{\#}$. In particular, $\mathscr{R}^{\#}$ is dense in $\mathscr{S}^{\#}$.

▌ PROOF The first statement follows from IV.3.3(2) for $n = 0$. The second statement follows from the first one because $\mathscr{R}^{\#}$ is an abundant set (as a consequence of Theorem IV.3.9). QED

IV.3.11 ***Theorem*** Let $\Gamma : T \to \mathscr{P}'(R)$ be μ-measurable and

$$\varDelta \triangleq \{\sigma \in \mathscr{S} \mid \sigma(t)\big(\overline{\Gamma(t)}\big) = 1 \ \mu\text{-a.e.}\}.$$

Then $\varDelta$ is convex and compact. In particular, $\mathscr{S}^{\#}$ is convex and compact.

▌PROOF The convexity of Δ is verified directly. By IV.2.1, $\mathscr{S}$ is compact and it suffices therefore to show that Δ is closed.

Let $\bar{\Gamma} : T \to \mathscr{P}'(R)$ be defined by the relation $\bar{\Gamma}(t) \triangleq \overline{\Gamma(t)}$ $(t \in T)$, (σ_j) be a sequence in Δ converging to some $\sigma \in \mathscr{S}$, $\eta > 0$, and $\epsilon > 0$. By I.7.1 and Lusin's theorem I.4.19, there exists a closed subset F_η of T such that $\mu(T \sim F_\eta) \leqslant \eta$ and $\Gamma \mid F_\eta$ is continuous; hence $\bar{\Gamma} \mid F_\eta$ is continuous. Let $\bar{t} \in F_\eta$,

$$G_1 \triangleq S(\bar{\Gamma}(\bar{t}), \epsilon) \qquad \text{and} \qquad G_{1/2} \triangleq S(\bar{\Gamma}(\bar{t}), \epsilon/2).$$

Then there exists $\delta > 0$ such that

$$\bar{\Gamma}(t) \subset G_{1/2} \qquad \text{and} \qquad \bar{\Gamma}(\bar{t}) \subset S(\bar{\Gamma}(t), \epsilon/2)$$

provided $t \in F_{\eta,\delta} \triangleq \{\tau \in F_\eta \mid d(\tau, \bar{t}) < \delta\}$. Now let $c(r) \triangleq 0$ if $G_1 = R$ and otherwise

$$c(r) \triangleq d[r, \bar{G}_{1/2}]/(d[r, \bar{G}_{1/2}] + d[r, R \sim G_1]) \qquad (r \in R),$$

and let χ be the characteristic function of $F_{\eta,\delta}$. Then c is continuous,

$$\int \chi(t)\, \mu(dt) \int c(r)\, \sigma_j(t)\, (dr) = 0 \qquad (j \in \mathbb{N})$$

and, by IV.1.11,

$$\begin{aligned} 0 &= \lim_j \int \chi(t)\, \mu(dt) \int c(r)\, \sigma_j(t)\, (dr) \\ &= \int \chi(t)\, \mu(dt) \int c(r)\, \sigma(t)\, (dr) \\ &\geqslant \int_{F_{\eta,\delta}} \sigma(t)(R \sim G_1)\, \mu(dt); \end{aligned}$$

hence $\sigma(t)(R \sim G_1) = 0$ or $\sigma(t)(G_1) = 1$ for μ-a.a. $t \in F_{\eta,\delta}$. Now $G_1 \subset S(\bar{\Gamma}(t), 3\epsilon/2)$ for all $t \in F_{\eta,\delta}$, and it follows that the compact set F_η can be covered by a finite collection of relatively open neighborhoods in each of which $\sigma(t)(S(\bar{\Gamma}(t), 3\epsilon/2)) = 1$ μ-a.e. Since ϵ is arbitrary, it follows from the dominated convergence theorem I.4.35 that $\sigma(t)(\bar{\Gamma}(t)) = 1$ for μ-a.a. $t \in F_\eta$. Now let $T' \triangleq \bigcup_{j=1}^{\infty} F_{1/j}$. Then $\mu(T \sim T') = 0$ and $\sigma(\bar{\Gamma}(t)) = 1$ for μ-a.a. $t \in T'$; hence $\sigma \in \Delta$.

Since, by Condition IV.3.1, $R^{\#}$ is μ-measurable, our conclusion remains valid with Δ replaced by $\mathscr{S}^{\#}$. QED

IV.3.12 ***Theorem*** Let $\phi^j \in \mathscr{B}$ $(j = 1,..., n)$, $\phi \triangleq (\phi^1,..., \phi^n)$, $\tilde{\sigma}(t) \in \operatorname{rpm}(\bar{R}^{\#}(t))$ for μ-a.a. $t \in T$, and assume that the function

$$t \to \psi(t) \triangleq \int \phi(t, r)\, \tilde{\sigma}(t)\, (dr) : T \to \mathbb{R}^n$$

is μ-measurable. Then there exists $\bar{\sigma} \in \mathscr{S}^{\#}$ such that

$$\psi(t) = \int \phi(t, r)\, \bar{\sigma}(t)\,(dr) \qquad \text{for} \quad \mu\text{-a.a. } t \in T.$$

▌PROOF By I.5.26 and Lusin's theorem I.4.19, for all $i \in \mathbb{N}$ there exists a closed subset F_i of T such that $\mu(T \sim F_i) \leqslant 1/i$ and $\psi \mid F_i$, $R^{\#} \mid F_i$, and $\phi \mid F_i \times R$ are continuous.

Now let $i \in \mathbb{N}$, and let $\delta > 0$ be such that

$$|\psi(t) - \psi(t')| \leqslant 1/i, \qquad |\phi(t, r) - \phi(t', r)| \leqslant 1/i \quad (r \in R),$$

and

$$R^{\#}(t) \subset S(R^{\#}(t'), 1/i)$$

provided $d(t, t') \leqslant \delta$, $t \in F_i$, and $t' \in F_i$. We can cover the closed (hence compact) subset F_i of the compact set T by a finite collection of open (relative to F_i) subsets $F_i^1,\ldots, F_i^{k_i}$ of F_i of diameter at most δ, and we may clearly assume that no set F_i^l is contained in the union of the others. Let

$$D_i^0 \triangleq T \sim F_i \qquad \text{and} \qquad D_i^l \triangleq F_i^l \sim \bigcup_{m=1}^{l-1} F_i^m \quad (l = 1,\ldots, k_i).$$

Then $D_i^1,\ldots, D_i^{k_i}$ are nonempty and form a partition of F_i. For each $l \in \{0, 1,\ldots, k_i\}$ we choose $t_i^l \in D_i^l$ (setting $t_i^0 \triangleq t_i^1$ if $D_i^0 = \varnothing$), denote by χ_i^l the characteristic function of D_i^l, and set

$$\sigma_i(t) \triangleq \sum_{l=0}^{k_i} \chi_i^l(t)\bar{\sigma}(t_i^l) \qquad \text{and} \qquad \psi_i(t) = \int \phi(t, r)\, \sigma_i(t)\,(dr) \quad (t \in T).$$

Then, for each $i \in \mathbb{N}$,

$$\sigma_i \in \mathscr{S}, \qquad \psi_i \text{ is } \mu\text{-measurable},$$
$$\sigma_i(t)(S(\bar{R}^{\#}(t), 1/i)) = 1 \qquad (t \in F_i),$$

and

$$|\psi(t) - \psi_i(t)| = |\psi(t) - \psi(t_i^l)| \leqslant 1/i \qquad \text{for} \quad t \in D_i^l \quad (l = 1,\ldots, k_i);$$

hence $|\psi(t) - \psi_i(t)| \leqslant 1/i$ for $t \in F_i$. We conclude that

$$\lim_i \psi_i = \psi \quad \text{in } \mu\text{-measure};$$

hence, by Egoroff's theorem I.4.18, there exists $J_1 \subset (1, 2,\ldots)$ such that $\lim_{i \in J_1} \psi_i(t) = \psi(t)$ μ-a.e.

By IV.2.1, there exist $J_2 \subset J_1$ and $\bar{\sigma} \in \mathscr{S}$ such that $\lim_{i \in J_2} \sigma_i = \bar{\sigma}$. We set

$$\bar{\psi}(t) \triangleq \int \phi(t, r)\, \bar{\sigma}(t)\,(dr) \qquad (t \in T).$$

We shall show that $\bar{\psi} = \psi$ μ-a.e. and $\bar{\sigma} \in \mathscr{S}^{\#}$ which will complete the proof of the theorem. We have

$$\text{(1)} \qquad \lim_{i \in J_2} \int_E \psi_i(t)\, \mu(dt) = \lim_{i \in J_2} \int \mu(dt) \int \chi_E(t)\, \phi(t, r)\, \sigma_i(t)\,(dr)$$

$$= \int_E \mu(dt) \int \phi(t, r)\, \bar{\sigma}(t)\,(dr)$$

$$= \int_E \bar{\psi}(t)\, \mu(dt) \qquad (E \in \Sigma).$$

Now $\lim_{i \in J_2} \psi_i = \psi$ μ-a.e., $|\psi_i(t)| \leqslant \sup_{r \in R} |\phi(t, r)|$ μ-a.e., and the function $t \to \sup_{r \in R} |\phi(t, r)|$ is μ-integrable because $\phi^j \in \mathscr{B}$ $(j = 1,..., n)$. It follows, by the dominated convergence theorem I.4.35, that

$$\lim_{i \in J_2} \int_E \psi_i(t)\, \mu(dt) = \int_E \psi(t)\, \mu(dt) \qquad (E \in \Sigma)$$

and, by (1), that

$$\int_E \bar{\psi}(t)\, \mu(dt) = \int_E \psi(t)\, \mu(dt) \qquad (E \in \Sigma);$$

hence, by I.4.34(7), that $\bar{\psi}(t) = \psi(t)$ μ-a.e.

Finally, we verify that $\bar{\sigma}(t)(\bar{R}^{\#}(t)) = 1$ μ-a.e. Indeed, let $\epsilon > 0$, and let $J \subset J_2$ be such that $\sum_{i \in J} 1/i \leqslant \epsilon$. We set

$$F^{\epsilon} \triangleq \bigcap_{i \in J} F_i, \qquad R_{\epsilon}^{\#}(t) \triangleq S(R^{\#}(t), \epsilon), \qquad \bar{R}_{\epsilon}^{\#}(t) = \overline{R_{\epsilon}^{\#}(t)} \quad (t \in F^{\epsilon})$$

and

$$\mathscr{S}_{\epsilon}^{\#} \triangleq \{\sigma \in \mathscr{S} \mid \sigma(t)(\bar{R}_{\epsilon}^{\#}(t)) = 1 \ \mu\text{-a.e. in } F^{\epsilon}\}.$$

Then $\bar{R}_{\epsilon}^{\#}(\cdot)$ is a continuous mapping on F^{ϵ} to $\mathscr{P}'(R)$ with compact values, $\mu(T \sim F^{\epsilon}) \leqslant \epsilon$, and

$$\sigma_i(t)(\bar{R}_{\epsilon}^{\#}(t)) = 1 \qquad (i \in J, \quad t \in F^{\epsilon}).$$

We may now apply IV.3.11, with F^{ϵ} replacing T and $\mathscr{S}_{\epsilon}$ replacing Δ, and conclude that

$$\bar{\sigma} = \lim_{i \in J} \sigma_i \in \mathscr{S}_{\epsilon}^{\#};$$

hence $\bar{\sigma}(t)(\bar{R}_{\epsilon}^{\#}(t)) = 1$ μ-a.e. in F^{ϵ}.

Now let $\eta > 0$. We shall show that

$$\mu(\{t \in T \mid \bar{\sigma}(t)(\bar{R}^{\#}(t)) \neq 1\}) < \eta$$

which will complete our argument. Let $i_0 \in \mathbb{N}$ be such that $\sum_{i=i_0+1}^{\infty} 2^{-i} < \eta$, and let $T_\eta \triangleq \bigcap_{i=i_0+1}^{\infty} F^{2^{-i}}$. Then $\mu(T \sim T_\eta) < \eta$ and

$$\bar{\sigma}(t)(\bar{R}^\#_{2^{-i}}(t)) = 1 \quad \mu\text{-a.e. in } T_\eta \quad (i > i_0).$$

Since $\bar{\sigma}(t) \in \text{rpm}(R)$ μ-a.e. and the characteristic function of $\bar{R}^\#_{2^{-i}}(t)$ converges for each $r \in R$ to the characteristic function of $\bar{R}^\#(t)$ as $i \to \infty$, it follows by the dominated convergence theorem I.4.35 that

$$\bar{\sigma}(t)(\bar{R}^\#(t)) = 1 \quad \mu\text{-a.e. in } T_\eta . \quad \text{QED}$$

IV.3.13 ***Theorem*** Theorem IV.3.12 remains valid if the assumption that $\phi^j \in \mathscr{B}$ $(j = 1,..., n)$ is replaced by the assumption that ϕ^j is a real-valued function on $T \times R$, $\phi^j(t, \cdot)$ is continuous, and $\phi^j(\cdot, r)$ is μ-measurable for all $(t, r) \in T \times R$.

▌PROOF For each $k \in \{0, 1, 2,...\}$ and $j \in \{1, 2,..., n\}$, let

$$T_k \triangleq \{\tau \in T \mid k \leqslant |\phi(\tau, \cdot)|_{\sup} < k + 1\},$$
$$\phi_k^j(t, r) \triangleq \phi^j(t, r) \quad (t \in T_k), \qquad \phi_k^j(t, r) \triangleq 0 \quad (t \in T \sim T_k),$$
$$\phi_k \triangleq (\phi_k^1,..., \phi_k^n).$$

By I.4.21, the set T_k is μ-measurable. Thus $\phi_k^j \in \mathscr{B}$ and

$$\psi(t) = \int \phi_k(t, r)\, \tilde{\sigma}(t)\,(dr) \qquad (t \in T_k ,\ k = 0, 1, 2,...,\ j = 1,..., n).$$

By IV.3.12, for each $k \in \{0, 1, 2,...\}$ there exists $\bar{\sigma}_k \in \mathscr{S}^\#$ such that

$$\text{(1)} \qquad \psi(t)\, \chi_k(t) = \int \phi_k(t, r)\, \bar{\sigma}_k(t)\,(dr) \qquad (t \in T,\ k = 0, 1, 2,...),$$

where χ_k is the characteristic function of T_k. We set $\bar{\sigma}(t) = \bar{\sigma}_k(t)$ for $t \in T_k$ and $k = 0, 1, 2,...$ and observe that $\bar{\sigma} \in \mathscr{S}^\#$ and, by (1),

$$\psi(t) = \int \phi(t, r)\, \bar{\sigma}(t)\,(dr) \qquad (t \in T). \quad \text{QED}$$

IV.3.14 ***Theorem*** Let $n \in \mathbb{N}$, $h : T \times T \times R \to \mathbb{R}^n$, $\nu \in \text{frm}^+(T)$, $\bar{\sigma} \in \mathscr{S}^\#$, $T' \subset T$, $\mu(T \sim T') = 0$, and assume that

(1) $R^\#(\tau)$ is closed for all $\tau \in T'$;
(2) $h \in \mathscr{B}(T \times T, \Sigma \otimes \Sigma, \nu \times \mu, R; \mathbb{R}^n)$;
and
(3) $\{h(\cdot, \tau, r) \mid r \in R^\#(\tau)\}$ is a convex subset of $\mathscr{F}(T, \Sigma, \nu, \mathbb{R}^n)$ for all $\tau \in T'$.

Then there exists $\bar{\rho} \in \mathscr{R}^{\#}$ such that

$$\int h(t, \tau, \bar{\sigma}(\tau))\, \mu(d\tau) = \int h(t, \tau, \bar{\rho}(\tau))\, \mu(d\tau) \qquad \text{for} \quad \nu\text{-a.a. } t \in T$$

[where $h(t, \tau, \bar{\sigma}(\tau)) \triangleq \int h(t, \tau, r)\, \bar{\sigma}(\tau)\,(dr)$].

▌PROOF *Step* 1 Since $h(\cdot, \cdot, r)$ is $\nu \times \mu$-measurable for all $r \in R$, and $(t, \tau) \to |h(t, \tau, \cdot)|_{\sup}$ is $\nu \times \mu$-integrable, it follows from I.4.48 and Fubini's theorem I.4.45 that there exists a set $T'' \subset T'$ such that $\mu(T \sim T'') = 0$ and both $h(\cdot, \tau, r)$ and $t \to |h(t, \tau, \cdot)|_{\sup}$ are ν-integrable for $(\tau, r) \in T'' \times R$. Now let $\tau \in T''$ be fixed, and let $\phi(r)$ denote, for all $r \in R$, the element $h(\cdot, \tau, r)$ of $L^1(T, \Sigma, \nu, \mathbb{R}^n)$. If $\lim_j r_j = \bar{r}$ in R, then, by (2) and the dominated convergence theorem I.4.35,

$$|\phi(\bar{r}) - \phi(r_j)| = \int |h(t, \tau, \bar{r}) - h(t, \tau, r_j)|\, \nu(dt) \underset{j}{\to} 0.$$

Thus $\phi : R \to L^1(T, \Sigma, \nu, \mathbb{R}^n)$ is continuous and, by I.2.9(1), $\phi(R^{\#}(\tau))$ is a compact subset of the separable Banach space $L^1(T, \Sigma, \nu, \mathbb{R}^n)$ (I.5.18). Since, by (3), $\phi(R^{\#}(\tau))$ is convex, it follows from I.6.13 that there exists $\tilde{\rho}(\tau) \in R^{\#}(\tau)$ such that

$$\int h(\cdot, \tau, r)\, \bar{\sigma}(\tau)\,(dr) = h(\cdot, \tau, \tilde{\rho}(\tau)) \qquad \text{in} \quad L^1(T, \Sigma, \nu, \mathbb{R}^n).$$

By I.5.26(3), $h(\cdot, \tau, \cdot)$ is $\nu \times \bar{\sigma}(\tau)$-measurable and therefore, by I.5.21,

$$\text{(4)} \qquad \int h(t, \tau, r)\, \bar{\sigma}(\tau)\,(dr) = h(t, \tau, \tilde{\rho}(\tau)) \qquad \text{for} \quad \nu\text{-a.a. } t \in T.$$

Step 2 Now let

$$\bar{h}(t, \tau, r) \triangleq (1 + |h(t, \tau, \cdot)|_{\sup})^{-1} h(t, \tau, r) \qquad (t, \tau \in T, \quad r \in R).$$

Then, by (2), I.4.21, and IV.2.8, $(t, \tau) \to \bar{h}(t, \tau, \bar{\sigma}(\tau))$ and $\bar{h}(\cdot, \cdot, r)$ are $\nu \times \mu$-measurable for all $r \in R$, and they are also $\nu \times \mu$-integrable since $|\bar{h}(t, \tau, r)| \leqslant 1$ and $|\bar{h}(t, \tau, \bar{\sigma}(\tau))| \leqslant 1$. Thus, by Fubini's theorem I.4.45(2) and I.4.48, there exists a set $T''' \subset T''$ such that $\mu(T \sim T''') = 0$, $t \to |\bar{h}(t, \tau, r) - \bar{h}(t, \tau, \bar{\sigma}(\tau))|$ is ν-integrable for all $\tau \in T'''$ and $r \in R$,

$$\text{(5)} \qquad \psi(\tau, r) \triangleq \int |\bar{h}(t, \tau, r) - \bar{h}(t, \tau, \bar{\sigma}(\tau))|\, \nu(dt)$$

exists for all $(\tau, r) \in T''' \times R$, and $\psi(\cdot, r)$ is μ-measurable. Furthermore, by the dominated convergence theorem I.4.35, we have

$$\lim_j \psi(\tau, r_j) = \psi(\tau, r) \quad (\tau \in T''') \qquad \text{whenever} \quad \lim_j r_j = r.$$

Thus $\psi(\tau, \cdot)$ is continuous on R for all $\tau \in T'''$.

By (4) and (5), $\psi(\tau, \tilde{\rho}(\tau)) = 0$ for all $\tau \in T'''$. We conclude, applying the Filippov–Castaing theorem I.7.10, that there exists $\bar{\rho} \in \mathscr{R}^{\#}$ such that $\psi(\tau, \bar{\rho}(\tau)) = 0$ $(\tau \in T''')$; hence, by (5) and I.4.27(6), for each $\tau \in T'''$ there exists a ν-null set Z_τ such that

$$(6) \qquad h(t, \tau, \bar{\sigma}(\tau)) = h(t, \tau, \bar{\rho}(\tau)) \qquad (\tau \in T''', t \in T \sim Z_\tau).$$

If we denote by Z the subset of $T \times T$ on which (6) fails to be valid, then Z is $\nu \times \mu$-mesurable and

$$\int \chi_Z(t, \tau)\, \nu(dt) \times \mu(d\tau) \leqslant \int \mu(d\tau) \int_{Z_\tau} \nu(dt) = 0.$$

Thus $\nu \times \mu(Z) = 0$ and $\mu(\{\tau \in T \mid (t, \tau) \in Z\}) = 0$ for ν-a.a. $t \in T$. Relation (6) and Fubini's theorem I.4.45(3) imply, therefore, that

$$\int h(t, \tau, \bar{\sigma}(\tau))\, \mu(d\tau) = \int h(t, \tau, \bar{\rho}(\tau))\, \mu(d\tau) \qquad \text{for} \quad \nu\text{-a.a } t \in T. \quad \text{QED}$$

Notes

The idea of embedding a set of "ordinary" functions (such as $\mathscr{R}$) in a larger set, whose elements need not be functions with the same domain, appears to have originated in connection with partial differential equations, and various constructions of "weak solutions" gave rise to Sobolev spaces (Sobolev [1]), distributions (Schwartz [1]), and generalized functions (e.g., Gel'fand and Shilov [1, 2]). Both distributions and other generalized functions are identified with continuous linear functionals on an appropriate topological vector space.

Within the context of the calculus of variations such a construction was first defined and investigated by Young [1, 2] in the form of "generalized curves" (see also Young [3] for a detailed discussion). Within the context of the optimal control theory of ordinary differential equations, weak solutions or similar constructions have been considered by Filippov [1] ("sliding regimes"), Warga [1] ("relaxed curves"), McShane [5] ("relaxed controls"), and Ghouila–Houri [1] ("commandes limites"). The sets $\mathscr{S}$ and $\mathscr{S}^{\#}$, as defined in Chapter IV, were introduced in Warga [8].

Theorem IV.1.8 is related to certain variants (Schwartz [2, Expose n° 4, p. 3]) of the Dunford–Pettis theorem (Dunford and Schwartz [1, Th. 6, p. 503]), but we prefer to treat the case of interest to us, namely $L^1(T, C(R))^*$, by itself. Theorems IV.2.1, IV.2.6, IV.3.9, IV.3.11, and IV.3.12 generalize results of Warga [8 and 12, Lemma 3.2, p. 366].

CHAPTER V

Control Problems Defined by Equations in Banach Spaces

V.0 Formulation of the Optimal Control Problem

We shall consider here the general control problem of Chapter III, defined by the sets $\mathscr{Y}$, $\mathscr{U}$, B, C_1, and $\mathscr{X}_1$ and functions

$$F : \mathscr{Y} \times \mathscr{U} \times B \to \mathscr{Y}, \qquad g_0 : \mathscr{Y} \times \mathscr{U} \times B \to \mathbb{R},$$
$$g_1 : \mathscr{Y} \times \mathscr{U} \times B \to \mathscr{X}_1 .$$

We restrict our attention to points $(y, u, b) \in \mathscr{Y} \times \mathscr{U} \times B$ that satisfy the equation of motion $y = F(y, u, b)$ and, among such points, consider those that "come close" to satisfying the restriction $g_1(y, u, b) \in C_1$ and, in so doing, tend to minimize $g_0(y, u, b)$.

We shall find it convenient to assume, without any loss of generality, that $\mathscr{X}_1$ is a topological vector space of the form $\mathbb{R}^m \times \mathscr{X}_2$ and that $C \triangleq C_1 \times C_2$, with $C_1 \subset \mathbb{R}^m$ and $C_2 \subset \mathscr{X}_2$. The reason for this assumption is that, in discussing necessary conditions for a minimum, we shall have to assume that C_2

is a convex set with a nonempty interior whereas C_1 can be an arbitrary convex set. Accordingly, we shall replace g_1 by a function $(g_1, g_2) : \mathscr{Y} \times \mathscr{U} \times B \to \mathbb{R}^m \times \mathscr{X}_2$. In this formulation, our problem is to determine "low" values of g_0 on the set

$$\mathscr{H}(\mathscr{U}) \triangleq \{(y, u, b) \in \mathscr{Y} \times \mathscr{U} \times B \mid y = F(y, u, b)\},$$

subject to "nearly" satisfying the restrictions $g_1(y, u, b) \in C_1$, $g_2(y, u, b) \in C_2$. It is also convenient to reduce our problem to a form corresponding to $C_1 = \{0\}$. We accomplish this by letting

$$\tilde{B} \triangleq B \times C_1, \qquad \tilde{b} \triangleq (b, c_1) \in \tilde{B},$$
$$\tilde{g}_0(y, u, \tilde{b}) \triangleq g_0(y, u, b), \qquad \tilde{g}_1(y, u, \tilde{b}) \triangleq g_1(y, u, b) - c_1,$$

and

$$\tilde{g}_2(y, u, \tilde{b}) \triangleq g_2(y, u, b) \qquad [(y, u, \tilde{b}) \in \mathscr{Y} \times \mathscr{U} \times \tilde{B}].$$

The old restriction $g_1(y, u, b) \in C_1$ is now equivalent to $\tilde{g}_1(y, u, \tilde{b}) = 0$. We shall assume, from now on, that this transformation has been carried out.

Formulation of the Problem

Let $m \in \mathbb{N}$, $\mathscr{Y}$ be a Banach space, and B a convex subset of a vector space. We refer to an element σ of $\mathscr{S}^{\#}$ as a *control function*, a point $b \in B$ as a *control parameter*, and a point $q \triangleq (\sigma, b) \in \mathscr{S}^{\#} \times B$ as a *control.* We assume given a topological vector space $\mathscr{X}_2$, a set $C_2 \subset \mathscr{X}_2$, functions $F : \mathscr{Y} \times \mathscr{S}^{\#} \times B \to \mathscr{Y}$ and $g \triangleq (g_0, g_1, g_2) : \mathscr{Y} \times \mathscr{S}^{\#} \times B \to \mathbb{R} \times \mathbb{R}^m \times \mathscr{X}_2$, and we set, for any $\mathscr{S}' \subset \mathscr{S}^{\#}$,

$$\mathscr{H}(\mathscr{S}') \triangleq \{(y, \sigma, b) \in \mathscr{Y} \times \mathscr{S}' \times B \mid y = F(y, \sigma, b)\}$$

and

$$\mathscr{A}(\mathscr{S}') \triangleq \{(y, \sigma, b) \in \mathscr{H}(\mathscr{S}') \mid g_1(y, \sigma, b) = 0, g_2(y, \sigma, b) \in C_2\}.$$

A point $(\bar{y}, \bar{q}) \triangleq (\bar{y}, \bar{\sigma}, \bar{b}) \in \mathscr{A}(\mathscr{S}')$ is a *minimizing $\mathscr{S}'$-solution* if $g_0(\bar{y}, \bar{q}) = \operatorname{Min} g_0(\mathscr{A}(\mathscr{S}'))$. A minimizing $\mathscr{S}^{\#}$-solution is a *minimizing relaxed solution* and a minimizing $\mathscr{U}$-solution is a *minimizing original solution* if $\mathscr{U} \subset \mathscr{R}^{\#}$. A sequence $((y_j, u_j, b_j))$ in $\mathscr{H}(\mathscr{U})$ is an *approximate $\mathscr{U}$-solution* if, for every neighborhood G of 0 in $\mathbb{R}^m \times \mathscr{X}_2$,

$$(g_1(y_j, u_j, b_j), g_2(y_j, u_j, b_j)) \in \{0\} \times C_2 + G$$

for all sufficiently large $j \in \mathbb{N}$. An approximate $\mathscr{U}$-solution $((\bar{y}_j, \bar{u}_j, \bar{b}_j))$ is a *minimizing approximate $\mathscr{U}$-solution* if $\lim_j g_0(\bar{y}_j, \bar{u}_j, \bar{b}_j)$ exists in $\mathbb{R}$ and

$$\lim_j g_0(\bar{y}_j, \bar{u}_j, \bar{b}_j) \leqslant \liminf_j g_0(y_j, u_j, b_j)$$

for every approximate $\mathscr{U}$-solution $((y_j, u_j, b_j))$.

V.1 Existence of Minimizing Relaxed and Approximate Solutions

V.1.1 *Theorem* Let C_2 be sequentially closed, B a topological space, and $\mathscr{A}(\mathscr{S}^{\#})$ nonempty, and let $\mathscr{A}_1$ denote either the set $\mathscr{A}(\mathscr{S}^{\#})$ or the set

$$\mathscr{A}' \triangleq \{(y, \sigma, b) \in \mathscr{A}(\mathscr{S}^{\#}) \mid g_0(y, \sigma, b) \leqslant g_0(y', \sigma', b') + \epsilon\}$$

for some $(y', \sigma', b') \in \mathscr{A}(\mathscr{S}^{\#})$ and $\epsilon > 0$. Assume that seq cl$(\mathscr{A}_1)$ is sequentially compact and that $F \mid \text{seq cl}(\mathscr{A}_1)$ and $g \mid \text{seq cl}(\mathscr{A}_1)$ are sequentially continuous. Then $\mathscr{A}_1$ is sequentially compact and there exists a minimizing relaxed solution $(\bar{y}, \bar{q}) \triangleq (\bar{y}, \bar{\sigma}, \bar{b})$.

▌PROOF Let $((y_j, q_j)) \triangleq ((y_j, \sigma_j, b_j))$ be a sequence in $\mathscr{A}_1$. Then there exist $J \subset (1, 2, \ldots)$ and $(\bar{y}, \bar{q}) \in \text{seq cl}(\mathscr{A}_1)$ such that

$$\lim_{j \in J}(y_j, q_j) = (\bar{y}, \bar{q});$$

hence

$$\bar{y} = \lim_{j \in J} y_j = \lim_{j \in J} F(y_j, q_j) = F(\bar{y}, \bar{q}),$$

$$g_1(\bar{y}, \bar{q}) = \lim_{j \in J} g_1(y_j, q_j) = 0,$$

$$g_2(\bar{y}, \bar{q}) = \lim_{j \in J} g_2(y_j, q_j) \in C_2,$$

and, if $\mathscr{A}_1 = \mathscr{A}'$, we have

$$g_0(\bar{y}, \bar{q}) = \lim_{j \in J} g_0(y_j, q_j) \leqslant g_0(y', \sigma', b') + \epsilon.$$

This shows that $(\bar{y}, \bar{q}) \in \mathscr{A}_1$ and $\mathscr{A}_1$ is therefore sequentially compact. It is clear that we can choose the sequence $((y_j, q_j))$ in $\mathscr{A}_1$ in such a manner that $\lim_j g_0(y_j, q_j) = \inf g_0(\mathscr{A}(\mathscr{S}^{\#}))$. Then

$$\lim_{j \in J} g_0(y_j, q_j) = g_0(\bar{y}, \bar{q}) = \inf g_0(\mathscr{A}(\mathscr{S}^{\#})),$$

showing that $(\bar{y}, \bar{q})$ is a minimizing relaxed solution. QED

V.1.2 *Theorem* Let C_2 be sequentially closed, B a topological space, $(\bar{y}, \bar{\sigma}, \bar{b})$ a minimizing relaxed solution, and $\mathscr{U}$ an abundant set. Furthermore, let $\mathscr{H}_1$ denote either the set $\mathscr{H}(\mathscr{S}^{\#})$ or the set

$$\mathscr{H}' \triangleq \{(y, \sigma, b) \in \mathscr{H}(\mathscr{S}^{\#}) \mid g_0(y, \sigma, b) \leqslant g_0(y_1, \sigma_1, b_1) + \epsilon, \mid g_1(y, \sigma, b) \mid \leqslant \epsilon, \\ g_2(y, \sigma, b) \in C_2 + U_2\}$$

for some $(y_1, \sigma_1, b_1) \in \mathscr{A}(\mathscr{S}^{\#})$, $\epsilon > 0$ and a sequentially closed neighborhood U_2 of the origin in $\mathscr{X}_2$. Assume that

(1) $\operatorname{seq\,cl}(\mathscr{H}_1)$ is sequentially compact, $F \mid \operatorname{seq\,cl}(\mathscr{H}_1)$ and $g \mid \operatorname{seq\,cl}(\mathscr{H}_1)$ are sequentially continuous, and

(2) $\bar{y}$ is the unique solution of the equation $y = F(y, \bar{\sigma}, \bar{b})$, and the equation $y = F(y, u, \bar{b})$ has a (not necessarily unique) solution y for all $u \in \mathscr{U}$ and in some neighborhood of $\bar{\sigma}$ in $\mathscr{S}^{\#}$.

Then $\mathscr{H}_1$ is sequentially compact and there exists a sequence $((\bar{y}_j, \bar{u}_j))$ in $\mathscr{Y} \times \mathscr{U}$ such that $((\bar{y}_j, \bar{u}_j, \bar{b}))$ is a minimizing approximate $\mathscr{U}$-solution and

$$\lim_j g(\bar{y}_j, \bar{u}_j, \bar{b}) = g(\bar{y}, \bar{\sigma}, \bar{b}).$$

▌ PROOF We first show that $\mathscr{H}_1$ is sequentially compact by the same argument that we used in V.1.1 to show that $\mathscr{A}_1$ was sequentially compact. Since, by IV.3.10, $\mathscr{U}$ is dense in the metric space $\mathscr{S}^{\#}$, there exists a sequence $(\bar{u}_j)$ in $\mathscr{U}$ converging to $\bar{\sigma}$. By assumption, the equation $y = F(y, \bar{u}_j, \bar{b})$ has a solution $\bar{y}_j$ for all sufficiently large $j \in \mathbb{N}$, and we shall assume that $(\bar{u}_j)$ was chosen so that $\bar{y}_j$ exists for all $j \in \mathbb{N}$. Since $\mathscr{H}_1$ is sequentially compact, so is the metric space $H \triangleq \operatorname{pr}_{\mathscr{Y} \times \mathscr{S}^{\#}}(\mathscr{H}_1)$, and there exist some $J \subset (1, 2, ...)$ and $(\tilde{y}, \tilde{\sigma}) \in H$ such that

$$\lim_{j \in J} (\bar{y}_j, \bar{u}_j) = (\tilde{y}, \tilde{\sigma}) = (\tilde{y}, \bar{\sigma}).$$

Therefore $\tilde{y} = F(\tilde{y}, \bar{\sigma}, \bar{b})$ and, by the uniqueness assumption, $\tilde{y} = \bar{y}$; hence

$$\lim_{j \in J} g(\bar{y}_j, \bar{u}_j, \bar{b}) = g(\bar{y}, \bar{\sigma}, \bar{b}) \in \mathbb{R} \times \{0\} \times C_2 .$$

This shows that $((\bar{y}_j, \bar{u}_j, \bar{b}))$ is an approximate $\mathscr{U}$-solution.

Now let $((y_j, u_j, b_j))$ be an approximate $\mathscr{U}$-solution, and assume that

$$\liminf_j g_0(y_j, u_j, b_j) < g_0(\bar{y}, \bar{\sigma}, \bar{b}).$$

We can determine $J_1 \subset (1, 2, ...)$ such that

$$\lim_{j \in J_1} g_0(y_j, u_j, b_j) = \liminf_j g_0(y_j, u_j, b_j)$$

and

$$(y_j, u_j, b_j) \in \mathscr{H}_1 \qquad (j \in J_1).$$

Since $\mathscr{H}_1$ is sequentially compact, there exist $J_2 \subset J_1$ and $(\tilde{y}, \tilde{\sigma}, \tilde{b}) \in \mathscr{H}_1$ such that $\lim_{j \in J_2}(y_j, u_j, b_j) = (\tilde{y}, \tilde{\sigma}, \tilde{b})$. Then

$$g_0(\tilde{y}, \tilde{\sigma}, \tilde{b}) = \lim_{j \in J_2} g_0(y_j, u_j, b_j) < g_0(\bar{y}, \bar{\sigma}, \bar{b})$$

and

$$g(\tilde{y}, \tilde{\sigma}, \tilde{b}) = \lim_{j \in J_2} g(y_j, u_j, b_j) \in \mathbb{R} \times \{0\} \times C_2;$$

hence

$$(\tilde{y}, \tilde{\sigma}, \tilde{b}) \in \mathscr{A}(\mathscr{S}^{\#}) \qquad \text{and} \qquad g_0(\tilde{y}, \tilde{\sigma}, \tilde{b}) < g_0(\bar{y}, \bar{\sigma}, \bar{b}),$$

contradicting the assumption that $(\bar{y}, \bar{\sigma}, \bar{b})$ is a minimizing relaxed solution. It follows that $((\bar{y}_j, \bar{u}_j, \bar{b}))_{j \in J}$ is a minimizing approximate $\mathscr{U}$-solution. QED

V.2 Necessary Conditions for a Relaxed Minimum

The basic result of this section, Theorem V.2.3, can also be derived as a special case of Theorem V.3.2. However, we prove V.2.3 independently because its derivation is simpler than that of V.3.2 and some readers may prefer to concentrate at first reading on "relaxed" problems.

The necessary conditions for a minimum that we study involve certain relations analogous to those known in the classical calculus of variations as "the Euler–Lagrange equations," "the Weierstrass E-condition," and "the transversality conditions." We combine these relations in defining an *extremal* of the optimal control problem. This expression is used in the calculus of variations to denote any solutions of the "Euler–Lagrange equations," and our definition is, therefore, more restrictive than the variational one. On the other hand, both definitions refer to necessary conditions for a minimum that often yield minimizing solutions when they are combined with admissibility conditions. We therefore borrow the term "extremal" in order to avoid coining new and unfamiliar expressions.

V.2.0 ***Definition.*** *Extremal* Let $(\bar{y}, \bar{\sigma}, \bar{b}) \in \mathscr{Y} \times \mathscr{S}^{\#} \times B$, $\bar{q} \triangleq (\bar{\sigma}, \bar{b})$, and $l \triangleq (l_0, l_1, l_2) \in [0, \infty) \times \mathbb{R}^m \times \mathscr{X}_2^*$. We refer to the point $(\bar{y}, \bar{\sigma}, \bar{b}, l)$ [also written $(\bar{y}, \bar{q}, l_0, l_1, l_2)$ or $(\bar{y}, \bar{\sigma}, \bar{b}, l_0, l_1, l_2)$] as an *extremal* of the optimal control problem if the expression

$$\begin{aligned}\chi(q) &\triangleq (\chi_0(q), \chi_1(q), \chi_2(q))\\ &\triangleq g_y(\bar{y}, \bar{q}) \circ [I - F_y(\bar{y}, \bar{q})]^{-1} D_2F(\bar{y}, \bar{q}; q - \bar{q}) + D_2 g(\bar{y}, \bar{q}; q - \bar{q})\end{aligned}$$

is defined in $\mathbb{R} \times \mathbb{R}^m \times \mathscr{X}_2$ for all $q \triangleq (\sigma, b) \in \mathscr{S}^\# \times B$ and we have

$$l \neq 0,$$

$$l_0\chi_0(q) + l_1 \cdot \chi_1(q) + l_2(\chi_2(q)) \geqslant 0 \qquad (q \in \mathscr{S}^\# \times B),$$

and

$$l_2(g_2(\bar{y}, \bar{q})) = \operatorname*{Max}_{c \in C_2} l_2(c).$$

The extremal $(\bar{y}, \bar{\sigma}, \bar{b}, l)$ is *admissible* if $(\bar{y}, \bar{\sigma}, \bar{b}) \in \mathscr{A}(\mathscr{S}^\#)$; it is *normal* if $l_0 \neq 0$ and *abnormal* if $l_0 = 0$. A point $(\bar{y}, \bar{\sigma}, \bar{b}) \in \mathscr{Y} \times \mathscr{S}^\# \times B$ is *extremal* if there exists some $l \in [0, \infty) \times \mathbb{R}^m \times \mathscr{X}_2{}^*$ such that $(\bar{y}, \bar{\sigma}, \bar{b}, l)$ is an extremal. An extremal point $(\bar{y}, \bar{\sigma}, \bar{b})$ is *admissible* if it belongs to $\mathscr{A}(\mathscr{S}^\#)$; it is *abnormal* if there exists an abnormal extremal $(\bar{y}, \bar{\sigma}, \bar{b}, l)$; and it is *normal* if it is not abnormal.

The relation

$$l_0\chi_0(q) + l_1 \cdot \chi_1(q) + l_2(\chi_2(q)) \geqslant 0 \qquad (q \in \mathscr{S}^\# \times B)$$

is a generalization of the Weierstrass E-condition and of Pontryagin's "maximum principle." The relation

$$l_2(g_2(\bar{y}, \bar{q})) = \operatorname*{Max}_{c \in C_2} l_2(c)$$

can be thought of as a generalization of the variational "transversality conditions."

We require two lemmas, the first of which, V.2.1, will also be used in V.3. We shall write $\eta_{0,1} \triangleq (\eta_0, \eta_1)$ and $\eta_{1,2} \triangleq (\eta_1, \eta_2)$ for

$$\eta \triangleq (\eta_0, \eta_1, \eta_2) \in \mathbb{R} \times \mathbb{R}^m \times \mathscr{X}_2 .$$

V.2.1 ***Lemma*** Let W be a convex subset of $\mathbb{R} \times \mathbb{R}^m \times \mathscr{X}_2$, C' an open convex subset of $\mathscr{X}_2$, $0 \in W$, and $0 \in \bar{C}'$. Then either there exists $l \triangleq (l_0, l_1, l_2) \in [0, \infty) \times \mathbb{R}^m \times \mathscr{X}_2{}^*$ such that $l \neq 0$,

$$l(w) = l_0 w_0 + l_1 \cdot w_1 + l_2(w_2) \geqslant 0 \qquad [w \triangleq (w_0, w_1, w_2) \in W],$$

and

$$l_2(c) \leqslant 0 \qquad (c \in \bar{C}'),$$

or there exist points $\xi^i \triangleq (\xi_0{}^i, \xi_1{}^i, \xi_2{}^i) \in W$ and numbers $\beta^i > 0$ $(i=0, 1,..., m)$ such that $\sum_{i=0}^m \beta^i = 1$, $\xi_2{}^i \in C'$, the set $\{\xi_{0,1}^0, ..., \xi_{0,1}^m\}$ is independent, $\xi_0{}^i < 0$ and $\sum_{i=0}^m \beta^i \xi_1{}^i = 0$.

▌PROOF We first consider the special case where W contains the point $(1, 0, 0) \in \mathbb{R} \times \mathbb{R}^m \times \mathscr{X}_2$. Let

$$W_{0,1} \triangleq \operatorname{pr}_{\mathbb{R}\times\mathbb{R}^m} W \triangleq \{\xi_{0,1} \in \mathbb{R} \times \mathbb{R}^m \mid \xi \triangleq (\xi_0, \xi_1, \xi_2) \in W\}.$$

Then either

(1) $0 \in \partial W_{0,1}$, or

(2) $0 \in W_{0,1}^\circ$ and there exists some $\bar{\xi} \triangleq (\bar{\xi}_0, \bar{\xi}_1, \bar{\xi}_2) \in W$ such that $\bar{\xi}_0 < 0$, $\bar{\xi}_1 = 0$, and $\bar{\xi}_2 \in C'$, or

(3) $0 \in W_{0,1}^\circ$ and every $\xi \triangleq (\xi_0, \xi_1, \xi_2) \in \mathbb{R} \times \mathbb{R}^m \times \mathscr{X}_2$ with $\xi_0 < 0$, $\xi_1 = 0$, and $\xi_2 \in C'$ is outside W.

If (1) holds, then, by I.6.11, the first alternative of the lemma is satisfied with $l_2 = 0$ and (l_0, l_1) an inward normal to $W_{0,1}$ at 0. [We have $l_0 \geqslant 0$ because $(1,0) \in W_{0,1}$.]

If (2) holds, then $0 \in W_{0,1}^\circ$ and there exist $\eta^i \in W$ and $\beta^i > 0$ $(i = 0, 1,..., m)$ such that $\sum_{i=0}^m \beta^i = 1$, $\eta_0{}^i < 0$, $\{\eta_{0,1}^0, ..., \eta_{0,1}^m\}$ is independent, and $\sum_{i=0}^m \beta^i \eta_1{}^i = 0$. For every $\alpha \in (0, 1]$, we have

$$\xi^i \triangleq \alpha\eta^i + (1-\alpha)\,\bar{\xi} \in W \quad (i = 0, 1,..., m),$$

$$\xi_0{}^i < 0, \qquad \text{and} \qquad \sum_{i=0}^m \beta^i \xi_1{}^i = 0.$$

Furthermore, by choosing α sufficiently close to 0, we can insure that $\xi_2{}^i \in C'$ and $(\xi_{0,1}^0, ..., \xi_{0,1}^m)$ is independent (the latter statement being verified by expressing $\bar{\xi}_{0,1}$ as a linear combination of $\eta_{0,1}^0, ..., \eta_{0,1}^m$).

Finally, we consider case (3). Let

$$W_2' \triangleq \{w_2 \mid w \triangleq (w_0, w_1, w_2) \in W,\ w_0 < 0,\ w_1 = 0\}.$$

Then W_2' is a nonempty convex subset of $\mathscr{X}_2$ and $W_2' \cap C' = \varnothing$. Since C' is an open convex set, there exist, by I.6.7, $l_2' \in \mathscr{X}_2{}^*$ and $\alpha \in \mathbb{R}$ such that $l_2' \neq 0$ and

(4) $$l_2'(w_2) \leqslant \alpha \leqslant l_2'(c) \qquad (w_2 \in W_2',\ c \in C').$$

Since $0 \in \bar{C}' \cap \overline{W}_2'$, we have $\alpha = 0$.

Now let $W^\# \triangleq \{(l_2'(w_2), w_0, w_1) \mid w \triangleq (w_0, w_1, w_2) \in W\}$, let 0_m represent the origin of $\mathbb{R}^m$, and let

$$H \triangleq (0, \infty) \times (-\infty, 0) \times \{0_m\}.$$

Then, by (4) (with $\alpha = 0$), we have $H \cap W^\# = \varnothing$. Since $0 \in W^\# \cap \overline{H}$ and both H and $W^\#$ are nonempty convex subsets of $\mathbb{R} \times \mathbb{R} \times \mathbb{R}^m$, it follows from I.6.10 that there exist $l_0', l_0 \in \mathbb{R}$ and $l_1 \in \mathbb{R}^m$ such that $|l_0'| + |l_0| + |l_1| \neq 0$ and

(5) $$l_0' l_2'(w_2) + l_0 w_0 + l_1 \cdot w_1 \geqslant 0 \geqslant l_0' \xi_0' + l_0 \xi_0$$
$$[w \triangleq (w_0, w_1, w_2) \in W, \ \xi_0' > 0, \ \xi_0 < 0].$$

It follows that $l_0' \leqslant 0$ and $l_0 \geqslant 0$. If we set $l_0' l_2' \triangleq l_2$, then (4) and (5) imply the first alternative of the lemma. Thus the lemma has been proven on the assumption that $(1, 0, 0) \in W$.

If $(1, 0, 0) \notin W$, we set $W' \triangleq \mathrm{co}(W \cup \{(1, 0, 0)\})$. Since $W \subset W'$, the first alternative of the lemma remains true for W if it is true for W'. If the second alternative is true for W', then we observe that, for each $i = 0, 1,..., m$, there exist $\alpha_i \in [0, 1]$ and $\tilde{\eta}^i \in W$ such that $\xi^i = (1 - \alpha_i)(1, 0, 0) + \alpha_i \tilde{\eta}^i$. Since W is convex and contains $(0, 0, 0)$, we have

$$\eta^i \triangleq \alpha_i \tilde{\eta}^i \in W \quad \text{and} \quad \xi^i = (1 - \alpha_i, 0, 0) + \eta^i \qquad (i = 0, 1,..., m).$$

Furthermore $\eta_2{}^i = \xi_2{}^i \in C'$, $\eta_0{}^i = \xi_0{}^i - (1 - \alpha_i) < 0$, and $\sum_{i=0}^m \beta^i \eta_1{}^i = \sum_{i=0}^m \beta^i \xi_1{}^i = 0$. We shall next verify that the set $\{\eta_{0,1}^0, ..., \eta_{0,1}^m\}$ is independent which will show that the set $\{\eta^0,..., \eta^m\}$ has all the properties listed for $\{\xi^0,..., \xi^m\}$ and will thus complete the proof of the lemma. Indeed, if $\sum_{i=0}^m \gamma^i \eta_{0,1}^i = 0$, then

$$\sum_{i=0}^m \gamma^i \eta_1{}^i = \sum_{i=0}^m \gamma^i \xi_1{}^i = 0,$$

and $\sum_{i=0}^m (\gamma^i + \alpha\beta^i)\, \xi_1{}^i = 0$ for all $\alpha \in \mathbb{R}$. Since $\sum_{i=0}^m \beta^i \xi_0{}^i < 0$, we can set

$$\bar{\alpha} \triangleq -\sum_{i=0}^m \gamma^i \xi_0{}^i \Big/ \sum_{i=0}^m \beta^i \xi_0{}^i$$

and observe that $\sum_{i=0}^m (\gamma^i + \bar{\alpha}\beta^i)\, \xi_{0,1}^i = 0$. Because the set $\{\xi_{0,1}^0, ..., \xi_{0,m}^m\}$ is independent, this implies that $\gamma^i = -\bar{\alpha}\beta^i$ for $i = 0, 1,..., m$. Thus

$$0 = \sum_{i=0}^m \gamma^i \eta_0{}^i = -\bar{\alpha} \sum_{i=0}^m \beta^i \xi_0{}^i + \bar{\alpha} \sum_{i=0}^m \beta^i (1 - \alpha_i).$$

Since $\sum_{i=0}^m \beta^i \xi_0{}^i < 0$ and $\sum_{i=0}^m \beta^i (1 - \alpha_i) \geqslant 0$, we must have $\bar{\alpha} = 0$; hence $(\gamma^0,..., \gamma^m) = (0,..., 0)$. QED

V.2.2 ***Lemma*** Let C', β^j and $\xi^j \triangleq (\xi_0{}^j, \xi_1{}^j, \xi_2{}^j) \in \mathbb{R} \times \mathbb{R}^m \times \mathscr{X}_2$ $(j = 0,..., m)$ have the properties described in V.2.1, $\mathscr{T}'$ be a neighborhood of 0 in $\mathscr{T}_{m+1}$, and $h : \mathscr{T}' \to \mathbb{R} \times \mathbb{R}^m \times \mathscr{X}_2$ be continuous and have a derivative $h'(0)$ such that

(1) $$h'(0)\, \Delta\theta = \sum_{j=0}^m \Delta\theta^j\, \xi^j \qquad (\Delta\theta \in \mathbb{R}^{m+1}).$$

Then there exist $\bar{\gamma} > 0$ and a function $\gamma \to \tilde{\theta}(\gamma) : (0, \bar{\gamma}] \to \mathscr{T}'$ such that $h_0(\tilde{\theta}(\gamma)) < h_0(0)$, $h_1(\tilde{\theta}(\gamma)) = h_1(0)$, and $h_2(\tilde{\theta}(\gamma)) - h_2(0) \in C'$ $(\gamma \in (0, \bar{\gamma}])$.

▌PROOF By assumption, h is differentiable at 0 and continuous in some neighborhood of 0, say for some $\alpha > 0$ and $\theta \in [0, \alpha]^{m+1}$. Let $\beta \triangleq (\beta^0,..., \beta^m)$, H be the element of $B(\mathbb{R}^{m+1}, \mathbb{R}^{m+1})$ defined by

$$H\,\Delta\theta \triangleq \sum_{j=0}^{m} \Delta\theta^j\, \xi_{0,1}^j \qquad [\Delta\theta \triangleq (\Delta\theta^0,..., \Delta\theta^m) \in \mathbb{R}^{m+1}],$$

and let

$$\psi(\theta) \triangleq h(\theta) - h(0) - h'(0)\theta \qquad (\theta \in \mathscr{T}').$$

By (1), we have $h'_{0,1}(0) = H$. Since, by assumption, the set $\{\xi_{0,1}^0,..., \xi_{0,1}^m\}$ is independent, it follows from I.3.3 that H has a bounded inverse. We conclude that, for any fixed $\gamma > 0$, the equations

$$h_{0,1}(\theta) - h_{0,1}(0) - \gamma h'_{0,1}(0)\,\beta = 0 \tag{2}$$

and

$$\theta = \gamma\beta - H^{-1}\psi_{0,1}(\theta) \tag{3}$$

are equivalent for $\theta \in \mathscr{T}'$.

We set

$$\beta_{\min} \triangleq \operatorname*{Min}_j \beta^j, \qquad \beta_{\max} \triangleq \operatorname*{Max}_j \beta^j, \qquad \text{and} \qquad \theta_{\max} \triangleq \operatorname*{Max}_j \theta^j.$$

The differentiability of h at 0 implies that $\psi_{0,1}(|\,\theta\,|) = o(|\,\theta\,|)$. There exists, therefore, some $\alpha' \in (0, \alpha]$ such that

$$|\,H^{-1}\psi_{0,1}(\theta)| \leqslant \frac{1}{3}\frac{\beta_{\min}}{\beta_{\max}}\,\theta_{\max} \qquad \text{if} \quad 0 \leqslant \theta^j \leqslant \alpha' \quad (j = 0,..., m).$$

Furthermore, for

$$\tilde{\gamma} \triangleq \frac{2}{3}\frac{\alpha'}{\beta_{\max}} \qquad \text{and} \qquad \gamma \in (0, \tilde{\gamma}],$$

the functions h and $\psi_{0,1}$ are continuous on the set

$$\mathscr{T}_\gamma' \triangleq \{\theta \in \mathbb{R}^{m+1} \mid |\,\theta^j - \gamma\beta^j\,| \leqslant \tfrac{1}{2}\gamma\beta_{\min}\ (j = 0,..., m)\} \subset [0, \alpha]^{m+1}.$$

Thus $\mathscr{T}_\gamma'$ is a bounded convex body in $\mathbb{R}^{m+1}$ and therefore, by Brouwer's fixed point theorem II.2.3, it has the fixed point property. Finally, the function

$$\theta \to \gamma\beta - H^{-1}\psi_{0,1}(\theta)$$

maps $\mathcal{T}_\gamma'$ into itself and is continuous on $\mathcal{T}_\gamma'$. Therefore the equivalent equations (2) and (3) admit a solution $\tilde{\theta}(\gamma) \in \mathcal{T}_\gamma'$ for every $\gamma \in (0, \tilde{\gamma}]$. Thus, for all $\gamma \in (0, \tilde{\gamma}]$,

$$h_0(\tilde{\theta}(\gamma)) = h_0(0) + \gamma h_0'(0)\,\beta = h_0(0) + \gamma \sum_{j=0}^{m} \beta^j \xi_0{}^j < h_0(0)$$

and

$$h_1(\tilde{\theta}(\gamma)) = h_1(0) + \gamma h_1'(0)\,\beta = h_1(0) + \gamma \sum_{j=0}^{m} \beta^j \xi_1{}^j = h_1(0).$$

We also observe that $\lim_{\gamma\to+0} \sup\{|\,\theta\,| \mid \theta \in \mathcal{T}_\gamma'\} = 0$ and consequently $\lim_{\gamma\to+0} \tilde{\theta}(\gamma) = 0$.

Now consider the point $h_2(\tilde{\theta}(\gamma)) \in \mathcal{X}_2$. We have

$$\lim_{\gamma\to+0} \tilde{\theta}(\gamma) = 0, \qquad \psi_{0,1}(\theta) = o(|\,\theta\,|),$$

and

$$\tilde{\theta}(\gamma) - \gamma\beta = -H^{-1}\psi_{0,1}(\tilde{\theta}(\gamma));$$

hence

$$\tilde{\theta}(\gamma) = \gamma\beta + o(\gamma).$$

Therefore

$$\lim_{\gamma\to+0} \gamma^{-1}[h_2(\tilde{\theta}(\gamma)) - h_2(0) - h_2'(0)\,\tilde{\theta}(\gamma)] = 0$$

and, by (1),

$$\lim_{\gamma\to+0} \gamma^{-1}[h_2(\tilde{\theta}(\gamma)) - h_2(0)] = \sum_{j=0}^{m} \beta^j \xi_2{}^j.$$

By assumption, $\xi_2{}^j \in C'$, $\beta^j > 0$, and $\sum_{j=0}^m \beta^j = 1$. It follows that $\sum_{j=0}^m \beta^j \xi_2{}^j \in C'$ and there exists an open neighborhood U of 0 in $\mathcal{X}_2$ such that $\sum_{j=0}^m \beta^j \xi_2{}^j + U \subset C'$; hence

$$\gamma^{-1}[h_2(\tilde{\theta}(\gamma)) - h_2(0)] \in C'$$

for all sufficiently small γ. Since $0 \in \bar{C}'$, we have $\gamma C' \subset C'$; thus $h_2(\tilde{\theta}(\gamma)) - h_2(0) \in C'$. This completes the proof of the lemma. QED

V.2.3 ***Theorem*** Let $Q \triangleq \mathcal{S}^\# \times B$, C_2 be convex, $C_2^\circ \neq \varnothing$, and $(\bar{y}, \bar{q}) \triangleq (\bar{y}, \bar{\sigma}, \bar{b}) \in \mathcal{A}(\mathcal{S}^\#)$. Assume that

(1) for every choice of a point $K \triangleq (q_0, ..., q_m) \in Q^{m+1}$ there exist neighborhoods Y_K of $\bar{y}$ in $\mathscr{Y}$ and $\mathscr{T}_K$ of 0 in $\mathscr{T}_{m+1}$ such that each of the functions

$$(y, \theta) \to \tilde{F}^K(y, \theta) \triangleq F\left(y, \bar{q} + \sum_{j=0}^{m} \theta^j(q_j - \bar{q})\right) : Y_K \times \mathscr{T}_K \to \mathscr{Y}$$

and

$$(y, \theta) \to \tilde{g}^K(y, \theta) \triangleq g\left(y, \bar{q} + \sum_{j=0}^{m} \theta^j(q_j - \bar{q})\right) : Y_K \times \mathscr{T}_K \to \mathbb{R} \times \mathbb{R}^m \times \mathscr{X}_2$$

is continuous and has a derivative at $(\bar{y}, 0)$, and $\tilde{F}^K$ has a continuous partial derivative $\tilde{F}_y{}^K$ on $Y_K \times \mathscr{T}_K$; and

(2) the mapping $I - F_y(\bar{y}, \bar{q})$ is a homeomorphism of $\mathscr{Y}$ onto $\mathscr{Y}$.

If $(\bar{y}, \bar{\sigma}, \bar{b})$ is a minimizing relaxed solution, then $(\bar{y}, \bar{\sigma}, \bar{b})$ is extremal.

▌PROOF Let $(\bar{y}, \bar{q})$ be a minimizing relaxed solution and

$$K \triangleq (q_0, ..., q_m) \in Q^{m+1}.$$

We consider the equation

$$(3) \qquad y - \tilde{F}^K(y, \theta) = 0 \qquad \text{for} \quad \theta \in \mathscr{T}_K.$$

Since $\bar{y} = F(\bar{y}, \bar{q}) = \tilde{F}^K(\bar{y}, 0)$, assumptions (1) and (2) insure that the implicit function theorem II.3.8(2) is applicable. It follows that there exists a neighborhood $\tilde{Y}_K \times \tilde{\mathscr{T}}_K$ of $(\bar{y}, 0)$ in $\mathscr{Y} \times \mathscr{T}_K$ such that equation (3) has a unique solution $\tilde{y}^K(\theta)$ in $\tilde{Y}_K$ for all $\theta \in \tilde{\mathscr{T}}_K$ and the function $(\theta^0, ..., \theta^m) \triangleq \theta \to \tilde{y}^K(\theta) : \tilde{\mathscr{T}}_K \to \tilde{Y}_K$ is continuous and has a derivative at 0 such that

$$\mathscr{D}\tilde{y}^K(0) = [I - \tilde{F}_y{}^K(\bar{y}, 0)]^{-1} \tilde{F}_\theta{}^K(\bar{y}, 0).$$

Thus $\tilde{y}^K(0) = \bar{y}$,

$$(4) \qquad \tilde{y}^K_{\theta^j}(0) = [I - F_y(\bar{y}, \bar{q})]^{-1} D_2F(\bar{y}, \bar{q}; q_j - \bar{q}) \qquad (j = 0, 1, ..., m)$$

and

$$\mathscr{D}\tilde{y}^K(0)\, \Delta\theta = \sum_{j=0}^{m} \Delta\theta^j\, \tilde{y}^K_{\theta^j}(0) \qquad [\Delta\theta \triangleq (\Delta\theta^0, ..., \Delta\theta^m) \in \mathbb{R}^{m+1}].$$

By (4) and the chain rule II.3.4, the function

$$\theta \to h^K(\theta) \triangleq g\left(\tilde{y}^K(\theta), \bar{q} + \sum_{j=0}^{m} \theta^j(q_j - \bar{q})\right)$$

$$= \tilde{g}^K(\tilde{y}^K(\theta), \theta) : \tilde{\mathscr{T}}_K \to \mathbb{R} \times \mathbb{R}^m \times \mathscr{X}_2$$

is continuous in some neighborhood of 0 in $\tilde{\mathscr{T}}_K$ and has a derivative at 0 satisfying the relation

$$\mathscr{D}h^K(0)\,\Delta\theta = [g_y(\bar{y}, \bar{q})\,\tilde{y}_\theta{}^K(0) + \tilde{g}_\theta{}^K(\bar{y}, 0)]\,\Delta\theta$$

$$= \sum_{j=0}^{m} \Delta\theta^j\,\chi(q_j) \qquad (\Delta\theta \in \mathbb{R}^{m+1}),$$

where

$$\chi(q) \triangleq g_y(\bar{y}, \bar{q}) \circ [I - F_y(\bar{y}, \bar{q})]^{-1}\,D_2F(\bar{y}, \bar{q}; q - \bar{q}) + D_2g(\bar{y}, \bar{q}; q - \bar{q}) \; (q \in Q)$$

Furthermore, in view of our assumptions, each of the functions

$$F(\bar{y}, \cdot) : Q \to \mathscr{Y} \qquad \text{and} \qquad g(\bar{y}, \cdot) : Q \to \mathbb{R} \times \mathbb{R}^m \times \mathscr{X}_2$$

is 2-differentiable at $\bar{q}$. It follows therefore from II.3.3 that the set $\{\chi(q) \mid q \in Q\}$ is convex.

Now let $W \triangleq \{\chi(q) \mid q \in Q\}$ and $C' \triangleq C_2{}^\circ - g_2(\bar{y}, \bar{q})$. Then Lemma V.2.1 is applicable and its first alternative yields the statement of the theorem. If the second alternative holds, then there exist $\tilde{q}_j \in Q$ $(j = 0,..., m)$ such that $\xi^j = \chi(\tilde{q}_j)$. Let $\tilde{K} \triangleq \{\tilde{q}_0, ..., \tilde{q}_m\}$ and $h(\theta) \triangleq h^{\tilde{K}}(\theta)$ for $\theta \in \mathscr{T}_{\tilde{K}}$. Then h and ξ^j satisfy the assumptions of V.2.2 and there exist $\bar{\gamma} > 0$ and a function $\gamma \to \tilde{\theta}(\gamma) : (0, \bar{\gamma}] \to \mathscr{T}_{m+1}$ such that, for $\tilde{q}(\theta) \triangleq \bar{q} + \sum_{j=0}^{m} \theta^j(\tilde{q}_j - \bar{q})$, we have

$$h_0(\tilde{\theta}(\gamma)) = g_0(\tilde{y}^{\tilde{K}}(\tilde{\theta}(\gamma)), \tilde{q}(\tilde{\theta}(\gamma))) < g_0(\bar{y}, \bar{q}) = h_0(0),$$
$$h_1(\tilde{\theta}(\gamma)) = g_1(\tilde{y}^{\tilde{K}}(\tilde{\theta}(\gamma)), \tilde{q}(\tilde{\theta}(\gamma))) = h_1(0) = 0,$$

and

$$h_2(\tilde{\theta}(\gamma)) - h_2(0) = g_2(\tilde{y}^{\tilde{K}}(\tilde{\theta}(\gamma)), \tilde{q}(\tilde{\theta}(\gamma))) - g_2(\bar{y}, \bar{q}) \in C_2{}^\circ - g_2(\bar{y}, \bar{q}).$$

This shows that $(\bar{y}, \bar{q})$ is not a minimizing relaxed solution, contrary to assumption. QED

We next consider *unilateral problems* characterized by the assumption that $\mathscr{X}_2 \triangleq C(P, \mathbb{R}^{m_2})$ and $C_2 \triangleq \{c \in C(P, \mathbb{R}^{m_2}) \mid c(p) \in A(p) \; (p \in P)\}$, where P is a compact metric space, $m_2 \in \{0, 1,...\}$, and A is a mapping on P to $\mathscr{P}'(\mathbb{R}^{m_2})$.

V.2.4 ***Lemma*** Let $m_2 \in \{0, 1, 2,...\}$, P be a compact metric space, and $A : P \to \mathscr{P}'(\mathbb{R}^{m_2})$ such that $A(p)$ is convex and $A(p)^\circ \neq \varnothing$ for all $p \in P$ and

$$G(A^\circ) \triangleq \{(p, v) \in P \times \mathbb{R}^{m_2} \mid v \in A(p)^\circ\}$$

is an open subset of $P \times \mathbb{R}^{m_2}$. If we set

$$C_2 \triangleq \{c \in C(P, \mathbb{R}^{m_2}) \mid c(p) \in A(p) \; (p \in P)\},$$

then C_2 is convex, $C_2^\circ \neq \varnothing$, and the set $\{c(p) \mid c \in C_2\}$ is dense in $A(p)$ for all $p \in P$.

▌PROOF *Step* 1 We shall first show that $C_2^\circ \neq \varnothing$, that is, there exist $\alpha > 0$ and a continuous $c_0 : P \to \mathbb{R}^{m_2}$ such that $g(p) \in A(p)$ $(p \in P)$ whenever $g \in C(P, \mathbb{R}^{m_2})$ and $|g - c_0|_{\sup} < \alpha$. Since $G(A^\circ)$ is open in $P \times \mathbb{R}^{m_2}$ and $\mathrm{pr}_P\, G(A^\circ) = P$, for every $\bar{p} \in P$ there exist $v(\bar{p}) \in A(\bar{p})^\circ$ and an open neighborhood $N(\bar{p})$ of $\bar{p}$ in P such that

$$v(\bar{p}) \in A(p)^\circ \qquad [p \in N(\bar{p})].$$

Since P is compact, a finite subcollection $\{N(p_1),..., N(p_k)\}$ of $\{N(\bar{p}) \mid \bar{p} \in P\}$ covers P, and we shall write N_i and v_i for $N(\bar{p}_i)$ and $v(\bar{p}_i)$, respectively. We set

$$\beta_j(p) \triangleq \begin{cases} d[p, P \sim N_j] & \text{if} \quad N_j \neq P, \\ 1 & \text{if} \quad N_j = P, \end{cases}$$

and

$$\gamma_j(p) \triangleq \beta_j(p) \Big/ \sum_{i=1}^{k} \beta_i(p) \qquad (j = 1, 2,..., k, \quad p \in P).$$

Then $0 \leqslant \gamma_j(p) \leqslant 1$, $\sum_{i=1}^k \gamma_i(p) = 1$ $(p \in P)$, $\gamma_j(p) = 0$ $(p \notin N_j)$, and γ_j is continuous for each $j \in \{1, 2,..., k\}$. We set

$$c_0(p) \triangleq \sum_{j=1}^{k} \gamma_j(p)\, v_j \qquad (p \in P)$$

and observe that $c_0 : P \to \mathbb{R}^{m_2}$ is continuous. If $\gamma_j(p) \neq 0$, then $p \in N_j$; hence $v_j \in A(p)^\circ$. Since $A(p)$ is convex, we conclude that $c_0(p) \in A(p)^\circ$.

The graph $\{(p, c_0(p)) \mid p \in P\}$ of the continuous function c_0 is a compact subset of the open set $G(A^\circ)$. It follows that there exists some $\alpha > 0$ such that $g(p) \in A(p)$ $(p \in P)$ whenever $|g - c_0|_{\sup} < \alpha$.

Step 2 Since $A(p)$ is convex for all $p \in P$, it follows directly that C_2 is convex.

Now let $\bar{p} \in P$ and $\bar{a} \in A(\bar{p})^\circ$. Then $(\bar{p}, \bar{a})$ belongs to the open subset $G(A^\circ)$ of $P \times \mathbb{R}^{m_2}$ and there exists, therefore, $\bar{\alpha} > 0$ such that $(p, \bar{a}) \in G(A^\circ)$ for all $p \in S(\bar{p}, \bar{\alpha})$; hence $\bar{a} \in A(p)^\circ$ for all $p \in S(\bar{p}, \bar{\alpha})$. We set $\alpha(p) \triangleq 1$ if $P = S(\bar{p}, \bar{\alpha})$ and otherwise

$$\alpha(p) \triangleq d[p, P \sim S(\bar{p}, \bar{\alpha})]/(d[p, P \sim S(\bar{p}, \bar{\alpha})] + d[p, S^F(\bar{p}, \bar{\alpha}/2)]) \qquad (p \in P),$$

and observe that $\alpha(\cdot)$ is continuous, $\alpha(p) = 1$ for $p \in S^F(\bar{p}, \bar{\alpha}/2)$, $\alpha(p) = 0$ for $p \notin S(\bar{p}, \bar{\alpha})$, and $0 \leqslant \alpha(p) \leqslant 1$. It follows that the function

$$p \to \bar{c}(p) \triangleq \alpha(p)\, \bar{a} + [1 - \alpha(p)]\, c_0(p) : P \to \mathbb{R}^{m_2}$$

is continuous and $\bar{c}(p) \in A(p)$ $(p \in P)$; hence $\bar{c} \in C_2$. Since $\bar{p}$ and $\bar{a}$ are arbitrary points of, respectively, P and $A(\bar{p})^\circ$ and, by I.6.3, $\overline{A(p)} = \overline{A(p)^\circ}$ $(p \in P)$, we conclude that

$$A(p)^\circ \subset \{c(p) \mid c \in C_2\} \subset A(p) \qquad (p \in P)$$

and $\{c(p) \mid c \in C_2\}$ is therefore dense in $A(p)$ for all $p \in P$. QED

V.2.5 ***Theorem*** Let m_2, P, $A(\cdot)$, and C_2 be defined as in Lemma V.2.4, $\mathscr{X}_2 \triangleq C(P, \mathbb{R}^{m_2})$, $l_0 \geqslant 0$, $l_1 \in \mathbb{R}^m$, $l_2 \in \mathscr{X}_2{}^*$, $l \triangleq (l_0, l_1, l_2) \neq 0$, and $\bar{c} \in C_2$. Then

$$l_2(\bar{c}) = \operatorname*{Max}_{c \in C_2} l_2(c)$$

if and only if there exist $\omega \in \text{frm}^+(P)$ and $\tilde{\omega} \in L^1(P, \Sigma_{\text{Borel}}(P), \omega, \mathbb{R}^{m_2})$ such that

(1) $$|\tilde{\omega}(p)| = 1 \quad (p \in P) \qquad \text{and} \qquad l_0 + |l_1| + \omega(P) > 0,$$

(2) $$l_2(c) = \int \tilde{\omega}(p) \cdot c(p)\, \omega(dp) \qquad (c \in \mathscr{X}_2),$$

and

(3) $$\tilde{\omega}(p) \cdot \bar{c}(p) = \operatorname*{Max}_{a \in A(p)} \tilde{\omega}(p) \cdot a \qquad \text{for} \quad \omega\text{-a.a.} \quad p \in P.$$

▌PROOF Assume that $l_2(\bar{c}) = \operatorname{Max}_{c \in C_2} l_2(c)$. By I.5.9, there exist $\omega \in \text{frm}^+(P)$ and $\tilde{\omega} \in L^1(P, \Sigma_{\text{Borel}}(P), \omega, \mathbb{R}^{m_2})$ such that $|\tilde{\omega}(p)| = 1$ $(p \in P)$ and

$$l_2(c) = \int \tilde{\omega}(p) \cdot c(p)\, \omega(dp) \qquad (c \in \mathscr{X}_2).$$

Since $l \triangleq (l_0, l_1, l_2) \neq 0$, we have $l_0 + |l_1| + \omega(P) > 0$. Thus (1) and (2) are valid.

By I.5.1 and I.2.17, the subset C_2 of $C(P, \mathbb{R}^{m_2})$ contains a dense denumerable subset $\{c_1, c_2, \ldots\}$. Since, by V.2.4, the set $\{c(p) \mid c \in C_2\}$ is dense in $A(p)$ for all $p \in P$, it follows that the set $\{c_1(p), c_2(p), \ldots\}$ is dense in $A(p)$ for all $p \in P$. For each $j \in \mathbb{N}$ and $\alpha \in C(P, [0, 1])$, we consider the continuous function

$$p \to k_j(p) \triangleq [1 - \alpha(p)]\, \bar{c}(p) + \alpha(p)\, c_j(p).$$

Since $A(p)$ is convex, we have $k_j(p) \in A(p)$ $(p \in P)$ and thus $k_j \in C_2$ $(j \in \mathbb{N})$. It follows that

$$\begin{aligned} l_2(\bar{c} - k_j) &= \int \tilde{\omega}(p)\, [\bar{c}(p) - k_j(p)]\, \omega(dp) \\ &= \int \alpha(p)\, \tilde{\omega}(p) \cdot [\bar{c}(p) - c_j(p)]\, \omega(dp) \geqslant 0 \qquad (j \in \mathbb{N}). \end{aligned}$$

Hence, by I.5.5(2), there exist sets P_j ($j \in \mathbb{N}$) such that $\omega(P \sim P_j) = 0$ and

$$\tilde{\omega}(p) \cdot [\bar{c}(p) - c_j(p)] \geqslant 0 \qquad (j \in \mathbb{N}, \quad p \in P_j).$$

Since $\{c_1(p), c_2(p),...\}$ is dense in $A(p)$ for all $p \in P$, this implies that

$$\begin{aligned}\tilde{\omega}(p) \cdot \bar{c}(p) &\geqslant \sup_{j \in \mathbb{N}} \tilde{\omega}(p) \cdot c_j(p) \\ &= \sup_{a \in A(p)} \tilde{\omega}(p) \cdot a \qquad \left(p \in \bigcap_{j=1}^{\infty} P_j\right).\end{aligned}$$

As $\bar{c}(p) \in A(p)$ ($p \in P$), relation (3) follows directly.

Now assume that there exist ω and $\tilde{\omega}$ as described and satisfying (1)–(3). Then, by (3), $\tilde{\omega}(p) \cdot \bar{c}(p) \geqslant \tilde{\omega}(p) \cdot c(p)$ ($p \in P$) for every $c \in C_2$; hence, by (2), $l_2(\bar{c}) \geqslant l_2(c)$. QED

Theorem V.2.5 shows that in unilateral problems the condition

$$l_2(g_2(\bar{y}, \bar{q})) \geqslant l_2(c) \qquad (c \in C_2)$$

is equivalent to the "local" condition

$$\tilde{\omega}(p) \cdot g_2(\bar{y}, \bar{q}) = \operatorname*{Max}_{a \in A(p)} \tilde{\omega}(p) \cdot a \qquad \text{for} \quad \omega\text{-a.a.} \quad p \in P.$$

V.3 Necessary Conditions for an Original Minimum

V.3.0 ***Summary*** If $\mathscr{U}$ is an abundant subset of $\mathscr{R}^{\#}$ and the optimal control problem admits a minimizing $\mathscr{U}$-solution $(\bar{y}, \bar{\sigma}, \bar{b})$, then $(\bar{y}, \bar{\sigma}, \bar{b})$ satisfies the same relations that were derived in V.2.3 for a minimizing relaxed solution but under somewhat stronger conditions (Theorems V.3.2 and V.3.3). An optimal control problem cannot admit a minimizing $\mathscr{U}$-solution that is not at the same time a minimizing relaxed solution except under very restrictive conditions that include the existence of admissible abnormal extremals (Theorem V.3.4).

V.3.1 ***Lemma*** Let C', β^j, and $\xi^j \in \mathbb{R} \times \mathbb{R}^m \times \mathscr{X}_2$ ($j = 0, 1,..., m$) be as described in Lemma V.2.1, $\mathscr{T}'$ and G be neighborhoods of 0 in $\mathscr{T}_{m+1}$ and $\mathscr{X}_2$, respectively, and $\beta \triangleq (\beta^0,..., \beta^m)$. For $i \in \mathbb{N}$, let $\psi \triangleq (\psi_0, \psi_1, \psi_2)$,

$$\bar{\eta} \triangleq (\bar{\eta}_0, \bar{\eta}_1, \bar{\eta}_2), \qquad \eta^i \triangleq (\eta_0^i, \eta_1^i, \eta_2^i), \qquad \text{and} \qquad e^i \triangleq (e_0^i, e_1^i, e_2^i)$$

be continuous functions on $\mathscr{T}'$ to $\mathbb{R} \times \mathbb{R}^m \times \mathscr{X}_2$ such that

$$\lim_i \eta^i(\theta) = \bar{\eta}(\theta) \quad \text{uniformly for} \quad \theta \in \mathscr{T}',$$
$$\lim_i e^i(\theta) = 0 \quad \text{uniformly for} \quad \theta \in \mathscr{T}',$$
$$\psi(\theta) = o(|\,\theta\,|) \qquad (\theta \to 0),$$

and

$$\eta^i(\theta) - \bar{\eta}(0) - \sum_{j=0}^{m} \theta^j \xi^j = \psi(\theta) + e^i(\theta) \qquad (\theta \in \mathscr{T}', \quad i \in \mathbb{N}). \tag{1}$$

Then there exist $\gamma' > 0$ and functions $\tilde{\theta}(\cdot) : (0, \gamma'] \to \mathscr{T}'$ and $\tilde{\imath}(\cdot) : (0, \gamma'] \to \mathbb{N}$ such that

$$\begin{aligned} &\psi(\tilde{\theta}(\gamma)) = o(\gamma), \qquad \tilde{\theta}(\gamma) - \gamma\beta = o(\gamma), \\ &e_{0,1}^{\tilde{\imath}(\gamma)}(\tilde{\theta}(\gamma)) = o(\gamma), \qquad \gamma^{-1} e_2^{\tilde{\imath}(\gamma)}(\tilde{\theta}(\gamma)) \in G \qquad (\gamma \in (0, \gamma']) \end{aligned} \tag{2}$$

and

$$\begin{aligned} &\eta_0^{\tilde{\imath}(\gamma)}(\tilde{\theta}(\gamma)) < \bar{\eta}_0(0), \qquad \eta_1^{\tilde{\imath}(\gamma)}(\tilde{\theta}(\gamma)) = \bar{\eta}_1(0), \\ &\eta_2^{\tilde{\imath}(\gamma)}(\tilde{\theta}(\gamma)) - \bar{\eta}_2(0) \in C' \qquad (\gamma \in (0, \gamma']). \end{aligned} \tag{3}$$

▌PROOF Since $\xi_2{}^j \in C'$, it follows that $\sum_{j=0}^m \beta^j \xi_2{}^j \in C'$. Thus there exists a neighborhood G_1 of 0 in $\mathscr{X}_2$ such that $\sum_{j=0}^m \beta^j \xi_2{}^j + G_1 \subset C'$. Since $\mathscr{X}_2$ is a topological vector space, we can determine a symmetric neighborhood G_2 of 0 in $\mathscr{X}_2$ such that $G_2 + G_2 \subset G_1$. We then set $G' \triangleq G \cap G_2$.

Let $M \in B(\mathbb{R}^{m+1}, \mathbb{R}^{m+1})$ be defined by

$$M\theta \triangleq \sum_{j=0}^{m} \theta^j \xi_{0,1}^j \qquad [\theta \triangleq (\theta^0, \ldots, \theta^m) \in \mathbb{R}^{m+1}].$$

By I.3.3, M has an inverse. Furthermore, relation (1) implies

$$\eta_{0,1}^i(\theta) - \bar{\eta}_{0,1}(0) = M\theta + \psi_{0,1}(\theta) + e_{0,1}^i(\theta) \qquad (i \in \mathbb{N}, \quad \theta \in \mathscr{T}'). \tag{4}$$

We shall now show that there exist $\bar{\gamma} > 0$ and functions $\tilde{\imath}(\cdot) : (0, \bar{\gamma}] \to \mathbb{N}$ and $\tilde{\theta}(\cdot) : (0, \bar{\gamma}] \to \mathscr{T}'$ satisfying the relations

$$\eta_{0,1}^{\tilde{\imath}(\gamma)}(\tilde{\theta}(\gamma)) - \bar{\eta}_{0,1}(0) = \gamma M\beta \qquad (\gamma \in (0, \bar{\gamma}]), \tag{5}$$

$$\gamma^{-1} e_2^{\tilde{\imath}(\gamma)}(\tilde{\theta}(\gamma)) \in G' \subset G \qquad (\gamma \in (0, \bar{\gamma}]), \quad \lim_{\gamma \to +0} \gamma^{-1} e_{0,1}^{\tilde{\imath}(\gamma)}(\tilde{\theta}(\gamma)) = 0, \tag{6}$$

$$\lim_{\gamma \to +0} \gamma^{-1}[\tilde{\theta}(\gamma) - \gamma\beta] = 0. \tag{7}$$

Indeed, since $\psi_{0,1}(\theta) = o(|\theta|)$ $(|\theta| \to 0)$, we can determine $h' \in (0, 1]$ such that

$$[0, h']^{m+1} \subset \mathscr{T}' \qquad \text{and} \qquad |M^{-1}\psi_{0,1}(\theta)| \leqslant \frac{1}{6}\frac{\beta_{\min}}{\beta_{\max}}\,\theta_{\max} \qquad (\theta \in [0, h']^{m+1}),$$

where $\beta_{\min} \triangleq \min_j \beta^j$, $\beta_{\max} \triangleq \max_j \beta^j$, and $\theta_{\max} \triangleq \max_j \theta^j$. We set $\bar{\gamma} \triangleq \frac{2}{3}h'/\beta_{\max}$ and choose, for every $\gamma \in (0, \bar{\gamma}]$, an integer $\tilde{\imath}(\gamma)$ sufficiently large so that

$$\text{(8)} \quad \gamma^{-1}e_2^{\tilde{\imath}(\gamma)}(\theta) \in G' \quad \text{and} \quad |M^{-1}e_{0,1}^{\tilde{\imath}(\gamma)}(\theta)| \leqslant \operatorname{Min}(\gamma\beta_{\min}/4, \gamma^2) \qquad (\theta \in \mathscr{T}').$$

We now observe that the set

$$\mathscr{T}''_\gamma \triangleq \{(\theta^0,\ldots, \theta^m) \in \mathbb{R}^{m+1} \mid |\theta^j - \gamma\beta^j| \leqslant \gamma\beta_{\min}/2 \ (j = 0, 1,\ldots, m)\}$$

is contained in $\mathscr{T}'$ and is a bounded convex body in $\mathbb{R}^{m+1}$. Furthermore, the function

$$\theta \to \gamma\beta - M^{-1}[\psi_{0,1}(\theta) + e_{0,1}^{\tilde{\imath}(\gamma)}(\theta)]$$

is a continuous mapping of $\mathscr{T}''_\gamma$ into itself. By Brouwer's fixed point theorem II.2.3, this function has a fixed point $\tilde{\theta}(\gamma)$ which, in view of (4), satisfies relation (5). Relation (6) is satisfied because of our choice of $\tilde{\imath}(\gamma)$ and (7) follows from $\psi(\theta) = o(|\theta|)$ $(|\theta| \to 0)$ and (8).

Relations (2) now follow from (6) and (7). Furthermore, relations (1), (2), and (6) imply that

$$\text{(9)} \qquad \eta_2^{\tilde{\imath}(\gamma)}(\tilde{\theta}(\gamma)) - \bar{\eta}_2(0) - \gamma \sum_{j=0}^{m} \beta^j \xi_2{}^j \in a(\gamma) + \gamma G',$$

where $a(\gamma) = o(\gamma)$. We choose γ' in $(0, \min(\bar{\gamma}, 1)]$ sufficiently close to 0 so that $\gamma^{-1}a(\gamma) \in G'$ $(0 < \gamma \leqslant \gamma')$. Then it follows from (9) that

$$\gamma^{-1}[\eta_2^{\tilde{\imath}(\gamma)}(\tilde{\theta}(\gamma)) - \bar{\eta}_2(0)] \in \sum_{j=0}^{m} \beta^j \xi_2{}^j + G_1 \subset C' \qquad (\gamma \in (0, \gamma']).$$

Since $0 \in \bar{C}'$ and C' is convex, we have therefore

$$\text{(10)} \qquad \eta_2^{\tilde{\imath}(\gamma)}(\tilde{\theta}(\gamma)) - \bar{\eta}_2(0) \in C'.$$

We recall that $M\beta = \sum_{j=0}^m \beta^j \xi_{0,1}^j$, $\sum_{j=0}^m \beta^j \xi_1{}^j = 0$, and $\xi_0{}^j < 0$ $(j = 0, 1,\ldots, m)$. Thus relations (5) and (10) yield (3). QED

V.3.2 *Theorem* Let $\mathscr{U}$ be an abundant subset of $\mathscr{R}^\#$, C_2 convex, $C_2{}^\circ \neq \varnothing$, and $(\bar{y}, \bar{q}) \triangleq (\bar{y}, \bar{\sigma}, \bar{b}) \in \mathscr{A}(\mathscr{S}^\#)$. For each choice of

$$K \triangleq ((\sigma_0, b_0),\ldots, (\sigma_m, b_m)) \in (\mathscr{S}^\# \times B)^{m+1},$$

set

$$\sigma^K(\theta) \triangleq \bar{\sigma} + \sum_{j=0}^{m} \theta^j(\sigma_j - \bar{\sigma})$$

and

$$b^K(\theta) \triangleq \bar{b} + \sum_{j=0}^{m} \theta^j(b_j - \bar{b}) \qquad (\theta \in \mathscr{T}_{m+1}),$$

and assume that there exist closed neighborhoods Y_K of $\bar{y}$ in $\mathscr{Y}$, $\mathscr{S}_K$ of $\bar{\sigma}$ in $\mathscr{S}^{\#}$, and $\mathscr{T}_K$ of 0 in $\mathscr{T}_{m+1}$ such that

(1) the functions

$$(y, \theta) \to \tilde{F}^K(y, \theta) \triangleq F\big(y, \sigma^K(\theta), b^K(\theta)\big) : Y_K \times \mathscr{T}_K \to \mathscr{Y}$$

and

$$(y, \theta) \to \tilde{g}^K(y, \theta) \triangleq g\big(y, \sigma^K(\theta), b^K(\theta)\big) : Y_K \times \mathscr{T}_K \to \mathbb{R} \times \mathbb{R}^m \times \mathscr{X}_2$$

are continuous and have derivatives at $(\bar{y}, 0)$, and $\tilde{F}^K$ has a continuous partial derivative $\tilde{F}_y{}^K$ on $Y_K \times \mathscr{T}_K$;

(2) the mapping $I - F_y(\bar{y}, \bar{\sigma}, \bar{b})$ is a homeomorphism of $\mathscr{Y}$ onto $\mathscr{Y}$;

(3) the functions

$$(y, \sigma, \theta) \to F^K(y, \sigma, \theta) \triangleq F\big(y, \sigma, b^K(\theta)\big) : Y_K \times \mathscr{S}_K \times \mathscr{T}_K \to \mathscr{Y}$$

and

$$(y, \sigma, \theta) \to g^K(y, \sigma, \theta) \triangleq g\big(y, \sigma, b^K(\theta)\big) : Y_K \times \mathscr{S}_K \times \mathscr{T}_K \to \mathbb{R} \times \mathbb{R}^m \times \mathscr{X}_2$$

are continuous;

(4) for each $(\sigma, \theta) \in \mathscr{S}_K \times \mathscr{T}_K$, the equation

$$y = F^K(y, \sigma, \theta)$$

has a unique solution $y = y^K(\sigma, \theta)$ in Y_K; and

(5) the set $y^K(\mathscr{S}_K \times \mathscr{T}_K)$ is conditionally compact in $\mathscr{Y}$.

Then either

(6) $(\bar{y}, \bar{\sigma}, \bar{b})$ is extremal,

or

(7) there exists a point $(y_1, u_1, b_1) \in \mathscr{A}(\mathscr{U})$ such that

$$g_0(y_1, u_1, b_1) < g_0(\bar{y}, \bar{\sigma}, \bar{b}).$$

In particular, if $(\bar{y}, \bar{\sigma}, \bar{b})$ is either a minimizing relaxed solution or a minimizing $\mathscr{U}$-solution, then $(\bar{y}, \bar{\sigma}, \bar{b})$ is extremal.

▌PROOF Let

$$\chi(q) \triangleq g_y(\bar{y}, \bar{q})[I - F_y(\bar{y}, \bar{q})]^{-1} D_2F(\bar{y}, \bar{q}; q - \bar{q}) + D_2 g(\bar{y}, \bar{q}; q - \bar{q})$$

for $q \in \mathscr{S}^{\#} \times B$, $W \triangleq \chi(\mathscr{S}^{\#} \times B)$, and $C' \triangleq C_2^{\circ} - g_2(\bar{y}, \bar{q})$. Since, by (1), the functions $q \to F(\bar{y}, q)$ and $q \to g(\bar{y}, q)$ are 2-differentiable at $\bar{q}$, it follows from II.3.3 that W is convex; furthermore $0 = \chi(\bar{q}) \in W$. Thus the assumptions of V.2.1 are satisfied and its first alternative yields statement (6) of the present theorem. We shall assume, therefore, that the second alternative of V.2.1 is valid and will show that it implies statement (7).

Let $\xi^i \in W$ $(i = 0, 1,..., m)$ be as described in V.2.1. Then there exist $q_i \triangleq (\sigma_i, b_i) \in \mathscr{S}^{\#} \times B$ $(i = 0, 1,..., m)$ such that $\xi^i = \chi(q_i)$, and we set $K \triangleq (q_0, q_1,..., q_m)$. Now let $((\sigma_j, \theta_j))$ be a sequence in $\mathscr{S}_K \times \mathscr{T}_K$ converging to some $(\tilde{\sigma}, \tilde{\theta}) \in \mathscr{S}_K \times \mathscr{T}_K$ and assume that there exist $\epsilon > 0$ and $J \subset (1, 2,...)$ such that

$$|\, y^K(\sigma_j, \theta_j) - y^K(\tilde{\sigma}, \tilde{\theta})| > \epsilon \qquad (j \in J). \tag{8}$$

Then, by (4) and (5), there exist $J_1 \subset J$ and $\tilde{y} \in \mathscr{Y}$ such that

$$\tilde{y} = \lim_{j \in J_1} y^K(\sigma_j, \theta_j) = \lim_{j \in J_1} F^K(y^K(\sigma_j, \theta_j), \sigma_j, \theta_j).$$

Since Y_K is closed and $F^K \mid Y_K \times \mathscr{S}_K \times \mathscr{T}_K$ continuous, this implies that $\tilde{y} = F^K(\tilde{y}, \tilde{\sigma}, \tilde{\theta}) \in Y_K$ and therefore $\tilde{y} = y^K(\tilde{\sigma}, \tilde{\theta})$, thus contradicting (8). We conclude that $y^K : \mathscr{S}_K \times \mathscr{T}_K \to Y_K$ is continuous.

It follows from (1), (2), and the implicit function theorem II.3.8 that there exist neighborhoods Y_K' of $\bar{y}$ in Y_K and $\mathscr{T}_K'$ of 0 in $\mathscr{T}_K$ and a unique function $\theta \to \tilde{y}^K(\theta) : \mathscr{T}_K' \to Y_K'$ such that

$$\tilde{y}^K(\theta) = \tilde{F}^K(\tilde{y}^K(\theta), \theta)$$

and

$$\begin{aligned} \mathscr{D}\tilde{y}^K(0)\, \Delta\theta &= [I - \tilde{F}_y{}^K(\bar{y}, 0)]^{-1} \circ \tilde{F}_\theta{}^K(\bar{y}, 0)\, \Delta\theta \\ &= [I - F_y(\bar{y}, \bar{q})]^{-1} \sum_{j=0}^{m} \Delta\theta^j\, D_2F(\bar{y}, \bar{q}; q_j - \bar{q}) \\ &\qquad [\Delta\theta \triangleq (\Delta\theta^0,..., \Delta\theta^m) \in \mathbb{R}^{m+1}]. \end{aligned} \tag{9}$$

We observe that $\tilde{y}^K(\theta) = y^K(\sigma^K(\theta), \theta)$ for $\theta \in \tilde{\mathscr{T}}_K \triangleq \{\theta \in \mathscr{T}_K' \mid \sigma^K(\theta) \in \mathscr{S}_K\}$. If we set

$$\bar{\eta}(\theta) \triangleq g(\tilde{y}^K(\theta), q^K(\theta)) \qquad (\theta \in \tilde{\mathscr{T}}_K),$$

then, by (9) and the chain rule II.3.4,

$$\bar{\eta}'(0)\,\Delta\theta = g_y(\bar{y}, \bar{q})\,\mathscr{D}\tilde{y}^K(0)\,\Delta\theta + \sum_{j=0}^{m} \Delta\theta^j\, D_2 g(\bar{y}, \bar{q}; q_j - \bar{q})$$

$$= \sum_{j=0}^{m} \Delta\theta^j\, \chi(q_j) \qquad [\Delta\theta \triangleq (\Delta\theta^0, \ldots, \Delta\theta^m) \in \mathbb{R}^{m+1}].$$

Since $\mathscr{U}$ is an abundant set, it follows from IV.3.3(2) that there exists, for each $\theta \in \mathscr{T}_{m+1}$, a sequence $(u_i(\theta))_i$ in $\mathscr{U}$ such that

$$\lim_i u_i(\theta) = \bar{\sigma} + \sum_{j=0}^{m} \theta^j(\sigma_j - \bar{\sigma}) \quad \text{uniformly for} \quad \theta \in \mathscr{T}_{m+1}$$

and

$$u_i(\cdot) : \mathscr{T}_{m+1} \to \mathscr{U} \quad \text{is continuous for all} \quad i \in \mathbb{N};$$

and we may assume that $u_i(\theta) \in \mathscr{S}_K$ for all i and all $\theta \in \tilde{\mathscr{T}}_K$ (replacing, if necessary, $\tilde{\mathscr{T}}_K$ by a smaller neighborhood of 0 in $\mathscr{T}_{m+1}$). We then set $G \triangleq \mathscr{X}_2$, $\mathscr{T}' \triangleq \tilde{\mathscr{T}}_K$,

$$\eta^i(\theta) \triangleq g\big(y^K(u_i(\theta), \theta), u_i(\theta), b^K(\theta)\big), \quad e^i(\theta) \triangleq \eta^i(\theta) - \bar{\eta}(\theta) \quad (i \in \mathbb{N}, \quad \theta \in \tilde{\mathscr{T}}_K)$$

and

$$\psi(\theta) \triangleq \bar{\eta}(\theta) - \bar{\eta}(0) - \sum_{j=0}^{m} \theta^j \chi(q_j) \qquad (\theta \in \tilde{\mathscr{T}}_K),$$

and verify that the assumptions of V.3.1 are satisfied. It follows now from V.3.1 that there exist $\gamma_1 > 0$, $\theta_1 \triangleq \tilde{\theta}(\gamma_1) \in \tilde{\mathscr{T}}_K$, $i_1 \triangleq \tilde{\imath}(\gamma_1) \in \mathbb{N}$, $u_1 \triangleq u_{i_1}(\theta_1) \in \mathscr{U}$, $b_1 \triangleq b^K(\theta_1) \in B$, and $y_1 \triangleq y^K(u_1, \theta_1) \in Y_K$ such that

$$\eta_0^{i_1}(\theta_1) = g_0{}^K(y_1, u_1, \theta_1) = g_0(y_1, u_1, b_1) < \bar{\eta}_0(0) = g_0(\bar{y}, \bar{\sigma}, \bar{b}),$$

$$\eta_1^{i_1}(\theta_1) = g_1(y_1, u_1, b_1) = \bar{\eta}_1(0) = 0,$$

$$\eta_2^{i_1}(\theta_1) = g_2(y_1, u_1, b_1) \in C' + g_2(\bar{y}, \bar{\sigma}, \bar{b}) \subset C.$$

This shows that statement (7) is valid.

If $(\bar{y}, \bar{\sigma}, \bar{b})$ is either a minimizing relaxed solution or a minimizing $\mathscr{U}$-solution, then statement (7) cannot hold, thus implying the validity of (6). QED

V.3.3 ***Theorem*** Let $m_2 \in \{0, 1, 2, \ldots\}$, P be a compact metric space, $A : P \to \mathscr{P}'(\mathbb{R}^{m_2})$, $\mathscr{X}_2 \triangleq C(P, \mathbb{R}^{m_2})$, and $C_2 \triangleq \{c \in \mathscr{X}_2 \mid c(p) \in A(p)\ (p \in P\}$. Assume that $A(p)$ is convex and $A(p)^\circ \neq \varnothing$ for all $p \in P$, the set

$$G(A^\circ) \triangleq \{(p, v) \in P \times \mathbb{R}^{m_2} \mid v \in A(p)^\circ\}$$

is open in $P \times \mathbb{R}^{m_2}$, and the conditions of Theorem V.3.2 are satisfied. If $(\bar{y}, \bar{\sigma}, \bar{b})$ is either a minimizing relaxed solution or a minimizing $\mathscr{U}$-solution and χ is defined as in V.3.2, then there exist $l_0 \geqslant 0$, $l_1 \in \mathbb{R}^m$, $\omega \in \mathrm{frm}^+(P)$, and $\tilde{\omega} \in L^1(P, \Sigma_{\mathrm{Borel}}(P), \omega, \mathbb{R}^{m_2})$ such that

$$(1) \qquad |\tilde{\omega}(p)| = 1 \quad (p \in P) \quad \text{and} \quad l_0 + |l_1| + \omega(P) > 0,$$

$$(2) \quad l_0\chi_0(q) + l_1 \cdot \chi_1(q) + \int \tilde{\omega}(p)\, \chi_2(q)(p)\, \omega(dp) \geqslant 0 \qquad (q \in \mathscr{S}^\# \times B),$$

and

$$(3) \quad \tilde{\omega}(p) \cdot g_2(\bar{y}, \bar{\sigma}, \bar{b})(p) = \operatorname*{Max}_{a \in A(p)} \tilde{\omega}(p) \cdot a \qquad \text{for} \quad \omega\text{-a.a.} \quad p \in P.$$

▌PROOF Our conclusions follow directly from V.3.2 and V.2.5. QED

Definition *Strict $\mathscr{U}$-Solution* A minimizing $\mathscr{U}$-solution $(\bar{y}, \bar{u}, \bar{b})$ is a *strict $\mathscr{U}$-solution* if it is not a minimizing relaxed solution.

V.3.4 ***Theorem*** Let $\mathscr{U}$ be an abundant subset of $\mathscr{R}^\#$,

$$(\bar{y}, \bar{u}, \bar{b}) \in \mathscr{A}(\mathscr{U}), \quad \mathscr{M} \triangleq \{(y, \sigma, b) \in \mathscr{A}(\mathscr{S}^\#) \mid g_0(y, \sigma, b) \leqslant g_0(\bar{y}, \bar{u}, \bar{b})\},$$

and assume that the conditions of V.3.2 are satisfied when $(\bar{y}, \bar{\sigma}, \bar{b})$ is replaced there by a point of $\mathscr{M}$. Then the following conditions are necessary for $(\bar{y}, \bar{u}, \bar{b})$ to be a strict $\mathscr{U}$-solution:

(1) $(\bar{y}, \bar{u}, \bar{b})$ is extremal;
(2) the set $\mathscr{M}^< \triangleq \{(y, \sigma, b) \in \mathscr{A}(\mathscr{S}^\#) \mid g_0(y, \sigma, b) < g_0(\bar{y}, \bar{u}, \bar{b})\}$ is nonempty; and
(3) every point (y, σ, b) in $\mathscr{M}^<$ is extremal and abnormal.

▌*Remark* It follows, under the conditions of V.3.4, that the optimal control problem has no strict $\mathscr{U}$-solutions if either (a) it has no abnormal admissible extremals, or (b) it has a minimizing relaxed solution $(\tilde{y}, \tilde{\sigma}, \tilde{b})$ (which, by V.3.2, is extremal) and $(\tilde{y}, \tilde{\sigma}, \tilde{b})$ is normal. Furthermore, if (b) is valid, then the problem admits a minimizing $\mathscr{U}$-solution if and only if there exists an extremal point $(y', u', b') \in \mathscr{A}(\mathscr{U})$ such that $g_0(y', u', b') = g_0(\tilde{y}, \tilde{\sigma}, \tilde{b})$.

▌PROOF Let $(\bar{y}, \bar{u}, \bar{b})$ be a strict $\mathscr{U}$-solution. Condition (1) follows from V.3.2 and condition (2) from the definition of a strict $\mathscr{U}$-solution.

Now let $(\tilde{y}, \tilde{\sigma}, \tilde{b}) \in \mathscr{M}^<$,

$$B^\# \triangleq B \times \mathbb{R}, \qquad \mathscr{X}_2^\# \triangleq \mathscr{X}_2 \times \mathbb{R}, \qquad C_2^\# \triangleq C_2 \times (-\infty, g_0(\bar{y}, \bar{u}, \bar{b})),$$

and, for $y \in \mathscr{Y}$, $\sigma \in \mathscr{S}^{\#}$, and $b^{\#} \triangleq (b, \alpha) \in B^{\#}$,

$$g_0^{\#}(y, \sigma, b^{\#}) \triangleq \alpha, \qquad g_1^{\#}(y, \sigma, b^{\#}) \triangleq g_1(y, \sigma, b),$$
$$g_2^{\#}(y, \sigma, b^{\#}) \triangleq (g_2(y, \sigma, b), g_0(y, \sigma, b))$$

and

$$F^{\#}(y, \sigma, b^{\#}) \triangleq F(y, \sigma, b)$$

We shall denote by P the optimal control problem we are discussing and by $P^{\#}$ the optimal control problem obtained by replacing B, $\mathscr{X}_2$, C_2, g, and F by $B^{\#}$, $\mathscr{X}_2^{\#}$, $C_2^{\#}$, $g^{\#}$, and $F^{\#}$, respectively. Then Theorem V.3.2 is applicable to $P^{\#}$ and $(\tilde{y}, \tilde{\sigma}, \tilde{b}, 0)$. We shall first show that the alternative V.3.2(7) cannot apply to $P^{\#}$. Indeed, this alternative implies the existence of points $y_1 \in \mathscr{Y}$, $u_1 \in \mathscr{U}$, and $(b_1, \alpha_1) \in B^{\#}$ such that $g_1(y_1, u_1, b_1) = 0$, $g_2(y_1, u_1, b_1) \in C_2$, and $g_0(y_1, u_1, b_1) \in (-\infty, g_0(\bar{y}, \bar{u}, \bar{b}))$, contradicting the assumption that $(\bar{y}, \bar{u}, \bar{b})$ is a minimizing $\mathscr{U}$-solution of P. Thus, by V.3.2(6), $(\tilde{y}, \tilde{\sigma}, \tilde{b}, 0)$ is extremal for $P^{\#}$, and there exist $\tilde{l}_0 \geq 0$, $\tilde{l}_1 \in \mathbb{R}^m$, and $\tilde{l}_2^{\#} \triangleq (\tilde{l}_2, l_3) \in \mathscr{X}_2^{\#*} = \mathscr{X}_2^* \times \mathbb{R}$ such that $(\tilde{y}, \tilde{\sigma}, (\tilde{b}, 0), \tilde{l}_0, \tilde{l}_1, \tilde{l}_2^{\#})$ is an extremal of $P^{\#}$. If we set $\tilde{q} \triangleq (\tilde{\sigma}, \tilde{b})$, $\tilde{q}^{\#} \triangleq (\tilde{\sigma}, \tilde{b}, 0) \triangleq (\tilde{q}, 0)$,

$$\chi(q) \triangleq g_y(\tilde{y}, \tilde{q}) \circ [I - F_y(\tilde{y}, \tilde{q})]^{-1} D_2F(\tilde{y}, \tilde{q}; q - \tilde{q})$$
$$+ D_2 g(\tilde{y}, \tilde{q}; q - \tilde{q}) \qquad (q \in \mathscr{S}^{\#} \times B),$$

and

$$\chi^{\#}(q^{\#}) \triangleq g_y^{\#}(\tilde{y}, \tilde{q}^{\#}) \circ [I - F_y^{\#}(\tilde{y}, \tilde{q}^{\#})]^{-1} D_2F^{\#}(\tilde{y}, \tilde{q}^{\#}; q^{\#} - \tilde{q}^{\#})$$
$$+ D_2 g^{\#}(\tilde{y}, \tilde{q}^{\#}; q^{\#} - \tilde{q}^{\#}) \qquad (q^{\#} \in \mathscr{S}^{\#} \times B^{\#})$$

then

$$\chi_0^{\#}(q^{\#}) = \alpha, \qquad \chi_1^{\#}(q^{\#}) = \chi_1(q), \qquad \chi_2^{\#}(q^{\#}) = (\chi_2(q), \chi_0(q))$$
$$(q^{\#} \triangleq (q, \alpha) \in \mathscr{S}^{\#} \times B \times \mathbb{R}),$$

and we have

$$\text{(4)} \qquad \tilde{l}_0\alpha + \tilde{l}_1 \cdot \chi_1(q) + \tilde{l}_2(\chi_2(q)) + l_3\chi_0(q) \geq 0 \qquad (q \in \mathscr{S}^{\#} \times B, \alpha \in \mathbb{R})$$

and

$$\text{(5)} \qquad \tilde{l}_2(g_2(\tilde{y}, \tilde{q})) + l_3 g_0(\tilde{y}, \tilde{q}) \geq \tilde{l}_2(c) + l_3\beta$$
$$[c \in C_2, \ \beta \in (-\infty, g_0(\bar{y}, \bar{u}, \bar{b}))].$$

For $q = \tilde{q}$, relation (4) implies $\tilde{l}_0\alpha \geq 0$ $(\alpha \in \mathbb{R})$; hence $\tilde{l}_0 = 0$. For $c = g_2(\tilde{y}, \tilde{q})$, relation (5) yields

$$l_3 g_0(\tilde{y}, \tilde{q}) \geq l_3\beta \qquad [\beta < g_0(\bar{y}, \bar{u}, \bar{b})];$$

hence $l_3 = 0$. It follows that $(0, \tilde{l}_1, \tilde{l}_2) \neq 0$ and (4) and (5) yield statement (3); that is, the conclusion that $(\tilde{y}, \tilde{\sigma}, \tilde{b}, 0, \tilde{l}_1, \tilde{l}_2)$ is an extremal of P. QED

V.4 Convex Cost Functionals

One of the assumptions recurring in the necessary conditions for both minimizing relaxed solutions and minimizing $\mathscr{U}$-solutions is the $(m+1)$-differentiability of the function $(\sigma, b) \to g_0(\bar{y}, \sigma, b)$. This assumption can be somewhat modified, replacing the $(m+1)$-differentiability of g_0 by certain convexity properties. We shall derive necessary conditions for both minimizing relaxed solutions and minimizing original solutions. As before, we write $\eta_{i,j} \triangleq (\eta_i, \eta_j)$ if $\eta \triangleq (\eta_0, \eta_1, \eta_2) \in \mathbb{R} \times \mathbb{R}^m \times \mathscr{X}_2$.

V.4.1 ***Theorem*** Let $\mathscr{U}$ be an abundant set, $(\bar{y}, \bar{q}) \triangleq (\bar{y}, \bar{u}, \bar{b})$ a minimizing $\mathscr{U}$-solution, C_2 a convex set with a nonempty interior, and

$$g_0(y, q) \triangleq \phi(g_1(y, q), g_2(y, q)) \qquad (y \in \mathscr{Y}, \quad q \in \mathscr{S}^\# \times B),$$

where $\phi : \mathbb{R}^m \times \mathscr{X}_2 \to \mathbb{R}$ is continuous and convex. For each choice of

$$K \triangleq (q_0, \ldots, q_m) \triangleq ((\sigma_0, b_0), \ldots, (\sigma_m, b_m)) \in (\mathscr{S}^\# \times B)^{m+1},$$

set

$$\sigma^K(\theta) \triangleq \bar{u} + \sum_{j=0}^{m} \theta^j(\sigma_j - \bar{u}), \qquad b^K(\theta) \triangleq \bar{b} + \sum_{j=0}^{m} \theta^j(b_j - \bar{b}),$$

and

$$q^K(\theta) \triangleq (\sigma^K(\theta), b^K(\theta)) \qquad (\theta \in \mathscr{T}_{m+1}),$$

and assume that there exist closed neighborhoods Y_K of $\bar{y}$ in $\mathscr{Y}$, $\mathscr{S}_K$ of $\bar{u}$ in $\mathscr{S}^\#$, and $\mathscr{T}_K$ of 0 in $\mathscr{T}_{m+1}$ such that

(1) the functions

$$(y, \theta) \to \tilde{F}^K(y, \theta) \triangleq F(y, q^K(\theta)) : Y_K \times \mathscr{T}_K \to \mathscr{Y}$$

and

$$(y, \theta) \to \tilde{g}^K_{1,2}(y, \theta) \triangleq g_{1,2}(y, q^K(\theta)) : Y_K \times \mathscr{T}_K \to \mathbb{R}^m \times \mathscr{X}_2$$

are continuous and have derivatives at $(\bar{y}, 0)$, and $\tilde{F}^K$ has a continuous partial derivative $\tilde{F}_y{}^K$ on $Y_K \times \mathscr{T}_K$;

(2) the mapping $I - F_y(\bar{y}, \bar{\sigma}, \bar{b})$ is a homeomorphism of $\mathscr{Y}$ onto $\mathscr{Y}$;

(3) the functions

$$(y, \sigma, \theta) \to F^K(y, \sigma, \theta) \triangleq F(y, \sigma, b^K(\theta)) : Y_K \times \mathscr{S}_K \times \mathscr{T}_K \to \mathscr{Y}$$

and

$$(y, \sigma, \theta) \to g_{1,2}^K(y, \sigma, \theta) \triangleq g_{1,2}(y, \sigma, b^K(\theta)) : Y_K \times \mathscr{S}_K \times \mathscr{T}_K \to \mathbb{R}^m \times \mathscr{X}_2$$

are continuous;

(4) for each $(\sigma, \theta) \in \mathscr{S}_K \times \mathscr{T}_K$, the equation

$$y = F^K(y, \sigma, \theta)$$

has a unique solution $y = y^K(\sigma, \theta)$ in Y_K; and

(5) the set $y^K(\mathscr{S}_K \times \mathscr{T}_K)$ is conditionally compact in $\mathscr{Y}$.

Let

$$\hat{\phi}(v_{1,2}) \triangleq \phi(g_{1,2}(\bar{y}, \bar{q}) + v_{1,2}) - \phi(g_{1,2}(\bar{y}, \bar{q})) \qquad (v_{1,2} \in \mathbb{R}^m \times \mathscr{X}_2)$$

and

$$\chi_i(q) \triangleq \mathscr{D}_1 g_i(\bar{y}, \bar{q}) \circ [I - F_y(\bar{y}, \bar{q})]^{-1} D_2 F(\bar{y}, \bar{q}; q - \bar{q}) + D_2 g_i(\bar{y}, \bar{q}; q - \bar{q}) \qquad [i = 1, 2, \quad q \triangleq (\sigma, b) \in \mathscr{S}^\# \times B].$$

Then there exist $l_0 \geqslant 0$, $l_1 \in \mathbb{R}^m$, and $l_2 \in \mathscr{X}_2^*$, not all 0, and such that

$$l_0 \hat{\phi}(\chi_1(q), \chi_2(q)) + l_1 \cdot \chi_1(q) + l_2(\chi_2(q)) \geqslant 0 \qquad (q \in \mathscr{S}^\# \times B)$$

and

$$l_2(c) \leqslant l_2(g_2(\bar{y}, \bar{q})) \qquad (c \in C_2).$$

▌PROOF Let

$$W \triangleq \{(\hat{\phi}(\chi_1(q), \chi_2(q)) + a, \chi_1(q), \chi_2(q)) \mid q \in \mathscr{S}^\# \times B, a \geqslant 0\}$$

and $C' \triangleq C_2^\circ - g_2(\bar{y}, \bar{q})$. By (1), the functions $q \to F(\bar{y}, q)$ and $q \to g_i(\bar{y}, q)$ $(i = 1, 2)$ are 2-differentiable at $\bar{q}$ and, by II.3.3, the set $\{(\chi_1(q), \chi_2(q)) \mid q \in \mathscr{S}^\# \times B\}$ is a convex subset of $\mathbb{R}^m \times \mathscr{X}_2$. Furthermore, $(\chi_1(\bar{q}), \chi_2(\bar{q})) = 0$. It follows that W is convex and contains the origin of $\mathbb{R} \times \mathbb{R}^m \times \mathscr{X}_2$. Thus we can apply Lemma V.2.1 and the first alternative of that lemma yields our conclusion. We shall now assume that the second alternative of V.2.1 holds and we shall show that it leads to a contradiction.

Let β^j and ξ^j be as defined in V.2.1. Then there exist $q_j \triangleq (\sigma_j, b_j) \in \mathscr{S}^\# \times B$ and $a_j \geqslant 0$ $(j = 0, 1, \ldots, m)$ such that

$$\xi^j \triangleq (\xi_0^j, \xi_1^j, \xi_2^j) = (\hat{\phi}(\chi_1(q_j), \chi_2(q_j)) + a_j, \chi_1(q_j), \chi_2(q_j));$$

hence

$$\hat{\phi}(\xi^j_{1,2}) \leqslant \xi_0{}^j < 0 \qquad (j = 0, 1,..., m)$$

and, since $\hat{\phi}$ is convex,

$$\hat{\phi}\left(\sum_{j=0}^{m} \beta^j \xi^j_{1,2}\right) \leqslant \sum_{j=0}^{m} \beta^j \hat{\phi}(\xi^j_{1,2}) < 0. \tag{6}$$

We let $K \triangleq (q_0, q_1, ..., q_m)$ and observe that the argument of Theorem V.3.2 can be repeated, unchanged, to show that there exists a neighborhood $\tilde{\mathscr{T}}_K$ of 0 in $\mathscr{T}_K$ such that the functions $(\sigma, \theta) \to y^K(\sigma, \theta) : \mathscr{S}_K \times \tilde{\mathscr{T}}_K \to \mathscr{Y}$ and $\theta \to \tilde{y}^K(\theta) \triangleq y^K(\sigma^K(\theta), \theta) : \tilde{\mathscr{T}}_K \to \mathscr{Y}$ are continuous and

$$\mathscr{D}\tilde{y}^K(0)\, \Delta\theta = [I - F_y(\bar{y}, \bar{q})]^{-1} \sum_{j=0}^{m} \Delta\theta^j\, D_2F(\bar{y}, \bar{q}; q_j - \bar{q})$$

$$[\Delta\theta \triangleq (\Delta\theta^0, ..., \Delta\theta^m) \in \mathbb{R}^{m+1}].$$

If we set

$$\bar{\eta}_{1,2}(\theta) \triangleq g_{1,2}(\tilde{y}^K(\theta), q^K(\theta)) \qquad (\theta \in \tilde{\mathscr{T}}_K),$$

then we derive, as in the proof of Theorem V.3.2, the relation

$$\bar{\eta}'_{1,2}(0)\, \Delta\theta = \sum_{j=0}^{m} \Delta\theta^j\, \chi_{1,2}(q_j) \qquad (\Delta\theta \in \mathbb{R}^{m+1}).$$

By IV.3.3(2) there exists a sequence $(u_i(\theta))$ in $\mathscr{U}$ converging to $\sigma^K(\theta)$ uniformly for $\theta \in \tilde{\mathscr{T}}_K$ and such that $\theta \to u_i(\theta) : \tilde{\mathscr{T}}_K \to \mathscr{U}$ is continuous for each i. We set, for $i \in \mathbb{N}$ and $\theta \in \tilde{\mathscr{T}}_K$,

$$\eta_0{}^i(\theta) \triangleq \sum_{j=0}^{m} \theta^j \xi_0{}^j, \quad \eta^i_{1,2}(\theta) \triangleq g^K_{1,2}(y^K(u_i(\theta), \theta), u_i(\theta), \theta), \quad \eta^i \triangleq (\eta_0{}^i, \eta_1{}^i, \eta_2{}^i),$$

$$\bar{\eta}_0(\theta) \triangleq \sum_{j=0}^{m} \theta^j \xi_0{}^j, \qquad \bar{\eta} \triangleq (\bar{\eta}_0, \bar{\eta}_1, \bar{\eta}_2) \triangleq (\bar{\eta}_0, \bar{\eta}_{1,2}),$$

$$e^i(\theta) \triangleq \eta^i(\theta) - \bar{\eta}(\theta),$$

and

$$\psi_0(\theta) \triangleq 0, \qquad \psi_{1,2}(\theta) \triangleq \bar{\eta}_{1,2}(\theta) - \bar{\eta}_{1,2}(0) - \sum_{j=0}^{m} \theta^j \chi_{1,2}(q_j)$$

In view of (6), we can determine $\epsilon > 0$ and neighborhoods G and $\tilde{G}$ of 0 in $\mathscr{X}_2$ such that $G + G \subset \tilde{G}$ and

$$\hat{\phi}\left(\sum_{j=0}^{m} \beta^j \xi^j_{1,2} + v_{1,2}\right) < 0 \qquad \text{if } |v_1| < \epsilon \quad \text{and} \quad v_2 \in \tilde{G}. \tag{7}$$

We set $\mathscr{T}' \triangleq \tilde{\mathscr{T}}_K$ and observe that the conditions of Lemma V.3.1 are satisfied. It follows that there exist $\gamma_1 > 0$ and $\bar{\imath}(\gamma) \in \mathbb{N}$, $\tilde{\theta}(\gamma) \in \tilde{\mathscr{T}}_K$, $u(\gamma) \triangleq u_{\bar{\imath}(\gamma)}(\tilde{\theta}(\gamma)) \in \mathscr{U}$, $b(\gamma) \triangleq b^K(\tilde{\theta}(\gamma))$, and $y(\gamma) \triangleq y^K(u(\gamma), \tilde{\theta}(\gamma)) \in \mathscr{Y}$ for $\gamma \in (0, \gamma_1]$ such that

$$\text{(8)} \quad \psi(\tilde{\theta}(\gamma)) = o(\gamma),\ e_1^{\bar{\imath}(\gamma)}(\tilde{\theta}(\gamma)) = o(\gamma),\ \gamma^{-1} e_2^{\bar{\imath}(\gamma)}(\tilde{\theta}(\gamma)) \in G,\ \tilde{\theta}(\gamma) - \gamma\beta = o(\gamma),$$

$$\text{(9)} \quad \eta^{\bar{\imath}(\gamma)}(\tilde{\theta}(\gamma)) - \bar{\eta}(0) - \textstyle\sum_{j=0}^{m} \tilde{\theta}^j(\gamma)\, \xi^j = \psi(\tilde{\theta}(\gamma)) + e^{\bar{\imath}(\gamma)}(\tilde{\theta}(\gamma)),$$

$$\text{(10)} \quad \eta_1^{\bar{\imath}(\gamma)}(\tilde{\theta}(\gamma)) = g_1(y(\gamma), u(\gamma), b(\gamma)) = \bar{\eta}_1(0) = g_1(\bar{y}, \bar{u}, \bar{b}) = 0,$$

and

$$\text{(11)} \quad \eta_2^{\bar{\imath}(\gamma)}(\tilde{\theta}(\gamma)) = g_2(y(\gamma), u(\gamma), b(\gamma)) \in C' + \bar{\eta}_2(0) = C' + g_2(\bar{y}, \bar{u}, \bar{b}) \subset C_2 .$$

Furthermore, by (8) and (9),

$$\text{(12)} \qquad \gamma^{-1}[\eta_{1,2}^{\bar{\imath}(\gamma)}(\tilde{\theta}(\gamma)) - g_{1,2}(\bar{y}, \bar{u}, \bar{b})] = \sum_{j=0}^{m} \beta^j \xi_{1,2}^j + o(1) + a_{1,2}(\gamma),$$

where $a_1(\gamma) = o(1)$ and $a_2(\gamma) \in G$.

Since $\hat{\phi}$ is convex and $\hat{\phi}(0) = 0$, it follows now from (7) and (12) that, for all sufficiently small positive γ,

$$\begin{aligned} &\gamma^{-1}\hat{\phi}(g_{1,2}(y(\gamma), u(\gamma), b(\gamma)) - g_{1,2}(\bar{y}, \bar{u}, \bar{b})) \\ &\quad \leqslant \hat{\phi}(\gamma^{-1}[g_{1,2}(y(\gamma), u(\gamma), b(\gamma)) - g_{1,2}(\bar{y}, \bar{u}, \bar{b})]) \\ &\quad = \hat{\phi}\left(\sum_{j=0}^{m} \beta^j \xi_{1,2}^j + o(1) + a_{1,2}(\gamma)\right) < 0; \end{aligned}$$

hence

$$\phi(g_{1,2}(y(\gamma), u(\gamma), b(\gamma))) < \phi(g_{1,2}(\bar{y}, \bar{u}, \bar{b})).$$

This last relation and relations (10) and (11) contradict the assumption that $(\bar{y}, \bar{u}, \bar{b})$ is a minimizing $\mathscr{U}$-solution. QED

We can derive necessary conditions for a minimizing relaxed solution as a corollary of the above theorem.

V.4.2 ***Theorem*** Let $(\bar{y}, \bar{q}) \triangleq (\bar{y}, \bar{\sigma}, \bar{b})$ be a minimizing relaxed solution, C_2 a convex set with a nonempty interior, and $g_0(y, q) \triangleq \phi(g_1(y, q), g_2(y, q))$ $(y \in \mathscr{Y},\ q \in \mathscr{S}^{\#} \times B)$, where $\phi : \mathbb{R}^m \times \mathscr{X}_2 \to \mathbb{R}$ is continuous and convex. For each choice of $K \triangleq (q_0, \ldots, q_m) \in (\mathscr{S}^{\#} \times B)^{m+1}$, set

$$q^K(\theta) \triangleq \bar{q} + \sum_{j=0}^{m} \theta^j (q_j - \bar{q}) \qquad (\theta \in \mathscr{T}_{m+1}),$$

and assume that

(1) there exist closed neighborhoods Y_K of $\bar{y}$ in $\mathscr{Y}$ and $\mathscr{T}_K$ of 0 in $\mathscr{T}_{m+1}$ such that each of the functions

$$(y, \theta) \to \tilde{F}^K(y, \theta) \triangleq F(y, q^K(\theta)) : Y_K \times \mathscr{T}_K \to \mathscr{Y}$$

and

$$(y, \theta) \to \tilde{g}_{1,2}^K(y, \theta) \triangleq g_{1,2}(y, q^K(\theta)) : Y_K \times \mathscr{T}_K \to \mathbb{R}^m \times \mathscr{X}_2$$

is continuous and has a derivative at $(\bar{y}, 0)$, and $\tilde{F}^K$ has a continuous partial derivative $\tilde{F}_y{}^K$ on $Y_K \times \mathscr{T}_K$; and

(2) the mapping $I - F_y(\bar{y}, \bar{q})$ is a homeomorphism of $\mathscr{Y}$ onto $\mathscr{Y}$.

Let

$$\hat{\phi}(v_{1,2}) \triangleq \phi(g_{1,2}(\bar{y}, \bar{q}) + v_{1,2}) - \phi(g_{1,2}(\bar{y}, \bar{q})) \qquad (v_{1,2} \in \mathbb{R}^m \times \mathscr{X}_2)$$

and

$$\chi_i(q) \triangleq \mathscr{D}_1 g_i(\bar{y}, \bar{q}) \circ [I - F_y(\bar{y}, \bar{q}]^{-1} D_2F(\bar{y}, \bar{q}; q - \bar{q}) \\ + D_2 g_i(\bar{y}, \bar{q}; q - \bar{q}) \qquad [i = 1, 2, \quad q \triangleq (\sigma, b) \in \mathscr{S}^\# \times B].$$

Then there exist $l_0 \geqslant 0$, $l_1 \in \mathbb{R}^m$, and $l_2 \in \mathscr{X}_2{}^*$, not all 0, and such that

$$l_0\hat{\phi}(\chi_1(q), \chi_2(q)) + l_1 \cdot \chi_1(q) + l_2(\chi_2(q)) \geqslant 0 \qquad (q \in \mathscr{S}^\# \times B)$$

and

$$l_2(c) \leqslant l_2(g_2(\bar{y}, \bar{q})) \qquad (c \in C_2).$$

▌PROOF Let $Q \triangleq \mathscr{S}^\# \times B$ and $K \triangleq (q_0, ..., q_m) \in Q^{m+1}$. We observe, as in the proof of V.2.3, that assumptions (1) and (2) and the implicit function theorem II.3.8 imply the existence of a neighborhood $\tilde{Y}_K \times \tilde{\mathscr{T}}_K$ of $(\bar{y}, 0)$ in $\mathscr{Y} \times \mathscr{T}_K$ such that the equation

$$y = \tilde{F}^K(y, \theta)$$

has a unique solution $\tilde{y}^K(\theta)$ in $\tilde{Y}_K$ for all $\theta \in \tilde{\mathscr{T}}_K$, and $\theta \to \tilde{y}^K(\theta) : \tilde{\mathscr{T}}_K \to \tilde{Y}_K$ is continuous. We may assume that both $\tilde{Y}_K$ and $\tilde{\mathscr{T}}_K$ are closed, and it follows then from I.2.9 that $\tilde{y}^K(\tilde{\mathscr{T}}_K)$ is compact.

Since $\mathscr{S}^\#$ is a convex subset of a normed vector space, it follows that $Q \triangleq \mathscr{S}^\# \times B$ is convex. We may, therefore, consider both F and $g_{1,2}$ as functions on $\mathscr{Y} \times Q$ and independent of σ, the set Q playing the role previously assigned to B. With this interpretation and with Y_K, $\mathscr{T}_K$ replaced by

$\tilde{Y}_K$, $\tilde{\mathscr{T}}_K$, respectively, our problem satisfies the conditions of V.4.1; assumptions V.4.1(1) and V.4.1(3) now coincide with assumption (1) and the function $y^K(\cdot, \cdot)$ of V.4.1(4) and V.4.1(5) can be identified with the function $\tilde{y}^K(\cdot)$ which we have defined above and which satisfies the conditions of V.4.1. Our conclusions now follow from V.4.1 QED

V.5 Weak Necessary Conditions for an Original Minimum

The necessary conditions for an original minimum that we have derived in Theorems V.3.2 and V.4.1 are based on the assumption that the functions

$$(y, \sigma, \theta) \to F^K(y, \sigma, \theta) \qquad \text{and} \qquad (y, \sigma, \theta) \to g^K(y, \sigma, \theta)$$

are continuous in an appropriate neighborhood of $(\bar{y}, \bar{\sigma}, 0)$ in $\mathscr{Y} \times \mathscr{S}^\# \times \mathscr{T}_{m+1}$. We can verify this assumption in the case of problems discussed in Chapters VI–VIII, and the required conditions do not unduly restrict the generality of our results. However, as Counterexample IX.2.2 demonstrates, there are simple conflicting control problems for which this continuity assumption is not satisfied and the necessary conditions of the type derived in V.3 are not valid. We may often, in such cases, verify some alternate assumptions that enable us to derive other necessary conditions for an original minimum. We refer to the latter conditions as *weak* because, whenever the assumptions of V.3.2 and the new ones are simultaneously satisfied, the results of V.3.2 imply that the new necessary conditions are true. These weak necessary conditions are described below.

V.5.1 ***Theorem*** Let $(\bar{y}, \bar{\rho}, \bar{b})$ be a minimizing $\mathscr{R}^\#$-solution, C_2 a convex subset of $\mathscr{X}_2$ with a nonempty interior, the set M both a subset of some vector space and a topological space, $M^\# : T \to \mathscr{P}'(M)$, and $\phi \in C(M, R)$. We denote by $\mathscr{M}^\#$ the set of μ-measurable selections of $M^\#$, and set

$$\hat{F}(y, \nu, b) \triangleq F(y, \phi \circ \nu, b)$$

and

$$\hat{g}(y, \nu, b) \triangleq g(y, \phi \circ \nu, b) \qquad (y \in \mathscr{Y}, \quad \nu \in \mathscr{M}^\#, \quad b \in B).$$

We assume that

(1) $M^\#(t)$ is convex and $\phi(M^\#(t)) = R^\#(t)$ for μ-a.a. $t \in T$;

(2) $\bar{\nu} \in \mathscr{M}^\#$ is such that $\bar{\rho} = \phi \circ \bar{\nu}$;

(3) for every choice of $K \triangleq ((\nu_0, b_0), \ldots, (\nu_m, b_m)) \in (\mathscr{M}^\# \times B)^{m+1}$ there

exist neighborhoods Y_K of $\bar{y}$ in $\mathscr{Y}$ and $\mathscr{T}_K$ of 0 in $\mathscr{T}_{m+1}$ such that each of the functions

$$(y, \theta) \to \hat{F}^K(y, \theta)$$
$$\triangleq \hat{F}\left(y, \bar{\nu} + \sum_{j=0}^{m} \theta^j(\nu_j - \bar{\nu}), \bar{b} + \sum_{j=0}^{m} \theta^j(b_j - \bar{b})\right): Y_K \times \mathscr{T}_K \to \mathscr{Y}$$

and

$$(y, \theta) \to \hat{g}^K(y, \theta)$$
$$\triangleq \hat{g}\left(y, \bar{\nu} + \sum_{j=0}^{m} \theta^j(\nu_j - \bar{\nu}), \bar{b} + \sum_{j=0}^{m} \theta^j(b_j - \bar{b})\right): Y_K \times \mathscr{T}_K \to \mathbb{R} \times \mathbb{R}^m \times \mathscr{X}_2$$

is continuous and has a derivative at $(\bar{y}, 0)$, and $\hat{F}^K$ has a continuous partial derivative $\hat{F}_y{}^K$ on $Y_K \times \mathscr{T}_K$; and

(4) the mapping $I - \hat{F}_y(\bar{y}, \bar{\nu}, \bar{b})$ is a homeomorphism of $\mathscr{Y}$ onto $\mathscr{Y}$.
Then

$$\begin{aligned} \chi(\nu, b) &\triangleq (\chi_0(\nu, b), \chi_1(\nu, b), \chi_2(\nu, b)) \\ &\triangleq \hat{g}_y(\bar{y}, \bar{\nu}, \bar{b}) \circ [I - \hat{F}_y(\bar{y}, \bar{\nu}, \bar{b})]^{-1} D_2\hat{F}(\bar{y}, (\bar{\nu}, \bar{b}); (\nu, b) - (\bar{\nu}, \bar{b})) \\ &\quad + D_2\hat{g}(\bar{y}, (\bar{\nu}, \bar{b}); (\nu, b) - (\bar{\nu}, \bar{b})) \end{aligned}$$

is defined for all $(\nu, b) \in \mathscr{M}^\# \times B$, and there exists

$$l \triangleq (l_0, l_1, l_2) \in [0, \infty) \times \mathbb{R}^m \times \mathscr{X}_2{}^*$$

such that $l \neq 0$,

$$(5) \qquad l_0\chi_0(\nu, b) + l_1 \cdot \chi_1(\nu, b) + l_2(\chi_2(\nu, b)) \geqslant 0 \qquad (\nu \in \mathscr{M}^\#, b \in B)$$

and

$$(6) \qquad l_2(\hat{g}_2(\bar{y}, \bar{\nu})) = \mathrm{Max}_{c \in C_2}\, l_2(c).$$

▌PROOF Since $M^\#(t)$ is convex μ-a.e., it follows that $\mathscr{M}^\#$ is convex. We set $\hat{B} \triangleq \mathscr{M}^\# \times B$, and consider the problem defined in V.0 but with B, F, and g replaced by $\hat{B}$, $\hat{F}$, and $\hat{g}$, respectively. Then the new problem does not involve any control functions and the corresponding definitions of a minimizing relaxed solution and of a minimizing $\mathscr{U}$-solution are identical. Since $(\bar{y}, \bar{\rho}, \bar{b})$ is a minimizing $\mathscr{R}^\#$-solution of the old problem and, by (1), $\phi \circ \nu \in \mathscr{R}^\#$ whenever $\nu \in \mathscr{M}^\#$, it follows that $(\bar{y}, \bar{\hat{b}})$, with $\bar{\hat{b}} \triangleq (\bar{\nu}, \bar{b})$, is a minimizing relaxed solution of the new problem. Our conclusions now follow from V.2.3. QED

V.6 An Illustration—A Class of Ordinary Differential Problems and Examples

V.6.0 ***Formulation of the Problem.*** In Chapters VI–VIII we shall apply the results of V.1–V.3 to the study of the optimal control of specific classes of functional equations. In the process, we shall constantly face the problem of choosing between greater simplicity and greater generality. While we shall try to be moderate about it, this conflict will quite often be resolved in favor of greater generality. As a consequence, some of our assumptions will tend to be lengthy and many of the arguments rather technical. This will have the obvious disadvantage of obscuring the basic concepts and methods which are essentially simple.

The present section is written in an effort to somewhat redress the balance. We consider a reasonably general class of optimal control problems defined by ordinary differential equations but in a streamlined form and with the frills removed. After we establish existence theorems and necessary conditions for relaxed and original minima, these results are illustrated by a few examples.

The optimal control problem that we now consider is presented in a form that does not involve any states, and we need not therefore define $\mathscr{Y}$ or F. We are given $n \in \mathbb{N}$, compact convex sets A_0 and A_1 in $\mathbb{R}^n$, a convex body $A \subset \mathbb{R}^n$, an interval $T \triangleq [t_0, t_1] \subset \mathbb{R}$, and a function $(v, r) \to f(v, r)$: $\mathbb{R}^n \times R \to \mathbb{R}^n$. We assume that f_v (that is, $\mathscr{D}_1 f$) exists, that both f and f_v are continuous, that the equation

$$(1) \qquad y(t) = a_0 + \int_{t_0}^{t} f(y(\tau), \sigma(\tau))\, d\tau$$

$$= a_0 + \int_{t_0}^{t} d\tau \int f(y(\tau), r)\, \sigma(\tau)\,(dr) \qquad (t \in T)$$

has a unique solution

$$\tilde{y}(\sigma, a_0) \triangleq (\tilde{y}^1(\sigma, a_0), \ldots, \tilde{y}^n(\sigma, a_0)) \in C(T, \mathbb{R}^n)$$

for every choice of $(\sigma, a_0) \in \mathscr{S} \times A_0$, and that there exists $c_1 < \infty$ such that

$$(2) \qquad |\tilde{y}(\sigma, a_0)| \leqslant c_1 \qquad (t \in T, \quad \sigma \in \mathscr{S}, \quad a_0 \in A_0).$$

[This would be the case, in particular, if f and f_v are continuous and there exists $c < \infty$ such that

$$|f(v, r)| \leqslant c(|v| + 1) \qquad (v \in \mathbb{R}^n, \quad r \in R),$$

the proof of this assertion following from II.4.3–II.4.5.] The problem we consider, in its original version, is to minimize $\tilde{y}^1(\sigma, a_0)(t_1)$ on the set

$$\{(\sigma, a_0) \in \mathscr{R} \times A_0 \mid \tilde{y}(\sigma, a_0)(t_1) \in A_1, \tilde{y}(\sigma, a_0)(T) \subset A\}$$

or, equivalently, to minimize $\tilde{y}^1(\sigma, a_0)(t_1)$ on the set

$$\{(\sigma, a_0, a_1) \in \mathscr{R} \times A_0 \times A_1 \mid \tilde{y}(\sigma, a_0)(t_1) - a_1 = 0, \tilde{y}(\sigma, a_0)(T) \subset A\}.$$

This problem is of the type defined at the beginning of this chapter, with

$$m \triangleq n, \qquad B \triangleq A_0 \times A_1, \qquad \mathscr{X}_2 \triangleq C(T, \mathbb{R}^n),$$

$$C_2 \triangleq \{w(\cdot) \in C(T, \mathbb{R}^n) \mid w(T) \subset A\}, \qquad R^{\#}(t) \triangleq R \quad (t \in T),$$

and

$$g_0(\sigma, a_0, a_1) \triangleq \tilde{y}^1(\sigma, a_0)(t_1), \qquad g_1(\sigma, a_0, a_1) \triangleq \tilde{y}(\sigma, a_0)(t_1) - a_1,$$

and

$$g_2(\sigma, a_0, a_1) \triangleq \tilde{y}(\sigma, a_0) \qquad (\sigma \in \mathscr{S}, \quad a_0 \in A_0, \quad a_1 \in A_1).$$

Since this problem involves no states, a minimizing relaxed solution is defined by some $(\bar{\sigma}, \bar{a}_0, \bar{a}_1) \in \mathscr{S} \times A_0 \times A_1$, a minimizing $\mathscr{R}$-solution by some $(\bar{\rho}, \bar{a}_0, \bar{a}_1) \in \mathscr{R} \times A_0 \times A_1$, and an approximate $\mathscr{R}$-solution by a sequence $((\bar{\rho}_j, a_0^j, a_1^j))$ in $\mathscr{R} \times A_0 \times A_1$. We summarize our conclusions concerning this problem in the next theorem.

V.6.1 ***Theorem*** Assume that there exists some $(\sigma', a_0') \in \mathscr{S} \times A_0$ such that $\tilde{y}(\sigma', a_0')(t_1) \in A_1$ and $\tilde{y}(\sigma', a_0')(T) \subset A$. Then there exists a minimizing relaxed solution $(\tilde{\sigma}, \tilde{a}_0, \tilde{a}_1)$ and a minimizing approximate $\mathscr{R}$-solution $((\bar{\rho}_j, \tilde{a}_0, \tilde{a}_1))$ such that

$$\lim_j \tilde{y}(\bar{\rho}_j, \tilde{a}_0)(t) = \tilde{y}(\tilde{\sigma}, \tilde{a}_0)(t) \qquad \text{uniformly for all } t \in T.$$

Furthermore, if $(\bar{\sigma}, \bar{a}_0, \bar{a}_1)$ is either a minimizing relaxed solution or a minimizing $\mathscr{R}$-solution and $\bar{y} \triangleq \tilde{y}(\bar{\sigma}, \bar{a}_0)$, then the matrix-differential equation

$$Z(t) = I + \int_t^{t_1} Z(\tau) f_v(\bar{y}(\tau), \bar{\sigma}(\tau)) \, d\tau \qquad (t \in T)$$

has a unique solution Z and there exist $l_0 \geqslant 0$, $l_1 \triangleq (l_1^1, ..., l_1^n) \in \mathbb{R}^n$, $\omega \in \text{frm}^+(T)$, and an ω-integrable $\tilde{\omega} : T \to \mathbb{R}^n$ such that

(1) $$l_0 + |l_1| + \omega(T) > 0 \quad \text{and} \quad |\tilde{\omega}(t)| = 1 \quad (t \in T);$$

(2) $$k(t)^T f(\bar{y}(t), \bar{\sigma}(t)) = \min_{r \in R} k(t)^T f(\bar{y}(t), r) \qquad \text{for a.a.} \quad t \in T,$$

where $e_1 \triangleq (1, 0, \ldots, 0) \in \mathbb{R}^n$ and

$$k(t)^T \triangleq \left[(l_0 e_1 + l_1)^T + \int_{[t,t_1]} \tilde{\omega}(\alpha)^T Z(\alpha)^{-1} \omega(d\alpha)\right] Z(t) \qquad (t \in T);$$

$$(3) \qquad k(t_0)^T \bar{y}(t_0) = \operatorname*{Min}_{a_0 \in A_0} k(t_0)^T a_0 , \qquad l_1{}^T \bar{y}(t_1) = \operatorname*{Max}_{a_1 \in A_1} l_1{}^T a_1 ;$$

and

$$(4) \qquad \tilde{\omega}(t)^T \bar{y}(t) = \operatorname*{Max}_{a \in A} \tilde{\omega}(t)^T a \qquad \text{for} \quad \omega\text{-a.a.} \quad t \in T.$$

Finally, if $(\bar{\sigma}, \bar{a}_0, \bar{a}_1)$ is a minimizing relaxed solution and conditions (1)–(4) imply $l_0 \neq 0$, then there exists no strict $\mathscr{U}$-solution.

▌PROOF Let c_1 be as defined in V.6.0(2), and let

$$c_2 \triangleq \sup\{|f_v(v, r)| \mid v \in S^F(0, c_1), r \in R\}.$$

Then $c_2 < \infty$ since f_v is continuous.

Step 1 We first show that the function

$$(\sigma, a_0) \to \tilde{y}(\sigma, a_0) : \mathscr{S} \times A_0 \to C(T, \mathbb{R}^n)$$

is continuous. Indeed, let

$$\lim_j \sigma_j = \sigma_0 , \qquad \lim_j a_0{}^j = a_0{}^0, \qquad \text{and} \qquad y_j \triangleq \tilde{y}(\sigma_j , a_0{}^j) \qquad (j = 0, 1, 2,\ldots).$$

Then

$$y_j(t) = a_0{}^j + \int_{t_0}^t f(y_j(\tau), \sigma_j(\tau))\, d\tau \qquad (t \in T, \quad j = 0, 1, 2,\ldots);$$

hence, by the mean value theorem II.3.6,

$$\begin{aligned}
(5) \quad & | y_j(t) - y_0(t)| \\
&\leqslant \left| \int_{t_0}^t [f(y_j(\tau), \sigma_j(\tau)) - f(y_0(\tau), \sigma_0(\tau))]\, d\tau \right| + | a_0{}^j - a_0{}^0 | \\
&\leqslant \int_{t_0}^t | f(y_j(\tau), \sigma_j(\tau)) - f(y_0(\tau), \sigma_j(\tau))|\, d\tau \\
&\quad + \left| \int_{t_0}^t f(y_0(\tau), \sigma_j(\tau) - \sigma_0(\tau))\, d\tau \right| + | a_0{}^j - a_0{}^0 | \\
&\leqslant c_2 \int_{t_0}^t | y_j(\tau) - y_0(\tau)|\, d\tau \\
&\quad + \left| \int_{t_0}^t f(y_0(\tau), \sigma_j(\tau) - \sigma_0(\tau))\, d\tau \right| + | a_0{}^j - a_0{}^0 | \\
&\qquad\qquad (t \in T, \quad j \in \mathbb{N}).
\end{aligned}$$

We set

$$h_j(t) \triangleq \int_{t_0}^{t} f(y_0(\tau), \sigma_j(\tau) - \sigma_0(\tau))\, d\tau$$

$$= \int_T \chi_{[t_0,t]}(\tau) f(y_0(\tau), \sigma_j(\tau) - \sigma_0(\tau))\, d\tau \qquad (t \in T, \quad j \in \mathbb{N}),$$

and observe that $\lim_j \sigma_j = \sigma_0$ implies $\lim_j h_j(t) = 0 \quad (t \in T)$. Since $|f(y_0(\tau), \sigma(\tau))| \leqslant c_1$ $(\tau \in T,\ \sigma \in \mathscr{S})$, it follows that $\{h_1, h_2, ...\}$ is bounded and equicontinuous and, by I.5.3, $\lim_j h_j(t) = 0$ uniformly for $t \in T$. Therefore, by (5) and Gronwall's inequality II.4.4, we have $\lim_j y_j = y_0$ in $C(T, \mathbb{R}^n)$, thus showing that $(\sigma, a_0) \to \tilde{y}(\sigma, a_0)$ is continuous.

Step 2 Since our problem does not involve any states, the assumptions of Theorems V.1.1 and V.1.2 require only the continuity of $(\sigma, a_0, a_1) \to g(\sigma, a_0, a_1)$ which follows from that of $(\sigma, a_0) \to \tilde{y}(\sigma, a_0)$. Thus, the existence of a minimizing relaxed solution $(\tilde{\sigma}, \tilde{a}_0, \tilde{a}_1)$ and of a minimizing approximate solution $((\bar{\rho}_j, \tilde{a}_0, \tilde{a}_1))$ as well as the relation $\lim_j \tilde{y}(\bar{\rho}_j, \tilde{a}_0) = \tilde{y}(\tilde{\sigma}, \tilde{a}_0)$ in $C(T, \mathbb{R}^n)$ follow from V.1.1 and V.1.2.

Now let $(\bar{\sigma}, \bar{a}_0, \bar{a}_1) \in \mathscr{S} \times A_0 \times A_1$, $(\sigma_j, a_0{}^j) \in \mathscr{S} \times A$ $(j = 0, 1, 2, ..., n)$ and

$$q_0(\theta) \triangleq (\bar{\sigma}, \bar{a}_0) + \sum_{j=0}^{n} \theta^j[(\sigma_j, a_0{}^j) - (\bar{\sigma}_j, \bar{a}_0)] \qquad (\theta \in \mathscr{T}_{n+1}).$$

We consider the function

$$\theta \triangleq (\theta^0, ..., \theta^n) \to \tilde{y}(q_0(\theta)) : \mathscr{T}_{n+1} \to C(T, \mathbb{R}^n).$$

Since $\tilde{y}(q_0(\theta))(\cdot)$ is the solution of the equation

$$y(t) = \bar{a}_0 + \sum_{j=0}^{n} \theta^j(a_0{}^j - \bar{a}_0)$$

$$+ \int_{t_0}^{t} \left(f(y(\tau), \sigma(\tau)) + \sum_{j=0}^{n} \theta^j f(y(\tau), \sigma_j(\tau) - \sigma(\tau))\right) d\tau \qquad (t \in T),$$

it follows from II.4.11 [with $V \triangleq S(0, c_1 + 1)$] and IV.2.7 that $\theta \to y^{\#}(\theta) \triangleq \tilde{y}(q_0(\theta))$ has a derivative at $\theta = 0$, and that $\bar{y} \triangleq \tilde{y}(q_0(0))$ and $p \triangleq \mathscr{D}y^{\#}(0)\, \Delta\theta$ satisfy the equation

$$p(t) = \int_{t_0}^{t} f_v(\bar{y}(\tau), \bar{\sigma}(\tau))\, p(\tau)\, d\tau \tag{6}$$

$$+ \sum_{j=0}^{n} \Delta\theta^j \left(a_0{}^j - \bar{a}_0 + \int_{t_0}^{t} f(\bar{y}(\tau), \sigma_j(\tau) - \sigma(\tau))\, d\tau\right)$$

$$[t \in T, \quad \Delta\theta \triangleq (\Delta\theta^0, ..., \Delta\theta^n) \in \mathbb{R}^{n+1}].$$

Thus $(\bar{\sigma}, \bar{a}_0, \bar{a}_1) \triangleq (\bar{q}_0, \bar{a}_1)$ satisfies the assumptions of V.3.2 and V.3.3 and therefore, if it is extremal, then there exist $l_0 \geqslant 0$, $l_1 \in \mathbb{R}^n$, $\omega \in \mathrm{frm}^+(T)$, and an ω-integrable $\tilde{\omega} : T \to \mathbb{R}^n$ satisfying statements (1), (4), and

$$(7) \qquad \sum_{i=0}^{2} l_i(Dg_i(\bar{q}_0 ; q_0 - \bar{q}_0))$$
$$= (l_0 e_1 + l_1)^T D\tilde{y}(\bar{q}_0 ; q_0 - \bar{q}_0)(t_1)$$
$$- l_1^T(a_1 - \bar{a}_1) + \int \tilde{\omega}(t) \cdot D\tilde{y}(\bar{q}_0 ; q_0 - \bar{q}_0)(t)\, \omega(dt)$$
$$\geqslant 0 \qquad (q_0 \in \mathscr{S} \times A_0, \quad a_1 \in A_1).$$

It is clear that if we choose any $(\sigma, a_0) \in \mathscr{S} \times A_0$ and set

$$\Delta\theta = (1, 0, ..., 0), \qquad (\sigma_j, a_0^j) = (\sigma, a_0) \triangleq q_0 \qquad (j = 0, 1, ..., n),$$

then

$$(8) \qquad D\tilde{y}(\bar{q}_0 ; q_0 - \bar{q}_0) = \mathscr{D}y^{\#}(0)\, \Delta\theta = p.$$

It follows, by (6) and II.4.8, that there exists a unique Z as defined in the statement of the theorem, and that

$$(9) \qquad D\tilde{y}(\bar{q}_0 ; q_0 - \bar{q}_0)(t)$$
$$= Z(t)^{-1} \left[Z(t_0)(a_0 - \bar{a}_0) + \int_{t_0}^{t} Z(\tau)\, f(\bar{y}(\tau), \sigma(\tau) - \bar{\sigma}(\tau))\, d\tau \right]$$
$$[t \in T, \quad q_0 \triangleq (\sigma, a_0) \in \mathscr{S} \times A_0].$$

We combine relation (9) with Fubini's theorem I.4.45 to obtain

$$(10) \quad \int \tilde{\omega}(t) \cdot D\tilde{y}(\bar{q}_0 ; q_0 - \bar{q}_0)(t)\, \omega(dt)$$
$$= \int \tilde{\omega}(\alpha)^T Z(\alpha)^{-1} Z(t_0)(a_0 - \bar{a}_0)\, \omega(d\alpha)$$
$$+ \int \omega(d\alpha) \int_{t_0}^{t_1} \chi_{[t_0, \alpha]}(\tau)\, \tilde{\omega}(\alpha)^T Z(\alpha)^{-1} Z(\tau)\, f(\bar{y}(\tau), \sigma(\tau) - \bar{\sigma}(\tau))\, d\tau$$
$$= \int \tilde{\omega}(\alpha)^T Z(\alpha)^{-1} Z(t_0)(a_0 - \bar{a}_0)\, \omega(d\alpha)$$
$$+ \int_{t_0}^{t_1} \left[\int_{[\tau, t_1]} \tilde{\omega}(\alpha)^T Z(\alpha)^{-1}\, \omega(d\alpha) \right] Z(\tau)\, f(\bar{y}(\tau), \sigma(\tau) - \sigma(\tau))\, d\tau;$$

hence, by (7) and (9),

$$\text{(11)}\quad k(t_0)^T (a_0 - \bar{a}_0) - l_1{}^T(a_1 - \bar{a}_1) + \int k(\tau)^T f(\bar{y}(\tau), \sigma(\tau) - \bar{\sigma}(\tau))\, d\tau \geqslant 0$$
$$(\sigma \in \mathscr{S}, \quad a_0 \in A_0, \quad a_1 \in A_1).$$

The compact metric space R has a dense subset $\{r_1, r_2, ...\}$. We choose $j \in \mathbb{N}$ and an arbitrary measurable $E \subset T$, and set $a_0 = \bar{a}_0$, $a_1 = \bar{a}_1$, $\sigma(t) = r_j$ $(t \in E)$, $\sigma(t) = \bar{\sigma}(t)$ $(t \notin E)$. Then (11) yields

$$\int_E k(\tau)^T [f(\bar{y}(\tau), r_j) - f(\bar{y}(\tau), \bar{\sigma}(\tau))]\, d\tau \geqslant 0$$

and, by I.4.34(7),

$$k(\tau)^T[f(\bar{y}(\tau), r_j) - f(\bar{y}(\tau), \bar{\sigma}(\tau))] \geqslant 0 \qquad \text{for a.a.} \quad \tau \in T,$$

say for $\tau \in T_j$. It follows that

$$k(\tau)^T f(\bar{y}(\tau), \bar{\sigma}(\tau)) \leqslant \inf_{j \in \mathbb{N}} k(\tau)^T f(\bar{y}(\tau), r_j) \qquad \Big(\tau \in T' \triangleq \bigcap_{j=1}^{\infty} T_j\Big).$$

Since f is continuous and $\bar{\sigma}(\tau)$ a probability measure for a.a. $\tau \in T$, it follows that

$$k(\tau)^T f(\bar{y}(\tau), \bar{\sigma}(\tau)) = \operatorname*{Min}_{r \in R} k(\tau)^T f(\bar{y}(\tau), r) \quad \text{a.e. in } T,$$

thus proving statement (2). If we set, in (11), $\sigma = \bar{\sigma}$ and alternately $a_1 = \bar{a}_1$ and $a_0 = \bar{a}_0$, then we obtain the two relations in (3).

If $(\bar{\sigma}, \bar{a}_0, \bar{a}_1)$ is either a minimizing relaxed solution or a minimizing $\mathscr{R}$-solution, then, by V.3.2, $(\bar{\sigma}, \bar{a}_0, \bar{a}_1)$ is extremal and statements (1)–(4) are therefore valid. Finally, if $(\bar{\sigma}, \bar{a}_0, \bar{a}_1)$ is a minimizing relaxed solution and statements (1)–(4) imply $l_0 \neq 0$, then, by V.3.2, $(\bar{\sigma}, \bar{a}_0, \bar{a}_1)$ is extremal, admissible, and normal, and our last conclusion follows from V.3.4. QED

V.6.2 ***Example I*** Our first illustrative problem has a minimizing relaxed solution $(\bar{\sigma}, \bar{a}_0, \bar{a}_1)$ which is simultaneously a minimizing $\mathscr{R}$-solution. Let $T \triangleq [0, 1]$, $R \triangleq [-1, 1]$, $n \triangleq 2$, $A_0 \triangleq \{(0, 0)\}$, $A_1 \triangleq \mathbb{R} \times \{1\}$, $A \triangleq \mathbb{R}^2$, $f \triangleq (f^1, f^2)$, and

$$f^1(v^1, v^2, r) \triangleq v^2 + \tfrac{1}{2}(r)^2, \quad f^2(v^1, v^2, r) \triangleq v^1 + r \qquad [(v^1, v^2) \in \mathbb{R}^2, \quad r \in R].$$

Here $(r)^2 \triangleq r \cdot r$, the parentheses being used to indicate that 2 is an exponent of a power and not a superscript. We shall exhibit later a point

$(\bar{\sigma}, \bar{a}_0) \in \mathscr{S} \times A_0$ such that $\tilde{y}(\bar{\sigma}, \bar{a}_0)(1) \in A_1$. Therefore the assumptions of V.6.0 and V.6.1 are satisfied because

$$|f(v, r)| \leqslant |v| + \tfrac{3}{2} \leqslant \tfrac{3}{2}(|v| + 1) \qquad (v \in \mathbb{R}^n, \quad r \in R).$$

Since $A = \mathbb{R}^n$, it follows from V.6.1(4) that $\tilde{\omega}(t) = 0$ ω-a.e. and therefore $\omega = 0$ [since, by V.6.1(1), $|\tilde{\omega}(t)| = 1$ for all $t \in T$]. Furthermore, by V.6.1(3),

$$l_1{}^1 \bar{y}^1(1) = \operatorname*{Max}_{\alpha \in \mathbb{R}} l_1{}^1 \alpha;$$

hence $l_1{}^1 = 0$. Thus

$$k(t)^T = (l_0, l_1{}^2)^T Z(t).$$

If we premultiply the equation that defines Z by $(l_0, l_1{}^2)^T$, then we conclude that

$$k(t)^T = (l_0, l_1{}^2)^T + \int_t^{t_1} k(\tau)^T f_v(\bar{y}(\tau), \bar{\sigma}(\tau))\, d\tau \qquad (t \in T).$$

By V.6.1(1), we have $l_0 + |l_1{}^2| > 0$; hence, by II.4.7, $k(t) \neq 0$ $(t \in T)$. It follows that $k \triangleq (k^1, k^2)$ and $\bar{y} \triangleq (\bar{y}^1, \bar{y}^2)$ satisfy a.e. in $[0, 1]$ the equations

$$\begin{aligned}
&\dot{\bar{y}}^1(t) = \bar{y}^2(t) + \tfrac{1}{2} \int (r)^2\, \bar{\sigma}(t)(dr), \qquad \bar{y}^1(0) = 0 \\
&\dot{\bar{y}}^2(t) = \bar{y}^1(t) + \int r\, \bar{\sigma}(t)(dr), \qquad \bar{y}^2(0) = 0, \qquad \bar{y}^2(1) = 1 \\
&\dot{k}^1(t) = -k^2(t), \qquad k^1(1) = l_0 \geqslant 0 \\
&\dot{k}^2(t) = -k^1(t) \\
&k(t) \neq 0 \qquad (t \in T).
\end{aligned} \tag{1}$$

Furthermore, by V.6.1(2), we have

$$\tfrac{1}{2}k^1(t) \int (r)^2\, \bar{\sigma}(t)(dr) + k^2(t) \int r\, \bar{\sigma}(t)(dr) = \operatorname*{Min}_{r \in [-1,1]} [\tfrac{1}{2}k^1(t)(r)^2 + k^2(t) r] \tag{2}$$

a.e. in $[0, 1]$.

It is easy to see that if $\alpha_1, \alpha_2 \in \mathbb{R}$ and $(\alpha_1, \alpha_2) \neq 0$, then the function

$$r \to \tfrac{1}{2}\alpha_1(r)^2 + \alpha_2 r : [0, 1] \to \mathbb{R}$$

achieves its minimum at

$$\begin{aligned}
&r = -\alpha_2/\alpha_1 && \text{if} \quad \alpha_1 > |\alpha_2|, \\
&r = -1 && \text{if} \quad \alpha_1 \leqslant \alpha_2 \quad \text{and} \quad \alpha_2 > 0, \\
&r = 1 && \text{if} \quad \alpha_1 \leqslant -\alpha_2 \quad \text{and} \quad \alpha_2 < 0, \\
&r = 1 \quad \text{and} \quad r = -1 && \text{if} \quad \alpha_1 < 0, \quad \alpha_2 = 0.
\end{aligned} \tag{3}$$

Thus (2) and (3) imply that $\bar{\sigma}(t)$ is concentrated at one point of R for a.a. values of t except when $k^1(t) < 0$, $k^2(t) = 0$.

The relations $\dot{k}^1(t) = -k^2(t)$, $\dot{k}^2(t) = -k^1(t)$ a.e. in $[0, 1]$ imply $k^1(t)\,\dot{k}^1(t) = k^2(t)\,\dot{k}^2(t)$ a.e. in $[0, 1]$; hence $[k^1(t)]^2 - [k^2(t)]^2 = \beta$ for some constant β. Figure 1 shows the sets

$$\{(v^1, v^2) \in \mathbb{R}^2 \mid (v^1)^2 - (v^2)^2 = \beta\}$$

for various choices of β. If $k^2(1) \geqslant k^1(1) \geqslant 0$, then, as Fig. 1 indicates,

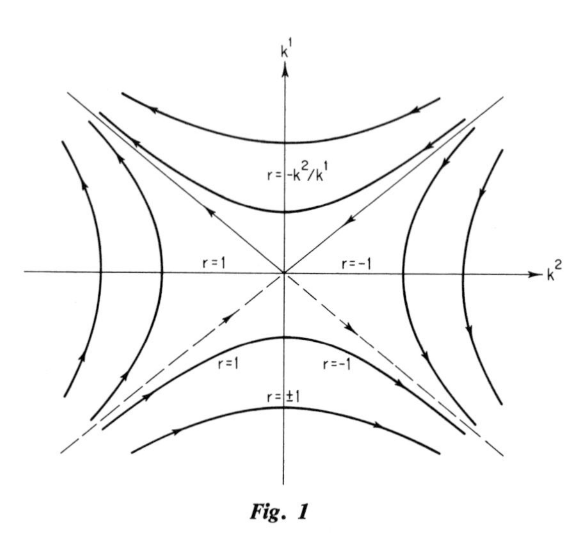

Fig. 1

$\bar{\sigma}(t)$ is a.e. concentrated at $r = -1$. Then the first two lines of (1) imply that for a.a. $t \in T$ we have

$$\dot{\bar{y}}^1(t) = \bar{y}^2(t) + \tfrac{1}{2}, \qquad \bar{y}^1(0) = 0,$$

$$\dot{\bar{y}}^2(t) = \bar{y}^1(t) - 1, \qquad \bar{y}^2(0) = 0,$$

and the (unique) solution of these equations is

$$\bar{y}^1(t) = 1 - \tfrac{1}{4}e^t - \tfrac{3}{4}e^{-t}, \qquad \bar{y}^2(t) = -\tfrac{1}{2} - \tfrac{1}{4}e^t + \tfrac{3}{4}e^{-t}.$$

This implies $\bar{y}^2(1) = -\frac{1}{2} - \frac{1}{4}e + \frac{3}{4}e^{-1} \neq 1$, thus contradicting (1). A similar argument rules out $-k^2(1) \geqslant k^1(1)$. Thus it follows (from Figure 1) that $k^1(t) > |\,k^2(t)\,| > 0$ for all $t \in T$; hence, by (3), $\bar{\sigma}(t)$ is concentrated at $r = -k^2(t)/k^1(t)$ a.e. in T. Thus $\bar{\sigma} \in \mathscr{R}$ and $\bar{\sigma}(t) = -k^2(t)/k^1(t)$ a.e. in T.

The equations $\dot{k}^1(t) = -k^2(t)$, $\dot{k}^2(t) = -k^1(t)$ a.e. in T, $k^1(t) > 0$ have solutions of the form

$$k^1(t) = \gamma\,\mathrm{Ch}(t - \delta), \qquad k^2(t) = -\gamma\,\mathrm{Sh}(t - \delta)$$

for some constants γ and δ, where

$$\mathrm{Ch}(x) \triangleq \tfrac{1}{2}(e^x + e^{-x}), \qquad \mathrm{Sh}(x) \triangleq \tfrac{1}{2}(e^x - e^{-x}) \qquad (x \in \mathbb{R}).$$

Thus

$$\bar{\sigma}(t) = -k^2(t)/k^1(t) = \mathrm{th}(t - \delta) \qquad (t \in T)$$

[where $\mathrm{th}(x) \triangleq \mathrm{Sh}(x)/\mathrm{Ch}(x)$ $(x \in \mathbb{R})$]. It follows, by (1), that for a.a. $t \in T$

$$\begin{aligned} \dot{\bar{y}}^1(t) &= \bar{y}^2(t) + \tfrac{1}{2}[\mathrm{th}(t - \delta)]^2, \qquad & y^1(0) &= 0, \\ \dot{\bar{y}}^2(t) &= \bar{y}^1(t) + \mathrm{th}(t - \delta), & \bar{y}^2(0) &= 0. \end{aligned}$$

These equations have unique solutions which are

$$\bar{y}^1(t) = \int_0^t \{\tfrac{1}{2}\,\mathrm{Ch}(t - \tau) \cdot [\mathrm{th}(\tau - \delta)]^2 + \mathrm{Sh}(t - \tau) \cdot \mathrm{th}(\tau - \delta)\}\, d\tau \qquad (t \in T),$$

$$\bar{y}^2(t) = \int_0^t \{\tfrac{1}{2}\,\mathrm{Sh}(t - \tau) \cdot [\mathrm{th}(\tau - \delta)]^2 + \mathrm{Ch}(t - \tau) \cdot \mathrm{th}(\tau - \delta)\}\, d\tau \qquad (t \in T).$$

We observe that, for fixed t and τ, the integrand in the expression for $\bar{y}^2$ is decreasing as a function of δ on $\mathbb{R}$. Indeed, the derivative of this function of δ is

$$\begin{aligned} &-[\mathrm{Ch}(\tau - \delta)]^{-2}[\mathrm{Sh}(t - \tau)\,\mathrm{th}(\tau - \delta) + \mathrm{Ch}(t - \tau)] \\ &\quad = -[\mathrm{Ch}(\tau - \delta)]^{-3}\,\mathrm{Ch}(t - \delta) < 0. \end{aligned}$$

As $\delta \to -\infty$ the integrand converges uniformly to

$$\tfrac{1}{2}\,\mathrm{Sh}(t - \tau) + \mathrm{Ch}(t - \tau) = \tfrac{3}{4}e^{t-\tau} + \tfrac{1}{4}e^{\tau - t};$$

hence, as $\delta \to -\infty$, the expression for $\bar{y}^2(1)$ converges to

$$\begin{aligned} \int_0^1 (\tfrac{3}{4}e^{1-\tau} + \tfrac{1}{4}e^{\tau-1})\, d\tau &= -\tfrac{3}{4} + \tfrac{3}{4}e + \tfrac{1}{4} - \tfrac{1}{4}e^{-1} \\ &= -\tfrac{1}{2} + \tfrac{3}{4}e - \tfrac{1}{4}e^{-1} \approx 1.448 \cdots. \end{aligned}$$

A similar computation shows that, as $\delta \to \infty$, the expression for $\bar{y}^2(1)$ converges to -0.904. Since the expression for $\bar{y}^2(1)$ is a decreasing function of δ, we conclude that there exists a unique value of δ which yields $\bar{y}^2(1) = 1$. This shows that there exists $(\bar{\sigma}, \bar{a}_0) \in \mathscr{S} \times A_0$ such that $\tilde{y}(\bar{\sigma}, \bar{a}_0)(1) \in A_1$.

Thus the problem has a unique minimizing relaxed solution $(\bar{\sigma}, \bar{a}_0, \bar{a}_1)$, $\bar{\sigma} \in \mathscr{R}$ and

$$\bar{\sigma}(t) = \mathrm{th}(t - \delta) \quad \text{a.e. in } T.$$

V.6.3 ***Example II*** We let $T \triangleq [0, 1]$, $R \triangleq [-1, 1]$, $n \triangleq 2$, $0 < \alpha < 1$, $A_0 \triangleq \{(0, \alpha)\}$, $A_1 \triangleq \mathbb{R} \times \{0\}$, $A \triangleq \mathbb{R}^2$, $f \triangleq (f^1, f^2)$,

$$f^1(v^1, v^2, r) \triangleq (v^2)^2 - (r)^2, \qquad f^2(v^1, v^2, r) \triangleq r \qquad [v \triangleq (v^1, v^2) \in \mathbb{R}^2, \quad r \in R].$$

The equations

$$\dot{\bar{y}}^1(t) = [\bar{y}^2(t)]^2 - \int (r)^2\, \sigma(t)\,(dr) \quad \text{a.e. in } T, \qquad \bar{y}^1(0) = 0,$$

$$\dot{\bar{y}}^2(t) = \int r\sigma(t)\,(dr) \quad \text{a.e. in } T, \qquad \bar{y}^2(0) = \alpha$$

have a unique solution for each $\sigma \in \mathscr{S}$, obtained by first evaluating $\bar{y}^2(t) = \alpha + \int_0^t d\tau \int r\sigma(\tau)\,(dr)$ and then substituting in the first equation. All these solutions are uniformly bounded for $\sigma \in \mathscr{S}$, since

$$\left| \int r\sigma(t)\,(dr) \right| \leqslant 1 \qquad \text{and} \qquad \left| \int (r)^2\, \sigma(t)\,(dr) \right| \leqslant 1 \qquad \text{a.e. in } T;$$

hence $|\bar{y}^2(t)| \leqslant \alpha + 1$, $|\bar{y}^1(t)| \leqslant (\alpha + 1)^2 + 1$ $(t \in T)$. We shall exhibit in what follows a function $\bar{\sigma} \in \mathscr{S}$ such that $\bar{y}(\bar{\sigma})(1) \in A_1$. Thus the assumptions of V.6.0 and V.6.1 are satisfied.

Since $A = \mathbb{R}^2$ and $A_1 = \mathbb{R} \times \{0\}$, we conclude, as in the previous example, that $\omega = 0$, $l_1{}^1 = 0$, and we have

$$\dot{\bar{y}}^1(t) = [\bar{y}^2(t)]^2 - \int (r)^2\, \bar{\sigma}(t)\,(dr) \quad \text{a.e. in } T, \qquad \bar{y}^1(0) = 0,$$

$$\dot{\bar{y}}^2(t) = \int r\bar{\sigma}(t)\,(dr) \quad \text{a.e. in } T, \qquad \bar{y}^2(0) = \alpha, \qquad \bar{y}^2(1) = 0,$$

$$\dot{k}^1(t) = 0 \quad \text{a.e. in } T, \qquad k^1(1) = l_0 \geqslant 0, \tag{1}$$

$$\dot{k}^2(t) = -2\bar{y}^2(t)\, k^1(t) \quad \text{a.e. in } T,$$

$$k(t) \neq 0 \qquad (t \in T),$$

and

$$-k^1(t) \int (r)^2\, \bar{\sigma}(t)\,(dr) + k^2(t) \int r\bar{\sigma}(t)\,(dr) = \operatorname*{Min}_{r \in [-1,1]} [-k^1(t)(r)^2 + k^2(t) r] \tag{2}$$

a.e. in T. Since $\dot{k}^1(t) = 0$ a.e., we have $k^1(t) = k^1(1) \geqslant 0$ $(t \in T)$, and we denote the constant $k^1(1)$ by k^1. Thus (2) implies that, for a.a. $t \in T$,

$$\bar{\sigma}(t)(\{-1\}) = 1 \qquad \text{if} \quad k^2(t) > 0,$$

$$\bar{\sigma}(t)(\{1\}) = 1 \qquad \text{if} \quad k^2(t) < 0, \tag{3}$$

$$\bar{\sigma}(t)(\{-1, 1\}) = 1 \qquad \text{if} \quad k^1 > 0 \quad \text{and} \quad k^2(t) = 0.$$

If $k^1 = 0$, then, by (1), $\dot{k}^2(t) = 0$ a.e. and $k^2(t) = k^2(1) \neq 0$ $(t \in T)$. If also $k^2(t) = k^2(1) > 0$, then $\bar{\sigma}(t) = -1$ a.e. and the second line in (1) yields $\bar{y}^2(t) = \alpha - t$ $(t \in T)$; hence $\bar{y}^2(1) = \alpha - 1 < 0$, contradicting $\bar{y}^2(1) = 0$. If $k^1 = 0$ and $k^2(t) = k^2(1) < 0$, then $\bar{\sigma}(t) = 1$ a.e. and we obtain $\bar{y}^2(t) = \alpha + t$ $(t \in T)$, thus contradicting $\bar{y}^2(1) = 0$. We conclude that $k^1 > 0$.

We now observe that the absolutely continuous functions $\bar{y}^2$ and k^2, the constant $k^1 > 0$, and the relaxed control function $\bar{\sigma}$ must satisfy relation (3) and [by (1)]

$$\dot{\bar{y}}^2(t) = \int r\bar{\sigma}(t)\,(dr), \qquad \dot{k}^2(t) = -2k^1\bar{y}^2(t) \quad \text{a.e. in } [0, 1]. \tag{4}$$

Among all the solutions of (2) and (3), they must also satisfy the conditions

$$\bar{y}^2(0) = \alpha, \qquad \bar{y}^2(1) = 0. \tag{5}$$

If ϕ is a function on T to some space Y and T_1 is a subinterval of T, let $\phi(T)$ be called the *trajectory* of ϕ and $\phi(T_1)$ an *arc of trajectory* of ϕ. Then we consider arcs of trajectories of the functions $(\bar{y}^2, k^2)$ that satisfy (3) and (4). If T^+ is a subinterval of T and $k^2(t) > 0$ $(t \in T^+)$, then

$$\dot{\bar{y}}^2(t) = -1 \qquad \text{and} \qquad \dot{k}^2(t) = -2k^1\bar{y}^2(t) \quad \text{a.e. in } T^+;$$

hence

$$\bar{y}^2(t) = -t + \beta_1 \qquad \text{and} \qquad k^2(t) = k^1(-t + \beta_1)^2 + \beta_2 \qquad (t \in T^+) \tag{6}$$

for some $\beta_1, \beta_2 \in \mathbb{R}$. Thus

$$k^2(t) = k^1[\bar{y}^2(t)]^2 + \beta_2 \qquad (t \in T^+). \tag{7}$$

We similarly verify that if T^- is a subinterval of T and $k(t) < 0$ $(t \in T^-)$, then, for some $\beta_2' \in \mathbb{R}$,

$$k^2(t) = -k^1[\bar{y}^2(t)]^2 + \beta_2' \qquad (t \in T^-).$$

If $k^2(t') = 0$ and $\bar{y}^2(t') > 0$, then $\dot{k}^2(t) < 0$ a.e. in some neighborhood of t' and therefore k^2 decreases from positive to negative values as t increases in that neighborhood; if $k^2(t'') = 0$ and $\bar{y}^2(t'') < 0$, then k^2 increases from negative to positive values in a neighborhood of t''.

Figure 2 summarizes this information, showing curves that contain possible arcs of trajectories of functions $(\bar{y}^2, k^2)$ satisfying (3) and (4). The arrows indicate the changes in the values of $(\bar{y}^2, k^2)$ as t increases. A function $(\bar{y}^2, k^2)$ satisfying (3)–(5) has a trajectory consisting of the arcs in Figure 2 except over the subset of T where $\bar{y}^2(t) = k^2(t) = 0$.

In order to satisfy relations (5), the trajectory of $(\bar{y}^2, k^2)$ must begin on the vertical line $\bar{y}^2 = \alpha$ and end at $\bar{y}^2 = 0$. Figure 2 shows that this is only

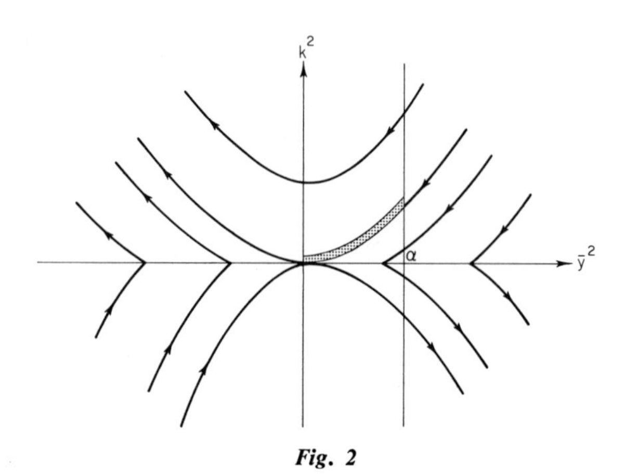

Fig. 2

possible if the trajectory contains the arc drawn in a heavy line and satisfying relation (7) for $\beta_2 = 0$. This yields

$$k^2(0) = k^1[\bar{y}^2(0)]^2 = k^1 \cdot (\alpha)^2.$$

Thus, by (6), $\beta_1 = \alpha$ and $k^2(t) = k^1 \cdot (\alpha - t)^2$ while $k^2(t) > 0$; hence

$$\bar{y}^2(t) = \alpha - t \qquad \text{and} \qquad k^2(t) = k^1 \cdot (\alpha - t)^2 \qquad \big(t \in [0, \alpha)\big).$$

Figure 2 shows that, for $t \in [\alpha, 1]$, we must have $\bar{y}^2(t) = k^2(t) = 0$. In view of (3) and (4), this implies that

$$-\bar{\sigma}(t)(\{-1\}) + \bar{\sigma}(t)(\{1\}) = 0 \quad \text{a.e. in } [\alpha, 1];$$

hence

$$\bar{\sigma}(t)(\{-1\}) = \bar{\sigma}(t)(\{1\}) = \tfrac{1}{2} \quad \text{a.e. in } [\alpha, 1].$$

This shows that there exist $\bar{a}_0 \triangleq (0, \alpha)$ and $\bar{\sigma} \in \mathscr{S}$ such that $\tilde{y}(\bar{\sigma}, \bar{a}_0)(1) \in A_1$ and that this function $\bar{\sigma}$, defined by

$$\bar{\sigma}(t)(\{-1\}) = 1 \quad \text{a.e. in } [0, \alpha)$$

and

$$\bar{\sigma}(t)(\{-1\}) = \bar{\sigma}(t)(\{1\}) = \tfrac{1}{2} \quad \text{a.e. in } [\alpha, 1]$$

is the only relaxed control function satisfying the conclusions of V.6.1. Thus, if we set $\bar{a}_1 = \tilde{y}(\bar{\sigma})(1)$, then $(\bar{\sigma}, \bar{a}_0, \bar{a}_1)$ is a minimizing relaxed solution.

We can obtain a minimizing approximate $\mathscr{R}$-solution $((\rho_j, \bar{a}_0, \bar{a}_1))$ by dividing, for each $j \in \mathbb{N}$, the interval $[\alpha, 1]$ into $2j$ subintervals of equal length and setting $\rho_j(t) \triangleq -1$ $(t \in [0, \alpha)]$, and $\rho_j(t) \triangleq +1$ and $\rho_j(t) \triangleq -1$ on alternate subintervals of $[\alpha, 1]$.

V.6.4 ***Example III*** Our final example involves a *unilateral restriction* $y(T) \subset A \neq \mathbb{R}^n$. The problem itself could be solved in a much simpler and more elegant manner than we do it here; but our purpose now is to illustrate the use of Theorem V.6.1. Assume that a train must travel between two stations that are a unit distance apart, and that its schedule calls for covering this unit distance in a unit time. The train starts from rest and makes a complete stop an arrival. The acceleration of the train can never exceed in absolute value a given number $\alpha \in (4, \infty)$. We wish to choose the acceleration as a measurable function of time into $[-\alpha, \alpha]$ so as to cover the distance between stations in the unit time and, in so doing, to minimize the maximum value of the velocity of the train.

Let the origin of time coincide with the instant of departure, $y^2(t)$ denote the distance of the train from the first station at time t, $y^3(t)$ the velocity, and $\rho(t)$ the acceleration. Then we must have

$$\text{(1)} \qquad \begin{aligned} &\dot{y}^2(t) = y^3(t), \qquad \dot{y}^3(t) = \rho(t) \qquad \text{a.e. in } [0, 1], \\ &y^2(0) = y^3(0) = 0, \qquad y^2(1) = 1, \qquad y^3(1) = 0. \end{aligned}$$

We wish to choose (y^2, y^3, ρ), with $(y^2, y^3) : [0, 1] \to \mathbb{R}^2$ absolutely continuous and $\rho : [0, 1] \to [-\alpha, \alpha]$ measurable, so as to minimize $\operatorname{Max}_{t\in[0,1]} y^3(t)$ subject to (1). This is equivalent, however, to choosing (y^2, y^3, ρ, β) so as to minimize β subject both to (1) and to the relation

$$\text{(2)} \qquad y^3(t) - \beta \leqslant 0 \qquad (t \in [0, 1]).$$

Furthermore, we observe that relations (1) imply that $|y^3(t)| \leqslant \alpha t \leqslant \alpha$ $(t \in [0, 1])$, and therefore we may restrict β to the set $[-2\alpha, 2\alpha]$.

We shall be able to transform our problem into the form described in V.6.0 by one additional step. We denote by y^1 the constant function on $[0, 1]$ with the value β. Then y^1 is absolutely continuous, $\dot{y}^1(t) = 0$ a.e. in $[0, 1]$, and $y^1(0) = \beta$. The problem we are now considering is to choose an absolutely continuous $y \triangleq (y^1, y^2, y^3) : [0, 1] \to \mathbb{R}^3$, a measurable $\rho : [0, 1] \to [-\alpha, \alpha]$, and $\beta \in [-2\alpha, 2\alpha]$ so as to minimize $y^1(1)$ subject to the conditions

$$\text{(3)} \qquad \begin{aligned} &\dot{y}^1(t) = 0, \qquad \dot{y}^2(t) = y^3(t), \qquad \dot{y}^3(t) = \rho(t) \quad \text{a.e. in } [0, 1], \\ &y^1(0) = \beta \in [-2\alpha, 2\alpha], \quad y^2(0) = y^3(0) = 0, \quad y^2(1) = 1, \quad y^3(1) = 0, \\ &y^3(t) - y^1(t) \leqslant 0 \qquad (t \in [0, 1]). \end{aligned}$$

This problem is now precisely in the form considered in V.6.0 and V.6.1, with $T \triangleq [0, 1]$, $R \triangleq [-\alpha, \alpha]$, $n \triangleq 3$, $A_0 \triangleq [-2\alpha, 2\alpha] \times \{0\} \times \{0\}$, $A_1 \triangleq \mathbb{R} \times \{1\} \times \{0\}$, $A \triangleq \{(v^1, v^2, v^3) \in \mathbb{R}^3 \mid v^3 - v^1 \leqslant 0\}$, $f \triangleq (f^1, f^2, f^3)$, and

$$f^1(v^1, v^2, v^3, r) \triangleq 0, \qquad f^2(v^1, v^2, v^3, r) \triangleq v^3, \qquad f^3(v^1, v^2, v^3, r) \triangleq r.$$

It is clear that the conditions of V.6.0 are satisfied, and we shall exhibit in what follows a point $(\bar{\sigma}, \bar{a}_0) \in \mathscr{S} \times A_0$ such that $\tilde{y}(\bar{\sigma}, \bar{a}_0)(1) \in A_1$ and $\tilde{y}(\bar{\sigma}, \bar{a}_0)(T) \subset A$. Thus we may apply Theorem V.6.1.

Relation V.6.1(3) implies that

$$l_1^T \bar{y}(1) = l_1^1 \bar{y}^1(1) + l_1^2 = \operatorname{Max}\{l_1^1 a_1^1 + l_1^2 \mid a_1^1 \in \mathbb{R}\};$$

hence

$$l_1^1 = 0. \tag{4}$$

By V.6.1(4), we have

$$\begin{aligned} \tilde{\omega}(t)^T \bar{y}(t) &= \operatorname{Max}\{\tilde{\omega}^1(t)\, a^1 + \tilde{\omega}^2(t)\, a^2 + \tilde{\omega}^3(t)\, a^3 \mid a^3 - a^1 \leqslant 0\} \\ &= \operatorname{Max}\{[\tilde{\omega}^1(t) + \tilde{\omega}^3(t)]\, a^1 + \tilde{\omega}^2(t)\, a^2 + \tilde{\omega}^3(t)(a^3 - a^1) \mid a^3 - a^1 \leqslant 0\} \end{aligned}$$

for ω-a.a. $t \in [0, 1]$. This relation implies that

$$\tilde{\omega}^1(t) + \tilde{\omega}^3(t) = \tilde{\omega}^2(t) = 0 \quad \omega\text{-a.e.},$$

$$\tilde{\omega}^3(t) \geqslant 0 \qquad \text{and} \qquad \tilde{\omega}^3(t)[\bar{y}^3(t) - \bar{y}^1(t)] = 0 \quad \omega\text{-a.e.},$$

and therefore, by (4) and V.6.1(1), that

$$\tilde{\omega}^2(t) = 0 \quad \omega\text{-a.e.}, \qquad \tilde{\omega}^1(t) = \tilde{\omega}^3(t) = 0 \quad \omega\text{-a.e. in } \{t \in T \mid \bar{y}^3(t) < \bar{y}^1(t)\},$$

$$\begin{gathered} \tilde{\omega}^1(t) = -\tilde{\omega}^3(t) = -\tfrac{1}{2} \quad \omega\text{-a.e. in } \{t \in T \mid \bar{y}^3(t) = \bar{y}^1(t)\}, \\ l_0 \geqslant 0, \qquad l_1^1 = 0, \qquad l_0 + |\, l_1 \,| + \omega([0, 1]) > 0. \end{gathered} \tag{5}$$

We now evaluate the solution of the equation

$$\dot{Z}(t) = -Z(t) f_v(\bar{y}(t), \bar{\sigma}(t)) \quad \text{a.e. in } [0, 1], \qquad Z(1) = I.$$

Each row $(z^1, z^2, z^3)^T$ of Z satisfies the equations

$$\dot{z}^1(t) = 0, \qquad \dot{z}^2(t) = 0, \qquad \dot{z}^3(t) = -z^2(t) \qquad \text{a.e. in } [0, 1],$$

and therefore z^1 and z^2 are constant and $z^3(t) = z^3(1) - z^2 \cdot (t - 1)$. It follows that

$$Z(t) = \begin{pmatrix} 1 & 0 & 0 \\ 0 & 1 & 1 - t \\ 0 & 0 & 1 \end{pmatrix} \qquad (t \in [0, 1]),$$

and we easily verify that

$$Z(t)^{-1} = \begin{pmatrix} 1 & 0 & 0 \\ 0 & 1 & t-1 \\ 0 & 0 & 1 \end{pmatrix}.$$

We set

$$\lambda(t) \triangleq \int_{[t,1]} \tilde{\omega}^3(\alpha)\, \omega(d\alpha) \qquad (t \in [0, 1]), \tag{6}$$

and observe that, by (5) and the definition of k in V.6.1(2), we have

$$\begin{aligned} k(t)^T &\triangleq \left(k^1(t), k^2(t), k^3(t)\right)^T \\ &= [(l_0\,, l_1^2, l_1^3)^T + (-\lambda(t), 0, \lambda(t))^T]\, Z(t) \\ &= (l_0 - \lambda(t), l_1^2, l_1^2(1-t) + \lambda(t) + l_1^3)^T \qquad (t \in [0, 1]). \end{aligned} \tag{7}$$

Furthermore, λ is nonnegative and nonincreasing and, by (5), λ is constant on any subinterval of [0, 1] where $\bar{y}^3(t) < \bar{y}^1(t)$.

We apply V.6.1(2) to obtain

$$[l_1^2(1-t) + \lambda(t) + l_1^3] \int r\bar{\sigma}(t)\,(dr) = \underset{r\in[-\alpha,\alpha]}{\text{Min}}\; [l_1^2(1-t) + \lambda(t) + l_1^3]\, r$$

for a.a. $t \in T$. This implies that, for a.a. $t \in [0, 1]$,

$$\begin{aligned} \bar{\sigma}(t)(\{-\alpha\}) &= 1 \qquad \text{if} \quad \tilde{\lambda}(t) \triangleq l_1^2(1-t) + \lambda(t) + l_1^3 > 0, \\ \bar{\sigma}(t)(\{\alpha\}) &= 1 \qquad \text{if} \quad \tilde{\lambda}(t) \triangleq l_1^2(1-t) + \lambda(t) + l_1^3 < 0. \end{aligned} \tag{8}$$

We next observe that, for all $y \triangleq (y^1, y^2, y^3)$ and $\sigma \in \mathscr{S}$ satisfying

$$\begin{aligned} &\left(y^1(t), y^2(t), y^3(t)\right) = \left(y^1(0), 0, 0\right) + \int_0^t \left(0, y^3(\tau), \int r\sigma(\tau)\,(dr)\right) d\tau, \\ &y^3(t) - y^1(t) \leqslant 0, \qquad y^2(1) = 1, \qquad y^3(1) = 0 \qquad (t \in [0, 1]), \end{aligned} \tag{9}$$

y^1 is a constant and we have

$$|\, y^3(t)| \leqslant \alpha t \qquad \text{and} \qquad y^2(t) \leqslant \int_0^t y^1(\tau)\, d\tau = y^1 \cdot t \qquad (t \in [0, 1]);$$

hence

$$\sup_{t\in T} |\, y^3(t)| \leqslant \alpha \qquad \text{and} \qquad y^1 \geqslant y^2(1) = 1. \tag{10}$$

It is clear that $\bar{y}^1 = \sup_{t \in [0,1]} \bar{y}^3(t)$ (since otherwise we could decrease $\bar{y}^1$ while satisfying all the conditions, thus contradicting the minimizing property of $\bar{y}^1$). It follows therefore from (10) that

$$\bar{y}^1 \in [1, \alpha]. \tag{11}$$

On the other hand, by (6), (7), and V.6.1(3), we have

$$\begin{aligned} k(0)^T \bar{y}(0) &= [l_0 - \lambda(0)]\, \bar{y}^1 \\ &= \operatorname{Min}\{[l_0 - \lambda(0)]\beta \mid \beta \in [-2\alpha, 2\alpha]\}; \end{aligned}$$

hence, by (11), $l_0 = \lambda(0)$.

The relation $l_0 = \lambda(0)$, together with (5) and (6), implies that

$$|\, l_1^2 \,| + |\, l_1^3 \,| + \lambda(0) > 0. \tag{12}$$

The set $L \triangleq \{t \in T \mid \bar{y}^3(t) < \bar{y}^1\}$ is relatively open in [0, 1], it contains 0 and 1, and it does not contain a point $t_{\max}$ where $\bar{y}^3(t_{\max}) = \operatorname{Max}_{t \in [0,1]} \bar{y}^3(t) = \bar{y}^1$. Thus L consists of intervals $[0, \gamma_1)$, $(\gamma_2, 1]$ (where $0 < \gamma_1 \leqslant \gamma_2 < 1$) and possibly some disjoint intervals of the form (t', t'') with $\gamma_1 \leqslant t' < t'' \leqslant \gamma_2$. Furthermore, λ is constant on each of these intervals and we have, in view of (11),

$$\bar{y}^3(\gamma_1) = \bar{y}^3(\gamma_2) = \bar{y}^3(t') = \bar{y}^3(t'') = \bar{y}^1 \in [1, \alpha].$$

These remarks, together with (3), (5), (8), (11), and (12), imply (by straightforward if somewhat tedious arguments) that $l_1^2 < 0$, $\tilde{\lambda}(t) < 0$ on $[0, \gamma_1)$, and $\tilde{\lambda}(t) > 0$ on $(\gamma_2, 1]$, and that $L = [0, \gamma_1) \cup (\gamma_2, 1]$. This, in turn, implies that

$$\bar{y}^3(t) = \begin{cases} \alpha t & (t \in [0, \gamma_1]), \\ \alpha\gamma_1 = \bar{y}^1 & (t \in [\gamma_1, \gamma_2]), \\ \alpha\gamma_1 - \alpha(t - \gamma_2) & (t \in [\gamma_2, 1]), \end{cases}$$

and therefore $\bar{y}^3(1) = 0 = \alpha\gamma_1 - \alpha(1 - \gamma_2)$; hence $\gamma_1 + \gamma_2 = 1$ and

$$1 = \bar{y}^2(1) = \int_0^1 \bar{y}^3(t)\, dt = \alpha\gamma_1(1 - \gamma_1).$$

We conclude that

$$\gamma_1 = \tfrac{1}{2}[1 - (1 - 4\alpha^{-1})^{1/2}], \qquad \gamma_2 = \tfrac{1}{2}[1 + (1 - 4\alpha^{-1})^{1/2}], \qquad \bar{y}^1 = \alpha\gamma_1 .$$

The control function $\bar{\rho} \in \mathscr{R}$, defined by

$$\bar{\rho}(t) = \alpha \quad (t \in [0, \gamma_1)), \quad \bar{\rho}(t) = 0 \quad (t \in [\gamma_1, \gamma_2]), \quad \bar{\rho}(t) = -\alpha \quad (t \in (\gamma_2, 1]),$$

yields the functions $\bar{y}^2$ and $\bar{y}^3$ as just described, and since $\bar{y}(0) \in A_0$, $\bar{y}(1) \in A_1$, and $\bar{y}(T) \subset A$, our use of Theorem V.6.1 was justified. Since $\bar{y}$, as just described, was shown to be the only function satisfying V.6.1, we conclude that $\bar{y}$ and $\bar{\rho}$ determine a minimizing $\mathcal{S}$-solution which is also a minimizing $\mathcal{R}$-solution.

V.7 State-Dependent Controls

V.7.1 ***An Example*** In the "original" version of the general optimal control problem, the control functions $u \in \mathcal{U}$, the control parameters $b \in B$, and the states $y \in \mathcal{Y}$ are subject to the "absolute" condition

$$y = F(y, u, b)$$

and to the "desirable" conditions

$$g_1(y, u, b) = 0, \qquad g_2(y, u, b) \in C_2 \subset \mathcal{X}_2 .$$

In this section we shall comment on certain problems in which the function g_2 does not satisfy the conditions of Theorems V.1.1, V.1.2, and V.3.2. Consider, as an example, an optimal control problem involving a space vehicle reentering the Earth's atmosphere with its trajectory modulated during a preassigned time interval $T \triangleq [t_0, t_1]$ by varying the "angle of attack" of a control surface. The ordinary differential equation of motion of such a vehicle involves a continuous state function $y \triangleq (y^1, ..., y^n) : T \to \mathbb{R}^n$ and a control function ρ; in particular, we denote by $y^2(t)$ the speed of the vehicle at time t and by $\rho(t)$ the angle of attack. In the "original" version of this problem the cost functional is $y^1(t_1)$, ρ is a measurable function on T to a preassigned interval $R \triangleq [-\alpha_{\max}, \alpha_{\max}]$, and (y, ρ) is subject to an additional restriction of the form

$$h(y, \rho)(t) \triangleq [y^2(t)]^2 \rho(t) \in [-g_{\max}, g_{\max}] \quad \text{a.e. in } T. \tag{1}$$

The number $g_{\max}$ is related to the largest "normal" acceleration that the astronauts can be subjected to.

Restriction (1) is of the form $g_2(y, \rho, b) \in C_2$ if we appropriately define the topological vector space $\mathcal{X}_2$. Since y^2 is continuous and ρ measurable and bounded, the function $h(y, \rho)(\cdot)$ is measurable and bounded, and can be considered an element of $L^p(T) \triangleq L^p(T, \mathcal{M}, m)$ for every $p \in [1, \infty]$ or of $\mathcal{F}(T) \triangleq \mathcal{F}(T, \mathcal{M}, m, \mathbb{R})$, where $\mathcal{M}$ and m refer to the Lebesgue measure. [It is not too hard to demonstrate that $(\mathcal{F}(T), |\cdot|_{\mathcal{F}})$ is a topological vector space.] If we define C_2 to be the subset of $L^p(T)$ respectively $\mathcal{F}(T)$ consisting of all measurable functions w such that $w(t) \in [-g_{\max}, g_{\max}]$ a.e. in T, then relation (1) is of the form $h(y, \rho) \in C_2$ and h defines a mapping on $C(T, \mathbb{R}^n) \times \mathscr{S}$ into $L^p(T)$ respectively $\mathscr{F}(T)$. However, we cannot apply Theorems V.1.1, V.1.2, and V.3.2 because h is not continuous. We therefore adopt a different approach.

Any real-valued measurable function ρ satisfying the restrictions

$$\rho(t) \in [-\alpha_{\max}, \alpha_{\max}] \quad \text{and} \quad [y^2(t)]^2 \rho(t) \in [-g_{\max}, g_{\max}] \quad \text{a.e. in } T$$

can be represented in the form

$$\rho(t) = u_1(t) \cdot \operatorname{Min}(\alpha_{\max}, g_{\max} \cdot [y^2(t)]^{-2}) \quad \text{a.e. in } T, \tag{2}$$

where u_1 is a measurable function on T to $[-1, 1]$. Thus, if the equation of motion of the space vehicle is

$$y(t) = v_0 + \int_{t_0}^{t} f(y(\tau), \rho(\tau))\, d\tau \qquad (t \in T),$$

then an equivalent equation of motion incorporating restriction (1) is

$$y(t) = v_0 + \int_{t_0}^{t} \tilde{f}(y(\tau), u_1(\tau))\, d\tau \qquad (t \in T),$$

where

$$\tilde{f}(v, \tilde{r}) \triangleq f(v, \tilde{r} \cdot \operatorname{Min}(\alpha_{\max}, g_{\max} \cdot [v^2]^{-2}))$$
$$(v \triangleq (v^1, \dots, v^n) \in \mathbb{R}^n,\ \tilde{r} \in [-1, 1]).$$

In this formulation, the "old" control functions ρ are replaced by "new" control functions u_1 that are measurable functions on T to $[-1, 1]$. If f is sufficiently "nice", then Theorems VI.1.1 and VI.1.3 (analogous to V.1.1 and V.1.2) ensure the existence of minimizing relaxed and approximate solutions. Unfortunately, in order to apply necessary conditions for a minimum, as in Theorem V.6.1, we must assume that $\tilde{f}(\cdot, \tilde{r})$ has a derivative $\tilde{f}_v(v, \tilde{r})$ for all $(v, \tilde{r}) \in \mathbb{R}^n \times [-1, 1]$, and this is not the case even if $f(\cdot, \cdot)$ is differentiable.

There is a way of avoiding this difficulty. We shall verify that statements (1) or (2) are equivalent to the statement that

$$\rho(t) = u(t)\, x(t) \quad \text{a.e. in } T \tag{3}$$

for some measurable $u : T \to [-1, 1]$ and some continuous $x : T \to \mathbb{R}$ that satisfies

$$x(t) \in [0, \alpha_{\max}], \qquad [y^2(t)]^2 \cdot x(t) - g_{\max} \leqslant 0 \qquad (t \in T). \tag{4}$$

Indeed, if $u : T \to [-1, 1]$ is measurable and $x : T \to \mathbb{R}$ is continuous and satisfies (4), then $\rho = u \cdot x$ is measurable and satisfies (1) and (2); conversely, if ρ satisfies (1) [or, equivalently, (2)], then (3) and (4) are valid when we set

$$x(t) \triangleq \mathrm{Min}(\alpha_{\max}, g_{\max} \cdot [y^2(t)]^{-2}) \quad (t \in T)$$

and

$$u(t) \triangleq \rho(t)/x(t) \qquad (t \in T).$$

Now our "reentry from space" problem is formulated in terms of a state function $y \triangleq (y^1,..., y^n) \in C(T, \mathbb{R}^n)$, a measurable original control function $u : T \to [-1, 1]$, a control parameter $x \in B \triangleq C(T)$, and the equation of motion

$$y(t) = v_0 + \int_{t_0}^{t} f(y(\tau), u(\tau)\, x(\tau))\, d\tau \qquad (t \in T), \tag{5}$$

with restriction (1) replaced by the "unilateral" restrictions (4). If we set $\mathscr{X}_2 \triangleq C(T, \mathbb{R}^2)$, $C_2 \triangleq \{w \in \mathscr{X}_2 \mid w(T) \subset [0, \alpha_{\max}] \times (-\infty, 0]\}$, $B \triangleq C(T)$, and

$$g_2(y, x)(t) \triangleq (x(t), [y^2(t)]^2 \cdot x(t) - g_{\max}) \qquad [y \in C(T, \mathbb{R}^n), \quad x \in B, \quad t \in T],$$

then our problem has the cost functional $y^1(t_1)$, the equation of motion (5), and the restriction

$$g_2(y, x) \in C_2 .$$

We have thus replaced the unmanageable restriction (1) by a reasonable restriction in $C(T, \mathbb{R}^2)$, at the cost of introducing a control parameter $x \in B \triangleq C(T)$. The new problem does not quite fit within the framework of V.6 but it is of the type that we shall discuss in Chapter VI. We also observe that, by our preceding remarks, this problem has a minimizing relaxed solution and a corresponding minimizing approximate $\mathscr{R}$-solution.

This new formulation also provides an alternate way of proving the existence of minimizing relaxed and approximate solutions. Indeed, if f is sufficiently "nice," then all the solutions y of Eq. (5) belong to some compact subset Y of $C(T, \mathbb{R}^n)$, and all the functions x such that, for some $y \in Y$,

$$x(t) \triangleq \mathrm{Min}(\alpha_{\max}, g_{\max}[y^2(t)]^{-2}) \qquad (t \in T),$$

belong to some compact subset X of $C(T)$ which we may assume to be convex. Our previous arguments remain unaffected if we restrict x to X and not, as before, to $C(T)$; that is, if we set $B \triangleq X$. Then Theorems VI.1.1 and

VI.1.3 (analogous to V.1.1 and V.1.2) are applicable to the newly formulated problem and ensure the existence of minimizing relaxed and approximate solutions.

We can also show that if we denote by $(\bar{y}, \bar{\sigma}, \bar{x})$ either a minimizing relaxed solution or a minimizing $\mathscr{R}$-solution, then we may assume that

$$\bar{x}(t) = \text{Min}(\alpha_{\max}, g_{\max} \cdot [\bar{y}^2(t)]^{-2}) \qquad (t \in T).$$

Indeed, the "relaxed" equation of motion is of the form

$$y(t) = v_0 + \int_{t_0}^{t} d\tau \int_{[-1,1]} f(y(\tau), x(\tau) \cdot r)\, \sigma(\tau)\,(dr) \qquad (t \in T).$$

If $(\bar{y}, \sigma_1, x_1)$ is a minimizing (relaxed or $\mathscr{R}$–) solution, then we set

$$\bar{x}(t) \triangleq \text{Min}(\alpha_{\max}, g_{\max} \cdot [\bar{y}^2(t)]^{-2}) \qquad (t \in T)$$

and define $\bar{\sigma}(t)$ as a measure concentrated on $[\bar{x}(t)]^{-1} x_1(t)[-1, 1]$ and such that

$$\sigma_1(t)(E) \triangleq \bar{\sigma}(t)([\bar{x}(t)]^{-1} x_1(t)E) \qquad [E \in \Sigma_{\text{Borel}}([-1, 1]), \quad t \in T].$$

It follows then that $(\bar{y}, \bar{\sigma}, \bar{x})$ also satisfies the equation of motion and the restriction $g_2(y, x) \in C_2$, and it is therefore a minimizing solution of the same kind as $(\bar{y}, \sigma_1, x_1)$.

V.7.2 ***More General Situations*** The example that we have just discussed illustrates a possible approach in more general situations. For the sake of concreteness, we consider the case where $\mathscr{U} = \mathscr{R}$ and $\mathscr{Y} = C(T, \mathbb{R}^n)$. Let $n, m_2 \in \mathbb{N}$, a closed $H \subset \mathbb{R}^{m_2}$, and $h \in C(\mathbb{R}^n \times R \times B, \mathbb{R}^{m_2})$ be given, and assume that $\mathscr{Y} = C(T, \mathbb{R}^n)$, $\mathscr{U} = \mathscr{R}$, and we consider the "original" problem of V.0 with an additional "desirable" restriction of the form

$$h(y(t), \rho(t), b) \in H \quad \mu\text{-a.e.}$$

Let us refer to this problem as Original Problem I.

In order to study this problem, we try the following approach. We seek to determine

(a) a compact metric space $\tilde{R}$ and a corresponding set $\tilde{\mathscr{R}}$ of μ-measurable functions on T to $\tilde{R}$,

(b) a compact convex subset X of $C(T, \mathbb{R}^k)$,

(c) a convex subset $\tilde{A}$ of some $\mathbb{R}^l$, and

(d) functions $\phi : \mathbb{R}^n \times \tilde{R} \times \mathbb{R}^k \to R$ and $\psi : \mathbb{R}^n \times \mathbb{R}^k \to \mathbb{R}^l$ such that for every $(y, \rho, b) \in \mathscr{Y} \times \mathscr{R} \times B$ satisfying

$$h^{\#}(y, \rho, b)(t) \triangleq h(y(t), \rho(t), b) \in H \quad \mu\text{-a.e.} \tag{1}$$

there exists $(\tilde{\rho}, x) \in \tilde{\mathscr{R}} \times X$ such that

$$\rho(t) = \phi(y(t), \tilde{\rho}(t), x(t)) \quad \mu\text{-a.e.} \quad \text{and} \quad \psi(y(t), x(t)) \in \tilde{A} \quad (t \in T); \tag{2}$$

and for every $(y, \tilde{\rho}, b, x) \in \mathscr{Y} \times \tilde{\mathscr{R}} \times B \times X$ satisfying

$$\psi^{\#}(y, x)(t) \triangleq \psi(y(t), x(t)) \in \tilde{A} \qquad (t \in T), \tag{3}$$

the function

$$t \to \rho(t) = \phi(y(t), \tilde{\rho}(t), x(t))$$

is μ-measurable and (y, ρ, b) satisfies (1).

If we are successful in determining such $\tilde{R}$, X, $\tilde{A}$, ϕ, and ψ, then we can replace restriction (1) by the manageable restriction (3). Specifically, we set

$$\tilde{B} \triangleq B \times X, \quad \tilde{\mathscr{X}}_2 \triangleq \mathscr{X}_2 \times C(T, \mathbb{R}^l), \quad \tilde{C}_2 \triangleq C_2 \times \{w \in C(T, \mathbb{R}^l) \mid w(T) \subset \tilde{A}\},$$

$$\tilde{g}_i(y, \tilde{\rho}, \tilde{b}) \triangleq g_i(y, \phi \circ (y, \tilde{\rho}, x), b) \qquad (i = 0, 1),$$

$$\tilde{g}_2(y, \tilde{\rho}, \tilde{b}) \triangleq (g_2(y, \phi \circ (y, \tilde{\rho}, x), b), \psi^{\#}(y, x)),$$

$$\tilde{F}(y, \tilde{\rho}, \tilde{b}) \triangleq F(y, \phi \circ (y, \tilde{\rho}, x), b) \qquad [y \in \mathscr{Y}, \quad \tilde{\rho} \in \tilde{\mathscr{R}}, \quad \tilde{b} \triangleq (x, b) \in \tilde{B}],$$

and consider Original Problem II defined as in V.0 but with B, $\mathscr{X}_2$, C_2, g, F replaced by $\tilde{B}$, $\tilde{\mathscr{X}}_2$, $\tilde{C}_2$, $\tilde{g}$, $\tilde{F}$, respectively. It is clear that Original Problem II is equivalent to Original Problem I, and we investigate questions of existence and necessary conditions for a (relaxed or an original) minimum using the formulation of Original Problem II and its relaxed version.

Notes

Sections V.1–V.4 generalize certain results of Warga [12] and [16]. Lemma V.2.1, based on the separation theorem for convex sets I.6.7, is analogous to similar applications of that theorem by Gamkrelidze [3] and Neustadt [3].

The convex set $\chi(\mathscr{S}^{\#} \times B)$ of Theorems V.2.3 and V.3.2 might be referred to as "a set of variations." The idea of "variations" goes back to Lagrange but that of a convex set of variations appears to have originated with

McShane [1, 2]. McShane's convex set of variations, in many forms and guises, has been a basic tool of the optimal control theory since its inception, and I am not aware of any necessary conditions for minimum in the optimal control theory that have been derived without relying on this concept.

General models of an optimal control problem (different from the model described in V.0) have been proposed and corresponding necessary conditions for an original minimum studied by Neustadt [3, 5] and by Gamkrelidze and Kharatishvili [1]. The necessary conditions for these models are expressed in terms of certain mappings which must be appropriately defined for particular problems. Necessary conditions related to still another model were studied in Warga [9], but the model of Warga [12] was found preferable, and it has been further generalized in V.0.

Necessary conditions for minimum with convexity assumptions replacing certain differentiability assumptions have been studied by R.T. Rockafellar [as mentioned in Neustadt [5]) and Neustadt [5]. This last paper inspired the consideration of the problem of Section V.4 (and, previously, of Theorem 2.3 of Warga [12, p. 364]).

Weak necessary conditions for an original minimum (Section V.5) generalize the well-known Euler–Lagrange equations of the calculus of variations.

Theorem V.6.1 deals with the unilateral problem defined by ordinary differential equations, also referred to in the terminology of Gamkrelidze [1, 2] as "the problem with restricted phase coordinates." Statements equivalent to the existence of a minimizing relaxed solution were proven, independently, by Filippov [1] in 1959, Roxin [1], Warga [1], and Ważewski [1, 2] in 1962, and Ghouila-Houri [1] in 1965. The existence of a corresponding minimizing approximate $\mathscr{R}$-solution (under slightly more restrictive conditions) and slightly less general necessary conditions for a relaxed minimum were established by Warga [1–3, 5]. Necessary conditions for an original minimum were first established by Gamkrelidze [1, 2] for those minimizing $\mathscr{R}$-solutions that also satisfy certain "regularity" assumptions. Similar conditions, not subject to the "regularity" assumption, were obtained by Dubovitskii and Milyutin [1] and Neustadt [4, 5] (who also refers to Gamkrelidze's oral communications on the subject). Finally, Theorem V.6.1 is a special case of Theorems 2.1, 2.2, and 3.2 of Warga [13].

The existence aspects of problems defined by ordinary differential equations and with state-dependent controls were discussed, for the Bolza problem, by McShane [4] and, in the context of optimal control, by Filippov [1], in several papers of Cesari (e.g. [1–3, 6]), and, in a very elegant manner, by McShane [5]. For certain problems defined by partial differential equations and with state-dependent controls, the existence aspects were studied by Cesari [4, 5]. Necessary conditions for the Bolza problem (with state-dependent controls) were studied by McShane [4] in those situations (customarily considered in the calculus of variations) where the implicit function theorem enables one to replace the counterparts of state-dependent control functions by differentiable combinations of other unrestricted control functions and state functions.

CHAPTER VI

Optimal Control of Ordinary Differential Equations

VI.0 Formulation of the "Standard" Problem

We have described in V.6 a class of problems defined by ordinary differential equations and have used it to illustrate the application of the general methods of Chapters IV and V. We shall now undertake a more detailed study of such problems based on somewhat weaker assumptions. The reason we concentrate at first on the optimal control of ordinary differential equations is twofold: first, because most applications of the optimal control theory involve ordinary differential equations and, secondly, because the reader may find this material somewhat simpler and therefore more suitable at the beginning. On the other hand, we pay for this choice by a certain amount of duplication, most of the results of VI.1 and VI.2 being in many respects generalizations of those in V.6 and special cases of those in VII.1 and VII.2.

We shall consider here certain unilateral problems defined by ordinary differential equations. These problems are of the general type discussed in V.1–V.3 for the case where $n \in \mathbb{N}$, $T \triangleq [t_0, t_1] \subset \mathbb{R}$, $\mathscr{Y} \triangleq C(T, \mathbb{R}^n)$, the equation of motion $y = F(y, \sigma, b)$ is an ordinary differential equation, and the condition $g_2(y, \sigma, b) \in C_2$ restricts $y(T)$ in a special manner. Specifically, let $m, n \in \mathbb{N}$, $m_2 \in \{0, 1, 2,...\}$, $v_0 \in V \subset \mathbb{R}^n$, $T = [t_0, t_1]$, μ be the Borel measure in T, and let

$$f : T \times V \times R \times B \to \mathbb{R}^n,$$

$$h_0 : V \times B \to \mathbb{R}, \qquad h_1 : V \times B \to \mathbb{R}^m,$$

$$h_2 : T \times V \times B \to \mathbb{R}^{m_2}, \qquad \text{and} \qquad A : T \to \mathscr{P}'(\mathbb{R}^{m_2})$$

be given functions. We set

$$\mathscr{Y} \triangleq C(T, \mathbb{R}^n), \quad \mathscr{X}_2 \triangleq C(T, \mathbb{R}^{m_2}), \quad C_2 \triangleq \{w(\cdot) \in \mathscr{X}_2 \mid w(t) \subset A(t)\ (t \in T)\},$$

and let

$$F(y, \sigma, b)(t) \triangleq v_0 + \int_{t_0}^{t} f(\tau, y(\tau), \sigma(\tau), b)\, d\tau \qquad (t \in T),$$

$$g_i(y, \sigma, b) \triangleq h_i(y(t_1), b) \qquad (i = 0, 1),$$

$$g_2(y, \sigma, b)(t) \triangleq h_2(t, y(t), b) \qquad (t \in T)$$

for all $(y, \sigma, b) \in \mathscr{Y} \times \mathscr{S}^\# \times B$ for which the expression

$$v_0 + \int_{t_0}^{t} f(\tau, y(\tau), \sigma(\tau), b)\, d\tau$$

defines a continuous function of t into V. For all other

$$(y, \sigma, b) \in \mathscr{Y} \times \mathscr{S}^\# \times B$$

we set

$$F(y, \sigma, b)(t) \triangleq y(t) + (1, 0,..., 0) \in \mathbb{R}^n \qquad (t \in T)$$

and

$$g_0 \triangleq 0, \qquad g_1 \triangleq 0, \qquad g_2 \triangleq 0.$$

We observe that the equation of motion $y = F(y, \sigma, b)$ in $\mathscr{Y} \times \mathscr{S}^\# \times B$ is equivalent to the ordinary differential equation

$$y(t) = v_0 + \int_{t_0}^{t} f(\tau, y(\tau), \sigma(\tau), b)\, d\tau \qquad (t \in T).$$

Problems of a somewhat different type will be discussed in VI.4, and certain more general problems will be reduced to the "standard" ones in VI.5.

VI.1 Existence of Minimizing Relaxed and Approximate Solutions

VI.1.1 ***Theorem*** Let B have a sequentially compact topology, $A(t)$ $(t \in T)$ be closed subsets of $\mathbb{R}^{m_2}$, $\mathscr{A}(\mathscr{S}^\#) \neq \varnothing$, and assume that $f(\cdot, v, r, b)$ is measurable and h_0, h_1, h_2 and $f(\tau, \cdot, \cdot, \cdot)$ continuous for all

$$(\tau, v, r, b) \in T \times V \times R \times B,$$

and

(1) there exist a closed $V_0 \subset V$ and an integrable $\psi : T \to \mathbb{R}$ such that

$$y(\tau) \in V_0, \qquad |f(\tau, y(\tau), r, b| \leqslant \psi(\tau)$$
$$[(\tau, r) \in T \times R, \quad (y, \sigma, b) \in \mathscr{A}(\mathscr{S}^\#)].$$

Then $\mathscr{A}(\mathscr{S}^\#)$ is sequentially compact, $F \mid \mathscr{A}(\mathscr{S}^\#)$ sequentially continuous, and there exists a minimizing relaxed solution.

▌PROOF By the definition of F, every point $(y, \sigma, b) \in \mathscr{Y} \times \mathscr{S}^\# \times B$ satisfying the equation $y = F(y, \sigma, b)$ is also a solution of the equation

$$y(t) = v_0 + \int_{t_0}^{t} f(\tau, y(\tau), \sigma(\tau), b)\, d\tau \qquad (t \in T).$$

The set $Y \triangleq \mathrm{pr}_y \mathscr{A}(\mathscr{S}^\#)$ is bounded [because $|y(t)| \leqslant \int_{t_0}^{t_1} \psi(\tau)\, d\tau + |v_0|$ for all $t \in T$ and $y \in Y$] and equicontinuous [because $|y(t'') - y(t')| \leqslant \int_{t'}^{t''} \psi(\tau)\, d\tau$ whenever $t_0 \leqslant t' \leqslant t'' \leqslant t_1$ and $y \in Y$]. Thus, by I.5.4, $\bar{Y}$ is sequentially compact and, by IV.3.11, $\mathscr{S}^\#$ is sequentially compact. It follows then easily that $\bar{Y} \times \mathscr{S}^\# \times B$ and its subset seq $\mathrm{cl}(\mathscr{A}(\mathscr{S}^\#))$ are sequentially compact. If $((y_j, \sigma_j, b_j))$ is a sequence in $\mathscr{A}(\mathscr{S}^\#)$ converging to some

$$(\bar{y}, \bar{\sigma}, \bar{b}) \in \mathscr{Y} \times \mathscr{S}^\# \times B,$$

then, by IV.2.9 (with $k = n$, $Z = B$, and T replaced by $[t_0, t]$), we have

$$\bar{y}(t) = \lim_j y_j(t) = v_0 + \lim_j \int_{t_0}^{t} f(\tau, y_j(\tau), \sigma_j(\tau), b_j)\, d\tau$$
$$= v_0 + \int_{t_0}^{t} f(\tau, \bar{y}(\tau), \bar{\sigma}(\tau), \bar{b})\, d\tau \qquad (t \in T).$$

Furthermore,

$$h_1(\bar{y}(t_1), \bar{b}) = \lim_j h_1(y_j(t_1), b_j) = 0$$

and

$$h_2(t, \bar{y}(t), \bar{b}) = \lim_j h_2(t, y_j(t), b_j) \in A(t) \qquad (t \in T).$$

Thus $(\bar{y}, \bar{\sigma}, \bar{b}) \in \mathscr{A}(\mathscr{S}^{\#})$, showing that $\mathscr{A}(\mathscr{S}^{\#})$ is sequentially compact and $F \mid \mathscr{A}(\mathscr{S}^{\#})$ sequentially continuous. The existence of a minimizing relaxed solution now follows from V.1.1. QED

VI.1.2 ***Theorem*** Condition (1) of Theorem VI.1.1 can be replaced by either of the following (stronger) conditions:

(1′) $V = \mathbb{R}^n$ and there exists an integrable $\psi_1 : T \to \mathbb{R}$ and an increasing $p : [0, \infty) \to [0, \infty)$ such that

$$| v \cdot f(t, v, r, b)| \leqslant \psi_1(t)(| v |_2^2 + 1)$$

and

$$| f(t, v, r, b)| \leqslant \psi_1(t)\, p(| v |)$$

for all $(t, v, r, b) \in T \times \mathbb{R}^n \times R \times B$, or

(1″) $V = \mathbb{R}^n$ and there exists an integrable $\psi_2 : T \to \mathbb{R}$ such that

$$| f(t, v, r, b)| \leqslant \psi_2(t)(| v | + 1) \qquad \text{for all} \quad (t, v, r, b) \in T \times \mathbb{R}^n \times R \times B.$$

▌PROOF Let $(y, \sigma, b) \in \mathscr{Y} \times \mathscr{S}^{\#} \times B$ satisfy the equation

$$y(t) = v_0 + \int_{t_0}^{t} f\big(\tau, y(\tau), \sigma(\tau), b\big)\, d\tau \qquad (t \in T),$$

and let (1′) be valid. Then

$$\begin{aligned}\left(\tfrac{1}{2} | y(\tau)|_2^2\right)^{\cdot} \leqslant | y(\tau) \cdot \dot{y}(\tau)| &= | y(\tau) \cdot f\big(\tau, y(\tau), \sigma(\tau), b\big)| \\ &\leqslant \psi_1(\tau)\big(| y(\tau)|_2^2 + 1\big) \quad \text{a.e. in } T;\end{aligned}$$

hence, setting $v(\tau) \triangleq \frac{1}{2} | y(\tau)|_2^2$ and integrating both sides over $[t_0, t]$, we obtain

$$v(t) - \tfrac{1}{2} | v_0 |^2 \leqslant 2 \int_{t_0}^{t} \psi_1(\tau)\, v(\tau)\, d\tau + \int_{t_0}^{t} \psi_1(\tau)\, d\tau \qquad (t \in T).$$

Gronwall's inequality II.4.4 guarantees the existence of a constant α_1 (depending on v_0, ψ_1, and T alone) such that

$$v(t) \leqslant \alpha_1 ;$$

hence

$$| y(t)| \leqslant n\, | y(t)|_2 \leqslant \alpha_2 \triangleq n[2\alpha_1]^{1/2} \qquad (t \in T).$$

Condition (1) of Theorem VI.1.1 is now satisfied, with

$$\psi(t) \triangleq \psi_1(t) p(\alpha_2) \qquad (t \in T).$$

If (1″) is valid, then (1′) is also valid, with $\psi_1 \triangleq 2\psi_2$ and $p(| v |) \triangleq | v | + 1$. QED

VI.1.3 ***Theorem*** Let the conditions of Theorem VI.1.1 be satisfied with $\mathscr{A}(\mathscr{S}^{\#})$ replaced by $\mathscr{H}(\mathscr{S}^{\#})$. Then $\mathscr{H}(\mathscr{S}^{\#})$ is sequentially compact and $F \mid \mathscr{H}(\mathscr{S}^{\#})$ sequentially continuous.

If $(\bar{y}, \bar{\sigma}, \bar{b})$ is a minimizing relaxed solution, $\mathscr{U}$ an abundant subset of $\mathscr{R}^{\#}$, $\bar{y}$ the unique solution of the equation $y = F(y, \bar{\sigma}, \bar{b})$, and there exists a neighborhood G of $\bar{\sigma}$ in $\mathscr{S}^{\#}$ such that the equation $y = F(y, u, \bar{b})$ has a (not necessarily unique) solution y for each $u \in \mathscr{U} \cap G$, then there exists a sequence $((y_j, u_j))$ in $\mathscr{Y} \times \mathscr{U}$ such that $((y_j, u_j, \bar{b}))$ is a minimizing approximate $\mathscr{U}$-solution and

$$\lim_j g(y_j, u_j, \bar{b}) = g(\bar{y}, \bar{\sigma}, \bar{b}).$$

▌PROOF With $\mathscr{A}(\mathscr{S}^{\#})$ replaced by $\mathscr{H}(\mathscr{S}^{\#})$, the proof of VI.1.1 shows that $\mathscr{H}(\mathscr{S}^{\#})$ is sequentially compact and $F \mid \mathscr{H}(\mathscr{S}^{\#})$ sequentially continuous. Our conclusion now follows from V.1.2. QED

We next consider certain alternate conditions that ensure the validity of the conclusions of Theorems VI.1.1 and VI.1.3. These results will be applied in VI.4. We first require a lemma.

VI.1.4 ***Lemma*** Let V be open and assume that for every bounded $W \subset V$ the function $f \mid T \times W \times R \times B$ belongs to $\mathscr{B}(T, W \times R \times B; \mathbb{R}^n)$ and there exists an integrable $\psi_W : T \to \mathbb{R}$ such that

$$\begin{aligned} &|f(\tau, v', r, b) - f(\tau, v'', r, b)| \\ &\qquad \leqslant \psi_W(\tau)|v' - v''| \qquad [(\tau, r, b) \in T \times R \times B; \quad v', v'' \in W]. \end{aligned}$$

Then for every abundant subset $\mathscr{U}$ of $\mathscr{R}^{\#}$, every $(\bar{y}, \bar{\sigma}, \bar{b}) \in \mathscr{H}(\mathscr{S}^{\#})$, and every sequence (u_j) in $\mathscr{U}$ converging to $\bar{\sigma}$ there exist $j_0 \in \mathbb{N}$ and a sequence $(y_j)_{j \geqslant j_0}$ in $\mathscr{Y}$ such that

$$(y_j, u_j, \bar{b}) \in \mathscr{H}(\mathscr{U}) \quad (j \geqslant j_0) \qquad \text{and} \qquad \lim_j y_j = \bar{y}.$$

▌PROOF Let $(\bar{y}, \bar{\sigma}, \bar{b}) \in \mathscr{H}(\mathscr{S}^{\#})$, and let (u_j) be a sequence in $\mathscr{U}$ converging to $\bar{\sigma}$. We shall show that there exists $j_0 \in \mathbb{N}$ such that the equation

$$(1) \qquad y(t) = v_0 + \int_{t_0}^{t} f(\tau, y(\tau), u_j(\tau), \bar{b})\, d\tau \qquad (t \in T)$$

has a solution y_j for every $j \geqslant j_0$. Indeed, let $d_1 > 0$ be such that

$$W \triangleq \{v \in \mathbb{R}^n \mid d[\bar{y}(T), v] < d_1\} \subset V,$$

and let $j \in \mathbb{N}$. Then W is an open neighborhood of $\bar{y}(T)$, $f \mid T \times W \times R \times B$ belongs to $\mathscr{B}(T, W \times R \times B; \mathbb{R}^n)$, and, by II.4.2, either there exists a solution $y_j \in C(T, W)$ of (1) or there exist $\alpha_j \in [t_0, t_1]$ and a function

$y_j : [t_0, \alpha_j) \to W$ satisfying the equation in (1) for $t \in [t_0, \alpha_j)$ and such that $\lim_{t \to \alpha_j} y_j(t) \in \partial W$. If the second alternative holds for all j in some (infinite) sequence J, then we combine the equations

$$\bar{y}(t) = v_0 + \int_{t_0}^{t} f(\tau, \bar{y}(\tau), \bar{\sigma}(\tau), \bar{b})\, d\tau \qquad (t \in [t_0, \alpha_j))$$

and

$$y_j(t) = v_0 + \int_{t_0}^{t} f(\tau, y_j(\tau), u_j(\tau), \bar{b})\, d\tau \qquad (t \in [t_0, \alpha_j))$$

to obtain

$$\begin{aligned} e_j(t) &\triangleq |y_j(t) - \bar{y}(t)| \\ &\leqslant \int_{t_0}^{t} |f(\tau, y_j(\tau), u_j(\tau), \bar{b}) - f(\tau, \bar{y}(\tau), u_j(\tau), \bar{b})|\, d\tau \\ &\quad + \left| \int_{t_0}^{t} f(\tau, \bar{y}(\tau), u_j(\tau) - \bar{\sigma}(\tau), \bar{b})\, d\tau \right| \\ &\leqslant \int_{t_0}^{t} \psi_W(\tau)\, e_j(\tau)\, d\tau + \left| \int_{t_0}^{t} f(\tau, \bar{y}(\tau), u_j(\tau) - \bar{\sigma}(\tau), \bar{b})\, d\tau \right| \\ &\qquad\qquad [j \in J, \quad t \in [t_0, \alpha_j)]. \end{aligned}$$

Since the functions $t \to \int_{t_0}^{t} f(\tau, \bar{y}(\tau), u_j(\tau) - \bar{\sigma}(\tau), \bar{b})\, d\tau : T \to \mathbb{R}$ converge to 0 for each $t \in T$ as $j \to \infty$ and are equicontinuous and uniformly bounded, it follows from Gronwall's inequality II.4.4 that

$$\lim_{j \in J} e_j(t) = 0 \quad \text{uniformly for } t \in [t_0, \alpha_j).$$

This contradicts the relations $d[\partial W, \bar{y}(T)] > 0$ and $\lim_{t \to \alpha_j} y_j(t) \in \partial W$. We conclude that there exists $j_0 \in \mathbb{N}$ such that (1) has a solution $y_j \in C(T, W)$ for all $j \geqslant j_0$; hence

$$(y_j, u_j, \bar{b}) \in \mathscr{H}(\mathscr{U}) \qquad (j \geqslant j_0).$$

The set $\{y_j \mid j \geqslant j_0\}$ is equicontinuous and bounded and therefore (by I.5.4) conditionally compact in $C(T, \mathbb{R}^n)$. If there exist a sequence J_1 and $\epsilon > 0$ such that

$$(2) \qquad |y_j - \bar{y}|_{\sup} > \epsilon \qquad (j \in J_1),$$

then the conditionally compact set $\{y_j \mid j \in J_1\}$ has a limit point $\tilde{y}$ which, by IV.2.9, is such that $(\tilde{y}, \bar{\sigma}, \bar{b}) \in \mathscr{H}(\mathscr{S}^{\#})$. By II.4.5, the equation

$$y(t) = v_0 + \int_{t_0}^{t} f(\tau, y(\tau), \bar{\sigma}(\tau), \bar{b})\, d\tau \qquad (t \in T)$$

has at most one solution, and therefore $\bar{y} = \tilde{y}$, contrary to (2). It follows that $\lim_j y_j = \bar{y}$ which completes the proof of the lemma. QED

VI.1.5 ***Theorem*** Let B be sequentially compact, V open, $\mathscr{U}$ an abundant subset of $\mathscr{R}^{\#}$, $A(t)$ $(t \in T)$ closed, and h_0, h_1, h_2 continuous. Assume, furthermore, that $\mathscr{A}(\mathscr{S}^{\#}) \neq \varnothing$; there exists a compact $V_0 \subset V$ such that

$$\{y(t) \mid t \in T,\ (y, \sigma, b) \in \mathscr{H}(\mathscr{S}^{\#})\} \subset V_0;$$

and for every bounded $W \subset V$ the function $f \mid T \times W \times R \times B$ belongs to $\mathscr{B}(T, W \times R \times B; \mathbb{R}^n)$ and there exists an integrable $\psi_W : T \to R$ such that

$$|f(\tau, v', r, b) - f(\tau, v'', r, b)| \leqslant \psi_W(\tau)|v' - v''| \qquad [(\tau, r, b) \in T \times R \times B,\quad v', v'' \in W].$$

Then

(1) there exists a minimizing relaxed solution, and
(2) for every minimizing relaxed solution $(\bar{y}, \bar{\sigma}, \bar{b})$ there exists a sequence $((y_j, u_j))$ in $\mathscr{Y} \times \mathscr{U}$ such that $((y_j, u_j, \bar{b}))$ is a minimizing approximate $\mathscr{U}$-solution and $\lim_j g(y_j, u_j, \bar{b}) = g(\bar{y}, \bar{\sigma}, \bar{b})$.

▌PROOF If we set

$$W = V_0$$

and

$$\psi(\tau) \triangleq \sup\{|f(\tau, v, r, b)| \mid (v, r, b) \in W \times R \times B\} \qquad (\tau \in T),$$

then ψ is integrable and we have

(3) $$|f(\tau, y(\tau), r, b)| \leqslant \psi(\tau) \qquad [(\tau, r) \in T \times R,\ (y, \sigma, b) \in \mathscr{H}(\mathscr{S}^{\#})].$$

It follows now from VI.1.1 that there exists a minimizing relaxed solution.

Now let $(\bar{y}, \bar{\sigma}, \bar{b})$ be a minimizing relaxed solution. Then, by VI.1.4, for every sequence (u_j) in $\mathscr{U}$ converging to $\bar{\sigma}$ there exist $j_0 \in \mathbb{N}$ and a sequence $(y_j)_{j \geqslant j_0}$ in $\mathscr{Y}$ such that

$$(y_j, u_j, \bar{b}) \in \mathscr{H}(\mathscr{U}) \quad (j \geqslant j_0) \qquad \text{and} \quad \lim_j y_j = \bar{y}.$$

Since $\mathscr{S}^{\#}$ is a metric space and $\mathscr{U}$ is dense in $\mathscr{S}^{\#}$, it follows that there exists a neighborhood G of $\bar{\sigma}$ in $\mathscr{S}^{\#}$ such that the equation $y = F(y, u, \bar{b})$ has a solution y for every $u \in \mathscr{U} \cap G$. This remark and (3) show that Theorem VI.1.3 is applicable, thus proving statement (2). QED

VI.2 Necessary Conditions for a Minimum

We investigate necessary conditions for a relaxed minimum and for an original minimum.

VI.2.1 ***Lemma*** Let $(\bar{y}, \bar{\sigma}, \bar{b}) \in C(T, V) \times \mathscr{S}^{\#} \times B$. Assume that V is open and for each point $L \triangleq (b_0, \dots, b_m) \in B^{m+1}$ there exist convex neighborhoods V^L of $\bar{y}(T)$ in V and $\mathscr{T}^L$ of 0 in $\mathscr{T}_{m+1}$ such that the function

$$(\tau, v, r, \theta) \to f^L(\tau, v, r, \theta)$$

$$\triangleq f\Big(\tau, v, r, \bar{b} + \sum_{j=0}^{m} \theta^j (b_j - \bar{b})\Big) : T \times V^L \times R \times \mathscr{T}^L \to \mathbb{R}^n$$

has the following properties:

(1) $f^L(\tau, \cdot, r, \cdot)$ has a derivative $f^L_{(v,\theta)}(\tau, v, r, \theta)$ for all

$$(\tau, v, r, \theta) \in T \times V^L \times R \times \mathscr{T}^L;$$

(2) $f^L \in \mathscr{B}(T, V^L \times R \times \mathscr{T}^L; \mathbb{R}^n)$; and

(3) $f^L_{(v,\theta)} \in \mathscr{B}\big(T, V^L \times R \times \mathscr{T}^L; B(\mathbb{R}^n \times \mathbb{R}^{m+1}, \mathbb{R}^n)\big)$.

Then

(4) for each $L \in B^{m+1}$ and $\sigma \in \mathscr{S}^{\#}$ the function

$$(\tau, v, \theta) \to f^{L,\sigma}(\tau, v, \theta) \triangleq f^L\big(\tau, v, \sigma(\tau), \theta\big) : T \times V^L \times \mathscr{T}^L \to \mathbb{R}^n$$

is such that

$$f^{L,\sigma} \in \mathscr{B}(T, V^L \times \mathscr{T}^L; \mathbb{R}^n), \qquad f^{L,\sigma}_{(v,\theta)} \in \mathscr{B}\big(T, V^L \times \mathscr{T}^L; B(\mathbb{R}^n \times \mathbb{R}^{m+1}, \mathbb{R}^n)\big),$$

and

$$f^{L,\sigma}_{(v,\theta)}(\tau, v, \theta)(\Delta v, \Delta\theta) = \int f^L_{(v,\theta)}(\tau, v, r, \theta)(\Delta v, \Delta\theta)\, \sigma(t)(dr)$$

$$[\Delta v \in \mathbb{R}^n, \quad \Delta\theta \in \mathbb{R}^{m+1}, \quad (\tau, v, \theta) \in T \times V^L \times \mathscr{T}^L];$$

and

(5) for each $L \in B^{m+1}$ there exist neighborhoods $\mathscr{S}'$ of $\bar{\sigma}$ in $\mathscr{S}^{\#}$ and $\mathscr{T}'$ of 0 in $\mathscr{T}_{m+1}$ such that, for every $(\sigma, \theta) \in \mathscr{S}' \times \mathscr{T}'$, the equation

$$y(t) = v_0 + \int_{t_0}^{t} f^L(\tau, y(\tau), \sigma(\tau), \theta)\, d\tau \qquad (t \in T)$$

has a unique solution $y^L(\sigma, \theta)(\cdot) \in C(T, V^L)$.

▌PROOF Let $L \in B^{m+1}$. Statement (4) is a direct consequence of IV.2.7. We next observe that if $(\sigma, \theta) \in \mathscr{S}^{\#} \times \mathscr{T}^L$ and

$$\psi^L(\tau) \triangleq \max\big(|f^L(\tau, \cdot, \cdot, \cdot)|_{\sup}, |f^L_{(v,\theta)}(\tau, \cdot, \cdot, \cdot)|_{\sup}\big) \qquad (\tau \in T),$$

then, by the mean value theorem II.3.6, we have

$$(6)\qquad |f^L(\tau, v_1, \sigma(\tau), \theta) - f^L(\tau, v_2, \sigma(\tau), \theta)| \leqslant \psi^L(\tau)|\, v_1 - v_2\,| \qquad (t \in T,\ v_1,\ \ v_2 \in V^L);$$

hence, by II.4.5, the equation in (5) has at most one solution y with $y(T) \subset V^L$. Thus, if statement (5) is not valid, then there exists a sequence $((\sigma_j, \theta_j))$ in $\mathscr{S}^\# \times \mathscr{T}^L$ converging to $(\bar{\sigma}, 0)$ and such that the equation in (5) has no solution y with range in V^L whenever $(\sigma, \theta) = (\sigma_j, \theta_j)$ $(j \in \mathbb{N})$. We set

$$\phi_j(t) \triangleq \left|\int_{t_0}^{t} f^L(\tau, \bar{y}(\tau), r, 0)(\sigma_j(\tau) - \bar{\sigma}(\tau))\,(dr)\right| + \theta_j(t - t_0) \qquad (t \in T,\quad j \in \mathbb{N})$$

and observe that, since V^L is a neighborhood of the compact set $\bar{y}(T)$, there exists $\alpha > 0$ such that $d[\bar{y}(T), \partial V^L] > \alpha$. By IV.2.9 and I.5.3, $\lim_j \phi_j(t) = 0$ uniformly for all $t \in T$ and thus there exists $j_0 \in \mathbb{N}$ such that

$$(7)\quad \phi_j(t) \leqslant \alpha[1 + \exp\left(\int_{t_0}^{t_1} \psi^L(\tau)\,d\tau\right) \cdot \int_{t_0}^{t_1} \psi^L(\tau)\,d\tau]^{-1} \qquad (j \geqslant j_0,\ t \in T).$$

Now let $j \in \{j_0, j_0 + 1, \ldots\}$ be fixed. Since, by assumption, the equation in (5) has no solution y with range in V^L for $(\sigma, \theta) = (\sigma_j, \theta_j)$, it follows from II.4.2 that there exist $t_j' \in (t_0, t_1]$ and a continuous function $y_j : [t_0, t_j') \to \mathbb{R}^n$ such that $\lim_{t \to t_j'} y_j(t) \in \partial V^L$ and

$$y_j(t) = v_0 + \int_{t_0}^{t} f^L(\tau, y_j(\tau), \sigma_j(\tau), \theta_j)\,d\tau \qquad (t \in [t_0, t_j')).$$

Since $(\bar{y}, \bar{\sigma})$ satisfies the equation

$$\bar{y}(t) = v_0 + \int_{t_0}^{t} f^L(\tau, \bar{y}(\tau), \bar{\sigma}(\tau), 0)\,d\tau \qquad (t \in [t_0, t_j')),$$

we can combine these last two equations to obtain

(8)

$$\begin{aligned}
|\,y_j(t) - \bar{y}(t)| &= \left|\int_{t_0}^{t} (f^L(\tau, y_j(\tau), \sigma_j(\tau), \theta_j) - f^L(\tau, \bar{y}(\tau), \bar{\sigma}(\tau), 0))\,d\tau\right| \\
&\leqslant \int_{t_0}^{t} |f^L(\tau, y_j(\tau), \sigma_j(\tau), \theta_j) - f^L(\tau, \bar{y}(\tau), \sigma_j(\tau), 0)|\,d\tau \\
&\quad + \left|\int_{t_0}^{t} d\tau \int f^L(\tau, \bar{y}(\tau), r, 0)(\sigma_j(\tau) - \bar{\sigma}(\tau))\,(dr)\right| \\
&\leqslant \int_{t_0}^{t} \psi^L(\tau)|\,y_j(\tau) - \bar{y}(\tau)|\,d\tau + \phi_j(t) \qquad (t \in [t_0, t_j')).
\end{aligned}$$

We apply Gronwall's inequality II.4.4 to relation (8) to show that

$$| y_j(t) - \bar{y}(t)| \leqslant \phi_j(t) + \exp\left[\int_{t_0}^{t_1} \psi^L(\tau)\, d\tau\right] \int_{t_0}^{t_1} \psi^L(\tau)\, \phi_j(\tau)\, d\tau \quad (t \in [t_0, t_j'))$$

which, together with (7), implies that

$$| y_j(t) - \bar{y}(t)| \leqslant \alpha \qquad (t \in [t_0, t_j')).$$

Thus

$$d[\bar{y}(T), \partial V^L] \leqslant \lim_{t \to t_j'} | y_j(t) - \bar{y}(t)| \leqslant \alpha,$$

contradicting the definition of α. This shows that statement (5) must be valid. QED

VI.2.2 ***Lemma*** Let $\mathscr{U}$ be an abundant subset of $\mathscr{R}^\#$, $A(t)$ convex and $A(t)^\circ \neq \varnothing$ for all $t \in T$,

$$G(A^\circ) \triangleq \{(t, v) \in T \times \mathbb{R}^{m_2} \mid v \in A(t)^\circ\}$$

open in $T \times \mathbb{R}^{m_2}$ and $(\bar{y}, \bar{q}) \triangleq (\bar{y}, \bar{\sigma}, \bar{b}) \in \mathscr{A}(\mathscr{S}^\#)$. If f and $(\bar{y}, \bar{\sigma}, \bar{b})$ satisfy the conditions of VI.2.1, the functions

$$h_0 : V \times B \to \mathbb{R}, \qquad h_1 : V \times B \to \mathbb{R}^m, \qquad \text{and} \qquad h_2 : T \times V \times B \to \mathbb{R}^{m_2}$$

are continuous and the partial derivatives $\mathscr{D}_1 h_0$, $\mathscr{D}_1 h_1$, and $\mathscr{D}_2 h_2$ exist and are continuous, then the conditions of Theorem V.3.2 are satisfied and we have

$$(1) \quad f_v(\tau, \bar{y}(\tau), \bar{\sigma}(\tau), \bar{b})\, \Delta v = \int f_v(\tau, \bar{y}(\tau), r, \bar{b})\, \Delta v\, \bar{\sigma}(\tau)\,(dr) \quad (\tau \in T,\ \Delta v \in \mathbb{R}^n),$$

$$(2) \quad (F_y(\bar{y}, \bar{\sigma}, \bar{b})\, \Delta y)(t) = \int_{t_0}^{t} f_v(\tau, \bar{y}(\tau), \bar{\sigma}(\tau), \bar{b})\, \Delta y(\tau)\, d\tau \qquad (t \in T, \Delta y \in \mathscr{Y})$$

and

$$(3) \qquad D_2 F(\bar{y}, \bar{q}; q - \bar{q})(t) = \int_{t_0}^{t} [f(\tau, \bar{y}(\tau), \sigma(\tau) - \bar{\sigma}(\tau), \bar{b}) + D_4 f(\tau, \bar{y}(\tau), \bar{\sigma}(\tau), \bar{b}; b - \bar{b})]\, d\tau$$

$$[t \in T, \quad q \triangleq (\bar{\sigma}, b) \in \mathscr{S}^\# \times B].$$

PROOF It follows from V.2.4 that C_2 is convex and $C_2^\circ \neq \varnothing$. Now let $K \triangleq (q_0, \ldots, q_m) \triangleq ((\sigma_0, b_0), \ldots, (\sigma_m, b_m)) \in (\mathscr{S}^\# \times B)^{m+1}$, $L \triangleq (b_0, \ldots, b_m)$, and $Y_K \triangleq C(T, V^L) \subset \mathscr{Y}$. It follows from VI.2.1 that relation (1) is satisfied and there exist neighborhoods $\mathscr{S}_K$ of $\bar{\sigma}$ in $\mathscr{S}^\#$ and $\mathscr{T}_K$ of 0 in $\mathscr{T}^L$ such that

$$F^K(y, \sigma, \theta)(t) \triangleq v_0 + \int_{t_0}^{t} f^L(\tau, y(\tau), \sigma(\tau), \theta)\, d\tau$$

is defined for all $(y, \sigma, \theta) \in Y_K \times \mathscr{S}_K \times \mathscr{T}_K$ and $t \in T$ and the equation $y = F^K(y, \sigma, \theta)$ has a unique solution $y^K(\sigma, \theta)$ in Y_K for all $(\sigma, \theta) \in \mathscr{S}_K \times \mathscr{T}_K$, thus satisfying V.3.2(4). By IV.2.9 and I.5.3, the function

$$F^K : Y_K \times \mathscr{S}_K \times \mathscr{T}_K \to \mathscr{Y}$$

is continuous and thus satisfies condition V.3.2(3). We next observe that VI.3.2(5) holds, that is, the set

$$\mathscr{F}^K \triangleq y^K(\mathscr{S}_K \times \mathscr{T}_K) = \{F^K(y^K(\sigma, \theta), \sigma, \theta) | \sigma \in \mathscr{S}_K, \theta \in \mathscr{T}_K\}$$

is conditionally compact. Indeed,

$$|f^L(\tau, y^K(\sigma, \theta)(\tau), \sigma(\tau), \theta)| \leqslant |f^L(\tau, \cdot, \cdot, \cdot)|_{\sup} \quad [\tau \in T, \ (\sigma, \theta) \in \mathscr{S}_K \times \mathscr{T}_K]$$

and therefore $\mathscr{F}^K$ is uniformly bounded and equicontinuous; hence, by I.5.4, $\bar{\mathscr{F}}^K$ is compact in $\mathscr{Y}$.

We set

$$\sigma^K(\theta) \triangleq \bar{\sigma} + \sum_{j=0}^{m} \theta^j(\sigma_j - \bar{\sigma}), \qquad b^K(\theta) \triangleq \bar{b} + \sum_{j=0}^{m} \theta^j(b_j - \bar{b}),$$

$$q^K(\theta) \triangleq (\sigma^K(\theta), b^K(\theta)) \qquad (\theta \in \mathscr{T}_{m+1}),$$

and

$$\tilde{f}^K(\tau, v, \theta) = f(\tau, v, \sigma^K(\theta)(\tau), b^K(\theta)) \qquad (\tau \in T, \quad v \in V^L, \quad \theta \in \mathscr{T}_{m+1}).$$

It is clear that $\mathscr{T}_K$ contains a convex neighborhood $\mathscr{T}_K'$ of 0 in $\mathscr{T}_K$ such that $\sigma^K(\theta) \in \mathscr{S}_K$ for $\theta \in \mathscr{T}_K'$. Thus

$$\tilde{F}^K(y, \theta) \triangleq F^K(y, \sigma^K(\theta), \theta) = F(y, \sigma^K(\theta), b^K(\theta))$$

is defined for all $(y, \theta) \in Y_K \times \mathscr{T}_K'$ and, by II.4.9, $\tilde{F}^K : Y_K \times \mathscr{T}_K' \to \mathscr{Y}$ is continuous and has a continuous derivative $\mathscr{D}\tilde{F}^K(\cdot, \cdot)$ such that

$$\mathscr{D}\tilde{F}^K(y, \theta)(\Delta y, \Delta\theta)(t) = \int_{t_0}^{t} \tilde{f}^K_{(v,\theta)}(\tau, y(\tau), \theta)\,(\Delta y(\tau), \Delta\theta)\, d\tau \tag{4}$$

$$(\Delta y \in \mathscr{Y}, \quad \Delta\theta \in \mathbb{R}^{m+1}, \quad t \in T).$$

This shows that $\tilde{F}^K$ satisfies Condition V.3.2(1). By II.4.10,

$$I - F_y(\bar{y}, \bar{\sigma}, \bar{b}) \ [= I - \tilde{F}_y{}^K(\bar{y}, 0)]$$

is a homeomorphism of $\mathscr{Y}$ onto $\mathscr{Y}$, thus satisfying condition V.3.2(2).

We observe that the function

$$g : \mathscr{Y} \times \mathscr{S}^{\#} \times B \to \mathbb{R} \times \mathbb{R}^m \times C(T, \mathbb{R}^{m_2}),$$

defined by

$$g_i(y, \sigma, b) \triangleq h_i(y(t_1), b) \ (i = 0, 1) \quad \text{and} \quad g_2(y, \sigma, b)(t) = h_2(t, y(t), b) \ (t \in T),$$

satisfies Conditions V.3.2(1) and V.3.2(3).

We have shown thus far that the conditions of V.3.2 are satisfied and relation (1) valid. If we set $\Delta\theta = 0$ and $K = ((\bar{\sigma}, \bar{b},..., (\bar{\sigma}, \bar{b}))$, then relation (2) follows from (4). Next we observe that, for $\Delta\theta = (1, 0,..., 0) \in \mathbb{R}^{m+1}$, $K = ((\sigma, b), (\bar{\sigma}, \bar{b}),..., (\bar{\sigma}, \bar{b})), \bar{q} = (\bar{\sigma}, \bar{b})$, and $q = (\sigma, b)$, we have

$$(5) \qquad (\tilde{F}_\theta{}^K(\bar{y}, 0)\, \Delta\theta)(t) = D_2F(\bar{y}, \bar{q}; q - \bar{q})(t) \qquad (t \in T)$$

and

$$\tilde{f}^K(\tau, \bar{y}(\tau), \theta) = f(\tau, \bar{y}(\tau), \bar{\sigma}(\tau), b^K(\theta)) + \theta^0 f(\tau, \bar{y}(\tau), \sigma(\tau) - \bar{\sigma}(\tau), b^K(\theta));$$

hence, by the chain rule II.3.4,

$$\tilde{f}_\theta{}^K(\tau, \bar{y}(\tau), 0)\, \Delta\theta = D_4 f(\tau, \bar{y}(\tau), \bar{\sigma}(\tau), \bar{b}; b - \bar{b}) + f(\tau, \bar{y}(\tau), \sigma(\tau) - \bar{\sigma}(\tau), \bar{b}).$$

This last relation together with (4) and (5) yields relation (3). QED

We observe that if the conditions of Lemma VI.2.1 are satisfied then, by Lemmas VI.2.2 and IV.2.7, the function $\tau \to f_v(\tau, \bar{y}(\tau), \bar{\sigma}(\tau), \bar{b})$ is integrable. Thus, by II.4.8, there exists a unique function $Z : T \to B(\mathbb{R}^n, \mathbb{R}^n)$ satisfying the equation

$$Z(t) = I + \int_t^{t_1} Z(\tau) f_v(\tau, \bar{y}(\tau), \bar{\sigma}(\tau), \bar{b})\, d\tau \qquad (t \in T).$$

VI.2.3 ***Theorem*** Let $\mathscr{U}$ be an abundant subset of $\mathscr{R}^\#$, $A(t)$ convex and $A(t)^\circ \neq \varnothing$ for all $t \in T$,

$$G(A^\circ) = \{(t, v) \in T \times \mathbb{R}^{m_2} \mid v \in A(t)^\circ\}$$

open in $T \times \mathbb{R}^{m_2}$, $(\bar{y}, \bar{\sigma}, \bar{b}) \in \mathscr{A}(\mathscr{S}^\#)$,

$$h_0 : V \times B \to \mathbb{R}, \qquad h_1 : V \times B \to \mathbb{R}^m \qquad \text{and} \qquad h_2 : T \times V \times B \to \mathbb{R}^{m_2}$$

continuous with continuous partial derivatives $\mathscr{D}_1 h_0$, $\mathscr{D}_1 h_1$, and $\mathscr{D}_2 h_2$, I the unit $n \times n$ matrix, Z the (unique) solution of the equation

$$Z(t) = I + \int_t^{t_1} Z(\tau) f_v(\tau, \bar{y}(\tau), \bar{\sigma}(\tau), \bar{b})\, d\tau \qquad (t \in T),$$

and let f and $(\bar{y}, \bar{\sigma}, \bar{b})$ satisfy the conditions of VI.2.1. Then

(1) $(\bar{y}, \bar{\sigma}, \bar{b})$ is extremal if and only if there exist $l_0 \geqslant 0$, $l_1 \in \mathbb{R}^m$, $\omega \in \mathrm{frm}^+(T)$, and an ω-integrable $\tilde{\omega} : T \to \mathbb{R}^{m_2}$ such that, setting

$$k(t)^T \triangleq [l_0 \mathscr{D}_1 h_0(\bar{y}(t_1), \bar{b}) + l_1{}^T \mathscr{D}_1 h_1(\bar{y}(t_1), \bar{b})$$

$$+ \int_{[t,t_1]} \tilde{\omega}(\alpha)^T \mathscr{D}_2 h_2(\alpha, \bar{y}(\alpha), \bar{b}) \, Z(\alpha)^{-1} \, \omega(d\alpha)] \, Z(t) \qquad (t \in T),$$

we have

(1a) $$l_0 + |l_1| + \omega(T) > 0 \quad \text{and} \quad |\tilde{\omega}(t)| = 1 \quad (t \in T),$$

(1b) $$k(t)^T f(t, \bar{y}(t), \bar{\sigma}(t), \bar{b}) = \min_{r \in \bar{R}^\#(t)} k(t)^T f(t, \bar{y}(t), r, \bar{b}) \quad \text{for a.a. } t \in T,$$

(1c) $$\int_{t_0}^{t_1} k(\tau)^T D_4 f(\tau, \bar{y}(\tau), \bar{\sigma}(\tau), \bar{b}; b - \bar{b}) \, d\tau$$

$$+ l_0 D_2 h_0(\bar{y}(t_1), \bar{b}; b - \bar{b}) + l_1{}^T D_2 h_1(\bar{y}(t_1), \bar{b}; b - \bar{b})$$

$$+ \int \tilde{\omega}(\alpha)^T D_3 h_2(\alpha, \bar{y}(\alpha), \bar{b}; b - \bar{b}) \, \omega(d\alpha) \geqslant 0 \qquad (b \in B),$$

(1d) $$\tilde{\omega}(t) \cdot h_2(t, \bar{y}(t), \bar{b}) = \max_{a \in A(t)} \tilde{\omega}(t) \cdot a \qquad \text{for } \omega\text{-a.a. } t \in T;$$

and

(2) either $(\bar{y}, \bar{\sigma}, \bar{b})$ is extremal or there exists a point $(y_1, u_1, b_1) \in \mathscr{A}(\mathscr{U})$ such that $h_0(y_1(t_1), \bar{b}) < h_0(\bar{y}(t_1), \bar{b})$.

In particular, if $(\bar{y}, \bar{\sigma}, \bar{b})$ is either a minimizing relaxed solution or a minimizing $\mathscr{U}$-solution, then $(\bar{y}, \bar{\sigma}, \bar{b})$ is extremal.

Finally, if $(y_1, u_1, b_1) \in \mathscr{A}(\mathscr{U})$, the conditions of VI.2.2 remain valid when $(\bar{y}, \bar{\sigma}, \bar{b})$ is replaced by an arbitrary point in

$$\{(y, \sigma, b) \in \mathscr{A}(\mathscr{S}^\#) \mid h_0(y(t_1), b) \leqslant h_0(y_1(t_1), b_1)\},$$

and there exists a minimizing relaxed solution which is normal, then the problem does not have any strict $\mathscr{U}$-solutions.

▌PROOF *Step* 1 We first assume that $(\bar{y}, \bar{\sigma}, \bar{b})$ is extremal and will derive relations (1a)–(1d) in Steps 1 and 2. We set

$$\chi(q) \triangleq (\chi_0(q), \chi_1(q), \chi_2(q))$$

$$\triangleq g_y(\bar{y}, \bar{q}) \circ [I - F_y(\bar{y}, \bar{q})]^{-1} D_2 F(\bar{y}, \bar{q}; q - \bar{q})$$

$$+ D_2 g(\bar{y}, \bar{q}; q - \bar{q}) \qquad [q \triangleq (\sigma, b) \in \mathscr{S}^\# \times B].$$

Then there exists $l \triangleq (l_0, l_1, l_2) \in [0, \infty) \times \mathbb{R}^m \times C(T, \mathbb{R}^{m_2})^*$ such that $l \neq 0$,

$$l_0\chi_0(q) + l_1 \cdot \chi_1(q) + l_2(\chi_2(q)) \geqslant 0 \qquad [q \triangleq (\sigma, b) \in \mathscr{S}^\# \times B] \tag{3}$$

and

$$l_2(g_2(\bar{y}, \bar{q})) = \operatorname{Max}_{c \in C_2} l_2(c). \tag{4}$$

Now let $q \triangleq (\sigma, b) \in \mathscr{S}^\# \times B$ and

$$p(t) \triangleq [(I - F_y(\bar{y}, \bar{q}))^{-1} D_2F(\bar{y}, \bar{q}; q - \bar{q})](t).$$

Then, by VI.2.2,

$$\begin{aligned}[(I - F_y(\bar{y}, \bar{q}))\, p](t) &= p(t) - \int_{t_0}^{t} f_v(\tau, \bar{y}(\tau), \bar{\sigma}(\tau), \bar{b})\, p(\tau)\, d\tau \\ &= \int_{t_0}^{t} [f(\tau, \bar{y}(\tau), \sigma(\tau) - \bar{\sigma}(\tau), \bar{b}) \\ &\qquad + D_4 f(\tau, \bar{y}(\tau), \bar{\sigma}(\tau), \bar{b}; b - \bar{b})]\, d\tau \\ &\qquad\qquad (t \in T);\end{aligned} \tag{5}$$

hence, by (5) and II.4.8(4),

$$p(t) = Z(t)^{-1} \int_{t_0}^{t} Z(\tau)\, m(\tau)\, d\tau \qquad (t \in T), \tag{6}$$

where

$$m(\tau) \triangleq f(\tau, \bar{y}(\tau), \sigma(\tau) - \bar{\sigma}(\tau), \bar{b}) + D_4 f(\tau, \bar{y}(\tau), \bar{\sigma}(\tau), \bar{b}; b - \bar{b}) \qquad (\tau \in T). \tag{7}$$

Since

$$g_i(y, \sigma, b) \triangleq h_i(y(t_1), b) \qquad (i = 0, 1)$$

and

$$g_2(y, \sigma, b)(t) = h_2(t, y(t), b) \qquad [t \in T, \quad (y, \sigma, b) \in \mathscr{Y} \times \mathscr{S}^\# \times B],$$

we have

$$\begin{aligned}\chi_i(q) &= \mathscr{D}_1 h_i(\bar{y}(t_1), \bar{b})\, p(t_1) + D_2 h_i(\bar{y}(t_1), \bar{b}; b - \bar{b}) \qquad (i = 0, 1), \\ \chi_2(q)(t) &= \mathscr{D}_2 h_2(t, \bar{y}(t), \bar{b})\, p(t) + D_3 h_2(t, \bar{y}(t), \bar{b}; b - \bar{b}) \qquad (t \in T).\end{aligned} \tag{8}$$

Thus, by (6) and (8),

$$\begin{aligned} l_0\chi_0(q) + l_1 \cdot \chi_1(q) &= \int_{t_0}^{t_1} [l_0 \mathscr{D}_1 h_0(\bar{y}(t_1), \bar{b}) \\ &\qquad + l_1^T \mathscr{D}_1 h_1(\bar{y}(t_1), \bar{b})]\, Z(\tau)\, m(\tau)\, d\tau \\ &\qquad + l_0 D_2 h_0(\bar{y}(t_1), \bar{b}; b - \bar{b}) + l_1^T D_2 h_1(\bar{y}(t_1), \bar{b}; b - \bar{b}).\end{aligned} \tag{9}$$

Now, by V.2.5, there exist $\omega \in \text{frm}^+(T)$ and an ω-integrable $\tilde{\omega} : T \to \mathbb{R}^{m_2}$ such that

$$(10) \quad |\tilde{\omega}(t)| = 1 \ (t \in T), \ l_2(c) = \int \tilde{\omega}(t) \cdot c(t)\, \omega(dt) \qquad [c \in C(T, \mathbb{R}^{m_2})]$$

and relation (1d) is satisfied.

It follows now from (6) and (8) that

$$l_2(\chi_2(q)) = \int \tilde{\omega}(t)^T \Big[\mathscr{D}_2 h_2(t, \bar{y}(t), \bar{b})\, Z(t)^{-1} \int_{t_0}^{t} Z(\tau)\, m(\tau)\, d\tau + D_3 h_2(t, \bar{y}(t), \bar{b}; b - \bar{b})\Big]\, \omega(dt) \qquad [q \triangleq (\sigma, b) \in \mathscr{S}^\# \times B].$$

We observe that, by VI.2.1((2) and (3)) [for $L = (\bar{b}, b, \ldots, b) \in B^{m+1}$], we have

$$|m(\tau)| \leqslant 2\, |f^L(\tau, \cdot, \cdot, \cdot)|_{\sup} + |f^L_{(v,\theta)}(\tau, \cdot, \cdot, \cdot)|_{\sup};$$

furthermore, the continuous functions $t \to Z(t)$ and $t \to \mathscr{D}_2 h_2(t, \bar{y}(t), \bar{b})$ are integrable on T. Thus we may apply Fubini's theorem I.4.45 and find that

$$l_2(\chi_2(q)) = \int_{t_0}^{t_1} \Big[\int_{[\tau, t_1]} \tilde{\omega}(t)^T\, \mathscr{D}_2 h_2(t, \bar{y}(t), \bar{b})\, Z(t)^{-1}\, \omega(dt)\Big]\, Z(\tau)\, m(\tau)\, d\tau + \int \tilde{\omega}(t)^T D_3 h_2(t, \bar{y}(t), \bar{b}; b - \bar{b})\, \omega(dt).$$

We combine this relation with (3), (7), and (9) to derive relation

$$\begin{aligned}(11) \quad & l_0 \chi_0(q) + l_1 \cdot \chi_1(q) + l_2(\chi_2(q)) \\ &= \int_{t_0}^{t_1} k(\tau)^T \,[f(\tau, \bar{y}(\tau), \sigma(\tau) - \bar{\sigma}(\tau), \bar{b}) \\ &\quad + D_4 f(\tau, \bar{y}(\tau), \bar{\sigma}(\tau), \bar{b}; b - \bar{b})]\, d\tau \\ &\quad + l_0 D_2 h_0(\bar{y}(t_1), \bar{b}; b - \bar{b}) + l_1{}^T D_2 h_1(\bar{y}(t_1), \bar{b}; b - \bar{b}) \\ &\quad + \int \tilde{\omega}(\alpha)^T D_3 h_2(\alpha, \bar{y}(\alpha), \bar{b}; b - \bar{b})\, \omega(d\alpha) \\ &\geqslant 0 \qquad [q \triangleq (\sigma, b) \in \mathscr{S}^\# \times B].\end{aligned}$$

Step 2 We next derive relations (1a)–(1c). Relation (1a) follows from (10) and $l \triangleq (l_0, l_1, l_2) \neq 0$. Relation (1c) is obtained from (11) by setting $\sigma = \bar{\sigma}$. To obtain relation (1b) we first set $b = \bar{b}$ in (11), yielding

$$(12) \qquad \int_{t_0}^{t_1} k(\tau)^T f(\tau, \bar{y}(\tau), \sigma(\tau) - \bar{\sigma}(\tau), \bar{b})\, d\tau \geqslant 0 \qquad (\sigma \in \mathscr{S}^\#).$$

By Condition IV.3.1, $\mathscr{R}^{\#}$ contains a denumerable subset $\{\rho_1, \rho_2, \ldots\}$ such that $\{\rho_1(\tau), \rho_2(\tau), \ldots\}$ is dense in $R^{\#}(\tau)$ for a.a. $\tau \in T$. If we set, for each $j \in \mathbb{N}$ and each measurable $E \subset T$,

$$\sigma(t) = \rho_j(t) \quad (t \in E), \qquad \sigma(t) = \bar{\sigma}(t) \quad (t \notin E),$$

then relation (12) yields

$$\int_E k(\tau)^T f(\tau, \bar{y}(\tau), \rho_j(\tau) - \bar{\sigma}(\tau), \bar{b})\, d\tau \geqslant 0 \qquad \text{for all measurable } E \subset T;$$

from which it follows, by I.4.34(7), that

$$k(t)^T f(t, \bar{y}(t), \rho_j(t) - \bar{\sigma}(t), \bar{b}) \geqslant 0 \quad \text{a.e. in } T,$$

say for $t \in T_j'$. This last relation is therefore valid for all $t \in T' \triangleq \bigcap_{j=1}^{\infty} T_j'$, and $T \sim T'$ is of measure 0. We recall that $\{\rho_1(t), \rho_2(t), \ldots\}$ is dense in $R^{\#}(t)$ for a.a. $t \in T$ and that the function $r \to k(t)^T f(t, \bar{y}(t), r, \bar{b})$ is continuous on R for all $t \in T$. This yields

$$(13) \quad k(t)^T f(t, \bar{y}(t), \bar{\sigma}(t), \bar{b}) \leqslant \mathrm{Min}_{r \in \bar{R}^{\#}(t)}\, k(t)^T f(t, \bar{y}(t), r, \bar{b}) \quad \text{a.e. in } T.$$

On the other hand, since $\bar{\sigma}(t)$ is supported on $\bar{R}^{\#}(t)$ a.e. in T, we have

$$\begin{aligned} k(t)^T f(t, \bar{y}(t), \bar{\sigma}(t), \bar{b}) &= \int_{\bar{R}^{\#}(t)} k(t)^T f(t, \bar{y}(t), r, \bar{b})\, \bar{\sigma}(t)\,(dr) \\ &\geqslant \underset{r \in \bar{R}^{\#}(t)}{\mathrm{Min}}\ k(t)^T f(t, \bar{y}(t), r, \bar{b}) \quad \text{a.e. in } T. \end{aligned}$$

This relation and (13) yield relation (1b).

Step 3 Next we assume that there exist $l_0 \geqslant 0$, $l_1 \in \mathbb{R}^m$, $\omega \in \mathrm{frm}^+(T)$, and an ω-integrable $\tilde{\omega} : T \to \mathbb{R}^{m_2}$ such that relations (1a)–(1d) are valid, and we shall deduce that $(\bar{y}, \bar{\sigma}, \bar{b})$ is extremal. We set

$$l_2(c) \triangleq \int \tilde{\omega}(t) \cdot c(t)\, \omega(dt) \qquad [c \in C(T, \mathbb{R}^{m_2})]$$

and observe that $l_2 \in C(T, \mathbb{R}^{m_2})^*$. The evaluation of

$$l_0 \chi_0(q) + l_1 \cdot \chi_1(q) + l_2(\chi_2(q))$$

that we carried out in Step 1 and that led to the equality in (11) remains valid. By (1a), $l \triangleq (l_0, l_1, l_2) \neq 0$. By (1b), we have

$$k(t)^T f(t, \bar{y}(t), \bar{\sigma}(t), \bar{b}) \leqslant k(t)^T f(t, \bar{y}(t), \sigma(t), \bar{b}) \qquad (\sigma \in \mathscr{S}^{\#}, \quad \text{a.a. } t \in T);$$

hence

$$\int_{t_0}^{t_1} k(\tau)^T [f(\tau, \bar{y}(\tau), \sigma(\tau) - \bar{\sigma}(\tau), \bar{b})] \geqslant 0 \qquad (\sigma \in \mathscr{S}^\#).$$

Combining this relation with (1c) and with the equality in (11) yields

$$l_0\chi_0(q) + l_1 \cdot \chi_1(q) + l_2(\chi_2(q)) \geqslant 0 \qquad [q \triangleq (\sigma, b) \in \mathscr{S}^\# \times B].$$

Finally, relation (1d) yields

$$\int \tilde{\omega}(t) \cdot h_2(t, \bar{y}(t), \bar{b})\, \omega(dt) \geqslant \int \tilde{\omega}(t) \cdot c(t)\, \omega(dt)$$
$$= l_2(c) \qquad (c \in C_2).$$

Thus $(\bar{y}, \bar{\sigma}, \bar{b}, l_0, l_1, l_2)$ is an extremal.

Step 4 Statement (2) follows from VI.2.2 and V.3.2. The last assertion of the theorem follows from VI.2.2 and V.3.4. QED

If $h : X \times Y \to Z$ is a given function and $h(\cdot, y)$ is a constant for all $y \in Y$, then we say that *h is independent of* $x \in X$ and write $h(y)$ for $h(x, y)$, treating h as a function on Y to Z.

VI.2.4 ***Theorem*** Let the conditions of Theorem VI.2.3 be satisfied and assume that h_2 is independent of $(t, b) \in T \times B$, $h_2(\cdot)$ has a continuous second derivative on V, $A(t) \triangleq (-\infty, 0]^{m_2}$ $(t \in T)$, and $(\bar{y}, \bar{\sigma}, \bar{b})$ is extremal. Then there exist $l_0 \geqslant 0$, $l_1 \in \mathbb{R}^m$, nonincreasing functions

$$\mu^j : T \to [0, \infty) \qquad (j = 1, 2,..., m_2),$$

and a continuous function $z : T \to \mathbb{R}^n$ such that, setting

$$\mu \triangleq (\mu^1,..., \mu^{m_2}),$$
$$\beta(t, v) \triangleq h_2'(v) f(t, v, \bar{\sigma}(t), \bar{b}) \qquad (t \in T, \quad v \in V),$$

and

$$k(t)^T \triangleq z(t)^T + \mu(t)^T h_2'(\bar{y}(t)) \qquad (t \in T),$$

we have

(1) $$z(t)^T = l_0 \mathscr{D}_1 h_0(\bar{y}(t_1), \bar{b}) + l_1 \cdot \mathscr{D}_1 h_1(\bar{y}(t_1), \bar{b}) + \int_t^{t_1} [z(\tau)^T f_v(\tau, \bar{y}(\tau), \bar{\sigma}(\tau), \bar{b}) + \mu(\tau)^T \beta_v(\tau, \bar{y}(\tau))]\, d\tau;$$

(2) $$l_0 + |l_1| + \sum_{j=1}^{m_2} \mu^j(t_0) > 0$$

and, for each $j = 1, 2,..., m_2$, $\mu^j |(t', t'')$ is constant if

$$h_2^j(\bar{y}(t)) < 0 \qquad (t' < t < t'');$$

(3) $$k(t)^T f(t, \bar{y}(t), \bar{\sigma}(t), \bar{b}) = \operatorname*{Min}_{r \in \bar{R}^\#(t)} k(t)^T f(t, \bar{y}(t), r, \bar{b}) \qquad \text{for a.a. } t \in T;$$

and

(4) $$\int_{t_0}^{t} k(\tau)^T D_4 f(\tau, \bar{y}(\tau), \bar{\sigma}(\tau), \bar{b}; b - \bar{b})\, d\tau$$
$$+ l_0 D_2 h_0(\bar{y}(t_1), \bar{b}; b - \bar{b}) + l_1 \cdot D_2 h_1(\bar{y}(t_1), \bar{b}; b - \bar{b}) \geqslant 0 \quad (b \in B).$$

▌PROOF *Step* 1 Let ω and $\tilde{\omega}$ be as in Theorem VI.2.3,

$$\mu^j(t) \triangleq \int_{[t,t_1]} \tilde{\omega}^j(\tau)\, \omega(d\tau) \qquad (t \in T, \quad j = 1, 2,..., m_2)$$

and $\mu \triangleq (\mu^1,..., \mu^{m_2})$. For each $j = 1, 2,..., m_2$, we have, by VI.2.3(1d),

$$\tilde{\omega}^j(t) \geqslant 0 \quad \omega\text{-a.e.} \qquad \text{and} \qquad \tilde{\omega}^j(t) = 0 \quad \omega\text{-a.e. in } \{t \in T \mid h_2^j(\bar{y}(t)) < 0\}.$$

These relations, together with VI.2.3(1a), show that μ^j is nonnegative and nonincreasing and statement (2) is valid.

Since each μ^j is nonincreasing, it is measurable. It follows then from II.4.6 and the continuity of h_2'' that there exists a continuous $z : T \to \mathbb{R}^n$ satisfying (1).

Step 2 Let $j \in \{1, 2,..., m_2\}$, $t \in T$, and $p \triangleq (p^1,..., p^{m_2}) : T \to \mathbb{R}^{m_2}$ be absolutely continuous. Then, by I.4.42 and Fubini's theorem I.4.45, we have

$$\begin{aligned}
\int_{[t,t_1]} \mu^j(\tau)\, \dot{p}^j(\tau)\, d\tau &= \int_{[t,t_1]} \dot{p}^j(\tau)\, d\tau \int_{[t,t_1]} \tilde{\omega}^j(\alpha)\, \chi_{[\tau,t_1]}(\alpha)\, \omega(d\alpha) \\
&= \int_{[t,t_1]} \tilde{\omega}^j(\alpha)\, \omega(d\alpha) \int_{[t,\alpha]} \dot{p}^j(\tau)\, d\tau \\
&= \int_{[t,t_1]} [p^j(\alpha) - p^j(t)]\, \tilde{\omega}^j(\alpha)\, \omega(d\alpha) \\
&= -p^j(t)\, \mu^j(t) + \int_{[t,t_1]} p^j(\alpha)\, \tilde{\omega}^j(\alpha)\, \omega(d\alpha);
\end{aligned}$$

hence

$$\int_{[t,t_1]} \tilde{\omega}^j(\alpha)\, p^j(\alpha)\, \omega(d\alpha) = \mu^j(t)\, p^j(t) + \int_{[t,t_1]} \mu^j(\alpha)\, \dot{p}^j(\alpha)\, d\alpha \qquad (t \in T).$$

(This is a form of the so-called *integration by parts formula.*) Thus, if $M : T \to B(\mathbb{R}^n, \mathbb{R}^{m_2})$ is absolutely continuous, then

$$(5)\quad \int_{[t,t_1]} \tilde{\omega}(\alpha)^T M(\alpha)\, \omega(d\alpha) = \mu(t)^T M(t) + \int_t^{t_1} \mu(\alpha)^T M'(\alpha)\, d\alpha \qquad (t \in T).$$

Step 3 Our next goal will be to show that the definition

$$k(t)^T \triangleq z(t)^T + \mu(t)^T h_2'(\bar{y}(t)) \qquad (t \in T)$$

is consistent with the definition of k in VI.2.3. To do so, we consider the function $Z : T \to B(\mathbb{R}^n, \mathbb{R}^n)$ that satisfies the equation

$$(6)\qquad Z(t) = I + \int_t^{t_1} Z(\tau) f_v(\tau, \bar{y}(\tau), \bar{\sigma}(\tau), \bar{b})\, d\tau \qquad (t \in T),$$

and observe that, by II.4.8, $W(t) \triangleq Z(t)^{-1}$ satisfies the equation

$$W(t) = I - \int_t^{t_1} f_v(\tau, \bar{y}(\tau), \bar{\sigma}(\tau), \bar{b})\, W(\tau)\, d\tau \qquad (t \in T).$$

If we set

$$V(t) \triangleq h_2'(\bar{y}(t)) \qquad (t \in T),$$

then it follows by II.3.12(5) and the chain rule II.3.4 that

$$V'(t) = h_2''(\bar{y}(t))\, \bar{y}'(t) = h_2''(\bar{y}(t))\, f(t, \bar{y}(t), \bar{\sigma}(t), \bar{b}) \qquad \text{for a.a. } t \in T$$

and

$$\begin{aligned} (V \cdot W)'(t) &= V'(t) \cdot W(t) + V(t) \cdot W'(t) \\ &= [h_2''(\bar{y}(t))\, f(t, \bar{y}(t), \bar{\sigma}(t), \bar{b}) \\ &\quad + h_2'(\bar{y}(t))\, f_v(t, \bar{y}(t), \bar{\sigma}(t), \bar{b})] \cdot W(t) \\ &= \beta_v(t, \bar{y}(t))\, W(t) \qquad \text{for a.a. } t \in T. \end{aligned}$$

We combine this relation with (5) to obtain

$$\begin{aligned} (7)\quad & \int_{[t,t_1]} \tilde{\omega}(\alpha)^T h_2'(\bar{y}(\alpha))\, Z(\alpha)^{-1}\, \omega(d\alpha) \\ &= \int_{[t,t_1]} \tilde{\omega}(\alpha)^T V \cdot W(\alpha)\, \omega(d\alpha) \\ &= \mu(t)^T V \cdot W(t) + \int_t^{t_1} \mu(\alpha)^T (V \cdot W)'(\alpha)\, d\alpha \\ &= \mu(t)^T h_2'(\bar{y}(t))\, Z(t)^{-1} + \int_t^{t_1} \mu(\alpha)^T \beta_v(\alpha, \bar{y}(\alpha))\, Z(\alpha)^{-1}\, d\alpha \quad (t \in T). \end{aligned}$$

It follows directly from (6) and II.4.8 that Eq. (1) has a unique solution z such that

$$z(t)^T = \Big[\int_t^{t_1} \mu(\alpha)^T \beta_v(\alpha, \bar{y}(\alpha))\, Z(\alpha)^{-1}\, d\alpha + l_0 \mathscr{D}_1 h_0(\bar{y}(t_1), \bar{b})$$

$$+ l_1 \cdot \mathscr{D}_1 h_1(\bar{y}(t_1), \bar{b})\Big]\, Z(t) \qquad (t \in T);$$

hence, by (7),

$$z(t)^T + \mu(t)^T h_2'(\bar{y}(t)) = \Big[l_0 \mathscr{D}_1 h_0(\bar{y}(t_1), \bar{b}) + l_1{}^T \mathscr{D}_1 h_1(\bar{y}(t_1), \bar{b})$$

$$+ \int_{[t,t_1]} \tilde{\omega}(\alpha)^T h_2'(\bar{y}(\alpha))\, Z(\alpha)^{-1}\, \omega(d\alpha)\Big]\, Z(t) \quad (t \in T).$$

This shows that our present definition of k coincides with that of VI.2.3. Thus relations (3) and (4) follow from VI.2.3(1b) and VI.2.3(1c), respectively. QED

We next consider the special case where the *unilateral* restriction

$$g_2(y, \sigma, b) \in C_2 \qquad [\text{i.e. } h_2(t, y(t), b) \in A(t) \quad \text{for all} \quad t \in T]$$

is discarded and $R^{\#}(\cdot)$ is constant. We can reduce the unilateral problem that we have been discussing in this chapter to this special case by setting $m_2 = 0$ (hence $\mathscr{X}_2 = \{0\}$). We also observe that our problem can be transformed (as in II.4.A) so that the function $(t, v, r, b) \to f(t, v, r, b)$ is independent of t. Indeed, any equation of the form

$$x(t) = v_0 + \int_{t_0}^{t} \phi(\tau, x(\tau))\, d\tau \qquad (t \in T)$$

is equivalent to $\hat{x}(t) = \hat{v}_0 + \int_{t_0}^{t} \hat{\phi}(\hat{x}(\tau))\, d\tau$ $(t \in T)$, where

$$\hat{x}(t) \triangleq (t, x(t)), \qquad \hat{\phi}(\hat{x}(t)) \triangleq (1, \phi(t, x(t))), \qquad \text{and} \qquad \hat{v}_0 \triangleq (t_0, v_0).$$

It must be borne in mind, however, that if this transformation is made then any assumptions about the dependence of the function $(t, v) \to \phi(t, v)$ on v must be extended to its dependence on (t, v).

VI.2.5 ***Theorem*** Let $(\bar{y}, \bar{\sigma}, \bar{b})$ be extremal,

$$m_2 = 0,\ R_1 \subset R,\ R^{\#}(t) \triangleq R_1 \qquad (t \in T),$$

V convex,

$$f(t, v, r, \bar{b}) \triangleq \hat{f}(v, r) \qquad [(t, v, r) \in T \times V \times R],$$

and the conditions of VI.2.3 satisfied. Then there exist $c_1 \in \mathbb{R}$, $l_0 \geqslant 0$, $l_1 \in \mathbb{R}^m$, and a continuous $k : T \to \mathbb{R}^n$ such that

$$l_0 + |\, l_1 \,| > 0,$$

(1) $$k(t)^T = l_0 \mathscr{D}_1 h_0(\bar{y}(t_1), \bar{b}) + l_1{}^T \mathscr{D}_1 h_1(\bar{y}(t_1), \bar{b}) + \int_t^{t_1} k(\tau)^T \hat{f}_v(\bar{y}(\tau), \bar{\sigma}(\tau))\, d\tau \qquad (t \in T);$$

(2) $$k(t)^T \hat{f}(\bar{y}(t), \bar{\sigma}(t)) = \operatorname*{Min}_{r \in \bar{R}_1} k(t)^T \hat{f}(\bar{y}(t), r) = c_1 \qquad \text{for a.a. } t \in T;$$

(3) $$\int_{t_0}^{t_1} k(\tau)^T D_4 f(\tau, \bar{y}(\tau), \bar{\sigma}(\tau), \bar{b}; b - \bar{b})\, d\tau + l_0 D_2 h_0(\bar{y}(t_1), \bar{b}; b - \bar{b}) + l_1{}^T D_2 h_1(\bar{y}(t_1), \bar{b}; b - \bar{b}) \geqslant 0 \qquad (b \in B);$$

and

(4) $$k(t)^T [\hat{f}_v(\bar{y}(t), r') \hat{f}(\bar{y}(t), \bar{\sigma}(t)) - \hat{f}_v(\bar{y}(t), \bar{\sigma}(t)) \hat{f}(\bar{y}(t), r')] = 0$$

for a.a. $t \in T$ and all

$$r' \in R'(t) \triangleq \{r \in R_1 \mid k(t)^T \hat{f}(\bar{y}(t), r) = k(t)^T \hat{f}(\bar{y}(t), \bar{\sigma}(t))\}.$$

▌PROOF We shall apply the results of Theorem VI.2.3. We have $\omega = 0$ because $l_2 \in \{0\}^* = \{0\}$; hence VI.2.3(1a) yields

(5) $$l_0 + |\, l_1 \,| > 0.$$

Relation VI.2.3(1c) yields relation (3), and VI.2.3(1b) yields the relation

$$k(t)^T \hat{f}(\bar{y}(t), \bar{\sigma}(t)) = \operatorname*{Min}_{r \in \bar{R}_1} k(t)^T \hat{f}(\bar{y}(t), r) \qquad \text{for a.a. } t \in T.$$

We also have

(6) $$k(\tau)^T = [l_0 \mathscr{D}_1 h_0(\bar{y}(t_1), \bar{b}) + l_1{}^T \mathscr{D}_1 h_1(\bar{y}(t_1), \bar{b})]\, Z(\tau) \qquad (\tau \in T)$$

and

(7) $$Z(t) = I + \int_t^{t_1} Z(\tau) f_v(\tau, \bar{y}(\tau), \bar{\sigma}(\tau), \bar{b})\, d\tau \qquad (t \in T).$$

We deduce statement (1) from (5)–(7) and II.4.8.

Now let

(8) $$\lambda(t, s) \triangleq k(t)^T \hat{f}(\bar{y}(t), s) = \int k(t)^T \hat{f}(\bar{y}(t), r)\, s(dr) \qquad [t \in T, \quad s \in \operatorname{rpm}(R)]$$

and

$$\lambda(t) \triangleq \lambda(t, \bar{\sigma}(t)) \qquad (t \in T). \tag{9}$$

If we set $S_1 \triangleq \text{rpm}(\bar{R}_1)$, then

$$\underset{r \in \bar{R}_1}{\text{Min}}\, k(t)^T \hat{f}(\bar{y}(t), r) = \underset{s \in S_1}{\text{Min}}\, k(t)^T \hat{f}(\bar{y}(t), s) \qquad \text{for} \quad \text{a.a. } t \in T$$

and therefore, by VI.2.3(1b), (8), and (9), we have

$$\lambda(t) = k(t)^T \hat{f}(\bar{y}(t), \bar{\sigma}(t)) = \underset{r \in \bar{R}_1}{\text{Min}}\, \lambda(t, r) = \underset{s \in S_1}{\text{Min}}\, \lambda(t, s) \quad \text{for} \quad \text{a.a. } t \in T. \tag{10}$$

We deduce from (1) and (8) that there exists $T' \subset T$ such that $\mu(T \sim T') = 0$, (10) is valid for all $t \in T'$, and

$$\begin{aligned} \lambda_t(t, s) &= \dot{k}(t)^T \hat{f}(\bar{y}(t), s) + k(t)^T \hat{f}_v(\bar{y}(t), s)\, \dot{\bar{y}}(t) \\ &= -\, k(t)^T \hat{f}_v(\bar{y}(t), \bar{\sigma}(t))\, \hat{f}(\bar{y}(t), s) \\ &\quad + k(t)^T \hat{f}_v(\bar{y}(t), s)\, f(\bar{y}(t), \bar{\sigma}(t)) \qquad (t \in T', \quad s \in S_1). \end{aligned} \tag{11}$$

For each $t \in T'$ we have, by (11),

$$\lim_{\substack{\alpha \to 0 \\ t+\alpha \in T'}} \alpha^{-1}[\lambda(t + \alpha, \bar{\sigma}(t)) - \lambda(t, \bar{\sigma}(t))] = \lambda_t(t, \bar{\sigma}(t)) = 0 \tag{12}$$

and, by (10),

$$\alpha^{-1}[\lambda(t + \alpha, \bar{\sigma}(t + \alpha)) - \lambda(t + \alpha, \bar{\sigma}(t))] \begin{cases} \leqslant 0 \text{ if } t + \alpha \in T' \text{ and } \alpha > 0, \\ \geqslant 0 \text{ if } t + \alpha \in T' \text{ and } \alpha < 0. \end{cases} \tag{13}$$

We observe that the functions $\hat{f}$ and $\hat{f}_v$ are continuous and therefore bounded on the compact set $V_1 \times R$, where V_1 is a compact and convex neighborhood of $\bar{y}(T)$. It follows that

$$|\hat{f}(v, s)| \leqslant c \qquad \text{and} \qquad |\hat{f}_v(v, s)| \leqslant c$$

for all $v \in V_1$ and $s \in \text{rpm}(R)$ and some $c \geqslant 0$. Similarly, $k(\cdot)$ is bounded on T and we may choose c so that $|k(t)| \leqslant c$ for all $t \in T$. Thus, by (11),

$$|\lambda_t(t, s)| \leqslant 2(c)^3 \triangleq c' \qquad (t \in T', \quad s \in S_1). \tag{14}$$

Furthermore, for all $s \in \text{rpm}(R)$ and $\tau_1, \tau_2 \in T'$, the second relation in (1), (8), the mean value theorem II.3.6, and VI.2.1(4) imply that

(15)

$$\begin{aligned}
&| \lambda(\tau_2, s) - \lambda(\tau_1, s)| \\
&\quad = | k(\tau_2)^T \hat{f}(\bar{y}(\tau_2), s) - k(\tau_1)^T \hat{f}(\bar{y}(\tau_1), s)| \\
&\quad \leqslant | k(\tau_2)| \, | \hat{f}(\bar{y}(\tau_2), s) - \hat{f}(\bar{y}(\tau_1), s)| + | k(\tau_2) - k(\tau_1)| \, | \hat{f}(\bar{y}(\tau_1), s)| \\
&\quad \leqslant c \, | \bar{y}(\tau_2) - \bar{y}(\tau_1)| \sup_{0 \leqslant \alpha \leqslant 1} | \hat{f}_v(\alpha \bar{y}(\tau_1) + (1 - \alpha) \bar{y}(\tau_2), s)| \\
&\qquad + c \, | k(\tau_2) - k(\tau_1)| \\
&\quad \leqslant 2(c)^3 \, | \tau_2 - \tau_1 |.
\end{aligned}$$

Thus, for each $s \in S_1$, the function $\lambda(\cdot, s)$ is absolutely continuous and we have

$$(16) \quad | \lambda(\tau_2, s) - \lambda(\tau_1, s)| = \left| \int_{\tau_1}^{\tau_2} \lambda_t(\tau, s) \, d\tau \right| \leqslant c' \, | \tau_2 - \tau_1 | \quad (\tau_1, \tau_2 \in T').$$

We now set

$$J_1(t, \tau) \triangleq (t - \tau)^{-1}[\lambda(t, \bar{\sigma}(t)) - \lambda(t, \bar{\sigma}(\tau))]$$

and

$$J_2(t, \tau) \triangleq (t - \tau)^{-1}[\lambda(t, \bar{\sigma}(\tau)) - \lambda(\tau, \bar{\sigma}(\tau))] \qquad (t, \tau \in T).$$

Then it follows from (13) that, for $t, \tau \in T'$,

$$(17) \quad (t - \tau)^{-1}[\lambda(t) - \lambda(\tau)] = J_1(t, \tau) + J_2(t, \tau) \begin{cases} \leqslant J_2(t, \tau) & \text{if } t > \tau, \\ \geqslant J_2(t, \tau) & \text{if } t < \tau, \end{cases}$$

and, by (16), that $-c' \leqslant J_2(t, \tau) \leqslant c'$ for all t and τ in T'; hence (17) implies that

$$| \lambda(t) - \lambda(\tau)| \leqslant c' \, | t - \tau | \qquad (t, \tau \in T').$$

Thus $\lambda(\cdot)$ is uniformly continuous on T' and has a uniformly continuous extension λ_1 to $\overline{T}' = T$, with

$$| \lambda_1(t) - \lambda_1(\tau)| \leqslant c' \, | t - \tau | \qquad (t, \tau \in T).$$

It follows that λ_1 is absolutely continuous and, by I.4.42, $\dot{\lambda}_1(t)$ exists a.e. in T, say for $t \in T'' \subset T'$. We can now deduce from (12) and (17) that

$$\lim_{\substack{\tau \to t \\ \tau \in T''}} [\lambda(t) - \lambda(\tau)]/(t - \tau) = \dot{\lambda}_1(t) = 0 \qquad (t \in T'');$$

therefore, by I.4.42, λ_1 is constant on T, hence λ is constant on T', and relation (2) follows from (10).

Finally, we must deduce relation (4). Let $\tau, t \in T'$ and $r' \in R'(t)$. Then $\lambda(t, r') = \lambda(t)$ and, by (2),

$$(18) \qquad (t-\tau)^{-1}[\lambda(t, r') - \lambda(\tau, r')] = (t-\tau)^{-1}[\lambda(t) - \lambda(\tau, r')]\begin{cases} \leqslant 0 & \text{if } t > \tau, \\ \geqslant 0 & \text{if } t < \tau. \end{cases}$$

By (11), the left side of (18) converges to a limit $\lambda_t(t, r')$ as $\tau \to t$, $\tau \in T'$, and thus (18) implies that $\lambda_t(t, r') = 0$. Relation (4) now follows from (11). QED

VI.3 Contingent Equations and Equivalent Control Functions

We continue our investigation of the control problem defined in VI.0 and discussed in VI.1 and VI.2, limiting our consideration of minimizing original solutions to the case $\mathscr{U} = \mathscr{R}^{\#}$. This control problem involves, in its original version, points $(y, \rho, b) \in \mathscr{Y} \times \mathscr{R}^{\#} \times B$ such that $y(t_0) = v_0$, y is absolutely continuous, and

$$(*) \qquad \dot{y}(t) = f\big(t, y(t), \rho(t), b\big) \quad \text{a.e. in } T;$$

and it involves, in its relaxed version, points $(y, \sigma, b) \in \mathscr{Y} \times \mathscr{S}^{\#} \times B$ such that $y(t_0) = v_0$, y is absolutely continuous, and

$$(**) \qquad \dot{y}(t) = f\big(t, y(t), \sigma(t), b\big) \quad \text{a.e. in } T.$$

We shall consider in this section relations equivalent to (*) and (**) that either do not involve any control functions (*contingent equations*) or are defined in terms of control functions other than elements of $\mathscr{R}^{\#}$ and $\mathscr{S}^{\#}$ (*equivalent control functions*).

VI.3.1 ***Theorem*** (Filippov–Castaing) Let $n \in \mathbb{N}$, V be an open subset of $\mathbb{R}^n$, $v_0 \in V$, and $\hat{f} : T \times V \times R \to \mathbb{R}^n$ be such that $\hat{f}(\cdot, v, r)$ is measurable and $\hat{f}(t, \cdot, \cdot)$ continuous for all $(t, v, r) \in T \times V \times R$. Let, furthermore, $y : T \to V$ be absolutely continuous, $y(t_0) = v_0$, and $R^{\#}(t)$ closed for a.a. $t \in T$. Then

(1) $\dot{y}(t) = \hat{f}\big(t, y(t), \rho(t)\big)$ a.e. in T for some $\rho \in \mathscr{R}^{\#}$ if and only if

$$\dot{y}(t) \in \hat{f}\big(t, y(t), R^{\#}(t)\big) \quad \text{a.e. in } T.$$

(2) If, furthermore, X is a compact metric space, $\mathscr{K}(X)$ the space of nonempty closed subsets of X with the Hausdorff metric, $\Gamma : T \to \mathscr{K}(X)$ measurable and $\phi : T \times V \times X \to \mathbb{R}^n$ such that $\phi(\cdot, v, x)$ is measurable, $\phi(t, \cdot, \cdot)$ continuous, and

$$\hat{f}(t, v, R^{\#}(t)) = \phi(t, v, \Gamma(t)) \qquad \text{for all} \quad (t, v, x) \in T \times V \times X,$$

then

$$\dot{y}(t) = \hat{f}(t, y(t), \rho(t)) \quad \text{a.e. in } T \qquad \text{for} \quad \text{some } \rho \in \mathscr{R}^{\#}$$

(that is, for some measurable selection ρ of $R^{\#}$) if and only if

$$\dot{y}(t) = \phi(t, y(t), \xi(t)) \quad \text{a.e. in } T \qquad \text{for} \quad \text{some measurable selection } \xi \text{ of } \Gamma.$$

▌PROOF If $\dot{y}(t) = \hat{f}(t, y(t), \rho(t))$ a.e. in T for some $\rho \in \mathscr{R}^{\#}$, then clearly $\dot{y}(t) \in \hat{f}(t, y(t), R^{\#}(t))$ a.e. in T. Conversely, if $\dot{y}(t) \in \hat{f}(t, y(t), R^{\#}(t))$ a.e. in T, then, by the Filippov–Castaing theorem I.7.10, there exists a measurable selection ρ of $R^{\#}$ such that $\dot{y}(t) = \hat{f}(t, y(t), \rho(t))$ a.e. in T. The same argument, applied to ϕ and Γ instead of $\hat{f}$ and $R^{\#}$, shows that $\dot{y}(t) = \phi(t, y(t), \xi(t))$ a.e. in T for some measurable selection ξ of Γ if and only if

$$\dot{y}(t) \in \phi(t, y(t), \Gamma(t)) = \hat{f}(t, y(t), R^{\#}(t)) \quad \text{a.e. in } T. \quad \text{QED}$$

VI.3.2 ***Theorem*** Let the conditions of Theorem VI.3.1 be satisfied except that $R^{\#}(t)$ need not be assumed closed. Then the following statements are equivalent:

(1a) $\qquad \dot{y}(t) = \hat{f}(t, y(t), \sigma(t))$ a.e. in T for some $\sigma \in \mathscr{S}^{\#}$;

(1b) $\qquad \dot{y}(t) \in \text{co}(\hat{f}(t, y(t), \bar{R}^{\#}(t)))$ a.e. in T;

and

(1c) $\qquad \dot{y}(t) = \sum_{j=0}^{n} \alpha^j(t) \hat{f}(t, y(t), \rho_j(t))$ a.e. in T

for some measurable $(\alpha^0, \ldots, \alpha^n) : T \to \mathscr{T}_n'$ and measurable $\rho_j : T \to R$ $(j = 0, 1, \ldots, n)$ such that $\rho_j(t) \in \bar{R}^{\#}(t)$ a.e. in T.

(2) If, furthermore, X is a compact metric space, $\mathscr{K}(X)$ the space of nonempty closed subsets of X with the Hausdorff metric, $\Gamma : T \to \mathscr{K}(X)$ measurable, and $\phi : T \times V \times X \to \mathbb{R}^n$ such that $\phi(\cdot, v, x)$ is measurable, $\phi(t, \cdot, \cdot)$ continuous, and

$$\phi(t, v, \Gamma(t)) = \text{co}(\hat{f}(t, v, \bar{R}^{\#}(t))) \qquad \text{for} \quad \text{all } (t, v, x) \in T \times V \times X,$$

then

$$\dot{y}(t) = \hat{f}(t, y(t), \sigma(t)) \quad \text{a.e. in } T \qquad \text{for} \quad \text{some } \sigma \in \mathscr{S}^{\#}$$

if and only if

$$\dot{y}(t) = \phi(t, y(t), \xi(t)) \quad \text{a.e. in } T \qquad \text{for some measurable selection } \xi \text{ of } \Gamma.$$

▌PROOF Let $K(t, v) \triangleq \operatorname{co}(\hat{f}(t, v, \bar{R}^{\#}(t)))$ for all $(t, v) \in T \times V$, $\sigma \in \mathscr{S}^{\#}$ and

$$\dot{y}(t) = \hat{f}(t, y(t), \sigma(t)) = \int f(t, y(t), r)\, \sigma(t)(dr) \tag{3}$$

$$\text{for a.a.} \quad t \in T, \quad \text{say for} \quad t \in T'.$$

Let $t \in T'$. Since $\hat{f}(t, y(t), r) \in K(t, y(t))$ $[r \in \bar{R}^{\#}(t)]$ and $\sigma(t)$ is a probability measure, it follows from (3) and I.6.13 that $\dot{y}(t) \in K(t, y(t))$ a.e. in T.

Now assume that

$$\dot{y}(t) \in K(t, y(t)) \qquad \text{for a.a. } t \in T,$$

say for $t \in T'$. Then, for each $t \in T'$, there exist, by Caratheodory's theorem I.6.2, $r_0, r_1, \ldots, r_n \in \bar{R}^{\#}(t)$ and $(\theta^0, \ldots, \theta^n) \in \mathscr{T}_n{}'$ such that

$$\dot{y}(t) = \sum_{j=0}^{n} \theta^j \hat{f}(t, y(t), r_j). \tag{4}$$

We define a regular probability measure $\tilde{\sigma}(t)$ by

$$\tilde{\sigma}(t)(R') = \sum_{r_j \in R'} \theta^j \qquad [R' \in \Sigma_{\text{Borel}}(R)].$$

Then, by (4),

$$\dot{y}(t) = \int \hat{f}(t, y(t), r)\, \tilde{\sigma}(t)(dr) = \hat{f}(t, y(t), \tilde{\sigma}(t)) \qquad \text{for} \quad \text{a.a. } t \in T,$$

and $\tilde{\sigma}(t)(\bar{R}^{\#}(t)) = 1$. It follows, by IV.3.13, that there exists $\sigma \in \mathscr{S}^{\#}$ such that

$$\dot{y}(t) = \hat{f}(t, y(t), \sigma(t)) \quad \text{a.e. in } T.$$

This proves that (1a) and (1b) are equivalent.

Now assume that the conditions of (2) are satisfied and $\dot{y}(t) = \hat{f}(t, y(t), \sigma(t))$ a.e. in T for some $\sigma \in \mathscr{S}^{\#}$. Then, by I.6.13,

$$\dot{y}(t) \in \operatorname{co}(\hat{f}(t, y(t), \bar{R}^{\#}(t))) = \phi(t, y(t), \Gamma(t)) \quad \text{a.e. in } T$$

and, by I.7.10, there exists a measurable selection ξ of Γ such that $\dot{y}(t) = \phi(t, y(t), \xi(t))$ a.e. in T. Conversely, if $\dot{y}(t) = \phi(t, y(t), \xi(t))$ a.e. in T for some measurable selection ξ of Γ, then (1b) is satisfied, whence (1a) follows. Thus statement (2) is valid.

Finally, we must show that statement (1c) is equivalent to (1b). Indeed, if

(1c) is satisfied, then, by I.6.1, (1b) also holds. Now let (1b) be satisfied, and therefore (1a) also hold, and let

$$X \triangleq \mathscr{T}_n' \times R^{n+1}, \qquad \Gamma(t) \triangleq \mathscr{T}_n' \times \bar{R}^{\#}(t)^{n+1} \quad (t \in T)$$

and

$$\phi(t, v, x) \triangleq \sum_{j=0}^{n} \theta^j \hat{f}(t, v, r_j) \quad [(t, v) \in T \times V,\ x \triangleq (\theta^0, \ldots, \theta^n, r_0, \ldots, r_n) \in X].$$

Then, by I.7.5, Γ is measurable, and (1c) follows from I.6.2, (1a), and statement (2). QED

VI.3.3 ***Theorem*** (Filippov) Let the conditions of VI.1.1 be satisfied, $R^{\#}(t)$ be closed for all $t \in T$, and $f(t, v, R^{\#}(t), b)$ convex for all

$$(t, v, b) \in T \times V \times B.$$

Then there exists a point $(\bar{y}, \bar{\rho}, \bar{b}) \in \mathscr{Y} \times \mathscr{R}^{\#} \times B$ that is both a minimizing relaxed solution and a minimizing $\mathscr{R}^{\#}$-solution.

▌PROOF By VI.1.1, there exists a minimizing relaxed solution $(\bar{y}, \bar{\sigma}, \bar{b})$. By VI.3.2(2) [with $X \triangleq R$, $\Gamma(t) \triangleq R^{\#}(t)$, and $\phi(t, v, r) \triangleq f(t, v, r, \bar{b})$], there exists a measurable selection $\bar{\rho}$ of $R^{\#}$ (that is, $\bar{\rho} \in \mathscr{R}^{\#}$) such that

$$\dot{\bar{y}}(t) = f(t, \bar{y}(t), \bar{\rho}(t), \bar{b}) \quad \text{a.e. in } T.$$

Since h_0, h_1, and h_2 depend only on t, y, and b, it follows that $(\bar{y}, \bar{\rho}, \bar{b})$ is a minimizing relaxed solution and, since $\bar{\rho} \in \mathscr{R}^{\#}$, also a minimizing $\mathscr{R}^{\#}$-solution. QED

VI.4 Unbounded Contingent Sets and Compactified Parametric Problems

VI.4.1 ***General Remarks*** The problems that we have so far considered in this chapter (that is, those defined in VI.0) have the property that the set $f(t, v, R^{\#}(t), b)$ (referred to in the terminology of Ważewski [1, 2] as the *contingent set*) is bounded for each fixed $(v, t) \in V \times T$. This is due to the fact that R is compact and $f(t, v, \cdot, b)$ is assumed continuous. In this section, we shall replace the compact metric space R by a topological space R^{top} (which need not be compact) and will consider original control functions with values in R^{top}. We shall investigate a class of problems for which the sets $f(t, v, R^{\text{top}}, b)$ need no longer be bounded and assumption VI.1.1(1) may be violated. If $\mathscr{H}$ and $\mathscr{A}$ are defined in a manner analogous to $\mathscr{H}(\mathscr{R})$ and $\mathscr{A}(\mathscr{R})$ (with R replaced by R^{top}), then it is no longer true that the set

$\{y \in C(T, \mathbb{R}^n) | (y, \rho, b) \in \mathscr{A}\}$ is equicontinuous and, in many cases, every minimizing approximate solution $((y_j, \rho_j, b_j))$ yields a set $\{y_1, y_2, ...\}$ in $C(T, \mathbb{R}^n)$ which is not conditionally compact. In fact, the sequence (y_j) may converge a.e. to a discontinuous function (see Example VI.4.7 below).

It is clear that, under such conditions, the existence theorems VI.1.1, VI.1.3 and VI.3.3 fail to be applicable and their conclusions are invalid. On the other hand, as we shall show in this section, it is possible to demonstrate under fairly general conditions the existence of a certain kind of minimizing approximate solutions. Thus our problem has a solution but the tools that we have employed so far for determining such a solution (relaxed controls) are by themselves no longer equal to the task.

We have discussed in Chapter III the reasons for considering minimizing approximate $\mathscr{U}$-solutions and have shown subsequently that it is appropriate to embed the original control functions in a larger set $\mathscr{S}^{\#}$ of relaxed control functions in order to "create" minimizing solutions of a new kind, whose knowledge enables us to determine a minimizing approximate $\mathscr{U}$-solution. In the present case, we must enlarge not only the set of control functions but also that of state functions. We accomplish the latter task by embedding our state functions in a set of "curve parametrizations." The latter are a standard tool for the study of geometric curves, that is, images of continuous mappings on a closed (finite) interval.

If $T \triangleq [t_0, t_1] \subset \mathbb{R}$, $y \triangleq (y^1, ..., y^n) : T \to \mathbb{R}^n$ is continuous, and

$$\tau(s) \triangleq t_0 + (t_1 - t_0)s \qquad (s \in [0, 1]),$$

then the function $\tilde{y} \triangleq (\tau, y \circ \tau)$ is also continuous and $(\tau, y \circ \tau)([0, 1])$ is the graph M_y of y. More generally, if $\tau : [0, 1] \to T$ is a continuous and nondecreasing surjection and $\tilde{y} \triangleq (\tau, y \circ \tau)$, then $\tilde{y}([0, 1])$ is also the graph of y. Thus the graph of y can be represented in many different ways as a continuous image of $[0, 1]$, and each such representation corresponding to a choice of a continuous function

$$\tilde{y} \triangleq (\tilde{y}^0, \tilde{y}^1, ..., \tilde{y}^n) \triangleq (\tau, \eta) : [0, 1] \to T \times \mathbb{R}^n$$

with $\tau : [0, 1] \to T$ a nondecreasing surjection is referred to as a *parametrization* of M_y. Now let $\mathscr{M}$ be the collection of subsets M of $\mathbb{R} \times \mathbb{R}^n$ such that $M = \tilde{y}([0, 1])$ for some continuous $\tilde{y} \triangleq (\tau, \eta) : [0, 1] \to \mathbb{R} \times \mathbb{R}^n$ with $\tau : [0, 1] \to \mathbb{R}$ a nondecreasing function. Then we refer to $\tilde{y}$ as a *parametrization* of M and our previous remarks show that $C(T, \mathbb{R}^n)$ can be embedded in $\mathscr{M}$ by identifying each $y \in C(T, \mathbb{R}^n)$ with its graph M_y. An element $M \in \mathscr{M}$ corresponds to some $y \in C(T, \mathbb{R}^n)$ (i.e., M is the graph of y) if and only if M has a parametrization $\tilde{y}_M \triangleq (\tau_M, \eta_M) : [0, 1] \to M$ such that $\tau_M([0, 1]) = T$ and $\tau_M : [0, 1] \to T$ is a bijection (and therefore increasing and a homeomorphism of $[0, 1]$ onto T); then we have $y = \eta_M \circ \tau_M^{-1}$. If an element $M \in \mathscr{M}$ has a parametrization $\tilde{y}_M \triangleq (\tau_M, \eta_M) : [0, 1] \to M$ such that τ_M is an injection, but $\tau_M([0, 1]) = [t_0^M, t_1^M] \neq T$, then M is the graph of some continuous function on $[t_0^M, t_1^M]$ to $\mathbb{R}^n$. There exist, however, elements $M \in \mathscr{M}$ that are not graphs of any function on a closed subinterval of $\mathbb{R}$ to $\mathbb{R}^n$; if M has a parametrization $\tilde{y}_M \triangleq (\tau_M, \eta_M)$ such that τ_M is constant on some subinterval $[\alpha, \beta]$ of $[0, 1]$, but η_M is not constant on $[\alpha, \beta]$, then M cannot be the graph of a function on a subset of $\mathbb{R}$ to $\mathbb{R}^n$.

Having embedded $C(T, \mathbb{R}^n)$ in the larger set $\mathscr{M}$, we introduce an appropriate concept of convergence in $\mathscr{M}$. We say that a sequence (M_j) in $\mathscr{M}$ *converges to* M if there exist parametrizations $\tilde{y}_j$ and $\tilde{y}$ of M_j and M, respectively, such that

$$\lim_j \tilde{y}_j(s) = \tilde{y}(s) \quad \text{uniformly for} \quad s \in [0, 1].$$

We extend the concept of an approximate solution by admitting into competition not only state functions y and control functions ρ defined on $T \triangleq [t_0, t_1]$ but also those defined on some interval $[t_0, \alpha]$. Then the requirement that $\alpha = t_1$ becomes an additional "desirable" condition [similar to $g_1(y, \rho, b) = 0$ and $g_2(y, \rho, b) \in C_2$] and treated in an analogous manner. Under certain conditions (that we discuss in Theorem VI.4.5), we are able to show the existence of a *regular* minimizing approximate solution, that is, one involving state functions and control functions defined on $T \triangleq [t_0, t_1]$ alone.

We shall consider problems for which there exists some $L < \infty$ such that $\int_{t_0}^{\alpha} |\dot{y}(t)|\, dt \leqslant L$ for all (y, ρ, b, α) that satisfy the preassigned equation of motion, that come sufficiently close to satisfying the "desirable" restrictions, and that render the cost functional sufficiently small. We shall define an appropriate parametrization of all such functions y and determine a "limit" of a minimizing sequence in terms of a minimizing relaxed solution of a related "compactified parametric problem." This parametrization is obtained by choosing an appropriate function $\phi : V \times R \times B \to \mathbb{R}$, setting

$$\theta_1(t) \triangleq \int_{t_0}^{t} \phi(y(\tau), \rho(\tau), b)\, d\tau \qquad (t \in [0, \alpha]),$$

and defining $\tau(\cdot)$ as the inverse function of $\theta(t_1)^{-1}\theta_1(\cdot)$. The function θ_1 is related to the length of the graph of y between $y(t_0)$ and $y(t)$.

We shall assume in this section that the ordinary differential equation that we wish to control is of the form

$$y(t) = v_0 + \int_{t_0}^{t} f(y(\alpha), \rho(\alpha), b)\, d\alpha \qquad (t \in T),$$

the argument t in $f(t, v, r, b)$ having been "transformed out" in the manner described in II.4.A. For the sake of greater notational simplicity we shall also assume that $T = [0, 1]$, which can be accomplished by another simple transformation. Finally, we consider a unilateral restriction of the form $h_2(y(t), b) \in A$ $(t \in T)$ instead of the more general unilateral restriction of the form $h_2(t, y(t), b) \in A(t)$ $(t \in T)$ considered in VI.1–VI.3.

VI.4.2 ***Formulation of the Problem and Definitions*** Let R^{top} be a topological space, $n, m, m_2 \in \mathbb{N}$, $v_0 \in \mathbb{R}^n$, A a closed subset of $\mathbb{R}^{m_2}$, B a sequentially compact space, and the functions f and g continuous, where

$$f : \mathbb{R}^n \times R^{\text{top}} \times B \to \mathbb{R}^n,$$

and

$$h \triangleq (h_0, h_1, h_2) : \mathbb{R}^n \times B \to \mathbb{R} \times \mathbb{R}^m \times \mathbb{R}^{m_2}.$$

We denote by $\mathscr{H}^{\text{top}}$ the collection of 4-tuples (y, ρ, b, α) such that $\alpha \in [0, \infty)$, $y \in C([0, \alpha], \mathbb{R}^n)$, $\rho : [0, \alpha] \to R^{\text{top}}$ is measurable, $b \in B$, and

$$y(t) = v_0 + \int_0^t f(y(\tau), \rho(\tau), b)\, d\tau \qquad (t \in [0, \alpha]).$$

We also write for all $\epsilon \geqslant 0$,

$$\mathscr{A}_\epsilon^{\text{top}} \triangleq \{(y, \rho, b, \alpha) \in \mathscr{H}^{\text{top}} \mid |\alpha - 1| + |h_1(y(\alpha), b)| \leqslant \epsilon,$$
$$d[h_2(y(t), b), A] \leqslant \epsilon \ (t \in [0, \alpha])\}.$$

A sequence $((y_j, \rho_j, b_j, \alpha_j))$ in $\mathscr{H}^{\text{top}}$ is an *approximate* R^{top}-solution if for every $\epsilon > 0$ there exists $j(\epsilon) \in \mathbb{N}$ such that $(y_j, \rho_j, b_j, \alpha_j) \in \mathscr{A}_\epsilon^{\text{top}}$ $[j \geqslant j(\epsilon)]$. An approximate R^{top}-solution $((y_j, \rho_j, b_j, \alpha_j))$ is *regular* if $\alpha_j = 1$ $(j \in \mathbb{N})$. An approximate R^{top}-solution $((\bar{y}_j, \bar{\rho}_j, \bar{b}_j, \bar{\alpha}_j))$ is a *minimizing approximate* R^{top}-solution if

$$\lim_j h_0(\bar{y}_j(\alpha_j), \bar{b}_j) \leqslant \liminf_j h_0(y_j(\alpha_j), b_j)$$

for every approximate R^{top}-solution $((y_j, \rho_j, b_j, \alpha_j))$. Finally, we say that an approximate R^{top}-solution $((y_j, \rho_j, b_j, \alpha_j))$ is an *exact* R^{top}*-solution* if $(y_j, \rho_j, b_j, \alpha_j) = (y, \rho, b, 1)$ $(j \in \mathbb{N})$ for some (y, ρ, b), and we identify this exact R^{top}-solution with the point (y, ρ, b).

We restrict ourselves to the consideration of problems satisfying the following assumption.

(1) *Assumption* There exist numbers $\gamma \in (0, \frac{1}{2}]$ and $L \in (0, \infty)$ and an approximate R^{top}-solution $((y_j^{\#}, \rho_j^{\#}, b_j^{\#}, \alpha_j^{\#}))$ such that

$$\int_0^\alpha |\dot{y}(t)|\, dt \leqslant L \qquad \text{for all } (y, \rho, b, \alpha) \in \mathscr{A}^0,$$

where

$$\mathscr{A}^0 \triangleq \{(y, \rho, b, \alpha) \in \mathscr{A}_\gamma^{\text{top}} \mid h_0(y(\alpha), b) \leqslant \gamma + \lim_j \inf h_0(y_j^{\#}(\alpha_j), b_j)\}.$$

THE FUNCTION ϕ AND THE SET $\tilde{\mathscr{A}}^0$ We next set $V_L \triangleq S(v_0, L+1)$ and choose a continuous function

$$\phi : V_L \times R^{\text{top}} \times B \to \mathbb{R}$$

and numbers $\beta_{\min}$ and $\beta_{\max}$ such that

$$(2) \quad 0 < 4\beta_{\min} \leqslant \phi(v, r, b)[|f(v, r, b)| + 1]^{-1} \leqslant \tfrac{1}{4}\beta_{\max}(L+1)^{-1} < \infty \qquad [v \in V_L, \quad (r, b) \in R^{\text{top}} \times B].$$

(We may choose, for example, $\phi(v, r, b) = |f(v, r, b)| + 1$, $\beta_{\min} = \frac{1}{4}$, and $\beta_{\max} = 4(L+1)$ but, as will be seen in Examples VI.4.6 and VI.4.7, it may be more convenient to choose ϕ differently.)

We denote by $\tilde{\mathscr{A}}^0$ the collection of all 5-tuples $(\tau, \eta, u, b, \beta)$ such that

$$u : [0, 1] \to R^{\text{top}} \text{ is measurable}, \qquad b \in B, \quad \beta \in (0, \infty),$$

$$h_0(\eta(1), b) \leqslant \gamma + \lim_j \inf h_0(y_j^{\#}(1), b_j^{\#}),$$

$$(3) \quad \tau(s) = \int_0^s \beta[\phi(\eta(\theta), u(\theta), b)]^{-1}\, d\theta \qquad (s \in [0, 1])$$

$$\eta(s) = v_0 + \int_0^s \beta[\phi(\eta(\theta), u(\theta), b)]^{-1} f(\eta(\theta), u(\theta), b)\, d\theta \qquad (s \in [0, 1]),$$

$$|\tau(1) - 1| + |h_1(\eta(1), b)| \leqslant \gamma, \qquad d[h_2(\eta(s), b), A] \leqslant \gamma \quad (s \in [0, 1]).$$

THE COMPACTIFIED PARAMETRIC PROBLEM Now let $(\tilde{R}, \tilde{R}', \Phi)$ be a metric compactification of R^{top},

$$(4) \qquad \tilde{v}_0 \triangleq (0, v_0), \qquad \tilde{B} \triangleq B \times [\beta_{\min}, \beta_{\max}], \qquad \tilde{V} \triangleq \mathbb{R} \times V_L,$$

and, for all $\tilde{v} \triangleq (v^0, v) \in \tilde{V}$, $\tilde{r} \in \tilde{R}'$, and $\tilde{b} \triangleq (b, \beta) \in \tilde{B}$, let

$$\tilde{f}(\tilde{v}, \tilde{r}, \tilde{b}) \triangleq \beta[\phi(v, \Phi^{-1}(\tilde{r}), b)]^{-1}\,(1, f(v, \Phi^{-1}(\tilde{r}), \tilde{b})),$$

$$(5) \qquad \tilde{h}_0(\tilde{v}, \tilde{b}) \triangleq h_0(v, b),$$

$$h_1(\tilde{v}, \tilde{b}) \triangleq (v^0 - 1, h_1(v, b)), \qquad \text{and} \qquad \tilde{h}_2(\tilde{v}, \tilde{b}) \triangleq h_2(v, b).$$

Assume, furthermore, that the function $\tilde{f} : \tilde{V} \times \tilde{R}' \times \tilde{B} \to \mathbb{R}^{n+1}$ has a continuous extension to $\tilde{V} \times \tilde{R} \times \tilde{B}$ which we continue to designate as $\tilde{f}$.

We set $\tilde{R}^{\#}(s) \triangleq \tilde{R}'$ $(s \in [0, 1])$ and define $\tilde{\mathscr{R}}^{\#}$ and $\tilde{\mathscr{S}}^{\#}$ exactly as we have defined $\mathscr{R}^{\#}$ and $\mathscr{S}^{\#}$ in Chapter IV but with T, R, and $R^{\#}$ replaced by $[0, 1]$,

$\tilde{R}$, and $\tilde{R}^{\#}$, respectively. Then we consider the problem defined in VI.0 but with T, R, $\mathscr{R}^{\#}$, $\mathscr{S}^{\#}$, V, v_0, B, f, and h replaced by $[0, 1]$, $\tilde{R}$, $\tilde{\mathscr{R}}^{\#}$, $\tilde{\mathscr{S}}^{\#}$, $\tilde{V}$, $\tilde{v}_0$, $\tilde{B}$, $\tilde{f}$, and $\tilde{h}$, respectively. We refer to this problem as the *compactified parametric problem.*

In order to prove Theorems VI.4.4 and VI.4.5 which contain the basic results of this section, we first require a lemma.

VI.4.3 ***Lemma*** Let the assumptions of VI.4.2 be satisfied. Then

(1) for each $(\tau, \eta, u, b, \beta) \in \tilde{\mathscr{A}}^0$, τ is a homeomorphism of $[0, 1]$ onto $[0, \tau(1)]$, τ^{-1} is absolutely continuous, $\eta([0, 1]) \subset S^F(v_0, L)$, and

$$\beta \in [2\beta_{\min}, \tfrac{1}{2}\beta_{\max}];$$

(2) there exists a bijection of $\mathscr{A}^0$ onto $\tilde{\mathscr{A}}^0$ such that the corresponding points $(y, \rho, b, \alpha) \in \mathscr{A}^0$ and $(\tau, \eta, u, b, \beta) \in \tilde{\mathscr{A}}^0$ satisfy the relations

$$\eta = y \circ \tau, \qquad u = \rho \circ \tau, \qquad \text{and} \qquad \tau(1) = \alpha;$$

and

(3) under the bijection of (2), the set of approximate R^{top}-solutions $((y_j, \rho_j, b_j, \alpha_j))$ in $\mathscr{A}^0$ is mapped onto the set of sequences

$$((\tau_j, \eta_j, u_j, b_j, \beta_j))$$

in $\mathscr{A}^0$ such that

$$\lim_j \tau_j(1) = 1, \qquad \lim_j h_1(\eta_j(1), b_j) = 0$$

and

$$\lim_j d[h_2(\eta_j(s), b_j), A] = 0 \quad \text{uniformly for} \quad s \in [0, 1].$$

▌PROOF *Step* 1 For each $(y, \rho, b, \alpha) \in \mathscr{A}^0$, we define

(4) $$\theta_1(t) \triangleq \int_0^t \phi(y(t'), \rho(t'), b)\, dt' \quad \text{and} \quad \theta(t) \triangleq \theta_1(\alpha)^{-1}\theta_1(t) \quad (t \in [0, \alpha]).$$

The function θ_1 is absolutely continuous and increasing because, by VI.4.2(2),

(5) $$4\beta_{\min} \leqslant \phi(y(t), \rho(t), b)[1 + |\dot{y}(t)|]^{-1} \leqslant \tfrac{1}{4}\beta_{\max}(L + 1)^{-1}.$$

It follows that θ is an increasing absolutely continuous bijection of $[0, \alpha]$ onto $[0, 1]$. We set

(6) $$\tau \triangleq \theta^{-1}, \qquad \eta \triangleq y \circ \tau, \qquad u \triangleq \rho \circ \tau,$$

and observe that, by I.4.43, τ is increasing and absolutely continuous, and $\tau^{-1}(E)$ measurable for every measurable $E \subset [0, \alpha]$. Therefore u is measurable and, by I.4.42(3), η is absolutely continuous. Since $\dot{\theta}(t)$ exists a.e. in $[0, \alpha]$, say for $t \in T'$, and $\dot{\theta}(t) > 0$ $(t \in T')$, it follows from (6), II.3.11, and the chain rule II.3.4 that $\theta(T')$ has measure $\theta(\alpha) - \theta(0) = 1$,

$$\text{(7)} \qquad \dot{\tau}(s) = [\dot{\theta}(\tau(s))]^{-1} \text{ and } \dot{\eta}(s) = [\dot{\theta}(\tau(s))]^{-1}\dot{y}(\tau(s)) \text{ a.e. in } [0, 1].$$

We set $\beta \triangleq \theta_1(\alpha)$ and observe that, by VI.4.2(1) and VI.4.2(2), we have $y([0, \alpha]) \subset S^F(v_0, L)$ and

$$|\phi(y(\tau), \rho(\tau), b)| \geqslant 4\beta_{\min};$$

hence

$$\beta \triangleq \theta_1(\alpha) \geqslant 4\beta_{\min} \cdot \alpha \geqslant 2\beta_{\min}.$$

Furthermore,

$$\text{(8)} \quad \dot{y}(t) = f(y(t), \rho(t), b) \quad \text{a.e. in } [0, \alpha] \qquad \text{and} \qquad \textstyle\int_0^\alpha |\dot{y}(t)|\, dt \leqslant L;$$

and therefore, by (5) and VI.4.2(2),

$$\begin{aligned} \beta \triangleq \theta_1(\alpha) &\leqslant \tfrac{1}{4}\beta_{\max} \cdot (L+1)^{-1} \int_0^\alpha (1 + |\dot{y}(t)|)\, dt \\ &\leqslant \tfrac{1}{4}\beta_{\max} \cdot (L+1)^{-1} (\tfrac{3}{2} + L) < \tfrac{1}{2}\beta_{\max}. \end{aligned}$$

Relations (4) and (6)–(8) now show that $(\tau, \eta, u, b, \beta)$ satisfies the relations

$$\text{(9)} \quad \begin{aligned} &\tau(s) = \int_0^s \beta[\phi(\eta(t'), u(t'), b)]^{-1}\, dt' \qquad (s \in [0, 1]), \\ &\eta(s) = v_0 + \int_0^s \beta[\phi(\eta(t'), u(t'), b)]^{-1} f(\eta(t'), u(t'), b)\, dt' \qquad (s \in [0, 1]), \\ &\beta \in [2\beta_{\min}, \tfrac{1}{2}\beta_{\max}), \\ &\tau(1) = \alpha, \qquad \eta(1) = y(\alpha), \qquad \eta([0, 1]) = y([0, \alpha]) \subset S^F(v_0, L). \end{aligned}$$

Thus $(\tau, \eta, u, b, \beta) \in \tilde{\mathscr{A}}^0$.

Step 2 Now let $(\tau, \eta, u, b, \beta) \in \tilde{\mathscr{A}}^0$. Then, by VI.4.2(2) and VI.4.2(3), we have $\dot{\tau}(s) > 0$ a.e. in $[0, 1]$, τ is increasing, and τ is a bijection of $[0, 1]$ onto $[0, \tau(1)]$. It follows, by I.4.42(3) and I.4.43, that τ^{-1} and $\eta \circ \tau^{-1}$ are absolutely continuous and $u \circ \tau^{-1}$ measurable. This proves the first two assertions of statement (1).

If we set

$$\text{(10)} \qquad \theta \triangleq \tau^{-1}, \qquad y \triangleq \eta \circ \theta, \qquad \rho \triangleq u \circ \theta, \qquad \text{and} \qquad \alpha \triangleq \tau(1),$$

then y is absolutely continuous and VI.4.2(3) (together with II.3.11 and II.3.4) implies that

$$\dot{y}(t) = \dot{\eta}(\theta(t))/\dot{\tau}(\theta(t)) = f(\eta \circ \theta(t), u \circ \theta(t), b)$$
$$= f(y(t), \rho(t), b) \quad \text{a.e. in } [0, \tau(1)],$$
$$y(0) = v_0, \qquad y(\alpha) = \eta(1).$$

Thus $(y, \rho, b, \alpha) \in \mathscr{A}^0$.

Finally, if (y, ρ, b, α) is chosen arbitrarily in $\mathscr{A}^0$ and the procedure of Step 1 is used to determine the corresponding point

$$(\tau, \eta, u, b, \beta) \in \tilde{\mathscr{A}}^0,$$

then the comparison of (6) and (10) shows that the latter point yields back the same (y, ρ, b, α). Thus the correspondence that we have established is one-to-one. This proves statement (2). Statement (3) now follows from relation (9) which also shows that $\eta([0, 1]) \subset S^F(v_0, L)$ and

$$\beta \in [2\beta_{\min}, \tfrac{1}{2}\beta_{\max}],$$

thus completing the proof of (1). QED

VI.4.4 ***Theorem*** Let the assumptions of VI.4.2 be satisfied, and assume furthermore that there exists $c \in \mathbb{R}$ such that

$$|\tilde{f}(\tilde{v}', \tilde{r}, \tilde{b}) - \tilde{f}(\tilde{v}'', \tilde{r}, \tilde{b})| \leqslant c\,|\tilde{v}' - \tilde{v}''| \qquad (\tilde{v}', \tilde{v}'' \in \tilde{V}, (\tilde{r}, \tilde{b}) \in \tilde{R} \times \tilde{B}).$$

Then the compactified parametric problem has a minimizing relaxed solution $(\bar{\tilde{y}}, \bar{\tilde{\sigma}}, \bar{\tilde{b}})$ and a corresponding minimizing approximate $\tilde{\mathscr{R}}^{\#}$-solution $((\tilde{y}_j, \tilde{\rho}_j, \bar{\tilde{b}}))$ such that

$$\bar{\tilde{b}} \triangleq (\bar{b}, \bar{\beta}), \qquad \bar{\beta} \in [2\beta_{\min}, \tfrac{1}{2}\beta_{\max}],$$

and

$$\lim_j (\tilde{y}_j, \tilde{\rho}_j) = (\bar{\tilde{y}}, \bar{\tilde{\sigma}}) \quad \text{in } C([0, 1], \mathbb{R}^{n+1}) \times \tilde{\mathscr{S}}^{\#}.$$

Furthermore, a minimizing approximate R^{top}-solution $((y_j, \rho_j, \bar{b}, \alpha_j))$ can be constructed out of $((\tilde{y}_j, \tilde{\rho}_j, \bar{\tilde{b}}))$ by setting

$$\tilde{y}_j \triangleq (\tilde{y}_j^0, \tilde{y}_j^1, \ldots, \tilde{y}_j^n),$$
$$u_j = \Phi^{-1} \circ \tilde{\rho}_j, \qquad \tau_j \triangleq \tilde{y}_j^0, \qquad \eta_j \triangleq (\tilde{y}_j^1, \ldots, \tilde{y}_j^n),$$
$$\rho_j \triangleq u_j \circ \tau_j^{-1}, \qquad y_j \triangleq \eta_j \circ \tau_j^{-1}, \qquad \alpha_j \triangleq \tau_j(1) \qquad (j \in \mathbb{N}).$$

▌PROOF *Step* 1 Our first objective will be to show that $\mathscr{A}(\tilde{\mathscr{S}}^{\#}) \neq \varnothing$. To do so, we consider the approximate R^{top}–solution $((y_j^{\#}, \rho_j^{\#}, b_j^{\#}, \alpha_j^{\#}))$

in $\mathscr{A}^0$ whose existence was postulated in VI.4.2(1). By VI.4.3, there exists a corresponding sequence $((\tau_j^\#, \eta_j^\#, u_j^\#, b_j^\#, \beta_j^\#))$ in $\tilde{\mathscr{A}}^0$ such that

$$\eta_j^\#([0,1]) \subset S^F(v_0, L), \qquad \lim_j \tau_j^\#(1) = 1, \qquad \lim_j h_1(\eta_j(1), b_j) = 0,$$
$$\lim_j d[h_2(\eta_j^\#(s), b_j^\#), A] = 0 \quad \text{uniformly} \qquad \text{for} \quad s \in [0,1].$$

If we set

$$\tilde{y}_j^\# \triangleq (\tau_j^\#, \eta_j^\#), \qquad \tilde{\rho}^\# \triangleq \Phi \circ u_j^\#, \qquad \tilde{b}_j^\# \triangleq (b_j^\#, \beta_j^\#) \qquad (j \in \mathbb{N}),$$

then we observe that $((\tilde{y}_j^\#, \tilde{\rho}_j^\#, \tilde{b}_j^\#))$ is an approximate $\tilde{\mathscr{R}}^\#$-solution of the compactified parametric problem. For all $j \in \mathbb{N}$ and a.a. $s \in [0,1]$, we have

$$\begin{aligned}
&\tilde{y}_j^\#([0,1]) \subset [0, 1+\gamma] \times S^F(v_0, L),\\
&\quad |\dot{\tilde{y}}_j^\#(s)| = |\tilde{f}(\tilde{y}_j^\#(s), \tilde{\rho}_j^\#(s), \tilde{b}_j^\#)|\\
&\qquad\qquad \leqslant \sup\{|\tilde{f}(\tilde{v}, \tilde{r}, \tilde{b})| \mid (\tilde{v}, \tilde{r}, \tilde{b}) \in [0, 1+\gamma] \times S^F(v_0, L)\} < \infty.
\end{aligned}$$

This shows that the set $\{\tilde{y}_j \mid j \in \mathbb{N}\}$ is bounded and equicontinuous in $C([0,1], \tilde{V})$ and therefore, by Ascoli's theorem I.5.4, there exist $J_1 \subset (1, 2, ...)$ and $\tilde{y}^\# \in C([0,1], \tilde{V})$ such that $\lim_{j \in J_1} y_j^\# = \tilde{y}^\#$. Since $\tilde{\mathscr{S}}^\#$ and $\tilde{B}$ are sequentially compact, there exist $\tilde{\sigma}^\# \in \tilde{\mathscr{S}}^\#$, $\tilde{b}^\# \in \tilde{B}$, and $J \subset J_1$ such that

$$\lim_{j \in J} (\tilde{y}_j^\#, \tilde{\rho}_j^\#, \tilde{b}_j^\#) = (\tilde{y}^\#, \tilde{\sigma}^\#, \tilde{b}^\#).$$

By VI.4.2(2), VI.4.2(5), and VI.4.3(1), we have

$$(1) \qquad |\tilde{f}(\tilde{v}, \tilde{r}, \tilde{b})| \leqslant \tfrac{1}{8}\beta_{\min}^{-1} \cdot \beta_{\max} \qquad ((\tilde{v}, \tilde{r}, \tilde{b}) \in \tilde{V} \times \tilde{R} \times \tilde{B}).$$

Thus it follows from IV.2.9 that $(\tilde{y}^\#, \tilde{\sigma}^\#, \tilde{b}^\#) \in \mathscr{H}(\mathscr{S}^\#)$. Finally, since $((\tilde{y}_j^\#, \tilde{\rho}_j^\#, \tilde{b}_j^\#))$ is an approximate $\tilde{\mathscr{R}}^\#$-solution, we conclude that $(\tilde{y}^\#, \tilde{\sigma}^\#, \tilde{b}^\#) \in \mathscr{A}(\mathscr{S}^\#)$ and therefore $\mathscr{A}(\mathscr{S}^\#) \neq \varnothing$.

Step 2 Let $(\tilde{y}, \tilde{\sigma}, (b, \beta)) \in \mathscr{A}(\tilde{\mathscr{S}}^\#)$. Then, by (1) and VI.1.4, there exists a sequence $((\tilde{y}_j, \tilde{\rho}_j, (b, \beta)))$ in $\mathscr{H}(\tilde{\mathscr{R}}^\#)$ such that $\lim_j(\tilde{y}_j, \tilde{\rho}_j) = (\tilde{y}, \tilde{\sigma})$, and therefore this sequence is an approximate $\tilde{\mathscr{R}}^\#$-solution. We verify that, for large j, $(\tilde{y}_j^0, (\tilde{y}_j^1, ..., \tilde{y}_j^n), \Phi^{-1} \circ \tilde{\rho}_j, b, \beta) \in \tilde{\mathscr{A}}^0$ and therefore, by VI.4.3(1), $\tilde{y}([0,1] \subset \mathbb{R} \times S^F(v_0, L) \subset \tilde{V}$. It follows now from (1) and VI.1.5 that there exist a minimizing relaxed solution $(\bar{\tilde{y}}, \bar{\tilde{\sigma}}, \bar{\tilde{b}})$ of the compactified parametric problem and a corresponding minimizing approximate $\tilde{\mathscr{R}}^\#$-solution $((\tilde{y}_j, \tilde{\rho}_j, \bar{\tilde{b}}))$ such that

$$\lim_j \tilde{h}_0(\tilde{y}_j(1), \bar{\tilde{b}}) = \tilde{h}_0(\bar{\tilde{y}}(1), \bar{\tilde{b}}).$$

Let

$$\tilde{y}_j \triangleq (\tilde{y}_j^0, \tilde{y}_j^1, \ldots, \tilde{y}_j^n), \qquad \tau_j \triangleq \tilde{y}_j^0, \qquad \eta_j \triangleq (\tilde{y}_j^1, \ldots, \tilde{y}_j^n),$$
$$u_j \triangleq \Phi^{-1} \circ \tilde{\rho}_j, \qquad \text{and} \qquad \bar{\tilde{b}} \triangleq (\bar{b}, \bar{\beta}) \qquad (j \in \mathbb{N}).$$

Then u_j are measurable functions on $[0, 1]$ to R^{top} $[= \Phi^{-1}(\tilde{R}')]$ and there exists $j_1 \in \mathbb{N}$ such that

$$h_0(\eta_j(1), \bar{b}) \leqslant \gamma + \lim_j \inf h_0(y_j^{\#}(\alpha_j), b_j^{\#}),$$

$$| \tau_j(1) - 1 | + | h_1(\eta_j(1), \bar{b})| \leqslant \gamma,$$

and

$$d[h_2(\eta_j(s), \bar{b}), A] \leqslant \gamma \qquad (j \geqslant j_1, \quad s \in [0, 1]).$$

We conclude that the sequence $((\tau_j, \eta_j, u_j, \bar{b}, \bar{\beta}))_{j \geqslant j_1}$ is in $\tilde{\mathscr{A}}^0$. It follows then from VI.4.3(1) that

$$\bar{\beta} \in [2\beta_{\min}, \tfrac{1}{2}\beta_{\max}].$$

Furthermore, the bijection referred to in VI.4.3(2) yields

$$\rho_j = u_j \circ \tau_j^{-1}, \qquad y_j = \eta_j \circ \tau_j^{-1}, \qquad \alpha_j = \tau_j(1) \qquad (j \in \mathbb{N}),$$

and by VI.4.3(3), $((y_j, \rho_j, \bar{b}, \alpha_j))$ is an approximate R^{top}–solution about which we assume that $\lim h_0(y_j(\alpha_j), \bar{b})$ exists [otherwise replacing

$$((y_j, \rho_j, \bar{b}, \alpha_j))$$

by an appropriate subsequence]. It follows then that

$$\text{(2)} \qquad \lim_j h_0(y_j(\alpha_j), \bar{b}) = \lim h_0(\eta_j(1), \bar{b}) = \tilde{h}_0(\bar{\tilde{y}}(1), \bar{\tilde{b}}).$$

To show that $((y_j, \rho_j, \bar{b}, \alpha_j))$ is a minimizing approximate R^{top}-solution, it suffices to compare $((y_j, \rho_j, \bar{b}, \alpha_j))$ with approximate R^{top}-solutions in $\mathscr{A}^0$. We consider therefore an arbitrary approximate R^{top}-solution $((\hat{y}_j, \hat{\rho}_j, \hat{b}_j, \hat{\alpha}_j))$ in $\mathscr{A}^0$ and the corresponding sequence $((\hat{\tau}_j, \hat{\eta}_j, \hat{u}_j, \hat{b}_j, \hat{\beta}_j))$ in $\tilde{\mathscr{A}}^0$ defined in VI.4.3(3). Then $((\hat{\tau}_j, \hat{\eta}_j), \Phi \circ \hat{u}_j, (\hat{b}_j, \hat{\beta}_j))$ is an approximate $\tilde{\mathscr{R}}^{\#}$-solution and therefore

$$\text{(3)} \qquad \lim_j h_0(\eta_j(1), \bar{b}) = \lim_j \tilde{h}_0(\tilde{y}_j(1), \bar{\tilde{b}})$$
$$\leqslant \lim_j \inf \tilde{h}_0((\hat{\tau}_j, \hat{\eta}_j)(1), (\hat{b}_j, \hat{\beta}_j))$$
$$= \lim_j \inf h_0(\hat{\eta}_j(1), \hat{b}_j) = \lim_j \inf h_0(\hat{y}_j(\hat{\alpha}_j), \hat{b}_j).$$

It follows now from (2) and (3) that

$$\lim_j h_0(y_j(\alpha_j), \bar{b}) \leqslant \lim_j \inf h_0(\hat{y}_j(\hat{\alpha}_j), \hat{b}_j),$$

thus showing that $((y_j, \rho_j, \bar{b}, \alpha_j))$ is a minimizing approximate R^{top}-solution. QED

VI.4.5 ***Theorem*** Let $((y_j, \rho_j, \bar{b}, \alpha_j))$ be a minimizing approximate R^{top}-solution, and assume that the conditions of VI.4.2 are satisfied and there exists $c_1 \in \mathbb{R}$ such that

(1) $|f(v', r, \bar{b}) - f(v'', r, \bar{b})| \leqslant c_1 |v' - v''| \quad [v', v'' \in S(v_0, L+1), r \in R^{\text{top}}].$

Then there exists a regular minimizing approximate R^{top}-solution $((\hat{y}_j, \hat{\rho}_j, \bar{b}, 1))$.

▌PROOF *Step* 1 We may assume that $((y_j, \rho_j, \bar{b}, \alpha_j))$ is in $\mathscr{A}^0$ and

(2) $$L(1 + c_1 e^{c_1}) \cdot |\alpha_j^{-1} - 1| \leqslant \tfrac{1}{2},$$

otherwise replacing $((y_j, \rho_j, \bar{b}, \alpha_j))$ by an appropriate subsequence

$$((y_j, \rho_j, \bar{b}, \alpha_j))_{j \geqslant j_0}.$$

Then, by VI.4.3,

(3) $$y_j([0, \alpha_j]) \subset S^F(v_0, L) \qquad (j \in \mathbb{N}).$$

Now let $j \in \mathbb{N}$ be fixed. We set

$$f_j(t, v) \triangleq f(v, \rho_j(t), \bar{b}) \qquad (t \in [0, \alpha_j], \quad v \in \mathbb{R}^n).$$

Since

(4) $$y_j(t) = v_0 + \int_0^t f(y_j(\tau), \rho_j(\tau), \bar{b})\, d\tau \qquad (t \in [0, \alpha_j]),$$

it follows that the function $\tau \to f_j(\tau, y_j(\tau))$ is integrable on $[0, \alpha_j]$ and, by (1) and (3),

(5) $$\begin{aligned} |f_j(t, v)| &\leqslant |f_j(t, y_j(t))| + c_1 |v - y_j(t)| \\ &\leqslant |f_j(t, y_j(t))| + c_1(2L + 1) \qquad (t \in [0, \alpha_j], \quad v \in S(v_0, L+1)); \end{aligned}$$

hence $\tau \to \sup\{|f_j(\tau, v)| \mid v \in S(v_0, L+1)\}$ is integrable on $[0, \alpha_j]$. Therefore, by II.4.2, either there exists a function $\hat{y}_j : [0, 1] \to S(v_0, L+1)$ such that

(6) $$\hat{y}_j(t) = v_0 + \int_0^t f_j(\alpha_j \tau, \hat{y}_j(\tau))\, d\tau$$

for all $t \in [0, 1]$, or there exist $t_j' \in (0, 1]$ and a function

$$\hat{y}_j : [0, t_j') \to S(v_0, L+1)$$

satisfying (6) for all $t \in [0, t_j')$ and with $\lim_{t \to t_j'} |\hat{y}_j(t) - v_0| = L + 1$.

Step 2 We shall show that the second of these alternatives leads to a contradiction. Assume, indeed, that it is true. If we set

$$\hat{t}_j \triangleq \alpha_j t_j' \qquad \text{and} \qquad x_j(s) \triangleq \hat{y}_j(s/\alpha_j) \quad (s \in [0, \hat{t}_j)),$$

then (by I.4.43) relation (6) for $t \in [0, t_j')$ is equivalent to

$$x_j(s) = v_0 + \alpha_j^{-1} \int_0^s f_j(\tau, x_j(\tau))\, d\tau \qquad (s \in [0, \hat{t}_j)).$$

Since $\hat{t}_j \leqslant \alpha_j$, we also have, by (4),

$$y_j(s) = v_0 + \int_0^s f_j(\tau, y_j(\tau))\, d\tau \qquad (s \in [0, \hat{t}_j)).$$

We set $e(s) \triangleq | x_j(s) - y_j(s)|$ and combine the last two relations with (1) and (3) to obtain

$$\begin{aligned} e(s) &\leqslant \alpha_j^{-1} c_1 \int_0^s e(\tau)\, d\tau + | 1 - \alpha_j^{-1} | \, | y_j(s) - v_0 | \\ &\leqslant \alpha_j^{-1} c_1 \int_0^s e(\tau)\, d\tau + | 1 - \alpha_j^{-1} | L \qquad (s \in [0, \hat{t}_j)). \end{aligned}$$

It follows, by Gronwall's inequality II.4.4, that

$$(7) \qquad e(s) \leqslant | 1 - \alpha_j^{-1} | L[1 + \alpha_j^{-1} c_1 \hat{t}_j \exp(\alpha_j^{-1} c_1 \hat{t}_j)] \qquad (s \in [0, \hat{t}_j));$$

hence, by (2), $e(s) \leqslant \frac{1}{2}$. Thus, by (3),

$$| x_j(s) - v_0 | \leqslant | y_j(s) - v_0 | + e(s) \leqslant L + \tfrac{1}{2} \qquad (s \in [0, \hat{t}_j)),$$

contradicting

$$\lim_{s \to \hat{t}_j} | x_j(s) - v_0 | = \lim_{t \to t_j'} | \hat{y}_j(t) - v_0 | = L + 1.$$

Step 3 We conclude that for all $j \in \mathbb{N}$ there exists

$$\hat{y}_j : [0, 1] \to S(v_0, L + 1)$$

satisfying (6) for all $t \in [0, 1]$. If we set

$$\hat{\rho}_j(t) \triangleq \rho_j(\alpha_j t) \qquad (t \in [0, 1], \quad j \in \mathbb{N}),$$

then it follows from (6) that

$$(8) \qquad \hat{y}_j(t) = v_0 + \int_0^t f(\hat{y}_j(\tau), \hat{\rho}_j(\tau), \bar{b})\, d\tau \qquad (t \in [0, 1], \quad j \in \mathbb{N}).$$

We may repeat the computations of Step 2, with $\hat{t}_j$ replaced by α_j, to show that (7) is valid for all $s \in [0, \alpha_j]$. It follows that

$$\lim_j |x_j(\alpha_j t) - y_j(\alpha_j t)| = \lim_j |\hat{y}_j(t) - y_j(\alpha_j t)|$$
$$= 0 \quad \text{uniformly for} \quad t \in [0, 1].$$

Since $((y_j, \rho_j, \bar{b}, \alpha_j))$ is a minimizing approximate R^{top}-solution, the last relation and (8) show that $((\hat{y}_j, \hat{\rho}_j, \bar{b}, 1))$ is a regular minimizing approximate R^{top}-solution. QED

We shall illustrate the application of Theorem VI.4.4 with two examples. The first one has a minimizing approximate R^{top}-solution which is an exact R^{top}-solution. The other example has a minimizing approximate solution with state functions y_j that converge to a discontinuous function.

VI.4.6 ***Example*** Let

$$n \triangleq 2, \quad m \triangleq 1, \quad v_0 \triangleq (0, 0), \quad 0 < L_1 < \infty, \quad R^{\text{top}} \triangleq [0, \infty),$$
$$h_0(v^1, v^2) \triangleq v^1, \quad h_1(v^1, v^2) \triangleq v^2 - L_1 \quad [v \triangleq (v^1, v^2) \in \mathbb{R}^2],$$

and

$$f^1(v^1, v^2, r) \triangleq (r)^2, \quad f^2(v^1, v^2, r) \triangleq v^1 r \quad [v \triangleq (v^1, v^2) \in \mathbb{R}^2, \quad r \in R^{\text{top}}].$$

The problem we are considering involves no control parameters and no restriction of the form $h_2(y(t), b) \in A$. We shall show that this problem has an exact R^{top}-solution (y, ρ) which is also a minimizing approximate R^{top}-solution, with

$$y \triangleq (y^1, y^2), \quad y^1(t) = [3L_1]^{1/3}, \quad y^2(t) = L_1 t, \quad \rho(t) = [L_1/3]^{1/3}(t)^{-1/3}.$$

We first show that Assumption VI.4.2(1) is satisfied. For all $\alpha \in (0, 3/2]$, the equation

$$y(t) = v_0 + \int_0^t f(y(\tau), \rho(\tau), b)\, d\tau \qquad (t \in [0, \alpha])$$

becomes

$$\text{(1)} \quad y^1(t) = \int_0^t [\rho(\tau)]^2\, d\tau, \qquad y^2(t) = \int_0^t y^1(\tau)\, \rho(\tau)\, d\tau \qquad (t \in [0, \alpha]).$$

We can determine a solution of these equations for any choice of a constant ρ. If $\rho^\#(t) = r^\#$ $(t \in [0, \alpha])$, then the corresponding solution $y^\#$ of (1) is defined by

$$y^{\#1}(t) = (r^\#)^2 \cdot t, \qquad y^{\#2}(t) = \tfrac{1}{2}(r^\#)^3(t)^2$$

and $h_1(y^{\#}(1)) = \frac{1}{2}(r^{\#})^3 - L_1$. Thus $(y^{\#}, \rho^{\#}) \in \mathscr{A}_0^{\text{top}}$ if we choose $r^{\#} = [2L_1]^{1/3}$. It follows then that $((y_j^{\#}, \rho_j^{\#}, 1))$ is an approximate R^{top}-solution when we set $r^{\#} \triangleq [2L_1]^{1/3}$, $y_j^{\#} \triangleq y^{\#}$, $\rho_j^{\#} \triangleq \rho^{\#}$ $(j \in \mathbb{N})$.

We define $\mathscr{A}^0$ as in VI.4.2(1), setting $\gamma \triangleq \frac{1}{2}$. It follows then that for every $(y, \rho, \alpha) \in \mathscr{A}^0$ we have

$$\begin{aligned}\int_0^{\alpha} |\dot{y}(t)|\, dt &= \int_0^{\alpha} ([\rho(t)]^2 + y^1(t)\,\rho(t))\, dt \\ &= y^1(\alpha) + y^2(\alpha) \leqslant y^1(\alpha) + L_1 + \tfrac{1}{2}.\end{aligned}$$

Since

$$y^1(\alpha) = h_0(y(\alpha)) \leqslant \gamma + h_0(y^{\#}(1)) = \tfrac{1}{2} + (r^{\#})^2 = \tfrac{1}{2} + [2L_1]^{2/3},$$

it follows that

$$\int_0^{\alpha} |\dot{y}(t)|\, dt \leqslant (2L_1)^{2/3} + L_1 + 1 \triangleq L.$$

Thus Assumption VI.4.2(1) is satisfied.

Next we set

$$\phi(v, r) \triangleq (r + 1)^2 \qquad [v \in S(0, L + 1) \subset \mathbb{R}^2, \quad r \in R^{\text{top}}]$$

and

$$\phi_1(v, r) \triangleq \phi(v, r)[|f(v, r)| + 1]^{-1} = (r + 1)^2[(r)^2 + |v^1|\, r + 1]^{-1},$$

and observe that

$$(L + 1)^{-1} \leqslant \phi_1(v, r) \leqslant 2.$$

This is consistent with VI.4.2(2) if we set

$$\beta_{\min} \triangleq \tfrac{1}{4}(L + 1)^{-1} \qquad \text{and} \qquad \beta_{\max} \triangleq 8(L + 1).$$

We define a metric compactification $(\tilde{R}, \tilde{R}', \Phi)$ of R^{top} $(\triangleq [0, \infty))$ by setting

$$\tilde{R} \triangleq [0, 1], \qquad \tilde{R}' \triangleq [0, 1), \qquad \text{and} \qquad \Phi(r) \triangleq r(r + 1)^{-1} \quad (r \in [0, \infty)).$$

We also set

$$\tilde{B} \triangleq [\beta_{\min}, \beta_{\max}] \qquad \text{and} \qquad \tilde{V} \triangleq \mathbb{R} \times S(0, L + 1) \subset \mathbb{R} \times \mathbb{R}^2.$$

Then

$$\Phi^{-1}(\tilde{r}) = \tilde{r}(1 - \tilde{r})^{-1} \qquad (\tilde{r} \in [0, 1))$$

and VI.4.2(5) yields for all $\tilde{v} \triangleq (v^0, v^1, v^2) \in \tilde{V}$, $\tilde{r} \in \tilde{R}$, and $\tilde{b} \triangleq \beta \in \tilde{B}$,

$$\tilde{f}^0(\tilde{v}, \tilde{r}, \tilde{b}) = \beta(1 - \tilde{r})^2, \qquad \tilde{f}^1(\tilde{v}, \tilde{r}, \tilde{b}) = \beta(\tilde{r})^2, \qquad \tilde{f}^2(\tilde{v}, \tilde{r}, \tilde{b}) = \beta v^1 \tilde{r}(1 - \tilde{r}),$$
$$\tilde{h}_0(\tilde{v}, \tilde{b}) = v^1, \qquad \text{and} \qquad h_1(\tilde{v}, \tilde{b}) = (v^0 - 1, v^2 - L_1).$$

It is now clear that the assumptions of Theorem VI.4.4 are satisfied.

Necessary Conditions for a Minimum

The compactified parametric problem thus reduces to the search for an absolutely continuous $\tilde{y} \triangleq (\tilde{y}^0, \tilde{y}^1, \tilde{y}^2) : [0, 1] \to \tilde{V}$, $\tilde{\sigma} \in \tilde{\mathscr{S}}^{\#}$ and

$$\bar{\beta} \in [\beta_{\min}, \beta_{\max}]$$

that minimize $\tilde{y}^1(1)$ subject to the restrictions

$$\dot{\tilde{y}}^0(s) = \bar{\beta} \int_0^1 (1 - \tilde{r})^2 \, \tilde{\sigma}(s)\,(d\tilde{r}), \qquad \dot{\tilde{y}}^1(s) = \bar{\beta} \int_0^1 (\tilde{r})^2 \, \tilde{\sigma}(s)\,(d\tilde{r}),$$

(1) $$\dot{\tilde{y}}^2(s) = \bar{\beta}\tilde{y}^1(s) \int_0^1 \tilde{r}(1 - \tilde{r}) \, \tilde{\sigma}(s)\,(d\tilde{r}) \qquad (s \in [0, 1]),$$

$$(\tilde{y}^0, \tilde{y}^1, \tilde{y}^2)(0) = (0, 0, 0), \qquad \tilde{y}^0(1) = 1, \qquad \tilde{y}^2(1) - L_1 = 0.$$

Theorem VI.4.4 guarantees that there exists a minimizing relaxed solution $(\bar{\tilde{y}}, \bar{\tilde{\sigma}}, \bar{\beta})$ which we denote by $(\tilde{y}, \tilde{\sigma}, \bar{\beta})$ to simplify notation.

We easily verify that the compactified parametric problem and its minimizing relaxed solution $(\tilde{y}, \tilde{\sigma}, \bar{\beta})$ satisfy the assumptions of Theorems VI.2.3 and VI.2.5. Then statements (1)–(3) of VI.2.5 imply that there exist

$$c_1 \in \mathbb{R}, \qquad l_0 \geqslant 0, \qquad l_1 \triangleq (l^1, l^2) \in \mathbb{R}^2,$$

and

$$k \triangleq (k^0, k^1, k^2) : [0, 1] \to \mathbb{R}^3$$

such that

$$l_0 + |l^1| + |l^2| > 0,$$

(2)

$$k^0(s) = l^1, \qquad k^1(s) = l_0 + \bar{\beta} \int_s^1 k^2(\theta)\, d\theta \int_0^1 \tilde{r}(1 - \tilde{r}) \, \tilde{\sigma}(\theta)(d\tilde{r}), \qquad k^2(s) = l^2$$
$$(s \in [0, 1]);$$

(3) if we denote by $\Gamma(s)$ the set of all $\tilde{r} \in [0, 1]$ that minimize

$$H(s, \tilde{r}) \triangleq k^0(s)(1 - \tilde{r})^2 + k^1(s)(\tilde{r})^2 + k^2(s)\tilde{y}^1(s)\,\tilde{r}(1 - \tilde{r})$$
$$= [l^1 + k^1(s) - l^2 \tilde{y}^1(s)](\tilde{r})^2 + [l^2 \tilde{y}^1(s) - 2l^1]\,\tilde{r} + l^1$$

over [0, 1], then

$$\tilde{\sigma}(s)(\Gamma(s)) = 1 \quad \text{and} \quad H_{\min}(s) \triangleq \underset{\tilde{r}\in\tilde{R}}{\operatorname{Min}}\ H(s, \tilde{r}) = c_1 \quad \text{for} \quad \text{a.a. } s \in [0, 1];$$

and

$$(4) \qquad (\beta - \bar{\beta}) \int_0^1 H_{\min}(s)\, ds \geqslant 0 \qquad (\beta \in [\beta_{\min}, \beta_{\max}]).$$

By VI.4.4, we have $\bar{\beta} \in [2\beta_{\min}, \frac{1}{2}\beta_{\max}]$ and therefore (3) and (4) imply that

$$(5) \quad H_{\min}(s) = \underset{\tilde{r}\in[0,1]}{\operatorname{Min}} \{[l^1 + k^1(s) - l^2\tilde{y}^1(s)](\tilde{r})^2 + [l^2\tilde{y}^1(s) - 2l^1]\,\tilde{r} + l^1\}$$
$$= \bar{\beta}^{-1}k(s)^T\,\dot{\tilde{y}}(s) = 0 \qquad \text{for} \quad \text{a.a. } s \in [0, 1].$$

This relation implies

$$(6) \quad H(s, 0) = k^0(s) = l^1 \geqslant 0 \quad \text{and} \quad H(s, 1) = k^1(s) \geqslant 0 \quad (s \in [0, 1]).$$

From (1), (2), and (5) we deduce that

$$\tilde{y}^1(s)\,\dot{k}^1(s) + l^2\dot{\tilde{y}}^2(s) = 0$$

and

$$l^1\dot{\tilde{y}}^0(s) + k^1(s)\,\dot{\tilde{y}}^1(s) + l^2\dot{\tilde{y}}^2(s) = 0 \qquad \text{a.e. in } [0, 1];$$

hence, after adding these relations,

$$l^1\dot{\tilde{y}}^0(s) + \big(k^1(s)\,\tilde{y}^1(s)\big)^{\cdot} + 2l^2\dot{\tilde{y}}^2(s) = 0 \qquad \text{a.e. in } [0, 1].$$

We integrate over [0, 1] and take account of the last line in (1) to obtain

$$(7) \qquad l^1 + k^1(1)\,\tilde{y}^1(1) + 2l^2L_1 = 0.$$

We also observe that, by (1), we have

$$(8) \qquad \tilde{y}^1(s) \geqslant 0 \qquad (s \in [0, 1]).$$

$l^1 > 0$ and $l^2 < 0$ We next prove that

$$(9) \qquad l^1 > 0 \quad \text{and} \quad l^2 < 0.$$

If $l^2 = 0$, then, by (6)–(8), we have $l^1 = k^1(1)\tilde{y}^1(1) = 0$; by (2), $k^1(s) = l_0 > 0$; and by (3) and (5), $\Gamma(s) = \{0\}$ for a.a. s. This implies, by (1), that $\tilde{y}^2(1) = \tilde{y}^2(0) = 0$, contrary to $\tilde{y}^2(1) - L_1 = 0$. Thus $l^2 \neq 0$. Now, by (2), we have $k^1(1) = l_0 \geqslant 0$ and, by (8), $\tilde{y}^1(1) \geqslant 0$; hence, by (6) and (7), $l^2 < 0$.

If $l^1 = 0$, then, by (5),

$$(10) \quad k^1(s)(\tilde{r})^2 + l^2\,\tilde{y}^1(s)[\tilde{r} - (\tilde{r})^2] \geqslant 0 \quad \text{for all } \tilde{r} \in [0, 1] \text{ and a.a. } s \in [0, 1];$$

hence $l^2 \tilde{y}^1(s) \geqslant 0$ $(s \in [0, 1])$, thus implying that $\tilde{y}^1(s) \leqslant 0$ $(s \in [0, 1])$. In view of (1), this is possible only if $\tilde{\sigma}(s)(\{0\}) = 1$ a.e., whence $\dot{\tilde{y}}^2(s) = 0$ a.e. Thus $\tilde{y}^2(1) = \tilde{y}^2(0) = 0$, contrary to $\tilde{y}^2(1) = L_1$. We conclude that $l^1 \neq 0$ and, in view of (6), $l^1 > 0$, thus completing the proof of (9).

Determination of the Minimizing Relaxed Solution. We now deduce from (6), (8), and (9) that

$$0 < l^1 - \tfrac{1}{2} l^2 \tilde{y}^1(s) \leqslant l^1 + k^1(s) - l^2 \tilde{y}^1(s) \qquad (s \in [0, 1]).$$

This relation, when combined with (5), shows that for a.a. $s \in [0, 1]$ the function $H(s, \cdot)$ achieves its minimum on $[0, 1]$ at the unique point

$$\tilde{\rho}(s) \triangleq [l^1 - \tfrac{1}{2} l^2 \tilde{y}^1(s)][l^1 + k^1(s) - l^2 \tilde{y}^1(s)]^{-1}. \tag{11}$$

Thus the relaxed control $\tilde{\sigma}$ belongs to $\tilde{\mathscr{R}}$ (and we shall show later on that it actually belongs to $\tilde{\mathscr{R}}^{\#}$ by proving that $\tilde{\rho}(s) \neq 1$ a.e.). Furthermore, since $H(s, \cdot)$ is a quadratic polynomial with a minimum value of 0, it must be a perfect square and therefore

$$[l^2 \tilde{y}^1(s) - 2l^1]^2 = 4l^1[l^1 + k^1(s) - l^2 \tilde{y}^1(s)].$$

This enables us to rewrite (11) in the form

$$\begin{aligned} \tilde{\rho}(s) &= 2l^1[2l^1 - l^2 \tilde{y}^1(s)]^{-1}, \\ 1 - \tilde{\rho}(s) &= -(l^2/2l^1)\, \tilde{y}^1(s)\, \tilde{\rho}(s) \qquad \text{for a.a. } s \in [0, 1]. \end{aligned} \tag{12}$$

We can now replace $\tilde{\sigma}$ by $\tilde{\rho}$ in (1) and deduce that

$$[\tilde{y}^1(s)]^2\, \dot{\tilde{y}}^1(s) = [2l^1/l^2]^2\, \dot{\tilde{y}}^0(s) \tag{13}$$

and

$$\dot{\tilde{y}}^2(s) = -(l^2/2l^1)[\tilde{y}^1(s)]^2\, \dot{\tilde{y}}^1(s) \quad \text{a.e. in } [0, 1].$$

Since $\tilde{y}^0(0) = \tilde{y}^1(0) = \tilde{y}^2(0) = 0$, these two relations imply that

$$\tfrac{1}{3}[\tilde{y}^1(s)]^3 = [2l^1/l^2]^2\, \tilde{y}^0(s) \tag{14}$$

and

$$\tilde{y}^2(s) = -\tfrac{1}{3}(l^2/2l^1)[\tilde{y}^1(s)]^3 = -(2l^1/l^2)\, \tilde{y}^0(s). \tag{15}$$

The conditions $\tilde{y}^2(1) = L_1$ and $\tilde{y}^0(1) = 1$ now yield

$$L_1 = -2l^1/l^2. \tag{16}$$

Minimizing Exact R^{top}-Solution We are now in a position to construct a minimizing approximate R^{top}-solution which is also an exact R^{top}-solution. We first show that $\tilde{\rho}(s) \in \tilde{R}' \triangleq [0, 1)$ for a.a. $s \in [0, 1]$. Indeed, by (9) and (12), if $\tilde{\rho}(\bar{s}) = 1$ for some $\bar{s} \in (0, 1]$, then $\tilde{y}^1(\bar{s}) = 0$ and therefore, by (1), $\tilde{\sigma}(s)(\{0\}) = 1$ a.e. in $(0, \bar{s}]$; hence $\tilde{\rho}(s) = 0$ a.e. in $[0, \bar{s}]$. It follows, by (1) and (2), that $\tilde{y}^1(s) = 0$ $(s \in [0, \bar{s}])$, implying [by (12)] that $\tilde{\rho}(s) = 1$ a.e. in $[0, \bar{s}]$, contrary to the previous conclusion. This shows that $\tilde{\rho}(s) < 1$ a.e. in $[0, 1]$, that is, $\tilde{\rho} \in \tilde{\mathscr{R}}^{\#}$.

Thus the compactified parametric problem has a minimizing relaxed solution $(\tilde{y}, \tilde{\rho}, \bar{\beta})$ with $\tilde{\rho} \in \tilde{\mathscr{R}}^{\#}$ and therefore the sequence with the repeated element $(\tilde{y}, \tilde{\rho}, \bar{\beta})$ is a minimizing approximate $\tilde{\mathscr{R}}^{\#}$-solution. We now find a corresponding element of $\tilde{\mathscr{A}}^0$ by setting

$$(17) \qquad \tau \triangleq \tilde{y}^0, \qquad \eta \triangleq (\tilde{y}^1, \tilde{y}^2), \qquad u \triangleq \Phi^{-1} \circ \tilde{\rho}, \qquad \text{and} \qquad \alpha \triangleq 1.$$

We then apply the bijection of VI.4.2 to determine the corresponding element $(y, \rho, 1) \in \mathscr{A}^0$ which satisfies the relations

$$(18) \qquad y = \eta \circ \tau^{-1} \qquad \text{and} \qquad \rho = u \circ \tau^{-1}.$$

Relations (12), (14)–(17) yield

$$\eta^1(s) = L_1^{2/3}[3\tau(s)]^{1/3}, \qquad \eta^2(s) = L_1\tau(s) \qquad (s \in [0, 1]),$$

$$\tilde{\rho}(s) = L_1[\tilde{y}^1(s) + L_1]^{-1} = L_1[\eta^1(s) + L_1]^{-1} \quad \text{a.e. in } [0, 1],$$

and

$$u(s) = \Phi^{-1}(\tilde{\rho}(s)) = \tilde{\rho}(s)[1 - \tilde{\rho}(s)]^{-1} = L_1[\tilde{y}^1(s)]^{-1} \quad \text{a.e. in } [0, 1].$$

It follows, by (17) and (18), that

$$y^1(t) = 3^{1/3}(L_1)^{2/3}(t)^{1/3}, \qquad y^2(t) = L_1 t \qquad (t \in [0, 1])$$

and

$$\rho(t) = u \circ \tau^{-1}(t) = L_1[y^1(t)]^{-1} = (L_1/3)^{1/3}(t)^{-1/3} \quad \text{a.e. in } [0, 1].$$

These relations now define a minimizing approximate R^{top}-solution which is the exact R^{top}-solution (y^1, y^2, ρ).

VI.4.7 ***Example*** The second example that we consider has a minimizing approximate solution with state functions that converge to a discontinuous function.

For $l \in \mathbb{N}$, we define the *euclidean norm* of $\mathbb{R}^l$ by $| a |_2 \triangleq (a \cdot a)^{1/2}$ $(a \in \mathbb{R}^l)$. We refer to a linear operator A in $\mathbb{R}^l$ as *positive definite* if $v^T A v > 0$ for every $v \in \mathbb{R}^l \sim \{0\}$. Now let $n \in \{2, 3, ...\}$, and let $p \triangleq (p^1, ..., p^{n-1}) : \mathbb{R}^{n-1} \to \mathbb{R}^{n-1}$ be bounded and continuous and have a bounded continuous derivative p' such that $p'(v^1,..., v^{n-1})$ is positive definite for all $(v^1,..., v^{n-1}) \in \mathbb{R}^{n-1}$. We set

$$m \triangleq 1, \qquad v_0 \triangleq 0 \in \mathbb{R}^n, \qquad 0 < L_1 < \infty, \qquad R^{\text{top}} \triangleq \mathbb{R}^{n-1},$$
$$f^n(v, r) \triangleq | r |_2 \qquad (v \in \mathbb{R}^n, \quad r \in R^{\text{top}}),$$
$$f(v, r) \triangleq \big(p(v^1,..., v^{n-1}) + r, f^n(v, r)\big) \qquad [v \triangleq (v^1,..., v^n) \in \mathbb{R}^n, \quad r \in R^{\text{top}}],$$

and

$$h_1(v) \triangleq v^n - L_1 .$$

We also assume that $h_0 : \mathbb{R}^n \to \mathbb{R}$ is continuous and has a continuous derivative, and h_0 is independent of v^n, that is,

$$h_0(v^1,..., v^n) = h_0(v^1,..., v^{n-1}) \qquad [(v^1,..., v^n) \in \mathbb{R}^n].$$

As in the previous example, our problem involves no control parameters and no restriction of the form $h_2(y(t), b) \in A$.

It is easy to verify that the problem has an exact R^{top}-solution $(y^\#, \rho^\#)$. We can obtain such a solution by setting

$$\rho^\#(t) \triangleq (L_1 , 0,..., 0) \in \mathbb{R}^{n-1} \qquad (t \in [0, 1]).$$

Then the equation

$$y(t) = \int_0^t f\big(y(\theta), \rho^\#(\theta)\big)\, d\theta \qquad (t \in [0, 1])$$

has a unique solution $y^\#$ (II.4.3 and II.4.5), and

$$h_1\big(y^\#(1)\big) = y^{\#n}(1) - L_1 = \int_0^1 | \rho^\#(\tau)|_2 \, d\tau - L_1 = 0.$$

Assumption VI.4.2(1) is satisfied if we choose $\gamma \triangleq \frac{1}{2}$ and

$$L \triangleq \tfrac{3}{2}[| p |_{\sup} + n(L_1 + \tfrac{1}{2})].$$

We set

$$\phi(v, r) \triangleq | r |_2 + 1 \qquad [v \in S(0, L + 1) \subset \mathbb{R}^n, \quad r \in R^{\text{top}}],$$
$$\beta_{\min} \triangleq \tfrac{1}{4}(| p |_{\sup} + n + 1)^{-1}, \qquad \beta_{\max} \triangleq 4(L + 1),$$

and verify that this choice of ϕ, $\beta_{\min}$, and $\beta_{\max}$ is consistent with VI.4.2(2).

We define a metric compactification $(\tilde{R}, \tilde{R}', \Phi)$ of R^{top} by setting

$$\tilde{R} \triangleq \{v \in \mathbb{R}^{n-1} \mid | v |_2 \leqslant 1\}, \qquad \tilde{R}' = \{v \in \mathbb{R}^{n-1} \mid | v |_2 < 1\}$$

and

$$\Phi(r) \triangleq (| r |_2 + 1)^{-1} r \qquad (r \in R^{\text{top}} \triangleq \mathbb{R}^{n-1}).$$

We also let

$$\tilde{B} \triangleq [\beta_{\min}, \beta_{\max}] \quad \text{and} \quad \tilde{V} \triangleq \mathbb{R} \times S(0, L+1) \subset \mathbb{R} \times \mathbb{R}^n.$$

Then

$$\Phi^{-1}(\tilde{r}) = (1 - |\tilde{r}|_2)^{-1}\tilde{r} \qquad (\tilde{r} \in \tilde{R}')$$

and VI.4.2(5) yields for all $\tilde{v} \triangleq (v^0, v) \in \tilde{V}$, $\tilde{r} \in \tilde{R}$, and $\beta \in \tilde{B}$,

$$\begin{aligned}
\tilde{f}^0(\tilde{v}, \tilde{r}, \tilde{b}) &= \beta(1 - |\tilde{r}|_2), \\
\tilde{f}^i(\tilde{v}, \tilde{r}, \tilde{b}) &= \beta(1 - |\tilde{r}|_2)\, p^i(v^1,..., v^{n-1}) + \beta\tilde{r}^i \qquad (i = 1,..., n-1), \\
\tilde{f}^n(\tilde{v}, \tilde{r}, \tilde{b}) &= \beta\, |\tilde{r}|_2, \\
\tilde{h}_0(\tilde{v}, \tilde{b}) &= h_0(v^1,..., v^{n-1}),
\end{aligned}$$

and

$$\tilde{h}_1(\tilde{v}, \tilde{b}) = (v^0 - 1, v^n - L_1).$$

The Minimizing Relaxed Solution

We conclude that the compactified parametric problem reduces to the search for an absolutely continuous

$$\tilde{y} \triangleq (\tilde{y}^0, \tilde{y}^1,..., \tilde{y}^n) : [0, 1] \to \tilde{V},$$

$\tilde{\sigma} \in \tilde{\mathscr{S}}$, and $\bar{\beta} \in [\beta_{\min}, \beta_{\max}]$ that minimize $h_0(\tilde{y}^1(1),..., \tilde{y}^{n-1}(1))$ subject to the conditions (in which $i = 1, 2,..., n-1$)

$$\begin{aligned}
\dot{\tilde{y}}^0(s) &= \bar{\beta} \int_{\tilde{R}} (1 - |\tilde{r}|_2)\, \tilde{\sigma}(s)\, (d\tilde{r}) \\
\dot{\tilde{y}}^i(s) &= \bar{\beta} p^i(\tilde{y}^1(s),..., \tilde{y}^{n-1}(s)) \int_{\tilde{R}} (1 - |\tilde{r}|_2)\, \tilde{\sigma}(s)\, (d\tilde{r}) + \bar{\beta} \int r^i \tilde{\sigma}(s)\, (d\tilde{r}), \\
\dot{\tilde{y}}^n(s) &= \bar{\beta} \int |\tilde{r}|_2\, \tilde{\sigma}(s)\, (d\tilde{r}) \qquad (s \in [0, 1]), \\
\tilde{y}(0) &= 0, \qquad \tilde{y}^0(1) = 1, \qquad \tilde{y}^n(1) - L_1 = 0.
\end{aligned} \tag{1}$$

Theorem VI.4.4 is applicable and guarantees that there exists a minimizing relaxed solution $(\bar{\tilde{y}}, \bar{\tilde{\sigma}}, \bar{\beta})$ which we shall denote by $(\tilde{y}, \tilde{\sigma}, \bar{\beta})$ to simplify notation. We set $\tau \triangleq \tilde{y}^0$, $\xi \triangleq (\tilde{y}^1,..., \tilde{y}^{n-1})$, and $\eta \triangleq (\xi, \tilde{y}^n)$. We shall show that either $h_0'(\xi(1)) = 0$ or $(\tilde{y}, \tilde{\sigma}, \bar{\beta}) \triangleq (\tau, \xi, \tilde{y}^n, \tilde{\sigma}, \bar{\beta})$ has the following properties: There exists $z : [0, 1] \to \mathbb{R}^{n-1} \sim \{0\}$ such that, if

$$s_1 \triangleq L_1(L_1 + 1)^{-1} \quad \text{and} \quad \lambda \triangleq -|z(s_1)|_2^{-1} z(s_1),$$

then

$$\bar{\beta} = L_1 + 1,$$

$$\tilde{\sigma}(s)(\{\lambda\}) = 1 \quad \text{a.e. in } [0, s_1], \qquad \tilde{\sigma}(s)(\{0\}) = 1 \quad \text{a.e. in } (s_1, 1],$$

(2)
$$\tau(s) = 0 \qquad \xi(s) = \bar{\beta} s \lambda, \qquad z(s) = z(s_1) \qquad (s \in [0, s_1]),$$

$$\tau(s) = (L_1 + 1)s - L_1, \qquad \xi(s) = L_1 \lambda + \bar{\beta} \int_{s_1}^{s} p(\xi(\theta))\, d\theta,$$

$$z(s)^T = h_0'(\xi(1)) + \bar{\beta} \int_s^1 z(\theta)^T p'(\xi(\theta))\, d\theta \qquad (s \in (s_1, 1]).$$

To prove these assertions we first observe that the compactified parametric problem satisfies the assumptions of VI.2.3 and VI.2.5. It follows from VI.2.5 that there exist $c_1 \in \mathbb{R}$, $l_0 \geqslant 0$, $l_1 \triangleq (l^0, l^1) \in \mathbb{R} \times \mathbb{R}$, and

$$k \triangleq (k^0, k^1, \ldots, k^n) \triangleq (k^0, \kappa, k^n) : [0, 1] \to \mathbb{R} \times \mathbb{R}^{n-1} \times \mathbb{R}$$

with the following properties:

(3)
$$l_0 + |l^0| + |l^1| > 0,$$

$$k^0(s) = l^0,$$

$$\kappa(s)^T = l_0 h_0'(\xi(1)) + \bar{\beta} \int_s^1 \kappa(\theta)^T p'(\xi(\theta))\, d\theta \int_{\tilde{R}} (1 - |\tilde{r}|_2)\, \tilde{\sigma}(\theta)\, (d\tilde{r}),$$

$$k^n(s) = l^1 \quad (s \in [0, 1]);$$

(4) if we denote by $\Gamma(s)$ the set of all $\tilde{r} \in \tilde{R}$ that minimize

$$H(s, \tilde{r}) \triangleq k^0(s)(1 - |\tilde{r}|_2) + \kappa(s) \cdot [p(\xi(s))(1 - |\tilde{r}|_2) + \tilde{r}] + k^n(s)|\tilde{r}|_2$$

over $\tilde{R}$, then

$$\tilde{\sigma}(s)(\Gamma(s)) = 1 \quad \text{and} \quad H_{\min}(s) \triangleq \operatorname*{Min}_{\tilde{r} \in \tilde{R}} H(s, \tilde{r}) = c_1 \quad \text{for a.a. } s \in [0, 1];$$

and

(5)
$$(\beta - \bar{\beta}) \int_0^1 H_{\min}(s)\, ds \geqslant 0 \qquad (\beta \in [\beta_{\min}, \beta_{\max}]).$$

By VI.4.4, we have $\bar{\beta} \in [2\beta_{\min}, \frac{1}{2}\beta_{\max}]$ and therefore (3)–(5) imply that

(6)
$$H_{\min}(s) = \operatorname*{Min}_{\tilde{r} \in \tilde{R}} \{[-l^0 - \kappa(s) \cdot p(\xi(s)) + l^1]|\tilde{r}|_2 + \kappa(s) \cdot \tilde{r}$$
$$+ l^0 + \kappa(s) \cdot p(\xi(s)) = 0 \qquad \text{for a.a. } s \in [0, 1].$$

We shall assume henceforth that $h_0'(\xi(1)) \neq 0$. If $l_0 = 0$, then, by (3) and II.4.6, $\kappa(\cdot) = 0$ and, by (6) and (3),

(7)
$$H_{\min}(s) = \operatorname*{Min}_{\tilde{r} \in \tilde{R}} \{(l^1 - l^0)|\tilde{r}|_2 + l^0\} = 0 \qquad \text{for a.a. } s \in [0, 1],$$

$$|l^0| + |l^1| > 0.$$

If also $l^1 - l^0 = 0$, then the first relation in (7) implies $l^0 = l^1 = 0$, contrary to the second relation. If $l^1 - l^0 > 0$, then, by (4) and (7), we have $\tilde{\sigma}(s)(\{0\}) = 1$ a.e.; hence, by (1), $\tilde{y}^n(1) = 0$ contradicting $\tilde{y}^n(1) - L_1 = 0$. If $l^1 - l^0 < 0$, then, by (4) and (7), we have $\int |\tilde{r}|_2 \tilde{\sigma}(s)(d\tilde{r}) = 1$ a.e. and, by (1), $\tilde{y}^0(1) = 0$, contrary to $y^0(1) = 1$. Thus the assumption that $l_0 = 0$ is inadmissible and therefore $l_0 > 0$.

Since we assume that $h_0'(\xi(1)) \neq 0$, it follows now from (3) and II.4.6 that $\kappa(s) \neq 0$ $(s \in [0, 1])$. If we set in (6)

$$\tilde{r} = r_0 \nu \qquad \text{where} \quad 0 \leqslant r_0 \leqslant 1 \quad \text{and} \quad |\nu|_2 = 1,$$

then we observe that $\operatorname{Min}_{|\nu|_2=1} \kappa(s) \cdot \nu = -|\kappa(s)|_2$ for all $s \in [0, 1]$, corresponding to $\nu = \bar{\nu}(s) \triangleq -|\kappa(s)|_2^{-1} \kappa(s)$; therefore

$$\text{(8)} \quad H_{\min}(s) = \operatorname*{Min}_{0 \leqslant r_0 \leqslant 1} \{[l^1 - l^0 - \kappa(s) \cdot p(\xi(s)) - |\kappa(s)|_2] r_0 + l^0 + \kappa(s) \cdot p(\xi(s))\}$$
$$= 0 \qquad \text{for} \quad \text{a.a. } s \in [0, 1].$$

We set

$$\zeta(s) \triangleq l^1 - l^0 - \kappa(s) \cdot p(\xi(s)) - |\kappa(s)|_2 \qquad (s \in [0, 1]),$$

and conclude that, for a.a. $s \in [0, 1]$ and all $\tilde{r} \in \Gamma(s)$,

$$\text{(9)} \qquad \tilde{r} = |\tilde{r}|_2 \bar{\nu}(s), \qquad \Gamma(s) = \begin{cases} \{0\} & \text{if} \quad \zeta(s) > 0, \\ \{\bar{\nu}(s)\} & \text{if} \quad \zeta(s) < 0. \end{cases}$$

We next show that $\zeta(\cdot)$ is an increasing function on [0, 1]. Indeed, by (1), (3), and (9), we have

$$\dot{\zeta}(s) = -\dot{\kappa}(s) \cdot p(\xi(s)) - \kappa(s)^T p'(\xi(s))[\bar{\beta} p(\xi(s)) \int_{\Gamma(s)} (1 - |\tilde{r}|_2) \tilde{\sigma}(s)(d\tilde{r})$$
$$+ \bar{\beta} \int_{\Gamma(s)} \tilde{r} \tilde{\sigma}(s)(d\tilde{r})] - ([\kappa(s) \cdot \kappa(s)]^{1/2})^{\cdot}$$
$$= -\bar{\beta} \kappa(s)^T p'(\xi(s)) \int_{\Gamma(s)} \tilde{r} \tilde{\sigma}(s)(d\tilde{r}) - |\kappa(s)|_2^{-1} \dot{\kappa}(s)^T \kappa(s)$$
$$= \bar{\beta} \kappa(s)^T p'(\xi(s)) \kappa(s) |\kappa(s)|_2^{-1} \int_{\Gamma(s)} |\tilde{r}|_2 \tilde{\sigma}(s)(d\tilde{r})$$
$$+ \bar{\beta} |\kappa(s)|_2^{-1} \kappa(s)^T p'(\xi(s)) \kappa(s) \int_{\Gamma(s)} (1 - |\tilde{r}|_2) \tilde{\sigma}(s)(d\tilde{r})$$
$$= \bar{\beta} |\kappa(s)|_2^{-1} \kappa(s)^T p'(\xi(s)) \kappa(s) \qquad \text{for} \quad \text{a.a. } s \in [0, 1].$$

Since $\kappa(s) \neq 0$ for all s and $p'(v^1, \ldots, v^{n-1})$ is assumed positive definite for all $(v^1, \ldots, v^{n-1}) \in \mathbb{R}^{n-1}$, we conclude that $\dot{\zeta}(s) > 0$ a.e. and the absolutely continuous function ζ is increasing.

It follows now from (9) that there exists $s_1 \in [0, 1]$ such that

$$\Gamma(s) = \begin{cases} \{-|\kappa(s)|_2^{-1}\kappa(s)\} & \text{for a.a. } s \in [0, s_1], \\ \{0\} & \text{for a.a. } s \in (s_1, 1). \end{cases} \tag{10}$$

Since $\tilde{\sigma}(s)(\Gamma(s)) = 1$ a.e., (1) implies that

$$\tau(s) \triangleq \tilde{y}^0(s) = 0 \quad (s \in [0, s_1]), \qquad \tau(s) \triangleq \tilde{y}^0(s) = \bar{\beta}(s - s_1) \quad (s \in (s_1, 1]),$$

$$\begin{aligned} &\xi(s) \triangleq (\tilde{y}^1(s), \ldots, \tilde{y}^{n-1}(s)) = -\bar{\beta}\int_0^s |\kappa(\theta)|_2^{-1}\kappa(\theta)\,d\theta \qquad (s \in [0, s_1]), \\ &\xi(s) = \xi(s_1) + \bar{\beta}\int_{s_1}^s p(\xi(\theta))\,d\theta \qquad (s \in (s_1, 1]), \\ &\tilde{y}^n(s) = \bar{\beta}s \quad (s \in [0, s_1]), \qquad \tilde{y}^n(s) = \bar{\beta}s_1 \quad (s \in (s_1, 1]). \end{aligned} \tag{11}$$

If we set $z \triangleq l_0^{-1}\kappa$, then, by (3),

$$z(s) = z(s_1) \qquad (s \in [0, s_1]) \tag{12}$$

and, by (1) and (11), we have

$$\tau(1) = 1 = \bar{\beta}(1 - s_1), \qquad \tilde{y}^n(1) = L_1 = \bar{\beta}s_1 . \tag{13}$$

Since $l_0 > 0$, all the relations in (2) now follow from (1), (3), and (11)–(13).

Minimizing Approximate R^{top}-Solution We can now apply the construction of VI.4.4 to determine a regular minimizing approximate R^{top}-solution (whose existence is guaranteed by VI.4.5). We begin by constructing a minimizing approximate $\tilde{\mathscr{R}}^{\#}$-solution $((\tilde{y}_j, \tilde{\rho}_j, \bar{\beta}))$. This we can accomplish (VI.1.3) by choosing a sequence $(\tilde{\rho}_j)$ in $\tilde{\mathscr{R}}^{\#}$ converging to $\tilde{\sigma}$ and determining $\tilde{y}_j$ as the (unique) solution of the equation

$$\tilde{y}_j(s) = \int_0^s \tilde{f}(\tilde{y}_j(\theta), \tilde{\rho}_j(\theta), \bar{\beta})\,d\theta \qquad (s \in [0, 1]).$$

We shall choose $\tilde{\rho}_j$ in such a manner that the corresponding function $\tilde{y}_j$ is characterized by $\tilde{y}_j{}^0(1) = 1$. We choose a sequence (ϵ_j) in $(0, 1] \cap (0, 1/L_1]$ converging to 0, and set for each $j \in \mathbb{N}$,

$$\tilde{\rho}_j(s) \triangleq (1 - \epsilon_j)\lambda \quad (s \in [0, s_1]), \qquad \tilde{\rho}_j(s) \triangleq L_1\epsilon_j\lambda \quad (s \in (s_1, 1]).$$

The corresponding function $\tilde{y}_j \triangleq (\tau_j, \xi_j, \tilde{y}_j{}^n) \triangleq (\tau_j, \eta_j)$ is such that

$$\tau_j(s) = \begin{cases} \bar{\beta}\epsilon_j s & (s \in [0, s_1]), \\ L_1\epsilon_j + \bar{\beta}(1 - L_1\epsilon_j)(s - s_1) & (s \in (s_1, 1]). \end{cases} \tag{14}$$

Next we set

$$u_j(s) \triangleq \Phi^{-1} \circ \tilde{\rho}_j(s) = [1 - |\tilde{\rho}_j(s)|_2]^{-1}\tilde{\rho}_j(s) \qquad (s \in [0, 1])$$

and deduce (remembering that $|\lambda|_2 = 1$) that

$$u_j(s) = \begin{cases} \epsilon_j^{-1}(1 - \epsilon_j)\,\lambda & (s \in [0, s_1]), \\ L_1\epsilon_j(1 - L_1\epsilon_j)^{-1}\,\lambda & (s \in (s_1, 1]). \end{cases} \tag{15}$$

It follows now from VI.4.4 that we obtain a minimizing approximate R^{top}-solution $(y_j, \rho_j, \bar{\beta}, \alpha_j)$ by setting

$$y_j \triangleq \eta_j \circ \tau_j^{-1}, \qquad \rho_j \triangleq u \circ \tau_j^{-1}, \qquad \text{and} \qquad \alpha_j \triangleq \tau_j(1) \qquad (j \in \mathbb{N}).$$

In view of (2), (14), and (15), we have, for each $j \in \mathbb{N}$,

$$\alpha_j \triangleq \tau_j(1) = 1,$$

$$\tau_j^{-1}(t) = \begin{cases} (\bar{\beta}\epsilon_j)^{-1}\,t & (t \in [0, L_1\epsilon_j]), \\ s_1 + [\bar{\beta}(1 - L_1\epsilon_j)]^{-1} \cdot (t - L_1\epsilon_j) & (t \in (L_1\epsilon_j, 1]), \end{cases}$$

$$\rho_j(t) = \begin{cases} \epsilon_j^{-1}(1 - \epsilon_j)\,\lambda & (t \in [0, L_1\epsilon_j]), \\ L_1\epsilon_j(1 - L_1\epsilon_j)^{-1}\,\lambda & (t \in (L_1\epsilon_j, 1]). \end{cases}$$

The function $y_j \triangleq (y_j^1, \ldots, y_j^n)$ is the solution of the equation

$$y_j(t) = \int_0^t f(y_j(\theta), \rho_j(\theta))\,d\theta \qquad (t \in [0, 1]),$$

that is,

$$y_j^i(t) = \int_0^t [p^i(y^1(\theta), \ldots, y^{n-1}(\theta)) + \rho_j^i(\theta)]\,d\theta$$
$$(i = 1, 2, \ldots, n-1, \quad t \in [0, 1])$$

$$y_j^n(t) = \int_0^t |\rho_j(\theta)|_2\,d\theta \qquad (t \in [0, 1]).$$

We may define a function $\bar{\xi} : [0, 1] \to \mathbb{R}^{n-1}$ by the equation

$$\bar{\xi}(t) = L_1\lambda + \int_0^t p(\bar{\xi}(\theta))\,d\theta \qquad (t \in [0, 1]).$$

Then we observe that $y_j(0) = 0$ $(j \in \mathbb{N})$ while

$$\lim_j (y_j^1, \ldots, y_j^{n-1})(t) = \bar{\xi}(t), \qquad \lim_j y_j^n(t) = L_1 \qquad (t \in (0, 1]),$$

thus showing that y_j converges pointwise to a discontinuous function.

VI.5 Variable Initial Conditions, Free Time, Infinite Time, Staging, Advance–Delay Differential Problems

We shall consider here certain problems defined by ordinary differential equations that appear to be more general than those defined in VI.0 but which can actually be transformed into the same form.

VI.5.1 ***Variable Initial Conditions*** The problem defined in VI.0 involves points $(y, \sigma, b) \in \mathscr{Y} \times \mathscr{S}^{\#} \times B$ that satisfy the equation

$$y(t) = v_0 + \int_{t_0}^{t} f(\tau, y(\tau), \sigma(\tau), b)\, d\tau \qquad (t \in T).$$

This implies, in particular, that $y(t_0)$ is fixed. A more general problem is one involving points $(y, \sigma, b) \in \mathscr{Y} \times \mathscr{S}^{\#} \times B$ satisfying the equation

$$(1) \qquad y(t) = \xi(b) + \int_{t_0}^{t} f(\tau, y(\tau), \sigma(\tau), b)\, d\tau \qquad (t \in T),$$

for a given $\xi : B \to V$ and with

$$(2) \quad h_1(y(t_1), b) = 0 \quad \text{and} \quad h_2(t, y(t), b) \in A(t) \quad \text{for all } t \in T.$$

This last problem can be transformed into that of VI.0 as follows: For every (y, σ, b) satisfying relations (1) and (2), we set

$$\tilde{y}(t) \triangleq y(t) - \xi(b) \qquad (t \in T).$$

Then $(\tilde{y}, \sigma, b) \in \mathscr{Y} \times \mathscr{S}^{\#} \times B$ and $(\tilde{y}, \sigma, b)$ satisfies the relations

$$\tilde{y}(t) = \int_{t_0}^{t} f(\tau, \tilde{y}(\tau) + \xi(b), \sigma(\tau), b)\, d\tau \qquad (t \in T),$$

$$h_1(\tilde{y}(t_1) + \xi(b), b) = 0 \quad \text{and} \quad h_2(t, \tilde{y}(t) + \xi(b), b) \in A(t) \qquad (t \in T).$$

These relations are of the type considered in VI.0.

VI.5.2 ***Free Time and Least Time Control Problems.*** Now we consider points $(y, \sigma, b) \in \mathscr{Y} \times \mathscr{S} \times B$ and $t_0', t_1' \in T$ such that

$$(1) \quad y(t) = v_0 + \int_{t_0'}^{t} f(\tau, y(\tau), \sigma(\tau), b)\, d\tau \qquad \text{for} \quad t \in T' \triangleq [t_0', t_1']$$

and

$$(2) \; h_1(y(t_1'), b, t_0', t_1') = 0 \quad \text{and} \quad h_2(t, y(t), b, t_0', t_1') \in A \qquad (t \in T').$$

[Observe that we no longer impose restrictions of the form $h_2(t, y(t), b) \in A(t)$ or $\rho(t) \in R^{\#}(t)$]. Among such points we wish to determine one that minimizes $h_0(y(t_1'), b, t_0', t_1')$. We refer to this problem as a *free time control problem* because the duration T' of the motion is not preassigned but must be chosen as a subinterval of T so that, together with $\sigma \in \mathscr{S}$ and $b \in B$, it minimizes $h_0(y(t_1'), b, t_0', t_1')$ subject to relations (2). The free time control problem becomes a *least time control problem* in the special case where

$$h_0(v, b, t_0', t_1') \triangleq t_1' - t_0'$$

for all (v, b, t_0', t_1') in the domain of definition of h_0.

We proceed as follows: We set

$$\tilde{T} \triangleq [0, 1], \qquad \tilde{B} \triangleq B \times \{(t_0', t_1') \in T \times T \mid t_0' \leqslant t_1'\}$$

and, for $(y, \sigma, (b, t_0', t_1')) \in \mathscr{Y} \times \mathscr{S} \times \tilde{B}$,

$$\text{(3)} \quad \begin{aligned} &\tilde{y}(s) \triangleq y(t_0' + [t_1' - t_0']s), \\ &\tilde{\sigma}(s) \triangleq \sigma(t_0' + [t_1' - t_0']s), \qquad \tilde{b} \triangleq (b, t_0', t_1') \qquad (s \in [0, 1]). \end{aligned}$$

We also denote by $\tilde{\mathscr{R}}$ the class of measurable functions on $[0, 1]$ to R and define $\tilde{\mathscr{S}}$ the same way as $\mathscr{S}$, but with T replaced by $[0, 1]$. Then

$$(\tilde{y}, \tilde{\sigma}, \tilde{b}) \in C([0, 1], \mathbb{R}^n) \times \tilde{\mathscr{S}} \times \tilde{B}$$

whenever $(y, \sigma, b) \in C(T, \mathbb{R}^n) \times \mathscr{S} \times B$. If, furthermore, (y, σ, b) satisfies (1) and (2), then I.4.43 yields

$$\text{(4)} \quad \tilde{y}(s) = v_0 + \int_0^s (t_1' - t_0') f(t_0' + [t_1' - t_0']\, \alpha, \tilde{y}(\alpha), \tilde{\sigma}(\alpha), b)\, d\alpha \qquad (s \in [0, 1])$$

and

$$\text{(5)} \quad h_1(\tilde{y}(1), \tilde{b}) = 0 \quad \text{and} \quad h_2(t_0' + [t_1' - t_0']\, s, \tilde{y}(s), \tilde{b}) \in A \qquad (s \in [0, 1]).$$

Conversely, every $(\tilde{y}, \tilde{\sigma}, \tilde{b}) \in C([0, 1], \mathbb{R}^n) \times \tilde{\mathscr{S}} \times \tilde{B}$ and satisfying (4) and (5) determines [through (3)] points $b \in B$ and $t_0', t_1' \in T$ and functions y and ρ on $T' \triangleq [t_0', t_1']$ that can be extended to all of T by setting

$$y(t) \triangleq y(t_0') \quad \text{and} \quad \sigma(t) \triangleq \sigma(t_0') \quad \text{for} \quad t < t_0'$$

and

$$y(t) \triangleq y(t_1') \quad \text{and} \quad \sigma(t) \triangleq \sigma(t_1') \quad \text{for} \quad t > t_1'.$$

Then $(y, \sigma, b) \in \mathscr{Y} \times \mathscr{S} \times B$ and (y, σ, b) satisfies relations (1) and (2).

Thus our present problem, involving y, σ, b, t_0', and t_1', is equivalent to the problem of minimizing $h_0(\tilde{y}(1), \tilde{b})$ subject to relations (4) and (5), which is of the form considered in VI.0.

Entirely similar arguments and transformations will be applicable if one of the points t_0' or t_1' is fixed or if t_0' and t_1' are subject to restrictions of the form $t_0' = \phi_0(c)$ and $t_1' = \phi_1(c)$, where c belongs to some set of parameters, $\phi_i(c) \in T$ $(i = 0, 1)$ and $\phi_0(c) \leqslant \phi_1(c)$ for all c.

VI.5.3 ***Infinite Time Problems.*** Another variant of the problem of VI.0 is obtained by replacing the finite interval T by $T_\infty \triangleq [t_0, \infty)$. For a given mapping $R_\infty{}^\#$ on T_∞ into $\mathscr{P}'(R)$, we let $\mathscr{R}_\infty{}^\#$ be the collection of measurable functions $\rho : T_\infty \to R$ such that $\rho(t) \in R_\infty{}^\#(t)$ a.e. in T_∞, and set $\mathscr{Y}_\infty \triangleq C(T_\infty, \mathbb{R}^n)$. We assume that $A(t) = A$ $(t \in T)$, and let

$$\mathscr{H}(\mathscr{R}_\infty{}^\#) \triangleq \Big\{(y, \rho, b) \in \mathscr{Y}_\infty \times \mathscr{R}_\infty{}^\# \times B \mid y(t) = v_0 + \int_{t_0}^{t} f(\tau, y(\tau), \rho(\tau), b)\, d\tau \ (t \in T_\infty)\Big\}$$

and, for $\gamma \geqslant 0$,

$$\mathscr{A}_\gamma(\mathscr{R}_\infty{}^\#) \triangleq \{(y, \rho, b) \in \mathscr{H}(\mathscr{R}_\infty{}^\#) \mid y(\infty) \triangleq \lim_{t\to\infty} y(t) \text{ exists}, \ |h_1(y(\infty), b)| \leqslant \gamma,\ d[h_2(t, y(t), b), A] \leqslant \gamma \ (t \in T_\infty)\}.$$

An approximate $\mathscr{R}_\infty{}^\#$-solution is a sequence $((y_j, \rho_j, b_j))$ in $\mathscr{H}(\mathscr{R}_\infty{}^\#)$ for which there exists a sequence (γ_j) in $[0, \infty)$ converging to 0 such that

$$(y_j, \rho_j, b_j) \in \mathscr{A}_{\gamma_j}(\mathscr{R}_\infty{}^\#) \qquad (j \in \mathbb{N}).$$

We say that a function $\psi : T_\infty \to [0, \infty)$ is integrable if $\psi \mid [t_0, \alpha]$ is integrable for all $\alpha \in T_\infty$ and

$$\int_{t_0}^{\infty} \psi(t)\, dt \triangleq \lim_{\alpha\to\infty} \int_{t_0}^{\alpha} \psi(t)\, dt < \infty.$$

We make the following assumption.

(1) *Assumption* There exist an integrable $\psi : T_\infty \to [0, \infty)$ and $\bar{\gamma} \in (0, \infty)$ such that

$$|f(t, y(t), \rho(t), b)| \leqslant \psi(t) \qquad [(y, \rho, b) \in \mathscr{A}_{\bar{\gamma}}(\mathscr{R}_\infty{}^\#)].$$

We may clearly assume that $\psi(t) > 0$ $(t \in T_\infty)$ [otherwise replacing $\psi(t)$ by $\psi(t) + e^{-t}$].

Now we set

$$\Psi(t) \triangleq \int_{t_0}^{t} \psi(\alpha)\, d\alpha \qquad (t \in T_\infty).$$

Then Ψ increases from 0 to $a \triangleq \int_{t_0}^{\infty} \psi(\alpha)\, d\alpha$ as t increases from t_0 to ∞, and Ψ has an inverse $\Phi \triangleq \Psi^{-1} : [0, a) \to T_\infty$. Now let $(y, \rho, b) \in \mathscr{H}(\mathscr{R}_\infty{}^{\#})$, and let

$$\tilde{y} \triangleq y \circ \Phi \qquad \text{and} \qquad \tilde{\rho} \triangleq \rho \circ \Phi.$$

We observe that, by I.4.42(3), I.4.43, and II.3.11, for every $\alpha \in [0, a)$ the functions $\Phi \,|[0, \alpha]$ and $\tilde{y}\,|[0, \alpha]$ are absolutely continuous, $\tilde{\rho}\,|[0, \alpha]$ is measurable, and

$$\dot{\Phi}(\beta) = [\psi \circ \Phi(\beta)]^{-1} \qquad \text{for} \quad \text{a.a. } \beta \in [0, \alpha].$$

We have

$$\text{(2)} \qquad y(t) = v_0 + \int_{t_0}^{t} f(\tau, y(\tau), \rho(\tau), b)\, d\tau \qquad (t \in T_\infty);$$

hence, setting $t = \Phi(\alpha)\,(\alpha \in [0, a))$ and applying I.4.43, we obtain

$$\text{(3)} \quad \tilde{y}(\alpha) = v_0 + \int_{0}^{\alpha} [\psi(\Phi(\beta))]^{-1} f(\Phi(\beta), \tilde{y}(\beta), \tilde{\rho}(\beta), b)\, d\beta \qquad (\alpha \in [0, a)).$$

Moreover, since

$$\tilde{y}(a) \triangleq \lim_{\alpha \to a} \tilde{y}(\alpha) = \lim_{t \to \infty} y(t) \triangleq y(\infty)$$

and, by (1),

$$|f(\Phi(\alpha), \tilde{y}(\alpha), b)| = |f(\Phi(\alpha), y(\Phi(\alpha)), b)| \leqslant \psi(\Phi(\alpha)) \qquad (\alpha \in [0, a)),$$

it follows that Eq. (3) remains valid for all $\alpha \in [0, a]$.

Now let $\tilde{T} \triangleq [0, a]$, $\tilde{R}^{\#} \triangleq R_\infty{}^{\#} \circ \Phi$, $\tilde{\mathscr{R}}^{\#}$ be the collection of measurable selections of $\tilde{R}^{\#}$, and the point $(\tilde{y}, \tilde{\rho}, b) \in C([0, a], \mathbb{R}^n) \times \tilde{\mathscr{R}}^{\#} \times B$ satisfy Eq. (3) for all $\alpha \in [0, a]$. Then we set

$$y \triangleq \tilde{y} \circ \Psi \qquad \text{and} \qquad \rho \triangleq \tilde{\rho} \circ \Psi$$

and apply I.4.43 to relation (3) to obtain relation (2). We also verify that

$$\lim_{t \to \infty} y(t) = \lim_{\alpha \to a} \tilde{y}(\alpha) = \tilde{y}(a)$$

exists. Thus the infinite time problem is actually equivalent to one of the type described in VI.0, but with T, $R^{\#}$, and $\mathscr{R}^{\#}$ replaced by $\tilde{T}$, $\tilde{R}^{\#}$, and $\tilde{\mathscr{R}}^{\#}$, respectively.

VI.5.4 ***Staging*** Most present-day space vehicles have multiple *stages*, that is, rocket engines mounted on top of one another with the payload (or personnel module) on top of the highest stage. The lowest stage is activated first and, when its fuel is used up, it is jettisoned and, when needed, the thrust is provided by the second stage, and then the following ones. The control of a k-stage rocket thus involves k additional parameters, namely the times when each of the rocket stages is discarded. The state of the vehicle at any time t is described by such quantities as its position and velocity components, its orientation (if important), and its mass (which varies with the time because of the expenditure of fuel and the jettisoning of stages). If

$$y(t) \triangleq \big(y^1(t),..., y^n(t)\big)$$

is the vector of data that describe the state of the vehicle at time t, $\rho(t)$ the setting of the (original) control function, and b the choice of control parameters, then y satisfies a system of ordinary differential equations of the form

$$\dot{y}(t) = f\big(t, y(t), \rho(t), b\big) \tag{1}$$

during any interval of time that does not include the jettisoning of the stages. If the j th stage is jettisoned at the time τ_j, then at least one of the components of y (the one describing the contribution of the j th stage and its fuel to the mass of the vehicle or the one describing the total instantaneous mass of the vehicle) is discontinuous at $t = \tau_j$ and the values

$$y(\tau_j -) \triangleq \lim_{\substack{t\to\tau_j \\ t<\tau_j}} y(t) \qquad \text{and} \qquad y(\tau_j +) \triangleq \lim_{\substack{t\to\tau_j \\ t>\tau_j}} y(t)$$

are related (the mass of the vehicle just before the jettisoning of the jth stage exceeding the mass after the jettisoning by the mass of the discarded stage).

A little reflection will show that this *rocket staging problem*, with an objective function and restrictions defined as in VI.0 but with the added staging features, is a special case of the following *mathematical staging problem*: We consider an interval

$$T' \triangleq [t_0', t_1'] \subset T \triangleq [t_0, t_1]$$

(with t_0' and t_1' either preassigned or left to choice), an open $V \subset \mathbb{R}^n$, $A \subset V$, functions $y : T' \to \mathbb{R}^n$ and $\rho : T' \to R$, and points

$$b \in B, \qquad \tau_0, \tau_1, \tau_2,..., \tau_{k+1} \in T', \qquad \text{and} \qquad a_0, a_1,..., a_k \in V$$

such that $t_0' \triangleq \tau_0 \leqslant \tau_1 \leqslant \cdots \leqslant \tau_k \leqslant \tau_{k+1} \triangleq t_1'$, y is continuous on each interval $[\tau_j, \tau_{j+1})$ ($j = 0,..., k$), and ρ is measurable on T'. We require that

$y(\tau_j-) \triangleq \lim_{t\to\tau_j, t<\tau_j} y(t)$ exist for $j = 1, 2,..., k + 1$ and that the following relations be satisfied:

(2) $$y(t) = a_j + \int_{\tau_j}^{t} f(\tau, y(\tau), \rho(\tau), b)\, d\tau \quad (j = 0, 1,..., k,\ t \in [\tau_j, \tau_{j+1})),$$

(3) $$\phi_j(\tau_j, a_j, y(\tau_j-), b) = 0 \quad (j = 1,..., k), \qquad h_1(\tau_{k+1}, y(\tau_{k+1}-), b) = 0,$$

and

(4) $$h_2(t, y(t), b) \in A \qquad (t \in [\tau_j, \tau_{j+1}),\ j = 0, 1,..., k),$$

where $f : T \times V \times R \times B \to \mathbb{R}^n$, $\phi_j : T \times V \times \mathbb{R}^n \times B \to \mathbb{R}^l$,

$$h_1 : T \times V \times B \to \mathbb{R}^m,\ A \subset \mathbb{R}^{m_2} \quad \text{and} \quad h_2 : T \times V \times B \to \mathbb{R}^{m_2}$$

are given. Here the first k relations in (3) correspond to the staging conditions relating the values of y immediately before and immediately after the jettisoning of the jth stage. The objective function is $h_0(\tau_{k+1}, y(\tau_{k+1}), b)$.

We next show how to transform this problem into the form of VI.0. We set

(5) $$\begin{aligned} y_j(s) &\triangleq y(\tau_j + s \cdot [\tau_{j+1} - \tau_j]), \\ \rho_j(s) &\triangleq \rho(\tau_j + s \cdot [\tau_{j+1} - \tau_j]) \qquad (s \in [0, 1),\ j = 0, 1,..., k) \end{aligned}$$

and

$$y_j(1) \triangleq y(\tau_{j+1}-) \qquad (j = 0, 1,..., k).$$

Then relations (2)–(4) can be rewritten as

(6) $$y_j(s) = a_j + \int_0^s (\tau_{j+1} - \tau_j) f(\tau_j + \alpha \cdot [\tau_{j+1} - \tau_j], y_j(\alpha), \rho_j(\alpha), b)\, d\alpha \qquad (j = 0, 1,..., k,\quad s \in [0, 1]),$$

(7) $$\phi_j(\tau_j, a_j, y_{j-1}(1), b) = 0 \quad (j = 1,..., k), \qquad h_1(\tau_{k+1}, y_k(1), b) = 0,$$

and

(8) $$h_2(\tau_j + s \cdot [\tau_{j+1} - \tau_j], y_j(s), b) \in A \qquad (s \in [0, 1],\ j = 0, 1,..., k).$$

We also set

$$\tilde{y}(s) \triangleq (y_0(s),..., y_k(s)) \quad \text{and} \quad \tilde{\rho}(s) \triangleq (\rho_0(s),..., \rho_k(s)) \quad \text{for} \quad s \in [0, 1],$$

$$\tilde{a} \triangleq (a_0, ..., a_k) \in V^{k+1},$$

$$\tilde{\tau} \triangleq (\tau_0, \tau_1, ..., \tau_{k+1}) \in \mathscr{T} \triangleq \{(\alpha_0, ..., \alpha_{k+1}) \in T^{k+2} \mid \alpha_0 \leqslant \cdots \leqslant \alpha_{k+1}\},$$

$$\tilde{v} \triangleq (v_0, ..., v_k) \in V^{k+1}, \qquad \tilde{A} \triangleq A^{k+1},$$

$$\tilde{B} \triangleq B \times V^{k+1} \times \mathscr{T}, \quad \tilde{R} \triangleq R^{k+1}, \quad \tilde{r} \triangleq (r_0, ..., r_k) \in \tilde{R}, \quad \tilde{b} \triangleq (b, \tilde{a}, \tilde{\tau}) \in \tilde{B},$$

and, for $s \in [0, 1]$, $\tilde{v} \in V^{k+1}$, $\tilde{r} \in \tilde{R}$, and $\tilde{b} \in \tilde{B}$,

$$\tilde{f}(s, \tilde{v}, \tilde{r}, \tilde{b}) \triangleq ((\tau_1 - \tau_0) f(\tau_0 + s \cdot [\tau_1 - \tau_0], v_0, r_0, b),\ldots, (\tau_{k+1} - \tau_k) f(\tau_k + s \cdot [\tau_{k+1} - \tau_k], v_k, r_k, b)),$$

$$\tilde{h}_0(\tilde{v}, \tilde{b}) \triangleq h_0(\tau_{k+1}, v_k, b),$$

$$\tilde{h}_1(\tilde{v}, \tilde{b}) \triangleq (\phi_1(\tau_1, a_1, v_0, b),\ldots, \phi_k(\tau_k, a_k, v_{k-1}, b), h_1(v_k, b)),$$

$$\tilde{h}_2(\alpha, \tilde{v}, \tilde{b}) \triangleq (h_2(\tau_0 + \alpha \cdot (\tau_1 - \tau_0), v_0, b),\ldots, h_2(\tau_k + \alpha \cdot (\tau_{k+1} - \tau_k), v_k, b)),$$

and

$$\xi(\tilde{b}) \triangleq \tilde{a}.$$

Then relations (6)–(8) become

$$\tilde{y}(s) = \tilde{\xi}(\tilde{b}) + \int_0^s \tilde{f}(\alpha, \tilde{y}(\alpha), \tilde{\rho}(\alpha), \tilde{b})\, d\alpha \qquad (s \in [0, 1]) \tag{9}$$

and

$$\tilde{h}_1(\tilde{y}(1), \tilde{b}) = 0, \qquad \tilde{h}_2(s, \tilde{y}(s), \tilde{b}) \in A \quad (s \in [0, 1]), \tag{10}$$

and the objective function is $\tilde{h}_0(\tilde{y}(1), \tilde{b})$.

This problem is now of the form considered in VI.5.1 (variable initial conditions).

VI.5.5 ***Advance–Delay Differential Equations*** An equation of the form

$$y(t) = v_0 + \int_{t_0}^{t} \phi(\tau, y(\tau - d_1), y(\tau - d_2),\ldots, y(\tau - d_k))\, d\tau \qquad (t \in T \triangleq [t_0, t_1]),$$

with $d_1,\ldots, d_k$ nonnegative numbers, is referred to as a "delay-differential equation," and this terminology continues to be used in the more general case where $d_1,\ldots, d_k$ are replaced by $d_1(\tau),\ldots, d_k(\tau)$ and each $d_i(\cdot)$ is a nonnegative function. If the d_i's contain both positive and negative numbers or functions, then we may refer to this equation as an "advance-delay differential equation."

One class of problems that arises in the optimal control of advance-delay differential equations is characterized, in its original version, by equations of motion of the form

$$y(t) = v_0 + \int_{t_0}^{t} f(\tau, y(\tau - d_1(\tau)),\ldots, y(\tau - d_k(\tau)), \rho(\tau), b)\, d\tau \qquad (t \in T \triangleq [t_0, t_1]).$$

Such problems are special cases of those that we shall study in Chapter VII. At the present time, we shall consider another class of problems with equations of motion that are in some respects more general and in other respects more particular than the equation above. These problems are more general because their equations of motion are of the form

$$y(t) = v_0 + \int_{t_0}^{t} f\big(\tau, y(\tau - d_1(\tau)),..., y(\tau - d_k(\tau)), \rho(\tau - d_1(\tau)), ...,$$
$$\rho(\tau - d_k(\tau)), b\big)\, d\tau \qquad (t \in T).$$

They are more particular because the "advances" and "delays" d_i define functions

$$\tau \to \tau - d_i(\tau) : T \to \mathbb{R} \qquad (\tau \in T, \quad i = 1, 2,..., k)$$

that are "iterates" of the same function. (In the special case where $d_1, d_2,..., d_k$ are constant and multiples of some number d, each function $\tau \to \tau - d_i$ is an iterate of $\tau \to \tau - d$.)

Specifically, we consider the problem of VI.0 except that the function $F : \mathscr{Y} \times \mathscr{R}^{\#} \times B \to \mathscr{Y}$ is defined in the following manner. Let $p(\cdot) : T \to \mathbb{R}$ be absolutely continuous,

$$\dot{p}(t) > 0 \quad \text{a.e. in } T, \qquad p(t) < t \quad (t \in T), \qquad \text{and} \qquad p(t_1) \geqslant t_0 .$$

Then p is increasing and injective, and there exists $c > 0$ such that

$$t - p(t) \leqslant c \qquad (t \in T).$$

We set

$$p_0(t) \triangleq t \qquad \text{and} \qquad p_{i+1}(t) \triangleq p \circ p_i(t) \qquad (t \in T)$$

for all $i \in \{0, 1, 2,...\}$ for which the composition is defined, and observe that, for some positive integer l not exceeding $| t_1 - t_0|/c$, we have

$$p_{l+1}(t_1) \leqslant t_0 < p_l(t_1),$$

the sets $p_i((p(t_1), t_1])$ $(i = 0, 1,..., l)$ are disjoint and cover $(t_0, t_1]$, and each $p_i(\cdot)$ $(i = 0, 1,..., l)$ is absolutely continuous on $[p(t_1), t_1]$.

We shall refer to a point $\tau' \in T$ as a p–translate of $\tau'' \in T$ if $\tau' = p_i(\tau'')$ or $\tau'' = p_i(\tau')$ for some $i = 0, 1, 2,...$. In the problem that we wish to consider, the relation $y = F(y, \rho, b)$ is a functional-differential equation of the form

$$(1) \qquad y(t) = v_0 + \int_{t_0}^{t} \phi(y, \rho, b)(\tau)\, d\tau \qquad (t \in T),$$

where $\phi(y, \rho, b)(\tau)$ depends on b, τ, and the values of y and ρ at the p–translates of τ. We assume, without loss of generality, that $t_0 = p_{l+1}(t_1)$; indeed, if this is not the case, we replace t_0 by $t_0' \triangleq p_{l+1}(t_1)$, T by $T' \triangleq [t_0', t_1]$, and set

$$\phi(y, \rho, b)(\tau) \triangleq 0 \qquad \text{for} \quad \tau \in [t_0', t_0) \quad \text{and} \quad \text{all } (y, \rho, b).$$

We shall reduce our problem to the form considered in VI.0 by partitioning $(t_0, t_1]$ into the disjoint intervals $p_j((p(t_1), t_1])$ $(j = 0, 1, \ldots, l)$ and then defining new state functions y_j and control functions ρ_j with the domain $[p(t_1), t_1]$ satisfying the relations

$$y_j(\alpha) = y \circ p_j(\alpha), \qquad \rho_j(\alpha) = \rho \circ p_j(\alpha) \qquad (\alpha \in [p(t_1), t_1], \quad j = 0, 1, \ldots, l).$$

In more precise terms, let $n, m, m_2 \in \mathbb{N}$, $T \triangleq [t_0, t_1] \subset \mathbb{R}$, $\mathscr{Y} = C(T, \mathbb{R}^n)$, V be an open subset of $\mathbb{R}^n$, $A \subset V$, $\tilde{V} \triangleq V^{l+1}$, $\tilde{R} \triangleq R^{l+1}$,

$$f : T \times \tilde{V} \times \tilde{R} \times B \to \mathbb{R}, \qquad (h_0, h_1) : V \times B \to \mathbb{R} \times \mathbb{R}^m,$$

and

$$h_2 : T \times V \times B \to \mathbb{R}^{m_2}.$$

With $p(\cdot)$ and l defined as before, we assume that $p_{l+1}(t_1) = t_0$. For all $(y, \rho, b) \in \mathscr{Y} \times \mathscr{R}^\# \times B$, we set

$$\begin{aligned}
&y_j(\alpha) \triangleq y(p_j(\alpha)), \qquad \rho_j(\alpha) \triangleq \rho(p_j(\alpha)) \qquad (j = 0, 1, \ldots, l, \\
&\qquad\qquad\qquad\qquad\qquad\qquad \alpha \in (p(t_1), t_1]), \\
&\tilde{y}(\alpha) \triangleq (y_0(\alpha), \ldots, y_l(\alpha)), \qquad \tilde{\rho}(\alpha) = (\rho_0(\alpha), \ldots, \rho_l(\alpha)) \\
&\qquad\qquad\qquad\qquad\qquad\qquad (\alpha \in (p(t_1), t_1]), \\
&\tilde{y}(p_i(\alpha)) \triangleq \tilde{y}(\alpha), \qquad \tilde{\rho}(p_i(\alpha)) \triangleq \tilde{\rho}(\alpha) \qquad (i = 1, \ldots, l, \quad \alpha \in (p(t_1), t_1]), \\
&\tilde{y}(t_0) \triangleq \tilde{y}(t_1), \qquad \tilde{\rho}(t_0) \triangleq \tilde{\rho}(t_1).
\end{aligned} \tag{2}$$

If the function $\tau \to f(\tau, \tilde{y}(\tau), \tilde{\rho}(\tau), b)$ is integrable on T, we set

$$F(y, \rho, b)(t) = v_0 + \int_{t_0}^{t} f(\tau, \tilde{y}(\tau), \tilde{\rho}(\tau), b)\, d\tau \qquad (t \in T)$$

and

$$g_i(y, \rho, b) \triangleq h_i(y(t_1), b) \quad (i = 0, 1), \qquad g_2(y, \rho, b)(t) \triangleq h_2(t, y(t), b) \qquad (t \in T);$$

otherwise we set

$$F(y, \rho, b)(t) \triangleq y(t) + (1, 0, \ldots, 0), \qquad g_0 = 0, \qquad g_1 = 0, \qquad g_2 = 0.$$

With $\mathscr{X}_2 \triangleq C(T, \mathbb{R}^{m_2})$ and $C_2 \triangleq \{c \in \mathscr{X}_2 \mid c(T) \subset A\}$, our original problem is as defined in V.0.

We transform the problem thus defined into the form discussed in VI.5.1 in the following manner: We observe that

$$\tilde{y}(\tau') = \tilde{y}(\tau'') \qquad \text{and} \qquad \tilde{\rho}(\tau') = \tilde{\rho}(\tau'')$$

if τ' is a p-translate of τ''. Thus we have

$$f(\tau, \tilde{y}(\tau), \tilde{\rho}(\tau), b) = f(p_i(\alpha), \tilde{y}(\alpha), \tilde{\rho}(\alpha), b) \tag{3}$$

for $\tau = p_i(\alpha)$ and $i = 0, 1,..., l$. We now set, for all $(y, \rho, b) \in \mathscr{Y} \times \mathscr{R}^\# \times B$ and $\alpha \in (p(t_1), t_1]$,

$$\begin{aligned} &f_i(\alpha, \tilde{y}(\alpha), \tilde{\rho}(\alpha), b) \triangleq f(p_i(\alpha), \tilde{y}(\alpha), \tilde{\rho}(\alpha), b)\, \dot{p}_i(\alpha) \qquad (i = 0, 1,..., l) \\ &\tilde{f}(\alpha, \tilde{y}(\alpha), \tilde{\rho}(\alpha), b) \triangleq (f_0(\alpha, \tilde{y}(\alpha), \tilde{\rho}(\alpha), b),..., f_l(\alpha, \tilde{y}(\alpha), \tilde{\rho}(\alpha), b)). \end{aligned} \tag{4}$$

We observe that, by I.4.43 and (2)–(4), the equation $y = F(y, \rho, b)$ is equivalent to the system

$$\begin{aligned} y(p_i(\alpha)) &= y(p_{i+1}(t_1)) + \int_{p_{i+1}(t_1)}^{p_i(\alpha)} f(\tau, \tilde{y}(\tau), \tilde{\rho}(\tau), b)\, d\tau \\ &= y(p_{i+1}(t_1)) + \int_{p(t_1)}^{\alpha} f_i(\beta, \tilde{y}(\beta), \tilde{\rho}(\beta), b)\, d\beta \\ &\qquad (\alpha \in [p(t_1), t_1], \quad i = 0, 1,..., l), \\ y(p_{l+1}(t_1)) &= v_0\,; \end{aligned}$$

hence, setting

$$a \triangleq (a_0, a_1,..., a_l) \triangleq (y \circ p_1(t_1), y \circ p_2(t_1),..., y \circ p_l(t_1), v_0),$$

we have

$$\tilde{y}(\alpha) = a + \int_{p(t_1)}^{\alpha} \tilde{f}(\beta, \tilde{y}(\beta), \tilde{\rho}(\beta), b)\, d\beta \qquad (\alpha \in [p(t_1), t_1]).$$

Finally, let

$$\begin{gathered} \tilde{R}^\#(\alpha) \triangleq R^\#(\alpha) \times R^\#(p_1(\alpha)) \times \cdots \times R^\#(p_l(\alpha)) \qquad (\alpha \in [p(t_1), t_1]), \\ \tilde{B} \triangleq B \times \mathbb{R}^{nl} \end{gathered}$$

and

$$\xi(\tilde{b}) \triangleq (a_0, a_1,..., a_{l-1}, v_0) \triangleq a \qquad [\tilde{b} \triangleq (b, a_0,..., a_{l-1}) \in \tilde{B}].$$

Then $\rho \in \mathscr{R}^\#$ if and only if $\tilde{\rho}$ is a measurable selection of $\tilde{R}^\#$.

Thus our problem is equivalent to that of VI.5.1 with T replaced by $[p(t_1), t_1]$, $R^{\#}$ by $\tilde{R}^{\#}$, B by $\tilde{B}$, f by $\tilde{f}$, and with the restrictions $h_1(y(t_1), b) = 0$ and $h_2(t, y(t), b) \in A$ $(t \in T)$ replaced by

$$\tilde{y}_{j+1}(t_1) - a_j = 0 \quad (j = 0, 1,..., l-1), \qquad h_1(\tilde{y}(t_1), b) = 0,$$

and

$$h_2(p_i(\alpha), \tilde{y}_i(\alpha), b) \in A \qquad (i = 0, 1,..., l, \quad \alpha \in [p(t_1), t_1]).$$

Notes

The results of VI.1 and VI.2 are (equivalent to) generalizations of Theorem V.6.1. The necessary conditions for a relaxed minimum in a form analogous to (but less general than) VI.2.4 were derived in Warga [3, 5] and similar conditions for an original minimum were obtained by Neustadt [3, 4]. A slightly less general form of Theorem VI.2.5 was derived in Warga [2].

Theorem VI.3.1 is based on the Filippov–Castaing theorem I.7.8 which was first established, under somewhat different conditions, by Filippov [1] and has been known as the "Filippov lemma." Theorem VI.3.2 is closely related to the ideas implicit in certain existence proofs in the optimal control of differential equations, and specifically in Filippov [1], Warga [1], and Ważewski [1, 2]. Theorem VI.3.2 essentially demonstrates that, in the optimal control of differential equations, whatever can be achieved with arbitrary selections of $\bar{R}^{\#}$ can also be achieved with measurable selections of $\bar{R}^{\#}$. Theorem VI.3.3 is equivalent to a slight generalization of the existence theorem first proven by Filippov [1].

Problems with unbounded contingent sets were studied by Cesari [1–3, 6] and McShane [1–5] under assumptions that precluded the need to consider discontinuous state functions. Problems of various types with possibly discontinuous original state functions were considered by Schmaedeke [1], Rishel [1], and Neustadt [2] and, under conditions analogous to those in VI.4.5, in Warga [7]. I wish to point out that the arguments in Warga [7] are incomplete, but they can be completed by the addition of Theorem VI.4.5.

Problems with staging and advance-delay differential equations were considered in Warga [10].

CHAPTER VII

Optimal Control of Functional-Integral Equations in $C(T, \mathbb{R}^n)$

VII.0 Formulation of the Problem

We shall consider in this chapter the control problem of Chapter V in the special case where $n \in \mathbb{N}$, $\mathscr{Y} = C(T, \mathbb{R}^n)$, and the function

$$F : \mathscr{Y} \times \mathscr{S}^{\#} \times B \to \mathscr{Y}$$

is defined by an expression of the type discussed in II.5. Specifically, we assume that we are given $k, n \in \mathbb{N}$, $W \subset \mathbb{R}^n$, $V \subset \mathbb{R}^k$, and functions

$$\xi : C(T, W) \to L^\infty(T, \Sigma, \mu, V) \qquad \text{and} \qquad f : T \times T \times V \times R \times B \to \mathbb{R}^n.$$

We set $\mathscr{Y} = C(T, \mathbb{R}^n)$ and

$$F(y, \sigma, b)(t) \triangleq \int f(t, \tau, \xi(y)(\tau), \sigma(\tau), b)\, \mu(d\tau) \qquad (t \in T)$$

for all $(y, \sigma, b) \in \mathscr{Y} \times \mathscr{S}^\# \times B$ for which the integral on the right defines a continuous function of t; for all other $(y, \sigma, b) \in \mathscr{Y} \times \mathscr{S}^\# \times B$ we set

$$F(y, \sigma, b)(t) \triangleq y(t) + (1, 0, \ldots, 0) \qquad (t \in T).$$

With $\mathscr{Y}$ and the function $F : \mathscr{Y} \times \mathscr{S}^\# \times B \to \mathscr{Y}$ as just defined, our problem is that of Chapter V. We observe that for $(y, \sigma, b) \in \mathscr{Y} \times \mathscr{S}^\# \times B$, we have $y = F(y, \sigma, b)$ if and only if

$$y(t) = \int f(t, \tau, \xi(y)(\tau), \sigma(\tau), b)\, \mu(d\tau) \qquad (t \in T).$$

As in Chapter V, we set

$$\mathscr{A}(\mathscr{S}^\#) \triangleq \{(y, \sigma, b) \in \mathscr{Y} \times \mathscr{S}^\# \times B \mid y = F(y, \sigma, b),\ g_1(y, \sigma, b) = 0,\ g_2(y, \sigma, b) \in C_2\}$$

and

$$\mathscr{H}(\mathscr{S}^\#) \triangleq \{(y, \sigma, b) \in \mathscr{Y} \times \mathscr{S}^\# \times B \mid y = F(y, \sigma, b)\}.$$

We shall refer to the control problem just described as a *unilateral problem* if $\mathscr{X}_2$, C_2, and g are defined as follows. We assume given $m_2 \in \{0, 1, 2, \ldots\}$, a compact metric space P, a mapping $A(\cdot) : P \to \mathscr{P}'(\mathbb{R}^{m_2})$ (the class of nonempty subsets of $\mathbb{R}^{m_2}$), and $\bar{g} : \mathscr{Y} \times B \to \mathbb{R} \times \mathbb{R}^m \times C(P, \mathbb{R}^{m_2})$; and we have $\mathscr{X}_2 \triangleq C(P, \mathbb{R}^{m_2})$, $C_2 \triangleq \{c \in \mathscr{X}_2 \mid c(p) \in A(p)\ (p \in P)\}$, and

$$g(y, \sigma, b) \triangleq \bar{g}(y, b) \qquad [(y, \sigma, b) \in \mathscr{Y} \times \mathscr{S}^\# \times B].$$

VII.1 Existence of Minimizing Solutions

VII.1.1 ***Theorem*** Let V and W be closed, and assume that

(1) B has a sequentially compact topology and C_2 is sequentially closed;

(2) $g \mid \text{seq cl}(\mathscr{A}(\mathscr{S}^\#))$ is sequentially continuous;

(3) $\mathscr{A}(\mathscr{S}^\#) \neq \varnothing$;

(4) $f \in C(T, \mathscr{B}(T, V \times R \times B; \mathbb{R}^n))$ and ξ is continuous.

Then $\mathscr{A}(\mathscr{S}^\#)$ is sequentially compact, $F \mid \mathscr{A}(\mathscr{S}^\#)$ is sequentially continuous, and there exists a minimizing relaxed solution.

▌PROOF By the definition of F, every point $(y, \sigma, b) \in \mathscr{Y} \times \mathscr{S}^\# \times B$ satisfying the equation $y = F(y, \sigma, b)$ is also a solution of the equation

$$y(t) = \int f(t, \tau, \xi(y)(\tau), \sigma(\tau), b)\, \mu(d\tau) \qquad (t \in T). \tag{5}$$

The set $Y \triangleq \mathrm{pr}_{\mathscr{Y}}\, \mathscr{A}(\mathscr{S}^{\#})$ is bounded and equicontinuous because, by (4) and (5),

$$|y(t)| \leqslant \sup_{t \in T} \int |f(t, \tau, \cdot, \cdot, \cdot)|_{\sup}\, \mu(d\tau) < \infty$$

and

$$|y(t) - y(t')| \leqslant |f(t, \cdot, \cdot, \cdot, \cdot) - f(t', \cdot, \cdot, \cdot, \cdot)| \underset{t' \to t}{\longrightarrow} 0 \qquad (t \in T, \quad y \in Y).$$

Thus, by I.5.4, $\bar{Y}$ is sequentially compact and, by IV.3.11, $\mathscr{S}^{\#}$ is sequentially compact. It follows then from (1) that $\bar{Y} \times \mathscr{S}^{\#} \times B$ and its subset seq $\mathrm{cl}(\mathscr{A}(\mathscr{S}^{\#}))$ are sequentially compact. If $((y_j, \sigma_j, b_j))$ is a sequence in $\mathscr{A}(\mathscr{S}^{\#})$ converging to some $(\bar{y}, \bar{\sigma}, \bar{b}) \in \mathscr{Y} \times \mathscr{S}^{\#} \times B$, then, by IV.2.9,

$$\bar{y}(t) = \lim_j y_j(t) = \lim_j \int f(t, \tau, \xi(y_j)(\tau), \sigma_j(\tau), b_j)\, \mu(d\tau)$$
$$= \int f(t, \tau, \xi(\bar{y})(\tau), \bar{\sigma}(\tau), \bar{b})\, \mu(d\tau) \qquad (t \in T).$$

Since Y is equicontinuous, it follows from I.5.3 that $\bar{y}$ is continuous and $\lim y_j(t) = \bar{y}(t)$ uniformly for $t \in T$. Thus $(\bar{y}, \bar{\sigma}, \bar{b}) \in \mathscr{A}(\mathscr{S}^{\#})$, showing that $\mathscr{A}(\mathscr{S}^{\#})$ is sequentially compact and $F \mid \mathscr{A}(\mathscr{S}^{\#})$ sequentially continuous. The existence of a minimizing relaxed solution now follows from V.1.1. QED

VII.1.2 ***Theorem*** Assume that the conditions of VII.1.1 are satisfied with $\mathscr{A}(\mathscr{S}^{\#})$ replaced by $\mathscr{H}(\mathscr{S}^{\#})$. Let $(\bar{y}, \bar{\sigma}, \bar{b})$ be a minimizing relaxed solution, $\mathscr{U}$ an abundant subset of $\mathscr{R}^{\#}$, and assume, furthermore, that $\bar{y}$ is the unique solution of the equation $y = F(y, \bar{\sigma}, \bar{b})$, and the equation $y = F(y, u, \bar{b})$ has a solution y for each $u \in \mathscr{U}$ and sufficiently close to $\bar{\sigma}$ (in the metric space $\mathscr{S}$). Then there exists a sequence $((y_j, u_j))$ in $\mathscr{Y} \times \mathscr{U}$ such that $((y_j, u_j, \bar{b}))$ is a minimizing approximate $\mathscr{U}$-solution and

$$\lim_j g(y_j, u_j, \bar{b}) = g(\bar{y}, \bar{\sigma}, \bar{b}).$$

▌PROOF With $\mathscr{A}(\mathscr{S}^{\#})$ replaced by $\mathscr{H}(\mathscr{S}^{\#})$, the proof of VII.1.1 shows that $\mathscr{H}(\mathscr{S}^{\#})$ is sequentially compact and $F \mid \mathscr{H}(\mathscr{S}^{\#})$ sequentially continuous. Our conclusion then follows from V.1.2. QED

We next observe that the preceding results are applicable to the unilateral problems defined by functional-integral equations in $C(T, \mathbb{R}^n)$.

VII.1.3 ***Theorem*** Let $m_2 \in \{0, 1, 2, \ldots\}$, P be a compact metric space, $A(p)$ a closed subet of $\mathbb{R}^{m_2}$ for each $p \in P$, and

$$\bar{g} : \mathscr{Y} \times B \to \mathbb{R} \times \mathbb{R}^m \times C(P, \mathbb{R}^{m_2}).$$

Then Theorems VII.1.1 and VII.1.2 remain applicable when

$$\mathscr{X}_2 \triangleq C(P, \mathbb{R}^{m_2}),\ C_2 \triangleq \{c \in \mathscr{X}_2 \mid c(p) \in A(p)\ (p \in P)\},$$

and $g(y, \sigma, b) \triangleq \bar{g}(y, b)$ for all $(y, \sigma, b) \in \mathscr{Y} \times \mathscr{S}^{\#} \times B$.

We now consider certain conditions that enable us to better characterize a minimizing relaxed control function or to ensure the existence of a point $(\bar{y}, \bar{\rho}, \bar{b}) \in \mathscr{Y} \times \mathscr{R}^{\#} \times B$ that is both a minimizing $\mathscr{R}^{\#}$-solution and a minimizing relaxed solution.

VII.1.4 ***Theorem*** Let the conditions of Theorem VII.1.1 be satisfied and $(\bar{y}, \bar{\sigma}, \bar{b})$ be any minimizing relaxed solution (whose existence is established in VII.1.1). Assume furthermore that

(1) $g(y, \sigma, b) = \bar{g}(y, b) \quad [(y, \sigma, b) \in \mathscr{Y} \times \mathscr{S}^{\#} \times B]$;

(2) $R^{\#}(\tau)$ is a closed set for all $\tau \in T$;

(3) $f(\cdot, \cdot, v, r, b)$ is $\Sigma \otimes \Sigma$-measurable for all $(v, r, b) \in V \times R \times B$;

and that there exists $\nu \in \text{frm}^+(T)$ such that

(4) $\bar{E} = T$ whenever $\nu(E) = \nu(T)$ [or, equivalently, that $\nu(A) > 0$ for every open $A \subset T$]; and

(5) $\{f(\cdot, \tau, v, r', b\} \mid r' \in R^{\#}(\tau)\}$ is a convex subset of $\mathscr{F}(T, \Sigma, \nu, \mathbb{R}^n)$ for all $(\tau, v, b) \in T \times V \times B$.

Then there exists $\bar{\rho} \in \mathscr{R}^{\#}$ such that $(\bar{y}, \bar{\rho}, \bar{b})$ is a minimizing relaxed solution (and, *a fortiori*, a minimizing $\mathscr{R}^{\#}$-solution).

▌PROOF Let

$$h(t, \tau, r) \triangleq f(t, \tau, \xi(\bar{y})(\tau), r, \bar{b}) \qquad (t, \tau \in T, \quad r \in R).$$

We observe that, by (3), I.4.22, and VII.1.1(4), $h(\cdot, \cdot, r)$ is $\nu \times \mu$-measurable and $h(t, \tau, \cdot)$ continuous; hence, by I.4.21, $(t, \tau) \to |h(t, \tau, \cdot)|_{\sup}$ is $\nu \times \mu$-measurable and, by Fubini's theorem I.4.45(3),

$$\int |h(t, \tau, \cdot)|_{\sup}\, \nu(dt) \times \mu(d\tau) \leqslant \int \nu(dt) \int |f(t, \tau, \cdot, \cdot, \cdot)|_{\sup}\, \mu(d\tau)$$
$$\leqslant \nu(T) \cdot |f| < \infty.$$

Thus, by IV.3.14, there exists $\bar{\rho} \in \mathscr{R}^{\#}$ such that

(6)
$$\bar{y}(t) = \int f(t, \tau, \xi(\bar{y})(\tau), \bar{\sigma}(\tau), \bar{b})\, \mu(d\tau)$$
$$= \int f(t, \tau, \xi(\bar{y})(\tau), \bar{\rho}(\tau), \bar{b})\, \mu(d\tau) \qquad \text{for } \nu\text{-a.a. } t \in T,$$

say for $t \in E$. Since, by VII.1.1(4), the right side of (6) is a uniformly continuous function of t, it follows from (4) and I.2.13 that (6) remains valid for all $t \in T$ [hence $\bar{y} = F(\bar{y}, \bar{\rho}, \bar{b})$] and, by (1), that $(\bar{y}, \bar{\rho}, \bar{b})$ is a minimizing relaxed solution. QED

VII.1.5 ***Theorem*** Let the conditions of VII.1.1 be satisfied, and assume furthermore that, for all

$$(t, \tau, v, r, b, y, \sigma) \in T \times T \times V \times R \times B \times \mathscr{Y} \times \mathscr{S}^{\#},$$

we have $g(y, \sigma, b) = \bar{g}(y, b)$, $R^{\#}(\tau)$ is closed, and

$$f(t, \tau, v, r, b) = \sum_{j=1}^{l} \alpha^j(t, \tau, v, b)\beta^j(\tau, v, r, b),$$

where $l \in \mathbb{N}$,

$$\alpha^j : T \times T \times V \times B \to \mathbb{R}, \beta^j : T \times V \times R \times B \to \mathbb{R}^n,$$

$\beta^j(\tau, \cdot, \cdot, \cdot)$ are continuous, and $\beta^j(\cdot, v, r, b)$ μ-measurable. Then there exists a minimizing relaxed solution $(\bar{y}, \bar{\sigma}, \bar{b})$ such that, for all $\tau \in T$, $\bar{\sigma}(\tau)$ is a measure supported on a set of at most $ln + 1$ points of $R^{\#}(\tau)$.

If, furthermore, for all $(\tau, v, b) \in T \times V \times B$,

$$\{(\beta^1(\tau, v, r, b),..., \beta^l(\tau, v, r, b)) | \, r \in R^{\#}(\tau)\}$$

is a convex subset of $\mathbb{R}^{ln}$, then there exists $\bar{\rho} \in \mathscr{R}^{\#}$ such that $(\bar{y}, \bar{\rho}, \bar{b})$ is a minimizing relaxed solution.

PROOF By VII.1.1, there exists a minimizing relaxed solution $(\bar{y}, \tilde{\sigma}, \bar{b})$. If we set

$$X \triangleq R^{nl+1} \times \mathscr{T}'_{nl},$$

$$\Gamma(\tau) \triangleq R^{\#}(\tau)^{nl+1} \times \mathscr{T}'_{nl} \qquad (\tau \in T),$$

$$\phi^j(\tau, x) \triangleq \sum_{i=0}^{nl} \theta^i \beta^j(\tau, \xi(\bar{y})(\tau), r^i, \bar{b})$$

$$[j = 1, 2,..., l, \quad \tau \in T, \quad x \triangleq (r^0,..., r^{nl}, \theta^0,..., \theta^{nl}) \in X],$$

$$\phi \triangleq (\phi^1,..., \phi^l), \qquad \text{and} \qquad \beta \triangleq (\beta^1,..., \beta^l),$$

then, for all $(\tau, x) \in T \times X$, $\phi(\tau, \cdot)$ is continuous and $\phi(\cdot, x)$ μ-measurable. By I.7.5, Γ is μ-measurable. Finally, by I.6.14, for each $\tau \in T$ there exists $\tilde{x}(\tau) \in \Gamma(\tau)$ such that

$$\phi(\tau, \tilde{x}(\tau)) = \int \beta(\tau, \xi(\bar{y})(\tau), r, \bar{b}) \, \tilde{\sigma}(\tau) \, (dr) \triangleq \beta(\tau, \xi(\bar{y})(\tau), \tilde{\sigma}(\tau), \bar{b}).$$

Since $\tau \to \beta(\tau, \xi(\bar{y})(\tau), \tilde{\sigma}(\tau), \bar{b})$ is μ-measurable, it follows from the Filippov–Castaing theorem I.7.10 that there exists a μ-measurable selection

$$\tau \to \bar{x}(\tau) \triangleq (\rho^0(\tau),..., \rho^{nl}(\tau), \theta^0(\tau),..., \theta^{nl}(\tau))$$

of Γ such that

$$\phi(\tau, \bar{x}(\tau)) = \sum_{i=0}^{nl} \theta^i(\tau)\, \beta(\tau, \xi(\bar{y})(\tau), \rho^i(\tau), \bar{b}) = \beta(\tau, \xi(\bar{y})(\tau), \tilde{\sigma}(\tau), \bar{b}).$$

If we define $\bar{\sigma}(\tau)$ to be the probability measure with support on

$$\{\rho^0(\tau),..., \rho^{nl}(\tau)\}$$

and such that $\bar{\sigma}(\tau)(\{\rho^i(\tau)\}) = \theta^i(\tau)$ $(i = 0,..., nl)$ then it follows that $\bar{\sigma} \in \mathscr{S}^{\#}$ and

$$\beta^j(\tau, \xi(\bar{y})(\tau), \tilde{\sigma}(\tau), \bar{b}) = \beta^j(\tau, \xi(\bar{y})(\tau), \bar{\sigma}(\tau), \bar{b}) \qquad (j = 1,..., l, \quad \tau \in T);$$

hence

$$f(t, \tau, \xi(\bar{y})(\tau), \tilde{\sigma}(\tau), \bar{b}) = f(t, \tau, \xi(\bar{y})(\tau), \bar{\sigma}(\tau), \bar{b}) \qquad (t, \tau \in T)$$

and

$$\bar{y}(t) = \int f(t, \tau, \xi(\bar{y})(\tau), \bar{\sigma}(\tau), \bar{b})\, \mu(d\tau) \qquad (t \in T).$$

This shows that $(\bar{y}, \bar{\sigma}, \bar{b})$ is a minimizing relaxed solution, thus proving our first conclusion.

Now assume that $\beta(\tau, v, R^{\#}(\tau), b)$ is a convex subset of $\mathbb{R}^{ln}$ for all

$$(\tau, v, b) \in T \times V \times B.$$

Then, by I.6.13,

$$\int \beta(\tau, \xi(\bar{y})(\tau), r, \bar{b})\, \tilde{\sigma}(\tau)\,(dr) \in \beta(\tau, \xi(\bar{y})(\tau), R^{\#}(\tau), \bar{b}) \qquad (\tau \in T)$$

and therefore, for each $\tau \in T$, there exists $\tilde{\rho}(\tau) \in R^{\#}(\tau)$ such that

$$\beta(\tau, \xi(\bar{y})(\tau), \tilde{\sigma}(\tau), \bar{b}) = \beta(\tau, \xi(\bar{y})(\tau), \tilde{\rho}(\tau), \bar{b}) \qquad (\tau \in T).$$

It follows again by the Filippov–Castaing theorem I.7.10 that there exists $\bar{\rho} \in \mathscr{R}^{\#}$ such that

$$\beta(\tau, \xi(\bar{y})(\tau), \tilde{\sigma}(\tau), \bar{b}) = \beta(\tau, \xi(\bar{y})(\tau), \bar{\rho}(\tau), \bar{b}) \qquad (\tau \in T)$$

and

$$\bar{y}(t) = \int f\big(t, \tau, \xi(\bar{y})(\tau), \bar{\rho}(\tau), \bar{b}\big)\, \mu(d\tau) \qquad (t \in T).$$

Since we assume that $g(y, \sigma, b) \triangleq \bar{g}(y, b)$, it follows that $(\bar{y}, \bar{\rho}, \bar{b})$ is a minimizing relaxed solution. QED

VII.2 Necessary Conditions for a Relaxed Minimum

We investigate necessary conditions for a point $(\bar{y}, \bar{\sigma}, b) \in \mathscr{Y} \times \mathscr{S}^{\#} \times B$ to be a minimizing relaxed solution.

VII.2.1 ***Lemma*** Let $(\bar{y}, \bar{\sigma}, \bar{b}) \triangleq (\bar{y}, \bar{q}) \in C(T, W) \times \mathscr{S}^{\#} \times B$. Assume that for each point $L \triangleq (b_0, \ldots, b_m) \in B^{m+1}$ there exist a closed convex neighborhood $V^L \subset V$ and a convex neighborhood $\mathscr{T}^L$ of 0 in $\mathscr{T}_{m+1}$ such that ξ and the function

$$(t, \tau, v, r, \theta) \to f^L(t, \tau, v, r, \theta)$$

$$\triangleq f\Big(t, \tau, v, r, \bar{b} + \sum_{j=0}^{m} \theta^j (b_j - \bar{b})\Big) : T \times T \times V^L \times R \times \mathscr{T}^L \to \mathbb{R}^n$$

have the following properties:

(1) $f^L(t, \tau, \cdot, r, \cdot)$ has a derivative $f^L_{(v,\theta)}(t, \tau, v, r, \theta)$ for all

$$(t, \tau, v, r, \theta) \in T \times T \times V^L \times R \times \mathscr{T}^L;$$

(2) $f^L \in C\big(T, \mathscr{B}(T, V^L \times R \times \mathscr{T}^L; \mathbb{R}^n)\big)$;

(3) $f^L_{(v,\theta)} \in C\big(T, \mathscr{B}(T, V^L \times R \times \mathscr{T}^L; B(\mathbb{R}^k \times \mathbb{R}^{m+1}, \mathbb{R}^n))\big)$; and

(4) ξ has a continuous derivative and

$$\xi(\bar{y})(\tau) \in V^L \qquad \text{for } \mu\text{-a.a. } \tau \in T.$$

Then

(5) for every choice of

$$K \triangleq \big((\sigma_0, b_0), \ldots, (\sigma_m, b_m)\big) \in (\mathscr{S}^{\#} \times B)^{m+1}$$

and for

$$\sigma^K(\theta) \triangleq \bar{\sigma} + \sum_{j=0}^{m} \theta^j (\sigma_j - \bar{\sigma}), \qquad b^K(\theta) \triangleq \bar{b} + \sum_{j=0}^{m} \theta^j (b_j - \bar{b}) \qquad (\theta \in \mathscr{T}_{m+1}),$$

$$L \triangleq (b_0, \ldots, b_m), \qquad \text{and} \qquad Y_K \triangleq \xi^{-1}\big(L^\infty(T, \Sigma, \mu, V^L)\big),$$

there exists a neighborhood $\mathscr{T}_K$ of 0 in $\mathscr{T}_{m+1}$ such that the functions

$$(y, \sigma, \theta) \to F^K(y, \sigma, \theta) \triangleq F(y, \sigma, b^K(\theta)) : Y_K \times \mathscr{S}^\# \times \mathscr{T}_K \to C(T, \mathbb{R}^n)$$

and

$$(y, \theta) \to \tilde{F}^K(y, \theta) \triangleq F(y, \sigma^K(\theta), b^K(\theta)) : Y_K \times \mathscr{T}_K \to C(T, \mathbb{R}^n)$$

are continuous and $\tilde{F}^K$ has a continuous derivative;

(6) $$(F_y(\bar{y}, \bar{\sigma}, \bar{b})\, \Delta y)(t) = \int f_v(t, \tau, \xi(\bar{y})(\tau), \bar{\sigma}(\tau), \bar{b})\, (\xi'(\bar{y})\, \Delta y)\, (\tau)\, \mu(d\tau)$$
$$[\Delta y \in C(T, \mathbb{R}^n),\ t \in T];$$

and

(7) $$D_2F(\bar{y}, \bar{q}; q - \bar{q})(t) = \int [f(t, \tau, \xi(\bar{y})(\tau), \sigma(\tau) - \bar{\sigma}(\tau), \bar{b})$$
$$+ D_5 f(t, \tau, \xi(\bar{y})(\tau), \bar{\sigma}(\tau), \bar{b}; b - \bar{b})]\, \mu(d\tau)$$
$$[t \in T,\ q \triangleq (\sigma, b) \in \mathscr{S}^\# \times B].$$

▌PROOF Let

$$K \triangleq (q_0, \ldots, q_m) \triangleq ((\sigma_0, b_0), \ldots, (\sigma_m, b_m)) \in (\mathscr{S}^\# \times B)^{m+1},$$

and $\mathscr{T}_K \triangleq \mathscr{T}^L$. The set Y_K is a closed neighborhood of $\bar{y}$ in $C(T, \mathbb{R}^n)$ because $L^\infty(T, \Sigma, \mu, V^L)$ is a closed neighborhood of $\xi(\bar{y})$ in the Banach space $L^\infty(T, \Sigma, \mu, \mathbb{R}^k)$ and ξ is continuous.

Step 1 We first prove that F^K is continuous. Let

$$\lim_j (y_j, \sigma_j, \theta_j) = (y, \sigma, \theta) \quad \text{in } Y_K \times \mathscr{S}^\# \times \mathscr{T}_K.$$

Then

$$\tau \to f^L(t, \tau, \xi(y_j)(\tau), \sigma_j(\tau), \theta_j) \qquad \text{and} \qquad \tau \to f^L(t, \tau, \xi(y)(\tau), \sigma(\tau), \theta)$$

are μ-integrable for each $t \in T$ and $j \in \mathbb{N}$ and, by IV.2.9,

(8) $$\lim_j \int f^L(t, \tau, \xi(y_j)(\tau), \sigma_j(\tau), \theta_j)\, \mu(d\tau) = \int f^L(t, \tau, \xi(y)(\tau), \sigma(\tau), \theta)\, \mu(d\tau).$$

Furthermore,

$$\left| \int [f^L(t, \tau, \xi(y_j)(\tau), \sigma_j(\tau), \theta_j) - f^L(t', \tau, \xi(y_j)(\tau), \sigma_j(\tau), \theta_j)]\, \mu(d\tau) \right|$$
$$\leqslant \int |f^L(t, \tau, \cdot, \cdot, \cdot) - f^L(t', \tau, \cdot, \cdot, \cdot)|_{\sup}\, \mu(d\tau) \qquad (t, t' \in T)$$

and, therefore, the functions $t \to \int f^L(t, \tau, \xi(y_j)(\tau), \sigma_j(\tau), \theta_j)\, \mu(d\tau)$ are equicontinuous. It follows, by I.5.3, that the convergence in (8) is uniform for all $t \in T$ and the right-hand side of (8) is a continuous function of t; therefore

$$F(y, \sigma, b^K(\theta))(t) \triangleq F^K(y, \sigma, \theta)(t) = \int f^L(t, \tau, \xi(y)(\tau), \sigma(\tau), \theta)\, \mu(d\tau)$$
$$[(y, \sigma, \theta) \in Y_K \times \mathscr{S}^\# \times \mathscr{T}_K]$$

and

$$\lim_j F^K(y_j, \sigma_j, \theta_j) = F^K(y, \sigma, \theta) \quad \text{in } C(T, \mathbb{R}^n).$$

Thus F^K is continuous.

Step 2 We next prove that $\tilde{F}^K$ is continuous and has a continuous derivative. Let

$$\text{(9)} \quad \tilde{f}^K(t, \tau, v, \theta) \triangleq f(t, \tau, v, \sigma^K(\tau), b^K(\theta)) \triangleq f^L(t, \tau, v, \bar{\sigma}(\tau), \theta) + \sum_{j=1}^{m} \theta^j f^L(t, \tau, v, \sigma_j(\tau) - \bar{\sigma}(\tau), \theta)$$
$$(t, \tau \in T,\ v \in V^L,\ \theta \in \mathscr{T}_K).$$

By IV.2.7, for each $\sigma \in \mathscr{S}^\#$ the function

$$(t, \tau, (v, \theta)) \to h(t, \tau, (v, \theta)) \triangleq f^L(t, \tau, v, \sigma(\tau), \theta) : T \times T \times (V^L \times \mathscr{T}_K) \to \mathbb{R}^n$$

is such that $h(t, \tau, \cdot)$ has a derivative $h_{(v,\theta)}(t, \tau, (v, \theta))$,

$$h \in C(T, \mathscr{B}(T, V^L \times \mathscr{T}^L; \mathbb{R}^n))$$

and

$$h_{(v,\theta)} \in C(T, \mathscr{B}(T, V^L \times \mathscr{T}^L; B(\mathbb{R}^k \times \mathbb{R}^{m+1}, \mathbb{R}^n))).$$

It follows, by (9), that $\tilde{f}^K$ has a derivative $\tilde{f}^K_{(v,\theta)}$,

$$\tilde{f}^K \in C(T, \mathscr{B}(T, V^L \times \mathscr{T}^L; \mathbb{R}^n))$$

and

$$\tilde{f}^K_{(v,\theta)} \in C(T, \mathscr{B}(T, V^L \times \mathscr{T}^L; B(\mathbb{R}^k \times \mathbb{R}^{m+1}, \mathbb{R}^n))).$$

Therefore, we may apply II.5.8 and conclude that

$$\tilde{F}(y, \theta)(t) = F(y, \sigma^K(\theta), b^K(\theta))(t) = \int \tilde{f}^K(t, \tau, \xi(y)(\tau), \theta)\, \mu(d\tau) \qquad (y \in Y_K,\quad \theta \in \mathscr{T}_K,\quad t \in T)$$

and $\tilde{F}^K$ has a continuous derivative such that

(10)

$$\mathscr{D}\tilde{F}^K(y, \theta)(\Delta y, \Delta\theta)(t) = \int \tilde{f}^K_{(v,\theta)}\,(t, \tau, \xi(y)(\tau), \theta)((\xi'(y)\,\Delta y)(\tau), \Delta\theta)\,\mu(d\tau)$$

$$[y \in Y_K,\ \theta \in \mathscr{T}_K,\ \Delta y \in C(T, \mathbb{R}^n),\ \Delta\theta \in \mathbb{R}^{m+1},\ t \in T].$$

This completes the proof of statement (5).

Step 3 Finally, we prove relations (6) and (7). If we set $\bar{K} \triangleq ((\bar{\sigma}, \bar{b}),..., (\bar{\sigma}, \bar{b}))$ and $\bar{L} \triangleq (\bar{b},..., \bar{b})$, then

$$\tilde{F}^{\bar{K}}(y, \theta) = F(y, \bar{\sigma}, \bar{b}) \qquad [y \in C(T, \mathbb{R}^n), \quad \theta \in \mathscr{T}_{\bar{K}}]$$

and

$$\tilde{f}^{\bar{K}}(t, \tau, v, \theta) = f(t, \tau, v, \bar{\sigma}(\tau), \bar{b}) \qquad (t, \tau \in T, \quad v \in V^L, \quad \theta \in \mathscr{T}_{\bar{K}}),$$

and relation (6) follows from (10) by setting $\Delta\theta = 0$. If, for an arbitrary $q \triangleq (\sigma, b) \in \mathscr{S}^{\#} \times B$, we set

$$K \triangleq ((\sigma, b), (\bar{\sigma}, \bar{b}),..., (\bar{\sigma}, \bar{b})) \qquad \text{and} \qquad L \triangleq (b, \bar{b},..., \bar{b}),$$

then

$$\tilde{F}^K(y, \theta) = F(y, \bar{\sigma} + \theta^0(\sigma - \bar{\sigma}), \bar{b} + \theta^0(b - \bar{b})) \ [y \in \mathscr{Y}, \theta \triangleq (\theta^0,..., \theta^m) \in \mathscr{T}_K]$$

and

$$\begin{aligned}\tilde{f}^K(t, \tau, v, \theta) = f(&t, \tau, v, \bar{\sigma}(\tau), \bar{b} + \theta^0(b - \bar{b})) \\ &+ \theta^0 f(t, \tau, v, \sigma(\tau) - \bar{\sigma}(\tau), \bar{b} + \theta^0(b - \bar{b})) \\ &\qquad (t, \tau \in T, \quad v \in V^L, \quad \theta \in \mathscr{T}_K);\end{aligned}$$

hence

$$D_2F(\bar{y}, \bar{q}; q - \bar{q}) = \tilde{F}^K_{\theta^0}(\bar{y}, 0)$$

and

$$\begin{aligned}\tilde{f}^K_{\theta^0}(t, \tau, \xi(\bar{y})(\tau), 0) = {} & D_5 f(t, \tau, \xi(\bar{y})(\tau), \bar{\sigma}(\tau), \bar{b}; b - \bar{b}) \\ &+ f(t, \tau, \xi(\bar{y})(\tau), \sigma(\tau) - \bar{\sigma}(\tau), \bar{b}) \qquad (t, \tau \in T).\end{aligned}$$

Relation (7) now follows from (10) by setting $\Delta y = 0$ and $\Delta\theta = (1, 0,..., 0)$. QED

VII.2.2 ***Lemma*** Under the conditions of Lemma VII.2.1, let the equation

$$\Delta y(t) = \int f_v(t, \tau, \xi(\bar{y})(\tau), \bar{\sigma}(\tau), \bar{b})(\xi'(\bar{y})\,\Delta y)(\tau)\,\mu(d\tau) \qquad (t \in T)$$

have $\Delta y = 0$ as its only solution in $\mathscr{Y}$. Then

(1) for each choice of $\lambda \in \mathrm{frm}^+(T)$, $F_y(\bar{y}, \bar{\sigma}, \bar{b})$ has a resolvent kernel (k^*, μ^*) such that k^* is $\lambda \times \mu^*$-measurable and

$$k^* \in C(T, L^1(T, \Sigma, \mu^*, B(\mathbb{R}^n, \mathbb{R}^n))).$$

(2) If, furthermore,

$$\lim_i \int f_v(t, \tau, \xi(\bar{y})(\tau), \bar{\sigma}(\tau), \bar{b})(\xi'(\bar{y})\, \Delta y_i)(\tau)\, \mu(d\tau) = 0$$

whenever $\Delta y_i \in \mathscr{Y}$, $| \Delta y_i | \leqslant 1$, and $\lim_i \mu(\{t \in T \mid \Delta y_i(t) \neq 0\}) = 0$, then we may assume that $\mu^* = \mu$.

▌PROOF Let $\mathscr{F} \triangleq F_y(\bar{y}, \bar{\sigma}, \bar{b})$, $\eta \triangleq \xi'(\bar{y})$, and

$$k(t, \tau) \triangleq f_v(t, \tau, \xi(\bar{y})(\tau), \bar{\sigma}(\tau), \bar{b}) \qquad (t, \tau \in T).$$

Then, by VII.2.1(6),

$$\mathscr{F}(\Delta y)(t) = \int k(t, \tau)(\eta\, \Delta y)(\tau)\, \mu(d\tau) \qquad (t \in T),$$

and we deduce from VII.2.1 [(1)–(3), with $L = (\bar{b}, \ldots, \bar{b})$], IV.2.7, and I.5.25, that

$$k \in C(T, L^1(T, \Sigma, \mu, B(\mathbb{R}^k, \mathbb{R}^n))).$$

Our conclusions now follow from II.5.5. QED

VII.2.3 ***Theorem*** Let C_2 be convex and have a nonempty interior, ξ have a continuous derivative, and $(\bar{y}, \bar{q}) \triangleq (\bar{y}, \bar{\sigma}, \bar{b}) \in \mathscr{Y} \times \mathscr{S}^\# \times B$. We assume that for each point

$$K \triangleq (q_0, \ldots, q_m) \triangleq ((\sigma_0, b_0), \ldots (\sigma_m, b_m)) \in (\mathscr{S}^\# \times B)^{m+1}$$

there exist a closed convex neighborhood $V_K \subset V$ containing $\xi(\bar{y})(\tau)$ for μ-a.a.$\tau \in T$ and a convex neighborhood $\mathscr{T}_K$ of 0 in $\mathscr{T}_{m+1}$ such that, setting

$$Y_K \triangleq \xi^{-1}(L^\infty(T, \Sigma, \mu, V_K)),$$

the functions

$$(t, \tau, v, r, \theta) \to f^K(t, \tau, v, r, \theta)$$

$$\triangleq f\left(t, \tau, v, r, \bar{b} + \sum_{j=0}^{m} \theta^j (b_j - b_m)\right) : T \times T \times V_K \times R \times \mathscr{T}_K \to \mathbb{R}^n$$

and

$$(y, \theta) \to \tilde{g}^K(y, \theta) \triangleq g\left(y, \bar{q} + \sum_{j=0}^{m} \theta^j(q_j - \bar{q})\right) : Y_K \times \mathcal{T}_K \to \mathbb{R} \times \mathbb{R}^m \times \mathcal{X}_2$$

have the following properties:

(1) for all $(t, \tau, v, r, \theta) \in T \times T \times V_K \times R \times \mathcal{T}_K$, the function $f^K(t, \tau, \cdot, r, \cdot)$ has a derivative $f^K_{(v,\theta)}(t, \tau, v, r, \theta)$;

(2) $f^K \in C(T, \mathcal{B}(T, V_K \times R \times \mathcal{T}_K ; \mathbb{R}^n))$;

(3) $f^K_{(v,\theta)} \in C(T, \mathcal{B}(T, V_K \times R \times \mathcal{T}_K ; B(\mathbb{R}^k \times \mathbb{R}^{m+1}, \mathbb{R}^n)))$; and

(4) $\tilde{g}^K(\cdot, \cdot)$ is continuous and has a derivative at $(\bar{y}, 0)$.

We assume, furthermore, that

(5) the equation

$$\Delta y(t) = \int f_v(t, \tau, \xi(\bar{y})(\tau), \bar{\sigma}(\tau), \bar{b})(\xi'(\bar{y})\, \Delta y)(\tau)\, \mu(d\tau) \qquad (t \in T)$$

has $\Delta y = 0$ as its only solution in $\mathcal{Y}$.

Then

(6) $\mathcal{F} \triangleq F_y(\bar{y}, \bar{\sigma}, \bar{b})$ has a resolvent kernel (k^*, μ^*) such that

$$k^* \in C(T, L^1(T, \Sigma, \mu^*, B(\mathbb{R}^n, \mathbb{R}^n)));$$

and

(7) $$D_2F(\bar{y}, \bar{q}; q - \bar{q})(t) = \int [f(t, \tau, \xi(\bar{y})(\tau), \sigma(\tau) - \bar{\sigma}(\tau), \bar{b}) + D_5 f(t, \tau, \xi(\bar{y})(\tau), \bar{\sigma}(\tau), \bar{b}; b - \bar{b})]\, \mu(d\tau)$$
$$[t \in T, q \triangleq (\sigma, b) \in \mathcal{S}^\# \times B].$$

(8) If, furthermore, $(\bar{y}, \bar{\sigma}, \bar{b})$ is a minimizing relaxed solution, then $(\bar{y}, \bar{\sigma}, \bar{b})$ is extremal.

▌PROOF This theorem follows directly from VII.2.1, VII.2.2, and V.2.3 [with

$$L = (b_0, \ldots, b_m) \qquad \text{if} \quad K = ((\sigma_0, b_0), \ldots, (\sigma_m, b_m))$$

and V^L, $\mathcal{T}^L$, and $\mathcal{S}_K$ replaced by V_K, $\mathcal{T}_K$, and $\mathcal{S}^\#$, respectively.] QED

VII.2.4 ***Theorem*** Under the conditions of VII.2.3, assume that

$$\lim_i \int f_v(t, \tau, \xi(\bar{y})(\tau), \bar{\sigma}(\tau), \bar{b})(\xi'(\bar{y})\, \Delta y_i)(\tau)\, \mu(d\tau) = 0$$

whenever $\Delta y_i \in C(T, \mathbb{R}^n)$, $|\Delta y_i| \leqslant 1$, and $\lim_i \mu(\{t \in T \mid \Delta y_i(t) \neq 0\}) = 0$. Then the measure μ^* in VII.2.3(6) may be replaced by μ.

▌PROOF This follows from VII.2.2(2). QED

VII.2.5 ***Theorem*** Let $p : T \to [0, 1]$ be μ-measurable, $\xi'(\bar{y})$ p-hereditary, $f_v(t, \tau, v, r, \bar{b}) = 0$ if $p(\tau) > p(t)$, and the function

$$\alpha \to \mu(p^{-1}([0, \alpha])) : [0, 1] \to \mathbb{R}$$

continuous. Then condition VII.2.3(5) follows from the other assumptions of VII.2.3.

▌PROOF If we set $K \triangleq ((\bar{\sigma}, \bar{b}), \ldots, (\bar{\sigma}, \bar{b})) \in (\mathscr{S}^{\#} \times B)^{m+1}$ and

$$k(t, \tau) \triangleq f_v(t, \tau, \xi(\bar{y})(\tau), \bar{\sigma}(\tau), \bar{b}),$$

then we deduce from IV.2.7, the continuity of ξ and conditions (1)–(3) of VII.2.3 that

$$k \in C(T, L^1(T, \Sigma, \mu, B(\mathbb{R}^k, \mathbb{R}^n))),$$

and it is clear from our assumptions that $k(t, \tau) = 0$ if $p(\tau) > p(t)$. Then it follows from II.5.6 that the equation

$$\Delta y(t) = \int f_v(t, \tau, \xi(\bar{y})(\tau), \bar{\sigma}(\tau), \bar{b})(\xi'(\bar{y})\, \Delta y)(\tau)\, \mu(d\tau) \qquad (t \in T)$$

has $\Delta y = 0$ as its only solution in $C(T, \mathbb{R}^n)$. QED

VII.3 Necessary Conditions for a Relaxed Minimum in Unilateral and Related Problems

We next proceed to apply our results to problems defined by functional-integral equations in $C(T, \mathbb{R}^n)$, where $\mathscr{X}_2$ is either

$$C(P, \mathbb{R}^{m_2}) \qquad \text{or} \qquad L^q(P, \Sigma_{\mathrm{Borel}}(P), \omega, \mathbb{R}^{m_2})$$

for some compact metric space P, $\omega \in \mathrm{frm}^+(P)$, and $q \in [1, \infty)$. These problems include, in particular, unilateral problems.

VII.3.1 ***Lemma*** Let $\zeta \in \mathrm{frm}^+(T)$, $\tilde{\zeta} \in L^1(T, \Sigma, \zeta, \mathbb{R}^n)$, and

$$\tilde{f} : T \times T \times R \to \mathbb{R}^n$$

be such that $\tilde{f}(\cdot, \cdot, r)$ is $\zeta \times \mu$-measurable for all $r \in R$, $\tilde{f}(t, \tau, \cdot)$ is continuous for all $t, \tau \in T$, $(t, \tau) \to |\tilde{\zeta}(t)| |\tilde{f}(t, \tau, \cdot)|_{\sup}$ is $\zeta \times \mu$-integrable and

$$(1) \quad \int \zeta(dt) \int \mu(d\tau) \int \tilde{\zeta}(t) \cdot \tilde{f}(t, \tau, r)(\sigma - \bar{\sigma})(\tau)\,(dr) \geqslant 0 \qquad (\sigma \in \mathscr{S}^{\#}).$$

Then

(2) there exists $\tilde{T} \subset T$ such that $\mu(T \sim \tilde{T}) = 0$ and the functions

$$t \to \int \tilde{\zeta}(t) \cdot \bar{f}(t, \tau, r)\, \sigma(\tau)\,(dr) \qquad \text{and} \qquad t \to |\,\tilde{\zeta}(t)\,|\,|\bar{f}(t, \tau, \cdot)|_{\sup}$$

are ζ-integrable for all $\sigma \in \mathscr{S}^{\#}$ and $\tau \in \tilde{T}$, and

$$(3) \qquad \int \zeta(dt) \int \tilde{\zeta}(t) \cdot \bar{f}(t, \tau, r)\, \bar{\sigma}(\tau)\,(dr) = \operatorname*{Min}_{r \in \bar{R}^{\#}(\tau)} \int \tilde{\zeta}(t) \cdot \bar{f}(t, \tau, r)\, \zeta(dt)$$

for μ-a.a. $\tau \in T$.

▌PROOF Let

$$\bar{f}(t, \tau, s) \triangleq \int \bar{f}(t, \tau, r)\, s(dr) \qquad [t, \tau \in T, \quad s \in \operatorname{frm}(R)].$$

Then $\bar{f}(t, \tau, \cdot)$ is continuous on $(\operatorname{frm}(R), |\cdot|_w)$ for all $t, \tau \in T$ and, by I.5.26(1) and I.5.27, $\bar{f}(\cdot, \cdot, s)$ is $\zeta \times \mu$-measurable for all $s \in \operatorname{frm}(R)$. It follows, by I.4.22, that $(t, \tau) \to \bar{f}(t, \tau, \sigma(\tau))$ is $\zeta \times \mu$-measurable for every $\sigma \in \mathscr{S}^{\#}$. Furthermore, by I.4.48, there exists $T' \subset T$ such that $\mu(T \sim T') = 0$ and $\bar{f}(\cdot, \tau, r)$ is ζ-measurable for all $(\tau, r) \in T' \times R$; hence, by I.4.22, $\bar{f}(\cdot, \tau, \sigma(\tau))$ is ζ-measurable for all $\tau \in T'$ and $\sigma \in \mathscr{S}^{\#}$.

Since $(t, \tau) \to |\tilde{\zeta}(t)||\bar{f}(t, \tau, \cdot)|_{\sup}$ is $\zeta \times \mu$-integrable, it follows from Fubini's theorem I.4.45 that there exists $\tilde{T} \subset T'$ such that $\mu(T \sim \tilde{T}) = 0$ and the functions

$$t \to |\tilde{\zeta}(t)||\bar{f}(t, \tau, \cdot)|_{\sup} \qquad \text{and} \qquad t \to \tilde{\zeta}(t) \cdot \bar{f}(t, \tau, \sigma(\tau))$$

are ζ-integrable for all $\tau \in \tilde{T}$ and $\sigma \in \mathscr{S}^{\#}$. This proves statement (2).

We may now apply Fubini's theorem I.4.45(3) to (1) to obtain

$$(4) \qquad \int \mu(d\tau) \int \tilde{\zeta}(t) \cdot \bar{f}(t, \tau, \sigma(\tau) - \bar{\sigma}(\tau))\, \zeta(dt) \geqslant 0 \qquad (\sigma \in \mathscr{S}^{\#}).$$

By Condition IV.3.1, the set $\mathscr{R}^{\#}$ contains a subset $\{\rho_1, \rho_2, \ldots\}$ such that $\{\rho_1(\tau), \rho_2(\tau), \ldots\}$ is dense in $R^{\#}(\tau)$ for μ-a.a. $\tau \in T$, say for $\tau \in T''$. For arbitrary $j \in \mathbb{N}$ and $E \in \Sigma$ we set, in (4), $\sigma(\tau) = \rho_j(\tau)$ $(\tau \in E)$ and $\sigma(\tau) = \bar{\sigma}(\tau)$ $(\tau \notin E)$ and obtain

$$\int_E \mu(d\tau) \int \tilde{\zeta}(t) \cdot \bar{f}(t, \tau, \rho_j(\tau) - \bar{\sigma}(\tau))\, \zeta(dt) \geqslant 0 \qquad (j \in \mathbb{N}, \quad E \in \Sigma).$$

This relation and I.4.34(7) imply that there exist sets T_j $(j \in \mathbb{N})$ such that $\mu(T \sim T_j) = 0$ and

$$(5) \qquad \int \tilde{\zeta}(t) \cdot \bar{f}(t, \tau, \rho_j(\tau) - \bar{\sigma}(\tau))\, \zeta(dt) \geqslant 0 \qquad (j \in \mathbb{N}, \quad \tau \in T_j).$$

Now let $T^{\#} \triangleq T'' \cap \tilde{T} \bigcap_{j=1}^{\infty} T_j$. For each $\tau \in T^{\#}$ and $r \in R$, $t \to \tilde{\zeta}(t) \cdot \bar{f}(t, \tau, r)$ is ζ-integrable and, for all $t \in T$ and $\tau \in T^{\#}$, $\tilde{\zeta}(t) \cdot \bar{f}(t, \tau, \cdot)$ is continuous on R. Thus, by I.5.26(3), the function $(t, r) \to \tilde{\zeta}(t) \cdot \bar{f}(t, \tau, r) : T \times R \to \mathbb{R}$ is $\zeta \times s$-measurable for every $s \in \text{rpm}(R)$ and $\tau \in T^{\#}$, and

$$\int |\tilde{\zeta}(t) \cdot \bar{f}(t, \tau, r)| \, \zeta(dt) \times s(dr) \leqslant \int |\tilde{\zeta}(t)| |\bar{f}(t, \tau, \cdot)|_{\sup} \zeta(dt) < \infty.$$

It follows, by Fubini's theorem I.4.45(3), that

$$(6) \quad \int \tilde{\zeta}(t) \cdot \bar{f}(t, \tau, \bar{\sigma}(\tau)) \, \zeta(dt) \triangleq \int \zeta(dt) \int \tilde{\zeta}(t) \cdot \bar{f}(t, \tau, r) \, \bar{\sigma}(\tau)(dr)$$
$$= \int \bar{\sigma}(\tau)(dr) \int \tilde{\zeta}(t) \cdot \bar{f}(t, \tau, r) \, \zeta(dt) \quad (\tau \in T^{\#}).$$

We next observe that if $\lim_j r_j = r$ in R, then

$$\lim_j \tilde{\zeta}(t) \cdot \bar{f}(t, \tau, r_j) = \tilde{\zeta}(t) \cdot \bar{f}(t, \tau, r) \qquad (t, \tau \in T)$$

and

$$|\tilde{\zeta}(t) \cdot \bar{f}(t, \tau, r_j)| \leqslant |\tilde{\zeta}(t)| \cdot |\bar{f}(t, \tau, \cdot)|_{\sup} \qquad (t, \tau \in T, \quad j \in \mathbb{N}).$$

It follows, by the dominated convergence theorem I.4.35, that

$$\lim_j \int \tilde{\zeta}(t) \cdot \bar{f}(t, \tau, r_j) \, \zeta(dt) = \int \tilde{\zeta}(t) \cdot \bar{f}(t, \tau, r) \, \zeta(dt),$$

showing that $r \to \int \tilde{\zeta}(t) \cdot f(t, \tau, r) \, \zeta(dt) : R \to \mathbb{R}$ is continuous for all $\tau \in T^{\#}$. Since $\{\rho_1(\tau), \rho_2(\tau), ...\}$ is dense in $R^{\#}(\tau)$ for all $\tau \in T^{\#}$, it follows now from (5) that

$$\int \tilde{\zeta}(t) \cdot \bar{f}(t, \tau, \bar{\sigma}(\tau)) \, \zeta(dt) \leqslant \min_{r \in \bar{R}^{\#}(\tau)} \int \tilde{\zeta}(t) \cdot \bar{f}(t, \tau, r) \, \zeta(dt) \qquad (\tau \in T^{\#})$$

On the other hand, by (6),

$$\int \tilde{\zeta}(t) \cdot \bar{f}(t, \tau, \bar{\sigma}(\tau)) \, \zeta(dt) \geqslant \int \bar{\sigma}(\tau)(dr') \min_{r \in \bar{R}^{\#}(\tau)} \int \tilde{\zeta}(t) \cdot \bar{f}(t, \tau, r) \, \zeta(dt)$$
$$= \min_{r \in \bar{R}^{\#}(\tau)} \int \tilde{\zeta}(t) \cdot \bar{f}(t, \tau, r) \, \zeta(dt) \qquad (\tau \in T^{\#})$$

These last two relations yield (3). QED

VII.3.2 ***Theorem*** Under the conditions of Theorem VII.2.3, let $(\bar{y}, \bar{\sigma}, \bar{b})$ be a minimizing relaxed solution, $\mathscr{F} \triangleq F_y(\bar{y}, \bar{\sigma}, \bar{b})$, $m_2 \in \{0, 1, 2...\}$, P be a compact metric space, $\mathscr{X}_2 = C(P, \mathbb{R}^{m_2})$, $g(y, \sigma, b) = \bar{g}(y, b)$ for all $(y, \sigma, b) \in \mathscr{Y} \times \mathscr{S}^* \times B$, and $f(\cdot, \cdot, v, r, b)$ be $\Sigma \otimes \Sigma$-measurable for all

$(v, r, b) \in V \times R \times B$. Then there exist $l_0 \geqslant 0$, $l_1 \in \mathbb{R}^m$, $\zeta \in \text{frm}^+(T)$, $\tilde{\zeta} \in L^1(T, \Sigma, \zeta, \mathbb{R}^n)$, $\omega \in \text{frm}^+(P)$, and $\tilde{\omega} \in L^1(P, \Sigma_{\text{Borel}}(P), \omega, \mathbb{R}^{m_2})$ such that

(1) $|\tilde{\zeta}(t)| = 1 \quad (t \in T), \quad |\tilde{\omega}(p)| = 1 \quad (p \in P), \quad l_0 + |l_1| + \omega(P) > 0$

and

$$\int \tilde{\zeta}(t) \cdot x(t)\, \zeta(dt) = \sum_{i=0}^{1} l_i \mathscr{D}_1 \bar{g}_i(\bar{y}, \bar{b})(I - \mathscr{F})^{-1} x$$
$$+ \int \tilde{\omega}(p) \cdot [\mathscr{D}_1 \bar{g}_2(\bar{y}, \bar{b}) \circ (I - \mathscr{F})^{-1} x](p)\, \omega(dp)$$
$$[x \in C(T, \mathbb{R}^n)];$$

(2) $$\int \tilde{\zeta}(t) \cdot f\big(t, \tau, \xi(\bar{y})(\tau), \bar{\sigma}(\tau), \bar{b}\big)\, \zeta(dt)$$
$$= \min_{r \in \bar{R}^{\#}(\tau)} \int \tilde{\zeta}(t) \cdot f\big(t, \tau, \xi(\bar{y})(\tau), r, \bar{b}\big)\, \zeta(dt) \qquad \text{for } \mu\text{-a.a. } \tau \in T;$$

(3) $$\int \zeta(dt) \int \tilde{\zeta}(t) \cdot D_5 f\big(t, \tau, \xi(\bar{y})(\tau), \bar{\sigma}(\tau), \bar{b}; b - \bar{b}\big)\, \mu(d\tau)$$
$$+ \sum_{i=0}^{1} l_i D_2 \bar{g}_i(\bar{y}, \bar{b}; b - \bar{b}) + \int \tilde{\omega}(p) \cdot D_2 \bar{g}_2(\bar{y}, \bar{b}; b - \bar{b})(p)\, \omega(dp) \geqslant 0$$
$$(b \in B);$$

and

(4) $$\int \tilde{\omega}(p) \cdot \bar{g}_2(\bar{y}, \bar{b})(p)\, \omega(dp) = \max_{c \in C_2} \int \tilde{\omega}(p) \cdot c(p)\, \omega(dp).$$

(5) If, furthermore, $A : P \to \mathscr{P}'(\mathbb{R}^{m_2})$ is such that $A(p)$ is a convex body in $\mathbb{R}^{m_2}$ for all $p \in P$, the set

$$G(A^\circ) \triangleq \{(p, v) \mid v \in A(p)^\circ\}$$

is open in $P \times \mathbb{R}^{m_2}$ and $C_2 = \{c \in C(P, \mathbb{R}^{m_2}) \mid c(p) \in A(p)\ (p \in P)\}$, then relation (4) can be replaced by

$$\tilde{\omega}(p) \cdot \bar{g}_2(\bar{y}, \bar{b})(p) = \max_{a \in A(p)} \tilde{\omega}(p) \cdot a \qquad \text{for } \omega\text{-a.a. } p \in P.$$

▌PROOF We shall use the notation of Theorem VII.2.3. We also set $\bar{q} \triangleq (\bar{\sigma}, \bar{b})$ and (as in the definition of an extremal)

$$\chi(q) \triangleq \big(\chi_0(q), \chi_1(q), \chi_2(q)\big)$$
$$\triangleq \bar{g}_y(\bar{y}, \bar{q}) \circ (I - \mathscr{F})^{-1} D_2 F(\bar{y}, \bar{q}; q - \bar{q})$$
$$+ D_2 \bar{g}(\bar{y}, \bar{q}; q - \bar{q}) \qquad [q \triangleq (\sigma, b) \in \mathscr{S}^{\#} \times B].$$

By VII.2.3(8), there exists $l \triangleq (l_0, l_1, l_2) \in [0, \infty) \times \mathbb{R}^m \times \mathscr{X}_2{}^*$ such that $(\bar{y}, \bar{q}, l)$ is an extremal. Since $l \circ \bar{g}_y(\bar{y}, \bar{b}) \circ (I - \mathscr{F})^{-1} \in C(T, \mathbb{R}^n)^*$ and $l_2 \in C(P, \mathbb{R}^{m_2})^*$, it follows from I.5.9 that there exist

$$\zeta \in \operatorname{frm}^+(T), \qquad \tilde{\zeta} \in L^1(T, \Sigma, \zeta, \mathbb{R}^n),$$

$\omega \in \operatorname{frm}^+(P)$, and $\tilde{\omega} \in L^1(P, \Sigma_{\text{Borel}}(P), \omega, \mathbb{R}^{m_2})$ such that

$$|\tilde{\zeta}(t)| = 1 \quad (t \in T), \qquad |\tilde{\omega}(p)| = 1 \quad (p \in P),$$

(6) $$l \circ \bar{g}_y(\bar{y}, \bar{b}) \circ (I - \mathscr{F})^{-1} x = \int \tilde{\zeta}(t) \cdot x(t)\, \zeta(dt) \qquad [x \in C(T, \mathbb{R}^n)],$$

and

(7) $$l_2 v = \int \tilde{\omega}(p) \cdot v(p)\, \omega(dp) \qquad [v \in C(P, \mathbb{R}^{m_2})].$$

The third relation of (1) is equivalent to $|l| \neq 0$ and the fourth relation of (1) is equivalent to (6). It follows now from VII.2.3(7), VII.2.3(8), (6), and (7) that

(8) $$\begin{aligned} l(\chi(q)) &= \int \tilde{\zeta}(t) \cdot D_2F(\bar{y}, \bar{q}; q - \bar{q})(t)\, \zeta(dt) + l(D_2\bar{g}(\bar{y}, \bar{b}; b - \bar{b})) \\ &= \int \zeta(dt) \int \tilde{\zeta}(t) \cdot [f(t, \tau, \xi(\bar{y})(\tau), \sigma(\tau) - \bar{\sigma}(\tau), \bar{b}) \\ &\quad + D_5 f(t, \tau, \xi(\bar{y})(\tau), \bar{\sigma}(\tau), \bar{b}; b - \bar{b})]\, \mu(d\tau) \\ &\quad + \sum_{i=0}^{1} l_i D_2 \bar{g}_i(\bar{y}, \bar{b}; b - \bar{b}) + \int \tilde{\omega}(p) \cdot D_2 \bar{g}_2(\bar{y}, \bar{b}; b - \bar{b})(p)\, \omega(dp) \\ &\geqslant 0 \qquad [q \triangleq (\sigma, b) \in \mathscr{S}^\# \times B] \end{aligned}$$

and

(9) $$\int \tilde{\omega}(p) \cdot \bar{g}_2(\bar{y}, \bar{b})(p)\, \omega(dp) \geqslant \int \tilde{\omega}(p) \cdot c(p)\, \omega(dp) \qquad (c \in C_2).$$

For $\sigma = \bar{\sigma}$, relation (8) yields (3). Relation (9) yields (4) because $\bar{g}_2(\bar{y}, \bar{b}) \in C_2$. It remains therefore to prove statements (2) and (5). For $b = \bar{b}$, relation (8) yields

(10) $$\int \zeta(dt) \int \tilde{\zeta}(t) \cdot f(t, \tau, \xi(\bar{y})(\tau), \sigma(\tau) - \bar{\sigma}(\tau), \bar{b})\, \mu(d\tau) \geqslant 0 \qquad (\sigma \in \mathscr{S}^\#).$$

We let

$$\bar{f}(t, \tau, r) \triangleq f(t, \tau, \xi(\bar{y})(\tau), r, \bar{b}) \qquad (t, \tau \in T, \quad r \in R),$$

and observe that $\bar{f}(t, \tau, \cdot)$ is continuous for all $t, \tau \in T$ and, by I.4.22, $\bar{f}(\cdot, \cdot, r)$ is $\zeta \times \mu$-measurable for all $r \in R$. If we set $\bar{K} = (\bar{q}, ..., \bar{q})$, then, by VII.2.3(2)

$$\int \zeta(dt) \int |\tilde{\zeta}(t)| |\bar{f}(t, \tau, \cdot)|_{\sup} \mu(d\tau) \leqslant \int |\tilde{\zeta}(t)| \zeta(dt) \cdot |f^{\bar{K}}| < \infty$$

and, by I.4.21, $(t, \tau) \to |\bar{f}(t, \tau, \cdot)|_{\sup}$ is $\zeta \times \mu$-measurable. Thus, by Fubini's theorem I.4.45, $(t, \tau) \to |\tilde{\zeta}(t)| |\bar{f}(t, \tau, \cdot)|_{\sup}$ is $\zeta \times \mu$-integrable. These remarks and relation (10) show that Lemma VII.3.1 is applicable, and VII.3.1(3) yields relation (2). Finally, for unilateral problems, relation (5) follows from (4) and V.2.5. QED

Analogous conclusions are reached when $\mathscr{X}_2 = L^q(P, \mathbb{R}^{m_2})$.

VII.3.3 ***Theorem*** Under the conditions of Theorem VII.2.3, let $(\bar{y}, \bar{\sigma}, \bar{b})$ be a minimizing relaxed solution, $m_2 \in \{0, 1, 2,...\}$, P be a compact metric space, $\omega \in \mathrm{frm}^+(P)$, $q \in [1, \infty)$, $\mathscr{X}_2 = L^q(P, \Sigma_{\mathrm{Borel}}(P), \omega, \mathbb{R}^{m_2})$, $g(y, \sigma, b) = \bar{g}(y, b)$ for $\sigma \in \mathscr{S}^\#$ and $(y, b) \in C(T, \mathbb{R}^n) \times B$, and $f(\cdot, \cdot, v, r, b)$ be $\Sigma \otimes \Sigma$-measurable for all $(v, r, b) \in V \times R \times B$. Then there exist $l_0 \geqslant 0$, $l_1 \in \mathbb{R}^m$, $\zeta \in \mathrm{frm}^+(T)$, $\tilde{\zeta} \in L^1(T, \Sigma, \zeta, \mathbb{R}^n)$, and $\tilde{\omega} \in L^{q/(q-1)}(P, \Sigma_{\mathrm{Borel}}(P), \omega, \mathbb{R}^{m_2})$ such that

$$|\tilde{\zeta}(t)| = 1 \quad (t \in T), \qquad l_0 + |l_1| + |\tilde{\omega}|_{q/(q-1)} > 0, \tag{1}$$

and

$$\int \tilde{\zeta}(t) \cdot x(t) \zeta(dt) = \sum_{i=0}^{1} l_i \mathscr{D}_1 \bar{g}_i(\bar{y}, \bar{b}) \circ (I - \mathscr{F})^{-1} x$$
$$+ \int \tilde{\omega}(p) \cdot [\mathscr{D}_1 \bar{g}_2(\bar{y}, \bar{b}) \circ (I - \mathscr{F})^{-1} x](p) \omega(dp) \qquad [x \in C(T, \mathbb{R}^n)];$$

$$\int \tilde{\zeta}(t) \cdot f(t, \tau, \xi(\bar{y})(\tau), \bar{\sigma}(\tau), \bar{b}) \zeta(dt) = \min_{r \in \bar{R}^\#(\tau)} \int \tilde{\zeta}(t) \cdot f(t, \tau, \xi(\bar{y})(\tau), r, \bar{b}) \zeta(dt) \tag{2}$$

for μ-a.a. $\tau \in T$;

$$\int \zeta(dt) \int \tilde{\zeta}(t) \cdot D_5 f(t, \tau, \xi(\bar{y})(\tau), \bar{\sigma}(\tau), \bar{b}; b - \bar{b}) \mu(d\tau) \tag{3}$$
$$+ \sum_{i=0}^{1} l_i D_2 \bar{g}_i(\bar{y}, \bar{b}; b - \bar{b})$$
$$+ \int \tilde{\omega}(p) \cdot D_2 \bar{g}_2(\bar{y}, \bar{b}; b - \bar{b})(p) \omega(dp) \geqslant 0 \qquad (b \in B);$$

and

(4) $$\int \tilde{\omega}(p) \cdot \bar{g}_2(\bar{y}, \bar{b})(p)\, \omega(dp) = \max_{c \in C_2} \int \tilde{\omega}(p) \cdot c(p)\, \omega(dp).$$

▌PROOF The proof is identical to that of VII.3.2 except that, in view of I.5.20, the element $l_2 \in \mathscr{X}_2{}^*$ is represented by the function

$$\tilde{\omega} \in L^{q/(q-1)}(P, \Sigma_{\text{Borel}}(P), \omega, \mathbb{R}^{m_2}). \quad \text{QED}$$

We also observe that Theorems VII.2.4 and VII.2.5 are directly applicable within the frameworks of VII.3.2 and VII.3.3.

VII.4 Necessary Conditions for an Original Minimum

We shall now investigate necessary conditions for a point $(\bar{y}, \bar{\sigma}, \bar{b})$ to be a minimizing $\mathscr{U}$-solution for the problem defined in VII.0 in general and for the corresponding unilateral problem in particular. Theorems VII.4.1 and VII.4.2 below are based on V.3.2, just as Theorems VII.2.3 and VII.3.2 were based on V.2.3.

VII.4.1 ***Theorem*** Let $\mathscr{U}$ be an abundant subset of $\mathscr{R}^{\#}$ and the conditions of VII.2.3 satisfied. We assume, furthermore, that for each choice of

$$K \triangleq ((\sigma_0, b_0), \ldots, (\sigma_m, b_m)) \in (\mathscr{S}^{\#} \times B)^{m+1},$$

the equation

$$y = F\left(y, \sigma, \bar{b} + \sum_{j=0}^{m} \theta^j (b_j - \bar{b})\right)$$

has a unique solution $\tilde{y}^K(\sigma, \theta)$ in Y_K for all $(\sigma, \theta) \in \mathscr{S}^{\#} \times \mathscr{T}_K$, and that the function

$$(y, \sigma, \theta) \to g\left(y, \sigma, \bar{b} + \sum_{j=0}^{m} \theta^j (b_j - \bar{b})\right) : Y_K \times \mathscr{S}^{\#} \times \mathscr{T}_K \to \mathbb{R} \times \mathbb{R}^m \times \mathscr{X}_2$$

is continuous.

Then $(\bar{y}, \bar{\sigma}, \bar{b})$ satisfies the conditions of V.3.2. If, furthermore, $(\bar{y}, \bar{\sigma}, \bar{b})$ is a minimizing $\mathscr{U}$-solution, then it is extremal and satisfies statements (6) and (7) of VII.2.3, and Theorems VII.2.4 and VII.2.5 are applicable.

▌PROOF We shall show that Theorem V.3.2 is applicable. Conditions V.3.2(1) and V.3.2(3) follow from Lemma VII.2.1 and our assumptions about g. Condition V.3.2(2) follows from VII.2.3(5). Finally, condition V.3.2(5) follows from I.5.4 because, for all $\sigma \in \mathscr{S}^{\#}$, $\theta \in \mathscr{T}_K$, and $t \in T$,

$$| y^K(\sigma, \theta)(t)| \leqslant \sup_{\beta \in T} \int | f^K(\beta, \tau, \cdot, \cdot, \cdot)|_{\sup} \mu(d\tau) = | f^K | < \infty$$

and

$$| y^K(\sigma, \theta)(t) - y^K(\sigma, \theta)(t')|$$
$$\leqslant \int | f^K(t, \tau, \cdot, \cdot, \cdot) - f^K(t', \tau, \cdot, \cdot, \cdot)|_{\sup} \mu(d\tau) \xrightarrow[t' \to t]{} 0.$$

It follows, by V.3.2, that $(\bar{y}, \bar{\sigma}, \bar{b})$ is extremal and, since $(\bar{y}, \bar{\sigma}, \bar{b})$ satisfies the conditions of VII.2.3, the remaining conclusions follow from VII.2.3–VII.2.5. QED

VII.4.2 ***Theorem*** Let $\mathscr{U}$ be an abundant subset of $\mathscr{R}^{\#}$ and the conditions of VII.3.2 (respectively VII.3.3) be satisfied except that $(\bar{y}, \bar{\sigma}, \bar{b})$ is assumed to be a minimizing $\mathscr{U}$-solution and not necessarily a minimizing relaxed solution. We assume, furthermore, that for each choice of $L \triangleq (b_0, \ldots, b_m) \in B^{m+1}$, the equation

$$y = F\left(y, \sigma, \bar{b} + \sum_{j=0}^{m} \theta^j (b_j - \bar{b})\right)$$

has a unique solution $\tilde{y}^L(\sigma, \theta)$ in $\xi^{-1}(L^\infty(T, V^L))$ for all $(\sigma, \theta) \in \mathscr{S}^{\#} \times \mathscr{T}^L$. Then the conclusions of VII.3.2 (respectively VII.3.3) are valid and Theorems VII.2.4 and VII.2.5 applicable.

▌PROOF The proof of VII.3.2 remains valid if the reference to VII.2.3 is replaced by a reference to VII.4.1. The proofs of VII.3.3 and VII.3.5 remain valid. QED

VII.4.3 ***Theorem*** Let $\mathscr{U}$ be an abundant subset of $\mathscr{R}^{\#}$, $(\bar{y}, \bar{u}, \bar{b}) \in \mathscr{A}(\mathscr{U})$,

$$\mathscr{M} \triangleq \{(y, \sigma, b) \in \mathscr{A}(\mathscr{S}^{\#}) | \ g_0(y, \sigma, b) \leqslant g_0(\bar{y}, \bar{u}, \bar{b})\},$$

and assume that the conditions of VII.2.3 remain satisfied when $(\bar{y}, \bar{\sigma}, \bar{b})$ is replaced by an arbitrary point of $\mathscr{M}$. Then the conditions of V.3.4 are satisfied and the conclusions of V.3.4 about strict $\mathscr{U}$-solutions remain valid.

▌PROOF The proof follows from VII.4.1. QED

VII.5 Problems with Pseudodelays

VII.5.1 ***Theorem*** Let $l \in \mathbb{N}$, $V \triangleq W^l$, $h_i : T \to T$ $(i = 1, 2, \ldots, l)$ be μ-measurable, and

$$\xi(y)(t) \triangleq (y \circ h_1(t), \ldots, y \circ h_l(t)) \qquad [y \in C(T, W), \quad t \in T].$$

Then $\xi \in B(C(T, W), L^\infty(T, W^l))$ and therefore ξ satisfies all the pertinent assumptions made in VII.1–VII.4.

If, furthermore, $p : T \to [0, 1]$ is μ-measurable,

$$p(h_i(\tau)) \leqslant p(\tau) \qquad (\tau \in T, \quad i = 1, 2,..., l),$$

the function $\alpha \to \mu(p^{-1}([0, \alpha])) : [0, 1] \to \mathbb{R}$ is continuous and

$$f_v(t, \tau, v, r, b) = 0 \qquad [(v, r, b) \in W^l \times R \times B, \quad p(\tau) > p(t)],$$

then assumption V.2.3(5) is satisfied.

▌PROOF The proof follows from II.5.10 and II.5.6. QED

Notes

In the special case where $T = [t_0, t_1] \subset \mathbb{R}$,

$$f(t, \tau, v, r, b) \triangleq \chi_{[t_0, t]}(\tau) \bar{f}(\tau, v, r, b),$$

$p(t) \triangleq t$, and ξ is p-hereditary, the equation of motion of VII.0 is referred to as a "functional-differential equation." If, furthermore, ξ is defined as in VII.5, then this equation of motion is a delay-differential equation. Necessary conditions for original minimum in the optimal control of delay-differential equations have been investigated, among others, by Oguztöreli [1], Banks [1], and Halanay [1]; and for functional-differential equations by Banks [2].

Necessary conditions for original minimum for problems with equations of motion of the form

$$y^i(t) = \int_{t_0}^{t} h^i(t - \tau) f^i(\tau, y(\tau), u(\tau))\, d\tau \qquad (i = 1, 2,..., n, \quad t \in [t_0, t_1])$$

were investigated by Friedman [1]. The problem of VII.0 in the case where ξ is the identity operator was studied in Warga [11, 13].

CHAPTER VIII

Optimal Control of Functional-Integral Equations in $L^p(T, \mathbb{R}^n)$

VIII.0 Formulation of the Problem

We consider in this chapter a problem formally resembling that of Chapter VII but defined entirely within the framework of L^p spaces. We assume given $k, n \in \mathbb{N}$, $p \in [1, \infty)$, $p_1 \in (1, \infty)$, $\xi : L^p(T, \Sigma, \mu, \mathbb{R}^n) \to L^{p_1}(T, \Sigma, \mu, \mathbb{R}^k)$, and $f : T \times T \times \mathbb{R}^k \times R \times B \to \mathbb{R}^n$. We set $\mathscr{Y} \triangleq L^p(T, \Sigma, \mu, \mathbb{R}^n)$ and, for all $(y, \sigma, b) \in \mathscr{Y} \times \mathscr{S}^{\#} \times B$,

$$F(y, \sigma, b)(t) \triangleq \int f(t, \tau, \xi(y)(\tau), \sigma(\tau), b)\, \mu(d\tau) \qquad \text{for} \quad \mu\text{-a.a.} \quad t \in T$$

if the integral on the right, as a function of t, belongs to $\mathscr{Y}$ and

$$F(y, \sigma, b)(t) \triangleq y(t) + (1, 0,..., 0) \in \mathbb{R}^n \qquad \mu\text{-a.e.},$$

otherwise. With $\mathscr{Y}$ and F as just defined, our problem is that of Chapter V.

We observe that, for $(y, \sigma, b) \in \mathscr{Y} \times \mathscr{S}^{\#} \times B$, we have $y = F(y, \sigma, b)$ if and only if

$$y(t) = \int f(t, \tau, \xi(y)(\tau), \sigma(\tau), b)\ \mu(d\tau) \quad \mu\text{-a.e.}$$

We define $\mathscr{H}(\mathscr{S}^{\#})$ and $\mathscr{A}(\mathscr{S}^{\#})$ as in Chapter V, that is,

$$\mathscr{H}(\mathscr{S}^{\#}) \triangleq \{(y, \sigma, b) \in \mathscr{Y} \times \mathscr{S}^{\#} \times B \mid y = F(y, \sigma, b)\}$$

and

$$\mathscr{A}(\mathscr{S}^{\#}) \triangleq \{(y, \sigma, b) \in \mathscr{H}(\mathscr{S}^{\#}) \mid g_1(y\ \sigma, b) = 0, g_2(y, \sigma, b) \in C_2\}.$$

As in II.6, we agree that $\infty/\infty \triangleq 1$ and $c/0 \triangleq \infty$ $(c > 0)$. We also write $L^q(T, \mathscr{Z}) \triangleq L^q(T, \Sigma, \mu, \mathscr{Z})$ for every separable Banach space $\mathscr{Z}$ and every $q \in [1, \infty]$.

VIII.1 Existence of Minimizing Solutions

VIII.1.1 *Theorem* Assume that

(1) $\mathscr{A}(\mathscr{S}^{\#}) \neq \varnothing$;

(2) $\xi : L^p(T, \mathbb{R}^n) \to L^{p_1}(T, \mathbb{R}^k)$ is continuous and

$$c_1 \triangleq \sup\{|\,\xi(y)|_{p_1} \mid (y, \sigma, b) \in \mathscr{A}(\mathscr{S}^{\#})\} < \infty;$$

(3) there exist a $\mu \times \mu$-measurable $\psi : T \times T \to \mathbb{R}$ and $\beta \in [0, p_1]$ such that

$$|\, t \to |\,\psi(t, \cdot)|_{p_1/(p_1-\beta)}\,|_p < \infty$$

and

$$|f(t, \tau, v, r, b)| \leqslant \psi(t, \tau)(1 + |\,v\,|^\beta) \qquad (t, \tau \in T, \quad v \in \mathbb{R}^k, \quad r \in R, \quad b \in B);$$

(4) there exists $\chi \in L^{p_1}(T)$ such that $|\,\xi(y)(\tau)| \leqslant \chi(\tau)$ μ-a.e. whenever $y \in \mathscr{Y}$ and $|\,y(t)| \leqslant c|\,\psi(t, \cdot)|_{p_1/(p_1-\beta)}$ μ-a.e., where $c \triangleq [\mu(T)]^{\beta/p_1} + (c_1)^\beta$;

(5) B is a compact metric space and C_2 sequentially closed;

(6) for all $(t, \tau, v, r, b) \in T \times T \times \mathbb{R}^k \times R \times B$, $f(\cdot, \cdot, v, r, b)$ is $\mu \times \mu$-measurable and $f(t, \tau, \cdot, \cdot, \cdot)$ continuous; and

(7) $g \mid \overline{\mathscr{A}(\mathscr{S}^{\#})}$ is continuous.

Then $\mathscr{A}(\mathscr{S}^{\#})$ is sequentially compact, $F \mid \mathscr{A}(\mathscr{S}^{\#})$ continuous, and there exists a minimizing relaxed solution.

▌PROOF Let S denote the topological space $\text{rpm}(R) \times B$ with the product topology corresponding to the weak norm topology of $\text{rpm}(R)$, $Q \triangleq \mathscr{S}^\# \times B$, and $q(t) \triangleq (\sigma(t), b)$ $[t \in T, (\sigma, b) \in Q]$. By (3) and Hölder's and Minkowski's inequalities I.5.13 and I.5.15, we have

$$(8)\quad \begin{aligned} |y(t)| &= |F(y, \sigma, b)(t)| \\ &\leqslant \int \psi(t, \tau)[1 + |\xi(y)(\tau)|^\beta]\, \mu(d\tau) \\ &\leqslant c\, |\psi(t, \cdot)|_{p_1/(p_1-\beta)} \qquad [\mu\text{-a.a. } t \in T, \quad (y, \sigma, b) \in \mathscr{A}(\mathscr{S}^\#)]. \end{aligned}$$

With $\tilde{f}$ replaced by f and $\tilde{F}$ by F, the conditions of II.6.3 are satisfied, and it follows that $F(K \times Q)$ is conditionally compact in $\mathscr{Y}$ for

$$K \triangleq \{y \in \mathscr{Y} \mid |y(t)| \leqslant c\, |\psi(t, \cdot)|_{p_1/(p_1-\beta)} \ \mu\text{-a.e.}\}.$$

Since, by (8), $y \in K$ whenever $(y, \sigma, b) \in \mathscr{A}(\mathscr{S}^\#)$, it follows that $\text{pr}_{\mathscr{Y}}\, \mathscr{A}(\mathscr{S}^\#) = F(\mathscr{A}(\mathscr{S}^\#))$ is conditionally compact. Now $\mathscr{S}^\#$ is sequentially compact (by IV.3.11) and therefore

$$\overline{\mathscr{A}(\mathscr{S}^\#)} \subset \overline{\text{pr}_{\mathscr{Y}}\, \mathscr{A}(\mathscr{S}^\#)} \times \mathscr{S}^\# \times B$$

is also sequentially compact.

Now let $((y_j, \sigma_j, b_j))$ be a sequence in $\mathscr{A}(\mathscr{S}^\#)$ converging to some $(\bar{y}, \bar{\sigma}, \bar{b}) \in \mathscr{Y} \times \mathscr{S}^\# \times B$. Then, by I.4.31, $\lim_j \xi(y_j) = \xi(\bar{y})$ in μ-measure and therefore, by (6) and I.2.15(2), for each $t \in T$ we have

$$\lim_j [\tau \to \sup_{r \in R} |f(t, \tau, \xi(y_j)(\tau), r, b_j) - f(t, \tau, \xi(\bar{y})(\tau), r, \bar{b})|] = 0$$

in μ-measure; hence

$$(9)\quad \lim_j [\tau \to |f(t, \tau, \xi(y_j)(\tau), \sigma_j(\tau), b_j) - f(t, \tau, \xi(\bar{y})(\tau), \sigma_j(\tau), \bar{b})|] = 0 \quad \text{in } \mu\text{-measure.}$$

By (3), (4), and (8), we have

$$(10)\qquad |f(t, \tau, \xi(y_j)(\tau), \sigma(\tau), b_j)| \leqslant \psi(t, \tau)[1 + |\chi(\tau)|^\beta] \qquad (\sigma \in \mathscr{S}, \tau \in T, \mu\text{-a.a. } t \in T)$$

and $\tau \to \psi(t, \tau)[1 + |\chi(\tau)|^\beta]$ is μ-integrable for μ-a.a. $t \in T$. Thus, by (9), (10), and the dominated convergence theorem I.4.35, we have

$$(11)\quad \lim_j \int |f(t, \tau, \xi(y_j)(\tau), \sigma_j(\tau), b_j) - f(t, \tau, \xi(\bar{y})(\tau), \sigma_j(\tau), \bar{b})|\, \mu(d\tau) = 0 \quad \text{for } \mu\text{-a.a. } t \in T.$$

On the other hand, for μ-a.a. $t \in T$, the function $(\tau, r) \to f(t, \tau, \xi(\bar{y})(\tau), r, \bar{b})$ is μ-measurable as a function of τ and continuous as a function of r on R; furthermore, by (3), (4), and (8),

$$|f(t, \tau, \xi(\bar{y})(\tau), r, \bar{b})| \leqslant \psi(t, \tau)[1 + |\chi(\tau)|^{\beta}] \qquad (\tau \in T, \quad r \in R).$$

It follows that $(\tau, r) \to f(t, \tau, \xi(\bar{y})(\tau), r, \bar{b})$ belongs to $L^1(T, C(R, \mathbb{R}^n))$ for μ-a.a. $t \in T$ and therefore

$$(12) \qquad \lim_j \int \mu(d\tau) \int f(t, \tau, \xi(\bar{y})(\tau), r, \bar{b})\, \sigma_j(\tau)\, (dr)$$
$$= \int \mu(d\tau) \int f(t, \tau, \xi(\bar{y})(\tau), r, \bar{b})\, \bar{\sigma}(\tau)\, (dr) \quad \mu\text{-a.e.}$$

By I.4.31 and Egoroff's theorem I.4.18, there exists $J \subset (1, 2, ...)$ such that $\lim_{j \in J} y_j(t) = \bar{y}(t)$ μ-a.e. Thus (11) and (12) yield

$$\bar{y}(t) = \lim_{j \in J} y_j(t) = \lim_{j \in J} \int f(t, \tau, \xi(y_j)(\tau), \sigma_j(\tau), b_j)\, \mu(d\tau)$$
$$= \int f(t, \tau, \xi(\bar{y})(\tau), \bar{\sigma}(\tau), \bar{b})\, \mu(d\tau) \quad \mu\text{-a.e.}$$

Since $g \mid \overline{\mathscr{A}(\mathscr{S}^{\#})}$ is continuous and C_2 is sequentially closed, this shows that $(\bar{y}, \bar{\sigma}, \bar{b}) \in \mathscr{A}(\mathscr{S}^{\#})$, hence $\mathscr{A}(\mathscr{S}^{\#})$ is closed, and that $F \mid \mathscr{A}(\mathscr{S}^{\#})$ is continuous. The existence of a minimizing relaxed solution now follows from V.1.1. QED

VIII.1.2 ***Theorem*** Assume that the conditions (2)–(6) of Theorem VIII.1.1 are satisfied and $g \mid \overline{\mathscr{H}(\mathscr{S}^{\#})}$ is sequentially continuous. Let $(\bar{y}, \bar{\sigma}, \bar{b})$ be a minimizing relaxed solution and $\mathscr{U}$ an abundant set and assume, furthermore, that $\bar{y}$ is the unique solution of the equation $y = F(y, \bar{\sigma}, \bar{b})$ and there exists a neighborhood G of $\bar{\sigma}$ in $\mathscr{S}^{\#}$ such that the equation $y = F(y, u, \bar{b})$ has a solution y for each $u \in \mathscr{U} \cap G$. Then there exists a sequence $((y_j, u_j))$ in $\mathscr{Y} \times \mathscr{U}$ such that $((y_j, u_j, \bar{b}))$ is a minimizing approximate $\mathscr{U}$-solution and

$$\lim_j g(y_j, u_j, \bar{b}) = g(\bar{y}, \bar{u}, \bar{b}).$$

▌PROOF With $\mathscr{A}(\mathscr{S}^{\#})$ replaced by $\mathscr{H}(\mathscr{S}^{\#})$, the proof of VIII.1.1 shows that $\mathscr{H}(\mathscr{S}^{\#})$ is sequentially compact and $F \mid \mathscr{H}(\mathscr{S}^{\#})$ is sequentially continuous. Our conclusion then follows from V.1.2. QED

VIII.1.3 ***Theorem*** Let the conditions of Theorem VIII.1.1 be satisfied and $(\bar{y}, \bar{\sigma}, \bar{b})$ be any minimizing relaxed solution (whose existence is established in VIII.1.1). Assume, furthermore, that

$$(1) \qquad g(y, \sigma, b) \triangleq \bar{g}(y, b) \qquad [(y, \sigma, b) \in \mathscr{Y} \times \mathscr{S}^{\#} \times B];$$

(2) $R^{\#}(\tau)$ is a closed set for all $\tau \in T$;

(3) $f(\cdot, \cdot, v, r, b)$ is $\Sigma \otimes \Sigma$-measurable for all $(v, r, b) \in \mathbb{R}^k \times R \times B$; and
(4) $\{f(\cdot, \tau, v, r', b) \mid r' \in R^\#(\tau)\}$ is a convex subset of $\mathscr{F}(T, \Sigma, \mu, \mathbb{R}^n)$ for all $(\tau, v, b) \in T \times \mathbb{R}^k \times B$.

Then there exists $\bar{\rho} \in \mathscr{R}^\#$ such that $(\bar{y}, \bar{\rho}, \bar{b})$ is a minimizing relaxed solution (and, therefore, also a minimizing $\mathscr{R}^\#$-solution).

PROOF Let

$$h(t, \tau, r) \triangleq f(t, \tau, \xi(\bar{y})(\tau), r, \bar{b}) \qquad (t, \tau \in T, \quad r \in R).$$

Then, by VIII.1.1(5) and I.4.22, $h(\cdot, \cdot, r)$ is $\mu \times \mu$-measurable and $h(t, \tau, \cdot)$ continuous for all $(t, \tau, r) \in T \times T \times R$. By VIII.1.1(3) and Hölder's and Minkowski's inequalities I.5.13 and I.5.15, we have

$$\begin{aligned}
&\int \mu(dt) \int |h(t, \tau, \cdot)|_{\sup}\, \mu(d\tau) \\
&\quad \leqslant \int \mu(dt) \int \psi(t, \tau)[1 + |\xi(\bar{y})(\tau)|^\beta]\, \mu(d\tau) \\
&\quad \leqslant \int |\psi(t, \cdot)|_{p_1/(p_1-\beta)}\, \mu(dt) \cdot [\mu(T)^{\beta/p_1} + |\xi(\bar{y})|_{p_1}^\beta] < \infty;
\end{aligned}$$

hence, by Fubini's theorem I.4.45, $(t, \tau) \to |h(t, \tau, \cdot)|_{\sup}$ is $\mu \times \mu$-integrable. Thus the conditions of IV.3.14 are satisfied with $\nu = \mu$ and there exists $\bar{\rho} \in \mathscr{R}^\#$ such that

$$\begin{aligned}
(6) \qquad \bar{y}(t) &= \int f(t, \tau, \xi(\bar{y})(\tau), \bar{\sigma}(\tau), \bar{b})\, \mu(d\tau) \\
&= \int f(t, \tau, \xi(\bar{y})(\tau), \bar{\rho}(\tau), \bar{b})\, \mu(d\tau) \qquad (\mu\text{-a.a., } t \in T).
\end{aligned}$$

Therefore, in view of (1), $(\bar{y}, \bar{\rho}, \bar{b})$ is a minimizing relaxed solution. QED

VIII.1.4 ***Theorem*** Let the conditions of VIII.1.1 be satisfied, and assume furthermore that $R^\#(\tau)$ is closed for all $\tau \in T$, $g(y, \sigma, b) \triangleq \bar{g}(y, b)$ for all $(y, \sigma, b) \in \mathscr{Y} \times \mathscr{S}^\# \times B$ and there exist $l \in \mathbb{N}$, $\alpha^j : T \times T \times \mathbb{R}^k \times B \to \mathbb{R}$ and $\beta^j : T \times \mathbb{R}^k \times R \times B \to \mathbb{R}^n$ $(j = 1, 2, ..., l)$ such that $\beta^j(\cdot, v, r, b)$ are μ-measurable, $\beta^j(\tau, \cdot, \cdot, \cdot)$ continuous and

$$f(t, \tau, v, r, b) \triangleq \sum_{j=1}^{l} \alpha^j(t, \tau, v, b)\, \beta^j(\tau, v, r, b)$$

$$[(t, \tau, v, r, b) \in T \times T \times \mathbb{R}^k \times R \times B].$$

Then there exists a minimizing relaxed solution $(\bar{y}, \bar{\sigma}, \bar{b})$ such that, for all $\tau \in T$, $\bar{\sigma}(\tau)$ is a measure supported on a set of $ln + 1$ points of $R^\#(\tau)$.

If, furthermore, for all $(\tau, v, b) \in T \times \mathbb{R}^k \times B$,

$$\{(\beta^1(\tau, v, r, b), \ldots, \beta^l(\tau, v, r, b)) \mid r \in R^{\#}(\tau)\}$$

is a convex subset of $\mathbb{R}^{ln}$, then there exists $\bar{\rho} \in \mathscr{R}^{\#}$ such that $(\bar{y}, \bar{\rho}, \bar{b})$ is a relaxed minimizing solution.

▌PROOF The proof is identical with that of VII.1.5, with references to VII.1.1 and VII.1.4(2) replaced by references to VIII.1.1 and VIII.1.3(1), respectively, and with V replaced by $\mathbb{R}^k$. QED

VIII.2 Necessary Conditions for a Relaxed Minimum

We investigate necessary conditions for a point $(\bar{y}, \bar{\sigma}, \bar{b})$ to be a minimizing relaxed solution.

VIII.2.1 ***Assumption*** Let $(\bar{y}, \bar{\sigma}, \bar{b}) \in \mathscr{Y} \times \mathscr{S}^{\#} \times B$. For each choice of a point $L \triangleq (b_0, \ldots, b^m) \in B^{m+1}$ there exists a convex neighborhood $\mathscr{T}^L$ of 0 in $\mathscr{T}_{m+1}$ such that ξ and the function

$$(t, \tau, v, r, \theta) \to f^L(t, \tau, v, r, \theta)$$

$$\triangleq f\left(t, \tau, v, r, \bar{b} + \sum_{j=0}^{m} \theta^j(b_j - \bar{b})\right) : T \times T \times \mathbb{R}^k \times R \times \mathscr{T}^L \to \mathbb{R}^n$$

have the following properties for all $(t, \tau, v, r, \theta) \in T \times T \times \mathbb{R}^k \times R \times \mathscr{T}^L$:

(1) $f^L(t, \tau, \cdot, r, \cdot)$ has a derivative $f^L_{(v,\theta)}(t, \tau, v, r, \theta)$ and both $f^L(t, \tau, \cdot, \cdot, \cdot)$ and $f^L_{(v,\theta)}(t, \tau, \cdot, \cdot, \cdot)$ are continuous;

(2) $f^L(\cdot, \cdot, v, r, \theta)$ is $\mu \times \mu$-measurable;

(3) there exist $\alpha(L), \beta(L) \in \mathbb{R}$,

$$\psi_0^L \in L^p(T, L^{p_1/(p_1-\beta(L))}(T)) \qquad \text{and} \qquad \psi_1^L \in L^p(T, L^{p_1/(p_1-\alpha(L)-1)}(T))$$

such that

$$0 \leqslant \alpha(L) \leqslant p_1 - 1, \qquad 0 \leqslant \beta(L) \leqslant p_1,$$
$$|f^L(t, \tau, v, r, \theta)| \leqslant (1 + |v|^{\beta(L)})\, \psi_0^L(t, \tau),$$
$$|f_\theta^L(t, \tau, v, r, \theta)| \leqslant (1 + |v|^{\beta(L)})\, \psi_0^L(t, \tau),$$

and

$$|f_v^L(t, \tau, v, r, \theta)| \leqslant (1 + |v|^{\alpha(L)})\, \psi_1^L(t, \tau);$$

and

(4) ξ has a continuous derivative.

VIII.2.2 ***Lemma*** Let Assumption VIII.2.1 be satisfied and

$$K \triangleq ((\sigma_0, b_0), ..., (\sigma_m, b_m)) \in (\mathscr{S}^\# \times B)^{m+1}.$$

We set $L \triangleq (b_0, ..., b_m)$ and, for all $\theta \in \mathscr{T}_{m+1}$,

$$\sigma^K(\theta) \triangleq \bar{\sigma} + \sum_{j=0}^{m} \theta^j(\sigma_j - \bar{\sigma}) \qquad \text{and} \qquad b^K(\theta) \triangleq \bar{b} + \sum_{j=0}^{m} \theta^j(b_j - \bar{b}).$$

Then

(1) the functions

$$(y, \sigma, \theta) \to F^K(y, \sigma, \theta) \triangleq F(y, \sigma, b^K(\theta)) : \mathscr{Y} \times \mathscr{S}^\# \times \mathscr{T}^L \to \mathscr{Y}$$

and

$$(y, \theta) \to \tilde{F}^K(y, \theta) \triangleq F(y, \sigma^K(\theta), b^K(\theta)) : \mathscr{Y} \times \mathscr{T}^L \to \mathscr{Y}$$

are continuous, and $\tilde{F}^K$ has a continuous derivative;

(2) for all $y \in \mathscr{Y}$ and $\sigma \in \mathscr{S}^\#$,

$$(t \to [\tau \to f(t, \tau, \xi(y)(\tau), \sigma(\tau), \bar{b})]) \in L^p(T, L^1(T, \mathbb{R}^n))$$

and

$$(t \to [\tau \to f_v(t, \tau, \xi(y)(\tau), \sigma(\tau), \bar{b})]) \in L^p(T, L^{p_1/(p_1-1)}(T, B(\mathbb{R}^n, \mathbb{R}^n)));$$

(3) for all $\Delta y \in \mathscr{Y}$ and μ-a.a. $t \in T$,

$$(F_y(\bar{y}, \bar{\sigma}, \bar{b})\, \Delta y)(t) = \int f_v(t, \tau, \xi(\bar{y}), \bar{\sigma}(\tau), \bar{b})(\xi'(\bar{y})\, \Delta y)(\tau)\, \mu(d\tau); \text{ and}$$

(4) for $\bar{q} \triangleq (\bar{\sigma}, \bar{b})$, all $q \triangleq (\sigma, b) \in \mathscr{S}^\# \times B$ and μ-a.a. $t \in T$,

$$D_2F(\bar{y}, \bar{q}; q - \bar{q})(t) = \int [f(t, \tau, \xi(\bar{y})(\tau), \sigma(\tau) - \bar{\sigma}(\tau), \bar{b})$$
$$+ D_5 f(t, \tau, \xi(\bar{y})(\tau), \bar{\sigma}(\tau), \bar{b}; b - \bar{b})]\, \mu(d\tau).$$

▌PROOF *Step* 1 We shall prove here that F^K is continuous. If

$$(y, \sigma, \theta) \in \mathscr{Y} \times \mathscr{S}^\# \times \mathscr{T}^L,$$

then, by I.4.22, the function $(t, \tau) \to f^L(t, \tau, \xi(y)(\tau), r, \theta)$ is $\mu \times \mu$-measurable for all $(r, \theta) \in R \times \mathscr{T}^L$. Furthermore, $f^L(t, \tau, \xi(y)(\tau), \cdot, \cdot)$ is continuous on $R \times \mathscr{T}^L$. It follows, by IV.2.8, that $(t, \tau) \to f^L(t, \tau, \xi(y)(\tau), \sigma(\tau), \theta)$ is $\mu \times \mu$-measurable for all $\theta \in \mathscr{T}^L$ and $f^L(t, \tau, \xi(\tau), \sigma(\tau), \cdot)$ is continuous for all $t, \tau \in T$.

We next observe that, for all $(y, \sigma, \theta) \in \mathscr{Y} \times \mathscr{S}^{\#} \times \mathscr{T}^L$, the function $t \to (\tau \to f^L(t, \tau, \xi(y)(\tau), \sigma(\tau), \theta))$ belongs to $L^p(T, L^1(T, \mathbb{R}^n))$. Indeed, by VIII.2.1(3) and Hölder's inequality I.5.13,

$$(5)\qquad \int | f^L(t, \tau, \xi(y)(\tau), \sigma(\tau), \theta) | \, \mu(d\tau) \leqslant | \psi_0{}^L(t, \cdot)|_{p_1/(p_1-\beta(L))} \, \big| \tau \to 1 + | \xi(y)(\tau)|^{\beta(L)} \big|_{p_1/\beta(L)} ,$$

and our assertion follows from I.4.45 and I.5.24. This implies, in particular, that $t \to \int f^L(t, \tau, \xi(y)(\tau), \sigma(\tau), \theta) \, \mu(d\tau)$ is an element of $L^p(T, \mathbb{R}^n)$. Thus

$$(6)\qquad F^K(y, \sigma, \theta)(t) = \int f^L(t, \tau, \xi(y)(\tau), \sigma(\tau), \theta) \, \mu(d\tau) \qquad (t \in T).$$

Now let $\lim_j (y_j, \sigma_j, \theta_j) = (y, \sigma, \theta)$ in $\mathscr{Y} \times \mathscr{S}^{\#} \times \mathscr{T}^L$ and

$$e_j(t, \tau, r) \triangleq | f^L(t, \tau, \xi(y_j)(\tau), r, \theta_j) - f^L(t, \tau, \xi(y)(\tau), r, \theta)| \qquad (j \in \mathbb{N},\ t, \tau \in T,\ r \in R).$$

By VIII.2.1 and II.6.7 [with $r_2 = r_1 = \beta(L)$, $\Gamma = R$ and $\phi(t, \tau, v, \theta, r) = f^L(t, \tau, v, r, \theta)$], we have

$$\lim_j \big| t \to | \tau \to \sup_{r \in R} e_j(t, \tau, r)|_1 \big|_p = 0$$

and, since

$$e_j(t, \tau, \sigma_j(\tau)) \leqslant \sup_{r \in R} e_j(t, \tau, r) \qquad (t, \tau \in T, \quad j \in \mathbb{N}),$$

it follows that

$$\lim_j \big| t \to | \tau \to e_j(t, \tau, \sigma_j(\tau))|_1 \big|_p = 0.$$

This relation and (6) show that

$$(7)\qquad \lim_j | F^K(y_j, \sigma_j, \theta_j) - F^K(y, \sigma_j, \theta)|_p = 0.$$

On the other hand, for μ-a.a. $t \in T$, the function $(\tau, r) \to f^L(t, \tau, \xi(y)(\tau), r, \theta)$ is continuous as a function of r, μ-measurable as a function of τ and, by VIII.2.1(3),

$$| f^L(t, \tau, \xi(y)(\tau), r, \theta)| \leqslant \psi_0{}^L(t, \tau)(1 + | \xi(y)(\tau)|^{\beta(L)}).$$

Thus, by I.5.25, $(\tau, r) \to f^L(t, \tau, \xi(y)(\tau), r, \theta)$ belongs to $L^1(T, C(R, \mathbb{R}^n))$ and therefore

$$\begin{aligned}\lim_j F^K(y, \sigma_j, \theta)(t) &= \lim_j \int f^L(t, \tau, \xi(y)(\tau), \sigma_j(\tau), \theta)\, \mu(d\tau) \\ &= \int f^L(t, \tau, \xi(y)(\tau), \sigma(\tau), \theta)\, \mu(d\tau) \\ &= F^K(y, \sigma, \theta)(t) \qquad (\mu\text{-a.a. } t \in T).\end{aligned}$$

This relation together with (5) and the dominated convergence theorem I.4.35 implies that

$$\lim_j F^K(y, \sigma_j, \theta) = F^K(y, \sigma, \theta) \quad \text{in } L^p(T, \mathbb{R}^n);$$

hence, by (7),

$$\begin{aligned}\lim_j |F^K(y_j, \sigma_j, \theta_j) - F^K(y, \sigma, \theta)|_p &\leqslant \lim_j |F^K(y_j, \sigma_j, \theta_j) - F^K(y, \sigma_j, \theta)|_p \\ &\quad + \lim_j |F^K(y, \sigma_j, \theta) - F^K(y, \sigma, \theta)|_p \\ &= 0.\end{aligned}$$

Thus F^K is continuous on $\mathscr{Y} \times \mathscr{S}^{\#} \times \mathscr{T}^L$.

Step 2 We next show that $\tilde{F}^K$ is continuous and has a continuous derivative. Let

$$\begin{aligned}\tilde{f}^K(t, \tau, v, \theta) &\triangleq f(t, \tau, v, \sigma^K(\theta), b^K(\theta)) \\ &= f^L(t, \tau, v, \bar{\sigma}(\tau), \theta) + \sum_{j=0}^{m} \theta^j f^L(t, \tau, v, \sigma_j(\tau) - \bar{\sigma}(\tau), \theta) \\ &\qquad (t, \tau \in T, \quad v \in \mathbb{R}^k, \quad \theta \in \mathscr{T}^L).\end{aligned}$$

By IV.2.8, for every $\sigma \in \mathscr{S}^{\#}$, the function

$$(t, \tau, v, \theta) \to h(t, \tau, v, \theta) \triangleq f^L(t, \tau, v, \sigma(\tau), \theta) : T \times T \times \mathbb{R}^k \times \mathscr{T}^L \to \mathbb{R}^n$$

is such that $h(t, \tau, \cdot, \cdot)$ has a derivative $h_{(v,\theta)}(t, \tau, v, \theta)$ for all $(t, \tau, v, \theta) \in T \times T \times \mathbb{R}^k \times \mathscr{T}^L$, $h(t, \tau, \cdot, \cdot)$ and $h_{(v,\theta)}(t, \tau, \cdot, \cdot)$ are continuous, and $h(\cdot, \cdot, v, \theta)$ is $\mu \times \mu$-measurable. In view of these properties and of Assumption VIII.2.1(3) we may apply II.6.8 and conclude that

$$\begin{aligned}\tilde{F}^K(y, \theta)(t) &= F(y, \sigma^K(\theta), b^K(\theta))(t) \\ &= \int \tilde{f}^K(t, \tau, \xi(y)(\tau), \theta)\, \mu(d\tau) \qquad (y \in \mathscr{Y}, \quad \theta \in \mathscr{T}^L, \quad t \in T)\end{aligned}$$

and $\tilde{F}^K$ has a continuous derivative such that

$$\mathscr{D}\tilde{F}^K(y, \theta)(\Delta y, \Delta\theta)(t) = \int \tilde{f}^K_{(v,\theta)}(t, \tau, \xi(y)(\tau), \theta)((\xi'(y)\, \Delta y)(\tau), \Delta\theta)\, \mu(d\tau)$$
$$(y, \Delta y \in \mathscr{Y},\ \theta \in \mathscr{T}^L,\ \Delta\theta \in \mathbb{R}^{m+1},\ t \in T).$$

This completes the proof of statement (1).

We observe that, for all $(t, \tau, y, \sigma, L) \in T \times T \times \mathscr{Y} \times \mathscr{S}^\# \times B^{m+1}$, we have $f^L(t, \tau, \xi(y)(\tau), \sigma(\tau), 0) = f(t, \tau, \xi(\bar{y})(\tau), \sigma(\tau), \bar{b})$ and, by Step 1,

$$(t \to [\tau \to f^L(t, \tau, \xi(y)(\tau), \sigma(\tau), 0)]) \in L^p(T, L^1(T, \mathbb{R}^n)).$$

This yields the first relation in (2). By VIII.2.1(3) and I.5.14,

$$\begin{aligned}
&|\, \tau \to f_v^L(t, \tau, \xi(y)(\tau), \sigma(\tau), \theta)|_{p_1/(p_1-1)} \\
&\quad \leqslant |\, \psi_1^L(t, \cdot)|_{p_1/(p_1-\alpha(L)-1)}\ |\, \tau \to 1 + |\, \xi(y)(\tau)|^{\alpha(L)}\, |_{p_1/\alpha(L)} \\
&\quad < \infty.
\end{aligned}$$

Since, as we have shown above, $(t, \tau) \to f^L(t, \tau, v, \sigma(\tau), 0)$ is $\mu \times \mu$-measurable, it follows from II.3.9 that $(t, \tau) \to f_v^L(t, \tau, \xi(y)(\tau), \sigma(\tau), 0)$ is $\mu \times \mu$-measurable and, therefore, Fubini's theorem I.4.45(1) and the above inequality yield the second relation of (2). Finally, relations (3) and (4) are derived exactly as in VII.2.1(Step 3). QED

We recall that, by VIII.2.2(3), Assumption VIII.2.1 implies that $\mathscr{F} \triangleq F_y(\bar{y}, \bar{\sigma}, \bar{b})$ exists and satisfies the relation

$$(\mathscr{F}\Delta y)(t) = \int f_v(t, \tau, \xi(\bar{y})(\tau), \bar{\sigma}(\tau), \bar{b})(\xi'(\bar{y})\, \Delta y)(\tau)\, \mu(d\tau)$$
$$(\Delta y \in \mathscr{Y},\ \mu\text{-a.a. } t \in T).$$

VIII.2.3 ***Theorem*** Let C_2 be convex and have a nonempty interior, and let $(\bar{y}, \bar{q}) \triangleq (\bar{y}, \bar{\sigma}, \bar{b})$, f and ξ satisfy Assumption VIII.2.1. Assume, furthermore, that for each $K \triangleq ((\sigma_0, b_0), \ldots, (\sigma_m, b_m)) \in (\mathscr{S}^\# \times B)^{m+1}$ and for $L \triangleq (b_0, \ldots, b_m)$ and $q_i \triangleq (\sigma_i, b_i)$ $(i = 0, 1, \ldots, m)$, the function

$$(y, \theta) \to \tilde{g}^K(y, \theta) = g\left(y, \bar{q} + \sum_{j=0}^{m} \theta^j(q_j - \bar{q})\right) : \mathscr{Y} \times \mathscr{T}^L \to \mathbb{R} \times \mathbb{R}^m \times \mathscr{X}_2$$

is continuous and has a derivative $\mathscr{D}g^K(\bar{y}, 0)$ at $(\bar{y}, 0)$, and that the equation

$$\begin{aligned}
\Delta y(t) &= (\mathscr{F}\Delta y)(t) \\
&= \int f_v(t, \tau, \xi(\bar{y})(\tau), \bar{\sigma}(\tau), \bar{b})(\xi'(\bar{y})\, \Delta y)(\tau)\, \mu(d\tau) \qquad (\mu\text{-a.a. } t \in T)
\end{aligned}$$

has $\Delta y = 0$ as its only solution in $\mathscr{Y}$.

Then

(1) $I - \mathscr{F}$ is a homeomorphism of $\mathscr{Y}$ onto itself and both $\mathscr{F}$ and $(I - \mathscr{F})^{-1} - I$ are compact operators in $\mathscr{Y}$.

If, furthermore, $(\bar{y}, \bar{\sigma}, \bar{b})$ is a minimizing relaxed solution, then there exists $l \triangleq (l_0, l_1, l_2) \in \mathbb{R} \times \mathbb{R}^m \times \mathscr{X}_2{}^*$ such that

(2) $l_0 \geqslant 0$ and $l_0 + |l_1| + |l_2| \neq 0$;

(3) for μ-a.a. $t \in T$ and all $q \triangleq (\sigma, b) \in \mathscr{S}^\# \times B$,

$$D_2F(\bar{y}, \bar{q}; q - \bar{q})(t) = \int [f(t, \tau, \xi(\bar{y})(\tau), \sigma(\tau) - \bar{\sigma}(\tau), \bar{b}) + D_5 f(t, \tau, \xi(\bar{y})(\tau), \bar{\sigma}(\tau), \bar{b}; b - \bar{b})]\, \mu(d\tau);$$

(4) $$\tilde{\zeta} \triangleq l \circ g_y(\bar{y}, \bar{q}) \circ (I - \mathscr{F})^{-1} \in L^{p/(p-1)}(T, \mathbb{R}^n)$$

[where $\tilde{\zeta}$ is interpreted both as an element of $L^p(T, \mathbb{R}^n)^*$ and the corresponding element of $L^{p/(p-1)}(T, \mathbb{R}^n)$];

(5) $$\int \tilde{\zeta}(t) \cdot D_2F(\bar{y}, \bar{q}; q - \bar{q})(t)\, \mu(dt) + l(D_2 g(\bar{y}, \bar{q}; q - \bar{q})) \geqslant 0 \qquad (q \in \mathscr{S}^\# \times B);$$

and

(6) $$l_2(g_2(\bar{y}, \bar{q})) \geqslant l_2(c) \qquad (c \in C_2).$$

▌PROOF By VIII.2.2(2), we have

$$(t \to [\tau \to f_v(t, \tau, \xi(\bar{y})(\tau), \bar{\sigma}(\tau), \bar{b})]) \in L^p(T, L^{p_1/(p_1-1)}(T, B(\mathbb{R}^n, \mathbb{R}^n))).$$

It follows, by VIII.2.1(4) and II.6.6, that $\mathscr{F}$ is a compact operator in $L^p(T, \mathbb{R}^n)$. Statement (1) now follows from Riesz's theorem I.3.13.

Now let $(\bar{y}, \bar{\sigma}, \bar{b})$ be a minimizing relaxed solution. In view of VIII.2.2(1) and (1), we may apply Theorem V.2.3. We set $\tilde{\zeta} \triangleq l \circ g_y(\bar{y}, \bar{q}) \circ (I - \mathscr{F})^{-1}$ and observe that $\tilde{\zeta}$ is a continuous linear functional on $L^p(T, \mathbb{R}^n)$ and may be identified therefore (I.5.20) with an element of $L^{p/(p-1)}(T, \mathbb{R}^n)$. Relation (2) follows from $l \neq 0$, relation (3) from VIII.2.2(4), and relations (5) and (6) now follow directly from V.2.3 and the definition of an extremal. QED

VIII.2.4 ***Theorem*** Let the conditions of Theorem VIII.2.3 be satisfied, $(\bar{y}, \bar{\sigma}, \bar{b})$ be a minimizing relaxed solution, and $g(y, \sigma, b) \triangleq \bar{g}(\bar{y}, \bar{b})$ for all $(y, \sigma, b) \in \mathscr{Y} \times \mathscr{S}^\# \times B$. Then

(1) $$\int \tilde{\zeta}(t) \cdot f(t, \tau, \xi(\bar{y})(\tau), \bar{\sigma}(\tau), \bar{b})\, \mu(dt) = \operatorname*{Min}_{r \in \bar{R}^\#(\tau)} \int \tilde{\zeta}(t) \cdot f(t, \tau, \xi(\bar{y})(\tau), r, \bar{b})\, \mu(dt) \qquad (\mu\text{-a.a. } \tau \in T)$$

and

(2) $$\int \mu(dt) \int \tilde{\zeta}(t) \cdot D_5 f\big(t, \tau, \xi(\bar{y})(\tau), \bar{\sigma}(\tau), \bar{b}; b - \bar{b}\big)\, \mu(d\tau)$$

$$+ \, l\big(D_2 \bar{g}(\bar{y}, \bar{b}; b - \bar{b})\big) \geqslant 0 \qquad (b \in B).$$

▌PROOF If we set $q = (\bar{\sigma}, b)$ in VIII.2.3(3) and VIII.2.3(5), then we obtain relation (2). If we set $q = (\sigma, \bar{b})$, then we obtain

(3) $$\int \mu(dt) \int \tilde{\zeta}(t) \cdot f\big(t, \tau, \xi(\bar{y})(\tau), \sigma(\tau) - \bar{\sigma}(\tau), \bar{b}\big)\, \mu(d\tau) \geqslant 0 \qquad (\sigma \in \mathscr{S}^{\#}).$$

Now let

$$\bar{f}(t, \tau, r) \triangleq f\big(t, \tau, \xi(\bar{y})(\tau), r, \bar{b}\big) \qquad (t, \tau \in T, \quad r \in R).$$

Then $\bar{f}(t, \tau, \cdot)$ is continuous and, by I.4.21 and I.4.22, the functions

$$(t, \tau) \to \bar{f}(t, \tau, r) \quad (r \in R) \qquad \text{and} \qquad (t, \tau) \to |\bar{f}(t, \tau, \cdot)|_{\sup}$$

are $\mu \times \mu$-measurable. Furthermore, by VIII.2.1(3) [with $L = (\bar{b}, ..., \bar{b}) \in B^{m+1}$] and Hölder's inequality I.5.13, we have

$$\int |\tilde{\zeta}(t)|\, \mu(dt) \int |\bar{f}(t, \tau, \cdot)|_{\sup}\, \mu(d\tau)$$

$$\leqslant \int |\tilde{\zeta}(t)|\, |\psi_0^L(t, \cdot)|_{p_1/(p_1 - \beta(L))}\, \mu(dt) \cdot |\tau \to 1 + |\xi(\bar{y})(\tau)|^{\beta(L)}|_{p_1/\beta(L)}$$

$$\leqslant |\tilde{\zeta}|_{p/(p-1)}\, |t \to |\psi_0^L(t, \cdot)|_{p_1/(p_1-\beta(L))}|_p\, |\tau \to 1 + |\xi(\bar{y})(\tau)|^{\beta(L)}|_{p_1/\beta(L)}$$

$$< \infty.$$

Thus, by Fubini's theorem I.4.45(3), $(t, \tau) \to |\tilde{\zeta}(t)|\, |\bar{f}(t, \tau, \cdot)|_{\sup}$ is $\mu \times \mu$-integrable.

This shows [taking into account relation (3)] that the conditions of Lemma VII.3.1 are satisfied with $\zeta = \mu$ and we conclude that

$$\int \mu(dt) \int \tilde{\zeta}(t) \cdot \bar{f}(t, \tau, r)\, \bar{\sigma}(\tau)\, (dr) = \operatorname*{Min}_{r \in \bar{R}^{\#}(\tau)} \int \tilde{\zeta}(t) \cdot \bar{f}(t, \tau, r)\, \zeta(dt)$$

for μ-a.a. $\tau \in T$, thus proving relation (1). QED

VIII.3 Necessary Conditions for an Original Minimum

VIII.3.1 ***Theorem*** Let $\mathscr{U}$ be an abundant subset of $\mathscr{R}^{\#}$, $(\bar{y}, \bar{\sigma}, \bar{b})$ a minimizing $\mathscr{U}$-solution, C_2 convex, $C_2{}^{\circ} \neq \varnothing$, and Assumption VIII.2.1 satisfied. Assume, furthermore, that for each

$$K \triangleq \big((\sigma_0, b_0), ..., (\sigma_m, b_m)\big) \in (\mathscr{S}^{\#} \times B)^{m+1}$$

and for $L \triangleq (b_0, ..., b_m)$ and $q_i \triangleq (\sigma_i, b_i)$ $(i = 0, 1, ..., m)$

(1) the equation

$$\Delta y(t) = (\mathscr{F} \Delta y)(t) = \int f_v(t, \tau, \xi(\bar{y})(\tau), \bar{\sigma}(\tau), \bar{b})(\xi'(\bar{y})\, \Delta y)(\tau)\, \mu(d\tau) \qquad (\mu\text{-a.a. } t \in T)$$

has $\Delta y = 0$ as its only solution in $\mathscr{Y}$;

(2) the equation

$$y = F\left(y, \sigma, \bar{b} + \sum_{j=0}^{m} \theta^j (b_j - \bar{b})\right)$$

has a unique solution $\tilde{y}^L(\sigma, \theta)$ in $\mathscr{Y}$ for all $(\sigma, \theta) \in \mathscr{S}^{\#} \times \mathscr{T}^L$;

(3) the function

$$(y, \sigma, \theta) \to g\left(y, \sigma, \bar{b} + \sum_{j=0}^{m} \theta^j (b_j - \bar{b})\right) : \mathscr{Y} \times \mathscr{S}^{\#} \times \mathscr{T}^L \to \mathbb{R} \times \mathbb{R}^m \times \mathscr{X}_2$$

is continuous, and the function

$$(y, \theta) \to \tilde{g}^K(y, \theta) = g\left(y, \bar{q} + \sum_{j=0}^{m} \theta^j (q_j - \bar{q})\right) : \mathscr{Y} \times \mathscr{T}^L \to \mathbb{R} \times \mathbb{R}^m \times \mathscr{X}_2$$

has a derivative $\mathscr{D}\tilde{g}^K(\bar{y}, 0)$ at $(\bar{y}, 0)$;

(4) $c_1 \triangleq \sup\{|\, \xi(\tilde{y}^L(\sigma, \theta))|_{p_1} \mid (\sigma, \theta) \in \mathscr{S}^{\#} \times \mathscr{T}^L\} < \infty$; and

(5) there exists $\chi \in L^{p_1}(T)$ such that $|\, \xi(y)(\tau)| \leqslant \chi(\tau)$ μ-a.e. whenever $y \in \mathscr{Y}$ and

$$|\, y(t)| \leqslant c\, |\, \psi_0{}^L(t, \cdot)|_{p_1/(p_1 - \beta(L))}, \qquad \text{where} \quad c \triangleq [\mu(T)]^{\beta(L)/p_1} + (c_1)^{\beta(L)}.$$

Then the conclusions of VIII.2.3 remain valid and VIII.2.4 is applicable with the minimizing relaxed solution replaced by the minimizing $\mathscr{U}$-solution $(\bar{y}, \bar{\sigma}, \bar{b})$.

▮ PROOF We first show that Theorem V.3.2 is applicable. Conditions V.3.2(1) and V.3.2(3) follow from VIII.2.2(1) and (3). Condition V.3.2(2) [stating that $I - F_y(\bar{y}, \bar{\sigma}, \bar{b})$ is a homeomorphism of $\mathscr{Y}$ onto $\mathscr{Y}$] follows from VIII.2.3(1). Condition V.2.3(4) is assumed in (2). Finally, condition V.3.2(5) follows from the first part of VIII.1.1 if we replace in it B and ψ by $\mathscr{T}^L$ and $\psi_0{}^L$, respectively. It follows, by V.3.2, that $(\bar{y}, \bar{\sigma}, \bar{b})$ is extremal and relations (2), (3), (5), and (6) of VIII.2.3 are derived exactly as in the proof of that theorem. Since the proof of VIII.2.4 depends only on relations VIII.2.3(3) and VIII.2.3(5), this theorem remains applicable in the present case. QED

VIII.3.2 ***Theorem*** Let $\mathscr{U}$ be an abundant subset of $\mathscr{R}^{\#}$, $(\bar{y}, \bar{u}, \bar{b}) \in \mathscr{A}(\mathscr{U})$,

$$\mathscr{M} \triangleq \{(y, \sigma, b) \in \mathscr{A}(\mathscr{S}^{\#}) \mid g_0(y, \sigma, b) \leqslant g_0(\bar{y}, \bar{u}, \bar{b})\},$$

and let the conditions of VIII.3.1 be satisfied whenever the minimizing $\mathscr{U}$-solution $(\bar{y}, \bar{\sigma}, \bar{b})$ is replaced by an arbitrary point of $\mathscr{M}$. Then the conditions of V.3.4 are satisfied and its conclusions about strict $\mathscr{U}$-solutions remain valid.

▌PROOF We verify, as in the proof of VIII.3.1, that the conditions of V.3.2 remain satisfied when $(\bar{y}, \bar{\sigma}, \bar{b})$ is replaced by a point of $\mathscr{M}$. QED

VIII.4 Problems with Pseudodelays

Let $l \in \mathbb{N}$, $k = ln$, $p_1 = p \in (1, \infty)$, $c \in \mathbb{R}$, and $h_i : T \to T$ $(i = 1, 2, \ldots, l)$ be such that $h_i^{-1}(E)$ is a μ-measurable set and $\mu(h_i^{-1}(E)) \leqslant c\mu(E)$ for each μ-measurable set E and $i = 1, 2, \ldots, l$. If we define ξ by

$$\xi(y)(t) \triangleq \big(y(h_1(t)), \ldots, y(h_l(t))\big) \qquad [y \in L^p(T, \mathbb{R}^n),\ t \in T],$$

then, by II.6.10,

$$\xi \in B\big(L^p(T, \mathbb{R}^n), L^p(T, \mathbb{R}^{ln})\big)$$

and ξ satisfies assumption VIII.1.1(4), with

$$\chi(t) \triangleq c \sum_{i=1}^{l} | \psi(h_i(t), \cdot)|_{p/(p-\beta)} \cdot$$

It follows that the results of VIII.1–VIII.3 remain applicable.

Notes

Existence and necessary conditions for a relaxed minimum were investigated in Warga [11] in the special case where ξ is the identity operator and $\mathscr{X}_2 = \{0\}$.

CHAPTER IX

Conflicting Control Problems with Relaxed Adverse Controls

Chapters IX and X are devoted to the study of "conflicting control problems" that involve both "friendly" and "adverse" controls. The "natural" method of "relaxing" the problem (which we shall study in the present chapter) consists in replacing the original controls, both friendly and adverse, by the corresponding relaxed controls. This approach yields results analogous to those of Chapters VI–VIII when the problems we consider are characterized by "additively coupled" conflicting controls. When the conflicting controls are not additively coupled, the minimizing relaxed solutions thus obtained may yield different results than the minimizing approximate $\mathscr{U}$-solutions. Thus the "naturally relaxed" problems with nonadditively coupled controls depict a different "physical" situation than the corresponding "original" problems. An appropriate "relaxed" model of the situation described by original controls will be discussed in Chapter X.

Sections IX.1 and IX.2 contain general results and the discussion of their significance. In Section IX.3 we discuss an important class of problems with additively coupled controls, and Section IX.4 is devoted to the derivation of a (known) generalization of the von Neumann minimax theorem which can be applied to zero-sum games with strategies that are relaxed controls.

The results of the present chapter can be applied to "naturally relaxed" problems with nonadditively coupled conflicting controls that are defined by differential or functional-integral equations of the type studied in Chapters VI–VIII, and at least one such application should have properly been included in the present chapter. However, in order to avoid the duplication of many definitions and arguments, we discuss such an application in Theorem X.3.7, after considering a similar problem within the model investigated in Chapter X.

IX.0 Formulation of the Problem

IX.0.1 ***An Example*** In this chapter we shall study a class of problems closely related to the unilateral problems first introduced in Chapter V. The latter are characterized by the assumption that $\mathscr{X}_2 \triangleq C(P, \mathbb{R}^{m_2})$ and $C_2 \triangleq \{c \in C(P, \mathbb{R}^{m_2}) \mid c(p) \in A(p) \ \ (p \in P)\}$ for some given space P, nonnegative integer m_2, and a mapping $A : P \to \mathscr{P}'(\mathbb{R}^{m_2})$. Our present considerations are motivated by the case where P is a collection of "adverse" controls. Specifically, if T_P, R_P, $R_P{}^\#$, $\mathscr{R}_P{}^\#$, $\mathscr{S}_P{}^\#$, and B_P are defined in a manner analogous to the definition of T, R, $R^\#$, $\mathscr{R}^\#$, $\mathscr{S}^\#$, and B, respectively, then we shall study the case where $P = \mathscr{S}_P{}^\# \times B_P$ or P is a subset of $\mathscr{S}_P{}^\# \times B_P$ (e.g., $P = \mathscr{U}_P \times B_P$, where $\mathscr{U}_P$ is an abundant subset of $\mathscr{R}_P{}^\#$).

Control problems of this type may arise, in particular, when F or g involve functions or parameters that are not accurately known or that are chosen with the intent to maximize the cost functional g_0 while the controls of the problem are chosen with the intent to minimize g_0. Let us consider the following example. If a drone airplane is to be programmed so as to minimize the fuel consumption while flying from a given point to some designated area in the time period $T \triangleq [t_0, t_1]$, then our problem is specified provided the equation of motion

$$y = F(y, \sigma, b) \triangleq F(y, q)$$

is known. However, this equation may involve the wind velocities in the flight corridor whose distribution is only imperfectly known. For example, we may only know that at any time t and point v the wind velocity is $p(t)w(t, v)$, where $w(\cdot, \cdot)$ is a known function and $p : [t_0, t_1] \to [0.8, 1.2]$ is an unknown measurable function. If this is the case, then the equation of motion is of the form

$$y = F(y, q, p),$$

and we must decide on our policy in the face of the uncertainty caused by our imperfect knowledge of p.

Assume (as it is always the case in a deterministic physical universe) that the equation of motion has a unique solution $\tilde{y}(q, p)$ for each choice of a control q and for each measurable $p : [t_0, t_1] \to [0.8, 1.2]$. Then the fuel consumption is a function $\tilde{x}_0(q, p)$ $[\triangleq g_0(\tilde{y}(q, p), q, p)]$ of q and p, and the restrictions concerning the area of destination can be written as

$$\tilde{y}(q, p)(t_1) \in A,$$

where A is a given set. A conservative approach to the problem would be to choose q so as to minimize $x_0(q) \triangleq \sup \tilde{x}_0(q, P)$ on the set $\{q \mid \tilde{y}(q, p)(t_1) \in A$ $(p \in P)\}$, where P is the collection of all measurable functions on $[t_0, t_1]$ to $[0.8, 1.2]$. Alternately, if we possess sufficient statistical data about the probable determinations of p, we may be able to define a probability measure space (P, Σ_P, ω) and decide to choose q so as to minimize the *expected value* $x_0(q) \triangleq \int \tilde{x}_0(q, p)\, \omega(dp)$ on the set $\{q \mid \tilde{y}(q, p)(t_1) \in A\ (p \in P)\}$.

Since the points p of P may have the nature of controls, we shall refer to q and p as *conflicting controls*, the search for a minimizing $\bar{q}$ being "hampered" by the condition $\tilde{y}(q, p)(t_1) \in A\ (p \in P)$ which involves the *adverse controls* p. The problem we have just discussed may serve as an example of the general *conflicting control problem* which we now proceed to study in its "naturally relaxed" version.

IX.0.2 ***Formulation of the Conflicting Control Problem*** Let Q and P be given sets, $\tilde{Q} \subset Q$, $m \in \mathbb{N}$, $m_2 \in \{0, 1, 2, \ldots\}$, $A : P \to \mathscr{P}'(\mathbb{R}^{m_2})$, $x_0 : Q \to \mathbb{R}$, $x_1 : Q \to \mathbb{R}^m$, and $x_2 : Q \times P \to \mathbb{R}^{m_2}$. When convenient, we write $x_2(q)(p)$ for $x_2(q, p)$ and $x(q)(p)$ for $(x_0(q), x_1(q), x_2(q, p))$, and set

$$\mathscr{A}(\tilde{Q}) \triangleq \{q \in \tilde{Q} \mid x_1(q) = 0,\ x_2(q)(p) \in A(p)\ (p \in P)\}.$$

A point $\bar{q} \in \mathscr{A}(\tilde{Q})$ is a *minimizing $\tilde{Q}$-control* if $x_0(\bar{q}) = \operatorname{Min}(\mathscr{A}(\tilde{Q}))$. A sequence (q_j) in $\tilde{Q}$ is an *approximate $\tilde{Q}$-control* if

$$\lim_j x_1(q_j) = 0 \qquad \text{and} \qquad \lim_j [\sup\{d[A(p), x_2(q_j)(p)] \mid p \in P\}] = 0.$$

An approximate $\tilde{Q}$-control $(\bar{q}_j)$ is a *minimizing approximate $\tilde{Q}$-control* if

$$\lim_j x_0(\bar{q}_j) \leqslant \lim_j \inf x_0(q_j)$$

for every approximate $\tilde{Q}$-control (q_j).

In the case where $Q = \mathscr{S}^{\#} \times B$, $\mathscr{S}' \subset \mathscr{S}^{\#}$ and $\tilde{Q} = \mathscr{S}' \times B$, we refer to an approximate $\tilde{Q}$-control as an *approximate $\mathscr{S}'$-control*.

IX.0.3 ***Original and Relaxed Adverse Controls*** As we have mentioned at the very beginning of this chapter, we are particularly interested in the case where $P \subset \mathscr{S}_P{}^\# \times B_P$, $\mathscr{S}_P{}^\#$ is a class of relaxed adverse control functions, and B_P is a given set of adverse control parameters. In an "original" statement of a conflicting control problem, the set P coincides with $\mathscr{U}_P \times B_P$, where $\mathscr{U}_P \subset \mathscr{R}_P{}^\#$. In the problems that we consider, the function

$$x_2(q, \cdot) : \mathscr{U}_P \times B_P \to \mathbb{R}^{m_2}$$

can be extended to a continuous function on $\mathscr{S}_P{}^\# \times B_P$, and $\mathscr{U}_P$ is an abundant subset of $\mathscr{R}_P{}^\#$. If, furthermore, A is a given closed subset of $\mathbb{R}^{m_2}$ and $A(p) = A$ $(p \in \mathscr{U}_P \times B_P)$, then the condition

$$x_2(q, p) \in A \qquad (p \in \mathscr{U}_P \times B_P)$$

is equivalent to the condition

$$x_2(q, p) \in A \qquad (p \in \mathscr{S}_P{}^\# \times B_P).$$

This shows that, whenever $A \subset \mathbb{R}^{m_2}$ is closed and $A(p) = A$ $(p \in \mathscr{U}_P \times B_P)$, the set $\mathscr{A}(\tilde{Q})$ coincides with the set

$$\{q \in \tilde{Q} \mid x_1(q) = 0,\ x_2(q)(\mathscr{S}_P{}^\# \times B_P) \subset A\},$$

and our conflicting control problem remains unchanged when $\mathscr{U}_P \times B_P$ is replaced by $\mathscr{S}_P{}^\# \times B_P$. The latter space is a compact metric space whenever B_P is a compact metric space.

It is in this context that we ought to interpret some of the ensuing results of this chapter in which P is assumed to be a compact metric space.

IX.1 Existence and Necessary Conditions for Optimal Controls

We shall find it convenient to apply a number of results, most of which were derived in Chapter V, after stating them or reformulating them in a form more appropriate for the conflicting control problem.

IX.1.1 ***Theorem*** Let P be an arbitrary set, Q a sequentially compact topological space, $\mathscr{A}(Q) \neq \varnothing$, and, for each $p \in P$, the function $q \to x(q)(p) : Q \to \mathbb{R} \times \mathbb{R}^m \times \mathbb{R}^{m_2}$ sequentially continuous and the set $A(p)$ closed. Then there exists a minimizing Q-control $\bar{q}$, and $(\bar{q}, \bar{q},...)$ is a minimizing approximate Q-control.

▌PROOF Let (q_j) be a sequence in $\mathscr{A}(Q)$ such that $\lim_j x_0(q_j) = \inf x_0(\mathscr{A}(Q))$ in $\bar{\mathbb{R}}$. Then there exist $J \subset (1, 2,...)$ and $\bar{q} \in Q$ such that $\lim_{j \in J} q_j = \bar{q}$. It follows that $\lim_{j\in J} x(q_j)(p) = x(\bar{q})(p)$ for each $p \in P$. Since $A(p)$ is closed, we have $x_2(\bar{q})(p) \in A(p)$. This shows that $\bar{q} \in \mathscr{A}(Q)$ and $x_0(\bar{q}) = \inf x_0(\mathscr{A}(Q))$.

Now let (q_j') be an approximate Q-control. Then there exists $J' \subset (1, 2, ...)$ such that $\lim_{j\in J'} x_0(q_j') = \liminf_j x_0(q_j')$, and our previous argument shows that there exist $J \subset J'$ and $\bar{q}' \in Q$ such that $\lim_{j\in J} q_j' = \bar{q}'$ and $\bar{q}' \in \mathscr{A}(Q)$; hence $x_0(\bar{q}) \leqslant x_0(\bar{q}') = \lim_{j\in J} x_0(q_j') = \liminf_j x_0(q_j')$, showing that $(\bar{q}, \bar{q}, ...)$ is a minimizing approximate Q-control. QED

IX.1.2 ***Theorem*** Let $\bar{q}$ be a minimizing Q-control, Q a convex subset of some vector space, P a compact metric space, $x_2(q)(\cdot) \in C(P, \mathbb{R}^{m_2})$ $(q \in Q)$, $A(p)$ convex and $A(p)^\circ \neq \varnothing$ for each $p \in P$, and

$$G(A^\circ) \triangleq \{(p, a) \in P \times \mathbb{R}^{m_2} \mid a \in A(p)^\circ\}$$

an open subset of $P \times \mathbb{R}^{m_2}$. Assume furthermore that, for each choice of $(q_0, ..., q_m) \in Q^{m+1}$, the function

$$\theta \to x\left(\bar{q} + \sum_{j=0}^{m} \theta^j(q_j - \bar{q})\right)(\cdot) : \mathscr{T}_{m+1} \to \mathbb{R} \times \mathbb{R}^m \times C(P, \mathbb{R}^{m_2})$$

is continuous in some neighborhood of 0 and has a derivative at 0. Then there exist $l_0 \geqslant 0$, $l_1 \in \mathbb{R}^m$, $\omega \in \text{frm}^+(P)$, and $\tilde{\omega} \in L^1(P, \Sigma_{\text{Borel}}(P), \omega, \mathbb{R}^{m_2})$ such that

$$(1) \qquad |\tilde{\omega}(p)| = 1 \quad (p \in P) \qquad \text{and} \qquad l_0 + |l_1| + \omega(P) > 0;$$

$$(2) \qquad l_0\, Dx_0(\bar{q}; q - \bar{q}) + l_1 \cdot Dx_1(\bar{q}; q - \bar{q}) + \int \tilde{\omega}(p) \cdot D_1x_2(\bar{q}, p; q - \bar{q})\, \omega(dp) \geqslant 0 \qquad (q \in Q);$$

and

$$(3) \qquad \tilde{\omega}(p) \cdot x_2(\bar{q}, p) = \operatorname*{Max}_{a\in A(p)} \tilde{\omega}(p) \cdot a \qquad \text{for} \quad \omega\text{-a.a. } p \in P.$$

▌PROOF We set $\mathscr{Y} = \mathbb{R}^0 \triangleq \{0\}$ [hence $F(y, \sigma, b) = 0$ for all $(y, \sigma, b) \in \mathscr{Y} \times \mathscr{S}^\# \times B$], $\mathscr{X}_2 = C(P, \mathbb{R}^{m_2})$, $C_2 = \{c \in \mathscr{X}_2 \mid c(p) \in A(p)\ (p \in P)\}$, $B = Q$, and $g(y, \sigma, b) = x(b)$ $(b \triangleq q \in Q)$ in V.2.3. We furthermore observe that the definitions of convergence in $C(P, \mathbb{R}^{m_2})$ and of a directional derivative imply that $Dx_2(\bar{q}; q - \bar{q})(p)$ [that is, the value at p of $Dx_2(\bar{q}; q - q)(\cdot)$] equals $D_1x_2(\bar{q}, p; q - \bar{q})$. Then our conclusions follow directly from V.2.3 and V.2.5. QED

We next consider the *minimax control problems* in which $x_0(q) \triangleq \sup \tilde{x}_0(q, P)$ $(q \in Q)$.

IX.1.3 ***Theorem*** Let $\bar{q}$ be a minimizing Q-control, Q a convex subset of some vector space, P a compact metric space, $\tilde{x}_0(q)(\cdot) \in C(P)$ and $x_2(q)(\cdot) \in C(P, \mathbb{R}^{m_2})$ $(q \in Q)$, $A(p)$ convex and $A(p)^\circ \neq \varnothing$ for each $p \in P$, and $G(A^\circ) \triangleq \{(p, a) \in P \times \mathbb{R}^{m_2} \mid a \in A(p)^\circ\}$ an open subset of $P \times \mathbb{R}^{m_2}$. Furthermore, let

$$x_0(q) \triangleq \sup \tilde{x}_0(q, P) \qquad (q \in Q),$$
$$\tilde{x}(q)(p) \triangleq \big(\tilde{x}_0(q, p), x_1(q), x_2(q, p)\big) \qquad (q \in Q, \quad p \in P),$$

and assume that, for each choice of $(q_0, \ldots, q_m) \in Q^{m+1}$, the function

$$\theta \to \tilde{x}\left(q + \sum_{j=0}^{m} \theta^j(q_j - \bar{q})\right)(\cdot) : \mathscr{T}_{m+1} \to C(P) \times \mathbb{R}^m \times C(P, \mathbb{R}^{m_2})$$

is continuous in some neighborhood of 0 and has a derivative at 0. Then there exist $l_1 \in \mathbb{R}^m$, $\omega \in \operatorname{frm}^+(P)$,

$$\tilde{\omega} \in L^1(P, \Sigma_{\text{Borel}}(P), \omega, \mathbb{R}^{m_2}), \qquad \text{and} \qquad \tilde{\omega}^0 \in L^1(P, \Sigma_{\text{Borel}}(P), \omega)$$

such that

(1)
$$\tilde{\omega}^0(p) \geqslant 0 \quad \omega\text{-a.e.}, \qquad \tilde{\omega}^0(p) + |\tilde{\omega}(p)| = 1 \quad \omega\text{-a.e.}, \qquad |l_1| + \omega(P) > 0,$$
$$\tilde{\omega}^0(p) = 0 \quad \omega\text{-a.e. in the set } \{p \in P \mid \tilde{x}_0(\bar{q}, p) < \sup \tilde{x}_0(\bar{q}, P)\};$$

(2)
$$l_1 \cdot Dx_1(\bar{q}; q - \bar{q}) + \int [\tilde{\omega}^0(p)\, D_1\tilde{x}_0(\bar{q}, p; q - \bar{q}) + \tilde{\omega}(p) \cdot D_1x_2(\bar{q}, p; q - \bar{q})]\, \omega(dp) \geqslant 0 \qquad (q \in Q);$$

and

(3)
$$\tilde{\omega}(p) \cdot x_2(\bar{q}, p) = \operatorname*{Max}_{a \in A(p)} \tilde{\omega}(p) \cdot a \qquad \text{for} \quad \omega\text{-a.a. } p \in P.$$

PROOF Let $Q' \triangleq Q \times \mathbb{R}$,

$$\xi_0(q') \triangleq q_0, \qquad \xi_1(q') \triangleq x_1(q),$$
$$\xi_2(q', p) = \big(\tilde{x}_0(q, p) - q_0, x_2(q, p)\big) \qquad [q' \triangleq (q, q_0) \in Q', \quad p \in P],$$

and

$$A'(p) \triangleq (-\infty, 0] \times A(p) \qquad (p \in P).$$

If we replace Q, m_2, A, and x by Q', $m_2 + 1$, A', and ξ, respectively, then the new conflicting control problem consists in the search for a point $\bar{q}' \in Q'$ that minimizes q_0 on the set

$$\begin{aligned}\mathscr{A}(Q') &\triangleq \{q' \in Q' \mid \xi_1(q') = 0, \xi_2(q') \in A'(p)\ (p \in P)\} \\ &= \{(q, q_0) \in Q \times \mathbb{R} \mid x_1(q) = 0, \tilde{x}_0(q, p) \leqslant q_0, \\ &\qquad \text{and } x_2(q, p) \in A(p)\ (p \in P)\}.\end{aligned}$$

It is clear that $\bar{q}' \triangleq (\bar{q}, \bar{q}_0)$ is a Q'-minimizing control if and only if $\bar{q}_0 = \sup \tilde{x}_0(\bar{q}, P)$ and $\bar{q}$ minimizes $\sup \tilde{x}_0(q, P)$ on the set $\mathscr{A}(Q) \triangleq \{q \in Q \mid x_1(q) = 0,\ x_2(q, p) \in A(p)\ (p \in P)\}$. Since the new problem satisfies the conditions of IX.1.2, it follows that there exist $l_0 \geqslant 0$, $l_1 \in \mathbb{R}^m$, $\omega \in \mathrm{frm}^+(P)$, and $\tilde{\omega}' \triangleq (\tilde{\omega}'^0, \ldots, \tilde{\omega}'^{m_2}) \in L^1(P, \Sigma_{\mathrm{Borel}}(P), \omega, \mathbb{R}^{m_2+1})$ such that

$$(4) \qquad |\tilde{\omega}'(p)| = 1 \quad (p \in P), \qquad l_0 + |l_1| + \omega(P) > 0,$$

$$(5) \qquad l_0\, D\xi_0(\bar{q}'; q' - \bar{q}') + l_1 \cdot D\xi_1(\bar{q}'; q' - \bar{q}') + \int \tilde{\omega}'(p) \cdot D_1\xi_2(\bar{q}', p; q' - \bar{q}')\, \omega(dp) \geqslant 0 \qquad (q' \in Q'),$$

and

$$(6) \qquad \tilde{\omega}'(p) \cdot \xi_2(\bar{q}', p) = \operatorname*{Max}_{a' \in A'(p)} \tilde{\omega}'(p) \cdot a' \qquad \text{for} \quad \omega\text{-a.a. } p \in P.$$

We now set

$$\tilde{\omega}^0 \triangleq \tilde{\omega}'^0 \qquad \text{and} \qquad \tilde{\omega} \triangleq (\tilde{\omega}'^1, \ldots, \tilde{\omega}'^{m_2}).$$

If we set $q' = (\bar{q}, q_0)$ in (5), then we obtain

$$\left[l_0 - \int \tilde{\omega}^0(p)\, \omega(dp)\right] (q_0 - \bar{q}_0) \geqslant 0 \qquad (q_0 \in \mathbb{R});$$

hence

$$(7) \qquad l_0 = \int \tilde{\omega}^0(p)\, \omega(dp).$$

Relation (6) yields relation (3) and the relation

$$\tilde{\omega}^0(p)\big(\tilde{x}_0(\bar{q}, p) - \bar{q}_0\big) = \operatorname*{Max}_{\alpha \in (-\infty, 0]} \tilde{\omega}^0(p) \cdot \alpha \qquad \text{for} \quad \omega\text{-a.a. } p \in P.$$

This last relation implies, in turn, that $\tilde{\omega}^0(p) \geqslant 0$ ω-a.e. and that $\tilde{\omega}^0(p) = 0$ ω-a.e. in $\{p \in P \mid \tilde{x}_0(\bar{q}, p) < \bar{q}_0 = \sup \tilde{x}_0(\bar{q}, P)\}$. It follows then from (4) and (7) that

$$|\tilde{\omega}'(p)| = \tilde{\omega}^0(p) + |\tilde{\omega}(p)| = 1 \quad \omega\text{-a.e.} \qquad \text{and} \qquad |l_1| + \omega(P) > 0.$$

Thus (1) is valid.

Finally, we set $q_0 = \bar{q}_0$ is relation (5) to obtain relation (2). QED

We now consider minimizing approximate and original controls. Theorem IX.1.4 deals with minimizing approximate $\mathscr{U}$-controls and IX.1.5 with strong necessary conditions for an original minimum.

IX.1.4 ***Theorem*** Let B be sequentially compact, $Q = \mathscr{S}^{\#} \times B$, P a compact metric space, $A(p)$ closed for each $p \in P$, and the function $q \to x(q)(\cdot)$

an element of $C(Q, \mathbb{R} \times \mathbb{R}^m \times C(P, \mathbb{R}^{m_2}))$. If $\bar{q} \triangleq (\bar{\sigma}, \bar{b})$ is a minimizing Q-control and $\mathscr{U}$ an abundant subset of $\mathscr{R}^{\#}$, then there exists a sequence (u_j) in $\mathscr{U}$ converging to $\bar{\sigma}$ and, for every choice of such a sequence (u_j), $((u_j, \bar{b}))$ is a minimizing approximate $\mathscr{U}$-control, and

$$\lim_j x(u_j, \bar{b})(\cdot) = x(\bar{\sigma}, \bar{b})(\cdot) \quad \text{in } \mathbb{R} \times \mathbb{R}^m \times C(P, \mathbb{R}^{m_2}).$$

▮ PROOF By IV.3.10, there exists a sequence (u_j) in $\mathscr{U}$ converging to $\bar{\sigma}$. Since $q \to x(q)(\cdot)$ is continuous, we have

$$\lim_j x(u_j, \bar{b})(\cdot) = x(\bar{\sigma}, \bar{b})(\cdot) \quad \text{in } \mathbb{R} \times \mathbb{R}^m \times C(P, \mathbb{R}^{m_2}).$$

Since $x_1(\bar{\sigma}, \bar{b}) = 0$ and $x_2(\bar{\sigma}, \bar{b}) \in C_2$, the sequence $((u_j, \bar{b}))$ is an approximate $\mathscr{U}$-control. Now assume, by way of contradiction, that there exists an approximate $\mathscr{U}$-control $((\tilde{u}_j, \tilde{b}_j))$ with $\liminf_j x_0(\tilde{u}_j, \tilde{b}_j) < x_0(\bar{\sigma}, \bar{b})$. Because $\mathscr{S}^{\#} \times B$ is sequentially compact, we may assume (extracting a subsequence, if necessary) that there exists $(\tilde{\sigma}, \tilde{b}) \in \mathscr{S}^{\#} \times B$ such that $\lim_j (\tilde{u}_j, \tilde{b}_j) = (\tilde{\sigma}, \tilde{b})$; hence $\lim_j x(\tilde{u}_j, \tilde{b}_j) = x(\tilde{\sigma}, \tilde{b})$ and, since

$$C_2 = \{c \in C(P, \mathbb{R}^{m_2}) \mid c(p) \in A(p) \ (p \in P)\}$$

is closed, we have $x_1(\tilde{\sigma}, \tilde{b}) = 0$ and $x_2(\tilde{\sigma}, \tilde{b}) \in C_2$. It follows that $(\tilde{\sigma}, \tilde{b}) \in \mathscr{A}(Q)$ and

$$x_0(\tilde{\sigma}, \tilde{b}) = \lim_j x_0(\tilde{u}_j, \tilde{b}_j) < x_0(\bar{\sigma}, \bar{b}),$$

contradicting the assumption that $(\bar{\sigma}, \bar{b})$ is a minimizing Q-control. QED

IX.1.5 ***Theorem*** Let $\mathscr{U}$ be an abundant subset of $\mathscr{R}^{\#}$, $(\bar{u}, \bar{b})$ a minimizing $\mathscr{U} \times B$-control, P a compact metric space, $x_2(\sigma, b)(\cdot) \in C(P, \mathbb{R}^{m_2})$ $[(\sigma, b) \in \mathscr{S}^{\#} \times B]$, $A(p)$ convex and $A(p)^\circ \neq \varnothing$ for each $p \in P$, and $G(A^\circ) \triangleq \{(p, a) \in P \times \mathbb{R}^{m_2} \mid a \in A(p)^\circ\}$ an open subset of $P \times \mathbb{R}^{m_2}$. Assume furthermore that, for each $((\sigma_0, b_0),\ldots, (\sigma_m, b_m)) \in (\mathscr{S}^{\#} \times B)^{m+1}$, the function

$$(\sigma, \theta) \to x\left(\sigma, \bar{b} + \sum_{j=0}^{m} \theta^j (b_j - \bar{b})\right) : \mathscr{S}^{\#} \times \mathscr{T}_{m+1} \to \mathbb{R} \times \mathbb{R}^m \times C(P, \mathbb{R}^{m_2})$$

is continuous and the function

$$\theta \to x\left(\bar{u} + \sum_{j=0}^{m} \theta^j(\sigma_j - \bar{u}), \bar{b} + \sum_{j=0}^{m} \theta^j (b_j - \bar{b})\right) : \mathscr{T}_{m+1} \to \mathbb{R} \times \mathbb{R}^m \times C(P, \mathbb{R}^{m_2})$$

has a derivative at 0. Then there exist $l_0 \geqslant 0$, $l_1 \in \mathbb{R}^m$, $\omega \in \text{frm}^+(P)$, and an ω-integrable $\tilde{\omega} : P \to \mathbb{R}^{m_2}$ such that

(1) $$|\tilde{\omega}(p)| = 1 \quad (p \in P) \qquad \text{and} \qquad l_0 + |l_1| + \omega(P) > 0;$$

(2) $$\sum_{i=0}^{1} l_i \cdot Dx_0\big((\bar{u}, \bar{b}); (\sigma, b) - (\bar{u}, \bar{b})\big)$$
$$+ \int \tilde{\omega}(p) \cdot D_1 x_2\big((\bar{u}, \bar{b}), p; (\sigma, b) - (\bar{u}, \bar{b})\big)\, \omega(dp) \geqslant 0$$
$$(\sigma \in \mathscr{S}^{\#}, b \in B);$$

and

(3) $$\tilde{\omega}(p) \cdot x_2(\bar{u}, \bar{b}, p) = \operatorname*{Max}_{a \in A(p)} \tilde{\omega}(p) \cdot a \qquad \text{for} \quad \omega\text{-a.a. } p \in P.$$

▌PROOF This follows from V.3.3 by setting $\mathscr{Y} = \mathbb{R}^0$, $\mathscr{X}_2 = C(P, \mathbb{R}^{m_2})$, $C_2 = \{c \in \mathscr{X}_2 \mid c(p) \in A(p)\ (p \in P)\}$, and $g(y, \sigma, b) \triangleq x(\sigma, b)$. QED

Finally, we consider weak necessary conditions for an original minimum. These weak conditions apply in many situations in which the function $(\sigma, \theta) \to x\big(\sigma, \bar{b} + \sum_{j=0}^{m} \theta^j(b_j - \bar{b})\big)$ is not continuous and Theorem IX.1.5 is therefore not applicable (and is, in fact, invalid as shown by Counterexample IX.2.2 below).

IX.1.6 ***Theorem*** Let $(\bar{\rho}, \bar{b})$ be a minimizing $\mathscr{R}^{\#} \times B$-control, P a compact metric space, $x_2(\rho, b)(\cdot) \in C(P, \mathbb{R}^{m_2})$ $[(\rho, b) \in \mathscr{R}^{\#} \times B)]$, $A(p)$ convex and $A(p)^{\circ} \neq \varnothing$ for each $p \in P$, and $G(A^{\circ}) \triangleq \{(p, a) \in P \times \mathbb{R}^{m_2} \mid a \in A(p)^{\circ}\}$ an open subset of $P \times \mathbb{R}^{m_2}$. Furthermore, let M be both a subset of some vector space and a topological space, $M^{\#} : T \to \mathscr{P}'(M)$, $\phi \in C(M, R)$, $\mathscr{M}^{\#}$ the collection of μ-measurable selections of $M^{\#}$, $\bar{\nu} \in \mathscr{M}^{\#}$, and

$$\hat{x}(\nu, b) \triangleq x(\phi \circ \nu, b) \qquad (\nu \in \mathscr{M}^{\#}, \quad b \in B).$$

We assume that

(1) $M^{\#}(t)$ is convex and $\phi\big(M^{\#}(t)\big) = R^{\#}(t)$ for μ-a.a. $t \in T$;

(2) $$\bar{\rho} = \phi \circ \bar{\nu};$$

and

(3) for each choice of $K \triangleq \big((\nu_0, b_0), \ldots, (\nu_m, b_m)\big) \in (\mathscr{M}^{\#} \times B)^{m+1}$, the function

$$\theta \to \hat{x}^K(\theta) \triangleq \hat{x}\left(\bar{\nu} + \sum_{j=0}^{m} \theta^j(\nu_j - \bar{\nu}), \bar{b} + \sum_{j=0}^{m} \theta^j(b_j - b)\right) :$$
$$\mathscr{T}_{m+1} \to \mathbb{R} \times \mathbb{R}^m \times C(P, \mathbb{R}^{m_2})$$

is continuous and has a derivative at 0.

Then there exist $l_0 \geqslant 0$, $l_1 \in \mathbb{R}^m$, $\omega \in \text{frm}^+(P)$, and an $\tilde{\omega}$-integrable $\tilde{\omega} : P \to \mathbb{R}^{m_2}$ such that

$$(4) \qquad |\tilde{\omega}(p)| = 1 \quad (p \in P) \qquad \text{and} \qquad l_0 + |l_1| + \omega(P) > 0;$$

$$(5) \qquad \sum_{i=0}^{1} l_i \cdot D\hat{x}_i((\bar{\nu}, \bar{b}); (\nu, b) - (\bar{\nu}, \bar{b}))$$
$$+ \int \tilde{\omega}(p) \cdot D_1\hat{x}_2((\bar{\nu}, \bar{b}), p; (\nu, b) - (\bar{\nu}, \bar{b}))\, \omega(dp) \geqslant 0$$
$$(\nu \in \mathscr{M}^{\#}, b \in B);$$

and

$$(6) \qquad \tilde{\omega}(p) \cdot \hat{x}_2(\bar{\nu}, \bar{b}) = \operatorname*{Max}_{a \in A(p)} \tilde{\omega}(p) \cdot a \qquad \text{for} \quad \omega\text{-a.a. } p \in P.$$

▌PROOF This theorem follows directly from V.5.1 and V.2.5 by setting $\mathscr{Y} = \mathbb{R}^0$, $\mathscr{X}_2 = C(P, \mathbb{R}^{m_2})$, $C_2 = \{c \in \mathscr{X}_2 \mid c(p) \in A(p) \quad (p \in P)\}$, and $g(y, \rho, b) = x(\rho, b)$. QED

IX.2 Conflicting Control Problems Defined by Functional Equations. Additively Coupled Conflicting Controls. A Counterexample

The example in IX.0 illustrates how the function $x \triangleq (x_0, x_1, x_2)$ that defines a conflicting control problem can arise from a functional equation. We shall now consider a general class of conflicting control problems defined by functional equations, and will describe this class in the following manner. Let $\mathscr{Y}_1$ and $\mathscr{Y}$ be Banach spaces, $Q \triangleq \mathscr{S}^{\#} \times B$, P be a given set, and $G : \mathscr{Y}_1 \times Q \to \mathscr{Y}_1$, $F : \mathscr{Y} \times Q \times P \to \mathscr{Y}$, $(g_0, g_1) : \mathscr{Y}_1 \times Q \to \mathbb{R} \times \mathbb{R}^m$, and $g_2 : \mathscr{Y} \times Q \times P \to \mathbb{R}^{m_2}$ given functions. If, for each choice of $(q, p) \in Q \times P$, the equations

$$y_1 = G(y_1, q) \quad \text{in } \mathscr{Y}_1 \qquad \text{and} \qquad y = F(y, q, p) \quad \text{in } \mathscr{Y}$$

have unique solutions $\tilde{y}_1(q)$ and $\tilde{y}(q, p)$, respectively, then we set

$$x_i(q) \triangleq g_i(\tilde{y}_1(q), q) \quad (i = 0, 1) \qquad \text{and} \qquad x_2(q, p) \triangleq g_2(\tilde{y}(q, p), q, p).$$

In order to apply the results of IX.1 to such a problem, we must establish conditions that ensure the appropriate continuity and "directional differentiability" properties of (x_0, x_1) and x_2, respectively. Those, in turn, depend on the properties of G, g_0, g_1 and of F, g_2, respectively. The case of (x_0, x_1) and G, g_0, g_1 is of the type that we have investigated in Chapters VI–VIII for various functional-integral equations, and it follows that x_0 and x_1 are continuous and have directional derivatives under most "reasonable" conditions. In the case of x_2, which depends both on q and p, the situation is more complicated. As we shall show in Counterexample IX.2.2 below, x_2 need not be continuous on $Q \times P$ even in some extremely simple and "smooth" cases for which P is a set of adverse controls. On the other hand, it is "usually" true that $x_2(\cdot, p)$ and $x_2(q, \cdot)$ are continuous for all $(q, p) \in Q \times P$. This is the case, in particular, if R_P and $R_P^\#$ satisfy the conditions imposed on R and $R^\#$, $\mathscr{R}_P^\#$ and $\mathscr{S}_P^\#$ are defined in the same manner as $\mathscr{R}^\#$ and $\mathscr{S}^\#$, B_P is a given metric space of adverse control parameters, $P = \mathscr{S}_P^\# \times B_P$, $x_2(q, p)$ is the solution $\tilde{y}(q, p)$ of the equation $y = F(y, q, p)$, and F is the functional-integral transformation

$$F(y, \sigma, b, \pi, b_P)(t) \triangleq \int \mu(d\tau) \int f(t, \tau, \xi(y)(\tau), r, b, r_P, b_P)\, \sigma(\tau)\,(dr) \times \pi(\tau)\,(dr_P),$$

where $(\sigma, b) \in Q$, $(\pi, b_P) \in P$, and f and ξ satisfy conditions analogous to those specified in Chapters VI–VIII.

We shall say that q and p are *additively coupled* in F if

$$F(y, q, p) = F_1(y, q) + F_2(y, p) \qquad (y \in \mathscr{Y}, \quad q \in Q, \quad p \in P)$$

for some functions $F_1 : \mathscr{Y} \times Q \to \mathscr{Y}$ and $F_2 : \mathscr{Y} \times P \to \mathscr{Y}$. While Counterexample IX.2.2 will show that it is not reasonable to expect $F : \mathscr{Y} \times Q \times P \to \mathscr{Y}$ to be sequentially continuous in the general case, the considerations of the three previous chapters show that functions like F_1 and F_2 are sequentially continuous under "usual" conditions. Under these conditions, F is also sequentially continuous if q and p are additively coupled in F. Furthermore, as we shall show below, the solution $\tilde{y}(q, p)$ of the equation $y = F(y, q, p)$ is "usually" a sequentially continuous function of (q, p) if F is sequentially continuous.

IX.2.1 ***Lemma*** Let $\mathscr{Y}$ be a Banach space, Q and P topological spaces, $F : \mathscr{Y} \times Q \times P \to \mathscr{Y}$ sequentially continuous, and assume that the equation $y = F(y, q, p)$ has a unique solution $\tilde{y}(q, p)$ in $\mathscr{Y}$ for all $(q, p) \in Q \times P$ and that the set $\tilde{y}(Q \times P)$ is conditionally compact in $\mathscr{Y}$. Then $\tilde{y}(\cdot, \cdot) : Q \times P \to \mathscr{Y}$ is sequentially continuous.

▌PROOF Assume, by way of contradiction, that $\tilde{y}(\cdot, \cdot)$ is not sequentially continuous at some $(\bar{q}, \bar{p}) \in Q \times P$. Then there exist a sequence $((q_j, p_j))$ in $Q \times P$ converging to $(\bar{q}, \bar{p})$ and some $\epsilon > 0$ such that

$$(1) \qquad |\tilde{y}(\bar{q}, \bar{p}) - \tilde{y}(q_j, p_j)| > \epsilon \qquad (j \in \mathbb{N}).$$

We may assume that there exists some $\bar{y} \in \mathscr{Y}$ such that $\lim_j \tilde{y}(q_j, p_j) = \bar{y}$, otherwise replacing $((q_j, p_j))$ by an appropriate subsequence. It follows that

$$\bar{y} = \lim_j \tilde{y}(q_j, p_j) = \lim_j F(\tilde{y}(q_j, p_j), q_j, p_j) = F(\bar{y}, \bar{q}, \bar{p}).$$

Since the equation $y = F(y, \bar{q}, \bar{p})$ has a unique solution, it follows that $\tilde{y}(\bar{q}, \bar{p}) = \bar{y} = \lim_j \tilde{y}(q_j, p_j)$, contrary to (1). QED

IX.2.2 ***A Counterexample*** We shall now define a very simple conflicting control problem with a function $x_2(\cdot, \cdot)$ that is not continuous. This problem satisfies the conditions of Theorems IX.1.1 and IX.1.2, and it possesses therefore a minimizing $\mathscr{S}^{\#} \times B$-control satisfying the necessary conditions of IX.1.2. However, the effects of this minimizing $\mathscr{S}^{\#} \times B$-control cannot be simulated by original controls. Furthermore, this problem also has a minimizing $\mathscr{U} \times B$-control which satisfies the weak necessary conditions of IX.1.6 but does not satisfy the "strong" conclusions of Theorem IX.1.5.

Let

$$T \triangleq [0, 1], \qquad R_P = R \triangleq \{r \triangleq (r^1, r^2) \in \mathbb{R}^2 \mid r \cdot r \triangleq (r^1)^2 + (r^2)^2 = 1\},$$

$$R^{\#}(t) \triangleq R \quad (t \in T), \qquad B \triangleq [-2, 2]^2, \qquad Q \triangleq \mathscr{S}^{\#} \times B,$$

$$P \triangleq \mathscr{S}_P^{\#} = \mathscr{S}^{\#} = \mathscr{S}, \qquad A \triangleq (-\infty, 0],$$

$$x_0(q) \triangleq x_0(\sigma, b^1, b^2) \triangleq b^1,$$

$$x_1(q) \triangleq x_1(\sigma, b^1, b_2) \triangleq b^2,$$

$$x_2(q, p) \triangleq x_2(\sigma, b^1, b^2, \pi) \triangleq \int_0^1 dt \int_{R \times R_P} \alpha \cdot \beta \, \sigma(t)(d\alpha) \times \pi(t)(d\beta) - b^1$$

$$[q \triangleq (\sigma, b^1, b^2) \in Q, \quad p \triangleq \pi \in \mathscr{S}_P^{\#}).$$

[Observe that x_2, as just defined, is an extension to $\mathscr{S}^{\#} \times B \times \mathscr{S}^{\#}$ of the function

$$(\rho, b^1, b^2, \rho_P) \to \int_0^1 \rho(t) \cdot \rho_P(t) \, dt - b^1 : \mathscr{R}^{\#} \times B \times \mathscr{R}^{\#} \to \mathbb{R}.]$$

We consider the problem of minimizing x_0 on the set

$$\mathscr{A}(Q) \triangleq \{(\sigma, b^1, b^2) \in \mathscr{S}^{\#} \times B \mid x_1(\sigma, b^1, b^2) \triangleq b^2 = 0,$$
$$x_2(\sigma, b^1, b^2, \pi) \in A \ (\pi \in \mathscr{S}^{\#})\}.$$

This problem clearly satisfies the conditions of IX.1.1 and IX.1.2.

Minimizing Relaxed Controls This problem has infinitely many minimizing Q-controls, and we can construct them as follows. Let $\bar{\rho}$ be an arbitrary element of $\mathscr{R}^{\#}$, that is, a measurable function on [0, 1] to R, and let $\bar{\sigma}$ be defined by

$$\bar{\sigma}(t)(\{\bar{\rho}(t)\}) = \bar{\sigma}(t)(\{-\bar{\rho}(t)\}) \triangleq \tfrac{1}{2} \qquad (t \in T).$$

Then we can easily verify that $(\bar{\sigma}, 0, 0)$ is a minimizing Q-control. Indeed, it is easy to verify that $(\bar{\sigma}, 0, 0) \in \mathscr{A}(Q)$. Now let $q \triangleq (\sigma, b^1, b^2) \in \mathscr{A}(Q)$. Then $x_1(q) \triangleq b^2 = 0$ and

$$x_2(\sigma, b^1, b^2, \pi) \triangleq \int_0^1 dt \int \alpha \cdot \beta\, \sigma(t)(d\alpha) \times \pi(t)(d\beta) - b^1 \leqslant 0 \qquad (\pi \in \mathscr{S}^{\#}).$$

If we set $\pi = \sigma$ and $s(t) \triangleq \int \alpha\, \sigma(t)(d\alpha)$, then the above relation and Fubini's theorem I.4.45 yield

$$\begin{aligned}\int_0^1 dt \int \alpha \cdot \left[\int \beta\, \pi(t)(d\beta)\right] \sigma(t)(d\alpha) - b^1 &= \int_0^1 dt \int \alpha \cdot s(t)\, \sigma(t)\,(d\alpha) - b^1 \\ &= \int_0^1 s(t) \cdot s(t)\, dt - b^1 \leqslant 0;\end{aligned}$$

hence $b^1 \geqslant \int_0^1 s(t) \cdot s(t)\, dt \geqslant 0$. Since $(\bar{\sigma}, 0, 0) \in \mathscr{A}(Q)$ and $x_0(\sigma, b^1, b^2) \triangleq b^1$, it follows that $(\bar{\sigma}, 0, 0)$ is a minimizing Q-control.

We next show that for every $(\rho, b^1, b^2) \in \mathscr{A}(\mathscr{R}^{\#} \times B)$ we have $b^1 \geqslant 1$. Indeed, we have

$$x_2(\rho, b^1, b^2, \pi) \triangleq \int_0^1 dt \int \rho(t) \cdot \beta\, \pi(t)(d\beta) - b^1 \leqslant 0 \qquad (\pi \in \mathscr{S}^{\#})$$

and, for $\pi(t) = \rho(t)$ $(t \in T)$, this yields

$$\int_0^1 \rho(t) \cdot \rho(t)\, dt - b^1 = \int_0^1 dt - b^1 = 1 - b^1 \leqslant 0.$$

This shows that no minimizing approximate $\mathscr{R}^{\#}$-control can simulate the effects of $(\bar{\sigma}, 0, 0)$.

THE FUNCTION x_2 IS NOT CONTINUOUS The above considerations also show that x_2 is not continuous. Indeed, let (ρ_j) be a sequence in $\mathscr{R}^{\#}$ converging to $\bar{\sigma}$. Then

$$x_2(\bar{\sigma}, b^1, b^2, \bar{\sigma}) = -b^1 \qquad [(b^1, b^2) \in B],$$

but

$$x_2(\rho_j, b^1, b^2, \rho_j) = 1 - b^1 \qquad [j \in \mathbb{N}, \quad (b^1, b^2) \in B].$$

MINIMIZING $\mathscr{R}^{\#} \times B$-CONTROLS Let $\bar{\rho}$ be an arbitrary measurable function on T to R. Then we shall show that $(\bar{\rho}, 1, 0)$ is a minimizing $\mathscr{R}^{\#} \times B$-control. Indeed, it follows from I.6.14 that

$$\int \beta\, \pi(t)(d\beta) \in \operatorname{co}(R) \qquad (\pi \in \mathscr{S}^{\#}, \quad t \in T), \tag{1}$$

and it is easy to verify that $| r_1 \cdot r_2 | \leqslant 1$ $[(r_1, r_2) \in R]$ and

$$\begin{aligned}\operatorname{co}(R) &= \{r \triangleq (r^1, r^2) \in \mathbb{R}^2 \mid (r^1)^2 + (r^2)^2 \leqslant 1\} \\ &= \{\gamma r \mid \gamma \in [0, 1], r \in R\}.\end{aligned}$$

Thus $| \alpha \cdot \int \beta\, \pi(t)(d\beta)| \leqslant 1$ $(\pi \in \mathscr{S}^{\#}, t \in T, \alpha \in R)$. Therefore

$$x_2(\bar{\rho}, 1, 0, \pi) = \int_0^1 dt \int \bar{\rho}(t) \cdot \beta\, \pi(t)(d\beta) - 1 \leqslant 0 \qquad (\pi \in \mathscr{S}^{\#}),$$

showing that $(\bar{\rho}, 1, 0) \in \mathscr{A}(\mathscr{R}^{\#} \times B)$ and $x_0(\bar{\rho}, 1, 0) = 1$. On the other hand, let $(\rho, b^1, b^2) \in \mathscr{A}(\mathscr{R}^{\#} \times B)$. Then

$$x_2(\rho, b^1, b^2, \rho) = \int_0^1 \rho(t) \cdot \rho(t)\, dt - b^1 = 1 - b^1 \leqslant 0,$$

showing that $x_0(\rho, b^1, b^2) = b^1 \geqslant 1$, which proves that $(\bar{\rho}, 1, 0)$ is a minimizing $\mathscr{R}^{\#} \times B$-control.

"STRONG" CONDITIONS FAIL We next show that the conclusions of Theorem IX.1.5 are not valid for any choice of an abundant subset $\mathscr{U}$ of $\mathscr{R}^{\#}$. Let, indeed, $\mathscr{U}$ be an abundant subset of $\mathscr{R}^{\#}$ and $\bar{u} \in \mathscr{U}$. Then, as we have just shown, $(\bar{u}, 1, 0)$ is a minimizing $\mathscr{R}^{\#} \times B$-control and, *a fortiori*, a minimizing $\mathscr{U} \times B$-control. If the conclusions of IX.1.5 are valid, then there exist $l_0 \geqslant 0$, $l_1 \in \mathbb{R}$, $\omega \in \operatorname{frm}^+(\mathscr{S}^{\#})$, and $\tilde{\omega} \in L^1(\mathscr{S}^{\#}, \Sigma_{\text{Borel}}(\mathscr{S}^{\#}), \omega)$ satisfying IX.1.5(1)–IX.1.5(3), with $A(\pi) \triangleq (-\infty, 0]$ $(\pi \in \mathscr{S}^{\#})$. We set $\bar{q} \triangleq (\bar{u}, 1, 0)$, $q \triangleq (\sigma, b^1, b^2)$, and denote by $\delta_{\bar{u}}(t)$ the Dirac measure at $\bar{u}(t)$. We have

$$Dx_0(\bar{q}; q - \bar{q}) = b^1 - 1, \tag{2}$$

$$Dx_1(\bar{q}; q - \bar{q}) = b^2, \tag{3}$$

and

$$\begin{aligned} &D_1 x_2(\bar{q}, \pi; q - \bar{q}) \\ &\quad = \int_0^1 dt \int \pi(t)\,(d\beta) \int \alpha \cdot \beta\,(\sigma(t) - \delta_{\bar{u}}(t))(d\alpha) - (b^1 - 1). \end{aligned} \tag{4}$$

It follows from IX.1.5(3) that

$$\tilde{\omega}(\pi)\, x_2(\bar{q}, \pi) = \operatorname*{Max}_{a \leqslant 0} \tilde{\omega}(\pi)\, a \qquad \text{for} \quad \omega\text{-a.a. } \pi \in \mathscr{S}^{\#};$$

hence $\tilde{\omega}(\pi) \geqslant 0$ ω-a.e. and $\tilde{\omega}(\pi)\, x_2(\bar{q}, \pi) = 0$ ω-a.e. Since, by IX.1.5(1), $|\tilde{\omega}(\pi)| = \tilde{\omega}(\pi) = 1$, we conclude that

$$\begin{aligned} x_2(\bar{q}, \pi) &= \int_0^1 dt \int \bar{u}(t) \cdot \beta\, \pi(t)(d\beta) - 1 \\ &= 0 \qquad \text{for} \quad \omega\text{-a.a. } \pi \in \mathscr{S}^{\#}. \end{aligned}$$

It is easy to verify that the above relation is satisfied only if $\pi(t) = \delta_{\bar{u}}(t)$ a.e., thus showing that ω is concentrated at $\{\delta_{\bar{u}}\}$.

Keeping this last remark in mind, we now apply relation IX.1.5(2) which yields, in view of (2)–(4),

$$\begin{aligned} (5) \qquad & [l_0 - \omega(\{\delta_{\bar{u}}\})](b^1 - 1) + l_1 b^2 \\ & + \omega(\{\delta_{\bar{u}}\}) \int_0^1 dt \int \alpha \cdot \bar{u}(t)\, (\sigma(t) - \delta_{\bar{u}}(t))\, (d\alpha) \geqslant 0 \\ & \qquad (\sigma \in \mathscr{S}^{\#}, \quad b^1, b^2 \in [-2, 2]). \end{aligned}$$

This implies that $l_1 = 0$ and $l_0 = \omega(\{\delta_{\bar{u}}\})$. By IX.1.5(1), we have $l_0 + |l_1| + \omega(\{\delta_{\bar{u}}\}) > 0$; hence $l_0 = \omega(\{\delta_{\bar{u}}\}) > 0$. Thus relation (5) yields, setting $s_\sigma(t) \triangleq \int \alpha\sigma(t)\, (d\alpha)$,

$$(6) \quad \int_0^1 dt \int \alpha \cdot \bar{u}(t)(\sigma(t) - \delta_{\bar{u}}(t))\, (d\alpha) = \int_0^1 \bar{u}(t) \cdot s_\sigma(t)\, dt - 1 \geqslant 0 \quad (\sigma \in \mathscr{S}^{\#}).$$

If we choose σ in such a manner that

$$\sigma(t)(\{\bar{u}(t)\}) = \sigma(t)(\{-\bar{u}(t)\}) = \tfrac{1}{2} \qquad (t \in T),$$

then relation (6) yields $-1 \geqslant 0$, a contradiction. We conclude that $(\bar{u}, 1, 0)$ does not satisfy the conclusions of IX.1.5.

WEAK CONDITIONS APPLY We have shown that for every choice of a measurable $\bar{\rho} : T \to R$ the point $(\bar{\rho}, 1, 0)$ is a minimizing $\mathscr{R}^{\#} \times B$-control. We shall now show that $(\bar{\rho}, 1, 0)$ satisfies the "weak" conditions of IX.1.6. Indeed, assuming familiarity with trigonometric functions and writing Π for 3.1415 ⋯ (to distinguish it from adverse controls), we set $M \triangleq [0, 2\Pi]$, $M^{\#}(t) \triangleq M$ $(t \in T)$,

$$\phi(m) \triangleq (\cos m, \sin m) \qquad (m \in M),$$

denote by $\mathscr{M}^{\#}$ the set of measurable functions on T to M, let

$$\hat{\mathscr{x}}_i(\nu, b^1, b^2) \triangleq b^i \qquad (\nu \in \mathscr{M}^{\#}, \quad (b^1, b^2) \in B \triangleq [-2, 2]^2, \quad i = 0, 1)$$

and

$$\hat{\mathscr{x}}_2(\nu, b^1, b^2, \pi) \triangleq \int_0^1 dt \int_{R_P} [\beta^1 \cos(\nu(t)) + \beta^2 \sin(\nu(t))]\, \pi(t)\, (d\beta) - b^1$$

$$[\nu \in \mathscr{M}^{\#}, \quad (b^1, b^2) \in B, \quad \pi \in \mathscr{S}_P^{\#}].$$

Then $\tilde{\phi} \triangleq \phi \mid (0, 2\Pi)$ is a homeomorphism of $(0, 2\Pi)$ onto $R \sim \{(1, 0)\}$ and the function $\bar{\nu} : T \to M$, defined by

$$\bar{\nu}(t) \triangleq \tilde{\phi}^{-1}(\bar{\rho}(t)) \qquad \text{if} \quad \bar{\rho}(t) \neq (1, 0),$$

$$\bar{\nu}(t) = 0 \qquad \text{if} \quad \bar{\rho}(t) = (1, 0)$$

is measurable. It is easy to verify that the conditions of IX.1.6 are satisfied and

$$D\hat{\mathscr{x}}_0((\bar{\nu}, 1, 0); (\nu, b^1, b^2) - (\bar{\nu}, 1, 0)) = b^1 - 1,$$

$$D\hat{\mathscr{x}}_1((\bar{\nu}, 1, 0); (\nu, b^1, b^2) - (\bar{\nu}, 1, 0)) = b^2,$$

and

$$D_1\hat{\mathscr{x}}_2((\bar{\nu}, 1, 0), \pi; (\nu, b^1, b^2) - (\bar{\nu}, 1, 0))$$

$$= \int_0^1 dt \int_{R_P} [-\beta^1 \sin(\nu(t)) + \beta^2 \cos(\nu(t))](\nu(t) - \bar{\nu}(t))\, \pi(t)\, (d\beta) - (b^1 - 1)$$

We can now verify that the conclusions of IX.1.6 are valid with ω concentrated at $\delta_{\bar{\rho}}$, $\tilde{\omega}(\pi) = 1$ $(\pi \in \mathscr{S}_P^{\#})$, $l_0 = \omega(\delta_{\bar{\rho}})$, and $l_1 = 0$, and yield the trivial relation

$$\sum_{i=0}^{1} l_i\, D\hat{\mathscr{x}}_i((\bar{\nu}, 1, 0); (\nu, b^1, b^2) - (\bar{\nu}, 1, 0))$$

$$+ \int \tilde{\omega}(\pi)^T D_1\hat{\mathscr{x}}_2((\bar{\nu}, 1, 0), \pi; (\nu, b^1, b^2) - (\bar{\nu}, 1, 0))\, \omega(d\pi) = 0 \geqslant 0$$

$$[\nu \in \mathscr{M}^{\#}, \quad (b^1, b^2) \in B].$$

IX.2.3 ***A "Physical" Interpretation of a Relaxed Conflicting Control Problem*** Counterexample IX.2.2 shows that a minimizing approximate $\mathscr{U}$-control of a conflicting control problem may yield entirely different results than a minimizing relaxed control as defined in this chapter. Since optimal control problems are intended to be mathematical models of physical problems, it is natural to inquire into the nature of physical conditions that give rise to Problems I and II; the first consisting in the search for a minimizing approximate $\mathscr{U}$-control and the second in the search for a minimizing relaxed control.

For the purposes of the present discussion we shall consider a minimax problem of the type discussed in IX.2.2. Let us assume, therefore, that we are given sets $\mathscr{R}^{\#}$ and $\mathscr{S}^{\#}$ of "friendly" control functions defined on some interval T of the time axis, an abundant subset $\mathscr{U}$ of $\mathscr{R}^{\#}$, sets $\mathscr{R}_P{}^{\#}$ and $\mathscr{S}_P{}^{\#}$ of "adverse" control functions with arguments in T, an abundant subset $\mathscr{U}_P$ of $\mathscr{R}_P{}^{\#}$, and a function $\tilde{x}_0 : \mathscr{S}^{\#} \times \mathscr{S}_P{}^{\#} \to \mathbb{R}$ such that $\tilde{x}_0(\cdot, \sigma_P)$ and $\tilde{x}_0(\sigma, \cdot)$ are continuous for each $\sigma \in \mathscr{S}^{\#}$ and $\sigma_P \in \mathscr{S}_P{}^{\#}$. We set

$$x_0(\sigma) \triangleq \sup\{\tilde{x}_0(\sigma, \sigma_P) \mid \sigma_P \in \mathscr{S}_P{}^{\#}\} \qquad (\sigma \in \mathscr{S}^{\#}),$$

and consider

(a) Problem I of determining a sequence (u_j) in $\mathscr{U}$ such that $\lim_j x_0(u_j) = \inf x_0(\mathscr{U})$, and

(b) Problem II of minimizing x_0 on $\mathscr{S}^{\#}$.

In physical terms, both these problems may refer to a "game" in which we choose a "friendly" control function in such a manner that our "opponent" can record our selection before deciding on his own choice of an "adverse" control. The "friendly" control functions u are chosen out of $\mathscr{U}$ and the "adverse" control functions u_P out of $\mathscr{U}_P$. Problem I will arise if our opponent's instruments can record every selection of u with sufficient precision. Indeed, since $\tilde{x}_0(u, \cdot)$ is continuous on $\mathscr{S}_P{}^{\#}$ and $\mathscr{U}_P$ is dense in $\mathscr{S}_P{}^{\#}$, our opponent can choose control functions u_P out of $\mathscr{U}_P$ in such a manner that $\tilde{x}_0(u, u_P)$ approximates $\sup \tilde{x}_0(u, \mathscr{S}_P{}^{\#})$ to an arbitrary degree of accuracy.

A different situation arises if our opponent's instruments yield a "blurred" recording of highly oscillatory functions $u \in \mathscr{U}$. Such a recording may yield fairly accurate data about the distribution of the values of u on small subintervals of T but even small errors in correlating the arguments and the corresponding values of u may yield large discrepancies between $u(t)$ and the recorded values $\tilde{u}(t)$ for most $t \in T$. As a result, the opponent's choice of u_P will "oppose" the recorded function $\tilde{u}$ instead of the true function u, and his effectiveness will be reduced.

Let us consider as an illustration the problem of Counterexample IX.2.2, with

$$R \triangleq \{r \in \mathbb{R}^2 \mid r \cdot r = 1\}, \qquad T \triangleq [0, 1], \qquad \mathscr{U} = \mathscr{U}_P = \mathscr{R},$$

and

$$\tilde{x}_0(u, u_P) \triangleq \int_0^1 u(t) \cdot u_P(t)\, dt \qquad (u \in \mathscr{U}, \quad u_P \in \mathscr{U}_P).$$

For every choice of u, the appropriate choice of an adverse control is $u_P = u$ which yields

$$\tilde{x}_0(u, u) = \sup \tilde{x}_0(u, \mathscr{S}) = 1.$$

Now let u be determined by choosing a point $r_1 \in R$ and assigning to $u(t)$ the values r_1 and $-r_1$ in a highly oscillatory and randomly distributed manner, with equal amounts of time spent at r_1 and $-r_1$. Then the recorded function $\tilde{u}$ will have a similar character but, with very high probability, the relations $\tilde{u}(t) = u(t)$ and $\tilde{u}(t) = -u(t)$ will be valid equally often. As a result, the combination of the friendly control function u and of the adverse control function $\tilde{u}$ will yield

$$x_0(u, \tilde{u}) = \int_0^1 u(t) \cdot \tilde{u}(t)\, dt$$

which will be nearly 0 instead of 1. Thus u will have an effect similar to that of the relaxed control σ with $\sigma(t)(\{r_1\}) = \sigma(t)(\{-r_1\} = \frac{1}{2}$ a.e. in T.

This is the type of a physical situation which seems to be best modeled by Problem II. Highly oscillatory functions $u \in \mathscr{U}$, with oscillations chosen by a random mechanism, have a net effect analogous to that of relaxed controls which they approximate in $\mathscr{S}$, so long as the opponent's recording capacity is inferior to our capacity of generating oscillatory functions.

The methods of the present chapter are suited to the investigation of all problems of the type (of Problem) II and of those problems of the type I in which the conflicting controls are additively coupled. In Chapter X we shall discuss an appropriately relaxed version of problems of the type I which is applicable even if the conflicting controls are not additively coupled.

IX.3 An Evasion Problem

We shall illustrate the application of some of our results with a discussion of an evasion problem, one of many problems with additively coupled conflicting controls that can be investigated by entirely analogous methods.

An *evader* moves in accordance with the equation

$$y(t) = v_0 + \int_0^t f(y(s), \rho(s))\, ds \qquad (t \in [0, t_1]), \tag{1}$$

where $v_0 \in \mathbb{R}^n$ and $t_1 > 0$ are given and an *escape control* ρ can be chosen arbitrarily as an element of the set $\mathscr{R}$ of measurable functions on $[0, t_1]$ to some compact metric space R. A swarm of *pursuers* move in accordance with the equation

$$(2) \qquad \tilde{y}(t) = \tilde{v}_0 + \int_0^t \tilde{f}(\tilde{y}(s), \tilde{\rho}(s))\, ds \qquad (t \in [0, t_1]),$$

each pursuer starting out from the same given point $\tilde{v}_0$ and each choosing a different *pursuit control* $\tilde{\rho} \in \mathscr{R}_P$, that is, a measurable function on $[0, t_1]$ to some compact metric space R_P. We assume that each of the equations, (1) and (2), has a unique solution $y(\rho)(\cdot)$ respectively $\tilde{y}(\tilde{\rho})(\cdot)$ for every choice of an evasion control ρ and a pursuit control $\tilde{\rho}$. The evader *evades capture before* a time $\beta \in [0, t_1]$ if his choice of ρ is such that

$$(3) \qquad h(y(\rho)(t), \tilde{y}(\tilde{\rho})(t)) \geqslant 0 \qquad (t \in [0, \beta], \quad \tilde{\rho} \in \mathscr{R}_P),$$

where h is a given continuous function and $h(v_0, \tilde{v}_0) \geqslant 0$. Typically,

$$h(w_1, w_2) = (w_1 - w_2) \cdot (w_1 - w_2) - (\delta)^2,$$

in which case condition (3) means that the evader remains at a (euclidean) distance of at least δ from each of the pursuers during the period $[0, \beta]$. The purpose of the evader is to choose an escape control ρ so as to evade capture as long as possible and be caught only inside a given set A_1. In mathematical terms, the evader desires to maximize β among all $(\rho, \beta) \in \mathscr{R} \times [0, t_1]$ for which

$$h(y(\rho)(t), \tilde{y}(\tilde{\rho})(t)) \geqslant 0 \qquad (t \in [0, \beta], \quad \tilde{\rho} \in \mathscr{R}_P)$$

and

$$y(\rho)(\beta) \in A_1 .$$

IX.3.1 ***A Transformation*** It is clear that the evader can replace the choice of a point $(\rho, \beta) \in \mathscr{R} \times [0, t_1]$ by the choice of $(\rho \mid [0, \beta], \beta)$, since the values of ρ for $t > \beta$ do not affect either the restrictions or the criterion of optimization. If we denote by $\mathscr{R}^\alpha$ and $\mathscr{R}_P{}^\alpha$ the classes of measurable functions on $[0, \alpha]$ to R and R_P, respectively, and set

$$y(\beta\tau) \triangleq \xi(\tau), \qquad \rho(\beta\tau) \triangleq u(\tau) \qquad (\tau \in [0, 1]),$$

$$\tilde{y}(\beta\tau) \triangleq \eta(\tau), \qquad \tilde{\rho}(\beta\tau) \triangleq \tilde{u}(\tau) \qquad (\tau \in [0, 1]),$$

then these relations establish a one-to-one correspondence between triplets

$$(\rho, y, \beta) \qquad \text{such that} \quad \beta \in (0, t_1] \quad \text{and} \quad (\rho, y) \in \mathscr{R}^\beta \times C([0, \beta], \mathbb{R}^n)$$

and triplets

$$(u, \xi, \beta) \in \mathscr{R}^1 \times C([0, 1], \mathbb{R}^n) \times (0, t_1],$$

and a similar correspondence between $(\tilde{\rho}, \tilde{y}, \beta)$ and $(\tilde{u}, \eta, \beta)$. If $\beta \in (0, t_1]$,

$$(\rho, y) \in \mathscr{R}^\beta \times C([0, \beta], \mathbb{R}^n)$$

and

$$(\tilde{\rho}, \tilde{y}) \in \mathscr{R}_P{}^\beta \times C([0, \beta], \mathbb{R}^n)$$

satisfy Eqs. (1) and (2) for $t \in [0, \beta]$ and correspond to (u, ξ, β) and $(\tilde{u}, \eta, \beta)$, respectively, then we have

$$\xi(\tau) = v_0 + \beta \int_0^\tau f(\xi(s), u(s))\, ds \qquad (\tau \in [0, 1]) \tag{4}$$

and

$$\eta(\tau) = \tilde{v}_0 + \beta \int_0^\tau \tilde{f}(\eta(s), \tilde{u}(s))\, ds \qquad (\tau \in [0, 1]). \tag{5}$$

Thus our evasion problem is mathematically equivalent to one of minimizing $-\beta$ on the set

$$\{(u, \beta) \in \mathscr{R}^1 \times [0, t_1] \mid \xi(u)(1) \in A_1,\ h(\xi(u)(\tau), \eta(\tilde{u})(\tau)) \geqslant 0 \qquad (\tilde{u} \in \mathscr{R}_P{}^1,\ \tau \in [0, 1])\},$$

where $\xi(u, \beta)(\cdot)$ and $\eta(\tilde{u}, \beta)(\cdot)$ are the solutions of Eqs. (4) and (5), respectively.

Let $\mathscr{S}_P{}^1$ be defined the same way as $\mathscr{S}$, with T and R replaced by $[0, 1]$ and R_P, respectively. We shall show in this section that, subject to our assumptions, the function

$$\tilde{u} \to \eta(\tilde{u}, \beta)(\cdot) : \mathscr{R}_P{}^1 \to C([0, 1], \mathbb{R}^n)$$

can be extended for each β to a continuous function

$$\tilde{\sigma} \to \eta(\tilde{\sigma}, \beta)(\cdot) : \mathscr{S}_P{}^1 \to C([0, 1], \mathbb{R}^n)$$

by defining $\eta(\tilde{\sigma}, \beta)(\cdot)$ to be the solution of the equation

$$\eta(\tau) = \tilde{v}_0 + \beta \int_0^\tau \tilde{f}(\eta(s), \tilde{\sigma}(s))\, ds \qquad (s \in [0, 1]).$$

It follows then that the condition

$$h(\xi(u, \beta)(\tau), \eta(\tilde{u}, \beta)(\tau)) \geqslant 0 \qquad (\tilde{u} \in \mathscr{R}_P{}^1, \quad \beta \in [0, t_1], \quad \tau \in [0, 1])$$

is equivalent to

$$h(\xi(u, \beta)(\tau), \eta(\tilde{\sigma}, \beta)(\tau)) \geqslant 0 \qquad (\tilde{\sigma} \in \mathscr{S}_P{}^1, \quad \beta \in [0, t_1], \quad \tau \in [0, 1]).$$

Thus, even if we restrict the escape controls u to be original, the problem remains unchanged when we replace pursuit controls by relaxed control functions.

Let $\mathscr{S}^1$ be defined the same way as $\mathscr{S}$ but with T replaced by $[0, 1]$. We obtain a relaxed version of the evasion problem by defining $\xi(\sigma, \beta)(\cdot)$ for $\sigma \in \mathscr{S}^1$ and $\beta \in [0, t_1]$ as the solution $\xi(\cdot)$ of the equation

$$\xi(\tau) = v_0 + \beta \int_0^\tau f(\xi(s), \sigma(s))\, ds \qquad (\tau \in [0, 1]),$$

and seeking to minimize $-\beta$ on the set

$$\{(\sigma, \beta) \in \mathscr{S}^1 \times [0, t_1] \mid \xi(\sigma, \beta)(1) \in A_1,\ h(\xi(\sigma, \beta)(\tau), \eta(\tilde{\sigma}, \beta)(\tau)) \geqslant 0 \quad (\tilde{\sigma} \in \mathscr{S}_P{}^1, \tau \in [0, 1])\}$$

THE MATHEMATICAL EVASION PROBLEM

We shall now state the precise conditions of the problem that we have been discussing above. Let R and R_P be compact metric spaces, $\mathscr{R}^1$, $\mathscr{S}^1$ and $\mathscr{R}_P{}^1$, $\mathscr{S}_P{}^1$ defined as before, $t_1 > 0$, $n \in \mathbb{N}$, $v_0, \tilde{v}_0 \in \mathbb{R}^n$, and $f : \mathbb{R}^n \times R \to \mathbb{R}^n$, $\tilde{f} : \mathbb{R}^n \times R_P \to \mathbb{R}^n$, and $h : \mathbb{R}^n \times \mathbb{R}^n \to \mathbb{R}$ given functions.

IX.3.2 ***Assumption***

(1) The functions f and $\tilde{f}$ have derivatives $\mathscr{D}_1 f$ and $\mathscr{D}_1 \tilde{f}$, and f, $\tilde{f}$, $\mathscr{D}_1 f$, and $\mathscr{D}_1 \tilde{f}$ are continuous;

(2) there exists $c \in \mathbb{R}$ such that

$$|f(v, r)| \leqslant c(|v| + 1) \quad \text{and} \quad |\tilde{f}(v, r_P)| \leqslant c(|v| + 1) \qquad (r \in R, \quad r_P \in R_P, \quad v \in \mathbb{R}^n);$$

(3) the function h has a continuous derivative $\mathscr{D}h$ and $h(v_0, \tilde{v}_0) \geqslant 0$; and

(4) $\sup_{r \in R} \mathscr{D}_1 h(w_1, w_2) f(w_1, r) + \inf_{r_P \in R_P} \mathscr{D}_2 h(w_1, w_2) \tilde{f}(w_2, r_P) < 0$ for all $w_1, w_2 \in \mathbb{R}^n$ such that $h(w_1, w_2) = 0$.

Item (4) may at first appear strange, but it has the following physical significance: if at some instant of time the evader is just barely evading capture [i.e., h(evader, some pursuer) $= 0$], then item (4) ensures that the evader is captured immediately thereafter because the pursuer can achieve a higher "speed." Item (4) in IX.3.2 is used only in the derivation of the necessary conditions. The latter can also be derived from Theorems IX.1.2

and IX.1.5 without assuming the validity of IX.3.2(4) but they become greatly simplified when IX.3.2(4) applies.

IX.3.3 ***Lemma*** Let Assumptions IX.3.2(1) and IX.3.2(2) be satisfied. Then each of the equations

(1) $$\xi(\tau) = v_0 + \beta \int_0^\tau f(\xi(s), \sigma(s))\, ds \qquad (\tau \in [0, 1])$$

and

(2) $$\eta(\tau) = \tilde{v}_0 + \beta \int_0^\tau \tilde{f}(\eta(s), \tilde{\sigma}(s))\, ds \qquad (\tau \in [0, 1])$$

has a unique solution $\xi(\sigma, \beta)(\cdot)$ respectively $\eta(\tilde{\sigma}, \beta)(\cdot)$ for every choice of $\beta \in [0, t_1]$, $\sigma \in \mathscr{S}^1$, and $\tilde{\sigma} \in \mathscr{S}_P{}^1$, and the functions

$$(\sigma, \beta) \to \xi(\sigma, \beta)(\cdot) : \mathscr{S}^1 \times [0, t_1] \to C([0, 1], \mathbb{R}^n)$$

and

$$(\tilde{\sigma}, \beta) \to \eta(\tilde{\sigma}, \beta)(\cdot) : \mathscr{S}_P{}^1 \times [0, t_1] \to C([0, 1], \mathbb{R}^n)$$

are continuous.

▌PROOF Let $\beta \in [0, t_1]$, $\sigma \in \mathscr{S}^1$, and $\tilde{\sigma} \in \mathscr{S}_P{}^1$ be fixed. By IX.3.2(1), IX.3.2(2), and II.4.3, there exist solutions ξ and η of (1) and (2) and, by Gronwall's inequality II.4.4, there exists $c_1 \in \mathbb{R}$ such that $|\xi(\tau)| \leqslant c_1$ and $|\eta(\tau)| \leqslant c_1$ $(\tau \in [0, 1])$ and c_1 is independent of the choice of β, σ, and $\tilde{\sigma}$. Since the functions f, $\tilde{f}$, $\mathscr{D}_1 f$, and $\mathscr{D}_1 \tilde{f}$ are continuous, it follows that the norms of their values are bounded by some c_2 on the compact sets $S^F(0, c_1) \times R$ respectively $S^F(0, c_1) \times R_P$ and, by the mean value theorem II.3.6, we have

(3) $$|f(v_1, \sigma(s)) - f(v_2, \sigma(s))| \leqslant c_2 |v_1 - v_2| \qquad (s \in [0, 1],\quad v_1, v_2 \in S^F(0, c_1),\quad \sigma \in \mathscr{S}^1)$$

and

(4) $$|\tilde{f}(v_1, \tilde{\sigma}(s)) - \tilde{f}(v_2, \tilde{\sigma}(s))| \leqslant c_2 |v_1 - v_2| \qquad (s \in [0, 1],\quad v_1, v_2 \in S^F(0, c_1),\quad \tilde{\sigma} \in \mathscr{S}_P{}^1).$$

Thus, by II.4.5, Eqs. (1) and (2) have unique solutions which we denote by $\xi(\sigma, \beta)(\cdot)$ and $\eta(\tilde{\sigma}, \beta)(\cdot)$, respectively.

Now let (σ_j, β_j) be a sequence in $\mathscr{S}^1 \times [0, t_1]$ converging to $(\bar{\sigma}, \bar{\beta})$, and let

$$\bar{\xi} \triangleq \xi(\bar{\sigma}, \bar{\beta}) \qquad \text{and} \qquad \xi_j \triangleq \xi(\sigma_j, \beta_j) \quad (j \in \mathbb{N}).$$

If (ξ_j) does not converge to $\bar{\xi}$, then there exist $J \subset (1, 2, ...)$ and $\epsilon > 0$ such that

(5) $$|\bar{\xi} - \xi_j|_{\sup} \geqslant \epsilon \qquad (j \in J).$$

Since $|\dot{\xi}_j(\tau)| \leqslant c_2 t_1$ a.e. in [0, 1] [because $|f(\xi_j(\tau), \sigma(\tau))| \leqslant c_2$], the sequence (ξ_j) is bounded and equicontinuous and, by Ascoli's theorem I.5.4, there exist $J_1 \subset J$ and $\xi_0 \in C([0, 1], \mathbb{R}^n)$ such that $\lim_{j \in J_1} |\xi_j - \xi_0|_{\sup} = 0$. It follows, by IV.2.9, that for each $\tau \in [0, 1]$ we have

$$\begin{aligned}\xi_0(\tau) = \lim_{j \in J_1} \xi_j(\tau) &= v_0 + \lim_{j \in J_1} \beta_j \int_0^\tau f(\xi_j(s), \sigma_j(s))\, ds \\ &= v_0 + \bar{\beta} \int_0^\tau f(\xi_0(s), \bar{\sigma}(s))\, ds,\end{aligned}$$

which implies that $\xi_0 = \xi(\bar{\sigma}, \bar{\beta}) \triangleq \bar{\xi}$ and contradicts (5). This shows that the function

$$(\sigma, \beta) \to \xi(\sigma, \beta) : \mathscr{S}^1 \times [0, t_1] \to C([0, 1], \mathbb{R}^n)$$

is continuous, and a similar argument applies to $(\tilde{\sigma}, \beta) \to \eta(\tilde{\sigma}, \beta)$. QED

IX.3.4 ***Definition*** Lemma IX.3.3 guarantees, in particular, that the set $\{\xi(\sigma, \beta)(\cdot) \mid \sigma \in \mathscr{S}^1, \beta \in [0, t_1]\}$ is bounded in $C([0, 1], \mathbb{R}^n)$ (being the image under a continuous mapping of the compact set $\mathscr{S}^1 \times [0, t_1]$) and therefore

$$|\xi(\sigma, \beta)(\tau)| \leqslant c_1 \qquad (\sigma \in \mathscr{S}^1, \quad \beta \in [0, t_1], \quad \tau \in [0, 1])$$

for some $c_1 < \infty$. Now let

$$A \triangleq [0, \infty), \qquad B \triangleq [0, t_1] \times A_1, \qquad P \triangleq \mathscr{S}_P^1 \times [0, 1],$$

and, for each $\sigma \in \mathscr{S}^1$, $b \triangleq (\beta, a_1) \in B$, and $(\tilde{\sigma}, \tau) \in P$, let

$$x_0(\sigma, b) \triangleq -\beta,$$

$$x_1(\sigma, b) \triangleq \xi(\sigma, \beta)(1) - a_1,$$

and

$$x_2(\sigma, b, \tilde{\sigma}, \tau) \triangleq h(\xi(\sigma, \beta)(\tau), \eta(\tilde{\sigma}, \beta)(\tau)).$$

It is clear that, with these definitions, the evasion problem coincides with the conflicting control problem of IX.0.2. As in IX.0.2, for any $\mathscr{S}' \subset \mathscr{S}^1$ we denote by $\mathscr{A}(\mathscr{S}' \times B)$ the set

$$\{(\sigma, b) \in \mathscr{S}' \times B \mid x_1(\sigma, b) = 0, x_2(\sigma, b, \tilde{\sigma}, \tau) \geqslant 0\ [(\tilde{\sigma}, \tau) \in P]\}.$$

We also write $\mathscr{R}^1$ for the collection of measurable functions on [0, 1] to R.

IX.3.5 ***Theorem*** Let Assumptions IX.3.2(1)–IX.3.2(3) be satisfied and $\mathscr{A}(\mathscr{S}^1 \times B) \neq \varnothing$. Then the evasion problem admits a minimizing $\mathscr{S}^1 \times B$-control $(\bar{\sigma}, \bar{\beta}, \bar{a}_1)$ and a minimizing approximate $\mathscr{R}^1$-control $((u_j, \bar{\beta}, \bar{a}_1))$, and we have

$$\lim_j x(u_j, \bar{\beta}, \bar{a}_1, \tilde{\sigma}, \tau) = x(\bar{\sigma}, \bar{\beta}, \bar{a}_1, \tilde{\sigma}, \tau) \tag{1}$$

uniformly for $\tilde{\sigma} \in \mathscr{S}_P{}^1$ and $\tau \in [0, 1]$ and

$$\lim_j \xi(u_j, \bar{\beta}) = \xi(\bar{\sigma}, \bar{\beta}) \quad \text{in } C([0, 1], \mathbb{R}^n). \tag{2}$$

▌PROOF Since $|\xi(\sigma, \beta)(\tau)| \leqslant c_1 < \infty$ $(\sigma \in \mathscr{S}^1, \beta \in [0, t_1], \tau \in [0, 1])$, it is clear that our problem remains unchanged if A_1 is replaced by the compact set $A_1' \triangleq A_1 \cap S^F(0, c_1)$ and B by $B' \triangleq [0, t_1] \times A_1'$. Now let $Q \triangleq \mathscr{S}^1 \times B'$. By IX.3.2(3) and Lemma IX.3.3, the function $q \to x(q)(\cdot) : Q \to C(P, \mathbb{R}^{m_2})$ is continuous. By IX.1.1, there exists a minimizing $\mathscr{S}^1 \times B'$-control $(\bar{\sigma}, \bar{\beta}, \bar{a}_1)$ and, by IX.1.4, $((u_j, \bar{\beta}, \bar{a}_1))$ is a minimizing approximate $\mathscr{R}^1$-control and satisfies (1) for any choice of a sequence (u_j) in $\mathscr{R}^1$ converging to $\bar{\sigma}$. Relation (2) follows from the continuity of $\xi(\cdot, \cdot)$ which was proven in IX.3.3. QED

We finally consider necessary conditions.

IX.3.6 ***Theorem*** Let $\mathscr{U}$ be either $\mathscr{S}^1$ or $\mathscr{R}^1$, $(\bar{\sigma}, \bar{\beta}, \bar{a}_1)$ be a minimizing $\mathscr{U} \times B$-control of the evasion problem, and Assumption IX.3.2 be satisfied. Let c_1 be as in Definition IX.3.4, and set

$$L \triangleq S^F(0, c_1) \subset \mathbb{R}^n,$$

$$\bar{\xi}(\cdot) \triangleq \xi(\bar{\sigma}, \bar{\beta})(\cdot), \qquad \text{and} \qquad \bar{\eta}(\tilde{\sigma})(\cdot) \triangleq \eta(\tilde{\sigma}, \bar{\beta})(\cdot) \qquad (\tilde{\sigma} \in \mathscr{S}_P{}^1).$$

Then either $\bar{\beta} = 0$, or $\bar{\beta} = t_1$, or there exist an absolutely continuous $z : [0, 1] \to \mathbb{R}^n$, $\nu \in \text{frm}^+(L)$, and $l_1 \in \mathbb{R}^n$ such that

$$h(\bar{\xi}(\tau), \bar{\eta}(\tilde{\sigma})(\tau)) > 0 \qquad (\tau \in [0, 1), \quad \tilde{\sigma} \in \mathscr{S}_P{}^1); \tag{1}$$

$$\dot{z}(\tau)^T = -\bar{\beta} z(\tau)^T \mathscr{D}_1 f(\bar{\xi}(\tau), \bar{\sigma}(\tau)) \quad \text{a.e. in } [0, 1]; \tag{2}$$

$$z(\tau)^T f(\bar{\xi}(\tau), \bar{\sigma}(\tau)) = \operatorname*{Min}_{r \in R} z(\tau)^T f(\bar{\xi}(\tau), r) \quad \text{a.e. in } [0, 1]; \tag{3}$$

(4) the measure ν is supported on the set

$$N \triangleq \{\bar{\eta}(\tilde{\sigma})(1) \mid \tilde{\sigma} \in \mathscr{S}_P{}^1, h(\bar{\xi}(1), \bar{\eta}(\tilde{\sigma})(1)) = 0\}$$

and $|l_1| + \nu(N) > 0$;

$$z(1)^T = l_1{}^T - \int \mathscr{D}_1 h(\bar{\xi}(1), w)\, \nu(dw); \tag{5}$$

and

$$l_1{}^T\bar{\xi}(1) = \operatorname*{Max}_{a_1 \in A_1} l_1{}^T a_1 . \tag{6}$$

▌PROOF We assume that $\bar{\beta} \neq 0$ and $\bar{\beta} \neq t_1$.

Step 1 We first prove statement (1). Since $(\bar{\sigma}, \bar{\beta}, \bar{a}_1) \in \mathscr{A}(\mathscr{U} \times B)$, we have

$$h(\bar{\xi}(\tau), \bar{\eta}(\tilde{\sigma})(\tau)) \geqslant 0 \qquad (\tau \in [0, 1], \quad \tilde{\sigma} \in \mathscr{S}_P{}^1).$$

Now assume, by way of contradiction, that there exist $\tau^* < 1$ and $\tilde{\sigma}^* \in \mathscr{S}_P{}^1$ such that

$$h(\bar{\xi}(\tau^*), \bar{\eta}(\tilde{\sigma}^*)(\tau^*)) = 0,$$

and let $r_P{}^*$ yield the minimum of the continuous function

$$r_p \to \mathscr{D}_2 h(\bar{\xi}(\tau^*), \bar{\eta}(\tilde{\sigma}^*)(\tau^*))\tilde{f}(\bar{\eta}(\tilde{\sigma}^*)(\tau^*), r_P) : R_P \to \mathbb{R}.$$

We set

$$\tilde{\sigma}^\#(\tau) \triangleq \tilde{\sigma}^*(\tau) \quad (\tau \in [0, \tau^*]), \qquad \tilde{\sigma}^\#(\tau) \triangleq r_P{}^* \quad (\tau \in (\tau^*, 1]).$$

Then $\tilde{\sigma}^\# \in \mathscr{S}_P{}^1$, $\bar{\eta}(\tilde{\sigma}^*)(\tau) = \bar{\eta}(\tilde{\sigma}^\#)(\tau)$ $(\tau \in [0, \tau^*])$, and

$$h(\bar{\xi}(\tau^*), \bar{\eta}(\tilde{\sigma}^\#)(\tau^*)) = 0.$$

We next refer to Assumption IX.3.2(4) which implies that there exists $\alpha < 0$ such that

$$\begin{aligned} &\mathscr{D}_1 h(\bar{\xi}(\tau^*), \bar{\eta}(\tilde{\sigma}^\#)(\tau^*)) f(\bar{\xi}(\tau^*), \bar{\sigma}(\tau)) \\ &\quad + \mathscr{D}_2 h(\bar{\xi}(\tau^*), \bar{\eta}(\tilde{\sigma}^\#)(\tau^*))\tilde{f}(\bar{\xi}(\tau^*), r_P{}^*) \leqslant \alpha \qquad (\tau \in [\tau^*, 1]). \end{aligned}$$

Let $de(\tau)/d\tau$ denote, as is customary, the derivative at τ of a function $\tau \to e(\tau)$. Since $\bar{\xi}(\cdot)$, $\bar{\eta}(\tilde{\sigma}^\#)(\cdot)$, and $h(\cdot, \cdot)$ are continuous and $\{f(\cdot, s) \mid s \in \operatorname{rpm}(R)\}$ and $\{\tilde{f}(\cdot, s) \mid s \in \operatorname{rpm}(R_P)\}$ are equicontinuous, it follows (applying the chain rule II.3.4) that there exists $\varDelta > 0$ such that $\tau^* + \varDelta < 1$ and

$$\begin{aligned} \frac{d}{d\tau} h(\bar{\xi}(\tau), \bar{\eta}(\tilde{\sigma}^\#)(\tau)) &= \bar{\beta}\mathscr{D}_1 h(\bar{\xi}(\tau), \bar{\eta}(\tilde{\sigma}^\#)(\tau)) f(\bar{\xi}(\tau), \bar{\sigma}(\tau)) \\ &\quad + \bar{\beta}\mathscr{D}_2 h(\bar{\xi}(\tau), \bar{\eta}(\tilde{\sigma}^\#)(\tau)) \tilde{f}(\eta(\tilde{\sigma}^\#)(\tau), r_P{}^*) \\ &\leqslant \tfrac{1}{2}\bar{\beta}\alpha < 0 \qquad \text{for} \quad \text{a.a. } \tau \in [\tau^*, \tau^* + \varDelta]. \end{aligned}$$

We easily verify that the function $\tau \to h(\bar{\xi}(\tau), \bar{\eta}(\tilde{\sigma}^\#)(\tau))$ is absolutely continuous [which follows from the chain rule II.3.4, the continuity of $\mathscr{D}h$

and the absolute continuity of $\bar{\xi}(\cdot)$ and $\bar{\eta}(\tilde{\sigma}^{\#})(\cdot)]$, and the relation above implies therefore that $h(\bar{\xi}(\tau), \bar{\eta}(\tilde{\sigma}^{\#})(\tau)) < 0$ $(\tau \in (\tau^*, \tau^* + \Delta])$, thus contradicting the assumption that $(\bar{\sigma}, \bar{\beta}, \bar{a}_1) \in \mathscr{A}(\mathscr{U} \times B)$.

Step 2 We next consider, for an arbitrary $\sigma \in \mathscr{S}^1$, the directional derivative $D_1\xi(\bar{\sigma}, \bar{\beta}; \sigma - \bar{\sigma})$ which is the derivative $\tilde{\xi}'(0)(\cdot)$ at $\alpha = 0$ of the function

$$\alpha \to \tilde{\xi}(\alpha)(\cdot) = \xi(\bar{\sigma} + \alpha(\sigma - \bar{\sigma}), \bar{\beta}) : [0, 1] \to C([0, 1], \mathbb{R}^n).$$

Since $\tilde{\xi}(\alpha)(\cdot)$ is the solution of the equation

$$\tilde{\xi}(\alpha)(\tau) = v_0 + \bar{\beta} \int_0^\tau f(\tilde{\xi}(\alpha)(s), \bar{\sigma}(s))\, ds$$
$$+ \alpha\bar{\beta} \int_0^\tau f(\tilde{\xi}(\alpha)(s), \sigma(s) - \bar{\sigma}(s))\, ds \qquad (\tau \in [0, 1])$$

and $\tilde{\xi}(0) = \bar{\xi}$, it follows from II.4.11 that

$$\tilde{\xi}'(0)(\tau) = \bar{\beta} \int_0^\tau \mathscr{D}_1 f(\bar{\xi}(s), \bar{\sigma}(s))\, \tilde{\xi}'(0)(s)\, ds$$
$$+ \bar{\beta} \int_0^\tau f(\bar{\xi}(s), \sigma(s) - \bar{\sigma}(s))\, ds \qquad (\tau \in [0, 1]).$$

This relation and II.4.8 imply that there exists $Z : [0, 1] \to B(\mathbb{R}^n, \mathbb{R}^n)$ satisfying

$$Z(\tau) = I + \bar{\beta} \int_\tau^1 Z(s)\, \mathscr{D}_1 f(\bar{\xi}(s), \bar{\sigma}(s))\, ds \qquad (\tau \in [0, 1]), \tag{7}$$

and we have

$$D_1\xi(\bar{\sigma}, \bar{\beta}; \sigma - \bar{\sigma})(\tau) = \tilde{\xi}'(0)(\tau) \tag{8}$$
$$= \bar{\beta} Z(\tau)^{-1} \int_0^\tau Z(s)\, f(\bar{\xi}(s), \sigma(s) - \bar{\sigma}(s))\, ds$$
$$(\tau \in [0, 1]).$$

Step 3 By IX.3.3, the functions $(\sigma, \beta) \to \xi(\sigma, \beta)$ and $(\tilde{\sigma}, \beta) \to \eta(\tilde{\sigma}, \beta)$ are continuous. By II.4.11, for every choice of $\tilde{\sigma} \in \mathscr{S}_P{}^1$, $\sigma_0, \sigma_1, \ldots, \sigma_n \in \mathscr{S}^1$, and $\beta_0, \beta_1, \ldots, \beta_n \in [0, t_1]$, the functions

$$\theta \to \xi\left(\bar{\sigma} + \sum_{j=0}^{n} \theta^j(\sigma_j - \bar{\sigma}), \bar{\beta} + \sum_{j=0}^{n} \theta^j(\beta_j - \bar{\beta})\right) : \mathscr{T}_{n+1} \to C([0, 1], \mathbb{R}^n)$$

and

$$\theta \to \eta\left(\tilde{\sigma}, \bar{\beta} + \sum_{j=0}^{n} \theta^j(\beta_j - \bar{\beta})\right) : \mathcal{T}_{n+1} \to C([0, 1], \mathbb{R}^n)$$

have a derivative at 0. Thus Theorems IX.1.2 or IX.1.5 are applicable (depending on whether $\mathscr{U} = \mathscr{S}^1$ or $\mathscr{U} = \mathscr{R}^1$), and it follows that there exist $l_0 \geqslant 0$, $l_1 \in \mathbb{R}^n$, $\omega \in \mathrm{frm}^+(P)$, and an ω-integrable $\tilde{\omega} : P \to \mathbb{R}$ (where $P \triangleq \mathscr{S}_P{}^1 \times [0, 1]$) such that

(9) $$|\tilde{\omega}(p)| = 1 \quad (p \in P) \qquad \text{and} \qquad l_0 + |l_1| + \omega(P) > 0,$$

(10) $$\sum_{i=0}^{1} l_i \cdot Dx_i\big((\tilde{\sigma}, \bar{\beta}, \bar{a}_1); (\sigma, \beta, a_1) - (\tilde{\sigma}, \bar{\beta}, \bar{a}_1)\big)$$
$$+ \int \tilde{\omega}(p) \cdot D_1x_2\big((\tilde{\sigma}, \bar{\beta}, \bar{a}_1), p; (\sigma, \beta, a_1) - (\tilde{\sigma}, \bar{\beta}, \bar{a}_1)\big)\, \omega(dp) \geqslant 0$$
$$(\sigma \in \mathscr{S}^1, \beta \in [0, t_1], a_1 \in A_1),$$

and

(11) $$\tilde{\omega}(p) \cdot x_2\big((\tilde{\sigma}, \bar{\beta}, \bar{a}_1), p\big) = \operatorname*{Max}_{a \geqslant 0} \tilde{\omega}(p) \cdot a \qquad \text{for} \quad \omega\text{-a.a. } p \in P.$$

Relations (9) and (11) imply that

(12) $$\tilde{\omega}(p) = -1 \quad \text{and} \quad h\big(\bar{\xi}(\tau), \bar{\eta}(\tilde{\sigma})(\tau)\big) = 0 \quad \text{for} \quad \omega\text{-a.a. } p \triangleq (\tilde{\sigma}, \tau) \in P.$$

In view of statement (1) (which we have proven in Step 1), this shows that ω is supported on the set

$$\Omega \triangleq \{(\tilde{\sigma}, 1) \mid h\big(\bar{\xi}(1), \bar{\eta}(\tilde{\sigma})(1)\big) = 0, \tilde{\sigma} \in \mathscr{S}_P{}^1\}.$$

If we set $\beta = \bar{\beta}$ and $a_1 = \bar{a}_1$ in (10) and refer to (12) and the definition of (x_0, x_1, x_2) in IX.3.4, then we obtain

(13) $$l_1 \cdot D_1\xi(\bar{\sigma}, \bar{\beta}; \sigma - \bar{\sigma})(1)$$
$$- \int_\Omega \mathscr{D}_1 h\big(\xi(\bar{\sigma}, \bar{\beta})(1), \eta(\tilde{\sigma}, \bar{\beta})(1)\big)\, D_1\xi(\bar{\sigma}, \bar{\beta}; \sigma - \bar{\sigma})(1)\, \omega\big(d(\tilde{\sigma}, \tau)\big)$$
$$\geqslant 0 \qquad (\sigma \in \mathscr{S}^1).$$

We now observe that N [as defined in (4)] is the image of the compact set Ω under the continuous mapping

$$(\tilde{\sigma}, \tau) \to \eta(\tilde{\sigma}, \bar{\beta})(1) : \mathscr{S}_P{}^1 \times [0, 1] \to \mathbb{R}^n.$$

It follows that N is compact, the expression

$$l_N(\phi) \triangleq \int_\Omega \phi(\tilde{\eta}(\tilde{\sigma}, \bar{\beta})(1))\, \omega(d(\tilde{\sigma}, \tau)) \qquad [\phi \in C(N)]$$

defines an element l_N of $C(N)^*$ and, by the Riesz representation theorem I.5.8, there exists $\nu \in \mathrm{frm}(N)$ such that

$$l_N(\phi) = \int \phi(w)\, \nu(dw). \tag{14}$$

The measure ν is positive because $l_N(\phi) \geqslant 0$ if $\phi(w) \geqslant 0$ $(w \in N)$. We may apply relation (14) to (13) and obtain

$$\Big[l_1{}^T - \int \mathscr{D}_1 h(\bar{\xi}(1), w)\, \nu(dw)\Big]\, D_1\xi(\bar{\sigma}, \bar{\beta}; \sigma - \bar{\sigma})(1) \geqslant 0 \qquad (\sigma \in \mathscr{S}^1). \tag{15}$$

If we let

$$z(\tau)^T \triangleq \Big[l_1{}^T - \int \mathscr{D}_1 h(\bar{\xi}(1), w)\, \nu(dw)\Big]\, Z(\tau),$$

then (5) holds and relation (2) follows from (7). Furthermore, (8) and (15) yield

$$\bar{\beta} \int_0^1 z(s)^T f(\bar{\xi}(s), \sigma(s) - \bar{\sigma}(s))\, ds \geqslant 0 \qquad (\sigma \in \mathscr{S}^1). \tag{16}$$

We now consider a dense subset $\{r_1, r_2, \ldots\}$ of R and choose arbitrary $j \in \mathbb{N}$ and measurable $E \subset [0, 1]$. If we denote by δ_j the Dirac measure at r_j, set

$$\sigma(s) \triangleq \delta_j \quad (s \in E) \qquad \text{and} \qquad \sigma(s) \triangleq \bar{\sigma}(s) \qquad (s \in [0, 1] \sim E)$$

and recall that $\bar{\beta} > 0$, then (16) yields

$$\int_E z(s)^T f(\bar{\xi}(s), \delta_j - \bar{\sigma}(s))\, ds \geqslant 0;$$

hence, by I.4.34(7), there exist $T_j \subset T$ $(j \in \mathbb{N})$ such that $T \sim T_j$ are null sets and

$$z(\tau)^T f(\bar{\xi}(\tau), r_j) \geqslant z(\tau)^T f(\bar{\xi}(\tau), \bar{\sigma}(\tau)) \qquad (\tau \in T_j).$$

It follows that

$$\underset{r \in R}{\mathrm{Min}}\; z(\tau)^T f(\bar{\xi}(\tau), r) \geqslant \int z(\tau)^T f(\bar{\xi}(\tau), r)\, \bar{\sigma}(\tau)(dr) \qquad \Big(\tau \in T' \triangleq \bigcap_{j \in \mathbb{N}} T_j\Big),$$

from which relation (3) follows directly.

It remains to prove that $|l_1| + \nu(N) > 0$ and that (6) is valid. If $|l_1| + \nu(N) = 0$, then it follows from (14) and the definition of l_N that $\omega(\Omega) = \omega(P) = 0$, and from (10) that

$$-l_0(\beta - \bar{\beta}) \geqslant 0 \qquad (\beta \in [0, t_1]).$$

This implies $l_0 = 0$ [since $\bar{\beta} \in (0, t_1)$], thus contradicting (9). Finally, if we set $\sigma = \bar{\sigma}$ and $\beta = \bar{\beta}$ in (10), then we obtain

$$\begin{aligned} l_1 \cdot Dx_1((\bar{\sigma}, \bar{\beta}, \bar{a}_1); (\bar{\sigma}, \bar{\beta}, a_1) - (\bar{\sigma}, \bar{\beta}, \bar{a}_1)) &= l_1 \cdot (\bar{a}_1 - a_1) \\ &= l_1 \cdot (\bar{\xi}(1) - a_1) \geqslant 0 \qquad (a_1 \in A), \end{aligned}$$

thus proving (6). QED

IX.3.7 ***Discussion*** We continue to use the notation and the assumptions of Theorems IX.3.5 and IX.3.6. It is easy to see that $\bar{\beta} = 0$ if and only if $h(v_0, \tilde{v}_0) = 0$. We shall assume henceforth that $\bar{\beta} \neq 0$ and $\bar{\beta} \neq t_1$.

Optimal Escape Controls It is conceivable that the evader will reach A_1 in an optimal way without being nearly captured, that is, in such a manner that

$$h(\bar{\xi}(1), \bar{\eta}(\tilde{\sigma})(1)) > 0 \qquad \text{for all } \tilde{\sigma} \in \mathscr{S}_P{}^1.$$

This would be the case if the evader is "faster" than all the pursuers, but his equation of motion [IX.3.3(1)] is such that $\bar{\beta}$ is the highest value of β for which $\xi(\sigma, \beta)(1) \in A_1$ for any $\sigma \in \mathscr{U}$.

If we rule out this last possibility, then it follows that $h(\bar{\xi}(1), \bar{\eta}(\tilde{\sigma})(1)) = 0$ for some $\tilde{\sigma}$, and thus the set N of IX.3.6(4) is nonempty. We then have two "qualitatively" different cases to consider. If $z(1) \neq 0$, then, by IX.3.6(2), we have $z(\tau) \neq 0$ $(\tau \in [0, 1])$, and the *optimal escape control* $\bar{\sigma}$ and the corresponding *escape trajectory* $\bar{\xi}$ satisfy condition IX.3.6(3) in a nontrivial manner. This would be the case if the evader's best policy is to "run for his life" toward the "shelter" A_1, and it is a very natural policy in many circumstances. There is, however, another possible case corresponding to $z(1) = 0$. Then conditions IX.3.6(5) and IX.3.6(6) yield

$$(1) \qquad l_1{}^T\bar{\xi}(1) = \underset{a_1 \in A_1}{\operatorname{Max}}\, l_1{}^T a_1, \qquad \text{where} \quad l_1{}^T = \int \mathscr{D}_1 h(\bar{\xi}(1), w)\, \nu(dw).$$

Consider, in particular, this last case in a situation where $A_1 = \mathbb{R}^n$ and

$$h(w_1, w_2) = (w_1 - w_2) \cdot (w_1 - w_2) - (\delta)^2$$

for some $\delta > 0$. Then (1) shows that $l_1 = 0$ and

(2) $$\int 2(\bar{\xi}(1) - w)\, \nu(dw) = 0.$$

Therefore, by IX.3.6(4), $\nu(N) > 0$ and (2) yields

$$\bar{\xi}(1) = \nu(N)^{-1} \int w\, \nu(dw).$$

Since $\nu^{-1}(N)\, \nu \in \mathrm{rpm}(N)$, it follows from I.6.14 that

$$\bar{\xi}(1) \in \mathrm{co}(N).$$

We can illustrate this last situation by the following example. Assume that the evader is on foot and the pursuers are driving jeeps in an open field with a marshy area in it. Then it is often to the advantage of the evader to get into the marshy area and to stay in one spot rather than to run away in a straight line [nontrivial condition IX.3.6(3)] and be caught soon after reaching dry land where the jeeps can drive at high speed. When the evader is finally captured (by those pursuers who reach the distance δ from him), his position will be in the convex hull of his captors.

Optimal Pursuit Controls We shall rule out the "trivial" situation where no pursuer ever catches up with the evader. Then N is nonempty and it appears reasonable to refer to a pursuit control $\tilde{\sigma}^{\#}$ as *optimal* if

$$\tilde{\sigma}^{\#} \in \Omega \triangleq \{\tilde{\sigma} \mid \bar{\eta}(\tilde{\sigma})(1) \in N\}.$$

We can easily characterize the optimal pursuit controls by observing that

$$h(\bar{\xi}(1), \eta(\tilde{\sigma}^{\#})(1)) = 0 \qquad (\tilde{\sigma}^{\#} \in \Omega)$$

while

$$h(\bar{\xi}(1), \bar{\eta}(\tilde{\sigma})(1)) \geqslant 0 \qquad (\tilde{\sigma} \in \mathscr{S}_P{}^1).$$

Thus any optimal pursuit control $\tilde{\sigma}^{\#}$ minimizes $h(\bar{\xi}(1), \eta(1))$ on the set of all $(\eta, \tilde{\sigma})$ such that

$$\eta(\tau) = \tilde{v}_0 + \bar{\beta} \int_0^{\tau} \tilde{f}(\eta(s), \tilde{\sigma}(s))\, ds \qquad (\tau \in [0, 1]).$$

[In particular, if $h(w_1, w_2) = (w_1 - w_2) \cdot (w_1 - w_2) - \delta^2$, then an optimal pursuer is one that comes closest to the final position $\bar{\xi}(1)$ of the evader.] Once $\bar{\xi}(1)$ is known, this last problem is a fairly simple case of problems investigated in V.6 and VI.2.

IX.4 Zero-Sum Games with Control Strategies

Let S_1 and S_2 be compact metric spaces and $M : S_1 \times S_2 \to \mathbb{R}$ a bounded function such that $M(\cdot, s_2)$ and $M(s_1, \cdot)$ are continuous for each choice of $s_1 \in S_1$ and $s_2 \in S_2$. In the terminology of the theory of games, the function M is referred to as the *payoff function*, (and, in particular, as the *payoff matrix* if both S_1 and S_2 are finite sets), and a choice of a point $s_1 \in S_1$ respectively $s_2 \in S_2$ as a *pure strategy* for *player* I respectively *player* II. These terms are intended to suggest a game in which each player chooses, unknown to the other, some pure strategy, say $\bar{s}_1$ and $\bar{s}_2$, and then player I pays to player II the amount of $M(\bar{s}_1, \bar{s}_2)$ dollars. This game is referred to as a *zero-sum* game because the amounts "paid" by the two players add up to 0 if we define "paying $-\alpha$ dollars" to mean "receiving α dollars."

Typical of such games is the game of "matching heads and tails" which corresponds to the case where a *stake* $c > 0$ is preassigned and

$$S_1 = S_2 \triangleq \{\text{heads, tails}\}, \qquad d(\text{heads, tails}) \triangleq 1,$$

$$M(\text{heads, heads}) = M(\text{tails, tails}) = c,$$

and

$$M(\text{heads, tails}) = M(\text{tails, heads}) = -c.$$

The considerable experience accumulated by persistent investigators matching heads and tails appears to indicate that no pure strategy is preferable for either player, and that the outcome is a matter of pure chance. However, this is no longer the case if the game is repeatedly played over and over again between the same players. If player I always chooses heads, then player II can take advantage of it by doing likewise and thus keep on winning. More generally, if any one player keeps choosing the same strategy or employs a detectable procedure for choosing strategies, then the other player may be able to take advantage of this information. On the other hand, if player I chooses his consecutive strategies in a random manner, with equal probability of choosing heads or tails, then on the long run he will have won and lost approximately equal numbers of games, and he is sure to keep his losses down to approximately 0. The essential feature of this policy is the random choice of consecutive strategies, since any predictable sequence of choices may possibly be taken advantage of by the other player.

These considerations appear to indicate that a mathematical zero-sum game ought to be a model of a sequence of encounters between player I and player II, and that each player be given a choice of a *mixed strategy* which is a probability measure on some appropriate σ-field in S_1 respectively S_2. It is natural in our present case to choose mixed strategies for player I out of rpm(S_1) and for player II out of rpm(S_2). In this formulation, whenever I and II choose mixed strategies μ^{I} and μ^{II}, respectively, then I pays to II the amount of

$$v(\mu^{\mathrm{I}}, \mu^{\mathrm{II}}) \triangleq \int M(s_1, s_2)\, \mu^{\mathrm{I}}(ds_1) \times \mu^{\mathrm{II}}(ds_2) \qquad \text{dollars.}$$

This is essentially von Neumann's [1] formulation of a two-person zero-sum game. The great advantage of this formulation is that it ensures the existence of mixed strategies $\bar{\mu}^{\mathrm{I}}$ and $\bar{\mu}^{\mathrm{II}}$ such that, if player I chooses $\bar{\mu}^{\mathrm{I}}$, he can never lose more than $v(\bar{\mu}^{\mathrm{I}}, \bar{\mu}^{\mathrm{II}})$ (called the *value* of the game) no matter what player II does; and player II, by choosing $\bar{\mu}^{\mathrm{II}}$, ensures that he will win at least $v(\bar{\mu}^{\mathrm{I}}, \bar{\mu}^{\mathrm{II}})$, no matter what player I does.

IX.4.1 ***Theorem*** (von Neumann *et al.*) Let S_1 and S_2 be compact metric spaces, and let $M : S_1 \times S_2 \to \mathbb{R}$ be a bounded function such that $M(s_1, \cdot)$ and $M(\cdot, s_2)$ are continuous for each choice of $s_1 \in S_1$ and $s_2 \in S_2$. Then the expression

$$v(\mu^{\mathrm{I}}, \mu^{\mathrm{II}}) \triangleq \int M(s_1, s_2)\, \mu^{\mathrm{I}}(ds_1) \times \mu^{\mathrm{II}}(ds_2)$$

defines a function $v : \mathrm{rpm}(S_1) \times \mathrm{rpm}(S_2) \to \mathbb{R}$, and there exist $\bar{\mu}^{\mathrm{I}} \in \mathrm{rpm}(S_1)$ and $\bar{\mu}^{\mathrm{II}} \in \mathrm{rpm}(S_2)$ such that

$$\text{(1)} \quad v(\bar{\mu}^{\mathrm{I}}, \mu^{\mathrm{II}}) \leqslant v(\bar{\mu}^{\mathrm{I}}, \bar{\mu}^{\mathrm{II}}) \leqslant v(\mu^{\mathrm{I}}, \bar{\mu}^{\mathrm{II}}) \qquad [\mu^{\mathrm{I}} \in \mathrm{rpm}(S_1), \quad \mu^{\mathrm{II}} \in \mathrm{rpm}(S_2)]$$

and

$$\text{(2)} \qquad \operatorname*{Min}_{\mu^{\mathrm{I}}} \operatorname*{Max}_{\mu^{\mathrm{II}}} v(\mu^{\mathrm{I}}, \mu^{\mathrm{II}}) = \operatorname*{Max}_{\mu^{\mathrm{II}}} \operatorname*{Min}_{\mu^{\mathrm{I}}} v(\mu^{\mathrm{I}}, \mu^{\mathrm{II}}) = v(\bar{\mu}^{\mathrm{I}}, \bar{\mu}^{\mathrm{II}}).$$

▌PROOF *Step* 1 By I.5.26(3), M is $\mu^{\mathrm{I}} \times \mu^{\mathrm{II}}$—measurable for all $\mu^{\mathrm{I}} \in \mathrm{rpm}(S_1)$ and $\mu^{\mathrm{II}} \in \mathrm{rpm}(S_2)$. Therefore the bounded function M is $\mu^{\mathrm{I}} \times \mu^{\mathrm{II}}$-integrable, and the expression

$$v(\mu^{\mathrm{I}}, \mu^{\mathrm{II}}) \triangleq \int M(s_1, s_2)\, \mu^{\mathrm{I}}(ds_1) \times \mu^{\mathrm{II}}(ds_2)$$

defines a bounded function $v(\cdot, \cdot) : \mathrm{rpm}(S_1) \times \mathrm{rpm}(S_2) \to \mathbb{R}$.

Now let M_0 be such that

$$|\, M(\cdot, \cdot)|_{\sup} \leqslant M_0 - 1;$$

hence

$$|\, v(\cdot, \cdot)|_{\sup} \leqslant M_0 - 1,$$

and let

$$Q \triangleq \mathrm{rpm}(S_1) \times [-M_0, M_0]^2, \qquad P \triangleq S_2, \qquad A(p) \triangleq (-\infty, 0] \quad (p \in P),$$

$$x_0(q) \triangleq \beta, \qquad x_1(q) \triangleq \gamma, \qquad x_2(q)(p) \triangleq \int M(s_1, p)\, \mu^{\mathrm{I}}(ds_1) - \beta$$

$$[q \triangleq (\mu^{\mathrm{I}}, \beta, \gamma) \in Q, p \in P].$$

We topologize Q by choosing the weak norm topology for $\mathrm{rpm}(S_1)$ [which we embed in $C(S_1)^*$, see I.3.11 and I.5.8]. Since S_1 is, like R, a compact metric space, it follows from IV.1.4 that $\mathrm{rpm}(S_1)$, and therefore also Q, are compact metric spaces, and the function $q \to x_2(q)(p)$ is continuous for each $p \in P$. Furthermore, since M is bounded and $M(s_1, \cdot)$ continuous for each s_1, it follows from the dominated convergence theorem I.4.35 that the function $p \to x_2(q)(p)$ is continuous for each $q \in Q$.

We may now apply Theorem IX.1.1 which implies that there exists $\bar{q} \triangleq (\bar{\mu}^{\mathrm{I}}, \bar{\beta}, \bar{\gamma}) \triangleq (\bar{\mu}^{\mathrm{I}}, \bar{\beta}, 0) \in Q$ that minimizes β on the set

$$\mathscr{A}(Q) \triangleq \{q \triangleq (\mu^{\mathrm{I}}, \beta, \gamma) \in Q \mid x_1(q) = 0, x_2(q)(p) \leqslant 0 \ (p \in P)\}.$$

Furthermore, the assumptions of Theorem IX.1.2 are also satisfied, and we have

$$Dx_0(\bar{q}; q - \bar{q}) = \beta - \bar{\beta}, \qquad Dx_1(\bar{q}; q - \bar{q}) = \gamma$$

and

$$Dx_2(\bar{q}; q - \bar{q})(p) = \int M(s_1, s_2)(\mu^{\mathrm{I}} - \bar{\mu}^{\mathrm{I}})(ds_1) - (\beta - \bar{\beta})$$

$$[q \triangleq (\mu^{\mathrm{I}}, \beta, \gamma) \in Q, \quad p \triangleq s_2 \in S_2].$$

It follows that there exist $l_0 \geqslant 0$, $l_1 \in \mathbb{R}$, $\omega \in \mathrm{frm}^+(S_2)$, and an ω-integrable $\tilde{\omega} : S_2 \to \mathbb{R}$ such that

$$(3) \qquad |\tilde{\omega}(s_2)| = 1 \quad (s_2 \in S_2) \qquad \text{and} \qquad l_0 + |l_1| + \omega(S_2) > 0,$$

$$(4) \qquad \left[l_0 - \int \tilde{\omega}(s_2)\, \omega(ds_2)\right] (\beta - \bar{\beta}) + l_1\gamma$$

$$+ \int \tilde{\omega}(s_2)\, \omega(ds_2) \int M(s_1, s_2)(\mu^{\mathrm{I}} - \bar{\mu}^{\mathrm{I}})(ds_1) \geqslant 0$$

$$(\beta, \gamma \in [-M_0, M_0], \quad \mu^{\mathrm{I}} \in \mathrm{rpm}(S_1)),$$

and

$$(5) \quad \tilde{\omega}(s_2) \left[\int M(s_1, s_2)\, \bar{\mu}^{\mathrm{I}}(ds_1) - \bar{\beta}\right] = \operatorname*{Max}_{a \leqslant 0} \tilde{\omega}(s_2)\, a \qquad \text{for} \quad \omega\text{-a.a. } s_2 \in S_2.$$

Since $\bar{\beta}$ is in the interior of $[-M_0, M_0]$ (because

$$\bar{\beta} = \sup_{p \in P} \int M(s_1, p)\, \bar{\mu}^{\text{I}}(ds_1) \qquad \text{and} \qquad |M(\cdot, \cdot)|_{\sup} \leqslant M_0 - 1),$$

relation (4) implies that

$$l_1 = 0 \qquad \text{and} \qquad \int \tilde{\omega}(s_2)\, \omega(ds_2) = l_0 \geqslant 0.$$

Thus, by (3) and (5), $\tilde{\omega}(s_2) = 1$ ω-a.e., $l_0 = \omega(S_2) > 0$, and

$$\int M(s_1, s_2)\, \bar{\mu}^{\text{I}}(ds_1) = \bar{\beta} \qquad \text{for} \quad \omega\text{-a.a. } s_2 \in S_2 . \tag{6}$$

Next we set $\bar{\mu}^{\text{II}} \triangleq \omega(S_2)^{-1}\omega$, and let $\beta = \bar{\beta}$ and $\gamma = 0$ in (4). We then obtain, applying Fubini's theorem I.4.45,

$$v(\bar{\mu}^{\text{I}}, \bar{\mu}^{\text{II}}) \leqslant v(\mu^{\text{I}}, \bar{\mu}^{\text{II}}) \qquad [\mu^{\text{I}} \in \text{rpm}(S_1)],$$

which proves the second inequality in (1). On the other hand, since $\bar{q} = (\bar{\mu}^{\text{I}}, \bar{\beta}, 0) \in \mathscr{A}(Q)$, we have

$$x_2(\bar{q})(s_2) = \int M(s_1, s_2)\, \bar{\mu}^{\text{I}}(ds_1) - \bar{\beta} \leqslant 0 \qquad (s_2 \in S_2).$$

It follows that

$$\begin{aligned} \int \mu^{\text{II}}(ds_2) \int M(s_1, s_2)\, \bar{\mu}^{\text{I}}(ds_1) - \bar{\beta} &= v(\bar{\mu}^{\text{I}}, \mu^{\text{II}}) - \bar{\beta} \\ &\leqslant 0 \qquad [\mu^{\text{II}} \in \text{rpm}(S_2)]. \end{aligned} \tag{7}$$

Similarly, relation (6) implies that

$$v(\bar{\mu}^{\text{II}}, \bar{\mu}^{\text{I}}) = \bar{\beta},$$

which together with (7) yields the first inequality in (1).

Step 2 We can now derive relation (2) from (1). We shall write $\inf_{\mu^{\text{I}}}$ respectively $\sup_{\mu^{\text{II}}}$ to mean "inf for all $\mu^{\text{I}} \in \text{rpm}(S_1)$" respectively "sup for all $\mu^{\text{II}} \in \text{rpm}(S_2)$." We observe that, by (1),

$$v(\bar{\mu}^{\text{I}}, \bar{\mu}^{\text{II}}) \leqslant \inf_{\mu^{\text{I}}} v(\mu^{\text{I}}, \bar{\mu}^{\text{II}}) \leqslant \sup_{\mu^{\text{II}}} \inf_{\mu^{\text{I}}} v(\mu^{\text{I}}, \mu^{\text{II}}) \tag{8}$$

and

$$v(\bar{\mu}^{\text{I}}, \bar{\mu}^{\text{II}}) \geqslant \sup_{\mu^{\text{II}}} v(\bar{\mu}^{\text{I}}, \mu^{\text{II}}) \geqslant \inf_{\mu^{\text{I}}} \sup_{\mu^{\text{II}}} v(\mu^{\text{I}}, \mu^{\text{II}}). \tag{9}$$

On the other hand, we have

$$\inf_{\mu^{\mathrm{I}}} v(\mu^{\mathrm{I}}, \mu^{\mathrm{II}}) \leqslant v(\tilde{\mu}^{\mathrm{I}}, \mu^{\mathrm{II}}) \qquad \text{for all } \tilde{\mu}^{\mathrm{I}} \text{ and } \mu^{\mathrm{II}};$$

hence

$$\sup_{\mu^{\mathrm{II}}} \inf_{\mu^{\mathrm{I}}} v(\mu^{\mathrm{I}}, \mu^{\mathrm{II}}) \leqslant \sup_{\mu^{\mathrm{II}}} v(\tilde{\mu}^{\mathrm{I}}, \mu^{\mathrm{II}}) \qquad \text{for all } \tilde{\mu}^{\mathrm{I}}$$

and therefore

$$\text{(10)} \qquad \sup_{\mu^{\mathrm{II}}} \inf_{\mu^{\mathrm{I}}} v(\mu^{\mathrm{I}}, \mu^{\mathrm{II}}) \leqslant \inf_{\mu^{\mathrm{I}}} \sup_{\mu^{\mathrm{II}}} v(\mu^{\mathrm{I}}, \mu^{\mathrm{II}}).$$

We have observed in Step 1 (when discussing the function x_2) that the function

$$s_2 \to \int M(s_1, s_2)\, \mu^{\mathrm{I}}(ds_1) : S_2 \to \mathbb{R}$$

is continuous for each μ^{I}. It follows then [by our choice of the topology for $\mathrm{rpm}(S_2)$] that the function

$$\mu^{\mathrm{II}} \to v(\mu^{\mathrm{I}}, \mu^{\mathrm{II}}) = \int \mu^{\mathrm{II}}(ds_2) \int M(s_1, s_2)\, \mu^{\mathrm{I}}(ds_1) : \mathrm{rpm}(S_2) \to \mathbb{R}$$

is continuous for each μ^{I}. The same argument shows that $\mu^{\mathrm{I}} \to v(\mu^{\mathrm{I}}, \mu^{\mathrm{II}})$ is continuous for each μ^{II}. Thus

$$\inf_{\mu^{\mathrm{I}}} v(\mu^{\mathrm{I}}, \mu^{\mathrm{II}}) = \operatorname*{Min}_{\mu^{\mathrm{I}}} v(\mu^{\mathrm{I}}, \mu^{\mathrm{II}}) \qquad \text{and} \qquad \sup_{\mu^{\mathrm{II}}} v(\mu^{\mathrm{I}}, \mu^{\mathrm{II}}) = \operatorname*{Max}_{\mu^{\mathrm{II}}} v(\mu^{\mathrm{I}}, \mu^{\mathrm{II}}).$$

We can therefore derive statement (2) from (8)–(10). QED

Theorem IX.4.1 is often referred to as the "Minimax–Maximin theorem" because of its relation (2). We can apply IX.4.1, in particular, whenever S_1 and S_2 are sets of conflicting controls; specifically, if $S_1 = \mathscr{S}^{\#} \times B$ for some choice of a compact metric space B and $S_2 = \mathscr{S}_P^{\#} \times B_P$, where B_P is compact and metric and $\mathscr{S}_P^{\#}$ is defined the same way as $\mathscr{S}^{\#}$ but with T, R, and $R^{\#}$ replaced by similarly defined T_P, R_P, and $R_P^{\#}$. We may then consider a situation where $M(\cdot, \cdot)$ is defined by a functional equation; for example, let $T = T_P \triangleq [t_0, t_1]$, and let $n \in \mathbb{N}$, $f: \mathbb{R}^n \times R \times B \times R_P \times B_P \to \mathbb{R}^n$, and $v_0 \in \mathbb{R}^n$ be given. If f and $\mathscr{D}_1 f$ exist and are continuous and bounded, then the ordinary differential equation

$$y(t) = v_0 + \int_t^t f\big(y(\tau), \sigma(\tau), b, \sigma_P(\tau), b_P\big)\, d\tau \qquad (t \in T)$$

has a unique solution $\tilde{y}(\sigma, b, \sigma_P, b_P)(\cdot)$ for each choice of $(\sigma, b) \in S_1 = \mathscr{S}^{\#} \times B$ and $(\sigma_P, b_P) \in S_2 = \mathscr{S}_P{}^{\#} \times B_P$. If $\tilde{y} \triangleq (\tilde{y}^1, ..., \tilde{y}^n)$, then we set

$$M(s_1, s_2) \triangleq \tilde{y}^1(\sigma, b, \sigma_P, b_P)(t_1) \qquad [s_1 \triangleq (\sigma, b) \in S_1, \quad s_2 \triangleq (\sigma_P, b_P) \in S_2],$$

and we can prove, using arguments analogous to those in Chapter VI and IX.1.3, that $M(\cdot, s_2)$ and $M(s_1, \cdot)$ are both continuous. Similar examples can be given in which M is defined by functional-integral equations in $C(T, \mathbb{R}^n)$ or in $L^p(T, \mathbb{R}^n)$, or the conditions on f and $\mathscr{D}_1 f$ are somewhat relaxed.

Notes

Existence and necessary conditions for a relaxed minimum were investigated in Warga [4] for minimax problems defined by differential equations, with adverse controls confined to control parameters. Most of the results of IX.1 and IX.2 were derived in Warga [15].

An evasion problem similar to the one of IX.3 (except that capture occurs when the evader and some pursuer are at exactly the same location) was proposed by Kelendzheridze [1] who derived necessary conditions for original minimum in the special case where the pursuers' motion is governed by a linear differential equation. An example was exhibited in Warga [6], showing that these results cannot be extended in general if the pursuers' equation is nonlinear. Linear evasion games in a Banach space were studied by Friedman [2]. The results of IX.3 were derived in Warga [14].

The theory of two person zero-sum games was initiated by von Neumann [1] who proved Theorem IX.4.1 in the special case where S_1 and S_2 are both finite sets. Von Neumann's theorem was generalized in many ways, and Theorem IX.4.1 is one such generalization.

CHAPTER X

Conflicting Control Problems with Hyperrelaxed Adverse Controls

X.0 Formulation of the Problem

X.0.1 ***General Remarks and Heuristic Considerations*** In the preceding chapter we have considered problems involving conflicting original controls and defined by functions

$$(x_0, x_1) : \mathscr{U} \times B \to \mathbb{R} \times \mathbb{R}^m$$

and

$$x_2 : \mathscr{U} \times B \times \mathscr{U}_P \times B_P \to \mathbb{R}^{m_2},$$

where $\mathscr{U} \times B$ and $\mathscr{U}_P \times B_P$ are sets of conflicting original controls. We have observed that, "in general," a problem of this type remains unaffected if the set $\mathscr{U}_P$ of original adverse control functions is replaced by the larger set $\mathscr{S}_P^{\#}$ of relaxed adverse control functions. If, furthermore, the conflicting controls are additively coupled, then we may extend x_2 to a continuous function on $\mathscr{S}^{\#} \times B \times \mathscr{S}_P^{\#} \times B_P$, and any minimizing relaxed solution $(\bar{\sigma}, \bar{b})$ is appropriately simulated by a sequence $((u_j, \bar{b}))$ of original controls that approach $(\bar{\sigma}, \bar{b})$ in the topology of $\mathscr{S}^{\#} \times B$. On the other hand, Counterexample IX.2.2 demonstrates that even simple problems with conflicting controls that are not additively coupled may not exhibit this "nice" pattern, and therefore the extension of controls to be relaxed may yield an unrelated problem.

In general, our "opponent" gains nothing by being allowed to employ relaxed controls, but "we" may gain a definite advantage if we can resort to relaxed controls. Our sense of honesty and fair play becomes particularly acute if our attempts to simulate relaxed controls (that is, to blur our opponent's picture by generating highly oscillatory original control functions as described in IX.2.3) are matched by his use of high-quality recording equipment. Under such conditions, that are modeled by Problem I of IX.2.3, we are denied the use of "friendly" relaxed controls in return for allowing a similar freedom to our opponent. In the process, we are thwarted in our mathematical endeavor to predict the outcome of the conflict because original controls lack the compactness properties of relaxed controls and there may exist no minimizing friendly original controls (which satisfy weak necessary conditions for a minimum.)

We shall show that, while it is not "fair" to replace all of the conflicting control functions by their relaxed versions, a fair trade-off will consist in allowing friendly control functions to be relaxed in exchange for granting the adverse control functions an even greater freedom and a larger set from which to operate. We shall refer to elements of this larger set of adverse control functions as "hyperrelaxed adverse control functions." In order to provide the motivation for introducing this new concept, we shall consider a variant of Counterexample IX.2.2.

Let $T \triangleq [0, 1]$, $R = R_P \triangleq \{r \in \mathbb{R}^2 \mid r \cdot r = |r|_2^2 = 1\}$, $\mathscr{U}$ be the set of measurable functions $u : T \to R$, and $\mathscr{U}_P = \mathscr{U}$. We consider the problem of determining

$$\inf_{u \in \mathscr{U}} \sup_{u_P \in \mathscr{U}_P} \left\{ \int_0^1 \left| \int_0^t u(\tau)\, d\tau \right|_2^2 dt + \int_0^1 u(t) \cdot u_P(t)\, dt \right\},$$

which is equivalent to the problem of determining

$$\inf \left\{ x_0(u, \alpha) \triangleq \int_0^1 \left| \int_0^t u(\tau)\, d\tau \right|_2^2 dt + \alpha \;\middle|\; u \in \mathscr{U},\ \alpha \in \mathbb{R}, \right.$$

$$\left. x_2(u, u_P, \alpha) \triangleq \int_0^1 u(t) \cdot u_P(t)\, dt - \alpha \leqslant 0 \ (u_P \in \mathscr{U}_P) \right\}.$$

It is easily seen that

$$\sup_{u_P \in \mathscr{U}} \int_0^1 u(t) \cdot u_P(t)\, dt = \int_0^1 u(t) \cdot u(t)\, dt = 1$$

for every $u \in \mathscr{U}$, and therefore the optimal choice of α is 1. Thus a minimizing sequence (u_j) of friendly original controls yields $\inf\{x_0(u, 1) \mid u \in \mathscr{U}\}$, and such a sequence can be obtained by choosing an arbitrary point $\bar{r} \in R$, dividing $[0, 1]$ into j equal subintervals, and setting $u_j(t) = \bar{r}$ and $u_j(t) = -\bar{r}$ in successive subintervals. This yields $1 \leqslant x_0(u_j, 1) \leqslant j^{-2} + 1$; hence

$$1 = \lim_j x_0(u_j, 1) = \inf_{u \in \mathscr{U}} x_0(u, 1).$$

If we denote by δ_r the Dirac measure at $r \in R$ and extend $x_0(\cdot, 1)$ to $\mathscr{S}$ in the usual manner [i.e., replace $\int_0^t u(\tau)\, d\tau$ by $\int_0^t d\tau \int r\, \sigma(\tau)(dr)$], then

$$\lim_j u_j = \bar{\sigma} \triangleq \tfrac{1}{2}\delta_{\bar{r}} + \tfrac{1}{2}\delta_{-\bar{r}} \quad \text{in } \mathscr{S}$$

and

$$\lim_j x_0(u_j, 1) = x_0(\bar{\sigma}, 1) = 1.$$

However, as it was the case with Counterexample IX.2.2, we have

$$x_2(\bar{\sigma}, \sigma_P, \alpha) \triangleq \int_0^1 dt \int r \cdot r_P(\tfrac{1}{2}\delta_{\bar{r}} + \tfrac{1}{2}\delta_{-\bar{r}})\,(dr) \times \sigma_P(t)\,(dr_P) - \alpha = -\alpha$$

for all relaxed adverse controls $\sigma_P \in \mathscr{S}_P$, while

$$x_2(u, u, \alpha) = 1 - \alpha \qquad \text{for} \quad u \in \mathscr{U}.$$

Our opponent was able to maximize $x_2(u, \cdot, \alpha)$ over $\mathscr{U}_P$ for any original control function u by choosing $u_P = u$ which yields $u(t) \cdot u_P(t) = 1$ $(t \in T)$. When we replaced u by the relaxed control function $\bar{\sigma}$, the individual point $u(t)$ was "smeared into a measure" and both $u_P(t)$ and its relaxed version "lost the grip" on the precise "location" of $u(t)$. In order to restore this lost advantage, we must allow our opponent the additional freedom of adjusting his control function "during" the averaging process represented by integrating with respect to $\bar{\sigma}(t)$, and this implies that $u_P(t)$ is replaced by a function on R into $\text{rpm}(R_P)$. Thus we replace an original $u_P : T \to R_P$ by a "hyperrelaxed adverse control function"

$$\pi : T \times R \to \text{rpm}(R_P),$$

which yields

$$x_2(\bar{\sigma}, \pi, \alpha) = \int_0^1 dt \int \bar{\sigma}(t)\,(dr) \int r \cdot r_P\, \pi(t, r)(dr_P) - \alpha.$$

It is now clear that $x_2(\bar{\sigma}, \cdot, \alpha)$ is maximized by choosing π so that

$$\pi(t, r) = \bar{\pi}(t, r) \triangleq \delta_r \qquad (t \in T, \quad r \in R),$$

yielding

$$\sup_{\pi} x_2(\bar{\sigma}, \pi, \alpha) = x_2(\bar{\sigma}, \bar{\pi}, \alpha) = \int_0^1 dt \int r \cdot r\, \bar{\sigma}(t)(dr) - \alpha = 1 - \alpha;$$

hence

$$\sup_{\pi} x_2(\bar{\sigma}, \pi, \alpha) = \lim_{j} \sup_{u_P \in \mathscr{U}_P} x_2(u_j, u_P, \alpha).$$

We now conjecture, and will prove later that, "in general," sequences (u_j) in $\mathscr{U}$ can be replaced in a fair manner by elements σ of $\mathscr{S}^{\#}$ provided our opponent has the freedom of choosing hyperrelaxed control functions instead of the original ones.

X.0.2 ***Formulation of the Problem*** Let R_P be a compact metric space and $R_P{}^{\#} : T \to \mathscr{P}'(R_P)$ a mapping satisfying Condition IV.3.1 (with R replaced by R_P). We denote by $\bar{R}_P{}^{\#}$ the mapping $t \to \overline{R_P{}^{\#}(t)} : T \to \mathscr{P}'(R_P)$. We refer to a $\Sigma_{\text{Borel}}(T \times R)$-measurable function $\pi : T \times R \to (\text{rpm}(R_P), |\cdot|_w)$ such that $\pi(t, r)(\bar{R}_P{}^{\#}(t)) = 1$ (μ-a.a. $t \in T$, $r \in R$) as a *hyperrelaxed control function*, and denote by $\mathscr{P}^{\#}$ the collection of all hyperrelaxed control functions in which we identify π_1 and π_2 whenever $\pi_1(t, \cdot) = \pi_2(t, \cdot)$ μ-a.e. We define the set $\mathscr{R}_P{}^{\#}$ of all μ-measurable selections of $R_P{}^{\#}$ as a subset of $\mathscr{P}^{\#}$ by identifying every $\rho \in \mathscr{R}_P{}^{\#}$ with the hyperrelaxed control function π_ρ defined by

$$\pi_\rho(t, r) \triangleq \delta_{\tilde{\rho}(t)} \qquad (t \in T, \quad r \in R),$$

where δ_r is the Dirac measure at $r \in R_P$ and $\tilde{\rho}$ is a Σ-measurable function that is μ-equivalent to ρ.

We assume given a set B_P (the set of *adverse control parameters*), $m \in N$, $m_2 \in \{0, 1, 2,...\}$, $A \subset \mathbb{R}^{m_2}$, and functions

$$x_0 : \mathscr{S}^{\#} \times B \to \mathbb{R}, \qquad x_1 : \mathscr{S}^{\#} \times B \to \mathbb{R}^m,$$

and

$$x_2 : \mathscr{S}^{\#} \times B \times \mathscr{P}^{\#} \times B_P \to \mathbb{R}^{m_2}.$$

We set, for all $\mathscr{S}' \subset \mathscr{S}^{\#}$,

$$\mathscr{A}(\mathscr{S}') \triangleq \{(\sigma, b) \in \mathscr{S}' \times B \mid x_1(\sigma, b) = 0,\ x_2(\sigma, b, \pi, b_P) \in A\ (\pi \in \mathscr{P}^{\#}, b \in B_P)\}.$$

A point $\bar{q} \triangleq (\bar{\sigma}, \bar{b}) \in \mathscr{A}(\mathscr{S}')$ is a *minimizing $\mathscr{S}'$-control* if

$$x_0(\bar{q}) = \text{Min}\, x_0(\mathscr{A}(\mathscr{S}')).$$

We refer to a minimizing $\mathscr{S}^{\#}$-control as a *minimizing relaxed control.* For any $\mathscr{U} \subset \mathscr{R}^{\#}$ and $\mathscr{U}_P \subset \mathscr{R}_P{}^{\#}$, a sequence $((u_j, b_j))$ in $\mathscr{U} \times B$ is an *approximate* $\mathscr{U}, \mathscr{U}_P$-control if

$$\lim_j x_1(u_j, b_j) = 0$$

and

$$\lim_j [\sup\{d[x_2(u_j, b_j, u_P, b_P), A] \mid u_P \in \mathscr{U}_P, b_P \in B_P\}] = 0.$$

An approximate $\mathscr{U}, \mathscr{U}_P$-control $((\bar{u}_j, \bar{b}_j))$ is a *minimizing approximate* $\mathscr{U}, \mathscr{U}_P$-control if

$$\lim_j x_0(\bar{u}_j, \bar{b}_j) \leqslant \lim_j \inf x_0(u_j, b_j)$$

for every approximate $\mathscr{U}, \mathscr{U}_P$-control $((u_j, b_j))$.

X.1 Existence of Minimizing Relaxed and Approximate Controls

Let $\hat{R}^{\#}(t) \triangleq R^{\#}(t) \times R_P{}^{\#}(t)$ $(t \in T)$. We can easily verify that the mapping $\hat{R}^{\#} : T \to \mathscr{P}'(R \times R_P)$ satisfies Condition IV.3.1. Indeed, an argument almost identical with that of I.7.5 shows that the mapping

$$t \to \bar{\hat{R}}(t) = \bar{R}^{\#}(t) \times \bar{R}_P{}^{\#}(t)$$

is μ-measurable and therefore $\hat{R}^{\#}$ is μ-measurable. Furthermore, if $\{\rho_1, \rho_2, ...\} \subset \mathscr{R}^{\#}$ and $\{\rho_{P1}, \rho_{P2}, ...\} \in \mathscr{R}_P{}^{\#}$ are such that $\{\rho_1(t), \rho_2(t), ...\}$ and $\{\rho_{P1}(t), \rho_{P2}(t), ...\}$ are dense in $R^{\#}(t)$ respectively $R_P{}^{\#}(t)$ for μ-a.a. $t \in T$, then $\{(\rho_i, \rho_{Pj})(t) \mid i, j \in \mathbb{N}\}$ is dense in $\hat{R}^{\#}(t)$ for μ-a.a. $t \in T$.

We shall denote by $\hat{\mathscr{S}}^{\#}$ the set, together with its algebraic structure and weak norm topology, that is defined exactly as $\mathscr{S}^{\#}$, but with R, $R^{\#}$ replaced by $R \times R_P$, $\hat{R}^{\#}$, respectively.

If S is a topological space and $\lambda \in \text{frm}^+(S)$, then we shall write $L^1(\lambda, \mathscr{X})$ for $L^1(S, \Sigma_{\text{Borel}}(S), \lambda, \mathscr{X})$ and $L^1(\lambda)$ for $L^1(\lambda, \mathbb{R})$. In this notation, the space previously denoted by $L^1(T, C(R))$ will now be denoted by $L^1(\mu, C(R))$.

The basic results of this section are contained in Theorem X.1.8.

X.1.1 ***Lemma*** For every $\sigma \in \mathscr{S}^{\#}$ there exists a unique $\zeta \in \text{frm}^+(T \times R)$ such that

$$\int \mu(dt) \int h(t, r)\, \sigma(t)\, (dr) = \int h(t, r)\, \zeta(d(t, r)) \qquad [h \in L^1(\zeta)].$$

▌PROOF *Step* 1 Let

$$l(c) \triangleq \int \mu(dt) \int c(t, r)\, \sigma(t)\, (dr) \qquad [c \in C(T \times R)].$$

Then $l \in C(T \times R)^*$ and, by the Riesz representation theorem I.5.8, there exists a unique $\zeta \in \text{frm}(T \times R)$ such that

$$l(c) = \int c(t, r)\, \zeta(d(t, r)) \qquad [c \in C(T \times R)].$$

By I.5.5, ζ is a positive measure because $l(c) \geqslant 0$ for all nonnegative c.

Step 2 Let F be a closed subset of $T \times R$. We can determine continuous functions $c_j : T \times R \to [0, 1]$ $(j \in \mathbb{N})$ such that $c_j(t, r) = 1$ if $(t, r) \in F$ and $c_j(t, r) = 0$ if $d[(t, r), F] \geqslant 1/j$. Then $\lim_j c_j(t, r) = \chi_F(t, r)$ for all $(t, r) \in T \times R$ and therefore (I.4.17 and I.4.35) the function

$$t \to \int \chi_F(t, r)\, \sigma(t)\,(dr) = \lim_j \int c_j(t, r)\, \sigma(t)\,(dr)$$

is μ-measurable and (by I.4.35)

$$\int \mu(dt) \int \chi_F(t, r)\, \sigma(t)\,(dr) = \int \chi_F(t, r)\, \zeta(d(t, r)) = \zeta(F). \tag{1}$$

If G is an open subset of $T \times R$, then $T \times R \sim G$ is closed and $\chi_G(t, r) = 1 - \chi_{T \times R \sim G}(t, r)$ for all $(t, r) \in T \times R$, and it follows that relation (1) is satisfied with F replaced by G.

Step 3 Now let Z be ζ-null. Then there exist open G_j $(j \in \mathbb{N})$ such that $G_j \supset Z$ and $\lim_j \zeta(G_j) = 0$. We may assume that $G_{j+1} \subset G_j$ for all j, otherwise replacing G_{j+1} by $G_j \cap G_{j+1}$. If we set $G' \triangleq \bigcap_{j=1}^{\infty} G_j$, then, by I.4.35,

$$\lim_j \int \chi_{G_j}(t, r)\, \sigma(t)\,(dr) = \int \chi_{G'}(t, r)\, \sigma(t)\,(dr) \qquad (t \in T)$$

and therefore, by (1) and I.4.35,

$$\int \mu(dt) \int \chi_{G'}(t, r)\, \sigma(t)\,(dr) = \lim_j \int \mu(dt) \int \chi_{G_j}(t, r)\, \sigma(t)\,(dr)$$

$$= \lim_j \zeta(G_j) = \zeta(G') = 0.$$

Thus $\int \chi_{G'}(t, r)\, \sigma(t)\,(dr) = 0$ μ-a.e. Since $\chi_Z(t, r) \leqslant \chi_{G'}(t, r)$ for all $(t, r) \in T \times R$, it follows that $\int \chi_Z(t, r)\, \sigma(t)\,(dr) = 0$ μ-a.e.

Step 4 Next let E be a ζ-measurable set, and let $F_j \subset E$ $(j \in \mathbb{N})$ be closed and such that $\lim_j \zeta(E \sim F_j) = 0$. We assume that $F_j \subset F_{j+1}$ for all j (replacing F_{j+1}, if necessary, by $F_j \cup F_{j+1}$), and set $F' \triangleq \bigcup_{j=1}^{\infty} F_j$. Then

$$\lim_j \chi_{F_j}(t, r) = \chi_{F'}(t, r) \qquad \text{for all } (t, r) \in T \times R$$

and therefore

(2) $$t \to \int \chi_{F'}(t, r)\, \sigma(t)\,(dr) = \lim_j \int \chi_{F_j}(t, r)\, \sigma(t)\,(dr)$$

is μ-measurable. Since $\zeta(E \sim F') = 0$, it follows from Step 3 that $\chi_{E \sim F'}(t, \cdot)$ is $\sigma(t)$-measurable for μ-a.a. $t \in T$ and $\int \chi_E(t, r)\, \sigma(t)\,(dr) = \int \chi_{F'}(t, r)\, \sigma(t)\,(dr)$ μ-a.e.; hence, by (1) and (2),

(3) $$\int \mu(dt) \int \chi_E(t, r)\, \sigma(t)\,(dr) = \lim_j \zeta(F_j) = \zeta(E).$$

Step 5 Now let $h \in L^1(\zeta)$ be nonnegative. Then we can determine (I.4.25) a sequence (h_j) of nonnegative ζ-simple functions such that $(h_j(t, r))_j$ increases to $h(t, r)$ for all $(t, r) \in T \times R$. Thus, by Step 4 and I.4.17,

$$t \to \int h(t, r)\, \sigma(t)\,(dr) = \lim_j \int h_j(t, r)\, \sigma(t)\,(dr)$$

is μ-measurable and, by (3),

(4) $$\begin{aligned}\int \mu(dt) \int h(t, r)\, \sigma(t)\,(dr) &= \lim_j \int \mu(dt) \int h_j(t, r)\, \sigma(t)\,(dr) \\ &= \lim_j \int h_j(t, r)\, \zeta(d(t, r)) \\ &= \int h(t, r)\, \zeta(d(t, r)).\end{aligned}$$

If h is an arbitrary element of $L^1(\zeta)$ then $h = h^+ - h^-$, where h^+ and h^- are nonnegative and ζ-integrable, and the conclusion of the theorem follows by applying (4) to h^+ and h^- and combining the results. QED

X.1.2 ***Lemma*** Let $\sigma \in \mathscr{S}^\#$ and let ζ be defined as in X.1.1. Then

$$L^1(\mu, C(R)) \subset L^1(\zeta) \qquad \text{and} \qquad L^1(\mu, C(R \times R_P)) \subset L^1(\zeta, C(R_P)).$$

▌PROOF Let $h \in L^1(\mu, C(R))$. Then, by I.5.26, for every $\epsilon > 0$ there exists a closed $F_\epsilon \subset T$ such that $\mu(T \sim F_\epsilon) \leqslant \epsilon$ and $h \mid F_\epsilon \times R$ is continuous and therefore ζ-integrable. Furthermore, by X.1.1,

$$\zeta(F_\epsilon \times R) = \int \chi_{F_\epsilon}(t)\, \mu(dt) \int \sigma(t)\,(dr) = \mu(F_\epsilon)$$

and

$$\zeta(T \times R \sim F_\epsilon \times R) = \mu(T \sim F_\epsilon) \leqslant \epsilon.$$

It follows that h is ζ-measurable. We have

$$\int \chi_{F_\epsilon \times R}(t, r) \mid h(t, r)\mid \zeta(d(t, r)) = \int_{F_\epsilon} \mu(dt) \int \mid h(t, r)\mid \sigma(t)(dr)$$
$$\leqslant \int \mid h(t, \cdot)\mid_{\sup} \mu(dt) < \infty,$$

from which we conclude that h is ζ-integrable. Thus $L^1(\mu, C(R)) \subset L^1(\zeta)$.

Now let $f \in L^1(\mu, C(R \times R_P))$. Then $f(t, r, \cdot)$ is continuous for all (t, r) and $f(\cdot, \cdot, r_P) \in L^1(\mu, C(R))$ for all r_P. It follows, by our previous argument, that $f(\cdot, \cdot, r_P) \in L^1(\zeta)$ for all r_P. Finally,

$$\mid f(t, r, \cdot)\mid_{\sup} \leqslant \mid f(t, \cdot, \cdot)\mid_{\sup}$$

for all (t, r) and therefore, by X.1.1,

$$\int \mid f(t, r, \cdot)\mid_{\sup} \zeta(d(t, r)) \leqslant \int \mid f(t, \cdot, \cdot)\mid_{\sup} \mu(dt) < \infty.$$

Thus, by I.5.25, $f \in L^1(\zeta, C(R_P))$. QED

X.1.3 ***Lemma*** Let $\hat{\sigma} \in \hat{\mathscr{S}}^{\#}$ and

$$\sigma(t)(E) = \hat{\sigma}(t)(E \times R_P) \qquad [t \in T, E \in \Sigma_{\text{Borel}}(R)].$$

Then $\sigma \in \mathscr{S}^{\#}$ and there exists (a not necessarily unique) $\pi \in \mathscr{P}^{\#}$ such that

$$\text{(1)} \qquad \int \mu(dt) \int f(t, r, r_P)\, \hat{\sigma}(t)\, (d(r, r_P))$$
$$= \int \mu(dt) \int \sigma(t)\, (dr) \int f(t, r, r_P)\, \pi(t, r)\, (dr_P)$$
$$[f \in L^1(\mu, C(R \times R_P))].$$

Conversely, for each $\sigma \in \mathscr{S}^{\#}$ and $\pi \in \mathscr{P}^{\#}$ there exists a unique $\hat{\sigma} \in \hat{\mathscr{S}}^{\#}$ satisfying relation (1).

▌PROOF *Step* 1 Let $\hat{\sigma} \in \hat{\mathscr{S}}^{\#}$ and $\sigma(t)(E) = \hat{\sigma}(t)(E \times R_P)$ $[t \in T, E \in \Sigma_{\text{Borel}}(R)]$. Then it is easily verified that $\sigma(t) \in \text{rpm}(R)$ and $\sigma(t)(\bar{R}^{\#}(t)) = 1$ μ-a.e. We shall now show that $\sigma \in \mathscr{S}^{\#}$ and

$$\text{(2)} \quad \int c(r)\, \sigma(t)\, (dr) = \int c(r)\, \hat{\sigma}(t)\, (d(r, r_P)) \qquad [\mu\text{-a.a. } t \in T, \quad c \in C(R)].$$

Indeed, for a fixed $t \in T$, let

$$\text{(3)} \qquad m(c) \triangleq \int c(r)\, \hat{\sigma}(t)\, (d(r, r_P)) \qquad [c \in C(R)].$$

Then $m \in C(R)^*$ and therefore (I.5.8) there exists $\sigma'(t) \in \text{frm}(R)$ such that

$$(4) \qquad m(c) = \int c(r)\, \sigma'(t)\,(dr) \qquad [c \in C(R)];$$

and we easily verify that $\sigma'(t) \in \text{rpm}(R)$. If we choose a closed $F \subset R$ and define c_j^F for $j \in \mathbb{N}$ as a continuous function on R to $[0, 1]$ that equals 1 on F and vanishes for $d[r, F] \geqslant 1/j$, then (3), (4), and the dominated convergence theorem I.4.35 imply that

$$\lim_j m(c_j^F) = \hat{\sigma}(t)(F \times R_P) = \sigma'(t)(F);$$

hence $\sigma(t)$ and $\sigma'(t)$ coincide on closed sets and, being regular, coincide on all of $\Sigma_{\text{Borel}}(R)$. Thus (3) and (4) imply (2), and it follows from (2) and IV.1.6 that $\sigma : T \to (\text{rpm}(R), |\cdot|_w)$ is μ-measurable. Thus $\sigma \in \mathscr{S}^\#$.

Step 2 Now let ζ be defined as in X.1.1, n_ζ denote the norm of $L^1(\zeta, C(R_P))$, and $f \in L^1(\mu, C(R \times R_P))$. Then, by X.1.1, X.1.2, and (2), $f \in L^1(\zeta, C(R_P))$ and

$$\begin{aligned}
&\left| \int \mu(dt) \int f(t, r, r_P)\, \hat{\sigma}(t)\,(d(r, r_P)) \right| \\
&\qquad \leqslant \int \mu(dt) \int |f(t, r, \cdot)|_{\sup}\, \hat{\sigma}(t)\,(d(r, r_P)) \\
&\qquad = \int \mu(dt) \int |f(t, r, \cdot)|_{\sup}\, \sigma(t)\,(dr) \\
&\qquad = \int |f(t, r, \cdot)|_{\sup}\, \zeta(d(t, r)) = n_\zeta(f).
\end{aligned}$$

Thus $f \to \int \mu(dt) \int f(t, r, r_P)\, \hat{\sigma}(t)(d(r, r_P))$ is a continuous linear functional on the normed vector space $(L^1(\mu, C(R \times R_P)), n_\zeta)$ (with two elements identified if they are identified in $L^1(\zeta, C(R_P))$) and, by the Hahn–Banach theorem I.3.8, can be extended to a continuous linear functional on $L^1(\zeta, C(R_P))$. The proof of Theorem IV.1.8 remains valid when $L^1(\mu, C(R))$ is replaced by $L^1(\zeta, C(R_P))$ (since the only properties of μ that are used in IV.1.8 are those of a positive Radon measure in a compact metric space). Therefore there exists a ζ-measurable function $\pi : T \times R \to (\text{frm}(R_P), |\cdot|_w)$ such that

$$\begin{aligned}
(5) \qquad &\int \mu(dt) \int f(t, r, r_P)\, \hat{\sigma}(t)\,(d(r, r_P)) \\
&\qquad = \int \zeta(d(t, r)) \int f(t, r, r_P)\, \pi(t, r)\,(dr_P) \qquad [f \in L^1(\mu, C(R \times R_p))],
\end{aligned}$$

and we easily verify that $\pi(t, r) \in \text{rpm}(R_P)$ and $\pi(t, r)(\bar{R}_P{}^{\#}(t)) = 1$ for ζ-a.a. $(t, r) \in T \times R$. By modifying π on a $\Sigma_{\text{Borel}}(T \times R)$-measurable set of ζ-measure 0, we can replace π by a $\Sigma_{\text{Borel}}(T \times R)$-measurable function such that $\pi(t, r) \in \text{rpm}(R_P)$ and

$$\pi(t, r)({}_P\bar{R}^{\#}(t)) = 1 \qquad \text{for all } (t, r) \in T \times R.$$

Thus $\pi \in \mathscr{P}^{\#}$, $(t, r) \to \int f(t, r, r_P)\, \pi(t, r)\,(dr_P)$ is ζ-integrable, and relation (1) follows from (5) and X.1.1.

Finally, if $\sigma \in \mathscr{S}^{\#}$ and $\pi \in \mathscr{P}^{\#}$, then the function

$$f \to \int \mu(dt) \int \sigma(t)\,(dr) \int f(t, r, r_P)\, \pi(t, r)\,(dr_P)$$

is a continuous linear functional on $L^1(\mu, C(R \times R_P))$ and therefore, by IV.1.8, there exists a unique μ-measurable $\hat{\sigma} : T \to (\text{frm}(R \times R_P), |\cdot|_w)$ satisfying relation (1), and it is easy to verify that $\hat{\sigma}(t) \in \text{rpm}(R \times R_P)$ and $\hat{\sigma}(t)(\overline{\hat{R}^{\#}(t)}) = 1$ for μ-a.a. $t \in T$; hence $\hat{\sigma} \in \hat{\mathscr{S}}^{\#}$. QED

X.1.4 ***Lemma*** Let $f \in L^1(\mu, C(R \times R_P))$,

$$h(t, r) \triangleq \min_{r_P \in \bar{R}_P{}^{\#}(t)} f(t, r, r_P) \qquad (t \in T, \quad r \in R)$$

and $\sigma_j \in \mathscr{S}^{\#}$ $(j = 0, 1, 2, \ldots)$. Then $h \in L^1(\mu, C(R))$ and the relation

$$\tilde{\sigma}(t)(E) \triangleq \sum_{j=0}^{\infty} 2^{-j-1} \sigma_j(t)(E) \qquad [t \in T, \quad E \in \Sigma_{\text{Borel}}(R)]$$

defines an element $\tilde{\sigma} \in \mathscr{S}^{\#}$. If $\tilde{\zeta}$ is defined as in X.1.1 in terms of $\tilde{\sigma}$, then there exists $\bar{\pi} \in \mathscr{P}^{\#}$ such that

$$h(t, r) = \int f(t, r, r_P)\, \bar{\pi}(t, r)\,(dr_P) \quad \tilde{\zeta}\text{-a.e.}$$

▌PROOF We first observe that the function h is defined because $f(t, r, \cdot)$ is continuous and $\bar{R}_P{}^{\#}(t)$ compact for all $t \in T$ and $r \in R$. Since $R_P{}^{\#}$ satisfies Condition IV.3.1, it follows from I.5.26 that for every $\epsilon > 0$ there exists a compact $F_\epsilon \subset T$ such that $\mu(T \sim F_\epsilon) \leqslant \epsilon$ and the functions $f \,|\, F_\epsilon \times R \times R_P$ and $\bar{R}_P{}^{\#} \,|\, F_\epsilon$ are continuous. It is then easy to show that $h \,|\, F_\epsilon \times R$ is continuous. Since ϵ is arbitrary, we conclude that $h(\cdot, r)$ is μ-measurable and $h(t, \cdot)$ is continuous for all $r \in R$ and μ-a.a. $t \in T$, and since $|h(t, \cdot)|_{\text{sup}} \leqslant |f(t, \cdot, \cdot)|_{\text{sup}}$ $(t \in T)$, we have (I.5.25) $h \in L^1(\mu, C(R))$.

Next we observe that, by I.4.9, $\tilde{\sigma}(t) \in \text{rpm}(R)$ μ-a.e. and it follows easily that $\tilde{\sigma} \in \mathscr{S}^{\#}$. We shall complete the proof of the lemma by exhibiting a $\Sigma_{\text{Borel}}(T \times R)$-measurable function $p : T \times R \to R_P$ such that $p(t, r)$ minimizes $f(t, r, \cdot)$ on $\bar{R}_P^{\#}(t)$ for $\tilde{\zeta}$-a.a. $(t, r) \in T \times R$. If we then define $\bar{\pi}(t, r)$ as the Dirac measure at $p(t, r)$, then $\bar{\pi} : T \times R \to (\text{rpm}(R_P), |\cdot|_w)$ is $\Sigma_{\text{Borel}}(T \times R)$-measurable and it is then clear that $\bar{\pi}$ satisfies the assertion of the lemma.

To prove the existence of such a function p, we observe that, by X.1.2, h and $f(\cdot, \cdot, r_P)$ are $\tilde{\zeta}$-measurable for each $r_P \in R_P$. It follows, by the Filippov–Castaing theorem I.7.10, that there exists a $\tilde{\zeta}$-measurable selection $p(\cdot, \cdot)$ of the $\tilde{\zeta}$-measurable mapping $(t, r) \to \bar{R}_P^{\#}(t)$ such that

$$f(t, r, p(t, r)) = h(t, r) \quad \tilde{\zeta}\text{-a.e.}$$

Finally, we may replace p by a $\Sigma_{\text{Borel}}(T \times R)$-measurable function that coincides with it $\tilde{\zeta}$-a.e. QED

▌ By X.1.3, for every $\sigma \in \mathscr{S}^{\#}$ and $\pi \in \mathscr{P}^{\#}$ there exists a unique $\hat{\sigma} \in \hat{\mathscr{S}}$ satisfying X.1.3(1). We shall henceforth denote this $\hat{\sigma}$ by $\sigma \otimes \pi$. It is clear that

$$[\alpha\sigma_1 + (1 - \alpha) \sigma_2] \otimes \pi = \alpha\sigma_1 \otimes \pi + (1 - \alpha) \sigma_2 \otimes \pi$$

for all $\alpha \in [0, 1]$, $\sigma_1, \sigma_2 \in \mathscr{S}^{\#}$, and $\pi \in \mathscr{P}^{\#}$.

X.1.5 ***Lemma*** For each $\sigma \in \mathscr{S}^{\#}$, the set $\{\sigma \otimes \pi \mid \pi \in \mathscr{P}^{\#}\}$ is a compact and sequentially compact subset of $\hat{\mathscr{S}}^{\#}$. If $\lim_j \sigma_j \otimes \pi_j = \sigma_0 \otimes \pi_0$ in $\hat{\mathscr{S}}^{\#}$, then $\lim_j \sigma_j = \sigma_0$ in $\mathscr{S}^{\#}$.

▌ PROOF Let $\sigma \in \mathscr{S}^{\#}$ and let (π_j) be a sequence in $\mathscr{P}^{\#}$. Then $(\sigma \otimes \pi_j)$ is a sequence in the compact metric space $\hat{\mathscr{S}}^{\#}$, and it has a subsequence $(\sigma \otimes \pi_j)_{j \in J}$ converging to some $\hat{\sigma} \in \hat{\mathscr{S}}^{\#}$. By X.1.3, there exist $\tilde{\sigma} \in \mathscr{S}^{\#}$ and $\tilde{\pi} \in \mathscr{P}^{\#}$ such that $\hat{\sigma} = \tilde{\sigma} \otimes \tilde{\pi}$. Therefore

$$\lim_{j \in J} \int \mu(dt) \int \sigma(t)\,(dr) \int f(t, r, r_P)\, \pi_j(t, r)\,(dr_P)$$

$$= \int \mu(dt) \int \tilde{\sigma}(t)\,(dr) \int f(t, r, r_P)\, \tilde{\pi}(t, r)\,(dr_P) \qquad [f \in L^1(\mu, C(R \times R_P))].$$

If we choose arbitrary f that are independent of r_P, then this relation implies that $\sigma = \tilde{\sigma}$. Thus

$$\lim_{j \in J} \sigma \otimes \pi_j = \sigma \otimes \tilde{\pi},$$

showing that $\{\sigma \otimes \pi \mid \pi \in \mathscr{P}^{\#}\}$ is a sequentially compact subset of the metric space $\hat{\mathscr{S}}^{\#}$, and therefore compact.

Now assume that $\lim_j \sigma_j \otimes \pi_j = \sigma_0 \otimes \pi_0$ in $\hat{\mathscr{S}}^\#$. Then

$$\lim_j \int \mu(dt) \int \sigma_j(t)(dr) \int f(t, r, r_P)\, \pi_j(t, r)(dr_P)$$

$$= \int \mu(dt) \int \sigma_0(t)(dr) \int f(t, r, r_P)\, \pi_0(t, r)(dr_P) \qquad [f \in L^1(\mu, C(R \times R_P))].$$

If we choose f that are independent of r_P, then we can conclude that $\lim_j \sigma_j = \sigma_0$ in $\mathscr{S}^\#$. QED

▌ *Remark* Lemma X.1.5 shows that if $(\sigma \otimes \pi_j)$ converges to some $\hat{\sigma}$ in $\hat{\mathscr{S}}^\#$, then $\hat{\sigma} = \sigma \otimes \tilde{\pi}$ for some $\tilde{\pi} \in \mathscr{P}^\#$. This raises the question whether an analogous result is valid if we associate a fixed $\pi \in \mathscr{P}^\#$ with a sequence (σ_j) in $\mathscr{S}^\#$; that is, if $(\sigma_j \otimes \pi)$ converges to some $\hat{\sigma}$ in $\hat{\mathscr{S}}^\#$, is it true that $\hat{\sigma} = \tilde{\sigma} \otimes \pi$ for some $\tilde{\sigma} \in \mathscr{S}^\#$? The answer to this question is in the negative as is shown by the following counterexample. Let $T = R = R_P = [0, 1]$, μ be the Borel measure in $[0, 1]$, δ_s represent the Dirac measure at s, and for all $t \in T$ and $j \in \mathbb{N}$, let

$$\bar{\pi}(t, r) = \delta_0 \quad \text{if} \quad r > 0, \qquad \bar{\pi}(t, 0) = \delta_1,$$

and

$$\sigma_j(t) = \frac{1}{j}\,\delta_0 + \frac{j-1}{j}\,\delta_{1/j}.$$

Then $\sigma_j \in \mathscr{S}^\#$, $\bar{\pi} \in \mathscr{P}^\#$,

$$\sigma_j \otimes \bar{\pi}(t)(\{(0, 1)\}) = \frac{1}{j}, \qquad \sigma_j \otimes \bar{\pi}(t)\left(\left\{\left(\frac{1}{j}, 0\right)\right\}\right) = \frac{j-1}{j},$$

and $\lim_j \sigma_j \otimes \bar{\pi} = \hat{\sigma}$, with $\hat{\sigma}(t) = \delta_{(0,0)}$. It follows that $\hat{\sigma} = \tilde{\sigma} \otimes \tilde{\pi}$ if and only if $\tilde{\sigma}(t) = \delta_0$ and $\tilde{\pi}(t, 0) = \delta_0 \neq \delta_1 = \bar{\pi}(t, 0)$.

This example also shows that even for simply defined functions x_2 independent of control parameters we cannot assert that $\sigma \to x_2(\sigma, \pi) : \mathscr{S}^\# \to \mathbb{R}^{m_2}$ is continuous for each $\pi \in \mathscr{P}^\#$. Indeed, let T, R, R_P, $\bar{\pi}$, and σ_j be defined as above, and let

$$x_2(\sigma, \pi) \triangleq \int_0^1 dt \int \sigma(t)(dr) \int (r + 1)\, r_P\, \pi(t, r)(dr_P) \qquad (\sigma \in \mathscr{S}^\#, \quad \pi \in \mathscr{P}^\#).$$

Then $\lim_j \sigma_j = \delta_0$ and $\lim_j x_2(\sigma_j, \bar{\pi}) = 0$ while $x_2(\delta_0, \bar{\pi}) = 1$.

X.1.6 ***Lemma*** Let $\sigma_j \in \mathscr{S}^\#$ $(j = 0, 1, 2, \ldots)$, $\pi_0 \in \mathscr{P}^\#$, and $\lim_j \sigma_j = \sigma_0$ in $\mathscr{S}^\#$. Then there exists a sequence $(\tilde{\pi}_j)$ in $\mathscr{P}^\#$ such that

$$\lim_j \sigma_j \otimes \tilde{\pi}_j = \sigma_0 \otimes \pi_0 \quad \text{in} \quad \hat{\mathscr{S}}^\#.$$

▌PROOF *Step* 1 Let $n \in \mathbb{N}$ and $\phi \in L^1(\mu, C(R \times R_P, \mathbb{R}^n))$. We set

$$\Phi(\sigma, \pi) \triangleq \int \mu(dt) \int \phi(t, r, r_P)\, \sigma \otimes \pi(t)(d(r, r_P)) \qquad (\sigma \in \mathscr{S}^\#, \quad \pi \in \mathscr{P}^\#)$$

and

$$K_j \triangleq \{\Phi(\sigma_j, \pi) \mid \pi \in \mathscr{P}^\#\} \qquad (j = 0, 1, 2,...).$$

We first show that for every $\bar{\lambda} \in \mathbb{R}^n$ there exist $x_j \in K_j$ $(j = 0, 1, 2,...)$ such that

$$(1) \quad \lim_j \bar{\lambda} \cdot x_j = \bar{\lambda} \cdot x_0 \qquad \text{and} \qquad \bar{\lambda} \cdot x_j = \min_{x \in K_j} \bar{\lambda} \cdot x \qquad (j = 0, 1, 2,...).$$

Indeed, if we set $f \triangleq \bar{\lambda} \cdot \phi$, let h, $\tilde{\sigma}$, $\tilde{\zeta}$ and $\bar{\pi}$ be defined as in X.1.4, and let

$$x_j \triangleq \Phi(\sigma_j, \bar{\pi}) \qquad (j = 0, 1, 2,...),$$

then, by X.1.4,

$$\bar{\lambda} \cdot x_j = \int \mu(dt) \int \sigma_j(t)\,(dr) \int f(t, r, r_P)\, \bar{\pi}(t, r)\,(dr_P)$$

$$= \int \mu(dt) \int h(t, r)\, \sigma_j(t)\,(dr) \qquad (j = 0, 1, 2,...),$$

and, since $h \in L^1(\mu, C(R))$ and $\lim_j \sigma_j = \sigma_0$ in $\mathscr{S}^\#$, we have $\lim_j \bar{\lambda} \cdot x_j = \bar{\lambda} \cdot x_0$. The second relation in (1) is valid because

$$h(t, r) = \int f(t, r, r_P)\, \bar{\pi}(t, r)\,(dr_P) \leqslant \int f(t, r, r_P)\, \pi(t, r)\,(dr_P) \qquad (\pi \in \mathscr{P}^\#)$$

for $\tilde{\zeta}$-a.a. $(t, r) \in T \times R$ and therefore (X.1.1) also for μ-a.a. $t \in T$ and $\sigma_j(t)$-a.a. $r \in R$.

Step 2 By X.1.5, the set $\{\sigma \otimes \pi \mid \pi \in \mathscr{P}^\#\}$ is sequentially compact in $\hat{\mathscr{S}}^\#$ for each $\sigma \in \mathscr{S}^\#$, and therefore each of the sets K_j $(j = 0, 1, 2,...)$ is compact. Furthermore, since $\mathscr{P}^\#$ is convex, so is every K_j.

We shall next prove that for every $\bar{x} \in K_0$ there exist $x_j \in K_j$ $(j \in \mathbb{N})$ such that

$$\lim_j x_j = \bar{x}.$$

Indeed, assume the contrary. Then there exist $J \subset (1, 2,...)$ and $\epsilon > 0$ such that $| x - \bar{x} |_2 \geqslant \epsilon$ for all $j \in J$ and $x \in K_j$. Let $j \in J$, and let a_j be a point in the compact and convex set K_j that minimizes $x \to | x - \bar{x} |_2^2 = (x - \bar{x}) \cdot (x - \bar{x})$. Then, for every $x \in K_j$, we have

$$| a_j + \theta(x - a_j) - \bar{x} |_2^2 \geqslant | a_j - \bar{x} |_2^2 \qquad (\theta \in [0, 1]);$$

hence

$$\theta^2 | x - a_j |_2^2 + 2\theta(x - a_j) \cdot (a_j - \bar{x}) \geqslant 0 \qquad (\theta \in [0, 1]),$$

implying that $(x - a_j) \cdot (a_j - \bar{x}) \geqslant 0$ $(x \in K_j)$. It follows, setting $\lambda_j \triangleq |a_j - \bar{x}|_2^{-1} (a_j - \bar{x})$, that

$$(2) \quad \lambda_j \cdot (x - \bar{x}) \geqslant \lambda_j \cdot (a_j - \bar{x}) = |a_j - \bar{x}|_2 \geqslant \epsilon \qquad (j \in J, \quad x \in K_j).$$

We may assume that $(\lambda_j)_{j \in J}$ converges to some $\bar{\lambda} \neq 0$, otherwise replacing J by a subsequence. By Step 1, there exist $x_j \in K_j$ $(j = 0, 1, 2,...)$ such that

$$(3) \qquad \lim_{j \in J} \bar{\lambda} \cdot x_j = \bar{\lambda} \cdot x_0 = \min_{x \in K_0} \bar{\lambda} \cdot x \leqslant \bar{\lambda} \cdot \bar{x}.$$

Since all K_j are contained in the closed ball of radius $|\phi|$ with center $0 \in \mathbb{R}^n$, it follows that

$$\lim_{j \in J} (\lambda_j - \bar{\lambda}) \cdot (x_j - \bar{x}) = 0;$$

hence, by (2) and (3),

$$(4) \qquad \epsilon \leqslant \lambda_j \cdot (x_j - \bar{x}) = \bar{\lambda} \cdot (x_j - \bar{x}) + (\lambda_j - \bar{\lambda}) \cdot (x_j - \bar{x})$$
$$\leqslant \epsilon/2 \qquad \text{for all sufficiently large } j \in J,$$

a contradiction.

Step 3 If we set $\bar{x} = \Phi(\sigma_0, \pi_0)$, then it follows from Step 2 that there exist $\pi_j \in \mathscr{P}^\#$ $(j \in \mathbb{N})$ such that $\lim_j \Phi(\sigma_j, \pi_j) = \Phi(\sigma_0, \pi_0)$. Now let $\{f_1, f_2, ...\}$ be a dense subset of $L^1(\mu, C(R \times R_P))$. For each $n \in \mathbb{N}$ we set $\phi \triangleq (f_1, f_2, ..., f_n)$ and denote by $(\pi_j{}^n)_j$ the corresponding sequence in $\mathscr{P}^\#$ such that

$$\lim_j \Phi(\sigma_j, \pi_j{}^n) = \Phi(\sigma_0, \pi_0).$$

Then we may choose a sufficiently large $j_n \in \mathbb{N}$ so that

$$|\Phi(\sigma_{j_n}, \pi_{j_n}^n) - \Phi(\sigma_0, \pi_0)| \leqslant 1/n.$$

We set $J \triangleq (j_1, j_2, ...)$ and $\tilde{\pi}_{j_n} \triangleq \pi_{j_n}^n$ $(n \in \mathbb{N})$. Then we have

$$(5) \quad \lim_{j \in J} \int \mu(dt) \int \sigma_j(t)(dr) \int f_i(t, r, r_P)\, \tilde{\pi}_j(t, r)(dr_P)$$
$$= \int \mu(dt) \int \sigma_0(t)(dr) \int f_i(t, r, r_P)\, \pi_0(t, r)(dr_P) \qquad (i \in \mathbb{N}).$$

Since every $f \in L^1(\mu, C(R \times R_P))$ can be approximated by a sequence in $\{f_1, f_2, ...\}$, we conclude that relation (5) remains valid with f replacing f_i. Therefore

$$(6) \qquad \lim_{j \in J} \sigma_j \otimes \tilde{\pi}_j = \sigma_0 \otimes \pi_0 \quad \text{in } \hat{\mathscr{S}}^\#.$$

Finally, assume that the conclusion of the lemma is not valid. Then there exist $\epsilon > 0$ and $J_1 \subset (1, 2,...)$ such that the distance in $\hat{\mathscr{S}}^{\#}$ from $\sigma_0 \otimes \pi_0$ to $\sigma_j \otimes \pi$ is at least ϵ for all $j \in J_1$ and $\pi \in \mathscr{P}^{\#}$. This is contradicted by relation (6) which must hold for some $J \subset J_1$ and a corresponding sequence $(\tilde{\pi}_j)_{j \in J}$.

QED

X.1.7 ***Theorem*** Let $\mathscr{U}$ and $\mathscr{U}_P$ be abundant subsets of $\mathscr{R}^{\#}$ and $\mathscr{R}_P{}^{\#}$, respectively, and

$$\lim_j x_2(\sigma_j, b, \pi_j, b_P) = x_2(\bar{\sigma}, b, \bar{\pi}, b_P) \qquad (b \in B, \quad b_P \in B_P)$$

whenever $\lim_j \sigma_j \otimes \pi_j = \bar{\sigma} \otimes \bar{\pi}$ in $\hat{\mathscr{S}}^{\#}$. Let $\mathscr{S}_P{}^{\#}$ be defined exactly as $\mathscr{S}^{\#}$ but with R, $R^{\#}$ replaced by R_P, $R_P{}^{\#}$, respectively, and let $\mathscr{S}_P{}^{\#}$ be embedded in $\mathscr{P}^{\#}$ by identifying $\sigma_P \in \mathscr{S}_P{}^{\#}$ with the hyperrelaxed adverse control function $(t, r) \to \sigma_P{}'(t)$, where $\sigma_P{}'$ is any Σ-measurable function μ-equivalent to σ_P. Then a sequence $((u_j, b_j))$ in $\mathscr{U} \times B$ is an approximate $\mathscr{U}$, $\mathscr{U}_P$-control if and only if

$$\lim_j x_1(u_j, b_j) = 0$$

and

$$\lim_j [\sup\{d[A, x_2(u_j, b_j, \pi, b_P)] \mid \pi \in \mathscr{P}^{\#}, b_P \in B_P\}] = 0$$

or

$$\lim_j x_1(u_j, b_j) = 0$$

and

$$\lim_j [\sup\{d[A, x_2(u_j, b_j, \sigma_P, b_P)] \mid \sigma_P \in \mathscr{S}_P{}^{\#}, b_P \in B_P\}] = 0.$$

PROOF For each $u \in \mathscr{U}$ we may select a μ-equivalent Σ-measurable function that is identified with u in $\mathscr{R}^{\#}$. Thus u may be assumed Σ-measurable and it is identified with the element $t \to \delta_{u(t)}$ of $\mathscr{S}^{\#}$, where δ_r is the Dirac measure at r. We have

$$(1) \qquad \int \mu(dt) \int \delta_{u(t)}(dr) \int f(t, r, r_P)\, \pi(t, r)\, (dr_P)$$

$$= \int \mu(dt) \int f(t, u(t), r_P)\, \pi(t, u(t))\, (dr_P)$$

$$[f \in L^1(\mu, C(R \times R_P)), \quad \pi \in \mathscr{P}^{\#}],$$

and therefore $u \otimes \pi = u \otimes \sigma_P$, where $\sigma_P(t) = \pi(t, u(t))$ μ-a.e. and σ_P is identified with the hyperrelaxed control function $(t, r) \to \sigma_P(t)$. If we choose in (1) f independent of t and r, then it follows from IV.1.6 that $\sigma_P : T \to (\mathrm{rpm}(R_P), |\cdot|_w)$ is μ-measurable, and we have $\sigma_P(t)(\bar{R}_P{}^{\#}(t)) = 1$ μ-a.e. Therefore σ_P is an element of the set $\mathscr{S}_P{}^{\#}$ of adverse relaxed controls that is defined exactly as $\mathscr{S}^{\#}$ but with R, $R^{\#}$ replaced by R_P, $R_P{}^{\#}$, respectively. It follows that for each $u \in \mathscr{U}$ we have

$$\{u \otimes \pi \mid \pi \in \mathscr{P}^{\#}\} = \{u \otimes \sigma_P \mid \sigma_P \in \mathscr{S}_P{}^{\#}\}. \tag{2}$$

Now let $\sigma_P \in \mathscr{S}_P{}^{\#}$, and let (u_{Pj}) be a sequence in the dense subset $\mathscr{U}_P$ of $\mathscr{S}_P{}^{\#}$ (IV.3.10) converging to σ_P. Then

$$\lim_j \int f(t, u(t), u_{Pj}(t))\, \mu(dt) = \int \mu(dt) \int f(t, u(t), r_P)\, \sigma_P(t)\,(dr_P)$$
$$[f \in L^1(\mu, C(R \times R_P))];$$

hence

$$\lim_j u \otimes u_{Pj} = u \otimes \sigma_P .$$

In view of (2) and our assumptions about x_2, this implies that, for each $u \in \mathscr{U}$ and $b \in B$, we have

$$\begin{aligned}
&\{x_2(u, b, \pi, b_P) \mid \pi \in \mathscr{P}^{\#}, b_P \in B\} \\
&\quad = \{x_2(u, b, \sigma_P, b_P) \mid \sigma_P \in \mathscr{S}_P{}^{\#}, b_P \in B_P\} \\
&\quad = \text{closure of } \{x_2(u, b, u_P, b_P) \mid u_P \in \mathscr{U}_P, b_P \in B_P\}.
\end{aligned}$$

Our conclusions now follow directly. QED

X.1.8 ***Theorem*** Let B and B_P have sequentially compact topologies, $(x_0, x_1) : \mathscr{S}^{\#} \times B \to \mathbb{R} \times \mathbb{R}^m$ be sequentially continuous, A closed, $\mathscr{A}(\mathscr{S}^{\#}) \neq \varnothing$, and

$$\lim_j x_2(\sigma_j, b_j, \pi_j, b_{Pj}) = x_2(\bar{\sigma}, \bar{b}, \bar{\pi}, \bar{b}_P)$$

whenever $\lim_j \sigma_j \otimes \pi_j = \bar{\sigma} \otimes \bar{\pi}$ in $\hat{\mathscr{S}}^{\#}$, $\lim_j b_j = \bar{b}$ in B, and $\lim_j b_{Pj} = \bar{b}_P$ in B_P. Then there exists a minimizing relaxed control $(\bar{\sigma}, \bar{b})$. Furthermore, for every choice of abundant subsets $\mathscr{U}$ of $\mathscr{R}^{\#}$ and $\mathscr{U}_P$ of $\mathscr{R}_P{}^{\#}$ there exists a minimizing approximate $\mathscr{U}, \mathscr{U}_P$-control $((\bar{u}_j, \bar{b}))$ which can be determined by selecting any sequence $(\bar{u}_j)$ in $\mathscr{U}$ converging to $\bar{\sigma}$.

PROOF *Step* 1 Since $\mathscr{A}(\mathscr{S}^{\#}) \neq \varnothing$ there exists a sequence $((\sigma_j, b_j))$ in $\mathscr{A}(\mathscr{S}^{\#})$ such that

$$\lim_j x_0(\sigma_j, b_j) = \inf x_0(\mathscr{A}(\mathscr{S}^{\#})).$$

Since $\mathscr{S}^{\#}$ and B are sequentially compact, we may assume that

$$\lim_j (\sigma_j, b_j) = (\bar{\sigma}, \bar{b}) \qquad \text{in } \mathscr{S}^{\#} \times B$$

and therefore

$$x_0(\bar{\sigma}, \bar{b}) = \lim_j x_0(\sigma_j, b_j) \qquad \text{and} \qquad x_1(\bar{\sigma}, \bar{b}) = \lim_j x_1(\sigma_j, b_j) = 0.$$

Now let $\pi \in \mathscr{P}^{\#}$ and $b_P \in B_P$. Then, by X.1.6, there exists a sequence $(\tilde{\pi}_j)$ in $\mathscr{P}^{\#}$ such that $\lim_j \sigma_j \otimes \tilde{\pi}_j = \bar{\sigma} \otimes \pi$. It follows that

$$x_2(\bar{\sigma}, \bar{b}, \pi, b_P) = \lim_j x_2(\sigma_j, b_j, \tilde{\pi}_j, b_P) \in A.$$

This shows that $(\bar{\sigma}, \bar{b}) \in \mathscr{A}(\mathscr{S}^{\#})$ and therefore $(\bar{\sigma}, \bar{b})$ is a minimizing relaxed control.

Step 2 Now let $\mathscr{U}$ and $\mathscr{U}_P$ be abundant subsets of $\mathscr{R}^{\#}$ and $\mathscr{R}_P{}^{\#}$, respectively, and let $(\bar{u}_j)$ be a sequence in the dense subset $\mathscr{U}$ of $\mathscr{S}^{\#}$ (IV.3.10) that converges to $\bar{\sigma}$ in $\mathscr{S}^{\#}$. Then we shall show that $((\bar{u}_j, \bar{b}))$ is an approximate $\mathscr{U}, \mathscr{U}_P$-control. Since $x_1(\bar{\sigma}, \bar{b}) = \lim_j x_1(\bar{u}_j, \bar{b}) = 0$, this conclusion will follow from X.1.7 if we prove that $\lim_j e_j = 0$, where

$$e_j \triangleq \sup\{d[A, x_2(\bar{u}_j, \bar{b}, \pi, b_P)] \mid \pi \in \mathscr{P}^{\#}, b_P \in B_P\} \qquad (j \in \mathbb{N}).$$

Indeed, by X.1.5, the set $\{\bar{u}_j \otimes \pi \mid \pi \in \mathscr{P}^{\#}\}$ is, for each $j \in \mathbb{N}$, a sequentially compact subset of $\hat{\mathscr{S}}^{\#}$ and there exists therefore a point $(\pi_j, b_{Pj}) \in \mathscr{P}^{\#} \times B_P$ such that

$$d[A, x_2(\bar{u}_j, \bar{b}, \pi_j, b_{Pj})] = e_j.$$

If (e_j) does not converge to 0, then there exists $J \subset (1, 2, \ldots)$ and $\epsilon > 0$ such that

$$e_j \geqslant \epsilon \qquad (j \in J). \tag{1}$$

We may assume that J was chosen so that the sequence $(b_{Pj})_{j \in J}$ converges in B_P to some $\bar{b}_P$ and the sequence $(\bar{u}_j \otimes \pi_j)_{j \in J}$ converges in $\hat{\mathscr{S}}^{\#}$ to some limit which, by X.1.3 and X.1.5, is of the form $\bar{\sigma} \otimes \bar{\pi}$. Then

$$\lim_{j \in J} x_2(\bar{u}_j, \bar{b}, \pi_j, b_{Pj}) = x_2(\bar{\sigma}, \bar{b}, \bar{\pi}, \bar{b}_P) \in A;$$

hence $\lim_{j \in J} e_j = 0$, contradicting (1).

Step 3 It remains to show that $((\bar{u}_j, \bar{b}))$ is a minimizing approximate $\mathscr{U}, \mathscr{U}_P$-control. Let $((u_j, b_j))$ be any other approximate $\mathscr{U}, \mathscr{U}_P$-control, that is,

$$\lim_j x_1(u_j, b_j) = 0$$

and

$$\lim_j [\sup\{d[x_2(u_j, b_j, u_P, b_P), A] \mid u_P \in \mathscr{U}_P, b_P \in B_P\}] = 0.$$

We observe that, by X.1.7,

$$\lim_j [\sup\{d[x_2(u_j, b_j, \pi, b_P), A] \mid \pi \in \mathscr{P}^\#, b_P \in B_P\}] = 0.$$

Now let $J \subset (1, 2, \ldots)$ be such that $\lim_{j\in J} x_0(u_j, b_j) = \liminf_j x_0(u_j, b_j)$ and the sequence $((u_j, b_j))_{j\in J}$ converges to some limit $(\tilde{\sigma}, \tilde{b})$ in $\mathscr{S}^\# \times B$. Then the argument of Step 1 shows that $(\tilde{\sigma}, \tilde{b}) \in \mathscr{A}(\mathscr{S}^\#)$. Since $(\bar{\sigma}, \bar{b})$ is a minimizing relaxed solution, we have

$$\lim_j x_0(\bar{u}_j, \bar{b}) = x_0(\bar{\sigma}, \bar{b}) \leqslant x_0(\tilde{\sigma}, \tilde{b}) = \lim_{j\in J} x_0(u_j, b_j) = \liminf_j x_0(u_j, b_j).$$

This shows that $((\bar{u}_j, \bar{b}))$ is a minimizing approximate $\mathscr{U}$, $\mathscr{U}_P$-control. QED

X.2 Necessary Conditions for a Relaxed Minimum

General Remarks

It is easily seen that for every choice of a Banach space $\mathscr{X}$ the definition (in I.5) of the Banach space $(BF(S, \mathscr{X}), |\cdot|_{\sup})$ extends immediately to the case where S is an arbitrary set (and not necessarily a topological space). We derive necessary conditions for a relaxed minimum by applying Theorem V.2.3, with $\mathscr{X}_2$ defined as the Banach space

$$(BF(\mathscr{P}^\# \times B_P, \mathbb{R}^{m_2}), |\cdot|_{\sup}),$$

$$C_2 = \{c \in \mathscr{X}_2 \mid c(\mathscr{P}^\# \times B_P) \subset A\}, \qquad (g_0, g_1)(y, \sigma, b) = (x_0, x_1)(\sigma, b),$$

$g_2(y, \sigma, b)$ defined as the function $(\pi, b_P) \to x_2(\sigma, b, \pi, b_P)$, and $\mathscr{Y}$ and F immaterial. These necessary conditions are based on assumptions that are in fact satisfied in most applications. Nevertheless, these necessary conditions suffer from a basic disadvantage: they involve an element $l_2 \in BF(\mathscr{P}^\# \times B_P, \mathbb{R}^{m_2})^*$, and so long as the set $\mathscr{P}^\#$ remains without any topological structure and l_2 is not restricted to a more "manageable" set of functions than arbitrary bounded functions, we cannot expect l_2 to exhibit many computationally useful properties.

With these considerations in mind and with the results of Section X.1 available, it would be natural to seek to define a compact metric topology in $\mathscr{P}^\#$ that (under realistic assumptions) renders the function

$$(\pi, b_P) \to x_2(\sigma, b, \pi, b_P) : \mathscr{P}^\# \times B_P \to \mathbb{R}^{m_2}$$

continuous for every $(\sigma, b) \in \mathscr{S}^\# \times B$. If this were accomplished and B_P was assumed to be a compact metric space, then l_2, when restricted to

$C(\mathscr{P}^{\#} \times B_P, \mathbb{R}^{m_2})$, could be represented in the manner described in I.5.9 by a finite regular Borel measure on $\mathscr{P}^{\#} \times B_P$ and a corresponding integrable function.

It would seem at first that a way of constructing a useful compact metric topology in $\mathscr{P}^{\#}$ is indicated in Lemma X.1.5 which asserts that, for each $\sigma \in \mathscr{S}^{\#}$, the set $\{\sigma \otimes \pi \mid \pi \in \mathscr{P}^{\#}\}$ is a compact subset of the metric space $\hat{\mathscr{S}}^{\#}$. Unfortunately, such a conclusion is not valid. Lemma X.1.5 enables us to construct, for each $\sigma \in \mathscr{S}^{\#}$, a compact metric topology in a collection $\mathscr{P}_{\sigma}{}^{\#}$ of equivalence classes of elements of $\mathscr{P}^{\#}$ (with π_1 and π_2 identified in $\mathscr{P}_{\sigma}{}^{\#}$ if $\sigma \otimes \pi_1 = \sigma \otimes \pi_2$); however, the collection $\mathscr{P}_{\sigma}{}^{\#}$ varies with σ and the same two elements π_1 and π_2 may be either identified or belong to different equivalence classes for different choices of σ. Furthermore, a sequence (π_j) that converges to a point $\bar{\pi}$ in $\mathscr{P}_{\sigma}{}^{\#}$ (in the sense that $\lim_j \sigma \otimes \pi_j = \sigma \otimes \bar{\pi}$) will generally not converge in $\mathscr{P}^{\#}_{\sigma'}$ for $\sigma' \neq \sigma$.

Since this attempt to define a "useful" compact metric topology in $\mathscr{P}^{\#}$ fails and no alternatives present themselves for consideration, we make do with less. For each minimizing relaxed control $(\bar{\sigma}, \bar{b})$ we select an appropriate denumerable subset $\{\sigma_1, \sigma_2, ...\}$ of $\mathscr{S}^{\#}$ and define a set $\tilde{\mathscr{P}}$ of equivalence classes in $\mathscr{P}^{\#}$ and a compact metric topology in $\tilde{\mathscr{P}}$ such that $\lim_j \pi_j = \bar{\pi}$ in $\tilde{\mathscr{P}}$ implies $\lim_j \sigma \otimes \pi_j = \sigma \otimes \bar{\pi}$ for $\sigma = \bar{\sigma}, \sigma_1, \sigma_2, ...$ By considering the functions

$$(\pi, b_P) \to x_2(\sigma, b, \pi, b_P) \quad \text{and} \quad (\pi, b_P) \to D_1x_2(\bar{\sigma}, \bar{b}), (\pi, b_P); (\sigma, b) - (\bar{\sigma}, \bar{b}))$$

for choices of $\sigma = \bar{\sigma}, \sigma_1, \sigma_2, ...$ only, we derive necessary conditions for a relaxed minimum which we shall later transform into the "pointwise" necessary conditions X.3.5 [and, in particular X.3.5(2)] in the special case of problems defined by ordinary differential equations. Similar "pointwise" conditions (analogous to those in VI.2.3, VII.3.2, and VIII.2.4) could be derived under appropriate conditions in the case where the functions x_0, x_1, and x_2 are defined by more general functional-integral equations.

X.2.1 ***Definition*** *The Space $\tilde{\mathscr{P}}$, the Element $\tilde{\sigma} \in \mathscr{S}^{\#}$, and the Measure $\tilde{\zeta}$* Let $\sigma_j \in \mathscr{S}^{\#}$ $(j = 0, 1, 2, ...)$. We have shown in X.1.4 that the relation

$$\tilde{\sigma}(t)(E) \triangleq \sum_{j=0}^{\infty} 2^{-j-1}\sigma_j(t)(E) \qquad [t \in T, \quad E \in \Sigma_{\text{Borel}}(R)]$$

defines an element $\tilde{\sigma} \in \mathscr{S}^{\#}$. In turn, if follows from X.1.1 that $\tilde{\sigma}$ determines a measure $\tilde{\zeta} \in \text{frm}^+(T \times R)$ such that

$$\int \mu(dt) \int h(t, r)\, \tilde{\sigma}(t)\,(dr) = \int h(t, r)\, \tilde{\zeta}(d(t, r)) \qquad [h \in L^1(\tilde{\zeta})].$$

We can verify that $\tilde{\zeta}$ is nonatomic. Indeed, let $E \in \Sigma_{\text{Borel}}(T \times R)$ and $\tilde{\zeta}(E) > 0$. By I.4.10, there exists a function $B : [0, 1] \to \Sigma$ such that $B(\alpha) \subset B(\beta)$ if $0 \leqslant \alpha \leqslant \beta \leqslant 1$, $B(0) = \varnothing$, $B(1) = T$, and $\mu(B(\alpha)) = \alpha\mu(T)$ for all $\alpha \in [0, 1]$. We set $h(\alpha) \triangleq \int_{B(\alpha)} \mu(dt) \int \chi_E(t, r)\, \tilde{\sigma}(t)(dr)$ and observe that $h : [0, 1] \to \mathbb{R}$ is continuous, $h(0) = 0$, and $h(1) = \tilde{\zeta}(E)$. It follows that $h(\bar{\alpha}) = \frac{1}{2}\tilde{\zeta}(E)$ for some $\bar{\alpha} \in (0, 1)$ and therefore $\tilde{\zeta}(E \cap [B(\bar{\alpha}) \times R]) = \frac{1}{2}\tilde{\zeta}(E)$, thus showing that $\tilde{\zeta}$ is nonatomic.

Thus the measure space $(T \times R, \Sigma_{\text{Borel}}(T \times R), \tilde{\zeta})$ has all the properties assumed for (T, Σ, μ) and $T \times R$ is a compact metric space just like T. Similarly, the mapping $(t, r) \to R_P{}^{\#}(t) : T \times R \to \mathscr{P}'(R_P)$ has all the properties assumed for $R^{\#}$. We conclude that all the results of Chapters IV and V remain applicable when $T, R, R^{\#}, \mu, \Sigma$ are replaced by $T \times R$, R_P, $R_P{}^{\#}$, $\tilde{\zeta}$, $\Sigma_{\text{Borel}}(T \times R)$, respectively, We now denote by $\tilde{\mathscr{P}}^{\#}$ the set defined just like $\mathscr{S}^{\#}$, that is, the set of $\tilde{\zeta}$-measurable functions $\pi : T \times R \to (\text{rpm}(R_P), |\cdot|_w)$ such that $\pi(t, r)(\bar{R}_P{}^{\#}(t)) = 1$ $\tilde{\zeta}$-a.e. We define $\tilde{\mathscr{P}}$ as the subset of $\tilde{\mathscr{P}}^{\#}$ whose elements are $\Sigma_{\text{Borel}}(T \times R)$-measurable, and it is clear that there is a one-to-one correspondence between $\tilde{\mathscr{P}}^{\#}$ and $\tilde{\mathscr{P}}$ (because all $\tilde{\zeta}$-equivalent elements of both are identified).

It follows now from IV.1.10 and IV.3.11 that $\tilde{\mathscr{P}}$ is a compact and convex subset of a separable normed vector space which can be identified with $(L^1(\tilde{\zeta}, C(R_P))^*, |\cdot|_w)$. The elements of this space are $\Sigma_{\text{Borel}}(T \times R)$-measurable functions $\nu : T \times R \to (\text{frm}(R_P), |\cdot|_w)$, with $\tilde{\zeta}$-ess sup $|\nu(t)|(R_P) < \infty$ and ν_1 identified with ν_2 if they coincide $\tilde{\zeta}$-a.e. We shall consider $\tilde{\mathscr{P}}$ as endowed with both the algebraic and the topological structures of $(L^1(\tilde{\zeta}, C(R_P))^*, |\cdot|_w)$.

X.2.2 ***Lemma*** Let σ_j, $\tilde{\sigma}$, $\tilde{\zeta}$, and $\tilde{\mathscr{P}}$ be as defined in X.2.1. Then $\lim_i \sigma_j \otimes \pi_i = \sigma_j \otimes \bar{\pi}$ in $\mathscr{S}^{\#}$ $(j = 0, 1, \ldots)$ if $\lim_i \pi_i = \bar{\pi}$ in $\tilde{\mathscr{P}}$. Furthermore, for each $j = 0, 1, \ldots$ and each $\tilde{\zeta}$-null set E, $\chi_E(t, \cdot) = 0$ $\sigma_j(t)$-a.e. for μ-a.a. $t \in T$.

▌PROOF Let $j \in \{0, 1, 2, \ldots\}$ and ζ_j be the unique measure corresponding to σ_j as in X.1.1. If $\tilde{\zeta}(E) = 0$, hence $\int \mu(dt) \int \chi_E(t, r)\, \tilde{\sigma}(t)(dr) = 0$, then

$$\int \chi_E(t, r)\, \tilde{\sigma}(t)\,(dr) = 0 \quad \mu\text{-a.e.},$$

say for $t \in T'$, and $\chi_E(t, \cdot) = 0$ $\tilde{\sigma}(t)$-a.e. for all $t \in T'$. It follows that $\chi_E(t, \cdot) = 0$ $\sigma_j(t)$-a.e. for $t \in T'$; hence

$$\zeta_j(E) = \int \mu(dt) \int \chi_E(t, r)\, \sigma_j(t)\,(dr) = 0.$$

Thus ζ_j is $\tilde{\zeta}$-continuous and, by the Radon–Nikodym theorem I.4.37, there exist $\tilde{\zeta}$-integrable $k_j : T \times R \to \mathbb{R}$ such that

$$(1) \quad \zeta_j(E) = \int_E k_j(t, r)\, \tilde{\zeta}(d(t, r)) \qquad [E \in \Sigma_{\text{Borel}}(T \times R), \quad j = 0, 1, 2, \ldots].$$

Now assume that $\lim_i \pi_i = \bar{\pi}$ in $\tilde{\mathscr{P}}$ and let $j \in \{0, 1, 2, ...\}$ be fixed. Then, by I.4.38 and X.1.2, the function $(t, r) \to k_j(t, r) \int \phi(t, r, r_P)\, \pi(t, r)(dr_P)$ is $\tilde{\zeta}$-integrable for each $\pi \in \tilde{\mathscr{P}}$ and $\phi \in L^1(\mu, C(R \times R_P))$, and

$$\int \tilde{\zeta}(d(t, r)) \int k_j(t, r)\, \phi(t, r, r_P)\, \pi(t, r)(dr_P)$$

$$= \int \zeta_j(d(t, r)) \int \phi(t, r, r_P)\, \pi(t, r)(dr_P).$$

Since $\lim_i \pi_i = \bar{\pi}$ in $\tilde{\mathscr{P}}$ and $k_j \phi \in L^1(\tilde{\zeta}, C(R_P))$, it follows that

$$\lim_i \int \zeta_j(d(t, r)) \int \phi(t, r, r_P)\, \pi_i(t, r)(dr_P)$$

$$= \int \zeta_j(d(t, r)) \int \phi(t, r, r_P)\, \bar{\pi}(t, r)(dr_P)$$

and, by X.1.1,

$$\lim_i \int \mu(dt) \int \sigma_j(t)\,(dr) \int \phi(t, r, r_P)\, \pi_i(t, r)(dr_P)$$

$$= \lim_i \int \mu(dt) \int \phi(t, r, r_P)\, \sigma_j \otimes \pi_i(t)(d(r, r_P))$$

$$= \int \mu(dt) \int \phi(t, r, r_P)\, \sigma_j \otimes \bar{\pi}(t)(d(r, r_P)).$$

Thus $\lim_i \sigma_j \otimes \pi_i = \sigma_j \otimes \bar{\pi}$ in $\hat{\mathscr{S}}^{\#}$. QED

X.2.3 ***Theorem*** Let $Q \triangleq \mathscr{S}^{\#} \times B$, $q \triangleq (\sigma, b)$, $p \triangleq (\pi, b_P)$, and $x_2(q)(p)$ denote $x_2(\sigma, b, \pi, b_P)$. Assume that A is a convex subset of $\mathbb{R}^{m_2}$, $A^\circ \neq \varnothing$, $\bar{q} \triangleq (\bar{\sigma}, \bar{b})$ is a minimizing relaxed control, $x_2(q)(\cdot)$ is bounded on $\mathscr{P}^{\#} \times B_P$ for every $q \in Q$, and, for every choice of $(q_0, q_1, ..., q_m) \in Q^{m+1}$, the functions

$$\theta \to (x_0, x_1)\left(\bar{q} + \sum_{j=0}^{m} \theta^j (q_j - \bar{q})\right) : \mathscr{T}_{m+1} \to \mathbb{R} \times \mathbb{R}^m$$

and

$$\theta \to x_2\left(\bar{q} + \sum_{j=0}^{m} \theta^j (q_j - \bar{q})\right)(\cdot) : \mathscr{T}_{m+1} \to BF(\mathscr{P}^{\#} \times B_P, \mathbb{R}^{m_2})$$

are continuous and have a derivative at $\theta = 0$. Then there exist $l_0 \geqslant 0$, $l_1 \in \mathbb{R}^m$, and $l_2 \in BF(\mathscr{P}^{\#} \times B_P, \mathbb{R}^{m_2})^*$ such that $(l_0, l_1, l_2) \neq 0$,

$$\text{(1)} \qquad \sum_{i=0}^{1} l_i \cdot Dx_i(\bar{q}; q - \bar{q}) + l_2(Dx_2(\bar{q}; q - \bar{q})) \geqslant 0 \qquad (q \in Q)$$

and

$$\text{(2)} \qquad l_2(x_2(\bar{q})) = \operatorname{Max}\{l_2(c) \mid c \in BF(\mathscr{P}^{\#} \times B_P, \mathbb{R}^{m_2}),\ c(\mathscr{P}^{\#} \times B_P) \subset A\}.$$

▌PROOF Follows directly from V.2.3 and the definition of an extremal. QED

X.2.4 ***Theorem*** Let the conditions of Theorem X.2.3 be satisfied, and let σ_j and $\tilde{\mathscr{P}}$ be defined as in X.2.1 with $\sigma_0 \triangleq \bar{\sigma}$. Assume, furthermore, that B_P is a compact metric space and that, for $\sigma = \bar{\sigma}$, σ_j $(j \in \mathbb{N})$ and $b \in B$, the functions $x_2((\sigma, b))(\cdot)$ and $Dx_2(\bar{q}; (\sigma, b) - \bar{q})(\cdot)$ on $\mathscr{P}^\# \times B$ are actually functions on $\tilde{\mathscr{P}} \times B_P$ [that is,

$$x_2((\sigma, b))(\pi_1, b_P) = x_2((\sigma, b))(\pi_2, b_P) \qquad \text{if} \quad \pi_1(t, r) = \pi_2(t, r) \quad \tilde{\zeta}\text{-a.e.}$$

and similarly for $Dx_2((\bar{\sigma}, \bar{b}); (\sigma, b) - (\bar{\sigma}, \bar{b}))(\cdot)$] and that they are continuous on $\tilde{\mathscr{P}} \times B_P$. Then there exist $l_0 \geqslant 0$, $l_1 \in \mathbb{R}^m$, $\omega \in \text{frm}^+(\tilde{\mathscr{P}} \times B_P)$ and an ω-integrable $\tilde{\omega} : \tilde{\mathscr{P}} \times B_P \to \mathbb{R}^{m_2}$ such that

$$\text{(1)} \qquad l_0 + |l_1| + \omega(\tilde{\mathscr{P}} \times B_P) > 0 \qquad \text{and} \qquad |\tilde{\omega}(p)| = 1 \quad \omega\text{-a.e.};$$

$$\text{(2)} \quad \sum_{i=0}^{1} l_i \cdot Dx_i(\bar{q}; (\sigma_j, b) - \bar{q}) + \int \tilde{\omega}(p) \cdot Dx_2(\bar{q}; (\sigma_j, b) - \bar{q})(p)\, \omega(dp) \geqslant 0$$
$$(j = 0, 1, 2, \ldots, b \in B);$$

and

$$\text{(3)} \qquad \tilde{\omega}(p) \cdot x_2(\bar{q})(p) = \operatorname*{Max}_{a \in A} \tilde{\omega}(p) \cdot a \qquad \text{for} \quad \omega\text{-a.a. } p \in \tilde{\mathscr{P}} \times B_P .$$

▌PROOF We can embed the Banach space $(C(\tilde{\mathscr{P}} \times B_P, \mathbb{R}^{m_2}), |\cdot|_{\sup})$ in $(BF(\mathscr{P}^\# \times B_P, \mathbb{R}^{m_2}), |\cdot|_{\sup})$ by identifying an element c of the former with the bounded function $f: \mathscr{P}^\# \times B_P \to \mathbb{R}^{m_2}$ such that

$$f(\pi, b_P) = c(\tilde{\pi}, b_P) \qquad \text{if} \quad \pi = \tilde{\pi} \quad \text{in } \tilde{\mathscr{P}}.$$

Now let l_0, l_1, and l_2 be as defined in X.2.3. By assumption,

$$x_2(\bar{q})(\cdot) \in C(\tilde{\mathscr{P}} \times B_P, \mathbb{R}^{m_2})$$

and therefore, by X.2.3(2),

$$l_2(x_2(\bar{q})) = \text{Max}\{l_2(c) \mid c \in C(\tilde{\mathscr{P}} \times B_P, \mathbb{R}^{m_2}),\ c(\tilde{\mathscr{P}} \times B_P) \subset A\}.$$

It follows, by V.2.5, that there exist $\omega \in \text{frm}^+(\tilde{\mathscr{P}} \times B_P)$ and $\tilde{\omega} \in L^1(\omega, \mathbb{R}^{m_2})$ satisfying relations (1) and (3) and such that

$$l_2(c) = \int \tilde{\omega}(p) \cdot c(p)\, \omega(dp) \qquad [c \in C(\tilde{\mathscr{P}} \times B_P, \mathbb{R}^{m_2})].$$

Since, by assumption, $Dx_2(\bar{q}; (\sigma_j, b) - \bar{q})(\cdot) \in C(\tilde{\mathscr{P}} \times B_P, \mathbb{R}^{m_2})$, relation (2) now follows from X.2.3(1). QED

Finally, we observe that if $(\bar{\rho}, \bar{b})$ is a minimizing $\mathscr{R}^{\#}$-control, then, in view of X.1.7, $(\bar{\rho}, \bar{b})$ is also a minimizing $\mathscr{R}^{\#} \times B$-control in the sense of Chapter IX. We can obtain therefore weak necessary conditions for an original minimum by applying Theorem IX.1.6, with P replaced by $\mathscr{S}_P{}^{\#} \times B_P$ and B_P assumed to be a compact metric space.

X.3 Hyperrelaxed and Relaxed Adverse Controls in Ordinary Differential Equations

As an illustration, we shall apply Theorems X.1.8 (on existence and approximation) and X.2.4 (on necessary conditions) to a conflicting control problem with hyperrelaxed adverse controls that is defined by ordinary differential equations of the type considered in Section V.6. The corresponding results are contained in Theorems X.3.1 and X.3.5. In Theorem X.3.7 we present analogous results for a similarly defined problem with relaxed adverse controls. (This last problem was not considered in Chapter IX where it properly belongs because the formulation of this problem and the proof of X.3.7 rely on the definitions and lemmas of the present section).

Let $\hat{R} \triangleq R \times R_P$, and let sets $A_0 \subset \mathbb{R}^m \times \mathbb{R}^{m_2}$ and $A_1 \subset \mathbb{R}^m$ be compact and convex and functions $f: \mathbb{R}^m \times R \to \mathbb{R}^m$ and $\hat{f}: \mathbb{R}^{m_2} \times \hat{R} \to \mathbb{R}^{m_2}$ be continuous and have continuous partial derivatives

$$\mathscr{D}_1 f: \mathbb{R}^m \times R \to B(\mathbb{R}^m, \mathbb{R}^m) \qquad \text{and} \qquad \mathscr{D}_1 \hat{f}: \mathbb{R}^{m_2} \times \hat{R} \to B(\mathbb{R}^{m_2}, \mathbb{R}^{m_2}).$$

We set $T = [t_0, t_1] \subset \mathbb{R}$, define μ as the Borel measure in T, and, for each $\sigma \in \mathscr{S}^{\#}$, $\hat{\sigma} \in \hat{\mathscr{S}}^{\#}$, and $b_0 \triangleq (a_0, \hat{a}_0) \in A_0$, consider the equations

$$y(t) = a_0 + \int_{t_0}^{t} f(y(\tau), \sigma(\tau))\, d\tau = \int_{t_0}^{t} d\tau \int f(y(\tau), r)\, \sigma(\tau)\, (dr) \qquad (t \in T)$$

and

$$\hat{y}(t) = \hat{a}_0 + \int_{t_0}^{t} \hat{f}(\hat{y}(\tau), \hat{\sigma}(\tau))\, d\tau = \int_{t_0}^{t} d\tau \int \hat{f}(y(\tau), \hat{r})\, \hat{\sigma}(\tau)\, (d\hat{r}) \qquad (t \in T).$$

As in Section V.6, we assume that for every choice of $(a_0, \hat{a}_0) \in A_0$, $\sigma \in \mathscr{S}^{\#}$, and $\hat{\sigma} \in \hat{\mathscr{S}}^{\#}$, these equations have unique solutions

$$y(\sigma, a_0)(\cdot) \triangleq (y^1(\sigma, a_0)(\cdot), \ldots, y^m(\sigma, a_0)(\cdot)) \qquad \text{and} \quad \hat{y}(\hat{\sigma}, \hat{a}_0)(\cdot)$$

and that there exists $c \in \mathbb{R}$ such that

$$|y(\sigma, a_0)(t)| \leqslant c \qquad \text{and} \qquad |\hat{y}(\hat{\sigma}, \hat{a}_0)(t)| \leqslant c$$
$$(\sigma \in \mathscr{S}^{\#}, \quad \hat{\sigma} \in \hat{\mathscr{S}}^{\#}, \quad (a_0, \hat{a}_0) \in A_0, \quad t \in T).$$

[This is the case, in particular, if

$$|f(v, r)| \leqslant c'(|v| + 1) \qquad \text{and} \qquad |\hat{f}(\hat{v}, \hat{r})| \leqslant c'(|\hat{v}| + 1)$$

for some $c' \in \mathbb{R}$ and all $v \in \mathbb{R}^m$, $\hat{v} \in \mathbb{R}^{m_2}$, $r \in R$, and $\hat{r} \in \hat{R}$.] We set

$$B \triangleq A_0 \times A_1, \qquad b \triangleq (b_0, b_1) \triangleq (a_0, \hat{a}_0, b_1),$$
$$x_0(\sigma, b) \triangleq y^1(\sigma, a_0)(t_1), \qquad x_1(\sigma, b) \triangleq y(\sigma, a_0)(t_1) - b_1,$$

and

$$x_2(\sigma \otimes \pi, b) \triangleq x_2(\sigma, b)(\pi) \triangleq \hat{y}(\sigma \otimes \pi, \hat{a}_0)(t_1) \qquad (\sigma \in \mathscr{S}^{\#}, \quad \pi \in \mathscr{P}^{\#}, \quad b \in B).$$

The problem thus defined does not involve any set B_P of "adverse" control parameters.

We first prove the existence of a minimizing relaxed control and a minimizing approximate $\mathscr{U}$, $\mathscr{U}_P$-control.

X.3.1 ***Theorem*** Let A be a closed subset of $\mathbb{R}^{m_2}$ and $\mathscr{A}(\mathscr{S}^{\#}) \neq \varnothing$. Then there exists a minimizing relaxed control $(\bar{\sigma}, \bar{b})$. Furthermore, for every choice of abundant subsets $\mathscr{U}$ of $\mathscr{R}^{\#}$ and $\mathscr{U}_P$ of $\mathscr{R}_P{}^{\#}$ there exists a minimizing approximate $\mathscr{U}$, $\mathscr{U}_P$-control $((\bar{u}_j, \bar{b}))$ which can be determined by selecting any sequence $(\bar{u}_j)$ in $\mathscr{U}$ converging to $\bar{\sigma}$.

▌PROOF The argument of V.6.1 (Step 1) shows that the functions

$$(\sigma, b) \to y(\sigma, a_0)(\cdot) : \mathscr{S}^{\#} \times B \to C(T, \mathbb{R}^m)$$

and

$$(\hat{\sigma}, b) \to \hat{y}(\hat{\sigma}, \hat{a}_0)(\cdot) : \hat{\mathscr{S}}^{\#} \times B \to C(T, \mathbb{R}^{m_2})$$

are continuous. It follows that the functions

$$(\sigma, b) \to (x_0, x_1)(\sigma, b) : \mathscr{S}^{\#} \times B \to \mathbb{R} \times \mathbb{R}^m$$

and

$$(\hat{\sigma}, b) \to x_2(\hat{\sigma}, b) = \hat{y}(\hat{\sigma}, \hat{a}_0)(t_1) : \hat{\mathscr{S}}^{\#} \times B \to \mathbb{R}^{m_2}$$

are also continuous. Thus the conditions of Theorem X.1.8 are satisfied and our conclusions follow from that theorem. QED

We now consider necessary conditions for a relaxed control $(\bar{\sigma}, \bar{b}) \triangleq (\bar{\sigma}, \bar{a}_0, \hat{\bar{a}}_0, \bar{b}_1)$ to yield a relaxed minimum. Since $|y(\sigma, a_0)(t)| \leqslant c$ and $|\hat{y}(\hat{\sigma}, \hat{a}_0)(t)| \leqslant c$ for all $t \in T$, $\sigma \in \mathscr{S}^{\#}$, $\hat{\sigma} \in \hat{\mathscr{S}}^{\#}$, and $(a_0, \hat{a}_0) \in A_0$, it follows that $\mathscr{D}_1 f(y(\sigma, a_0)(t), \sigma(t))$ and $\mathscr{D}_1 \hat{f}(\hat{y}(\hat{\sigma}, \hat{a}_0)(t), \hat{\sigma}(t))$ are both uniformly bounded for all $t \in T$, $(a_0, \hat{a}_0) \in A_0$, $\sigma \in \mathscr{S}^{\#}$, and $\hat{\sigma} \in \hat{\mathscr{S}}^{\#}$. It follows, by II.4.8, that the equations

$$Z(t) = I_m + \int_t^{t_1} Z(\tau)\, d\tau \int \mathscr{D}_1 f(y(\bar{\sigma}, \bar{a}_0)(\tau), r)\, \bar{\sigma}(\tau)\, (dr) \qquad (t \in T)$$

and

$$\hat{Z}(t) = I_{m_2} + \int_t^{t_1} \hat{Z}(\tau)\, d\tau \int \bar{\sigma}(\tau)\,(dr) \int \mathscr{D}_1 \hat{f}\big(\hat{y}(\bar{\sigma} \otimes \pi, \hat{a}_0)(\tau), r, r_P\big)\, \pi(\tau, r)\,(dr_P) \qquad (t \in T)$$

(where I_k denotes the unit $k \times k$ matrix) have unique solutions

$$Z : T \to B(\mathbb{R}^m, \mathbb{R}^m) \qquad \text{and} \qquad \hat{Z}(\pi) : T \to B(\mathbb{R}^{m_2}, \mathbb{R}^{m_2})$$

for each $\pi \in \mathscr{P}^\#$.

Next we define an appropriate set $\tilde{\mathscr{P}}$. By Condition IV.3.1, there exists a denumerable subset $\mathscr{R}_\infty{}^\#$ of $\mathscr{R}^\#$ such that $\{\rho(t) \mid \rho \in \mathscr{R}_\infty{}^\#\}$ is dense in $\bar{R}^\#(t)$ for a.a. $t \in T$. If we denote by $\mathscr{I}_\infty$ the collection of all closed subintervals of T with rational endpoints then $\mathscr{R}_\infty{}^\# \times \mathscr{I}_\infty$ is denumerable and of the form $\{(\rho_j, T_j) \mid j \in \mathbb{N}\}$. We denote by δ_r the Dirac measure at r and set $\sigma_0 \triangleq \bar{\sigma}$ and, for all $j \in \mathbb{N}$,

$$\sigma_j(t) \triangleq \delta_{\rho_j(t)} \quad (t \in T_j) \qquad \text{and} \qquad \sigma_j(t) \triangleq \bar{\sigma}(t) \quad (t \in T \sim T_j).$$

Then $\tilde{\sigma}$, $\tilde{\zeta}$, and $\tilde{\mathscr{P}}$ are defined accordingly as in Definition X.2.1.

X.3.2 ***Lemma*** Let $\hat{A}_0 \triangleq \{\hat{a}_0 \mid (a_0, \hat{a}_0) \in A_0\}$. For every choice of $(\hat{q}, \hat{q}_0, \hat{q}_1, ..., \hat{q}_m) \in (\hat{\mathscr{S}}^\# \times \hat{A}_0)^{m+2}$, the function

$$\theta \to \eta(\theta) \triangleq \hat{y}\left(\hat{q} + \sum_{j=0}^{m} \theta^j(\hat{q}_j - \hat{q})\right) : \mathscr{T}_{m+1} \to C(T, \mathbb{R}^{m_2})$$

has a derivative η', and $\lim |\theta_1 - \theta|^{-1}[\eta(\theta_1) - \eta(\theta) - \eta'(\theta)(\theta_1 - \theta)] = 0$ as $\theta_1 \to \theta$, $\theta_1 \in \mathscr{T}_{m+1} \sim \{\theta\}$, uniformly for all

$$\theta \in \mathscr{T}_{m+1} \qquad \text{and} \qquad (\hat{q}, \hat{q}_0, ..., \hat{q}_m) \in (\hat{\mathscr{S}}^\# \times \hat{A}_0)^{m+2}.$$

Furthermore, for

$$\pi \in \mathscr{P}^\#, \quad \hat{\bar{q}} \triangleq (\bar{\sigma} \otimes \pi, \hat{a}_0) \in \hat{\mathscr{S}}^\# \times \hat{A}_0, \quad \text{and} \quad \hat{q}' \triangleq (\hat{\sigma}', \hat{a}_0{}') \in \hat{\mathscr{S}}^\# \times \hat{A}_0,$$

we have

$$(1) \quad D\hat{y}(\hat{\bar{q}}; \hat{q}' - \hat{\bar{q}})(t_1) = \hat{Z}(\pi)(t_0)(\hat{a}_0{}' - \hat{a}_0) + \int_{t_0}^{t_1} \hat{Z}(\pi)(\tau) \hat{f}\big(\hat{y}(\hat{\bar{q}})(\tau), \hat{\sigma}'(\tau) - \bar{\sigma} \otimes \pi(\tau)\big)\, d\tau.$$

▌PROOF Let $\hat{q}, \hat{q}_0, ..., \hat{q}_m \in \hat{\mathscr{S}}^\# \times \hat{A}_0$. For all $\theta \in \mathscr{T}_{m+1}$ and $t \in T$, we have

$$\eta(\theta)(t) = \hat{a}_0 + \sum_{j=0}^{m} \theta^j(\hat{a}_{0,j} - \hat{a}_0) + \int_{t_0}^{t} \hat{f}\big(\eta(\theta)(\tau), \hat{\sigma}(\tau)\big)\, d\tau$$

$$+ \sum_{j=0}^{m} \theta^j \int_{t_0}^{t} \hat{f}\big(\eta(\theta)(\tau), \hat{\sigma}_j(\tau) - \hat{\sigma}(\tau)\big)\, d\tau$$

and, by II.4.11, $\eta : \mathscr{T}_{m+1} \to C(T, \mathbb{R}^{m_2})$ has a derivative such that

$$(2)\qquad (\eta'(\theta)\,\Delta\theta)(t) = \sum_{j=0}^{m} \Delta\theta^j(\hat{a}_{0,j} - \hat{a}_0) + \int_{t_0}^{t} \mathscr{D}_1\hat{f}\Big(\eta(\theta)(\tau), \hat{\sigma}(\tau) + \sum_{j=0}^{m} \theta^j[\hat{\sigma}_j(\tau) - \hat{\sigma}(\tau)]\Big)(\eta'(\theta)\,\Delta\theta)(\tau)\, d\tau + \sum_{j=0}^{m} \Delta\theta^j \int_{t_0}^{t} \hat{f}\big(\eta(\theta)(\tau), \hat{\sigma}_j(\tau) - \hat{\sigma}(\tau)\big)\, d\tau$$

$$(\theta \in \mathscr{T}_{m+1}\,,\ \Delta\theta \in \mathbb{R}^{m+1},\ t \in T).$$

Relation (1) follows from (2) by setting $\hat{a}_0' = \hat{a}_{0,0}\,, \hat{\sigma}' = \hat{\sigma}_0\,, \Delta\theta = (1, 0,\ldots, 0)$, and applying Theorem II.4.8.

Since $\hat{f}$ and $\mathscr{D}_1\hat{f}$ are continuous and $\{\hat{y}(\hat{q}) \mid \hat{q} \in \hat{\mathscr{S}}^{\#} \times \hat{A}_0\}$ bounded, it follows from (2) that there exist $c_1\,, c_2 \in \mathbb{R}$ such that

$$|(\eta'(\theta)\,\Delta\theta)(t)| \leqslant c_1 \int_{t_0}^{t} |(\eta'(\theta)\,\Delta\theta)(\tau)|\, d\tau + c_2$$

for all $\theta, \hat{q}, \hat{q}_0\,,\ldots, \hat{q}_m$ and t provided $|\,\Delta\theta\,| \leqslant 1$. It follows, by Gronwall's inequality II.4.4, that there exists $c_3 \in \mathbb{R}$ such that $|\,\eta'(\theta)| \leqslant c_3$ for all $\hat{q}, \hat{q}_0\,,\ldots, \hat{q}_m$ and θ, and we may assume that $|\,\mathscr{D}_1 f(\hat{y}(\hat{\sigma})(\tau), \hat{r})| \leqslant c_3$ for all $\hat{\sigma} \in \hat{\mathscr{S}}^{\#}$ and $\hat{r} \in \hat{R}$. Thus, if we set $\hat{\sigma} + \sum_{j=0}^{m} \theta^j(\hat{\sigma}_j - \hat{\sigma}) \triangleq \hat{\sigma}(\theta)$, then we can deduce from (2) that

$$(3)\qquad |\,\eta'(\theta_1)(t) - \eta'(\theta)(t)| \leqslant c_3 \int_{t_0}^{t} |\,\eta'(\theta_1)(\tau) - \eta'(\theta)(\tau)|\, d\tau + c_3 \int_{t_0}^{t} |\,\mathscr{D}_1\hat{f}(\eta(\theta_1)(\tau), \hat{\sigma}(\theta_1)(\tau)) - \mathscr{D}_1\hat{f}(\eta(\theta)(\tau), \hat{\sigma}(\theta)(\tau))|\, d\tau + 2 \int_{t_0}^{t} |\,\hat{f}(\eta(\theta_1)(\tau), \cdot) - \hat{f}(\eta(\theta)(\tau), \cdot)|_{\sup}\, d\tau.$$

We have $|\,\eta(\theta_1) - \eta(\theta)| \leqslant c_3\,|\,\theta_1 - \theta\,|$ and, since $\hat{R}$ is compact and $\hat{f}$ and $\mathscr{D}_1\,\hat{f}$ continuous, we conclude that the last two terms on the right of (3) converge to 0 as $\theta_1 \to \theta$, uniformly for all $\theta, \hat{q},\ldots, \hat{q}_m$ and t; hence, by (3) and Gronwall's inequality II.4.4, $\lim_{\theta_1 \to \theta} |\eta'(\theta_1) - \eta'(\theta)| = 0$ uniformly for all $\theta, \hat{q}, \ldots, \hat{q}_m$. By the mean value theorem II.3.6.

$$\begin{aligned} &|\theta_1 - \theta|^{-1} [\eta(\theta_1) - \eta(\theta) - \eta'(\theta)(\theta_1 - \theta)| \\ &\qquad \leqslant \sup_{0 \leqslant \alpha \leqslant 1} |\eta'(\theta + \alpha[\theta_1 - \theta]) - \eta'(\theta)|, \end{aligned}$$

from which our remaining conclusion follows directly. QED

X.3.3 ***Lemma*** For $\sigma = \bar{\sigma}, \sigma_1, \sigma_2, \ldots$ and $(a_0, \hat{a}_0, b_1) \in B$, the functions

$$\pi \to \hat{y}(\sigma \otimes \pi, \hat{a}_0) : \tilde{\mathscr{P}} \to C(T, \mathbb{R}^{m_2}) \quad \text{and} \quad \pi \to \hat{Z}(\pi) : \tilde{\mathscr{P}} \to C(T, B(\mathbb{R}^{m_2}, \mathbb{R}^{m_2}))$$

are continuous.

▌PROOF We have observed (in the proof of X.3.1) that the function $(\hat{\sigma}, b) \to \hat{y}(\hat{\sigma}, \hat{a}_0) : \hat{\mathscr{S}}^{\#} \times B \to C(T, \mathbb{R}^{m_2})$ is continuous and it follows, by X.2.2, that $\pi \to \hat{y}(\sigma \otimes \pi, \hat{a}_0) : \tilde{\mathscr{P}} \to C(T, \mathbb{R}^{m_2})$ is also continuous for $\sigma = \bar{\sigma}, \sigma_1, \sigma_2, \ldots$ and $(a_0, \hat{a}_0) \in A_0$. Therefore the function

$$(\pi, t, \hat{r}) \to \mathscr{D}_1 f(\hat{y}(\bar{\sigma} \otimes \pi, \hat{a}_0)(t), \hat{r})$$

on the compact set $\tilde{\mathscr{P}} \times T \times \hat{R}$ is uniformly continuous. It follows, by an argument analogous to that of V.6.1 (Step 1) (with σ, y replaced by $\bar{\sigma} \otimes \pi, \hat{Z}$), that the function

$$\pi \to \hat{Z}(\pi) : \tilde{\mathscr{P}} \to C(T, B(\mathbb{R}^{m_2}, \mathbb{R}^{m_2}))$$

is also continuous. QED

X.3.4 ***Lemma*** For every choice of $(\sigma_j', b_j) \in \mathscr{S}^{\#} \times B$ $(j = 0, 1, \ldots, m)$, the function

$$\theta \to x_2\big((\bar{\sigma}, \bar{b}) + \sum_{j=0}^{m} \theta^j [(\sigma_j', b_j) - (\bar{\sigma}, \bar{b})]\big) : \mathscr{T}_{m+1} \to BF(\mathscr{P}^{\#}, \mathbb{R}^{m_2})$$

is continuous and has a derivative at 0, and the function

$$\pi \to Dx_2\big((\bar{\sigma}, \bar{b}); (\sigma, b) - (\bar{\sigma}, \bar{b})\big)(\pi) : \tilde{\mathscr{P}} \to \mathbb{R}^{m_2}$$

is continuous for $\sigma = \bar{\sigma}, \sigma_1, \sigma_2, \ldots$ and $b \in B$.

▌PROOF By X.3.2, the function

$$\theta \to x_2\big((\bar{\sigma}, \bar{b}) + \sum_{j=0}^{m} \theta^j [(\sigma_j', b_j) - (\bar{\sigma}, \bar{b})]\big)(\cdot) : \mathscr{T}_{m+1} \to BF(\mathscr{P}^{\#}, \mathbb{R}^{m_2})$$

is continuous and differentiable, and

(1)
$$
\begin{aligned}
&Dx_2\big((\bar{\sigma}, \bar{b}); (\sigma, b) - (\bar{\sigma}, \bar{b})\big)(\pi) \\
&\quad = \hat{Z}(\pi)(t_0)(\hat{a}_0 - \hat{a}_0) \\
&\quad\quad + \int_{t_0}^{t_1} \hat{Z}(\pi)(\tau) \hat{f}\big(\hat{y}(\bar{\sigma} \otimes \pi, \hat{a}_0)(\tau), \sigma \otimes \pi(\tau) - \bar{\sigma} \otimes \pi(\tau)\big)\, d\tau \\
&\qquad\qquad\qquad (\sigma \in \mathscr{S}^{\#}, \quad b \in B, \quad \pi \in \mathscr{P}^{\#}).
\end{aligned}
$$

By X.3.3, $\pi \to \hat{y}(\bar{\sigma} \otimes \pi, \hat{a}_0)$ and $\pi \to \hat{Z}(\pi)$ are continuous and, by X.2.2, $\pi \to \sigma \otimes \pi : \tilde{\mathscr{P}} \to \hat{\mathscr{S}}^{\#}$ is continuous for $\sigma = \bar{\sigma}, \sigma_1, \sigma_2, \ldots$. Therefore our last conclusion follows from (1). QED

X.3.5 ***Theorem*** Let $(\bar{\sigma}, \bar{b})$ be a minimizing relaxed control, A convex, and $A^\circ \neq \varnothing$. Then there exist $l_0 \geqslant 0$, $l_1 \in \mathbb{R}^m$, $\omega \in \mathrm{frm}^+(\tilde{\mathscr{P}})$, and $\tilde{\omega} \in L^1(\omega, \mathbb{R}^{m_2})$ such that

(1) $$l_0 + |l_1| + \omega(\tilde{\mathscr{P}}) > 0 \quad \text{and} \quad |\tilde{\omega}(\pi)| = 1 \quad \omega\text{-a.e.};$$

(2) if we set

$$h(\pi, t, r) \triangleq \operatorname*{Max}_{r_P \in \bar{R}_P^{\#}(t)} \tilde{\omega}(\pi)^T \hat{Z}(\pi)(t) \hat{f}\big(\hat{y}(\bar{\sigma} \otimes \pi, \hat{a}_0)(t), r, r_P\big) \qquad (\pi \in \tilde{\mathscr{P}}, \quad t \in T, \quad r \in R),$$

$$e_1 \triangleq (1, 0, \ldots, 0) \in \mathbb{R}^m$$

and

$$H(t, r) \triangleq (l_0 e_1 + l_1)^T Z(t) f\big(y(\bar{\sigma}, \bar{a}_0)(t), r\big) + \int h(\pi, t, r)\, \omega(d\pi) \qquad (t \in T, \quad r \in R),$$

then $H \in L^1(\mu, C(R))$ and

$$\int H(t, r)\, \bar{\sigma}(t)\,(dr) = \operatorname*{Min}_{r \in \bar{R}^{\#}(t)} H(t, r) \qquad \text{for a.a. } t \in T;$$

(3) $$\tilde{\omega}(\pi) \cdot \hat{y}(\bar{\sigma} \otimes \pi, \hat{a}_0)(t_1) = \operatorname*{Max}_{a \in A} \tilde{\omega}(\pi) \cdot a \qquad \text{for } \omega\text{-a.a. } \pi \in \tilde{\mathscr{P}};$$

(4) for ω-a.a. $\pi \in \tilde{\mathscr{P}}$, we have

$$\int \tilde{\omega}(\pi)^T \hat{Z}(\pi)(t) \hat{f}\big(\hat{y}(\bar{\sigma} \otimes \pi, \hat{a}_0)(t), r, r_P\big) \pi(t, r)(dr) = h(\pi, t, r)$$

for a.a. $t \in T$ and $\bar{\sigma}(t)$-a.a. $r \in R$; and

(5) if we set $\lambda^T \triangleq \int \tilde{\omega}(\pi)^T \hat{Z}(\pi)(t_0)\, \omega(d\pi)$, then

$$(l_0 e_1 + l_1)^T Z(t_0)\, \bar{a}_0 + \lambda^T \hat{\bar{a}}_0 = \operatorname*{Min}_{(a_0, \hat{a}_0) \in A_0} [l_0 e_1 + l_1)^T Z(t_0)\, a_0 + \lambda^T \hat{a}_0]$$

and

$$l_1{}^T \bar{b}_1 = \operatorname*{Max}_{b_1 \in A_1} l_1{}^T b_1 .$$

▌PROOF *Step* 1 By X.3.4, the function $x_2 : \mathscr{S}^\# \times B \to BF(\mathscr{P}^\#, \mathbb{R}^{m_2})$ satisfies the conditions of Theorem X.2.3, and it was shown in the proof of V.6.1 that x_0 and x_1 also satisfy these conditions. By X.3.3 and X.3.4, the functions $\pi \to x_2(\sigma, b)(\pi)$ and $\pi \to Dx_2((\bar{\sigma}, \bar{b}); (\sigma, b) - (\bar{\sigma}, \bar{b}))(\pi)$ are continuous on $\tilde{\mathscr{P}}$ whenever $\sigma = \bar{\sigma}, \sigma_1, \sigma_2, ...$ and $b \in B$. Thus the conditions of Theorem X.2.4 are satisfied, and relations (1) and (3) follow from X.2.4(1) and X.2.4(3).

Step 2 We shall next prove relation (4). Since $x_2(\bar{\sigma}, \bar{b})(\pi) \in A$ $(\pi \in \tilde{\mathscr{P}})$, it follows from (3) that there exists a set $\tilde{\mathscr{P}}' \subset \tilde{\mathscr{P}}$ such that $\omega(\tilde{\mathscr{P}} \sim \tilde{\mathscr{P}}') = 0$ and

$$\tilde{\omega}(\bar{\pi}) \cdot \hat{y}(\bar{\sigma} \otimes \bar{\pi}, \hat{\bar{a}}_0)(t_1) = \operatorname*{Max}_{\pi \in \tilde{\mathscr{P}}} \tilde{\omega}(\bar{\pi}) \cdot \hat{y}(\bar{\sigma} \otimes \pi, \hat{\bar{a}}_0)(t_1) \qquad (\bar{\pi} \in \tilde{\mathscr{P}}').$$

This relation and X.3.2(1) imply that

$$(6)\quad \tilde{\omega}(\bar{\pi}) \cdot D_1 \hat{y}(\bar{\sigma} \otimes \bar{\pi}, \hat{\bar{a}}_0 ; \bar{\sigma} \otimes \pi - \bar{\sigma} \otimes \bar{\pi})(t_1)$$

$$= \int_{t_0}^{t_1} \tilde{\omega}(\bar{\pi})^T \hat{Z}(\bar{\pi})(\tau) \hat{f}(\hat{y}(\bar{\sigma} \otimes \bar{\pi}, \hat{\bar{a}}_0)(\tau), \bar{\sigma} \otimes \pi(\tau) - \bar{\sigma} \otimes \bar{\pi}(\tau))\, d\tau$$

$$\leqslant 0 \qquad (\pi \in \tilde{\mathscr{P}}, \quad \bar{\pi} \in \tilde{\mathscr{P}}').$$

By X.1.4, for each $\bar{\pi} \in \tilde{\mathscr{P}}'$ we have $h(\bar{\pi}, \cdot, \cdot) \in L^1(\mu, C(R))$, and there exists $\bar{\pi}' \in \tilde{\mathscr{P}}$ such that

$$h(\bar{\pi}, \tau, r) = \int \tilde{\omega}(\bar{\pi})^T \hat{Z}(\bar{\pi})(\tau) \hat{f}(\hat{y}(\bar{\sigma} \otimes \bar{\pi}, \hat{\bar{a}}_0)(\tau), r, r_P)\, \bar{\pi}'(\tau, r)\, (dr_P)$$

for ζ-a.a. $(\tau, r) \in T \times R$; hence, by X.2.2, for a.a. $\tau \in T$ and $\bar{\sigma}(\tau)$-a.a. $r \in R$. It follows therefore from (6) and the definition of h that, for all $\bar{\pi} \in \tilde{\mathscr{P}}'$, we have

$$(7)\quad \int_{t_0}^{t_1} d\tau \int [h(\bar{\pi}, \tau, r)$$

$$- \int \tilde{\omega}(\bar{\pi})^T \hat{Z}(\bar{\pi})(\tau) f(\hat{y}(\bar{\sigma} \otimes \bar{\pi}, \hat{\bar{a}}_0)(\tau), r, r_P)\, \bar{\pi}(\tau, r)(dr_P)]\, \bar{\sigma}(\tau)(dr) = 0.$$

Furthermore, the coefficient of $\bar{\sigma}(\tau)\,(dr)$ is nonnegative (again, by the definition of h), and relation (4) follows therefore from (7) and I.4.34(7).

Step 3 As in the proof of V.6.1, we derive the relation

$$Dy\big((\bar\sigma, \bar a_0); (\sigma, a_0) - (\bar\sigma, \bar a_0)\big)(t_1)$$

$$= Z(t_0)(a_0 - \bar a_0) + \int_{t_0}^{t_1} Z(\tau)\, f\big(y(\bar\sigma, \bar a_0)(\tau), \sigma(\tau) - \bar\sigma(\tau)\big)\, d\tau$$

$$[\sigma \in \mathscr{S}^{\#},\ (a_0, \hat a_0) \in A_0].$$

This relation, together with X.2.4(2) and X.3.2(1), yields

$$\begin{aligned}
(8)\quad & \sum_{i=0}^{1} l_i \cdot Dx_i\big((\bar\sigma, \bar b); (\sigma_j, b) - (\bar\sigma, \bar b)\big) \\
&\quad + \int \tilde\omega(\pi) \cdot Dx_2\big((\bar\sigma, \bar b); (\sigma_j, b) - (\bar\sigma, \bar b)\big)(\pi)\, \omega(d\pi) \\
&= (l_0 e_1 + l_1)^T Z(t_0)(a_0 - \bar a_0) \\
&\quad + \int \tilde\omega(\pi)^T \hat Z(\pi)(t_0)\, \omega(d\pi) \cdot (\hat a_0 - \hat{\bar a}_0) - l_1{}^T(b_1 - \bar b_1) \\
&\quad + \int_{t_0}^{t_1} (l_0 e_1 + l_1)^T Z(\tau)\, f\big(y(\bar\sigma, \bar a_0)(\tau), \sigma_j(\tau) - \bar\sigma(\tau)\big)\, d\tau \\
&\quad + \int \omega(d\pi) \int_{t_0}^{t_1} \tilde\omega(\pi)^T \hat Z(\pi)(\tau)\, \hat f\big(\hat y(\bar\sigma \otimes \pi, \hat{\bar a}_0)(\tau), \sigma_j \otimes \pi(\tau)\big)\, d\tau \\
&\quad - \int \omega(d\pi) \int_{t_0}^{t_1} \tilde\omega(\pi)^T \hat Z(\pi)(\tau)\, \hat f\big(\hat y(\bar\sigma \otimes \pi, \hat{\bar a}_0)(\tau), \bar\sigma \otimes \pi(\tau)\big)\, d\tau \\
&\triangleq I_1 + I_2 + \cdots + I_5 - I_6 \\
&\geqslant 0 \qquad [j = 0, 1, 2, \ldots,\ \ (a_0, \hat a_0, b_1) \in B].
\end{aligned}$$

Statement (5) follows by setting $j = 0$ and recalling that $\sigma_0 = \bar\sigma$.

We next observe that, by X.3.3, the function

$$(\pi, t, r, r_P) \to \hat Z(\pi)(t)\, \hat f\big(\hat y(\bar\sigma \otimes \pi, \hat{\bar a}_0)(t), r, r_P\big) : \tilde{\mathscr{P}} \times T \times R \times R_P \to \mathbb{R}^{m_2}$$

is continuous and therefore the function

$$(\pi, t, r, r_P) \to \tilde\omega(\pi)^T\, \hat Z(\pi)(t)\, \hat f\big(\hat y(\bar\sigma \otimes \pi, \hat{\bar a}_0)(t), r, r_P\big)$$

belongs to

$$L^1(\omega, C(T \times R \times R_P)) \subset \bigcap_{j=0}^{\infty} L^1(\omega \times \zeta_j, C(R_P)) \cap L^1(\omega \times \mu, C(R \times R_P)).$$

It follows, as in X.1.4, that $h \in \bigcap_{j=0}^{\infty} L^1(\omega \times \zeta_j) \cap L^1(\omega \times \mu, C(R))$. We may therefore replace in (8) the term I_5 by $\int \omega(d\pi) \int_{t_0}^{t_1} d\tau \int h(\pi, \tau, r)\, \sigma_j(\tau)\,(dr)$ which is at least as large. We may also replace [in view of (7)] the term I_6 by $\int \omega(d\pi) \int_{t_0}^{t_1} d\tau \int h(\pi, \tau, r)\, \bar{\sigma}(\tau)\,(dr)$. Since h is $\omega \times \zeta_j$-integrable for $j = 0, 1, 2, \ldots$, we may apply X.1.1 and Fubini's theorem I.4.45 to these new "replacement" terms and, after setting $a_0 = \bar{a}_0$, $\hat{a}_0 = \hat{\bar{a}}_0$, and $b_1 = \bar{b}_1$, obtain from (8) the relation

$$\begin{aligned}(9) \qquad & \int_{t_0}^{t_1} d\tau \int [(l_0 e_1 + l_1)^T Z(\tau) f(y(\bar{\sigma}, \bar{a}_0)(\tau), r) \\ & \qquad + \int h(\pi, \tau, r)\, \omega(d\pi)](\sigma_j(\tau) - \bar{\sigma}(\tau))\,(dr) \\ & = \int_{t_0}^{t_1} d\tau \int H(\tau, r)(\sigma_j(\tau) - \bar{\sigma}(\tau))\,(dr) \geqslant 0 \qquad (j \in \mathbb{N}).\end{aligned}$$

Step 4 We now recall that the functions σ_j were constructed in such a manner that, for every element $\rho \in \mathscr{R}_\infty{}^\#$ and an arbitrary subinterval T' of T with rational endpoints, there exists $j \in \mathbb{N}$ such that

$$\sigma_j(t) = \delta_{\rho(t)} \quad (t \in T') \qquad \text{and} \qquad \sigma_j(t) = \bar{\sigma}(t) \quad (t \notin T').$$

It is also easy to verify that $\psi(t) \geqslant 0$ a.e. if ψ is integrable and $\int_\alpha^\beta \psi(\tau)\, d\tau \geqslant 0$ for all rational α and β. Thus it follows from (9) that for every $\rho \in \mathscr{R}_\infty{}^\#$ there exists a null set $Z_\rho \subset T$ such that

$$(10) \qquad H(t, \rho(t)) - \int H(t, r)\, \bar{\sigma}(t)\,(dr) \geqslant 0$$

for all $t \in T \sim Z_\rho$, and therefore relation (10) holds for all $\rho \in \mathscr{R}_\infty{}^\#$ and all $t \in T \sim \bigcup_{\rho \in \mathscr{R}_\infty{}^\#} Z_\rho \triangleq T \sim Z$. The set Z is a null set because $\mathscr{R}_\infty{}^\#$ is denumerable.

Finally, we recall that $h \in L^1(\omega \times \mu, C(R))$ and therefore (by the dominated convergence theorem I.4.35) the function $(\tau, r) \to \int h(\pi, \tau, r)\, \omega(d\pi)$ belongs to $L^1(\mu, C(R))$. Since $(\tau, r) \to Z(\tau) f(y(\bar{\sigma}, \bar{a}_0)(\tau), r)$ is continuous, we conclude that $H \in L^1(\mu, C(R))$. Since $\{\rho(t) \mid \mathscr{R}_\infty{}^\#\}$ is dense in $\bar{R}^\#(t)$ for a.a. $t \in T$, we can now deduce from (10) that, for a.a. $t \in T$,

$$\int H(t, r)\, \bar{\sigma}(t)\,(dr) = \inf_{\rho \in \mathscr{R}_\infty{}^\#} H(t, \rho(t)) = \operatorname*{Min}_{r \in \bar{R}^\#(t)} H(t, r).$$

This completes the proof of statement (2) and of the theorem. QED

X.3.6 ***An Example*** We shall apply Theorems X.3.1 and X.3.5 to the simple illustrative problem considered in X.0.1. This will demonstrate, in particular, that the necessary conditions of X.3.5 are "constructive" in the sense that, for some problems at least, they determine a minimizing relaxed control and a corresponding minimizing approximate $\mathscr{U}$, $\mathscr{U}_P$-control.

It is easily seen that the problem of X.0.1 is a special case of the problem discussed in this section, and corresponds to the case where

$$T = [0, 1], \qquad R = R_P = \{r \in \mathbb{R}^2 \mid r \cdot r = |r|_2^2 = 1\},$$

$$\mathscr{R}^{\#} = \mathscr{R}_P = \mathscr{R}, \qquad m = 3, \qquad m_2 = 1,$$

$$A_0 = \{(\alpha, 0, 0, -\alpha) \mid \alpha \in [-2, 2]\}, \qquad A_1 = \mathbb{R}^3, \qquad A = (-\infty, 0],$$

$$f^1(v, r) = [v^2]^2 + [v^3]^2, \qquad (f^2, f^3)(v, r) = r$$

$$[v \triangleq (v^1, v^2, v^3) \in \mathbb{R}^3, \quad r \in R],$$

and

$$\hat{f}(w, r, r_P) = r \cdot r_P \qquad (w \in \mathbb{R}, \quad r, r_P \in R).$$

Actually the set A_0 should be expressed as $\{(\alpha, 0, 0, -\alpha) \mid \alpha \in \mathbb{R}\}$, but it is clear that $-1 \leqslant \int_0^1 u(t) \cdot u_P(t)\, dt \leqslant 1$ for all $u, u_P \in \mathscr{R}$ and we may restrict α to any interval containing $[-1, 1]$. For the sake of future convenience we restrict α to a larger interval such as $[-2, 2]$.

It is clear that the equations

$$y^1(t) = \alpha + \int_0^t ([y^2(\tau)]^2 + [y^3(\tau)]^2)\, d\tau, \qquad (y^2, y^3)(t) = \int_0^t d\tau \int r\, \sigma(\tau)(dr),$$

$$\hat{y}(t) = -\alpha + \int_0^t d\tau \int \hat{r}\, \hat{\sigma}(\tau)(d\hat{r}) \qquad (t \in T)$$

have unique uniformly bounded solutions for all $\alpha \in [-2, 2]$, $\sigma \in \mathscr{S}^{\#}$, and $\hat{\sigma} \in \hat{\mathscr{S}}^{\#}$, and thus the assumptions of this section are satisfied. Therefore, by X.3.1, there exists a minimizing relaxed control $(\bar{\sigma}, \bar{b}) \triangleq (\bar{\sigma}, \bar{\alpha}, 0, 0, -\bar{\alpha}, \bar{b}_1)$ and a corresponding minimizing $\mathscr{R}$, $\mathscr{R}$-control $((u_j, \bar{b}))$.

Now we apply the necessary conditions of X.3.5. We observe that $\hat{Z}(\pi)(t) = 1$ for all $\pi \in \tilde{\mathscr{P}}$ and $t \in T$, and conditions X.3.5(1) and X.3.5(3) show that $\tilde{\omega}(\pi) = 1$ ω-a.e. Thus

$$h(\pi, t, r) = \operatorname*{Max}_{r_P \in R} r \cdot r_P = 1 \qquad (\pi \in \tilde{\mathscr{P}}, \quad t \in T, \quad r \in R)$$

and, by X.3.5(4), $\pi(t, r) = \delta_r$ for ω-a.a. $\pi \in \tilde{\mathscr{P}}$, a.a. $t \in T$, and $\bar{\sigma}(t)$-a.a. $r \in R$. Thus, for ω-a.a. $\pi \in \tilde{\mathscr{P}}$, we have

$$\hat{y}(\bar{\sigma} \otimes \pi, -\bar{\alpha})(1) = -\bar{\alpha} + \int_0^1 d\tau \int \bar{\sigma}(\tau)\, (dr) \int r \cdot r_P \delta_r(dr_P) = 1 - \bar{\alpha}$$

and therefore, by X.3.5(3),

$$\bar{\alpha} = 1. \tag{1}$$

Since $A_1 = \mathbb{R}^3$, X.3.5(5) implies that $l_1 = 0$. We must have $l_0 > 0$ since otherwise X.3.5(1) implies $\omega(\tilde{\mathscr{P}}) > 0$ and then the first relation of X.3.5(5) yields $\omega(\tilde{\mathscr{P}}) \cdot (-1) = \mathrm{Min}_{\alpha\in[-2,2]}\, \omega(\tilde{\mathscr{P}})\, \alpha$, a contradiction.

Now let

$$k \triangleq (k^1, k^2, k^3), \quad k(t)^T \triangleq l_0 e_1{}^T Z(t), \quad \text{and} \quad \bar{y}(t) \triangleq y(\bar{\sigma}, \bar{\alpha}, 0, 0)(t) \quad (t \in T)$$

Then we have

$$\begin{aligned} \bar{y}(t) &= (\bar{\alpha}, 0, 0) + \int_0^t f(\bar{y}(\tau), \bar{\sigma}(\tau))\, d\tau \qquad (t \in T), \\ k(t)^T &= l_0 e_1{}^T + \int_t^1 k(\tau)^T \mathscr{D}_1 f(\bar{y}(\tau), \bar{\sigma}(\tau))\, d\tau, \end{aligned} \tag{2}$$

and, by X.3.5(2),

$$k(t)^T f(\bar{y}(t), \bar{\sigma}(t)) = \operatorname*{Min}_{r\in R} k(t)^T f(\bar{y}(t), r) \quad \text{a.e. in } T.$$

Then it follows, exactly as in the proof of VI.2.5, that there exists $c_1 \in \mathbb{R}$ such that

$$\begin{aligned} k(t)^T f(\bar{y}(t), \bar{\sigma}(t)) &= \operatorname*{Min}_{r\in R}(k^1(t)([y^2(t)]^2 + [y^3(t)]^2) + (k^2(t), k^3(t)) \cdot r) \\ &= k^1(t)([y^2(t)]^2 + [y^3(t)]^2) - |(k^2(t), k^3(t))|_2 \\ &= c_1 \quad \text{a.e. in } T. \end{aligned} \tag{3}$$

Since the first column of $\mathscr{D}_1 f(\bar{y}(t), \bar{\sigma}(t))$ has all its coefficients 0, it follows that the first column of $Z(t)$ is constant and thus $k^1(t) = l_0$ $(t \in T)$. Furthermore, we have $y^2(0) = y^3(0) = 0$ and, by (2), $k(1) = l_0 e_1 = (l_0, 0, 0)$. We can deduce therefore from (3) that

$$-|(k^2(0), k^3(0))|_2 = l_0([y^2(1)]^2 + [y^3(1)]^2) = c_1\,;$$

hence $c_1 = 0$ and

$$l_0([y^2(t)]^2 + [y^3(t)]^2) = |(k^2(t), k^3(t))|_2 \qquad (t \in T). \tag{4}$$

If we set $(y^2, y^3) \triangleq \eta$ and $(k^2, k^3) \triangleq \kappa$ then Eqs. (2) yield

$$\dot{\kappa}(t) = -2 l_0 \eta(t) \qquad \text{and} \qquad \dot{\eta}(t) = \int r\, \bar{\sigma}(t)(dr) \qquad \text{a.e. in } T;$$

hence $\dot{\kappa}(t) \cdot \eta(t) = -2l_0\eta(t) \cdot \eta(t)$ a.e. and, by (3),

$$\kappa(t) \cdot \dot{\eta}(t) = \operatorname*{Min}_{r \in R} \kappa(t) \cdot r = - |\kappa(t)|_2 \quad \text{a.e.}$$

We combine these last two relations with (4) to obtain

$$\left(\kappa(t) \cdot \eta(t)\right)^{\cdot} = -2l_0 |\eta(t)|_2^2 - |\kappa(t)|_2 = -3l_0 |\eta(t)|_2^2 \quad \text{a.e.}$$

and then integrate both sides to find that

$$0 = \kappa(1) \cdot \eta(1) - \kappa(0) \cdot \eta(0) = -3l_0 \int_0^1 |\eta(t)|_2^2 \, dt.$$

Thus $\eta(t) = 0$ $(t \in T)$ and therefore

$$\dot{\eta}(t) = \int r \, \bar{\sigma}(t)(dr) = 0 \quad \text{a.e. in } T. \tag{5}$$

We thus conclude that any minimizing relaxed control $(\bar{\sigma}, \bar{b})$ satisfies relation (5) and we verify that, by (1) and (2), every relaxed control function satisfying (5) yields the same function $\bar{y}$, namely $(1, 0, 0)$.

To obtain a minimizing $\mathscr{R}, \mathscr{R}$-control we may choose an arbitrary $\bar{\sigma}$ satisfying (5) and a corresponding sequence (u_j) in $\mathscr{R}$ converging to $\bar{\sigma}$; for example, we may choose some $\bar{r} \in R$, set $\bar{\sigma}(t) = \frac{1}{2}\,\delta_{\bar{r}} + \frac{1}{2}\,\delta_{-\bar{r}}$, and then define (u_j) as in X.0.1.

▌Finally, we derive results analogous to Theorems X.3.1 and X.3.5 that apply to minimizing relaxed controls as defined in Chapter IX. Specifically; we define the functions

$$y : \mathscr{S}^{\#} \times A_0 \to C(T, \mathbb{R}^m), \qquad \hat{y} : \hat{\mathscr{S}}^{\#} \times A_0 \to C(T, \mathbb{R}^{m_2}),$$

$$x_0 : \mathscr{S}^{\#} \times B \to \mathbb{R}, \qquad \text{and} \qquad x_1 : \mathscr{S}^{\#} \times B \to \mathbb{R}^m$$

as at the beginning of this section; define $\mathscr{S}_P{}^{\#}$ exactly as $\mathscr{S}^{\#}$ but with R, $R^{\#}$ replaced by R_P, $R_P{}^{\#}$, respectively; define $\sigma \otimes \sigma_P$ $(\sigma \in \mathscr{S}^{\#},\ \sigma_P \in \mathscr{S}_P{}^{\#})$ by the relation

$$\int_{t_0}^{t_1} dt \int \phi(t, r, r_P)\, \sigma \otimes \sigma_P(t)\, \big(d(r, r_P)\big)$$

$$\triangleq \int_{t_0}^{t_1} dt \int \phi(t, r, r_P)\, \sigma(t)\, (dr) \times \sigma_P(t)\, (dr_P) \qquad [\phi \in L^1(\mu, C(R \times R_P)];$$

and set

$$x_2(\sigma, b)(\sigma_P) \triangleq \hat{y}(\sigma \otimes \sigma_P\,, \hat{a}_0)(t_1) \quad [\sigma \in \mathscr{S}^{\#},\ \sigma_P \in \mathscr{S}_P{}^{\#},\ b \triangleq (a_0\,, \hat{a}_0\,, b_1) \in B].$$

We recall that a point $(\bar{\sigma}, \bar{b}) \in \mathscr{S}^{\#} \times B$ is a minimizing $\mathscr{S}^{\#} \times B$-control in the sense of Chapter IX if $x_1(\bar{\sigma}, \bar{b}) = 0$, $x_2(\bar{\sigma}, \bar{b})(\mathscr{S}_P{}^{\#}) \subset A$, and

$$x_0(\bar{\sigma}, \bar{b}) = \inf\{x_0(\sigma, b) \mid \sigma \in \mathscr{S}^{\#}, b \in B, x_1(\sigma, b) = 0, x_2(\sigma, b)(\mathscr{S}_P{}^{\#}) \subset A\}.$$

In order to simplify our arguments, we assume that $R^{\#}(t) = R$ $(t \in T)$.

X.3.7 ***Theorem*** Let A be a convex body in $\mathbb{R}^{m_2}$ and

$$\mathscr{A}(\mathscr{S}) \triangleq \{(\sigma, b) \in \mathscr{S} \times B \mid x_1(\sigma, b) = 0, x_2(\sigma, b)(\mathscr{S}_P{}^{\#}) \subset A\} \neq \varnothing.$$

Then there exists a minimizing $\mathscr{S} \times B$-control $(\bar{\sigma}, \bar{b}) \triangleq (\bar{\sigma}, \bar{a}_0, \hat{a}_0, \bar{b}_1)$ and $((\bar{\sigma}, \bar{b}), (\bar{\sigma}, \bar{b}), \ldots)$ is a minimizing approximate $\mathscr{S}$-control (both in the sense of Chapter IX). Furthermore, the equations

$$Z(t) = I_m + \int_t^{t_1} Z(\tau)\, d\tau \int \mathscr{D}_1 f(y(\bar{\sigma}, \bar{a}_0)(\tau), r)\, \bar{\sigma}(\tau)\,(dr) \qquad (t \in T)$$

and

$$\hat{Z}(t) = I_{m_2} + \int_t^{t_1} \hat{Z}(\tau)\, d\tau \int \mathscr{D}_1 \hat{f}(\hat{y}(\bar{\sigma} \otimes \sigma_P, \hat{a}_0)(\tau), r, r_P)\, \bar{\sigma}(\tau)(dr) \times \sigma_P(\tau)(dr_P)$$
$$(t \in T)$$

have unique solutions $Z: T \to B(\mathbb{R}^m, \mathbb{R}^m)$ and $\hat{Z}(\sigma_P): T \to B(\mathbb{R}^{m_2}, \mathbb{R}^{m_2})$ for each $\sigma_P \in \mathscr{S}_P{}^{\#}$, and there exist $l_0 \geqslant 0$, $l_1 \in \mathbb{R}^m$, $\omega \in \mathrm{frm}^+(\mathscr{S}_P{}^{\#})$, and $\tilde{\omega} \in L^1(\omega, \mathbb{R}^{m_2})$ such that

(1) $\quad l_0 + |l_1| + \omega(\mathscr{S}_P{}^{\#}) > 0 \quad$ and $\quad |\tilde{\omega}(\sigma_P)| = 1 \quad \omega$-a.e.;

(2) if we set

$$h(\sigma_P, t, s) \triangleq \sup_{r_P \in \bar{R}_P{}^{\#}(t)} \tilde{\omega}(\sigma_P)^T\, \hat{Z}(\sigma_P)(t) \int \hat{f}(\hat{y}(\bar{\sigma} \otimes \sigma_P, \hat{a}_0)(t), r, r_P)\, s(dr)$$
$$[\sigma_P \in \mathscr{S}_P{}^{\#}, t \in T, s \in \mathrm{rpm}(R)],$$

$$e_1 \triangleq (1, 0, \ldots, 0) \in \mathbb{R}^m,$$

and

$$H(t, s) \triangleq (l_0 e_1 + l_1)^T Z(t) \int f(y(\bar{\sigma}, \bar{a}_0)(t), r)\, s(dr) + \int h(\sigma_P, t, s)\, \omega(d\sigma_P)$$
$$[t \in T, s \in \mathrm{rpm}(R)],$$

then $H \in L^1(\mu, C(\mathrm{rpm}(R), |\cdot|_w))$ and

$$H(t, \bar{\sigma}(t)) = \inf\{H(t, s) \mid s \in \mathrm{rpm}(R)\} \qquad \text{for a.a. } t \in T;$$

(3) $\quad \tilde{\omega}(\sigma_P) \cdot \hat{y}(\bar{\sigma} \otimes \sigma_P, \hat{a}_0)(t_1) = \operatorname*{Max}_{a \in A} \tilde{\omega}(\sigma_P) \cdot a \qquad$ for ω-a.a. $\sigma_P \in \mathscr{S}_P{}^{\#}$;

(4) for ω-a.a. $\sigma_P \in \mathscr{S}_P{}^{\#}$, we have

$$\int \tilde{\omega}(\sigma_P)^T \hat{Z}(\sigma_P)(t) \hat{f}(\hat{y}(\bar{\sigma} \otimes \pi, \hat{a}_0)(t), \bar{\sigma}(t), r_P) \, \sigma_P(t)(dr_P) = h(\sigma_P, t, \bar{\sigma}(t))$$

for a.a. $t \in T$; and

(5) if we set $\lambda^T \triangleq \int \tilde{\omega}(\sigma_P)^T \hat{Z}(\sigma_P)(t_0) \, \omega(d\sigma_P)$, then

$$(l_0 e_1 + l_1)^T Z(t_0) \, \bar{a}_0 + \lambda^T \hat{\bar{a}}_0 = \underset{(a_0, \hat{a}_0) \in A_0}{\operatorname{Min}} [(l_0 e_1 + l_1)^T Z(t_0) \, a_0 + \lambda^T \hat{a}_0]$$

and

$$l_1{}^T \bar{b}_1 = \underset{b_1 \in A_1}{\operatorname{Max}} \, l_1{}^T b_1 .$$

▌PROOF *Step* 1 We easily verify that for all $\phi \in L^1(\mu, C(R \times R_P))$ and $\sigma_P \in \mathscr{S}_P{}^{\#}$, the function

$$(t, r) \to \int \phi(t, r, r_P) \, \sigma_P(t)(dr)$$

belongs to $L^1(\mu, C(R))$. It follows that

$$\lim_j \sigma_j \otimes \sigma_P = \sigma \otimes \sigma_P \quad \text{in } \hat{\mathscr{S}}^{\#} \qquad (\sigma_P \in \mathscr{S}_P{}^{\#})$$

if $\lim_j \sigma_j = \sigma$ in $\mathscr{S}$.

We can now show that there exist a minimizing $\mathscr{S} \times B$-control $(\bar{\sigma}, \bar{b})$ and a corresponding minimizing approximate $\mathscr{S}$-control $((\bar{\sigma}, \bar{b}), (\bar{\sigma}, \bar{b}), \ldots)$ in the sense of Chapter IX. Indeed, let $\sigma_P \in \mathscr{S}_P{}^{\#}$ be fixed,

$$b_j \triangleq (a_{0,j}, \hat{a}_{0,j}, b_{1,j}) \in B \quad (j \in \mathbb{N}), \qquad b \triangleq (a_0, \hat{a}_0, b_1) \in B,$$

and

$$\lim_j (\sigma_j, b_j) = (\sigma, b) \quad \text{in } \mathscr{S} \times B.$$

Then the argument of V.6.1(Step 1) can be applied with minor changes to show that

$$\lim_j y(\sigma_j, a_{0,j})(t) = y(\sigma, a_0)(t)$$

and

$$\lim_j \hat{y}(\sigma_j \otimes \sigma_P, \hat{a}_{0,j})(t) = \hat{y}(\sigma \otimes \sigma_P, \hat{a}_0)(t)$$

uniformly for all $t \in T$; hence

$$\lim_j x_i(\sigma_j, b_j) = x_i(\sigma, b) \quad (i = 0, 1)$$

and

$$\lim_j x_2(\sigma_j, b_j)(\sigma_P) = x_2(\sigma, b)(\sigma_P).$$

Thus the assumptions of IX.1.1 are satisfied and it follows that there exists a minimizing $\mathscr{S} \times B$-control $(\bar{\sigma}, \bar{b}) \triangleq (\bar{\sigma}, \bar{a}_0, \hat{\bar{a}}_0, \bar{b}_1)$ such that $((\bar{\sigma}, \bar{b}), (\bar{\sigma}, \bar{b}),...)$ is a minimizing approximate $\mathscr{S}$-control (both in the sense of Chapter IX). The existence of unique Z and $\hat{Z}(\sigma_P)$ is assured by II.4.8.

Step 2 Arguments quite analogous to those of X.3.3 and X.3.4 show that (x_0, x_1, x_2) is a function on $\mathscr{S} \times B$ to $\mathbb{R} \times \mathbb{R}^m \times C(\mathscr{S}_P{}^{\#}, \mathbb{R}^{m_2})$ that has a directional derivative at $(\bar{\sigma}, \bar{b})$ in the direction of every $(\sigma, b) \in \mathscr{S} \times B$. Therefore we may apply Theorem IX.1.2 which guarantees the existence of $l_0 \geqslant 0$, $l_1 \in \mathbb{R}^m$, $\omega \in \text{frm}^+(\mathscr{S}_P{}^{\#})$, and $\tilde{\omega} \in L^1(\omega, \mathbb{R}^{m_2})$ satisfying relations (1) and (3) and such that

$$(6) \qquad \sum_{i=0}^{1} l_i \cdot Dx_i((\bar{\sigma}, \bar{b}); (\sigma, b) - (\bar{\sigma}, \bar{b})) + \int \tilde{\omega}(\sigma_P) \cdot Dx_2((\bar{\sigma}, \bar{b}); (\sigma, b) - (\bar{\sigma}, \bar{b}))(\sigma_P)\, \omega(d\sigma_P) \geqslant 0 \qquad (\sigma \in \mathscr{S}, \quad b \in B).$$

We then deduce from (3), as in X.3.5 (Step 2), that there exists $\mathscr{S}_P{}' \subset \mathscr{S}_P{}^{\#}$ such that $\omega(\mathscr{S}_P{}^{\#} \sim \mathscr{S}_P{}') = 0$ and

$$(7) \quad \tilde{\omega}(\bar{\sigma}_P)^T D_1 \hat{y}(\bar{\sigma} \otimes \bar{\sigma}_P, \hat{\bar{a}}_0; \bar{\sigma} \otimes \sigma_P - \bar{\sigma} \otimes \bar{\sigma}_P)(t_1) = \int_{t_0}^{t_1} \tilde{\omega}(\bar{\sigma}_P)^T \hat{Z}(\bar{\sigma}_P)(\tau) \hat{f}(\hat{y}(\bar{\sigma} \otimes \bar{\sigma}_P, \hat{\bar{a}}_0)(\tau), \bar{\sigma} \otimes \sigma_P(\tau) - \bar{\sigma} \otimes \bar{\sigma}_P(\tau))\, d\tau \leqslant 0 \qquad (\sigma_P \in \mathscr{S}_P{}^{\#}, \quad \bar{\sigma}_P \in \mathscr{S}_P{}').$$

An argument similar to that of X.1.4 shows that for each $\bar{\sigma}_P \in \mathscr{S}_P{}'$ there exists $\bar{\sigma}_P{}' \in \mathscr{S}_P{}^{\#}$ such that

$$h(\bar{\sigma}_P, \tau, \bar{\sigma}(\tau)) = \tilde{\omega}(\bar{\sigma}_P)^T \hat{Z}(\bar{\sigma}_P)(\tau) \int \hat{f}(\hat{y}(\bar{\sigma} \otimes \bar{\sigma}_P, \hat{\bar{a}}_0)(\tau), r, r_P)\, \bar{\sigma}(\tau)(dr) \times \bar{\sigma}_P{}'(dr_P)$$

for a.a. $\tau \in T$. Thus, if we set

$$I(\bar{\sigma}_P) \triangleq \int_{t_0}^{t_1} [h(\bar{\sigma}_P, \tau, \bar{\sigma}(\tau)) - \int \bar{\sigma}_P(\tau)(dr_P) \int \tilde{\omega}(\bar{\sigma}_P)^T \hat{Z}(\bar{\sigma}_P)(\tau) \hat{f}(\hat{y}(\bar{\sigma} \otimes \bar{\sigma}_P, \hat{\bar{a}}_0)(\tau), r, r_P)\, \bar{\sigma}(\tau)(dr)]\, d\tau \qquad (\bar{\sigma}_P \in \mathscr{S}_P{}'),$$

then it follows from (7) and Fubini's theorem I.4.45 that $I(\bar{\sigma}_P) \leqslant 0$ $(\bar{\sigma}_P \in \mathscr{S}_P{}')$.

By the definition of h, the integrand of $I(\bar{\sigma}_P)$ is nonnegative and therefore relation (4) follows from I.4.34(7).

As in X.3.5(Step 2), we transform relation (6) into

$$
\begin{aligned}
(8)\quad & (l_0 e_0 + l_1)^T Z(t_0)(a_0 - \bar{a}_0) \\
& + \int \tilde{\omega}(\sigma_P)^T \hat{Z}(\sigma_P)(t_0)\, \omega(d\sigma_P) \cdot (\hat{a}_0 - \hat{\bar{a}}_0) - l_1{}^T(b_1 - \bar{b}_1) \\
& + \int_{t_0}^{t_1} (l_0 e_0 + l_1)^T Z(\tau) f\big(y(\bar{\sigma}, \bar{a}_0)(\tau), \sigma(\tau) - \bar{\sigma}(\tau)\big)\, d\tau \\
& + \int \omega(d\sigma_P) \int_{t_0}^{t_1} \tilde{\omega}(\sigma_P)^T \hat{Z}(\sigma_P)(\tau) \hat{f}\big(\hat{y}(\bar{\sigma} \otimes \sigma_P, \hat{\bar{a}}_0)(\tau), \sigma \otimes \sigma_P(\tau)\big)\, d\tau \\
& - \int \omega(d\sigma_P \int_{t_0}^{t_1} \tilde{\omega}(\sigma_P)^T \hat{Z}(\sigma_P)(\tau) \hat{f}\big(\hat{y}(\bar{\sigma} \otimes \sigma_P, \hat{\bar{a}}_0)(\tau), \bar{\sigma} \otimes \sigma_P(\tau)\big)\, d\tau \\
& \triangleq I_1 + \cdots + I_5 - I_6 \geqslant 0 \qquad [\sigma \in \mathscr{S},\quad (a_0, \hat{a}_0, b_1) \in B].
\end{aligned}
$$

Relation (5) follows by setting $\sigma = \bar{\sigma}$. Next we observe that (by I.5.26, I.4.21 and the definition of h) $(\sigma_P, \tau) \to h(\sigma_P, \tau, \sigma(\tau))$ is $\omega \times \mu$-integrable for each $\sigma \in \mathscr{S}$ and therefore, by the definition of h, (4), and Fubini's theorem I.4.45, we have

$$
I_5 \leqslant \int \omega(d\sigma_P) \int_{t_0}^{t_1} h(\sigma_P, \tau, \sigma(\tau))\, d\tau = \int_{t_0}^{t_1} d\tau \int h(\sigma_P, \tau, \sigma(\tau))\, \omega(d\sigma_P)
$$

and

$$
I_6 = \int \omega(d\sigma_P) \int_{t_0}^{t_1} h(\sigma_P, \tau, \bar{\sigma}(\tau))\, d\tau = \int_{t_0}^{t_1} d\tau \int h(\sigma_P, \tau, \bar{\sigma}(\tau))\, \omega(d\sigma_P).
$$

We set $(a_0, \hat{a}_0, b_1) = (\bar{a}_0, \hat{\bar{a}}_0, \bar{b}_1)$ and combine these last relations with (8) to obtain

$$
(9)\qquad \int_{t_0}^{t_1} [H(\tau, \sigma(\tau)) - H(\tau, \bar{\sigma}(\tau))]\, d\tau \geqslant 0 \qquad (\sigma \in \mathscr{S}).
$$

By IV.1.4, $(\text{rpm}(R), |\cdot|_w)$ is separable, and it is easy to verify that $h \in L^1(\omega \times \mu, C(\text{rpm}(R), |\cdot|_w))$; from which we deduce that H is bounded and, for all $t \in T$ and $s \in \text{rpm}(R)$, $H(\cdot, s)$ is measurable and $H(t, \cdot)$ continuous on $(\text{rpm}(R), |\cdot|_w)$. We select a dense subset $\{s_1, s_2, \ldots\}$ of $(\text{rpm}(R), |\cdot|_w)$ and, for each $j \in \mathbb{N}$ and $E \in \Sigma_{\text{Borel}}(T)$, set

$$
\sigma(\tau) = s_j \quad (\tau \in E), \qquad \sigma(\tau) = \bar{\sigma}(\tau) \quad (\tau \notin E).
$$

Then it follows that $H(\tau, s_j) \geqslant H(\tau, \bar{\sigma}(\tau))$ a.e. for all $j \in \mathbb{N}$ and therefore

$$H(\tau, \bar{\sigma}(\tau)) = \inf_j H(\tau, s_j) = \operatorname*{Min}_{s \in \mathrm{rpm}(R)} H(\tau, s) \qquad \text{for a.a. } \tau \in T.$$

This proves relation (2) and completes the proof of the theorem. QED

Notes

The results of this chapter were derived during November and December of 1971, and have hitherto been unpublished. The reader familiar with the theory of probability will observe that, for given $\hat{\sigma} \triangleq \sigma \otimes \pi \in \hat{\mathscr{S}}^{\#}$, $t \in T$, and $E \in \Sigma_{\mathrm{Borel}}(R_P)$, $\sigma(t)$ is a marginal probability measure of $\hat{\sigma}(t)$ and $\pi(t, r)(E)$ is the conditional probability of E, given r.

A theorem related to X.3.7 appears in Warga [15, Th. 6.3, p. 665]; I must point out, however, that the statement of this last theorem is imprecise and its derivation incorrect because, in the language of X.3.7, the function $(\tau, \sigma_P) \to \sigma_P(\tau) : T \times \mathscr{S}_P^{\#} \to (\mathrm{rpm}(R_P), |\cdot|_w)$ is tacitly treated there as a $\mu \times \omega$-measurable function (which cannot be justified, at least without choosing for each σ_P an appropriate element of its μ-equivalence class). Nevertheless, the theorem [15, Th. 6.3, p. 665] can be made precise and its proof made valid by appropriately redefining the meaning of "integrals" of the form $\int \omega(d\sigma_P) \int a(\sigma_P, \tau, r_P)\, \sigma_P(\tau)\, (dr_P)$, where $(\tau, \sigma_P, r_P) \to a(\sigma_P, \tau, r_P)$ belongs to $L^1(\mu, C(\mathscr{S}_P^{\#} \times R_P))$. I have not followed this procedure in this book because Theorem X.3.7 appears to me both theoretically and computationally superior to a "validated" form of the theorem [15, Th. 6.3, p. 665].

References

Banks, H. T.

[1] Necessary conditions for control problems with variable time lags, *SIAM J. Control* **6** (1968), 9–47.

[2] Variational problems involving functional differential equations, *SIAM J. Control* **7** (1969), 1–17.

Bishop, E.

[1] "Foundations of Constructive Analysis." McGraw-Hill, New York, 1967.

Castaing, C.

[1] Sur les multi-applications mesurables, *Rev. Française Informat. Recherche Opérationnelle* (1967), No. 1.

Cesari, L.

[1] An existence theorem in problems of optimal control, *J. Soc. Indust. Appl. Math., Ser. A, Control* **3** (1965), 7–22.

[2] Existence theorems for optimal solutions in Pontryagin and Lagrange problems, *J. Soc. Indust. Appl. Math., Ser. A, Control* **3** (1965), 475–498.

[3] Existence theorems for optimal controls of the Mayer type, *SIAM J. Control* **6** (1968), 517–552.

[4] Existence theorems for multidimensional Lagrange problems, *J. Optimization Theory Appl.* **1** (1967), 87–112.

[5] Multidimensional Lagrange problems of optimization in a fixed domain and an application to a problem of magneto-hydrodynamics, *Arch. Rational Mech. Anal.* **29** (1968), 81–104.

[6] Closure, lower closure, and semicontinuity theorems in optimal control, *SIAM J. Control* **9** (1971), 287–315.

Dieudonné, J.
[1] "Foundations of Modern Analysis" (enlarged and corrected printing). Academic Press, New York, 1969.

Dubovitskii, A. Ya., and Milyutin, A. A.
[1] Extremum problems in the presence of constraints, *Z. Wycisl. Mat. i Mat. Fiz.* **5** (1965), 395–453.

Dunford, N., and Schwartz, J. T.
[1] "Linear Operators," Part I. Wiley (Interscience), New York, 1967.

Fan, K.
[1] "Convex Sets and Their Applications" (Summer Lectures 1959). Argonne Nat. Lab., Appl. Math. Div. (bound mimeographed notes).

Filippov, A. F.
[1] On certain questions in the theory of optimal control, *J. Soc. Indust. Appl. Math., Ser. A, Control* **1** (1962), 76–84; *Vestnik Moskov. Univ. Ser. Mat. Mech. Astronom.* **2** (1959), 25–32.

Friedman, A.
[1] Optimal control for hereditary processes, *Arch. Rational Mech. Anal.* **15** (1964), 396–416.
[2] Differential games of pursuit in Banach space, *J. Math. Anal. Appl.* **25** (1969), 93–113

Gamkrelidze, R. V.
[1] Time-optimal processes with restricted phase coordinates, *Dokl. Akad. Nauk SSSR* **125** (1959), 475–478.
[2] Optimal control processes with restricted phase coordinates, *Izv. Akad. Nauk SSSR, Ser. Mat.* **24** (1960), 315–356.
[3] On some extremal problems in the theory of differential equations with applications to the theory of optimal control, *J. Soc. Indust. Appl. Math., Ser. A, Control* **3** (1965), 106–128.

Gamkrelidze, R. V., and Kharatishvili, G. L.
[1] Extremal problems in linear topological spaces I, *Math. Systems Theory* **3** (1967), 229–256.

Gel'fand, I. M., and Shilov, G. E.
[1] "Generalized Functions," Vol. 1. Academic Press, New York, 1964.
[2] "Generalized Functions," Vol. 2. Academic Press, New York, 1968.

Ghouila-Houri, A.
[1] Sur la généralisation de la notion de commande d'un système guidable, *Rev. Française Informat. Recherche Operationnelle* (1967), No. 4, 7–32.

Goursat, E.
[1] "Cours d'Analyse Mathematique," Tome III. Gauthier-Villars, Paris, 1942.

Graves, L. M.
[1] "The Theory of Functions of Real Variables." McGraw-Hill, New York, 1946.

Halanay, A.
[1] Optimal controls for systems with time-lag, *SIAM J. Control* **6** (1968), 215–234.

Kelendzeridze, D. L.
[1] On the theory of optimal pursuit, *Soviet Math. Dokl.* **2** (1961), 654–656.

Köthe, G.
[1] "Topological Vector Spaces," Vol. I. Springer-Verlag, New York, 1969.

Krasnoselskii, M. A.
[1] "Positive Solutions of Operator Equations." Noordhoff, Groningen, 1964.

McShane, E. J.
[1] On multipliers for Lagrange problems, *Amer. J. Math.* **61** (1939), 809–819.
[2] Generalized curves, *Duke Math. J.* **6** (1940), 513–536.
[3] Necessary conditions in generalized-curve problems of the calculus of variations, *Duke Math. J.* **7** (1940), 1–27.
[4] Existence theorems for Bolza problems in the calculus of variations, *Duke Math. J.* **7** (1940), 28–61.
[5] Relaxed controls and variational problems, *SIAM J. Control* **5** (1967), 438–485.

Munroe, M. E.
[1] "Introduction to Measure and Integration." Addison-Wesley, Cambridge, Massachusetts, 1953.

Neustadt, L. W.
[1] The existence of optimal controls in the absence of convexity conditions, *J. Math. Anal. Appl.* **7** (1963), 110–117.
[2] A general theory of minimum-fuel space trajectories, *J. Soc. Indust. Appl. Math., Ser A., Control* **3** (1965), 317–356.
[3] An abstract variational theory with applications to a broad class of optimization problems I, General Theory, *SIAM J. Control* **4** (1966), 505–527.
[4] An abstract variational theory with applications to a broad class of optimization problems II. Applications, *SIAM J. Control* **5** (1967), 90–137.
[5] A general theory of extremals, *J. Comput. System Sci.* **3** (1969), 57–92.

Oguztöreli, M. N.
[1] "Time-Lag Control Systems." Academic Press, New York, 1966.

Pontryagin, L. S., Boltyanskii, V. G., Gamkrelidze, R. V., and Mishchenko, E. F.
[1] "The Mathematical Theory of Optimal Processes." Wiley (Interscience), New York, 1962.

Riesz, F., and Sz.-Nagy, B.
[1] "Functional Analysis." Ungar, New York, 1955.

Rishel, R. W.
[1] An extended Pontryagin principle for control systems whose control laws contain measures, *SIAM J. Control* **3** (1965), 191–205.

Roxin, E.
[1] The existence of optimal controls, *Michigan Math. J.* **9** (1962), 109–119.

Rudin, W.
[1] "Real and Complex Analysis." McGraw-Hill, New York, 1966.

Saeks, R.
[1] Causality in Hilbert space, *SIAM Rev.* **12** (1970), 357–383.

Schmaedeke, W. W.
[1] Optimal control theory for nonlinear vector differential equations containing measures, *SIAM J. Control* **3** (1965), 231–280.

Schwartz, L.
[1] "Théorie des Distributions." Hermann, Paris, 1966.
[2] "Séminaire Schwartz," 1953–1954. Faculté des Sciences de Paris.

Sobolev, S. L.
[1] "Applications of Functional Analysis in Mathematical Physics" (Transl. Math. Monogr., Vol. 7). Amer. Math. Soc., Providence, Rhode Island, 1963.

Volterra, V., and Pérès, J.
[1] "Théorie Générale des Fonctionnelles." Gauthier-Villars, Paris, 1936.

von Neumann, J.
[1] Zur Theorie der Geselschaftsspiele, *Math. Ann.* **100** (1928), 295–320.

Warga, J.
[1] Relaxed variational problems, *J. Math. Anal. Appl.* **4** (1962), 111–128.
[2] Necessary conditions for minimum in relaxed variational problems, *J. Math. Anal. Appl.* **4** (1962), 129–145.
[3] Minimizing variational curves restricted to a preassigned set, *Trans. Amer. Math. Soc.* **112** (1964), 432–455.
[4] On a class of minimax problems in the calculus of variations, *Michigan Math. J.* **12** (1965), 289–311.
[5] Unilateral variational problems with several inequalities, *Michigan Math. J.* **12** (1965), 449–480.
[6] Minimax problems and unilateral curves in the calculus of variations, *J. Soc. Indust. Appl. Math., Ser. A, Control* **3** (1965), 91–105.
[7] Variational problems with unbounded controls, *J. Soc. Indust. Appl. Math., Ser. A, Control* **3** (1965), 424–438.
[8] Functions of relaxed controls, *SIAM J. Control* **5** (1967), 628–641.
[9] Restricted minima of functions of controls, *SIAM J. Control* **5** (1967), 642–656.
[10] The reduction of certain control problems to an "ordinary differential" type, *SIAM Rev.* **10** (1968), 219–222.
[11] Relaxed controls for functional equations, *J. Functional Analysis* **5** (1970), 71–93.
[12] Control problems with functional restrictions, *SIAM J. Control* **8** (1970), 360–371.
[13] Unilateral and minimax control problems defined by integral equations, *SIAM J. Control* **8** (1970), 372–382.
[14] On a class of pursuit and evasion problems, *J. Differential Equations* **9** (1971), 155–167.
[15] Conflicting and minimax controls, *J. Math. Anal. Appl.* **33** (1971), 655–673.
[16] Normal control problems have no minimizing strictly original solutions, *Bull. Amer. Math. Soc.* **77** (1971), 625–628.

Ważewski, T.
[1] Sur une généralisation de la notion des solutions d'une equation au contingent, *Bull. Acad. Polon. Sci., Ser. Sci. Math. Astronom. Phys.* **10** (1962), 11–15.
[2] Sur les systèmes de commande non linéaires dont le contredomaine de commande n'est pas forcément convexe, *Bull. Acad. Polon. Sci., Ser. Sci. Math. Astronom. Phys.* **10** (1962), 17–21.

Young, L. C.
[1] Generalized curves and the existence of an attained absolute minimum in the calculus of variations, *C. R. Sci. Lettres Varsovie, C III* **30** (1937), 212–234.
[2] Necessary conditions in the calculus of variations, *Acta Math.* **69** (1938), 239–258.
[3] "Lectures on the Calculus of Variations and Optimal Control Theory." Saunders, Philadelphia, 1969.

Notation Index

General

Specific to Part Two

Subject Index

G

H

I

J

K

L

P

R

S